Lecture Notes in Computer Science 5888

Commenced Publication in 1973
Founding and Former Series Editors:
Gerhard Goos, Juris Hartmanis, and Jan van Leeuwen

Juan A. Garay   Atsuko Miyaji
Akira Otsuka (Eds.)

Cryptology
and Network
Security

8th International Conference, CANS 2009
Kanazawa, Japan, December 12-14, 2009
Proceedings

 Springer

Volume Editors

Juan A. Garay
AT&T Labs Research
Florham Park, NJ, USA
E-mail: garay@research.att.com

Atsuko Miyaji
Japan Advanced Institute of Science and Technology (JAIST)
Nomi, Ishikawa, Japan
E-mail: miyaji@jaist.ac.jp

Akira Otsuka
National Institute of Advanced Industrial Science and Technoloy (AIST)
Tokyo, Japan
E-mail: a-otsuka@aist.go.jp

Library of Congress Control Number: 2009939618

CR Subject Classification (1998): E.3, G.2, D.4.6, K.6.5, C.2, E.4, K.4.1

LNCS Sublibrary: SL 4 – Security and Cryptology

ISSN 0302-9743

ISBN 978-3-642-10432-9 Springer Berlin Heidelberg New York

springer.com

© Springer-Verlag Berlin Heidelberg 2009

Typesetting: Camera-ready by author, data conversion by Scientific Publishing Services, Chennai, India
Printed on acid-free paper   SPIN: 12799756   06/3180   5 4 3 2 1 0

Preface

The 8th International Conference on Cryptology and Network Security (CANS 2009) was held at the Ishikawa Prefectural Museum of Art in Kanazawa, Japan, during December 12–14, 2009. The conference was jointly co-organized by the National Institute of Advanced Industrial Science and Technology (AIST), Japan, and the Japan Advanced Institute of Science and Technology (JAIST). In addition, the event was supported by the Special Interest Group on Computer Security (CSEC), IPSJ, Japan, the Japan Technical Group on Information Security (ISEC), IEICE, the Japan Technical Committee on Information and Communication System Security(ICSS), IEICE, and the Society of Information Theory and its Applications (SITA), Japan, and co-sponsored by the National Institute of Information and Communications Technology, Japan, ComWorth Co., LTD, Japan, Hitachi, Ltd., Hokuriku Telecommunication Network Co.,Inc., and Internet Initiative Japan Inc.

The conference received 109 submissions from 24 countries, out of which 32 were accepted for publication in these proceedings. At least three Program Committee (PC) members reviewed each submitted paper, while submissions co-authored by a PC member were submitted to the more stringent evaluation of five PC members. In addition to the PC members, many external reviewers joined the review process in their particular areas of expertise. We were fortunate to have this energetic team of experts, and are deeply grateful to all of them for their hard work, which included a very active discussion phase—almost as long as the initial individual reviewing period. The paper submission, review and discussion processes were effectively and efficiently made possible by the Web-based system iChair.

The main goal of CANS as a conference is to promote research on all aspects of network security, as well as to build a bridge between research on cryptology and on network security. The broad range of areas covered by the high-quality accepted papers to the current edition, which include novel cryptographic and number-theoretic constructs enabling new functionalities, the various security aspects of protocols and wireless and sensor networks, new cipher designs, advanced cryptanalytic techniques without which soundness would be unattainable, and the treatment of privacy in multifarious settings, from privacy-preserving computation to privacy in the ever-popular social networks, attests—exceedingly—to the fulfillment of that goal. In addition, the conference featured three invited speakers: Craig Gentry from IBM Research, Adrian Perrig from Carnegie Mellon University, and Adam Smith from The Pennsylvania State University, whose lectures on cutting-edge research areas— "Computing on Encrypted Data," "Building Secure Networked Systems with Code Attestation," and "A Cryptographer's-Eye View of Privacy in Statistical Databases," respectively—contributed in no small part to the richness of the program.

Finally, we thank all the authors who submitted papers to this conference; the Organizing Committee members, colleagues and student helpers for their valuable time and effort; and all the conference attendees who made this event a truly intellectually stimulating one through their active participation.

December 2009   Juan A. Garay
                                                   Atsuko Miyaji
                                                   Akira Otsuka

CANS 2009
8th International Conference on Cryptology and Network Security

Jointly organized by

National Institute of Advanced Industrial Science and Technology (AIST)
and
Japan Advanced Institute of Science and Technology (JAIST).

General Chair

Akira Otsuka (AIST, Japan)

Program Co-chairs

Juan A. Garay (AT&T Labs - Research, USA)
Atsuko Miyaji (JAIST, Japan)

Organizing Committee

Local Arrangements
  Co-chairs   Kazumasa Omote (JAIST, Japan)
                  Shoichi Hirose (University of Fukui, Japan)
                  Yuji Suga (IIJ, Japan)
                  Natsume Matsuzaki (Panasonic, Japan)
Finance Co-chairs   Katsuyuki Takashima (Mitsubishi Electric, Japan)
                  Shoko Yonezawa (AIST, Japan)
Publicity Co-chairs   Tomoyuki Asano (Sony, Japan)
                  Daisuke Inoue (NICT, Japan)
                  Kensuke Shibata (NTT Labs, Japan)
Liaison Co-chairs   Keisuke Takemori (KDDI R&D Labs., Japan)
                  Shinichiro Matsuo (NICT, Japan)
System Co-chairs   Atsuo Inomata (NAIST, Japan)
                  Yutaka Oiwa (AIST, Japan)
                  Takao Ogura (Fujitsu Laboratories, Japan)
                  Toshihiro Tabata (Okayama University, Japan)
Publication Co-chairs Isao Echizen (NII, Japan)
                  Takeshi Okamoto (Tsukuba University of Technology, Japan)
Registration Co-chairs Tatsuyuki Matsushita (Toshiba, Japan)
                  Kunihiko Miyazaki (Hitachi, Japan)

Program Committee

Jean-Luc Beuchat	University of Tsukuba, Japan
Alexandra Boldyreva	Georgia Institute of Technology, USA
Colin Boyd	Queensland University of Technology, Australia
Emmanuel Bresson	DCSSI Crypto Lab, France
Mike Burmester	FSU, USA
Koji Chida	NTT Labs, Japan
Ee-Chien Chang	National University of Singapore, Singapore
Rosario Gennaro	IBM T.J. Watson Research Center, USA
Trevor Jim	AT&T Labs - Research, USA
Charanjit Jutla	IBM T.J. Watson Research Center, USA
Seny Kamara	Microsoft Research, USA
Aggelos Kiayias	UConn, USA
Vladimir Kolesnikov	Bell Labs, USA
Tanja Lange	Technical University Eindhoven, The Netherlands
Dong Hoon Lee	Korea University, Korea
Pil Joong Lee	Pohang University of Science and Technology, Korea
Benoit Libert	UCL, Belgium
Helger Lipmaa	Cybernetica AS, Estonia
Subhamoy Maitra	Indian Statistical Institute, India
Mark Manulis	TU Darmstadt, Germany
David Nowak	AIST, Japan
Wakaha Ogata	Tokyo Institute of Technology, Japan
C. Pandu Rangan	IIT Madras, India
Kenny Paterson	Royal Holloway, University of London, UK
Rene Peralta	NIST, USA
Josef Pieprzyk	Macquarie University, Australia
German Saez	UPC, Spain
Taizo Shirai	Sony, Japan
Adam Smith	Penn State University, USA
Jessica Staddon	PARC, USA
Douglas Stinson	University of Waterloo, Canada
Willy Susilo	University of Wollongong, Australia
Tsuyoshi Takagi	Future University - Hakodate, Japan
Gene Tsudik	UC Irvine, USA
Ivan Visconti	University of Salerno, Italy
Xiaoyun Wang	Shandong University, China
Duncan S. Wong	City University of Hong Kong, China
Sung-Ming Yen	National Central University, Taiwan
Hong-Sheng Zhou	UConn, USA
Jianying Zhou	I2R, Singapore

External Reviewers

Man Ho Au, Toru Akishita, Franz Baader, Lucas Ballard, Paulo Barreto, Rebecca Braynard, Luigi Catuogno, Nishanth Chandran, Chien-Ning Chen, Joo Yeon Cho, Kwan Tae Cho, Sherman Chow, Cheng-Kang Chu, Ji Young Chun, Baudoin Collard, Emiliano De Cristofaro, Giovanni Di Crescenzo, Vanesa Daza, Karim El Defrawy, Yvo Desmedt, Jérémie Detrey, Xuhua Ding, Orr Dunkelman, Lei Fan, Pierre-Alain Fouque, Thomas Fuhr, David Galindo, Hossein Ghodosi, Jon Giffin, Juan Gonzalez, Choudary Gorantla, Matthew Green, Gerhard Hancke, Shoichi Hirose, Jin Hong, Xinyi Huang, Dai Ikarashi, Yuto Kawahara, Jun Kurihara, Bum Han Kim, Jihye Kim, Ki Tak Kim, Sun Young Kim, Woo Kwon Koo, Simon Kramer, Jun Kurihara, Jun Ho Lee, Wei-Chih Lien, Hsi-Chung Lin, Joseph K. Liu, Yanbin Lu, Vadim Lyubashevsky, Di Ma, Francesco Matarazzo, Daniele Micciancio, Marine Minier, Chris J. Mitchell, Shiho Moriai, Paz Morillo, Benjamin Morin, Mridul Nandi, Ching Yu Ng, Yasuyuki Nogami, Mehrdad Nojoumian, Adam O'Neill, Jung Ha Paik, Prabal Paul, Souradyuti Paul, Sean Peisert, Olivier Pereira, Ludovic Perret, Fabio Petagna, Le Trieu Phong, Benny Pinkas, Bertram Poettering, Axel Poschmann, Carla Rafols, Dominik Raub, Mariana Raykova, Mathieu Renauld, Hyun Sook Rhee, Francisco Rodríguez-Henríquez, Arnab Roy, Markus Ruckert, Alexandre Ruiz, Koichi Sakumoto, Palash Sarkar, Dominique Schroeder, Jae Woo Seo, Elaine Shi, Kyoji Shibutani, Masaaki Shirase, Jason Smith, Francois-Xavier Standaert, Hidema Tanaka, Toshiaki Tanaka, Ersin Uzun, Frederik Vercauteren, Marion Videau, Nguyen Vo, Christian Wachsmann, Bogdan Warinschi, Dai Watanabe, Hao-Chi Wong, Xiaokang Xiong, Yeon-Hyeong Yang, Yanjiang Yang, Tsz Hon Yuen, Greg Zaverucha, Fan Zhang.

Table of Contents

Network Security

Privacy and Anonymity

Functional and Searchable Encryption

Invited Talk 2

Authentication

Block Cipher Design

Cryptanalysis II

Algebraic and Number-Theoretic Schemes

Wireless and Sensor Network Security II

Invited Talk 3

Cryptographic Protocol and Schemes II

Improved Garbled Circuit Building Blocks and Applications to Auctions and Computing Minima[*]

Vladimir Kolesnikov[1], Ahmad-Reza Sadeghi[2], and Thomas Schneider[2]

[1] Bell Laboratories, 600 Mountain Ave. Murray Hill, NJ 07974, USA
kolesnikov@research.bell-labs.com
[2] Horst Görtz Institute for IT-Security, Ruhr-University Bochum, Germany
{ahmad.sadeghi,thomas.schneider}@trust.rub.de

Abstract. We consider generic Garbled Circuit (GC)-based techniques for Secure Function Evaluation (SFE) in the semi-honest model.

We describe efficient GC constructions for addition, subtraction, multiplication, and comparison functions. Our circuits for subtraction and comparison are approximately two times smaller (in terms of garbled tables) than previous constructions. This implies corresponding computation and communication improvements in SFE of functions using our efficient building blocks. The techniques rely on recently proposed "free XOR" GC technique.

Further, we present concrete and detailed improved GC protocols for the problem of secure integer comparison, and related problems of auctions, minimum selection, and minimal distance. Performance improvement comes both from building on our efficient basic blocks and several problem-specific GC optimizations. We provide precise cost evaluation of our constructions, which serves as a baseline for future protocols.

Keywords: Secure Computation, Garbled Circuit, Millionaires Problem, Auctions, Minimum Distance.

1 Introduction

We are motivated by secure function evaluation (SFE) of integer comparison, and related problems such as auctions and biometric authentication. For this, we propose new, more efficient SFE protocols for these functions. More specifically, we propose improved constructions for subtraction, and comparison functions, and demonstrate their advantages on the example of our motivating applications.

Comparison is a widely used basic primitive. In particular, it plays an especially important role in financial transactions, biometric authentication, database mining applications, etc.

---

[*] Supported by EU FP6 project SPEED, EU FP7 project CACE and ECRYPT II.

J.A. Garay, A. Miyaji, and A. Otsuka (Eds.): CANS 2009, LNCS 5888, pp. 1–20, 2009.

Auctions. With the growth of the Internet and its widespread acceptance as the medium for electronic commerce, online auctions continue to grow in popularity. Additionally, many sellers consider the "name your price" model. For example, sites such as priceline.com ask a buyer for a price he is willing to pay for a product, and the deal is committed to if that price is greater than a certain (secret) threshold. In many such situations, it is vital to maintain the privacy of bids of the players. Indeed, revealing an item's worth can result in artificially high prices or low bids, specifically targeted for a particular buyer or seller. While a winning bid or a committed deal may necessarily reveal the cost of the transaction, it is highly desirable to keep all other information (e.g., unsuccessful bids) secret. There has been a large stream of work dedicated to ensuring privacy and security of online auctions and haggling (e.g., [13,7,39]). Our work complements, extends, and builds on it.

Biometric authentication. Widespread adoption of biometric authentication (e.g., fingerprint or face recognition) is causing strong concerns of privacy violations. Indeed, improper use of biometric information has far more implications than "simple" collection of personal information. Adoption of privacy-preserving biometric authentication is highly desired and will benefit the users and the administrators of the systems alike. Because biometric images are never scanned perfectly, the identity of the user is determined by proximity of the scanned and stored biometrics. It is natural, therefore, that threshold comparisons are frequently employed in such identification systems. Further, in some multi-user systems, it may be desired to simply find the closest match in the database. In such systems, secure comparison would be also extensively used.

State of the art for secure comparison and related algorithms. Starting with the original paper of Yao [45], secure comparison, also referred to as the "two Millionaires problem", has attracted much attention [46,20,37,31]. A variety of techniques are employed in these solutions – homomorphic encryption, evaluation of branching programs, Garbled Circuit (GC).

Today, in the standard computational setting, the most efficient protocol is the simple evaluation of the generic GC. Indeed, the size of the comparison circuit is quite small (linear in the size of the inputs), and its secure evaluation is rather efficient (linear number of Oblivious Transfers (OT) and evaluations of a cryptographic hash function, such as SHA-256).

Most popular alternative solutions are based on homomorphic encryptions. For comparison, they offer a similar complexity compared to GC, as they still must perform a linear (in the input) number of public key operations by both players. However, GC offers more flexible and cheap programming possibilities, due to its low cost of manipulation of boolean values. In contrast, homomorphic encryptions are not suitable, e.g., for branching based on the encrypted value which can be achieved only with much more expensive techniques than GC).

In sum, GC approach is a clear choice for integer comparison, its extensions, such as auctions, simple integer manipulations (addition and even multiplications) and a variety of other problems that have small circuit representation. We build our solutions in this framework.

Our contributions. As justified above, our work is based on GC. We advance the state of the art of SFE for subtraction and comparison functions, by constructing their more efficient GC representations. We work in the semi-honest model which is appropriate for many application scenarios.

More specifically, our optimizations take advantage of the recently proposed method of GC construction [33], where XOR gates are evaluated essentially for free (one XOR operation on keys, and no garbled table entries to generate or transfer). We show how to compute comparison and other basic functions with circuits consisting mostly of XOR gates. This results in reduction of the size of GC (i.e., the size of garbled tables) by approximately half (see Table 2 for detailed comparison). We note that the method of [33] (and thus our work) requires the use of a weak form of Random Oracle, namely of correlation-robust functions [26].

As further contribution, we then follow through, and discuss in detail GC-based constructions for the Millionaires problem, computing first-price auctions and minimum Hamming- or Euclidian distance. In addition to improvements due to our new building blocks, our protocols benefit from a number of GC-based optimizations. In addition to establishing a new performance baseline for these problems, we aim to promote GC as a very efficient solution, and prevent its frequent unfair dismissal as an "impractical generic approach".

Related work. SFE (and in particular GC), and secure comparison has received much attention in the literature, all of which we cannot possibly include here. In this section we summarize relevant work to give the reader a perspective on our results. We discuss additional related work (on which we improve) in the individual sections of the paper.

Circuit-Based Secure Function Evaluation. GC technique of SFE was introduced by Yao [46], with a formal proof of security (in the semi-honest model) given in [34]. Extensions of Yao's garbled circuit protocol to security against covert players were given in [1,25], and against malicious players in [27,35,40]. Our constructions rely on the recent work of [33], where a GC technique is proposed that allows evaluation of XOR gates "for free", i.e., with no communication and negligible computation costs. In [33] improved circuit constructions for multiplexer, addition and (in-)equality testing are presented. Our main contribution – the building block constructions – further improve their proposals (e.g., subtraction and comparison are improved by a factor of two).

Secure Two-Party Comparison. The first secure two-party comparison protocol was proposed in [45], and today GC [46] is the most efficient solution to this problem as shown in this paper: our solution for comparing two ℓ-bit numbers requires $16\ell t$ bit offline communication and $3\ell t$ bit online communication, where t is a symmetric security parameter (i.e., length of a symmetric key).

Homomorphic encryption is another popular tool for comparison. The protocol of Fischlin [20] uses the Goldwasser-Micali XOR-homomorphic encryption scheme [24] and has communication complexity $\ell T(\kappa + 1)$, where κ is a statistical correctness parameter (e.g., $\kappa = 40$) and T is an asymmetric security parameter (i.e., size of an RSA modulus). The comparison protocol of [6] uses bitwise Paillier encryption and has communication complexity $4\ell T$. This protocol was improved in [14,16,15] to communication complexity $2\ell T$ by using a new homomorphic encryption scheme with smaller ciphertext size. These two-party protocols were extended to comparisons in the multi-party setting with logarithmic and linear round complexity in [21].

Minimum Selection. A two-party protocol for finding k-Nearest Neighbors was given in [44], and improved from quadratic to linear communication complexity in [43]. Our protocol for finding the nearest neighbor is a more efficient protocol for the special case $k = 1$. A simple protocol to select the minimum of homomorphically encrypted values based on the multiplicative hiding assumption was given in [30] in the context of privacy-preserving benchmarking. However, multiplicative blinding reveals some information about the magnitude of the blinded value. Our minimum selection protocol can be used as a provably secure replacement of this protocol. Finally, we note that in our minimum Hamming distance protocol we use several steps of the Hamming distance protocol of [28].

Efficient circuits for addition and multiplication. Boyar et al. [10,11,9] considered multiplicative complexity[1] of symmetric functions (i.e., functions only dependent on the hamming weight of their inputs). As a corollary, Boyar et al. describe efficient circuits for addition (and thus multiplication, via Karatsuba-Ofman method [29]). Our subtraction and comparison building blocks are extensions of their construction.

2   Preliminaries

In this section, we summarize our conventions and setting in §2.1 and cryptographic tools used in our constructions: oblivious transfer (OT) in §2.3, garbled circuits (GC) with free XOR in §2.4, and additively homomorphic encryption (HE) in §2.2. Reader familiar with the prerequisites may safely skip to §3.

2.1   Parameters, Notation and Model

We denote symmetric security parameter by t and the asymmetric security parameter, i.e., bitlength of RSA moduli, by T. Recommended parameters for short-term security (until 2010) are for example $t = 80$ and $T = 1024$ [23]. The bitlength of a garbled value is $t' := t + 1$ (cf. §2.4 for details). The statistical correctness parameter is denoted with κ (the probability of a protocol failure is bounded by $2^{-\kappa}$) and the statistical security parameter with σ. In practice, one can choose $\kappa = \sigma = 80$. The bitlength of protocol inputs is denoted with ℓ and the number of inputs with n. We write $\mathbf{x}^\ell$ to denote ℓ-bit value x.

We work in the semi-honest model. We note that the method of [33] (and thus our work) requires the use of a weak form of Random Oracle, namely of correlation-robust functions [26].

---

[1] Multiplicative complexity of a function measures the number of AND gates in its circuit (and gives NOT and XOR gates for free).

2.2 Homomorphic Encryption (HE)

Some of our constructions make black-box usage of a semantically secure homomorphic encryption scheme with plaintext space $(P, +, 0)$, ciphertext space $(C, *, 1)$, and probabilistic polynomial-time algorithms (Gen, Enc, Dec).

An *additively homomorphic encryption* scheme allows addition under encryption as it satisfies $\forall x, y \in P : \mathsf{Dec}(\mathsf{Enc}(x) * \mathsf{Enc}(y)) = x + y$. It can be instantiated with a variety of cryptosystems including [41,17], or the cryptosystem of [14,16,15] which is restricted to small plaintext space P - just to name a few.

For the sake of completeness we mention, that the cryptosystem of [8] allows for an arbitrary number of additions and one multiplication and *fully homomorphic encryption* schemes allow to evaluate an arbitrary number of additions and multiplications on ciphertexts. Possible candidates are the cryptosystem of [3] (size of ciphertexts grows exponentially in the number of multiplications) or the recently proposed scheme without such a restriction [22]. However, the size of ciphertexts in these schemes is substantially larger than that of the purely additively homomorphic schemes.

2.3 Oblivious Transfer (OT)

Parallel 1-out-of-2 Oblivious Transfer of m ℓ-bit strings, denoted as OT_ℓ^m, is a two-party protocol run between a chooser C and a sender S. For $i = 1, \ldots, m$, S inputs a pair of ℓ-bit strings $s_i^0, s_i^1 \in \{0,1\}^\ell$ and C inputs m choice bits $b_i \in \{0,1\}$. At the end of the protocol, C learns the chosen strings $s_i^{b_i}$, but nothing about the unchosen strings $s_i^{1-b_i}$ whereas S learns nothing about the choices b_i.

Efficient OT protocols. We use OT_ℓ^m as a black-box primitive which can be instantiated efficiently with different protocols [38,2,36,26]. For example the protocol of [2] implemented over a suitably chosen elliptic curve has communication complexity $m(6(2t + 1)) + (2t + 1) \sim 12mt$ bits and is secure against malicious C and semi-honest S in the standard model as described in the full version of this paper [32]. Similarly, the protocol of [38] implemented over a suitably chosen elliptic curve has communication complexity $m(2(2t + 1) + 2\ell)$ bits and is secure against malicious C and semi-honest S in the random oracle model. Both protocols require $\mathcal{O}(m)$ scalar point multiplications.

Extending OT efficiently. The extensions of [26] can be used to efficiently reduce the number of computationally expensive public-key operations of OT_ℓ^m to be independent of m. Their transformation for semi-honest receiver reduces OT_ℓ^m to OT_ℓ^t and a small additional overhead: one additional message, $2m(\ell + t)$ bits additional communication, and $\mathcal{O}(m)$ invocations of a correlation robust hash function ($2m$ for S and m for C) which is substantially cheaper than $\mathcal{O}(m)$ asymmetric operations. Also a slightly less efficient extension for malicious receiver is given in their paper.

2.4 Garbled Circuits (GC)

The most efficient method for secure evaluation of a boolean circuit C for computationally bounded players is Yao's garbled circuit approach [46,34]. We briefly summarize the main ideas of this protocol in the following. The circuit *constructor* (server $\mathcal{S}$) creates a *garbled circuit* $\widetilde{C}$ with algorithm CreateGC: for each wire W_i of the circuit, he randomly chooses two garbled values $\widetilde{w}_i^0, \widetilde{w}_i^1$, where $\widetilde{w}_i^j$ is the *garbled value* of W_i's value j. (Note: $\widetilde{w}_i^j$ does not reveal j.) Further, for each gate G_i, $\mathcal{S}$ creates a *garbled table* $\widetilde{T}_i$ with the following property: given a set of garbled values of G_i's inputs, $\widetilde{T}_i$ allows to recover the garbled value of the corresponding G_i's output, but nothing else. $\mathcal{S}$ sends these garbled tables, called *garbled circuit* $\widetilde{C}$ to the *evaluator* (client $\mathcal{C}$). Additionally, $\mathcal{C}$ obliviously obtains the *garbled inputs* $\widetilde{w}_i$ corresponding to inputs of both parties (details on how this can be done later in §2.4). Now, $\mathcal{C}$ can evaluate the garbled circuit $\widetilde{C}$ on the garbled inputs with algorithm EvalGC to obtain the *garbled outputs* simply by evaluating the garbled circuit gate by gate, using the garbled tables $\widetilde{T}_i$. Finally, $\mathcal{C}$ translates the garbled outputs into output values given for the respective players (details below in §2.4). Correctness of GC follows from the method of how garbled tables $\widetilde{T}_i$ are constructed.

Improved Garbled Circuit with free XOR [33]. An efficient method for creating garbled circuits which allows "free" evaluation of XOR gates was presented in [33]. More specifically, a garbled XOR gate has no garbled table (*no communication*) and its evaluation consists of XOR-ing its garbled input values (*negligible computation*). The other gates, called *non-XOR gates*, are evaluated as in Yao's GC construction [46] with a *point-and-permute technique* (as used in [37]) to speed up the implementation of the GC protocol: the garbled values $\widetilde{w}_i = \langle k_i, \pi_i \rangle \in \{0,1\}^{t'}$ consist of a symmetric key $k_i \in \{0,1\}^t$ and a random permutation bit $\pi_i \in \{0,1\}$ and hence have length $t' = t + 1$ bits. The permutation bit π_i is used to select the right table entry for decryption with the t-bit key k_i (recall, t is the symmetric security parameter). The encryption of the garbled table entries is done with the symmetric encryption function $\mathsf{Enc}_{k_1,\ldots,k_d}^s(m) = m \oplus H(k_1||\ldots||k_d||s)$, where d is the number of inputs of the gate, s is a unique identifier for the specific row in the gate's garbled table used once, and H is a suitably chosen cryptographic hash function. Hence, creation of the garbled table of a non-XOR d-input gate requires 2^d invocations of H and its evaluation needs one invocation, while XOR gates are "for free".

The main observation of [33] is, that the constructor $\mathcal{S}$ chooses a global key difference $\Delta \in_R \{0,1\}^t$ which remains unknown to evaluator $\mathcal{C}$ and relates the garbled values as $k_i^0 = k_i^1 \oplus \Delta$. (This technique was subsequently extended in the LEGO paper [40] which allows to compose garbled circuits dynamically with security against malicious circuit constructor). Clearly, the usage of such garbled values allows for *free evaluation of XOR gates* with input wires W_1, W_2 and output wire W_3 by computing $\widetilde{w}_3 = \widetilde{w}_1 \oplus \widetilde{w}_2$ (no communication and negligible computation). However, using related garbled values requires that the hash function H used to create the garbled tables of non-XOR gates has to be modeled to be correlation robust (as defined in [26]) which is stronger than modeling H as a key-derivation function (standard model) but weaker than modeling H as a random-oracle (ROM). In practice, H can be chosen from the SHA-2 family.

Input/Output Conversion. In secure two-party computation protocols executed between circuit constructor S and circuit evaluator C, each of the inputs and outputs of the securely computed functionality can be given in different forms depending on the application scenario: privately known to one party (§A.1), secret-shared between both parties (§A.2), or homomorphically encrypted under the public key of the other party (§A.3). These inputs can be converted from different forms to garbled inputs given to C. Afterwards, C evaluates the garbled circuit, obtains the garbled outputs, and converts them into outputs in the needed form.

The resulting communication complexities of these input and output conversion protocols for semi-honest parties are summarized in Table 1 and a detailed description of these known techniques is given in §A.

Table 1. Communication complexity for converting ℓ-bit inputs/outputs in different forms to inputs/outputs of a garbled circuit (parameters defined in §2.1). SS: Secret-Shared, HE: Homomorphically Encrypted.

	Input	Output
Private S (§A.1)	$\ell t'$ bits	ℓ bits
Private C (§A.1)	$\mathrm{OT}^{\ell}_{t'}$	ℓ bits
SS (§A.2)	$\mathrm{OT}^{\ell}_{t'}$	ℓ bits
HE (§A.3)	1 ciphertext $+ 5\ell t'$ bits $+ \mathrm{OT}^{\ell}_{t'}$	1 ciphertext $+ (\ell + \sigma)(5t' + 1)$ bits

3   Building Blocks for GC

In this section we present our basic contribution – improved circuit constructions for several frequently used primitives, such as integer subtraction (§3.1), comparison (§3.2), and selection of the minimum value and index (§3.3)[2]. As summarized in Table 2, our improved circuit constructions are smaller than previous solutions by 33% to 50% when used with the GC of [33]. This reduction in size immediately translates into a corresponding improvement in communication and computation complexity of any GC protocol built from these blocks. The efficiency improvements are achieved by modifying the underlying circuits, i.e., by carefully replacing larger (more costly) non-XOR gates (e.g., a 3-input gate) with smaller non-XOR gates (e.g., a 2-input gate) and (free) XOR gates.

Multiplexer Circuit (MUX). Our constructions use ℓ-bit multiplexer circuits MUX to select one of the ℓ-bit inputs $\mathbf{x}^\ell$ or $\mathbf{y}^\ell$ as output $\mathbf{z}^\ell$, depending on the selection bit c. We use the construction of [33] with ℓ non-XOR gates.

---

[2] As noted in §1, Boyar et al. [11,9] had previously proposed improved circuits for addition and multiplication. Further, the circuits for subtraction and comparison can be relatively naturally derived from the same ideas. We leave these building blocks in our presentation for completeness and readability.

Table 2. Size of efficient circuit constructions for ℓ-bit values and computing the minimum value and index of n ℓ-bit values (in table entries)

Circuit	Standard GC	[33]	This Work (Improvement)	
Multiplexer	8ℓ	4ℓ		
Addition/Subtraction (§3.1)	16ℓ	8ℓ	4ℓ	(50%)
Multiplication (§3.1)	$20\ell^2 - 16\ell$	$12\ell^2 - 8\ell$	$8\ell^2 - 4\ell$	(33%)
Equality Test (§3.2)	8ℓ	4ℓ		
Comparison (§3.2)	8ℓ		4ℓ	(50%)
Minimum Value + Index (§3.3)	$\approx 15\ell n$ [39]		$8\ell(n-1) + 4(n+1)$	(47%)

3.1 Integer Addition, Subtraction and Multiplication

Addition circuits (ADD) to add two unsigned integer values $\mathbf{x}^\ell, \mathbf{y}^\ell$ can be efficiently composed from a chain of 1-bit adders (+), often called full-adders, as shown in Fig. 1. (The first 1-bit adder has constant input $c_1 = 0$ and can be replaced by a smaller half-adder). Each 1-bit adder has as inputs the carry-in bit c_i from the previous 1-bit adder and the two input bits x_i, y_i. The outputs are the carry-out bit $c_{i+1} = (x_i \wedge y_i) \vee (x_i \wedge c_i) \vee (y_i \wedge c_i)$ and the sum bit $s_i = x_i \oplus y_i \oplus c_i$ (the latter can be computed "for free" using "free XOR" [33]). The efficient construction of a 1-bit adder shown in Fig. 2 computes the carry-out bit as $c_{i+1} = c_i \oplus ((x_i \oplus c_i) \wedge (y_i \oplus c_i))$. Overall, the efficient construction for a 1-bit adder consists of four free XOR gates and a single 2-input AND gate which has size $2^2 = 4$ table entries. The overall size of the efficient addition circuit is $|\mathsf{ADD}^\ell| = \ell \cdot |+| = 4\ell$ table entries.

Subtraction in two's complement representation is defined as $\mathbf{x}^\ell - \mathbf{y}^\ell = \mathbf{x}^\ell + \bar{\mathbf{y}}^\ell + 1$. Hence, a subtraction circuit (SUB) can be constructed analogously to the addition circuit from 1-bit subtractors (−) as shown in Fig. 3. Each 1-bit subtractor computes the carry-out bit $c_{i+1} = (x_i \wedge \bar{y}_i) \vee (x_i \wedge c_i) \vee (\bar{y}_i \wedge c_i)$ and the difference bit $d_i = x_i \oplus \bar{y}_i \oplus c_i$. We instantiate the 1-bit subtractor efficiently as shown in Fig. 4 to compute $c_{i+1} = x_i \oplus ((x_i \oplus c_i) \wedge (y_i \oplus c_i))$ with the same size as the 1-bit adder.

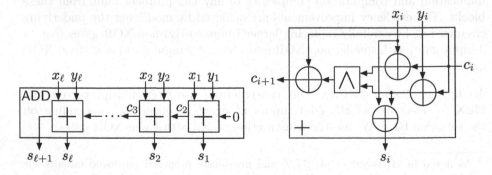

Fig. 1. Addition Circuit (ADD)   **Fig. 2.** Improved 1-bit Adder (+)

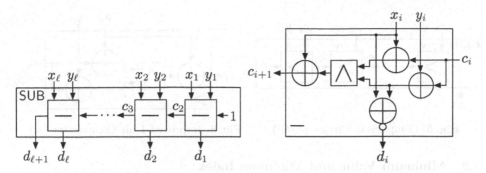

Fig. 3. Subtraction Circuit (SUB)   **Fig. 4.** Improved 1-bit Subtractor $(-)$

Multiplication circuits (MUL) to multiply two ℓ-bit integers $\mathbf{x}^\ell, \mathbf{y}^\ell$ can be constructed according to the "school method" for multiplication, i.e., adding up the bitwise multiplications of y_i and $\mathbf{x}^\ell$ shifted corresponding to the position: $\mathbf{x}^\ell \cdot \mathbf{y}^\ell = \sum_{i=1}^{\ell} 2^{i-1}(y_i \cdot \mathbf{x}^\ell)$. This circuit is composed from ℓ^2 of 1-bit multipliers (2-input AND gates) and $(\ell - 1)$ of ℓ-bit adders. Using the efficient implementation for adders, the size of the multiplication circuit is improved to $4\ell^2 + 4\ell(\ell - 1) = 8\ell^2 - 4\ell$ table entries. Alternatively, for multiplication of large ℓ-bit numbers, a circuit based on Karatsuba-Ofman multiplication [29] of size approximately $\mathcal{O}(\ell^{1.6})$ is more efficient.

3.2 Integer Comparison

We present improved circuit constructions for comparison of two ℓ-bit integers $\mathbf{x}^\ell$ and $\mathbf{y}^\ell$, i.e.,

$$z = [\mathbf{x}^\ell > \mathbf{y}^\ell] := \begin{cases} 1 & \text{if } \mathbf{x}^\ell > \mathbf{y}^\ell, \\ 0 & \text{else.} \end{cases}$$

Note that this functionality is more general than checking equality of ℓ-bit integers $\mathbf{x}^\ell$ and $\mathbf{y}^\ell$, i.e., $z = [\mathbf{x}^\ell = \mathbf{y}^\ell]$, for which an improved construction was given in [33].

As shown in Fig. 5, a comparison circuit (CMP) can be composed from ℓ sequential 1-bit comparators (>). (The first 1-bit comparator has constant input $c_1 = 0$ and can be replaced by a smaller gate). Our improved instantiation for a 1-bit comparator shown in Fig. 6 uses one 2-input AND gate with 4 table entries and three free XOR gates. Note, this improved bit comparator is exactly the improved bit subtractor shown in Fig. 4 restricted to the carry output: $[\mathbf{x}^\ell > \mathbf{y}^\ell] \Leftrightarrow [\mathbf{x}^\ell - \mathbf{y}^\ell - 1 \geq 0]$ which coincides with an underflow in the corresponding subtraction denoted by subtractor's most significant output bit $d_{\ell+1}$. The size of this comparison circuit is $\left|\mathsf{CMP}^\ell\right| = \ell \cdot |>| = 4\ell$ table entries.

Improved comparison circuits for $[\mathbf{x}^\ell < \mathbf{y}^\ell]$, $[\mathbf{x}^\ell \geq \mathbf{y}^\ell]$, or $[\mathbf{x}^\ell \leq \mathbf{y}^\ell]$ can be obtained from the improved circuit for $[\mathbf{x}^\ell > \mathbf{y}^\ell]$ by interchanging $\mathbf{x}^\ell$ with $\mathbf{y}^\ell$ and/or setting the initial carry to $c_1 = 1$.

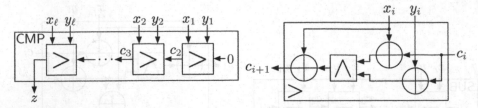

Fig. 5. Comparison Circuit (CMP)   **Fig. 6.** Improved 1-bit Comparator ($>$)

3.3 Minimum Value and Minimum Index

Finally, we show how the improved blocks presented above can be combined to obtain an improved minimum circuit (MIN) which selects the minimum value $\mathbf{m}^\ell$ and minimum index i of a list of n ℓ-bit values $\mathbf{x}_0^\ell, \ldots, \mathbf{x}_{n-1}^\ell$, i.e., $\forall j \in \{0, \ldots, n-1\} : (\mathbf{m}^\ell < \mathbf{x}_j^\ell) \vee (\mathbf{m}^\ell = \mathbf{x}_j^\ell \wedge i \leq j)$. E.g., for the list $3, 2, 5, 2$ the outputs would be $\mathbf{m}^\ell = 2$ and $i = 1$ as the leftmost minimum value of 2 is at position 1. W.l.o.g. we assume that n is a power of two, so the minimum index can be represented with $\log n$ bits.

Performance improvement of MIN mainly comes from the improved building blocks for integer comparison. We shave off an additive factor by carefully arranging tournament-style circuit so that some of the index wires can be reused and eliminated. That is, at depth d of the resulting tree we keep track of the ℓ-bit minimum value $\mathbf{m}^\ell$ of the sub-tree containing 2^d values but store and propagate only the d least significant bits $\mathbf{i}_d^d$ of the minimum index.

More specifically, the minimum value and minimum index are selected pairwise in a tournament-like way using a tree of minimum blocks (min) as shown in Fig. 7. As shown in Fig. 8, each minimum block at depth d gets as inputs the minimum ℓ-bit values $\mathbf{m}_{d,L}^\ell$ and $\mathbf{m}_{d,R}^\ell$ of its left and right subtrees T_L, T_R and the d least significant bits of their minimum indices $\mathbf{i}_{d,L}^d$ and $\mathbf{i}_{d,R}^d$, and outputs the minimum ℓ-bit value $\mathbf{m}_{d+1}^\ell$ and $(d+1)$-bit minimum index $\mathbf{i}_{d+1}^{d+1}$ of the tree. First, the two minimum values are compared with a comparison circuit (cf. §3.2). If

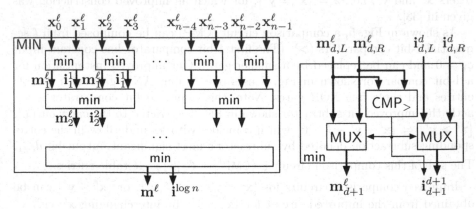

Fig. 7. Minimum Circuit (MIN)   **Fig. 8.** Minimum Block (min)

the minimum value of T_L is bigger than that of T_R (in this case, the comparison circuit outputs value 1), $\mathbf{m}_{d+1}^{\ell}$ is chosen to be the value of T_R with an ℓ-bit multiplexer block [33]. In this case, the minimum index $\mathbf{i}_{d+1}^{d+1}$ is set to 1 concatenated with the minimum index of T_R using another d-bit multiplexer. Alternatively, if the comparison yields 0, the minimum value of T_L and the value 0 concatenated with the minimum index of T_L are output. Overall, the size of the efficient minimum circuit is $\left|\mathsf{MIN}^{\ell,n}\right| = (n-1) \cdot (\left|\mathsf{CMP}^{\ell}\right| + \left|\mathsf{MUX}^{\ell}\right|) + \sum_{j=1}^{\log n} \frac{n}{2^j} \left|\mathsf{MUX}^{j-1}\right| = 8\ell(n-1) + 4n\sum_{j=1}^{\log n} \frac{j-1}{2^j} < 8\ell(n-1) + 4n(1+\frac{1}{n}) = 8\ell(n-1) + 4(n+1)$.

Our method of putting the minimum blocks together in a tree (cf. Fig. 7) is non-trivial: If the minimum blocks would have been arranged sequentially (according to the standard selection algorithm to find the minimum), the size of the circuit would have been $(n-1) \cdot (\left|\mathsf{CMP}^{\ell}\right| + \left|\mathsf{MUX}^{\ell}\right| + \left|\mathsf{MUX}^{\log n}\right|) = 8\ell(n-1) + 4(n-1)\log n$ table entries which is less efficient than the tree.

In previous work [39], a circuit for computing first-price auctions (which is functionally equivalent to computing the maximum value and index) with a size of approximately $15\ell n$ table entries is mentioned over which our explicit construction improves by a factor of approximately $\frac{15}{8}$.

4   Applications

We now describe how our efficient circuit constructions (§3) can be applied to improve previous solutions for several applications. We note that constructions of this section are not novel and may be folklore knowledge. We explicate them for concreteness, and use them to demonstrate the usefulness of our building blocks and to arrive at performance estimates to form a baseline for future protocols.

4.1   Integer Comparison (Millionaires Problem)

The "Millionaires problem" was introduced by Yao in [45] as motivation for secure compuation: two millionaires want to securely compare their respective private input values (e.g., their amount of money) without revealing more information than the outcome of the comparison to the other party. More concretely, client C holds a private ℓ-bit value x^{ℓ} and server S holds a private ℓ-bit value y^{ℓ}. The output bit $z = [x^{\ell} > y^{\ell}]$ should be revealed to both.

We obtain an efficient comparison protocol by evaluating the comparison circuit of §3.2 with the GC protocol of [33] and an efficient OT protocol. Our protocol, when executed without precomputation has asymptotic communication complexity $5\ell t + \mathsf{OT}_t^{\ell}$ bit with symmetric security parameter t (cf. §2.1).

In many practical application scenarios it is beneficial to shift as much of the computation and communication cost of a protocol into a setup (precomputation) phase, which is executed before the parties' inputs are known, while the parties' workload is low. In the following we apply a folklore technique, which demonstrates that GC protocols are ideally suited for precomputation as (in contrast to many protocols based on homomorphic encryption) almost their entire cost can be shifted into the setup phase.

Millionaires with setup. GC protocols allow to move all expensive operations (i.e., computationally expensive OT and creation of GC, as well as the transfer of GC which dominates the communication complexity) into the setup phase. The idea is to create and transfer the garbled circuit in the setup phase, and pre-compute the OTs [5]: for this, the parallel OT protocol is run on randomly chosen (by C and S) values of corresponding sizes (instead of private inputs of C and pairs of garbled input values of S). Then, in the online phase, C uses its randomly chosen value to mask his private inputs, and sends them to S. S replies with encryptions of wire's garbled inputs using his random values from the setup phase. Which garbled input is masked with which random value is determined by C's message. Finally, C can use the masks he received from the OT protocol in the setup phase to exactly decrypt the correct garbled input value.

More precisely, the **setup phase** works as follows: for $i = 1, \ldots, \ell$, C chooses random bits $r_i \in_R \{0,1\}$ and S chooses random masks $m_i^0, m_i^1 \in_R \{0,1\}^{t'}$ (recall, $t' = t+1$ is the bitlength of garbled values). Both parties run a $\mathrm{OT}_{t'}^{\ell}$ protocol on these randomly chosen values, where S inputs the pairs m_i^0, m_i^1 and C inputs r_i and C obliviously obtains the mask $m_i = m_i^{r_i}$. Additionally, S creates a garbled circuit $\widetilde{C}$ with garbled inputs $\widetilde{x}_i^0, \widetilde{x}_i^1$ and $\widetilde{y}_i^0, \widetilde{y}_i^1$ and sends $\widetilde{C}$ together with the output decryption table to C. This message has the size $4\ell t' + 1 \sim 4\ell t$ bits. Overall, the setup phase has a slightly smaller communication complexity than the Millionaires protocol without setup described above.

In the **online phase**, S sends the garbled values $\widetilde{\mathbf{y}}^{\ell}$ corresponding to his input $\mathbf{y}^{\ell}$ to C and the online part of the OT protocol is executed: for each $i = 1, \ldots, \ell$, C masks its input bits x_i with r_i as $X_i = x_i \oplus r_i$ and sends these masked bits to S. S responds with the masked pair of t'-bit strings $\langle M_i^0, M_i^1 \rangle = \langle m_i^0 \oplus \widetilde{x}_i^0, m_i^1 \oplus \widetilde{x}_i^1 \rangle$ if $X_i = 0$ or $\langle M_i^0, M_i^1 \rangle = \langle m_i^0 \oplus \widetilde{x}_i^1, m_i^1 \oplus \widetilde{x}_i^0 \rangle$ otherwise. C obtains $\langle M_i^0, M_i^1 \rangle$ and decrypts $\widetilde{x}_i = M_i^{r_i} \oplus m_i$. Using the garbled inputs $\widetilde{\mathbf{x}}^{\ell}, \widetilde{\mathbf{y}}^{\ell}$, C evaluates the garbled circuit $\widetilde{C}$, obtains the result from the output decryption table and sends it back to S. Overall, in the online phase $\ell t' + 2\ell t' + 1 \sim 3\ell t$ bits are sent.

Cost Evaluation. In the following we show how the GC-based comparison protocol outperforms those based on homomorphic encryption:

Computation Complexity. As our improved GC for integer comparison consists of no more than ℓ non-XOR 2-to-1 gates (cf. comparison circuit in §3.2), C needs to invoke the underlying cryptographic hash-function (e.g., SHA-256 for $t = 128$ bit symmetric security) exactly ℓ times to evaluate the GC (cf. §2.4). All other operations are negligible (XOR of t-bit strings). Hence, the computational complexity of the online phase of our protocol is negligible as compared to that of protocols based on homomorphic encryption. Even with an additional setup phase, those protocols need to invoke a few modular operations for each input bit which are usually by several orders of magnitude more expensive than the evaluation of a cryptographic hash function used in our protocols. Further the computational complexity of the setup phase in our protocol is more efficient than in protocols based on homomorphic encryption when using efficient OT protocols implemented over elliptic curves and efficient extensions of OT for a large number of inputs (cf. §2.3).

Communication Complexity. Table 3 shows that also the communication complexity of our protocol is much lower than that of previous protocols which are based on homomorphic encryption. As underlying $OT_{t'}^{\ell}$ protocol we use the protocol of [2] implemented over a suitably chosen elliptic curve and using point compression (see the full version of this paper [32] for details). This protocol has asymptotic communication complexity $12\ell t$ bits and is secure in the standard model. (Using the protocol of [38] which is secure in the random oracle model would result in communication complexity $6\ell t$ bits and much lower computation complexity.) The chosen values for the security parameters correspond to standard recommendations for short-term (upto 2010), medium-term (upto 2030) and long-term security (after 2030) [23].

Table 3. Asymptotic communication complexity of comparison protocols on ℓ-bit values. Parameters defined in §2.1: $\ell = 16, \kappa = 40$, short-term security: $t = 80, T = 1024$, medium-term security: $t = 112, T = 2048$, long-term security: $t = 128, T = 3082$.

Communication Complexity	Previous Work			This Work		
	[20]	[6]	[14]	Setup Phase	Online Phase	Total
Asymptotic	$(\kappa+1)\ell T$	$4\ell T$	$2\ell T$	$16\ell t$	$3\ell t$	$19\ell t$
short-term	82 kByte	8 kByte	4 kByte	2.5 kByte	0.5 kByte	3.0 kByte
medium-term	164 kByte	16 kByte	8 kByte	3.5 kByte	0.7 kByte	4.2 kByte
long-term	246 kByte	24 kByte	12 kByte	4.0 kByte	0.8 kByte	4.8 kByte

4.2 First-Price Auctions

In standard auction systems such as ebay, the auctioneer learns the inputs of all bidders and hence can deduce valuable information about the bidding behavior of unsuccessful bidders or cheat while computing the auction function depending on bidders' input values. To overcome this, a secure protocol can be used instead. Bidders provide their bids in a "smartly" encrypted form to the protocol which allows the auctioneer to compute the auction function without learning the bids. In the following we show how our constructions can be used to improve two previously proposed secure auction systems: one in which all bids are collected before the auction function is computed (Offline Auctions), and another one where bids are input dynamically and the current highest bid is published (Online Auctions).

Offline Auctions. In the offline auction system of [39], the auction function is computed by two parties, an *auction issuer* and the *auctioneer*, who are assumed not to collude. The auction issuer creates a garbled circuit which computes the auction function and sends it to the auctioneer. For each of the bidders' input bits b, a proxy-OT protocol is run, where the auction issuer inputs the two complementary garbled input values $\widetilde{b}^0, \widetilde{b}^1$ of the garbled circuit, the bidder inputs b and the auctioneer obtains the corresponding garbled value $\tilde{b}$. Then, the auctioneer evaluates the garbled circuit on the garbled inputs and obtains the outcome of the auction as output.

In order to run a first-price auction which outputs the maximum bid and the index of the maximum bidder, our improved minimum circuit of §3.3 can be used. This circuit is substantially smaller and hence the resulting protocol is more efficient than the circuit used in [39] as shown in Table 2.

Online Auctions. In the following we show that our GC-based comparison protocol outperforms the comparison protocol of Damgård, Geisler and Kroigård presented in [14] for the online auction scenario.

The auction system proposed in [14,15,16] extends the idea of splitting the computation of the auction function between two parties, the auctioneer (called *server*) and another party (called *assisting server*) who are assumed not to collude. Each bidder can submit a maximum bid b which he secret-shares between server and assisting server over respective secure channels. Afterwards, the bidder can go offline, while the server and assisting server run a secure comparison protocol to compare the secret-shared maximum bid with the publicly known value of the currently highest bid to keep track which bidder is still "in the game". A detailed description of the scenario can be found in [16].

Our protocol uses the efficient comparison protocol of §4.1 with inputs given in different forms: the bid is secret-shared between both players (cf. §A.2 for simple folklore technique to use such inputs in GC) and the other input is publicly known to both parties (e.g., can be treated as a private input of circuit constructor $\mathcal{S}$). The resulting circuit-based protocol for online auctions has exactly the same performance as our solution for the Millionaires problem described in §4.1 with the same efficiency improvements over previous solutions. In particular, the possibility to move all expensive operations into the setup phase, which can be executed during idle times (whenever no new bids are received), is very beneficial for this application as this enables the bidders to instantly see if their current bid was successful or if another bidder meanwhile gave a higher bid. This feature is important towards the end of the auction, where the frequency of bids is high. We further recall that the workload of the setup phases can be reduced by extending OTs efficiently (cf. §2.3).

4.3 Minimum Distance

Finally, we give an efficient protocol for secure computation of the *minimum distance* (or *nearest neighbor*) between a private query point Q, held by client $\mathcal{C}$, and an ordered list of private points $P_0, \ldots, P_{n-1}$ (called *database*), held by server $\mathcal{S}$. The protocol consists of two sub-protocols: the first sub-protocol computes for $i = 1, \ldots, n$ the encrypted distance $[\![\delta_i]\!]$ of the query point Q to each point P_i in the database, using a suitably chosen homomorphic encryption scheme, and outputs these encrypted distances to $\mathcal{S}$. The second sub-protocol securely selects the minimum value and index of these encrypted distances and outputs the minimum distance δ_{min} and minimum index i_{min} to $\mathcal{C}$.

Distance Computation. We sketch the sub-protocols to securely compute the distance $[\![\delta_i]\!]$ between query point Q and points P_i in the database next.

Hamming Distance. The Hamming distance between two points $P = (p_1, \ldots, p_m)$ and $Q = (q_1, \ldots, q_m)$ with $p_j, q_j \in \{0, 1\}$ is defined as $d_H(P, Q) := \sum_{j=1}^{m} p_j \oplus q_j = \sum_{j=1}^{m} (1 - p_j) q_j + p_j (1 - q_j)$. With an additively homomorphic cryptosystem, the Hamming distance can be computed as follows: $\mathcal{C}$ generates a public-key pk and corresponding secret-key and sends (the verifiably correct) pk and bitwise homomorphic encryptions of Q, $[\![q_1]\!], \ldots, [\![q_m]\!]$, to $\mathcal{S}$. As computing the Hamming distance is a linear operation, $\mathcal{S}$ can compute the encrypted Hamming distance to each point $P = P_i$ in its database as $[\![\delta_i]\!] = [\![d_H(P, Q)]\!]$ from $[\![q_j]\!]$ and p_j using standard techniques as proposed in [28].

Euclidean Distance. The Euclidean distance can be seen as an extension of the Hamming distance from 1-bit coordinates to ℓ-bit coordinates, i.e., for $j = 1, \ldots, m : p_j, q_j \in \{0, 1\}^\ell$. The Euclidean distance is then defined as $d_E(P, Q) := \sqrt{\sum_{j=1}^{m} (p_j - q_j)^2}$. As the Euclidean distance is not negative, it is sufficient to compute the square of the Euclidean distance instead, in order to find the minimum (or maximum) Euclidean distance: $d_E(P, Q)^2 = \sum_{j=1}^{m} (p_j - q_j)^2$. The encryption of the square of the Euclidean distance $[\![\delta_i^2]\!] = [\![d_E(P_i, Q)^2]\!]$ can be computed analogously to the protocol for the Hamming distance by using additively homomorphic encryption which allows for at least one multiplication (cf. §2.2). Alternatively, when using an additively homomorphic encryption scheme, one can run an additional round for multiplication as used in [18].

Minimum Selection. After having securely computed the homomorphically encrypted distances $[\![\delta_i]\!]$ held by $\mathcal{S}$, the minimum and minimum index of these values can be selected by converting these homomorphically encrypted values to garbled values as described in §A.3 and securely evaluating the minimum circuit of §3.3. The asymptotic communication complexity of this minimum selection protocol is $13\ell nt$ bits for the garbled circuits (when GCs are pre-computed), n homomorphic ciphertexts, and $\text{OT}_{t'}^{n\ell}$. The number of homomorphic ciphertexts can be further reduced using packing (§A.3), and the number of OTs can be reduced to a constant number of OTs (§2.3). As for the other application scenarios described before, all expensive operations can be moved into a setup phase and the entire protocol has a constant number of rounds.

Our minimum selection protocol can also be used as a provably secure[3] replacement for the minimum selection protocol of [30], which was used in the context of privacy-preserving benchmarking. In this scenario, mutually distrusting companies want to compare their key performance indicators (KPI) with the statistics of their peer group using an untrusted central server.

---

[3] The minimum selection protocol of [30] requires multiplicative-blinding of an additively homomorphically encrypted value which reveals some information about the magnitude of the blinded value.

Acknowledgements. We thank anonymous reviewers of CANS 2009 for helpful comments, in particular for pointing out previous work by Boyar et al.

References

1. Ahn, L.v., Hopper, N.J., Langford, J.: Covert two-party computation. In: ACM Symposium on Theory of Computing (STOC 2005), pp. 513–522. ACM, New York (2005)
2. Aiello, W., Ishai, Y., Reingold, O.: Priced oblivious transfer: How to sell digital goods. In: Pfitzmann, B. (ed.) EUROCRYPT 2001. LNCS, vol. 2045, pp. 119–135. Springer, Heidelberg (2001)
3. Armknecht, F., Sadeghi, A.-R.: A new approach for algebraically homomorphic encryption. Cryptology ePrint Archive, Report 2008/422 (2008)
4. Barni, M., Failla, P., Kolesnikov, V., Lazzeretti, R., Sadeghi, A.-R., Schneider, T.: Secure evaluation of private linear branching programs with medical applications. In: Backes, M., Ning, P. (eds.) ESORICS 2009. LNCS, vol. 5789, pp. 424–439. Springer, Heidelberg (2009), http://eprint.iacr.org/2009/195
5. Beaver, D.: Precomputing oblivious transfer. In: Coppersmith, D. (ed.) CRYPTO 1995. LNCS, vol. 963, pp. 97–109. Springer, Heidelberg (1995)
6. Blake, I.F., Kolesnikov, V.: Strong conditional oblivious transfer and computing on intervals. In: Lee, P.J. (ed.) ASIACRYPT 2004. LNCS, vol. 3329, pp. 515–529. Springer, Heidelberg (2004)
7. Blake, I.F., Kolesnikov, V.: Conditional encrypted mapping and comparing encrypted numbers. In: Di Crescenzo, G., Rubin, A. (eds.) FC 2006. LNCS, vol. 4107, pp. 206–220. Springer, Heidelberg (2006)
8. Boneh, D., Goh, E.-J., Nissim, K.: Evaluating 2-dnf formulas on ciphertexts. In: Kilian, J. (ed.) TCC 2005. LNCS, vol. 3378, pp. 325–341. Springer, Heidelberg (2005)
9. Boyar, J., Damgård, I., Peralta, R.: Short non-interactive cryptographic proofs. Journal of Cryptology 13(4), 449–472 (2000)
10. Boyar, J., Peralta, R.: Short discreet proofs. In: Maurer, U.M. (ed.) EUROCRYPT 1996. LNCS, vol. 1070, pp. 131–142. Springer, Heidelberg (1996)
11. Boyar, J., Peralta, R., Pochuev, D.: On the multiplicative complexity of boolean functions over the basis $(\wedge, \oplus, 1)$. Theor. Comput. Sci. 235(1), 43–57 (2000)
12. Brickell, J., Porter, D.E., Shmatikov, V., Witchel, E.: Privacy-preserving remote diagnostics. In: ACM Conference on Computer and Communications Security (CCS 2007), pp. 498–507. ACM, New York (2007)
13. Di Crescenzo, G.: Private selective payment protocols. In: Frankel, Y. (ed.) FC 2000. LNCS, vol. 1962, pp. 72–89. Springer, Heidelberg (2001)
14. Damgård, I., Geisler, M., Krøigaard, M.: Efficient and secure comparison for on-line auctions. In: Pieprzyk, J., Ghodosi, H., Dawson, E. (eds.) ACISP 2007. LNCS, vol. 4586, pp. 416–430. Springer, Heidelberg (2007)
15. Damgård, I., Geisler, M., Krøigaard, M.: A correction to efficient and secure comparison for on-line auctions. Cryptology ePrint Archive, Report 2008/321 (2008)
16. Damgård, I., Geisler, M., Krøigaard, M.: Homomorphic encryption and secure comparison. Journal of Applied Cryptology 1(1), 22–31 (2008)
17. Damgård, I., Jurik, M.: A generalisation, a simplification and some applications of paillier's probabilistic public-key system. In: Kim, K.-c. (ed.) PKC 2001. LNCS, vol. 1992, pp. 119–136. Springer, Heidelberg (2001)

18. Erkin, Z., Franz, M., Guajardo, J., Katzenbeisser, Lagendijk, I., Toft, T.: Privacy-preserving face recognition. In: Privacy Enhancing Technologies (PET 2009). LNCS, vol. 5672, pp. 235–253. Springer, Heidelberg (2009)
19. Feigenbaum, J., Pinkas, B., Ryger, R.S., Saint-Jean, F.: Secure computation of surveys. In: EU Workshop on Secure Multiparty Protocols (SMP). ECRYPT (2004)
20. Fischlin, M.: A cost-effective pay-per-multiplication comparison method for millionaires. In: Naccache, D. (ed.) CT-RSA 2001. LNCS, vol. 2020, pp. 457–472. Springer, Heidelberg (2001)
21. Garay, J.A., Schoenmakers, B., Villegas, J.: Practical and secure solutions for integer comparison. In: Okamoto, T., Wang, X. (eds.) PKC 2007. LNCS, vol. 4450, pp. 330–342. Springer, Heidelberg (2007)
22. Gentry, C.: Fully homomorphic encryption using ideal lattices. In: ACM Symposium on Theory of Computing (STOC 2009), pp. 169–178. ACM, New York (2009)
23. Giry, D., Quisquater, J.-J.: Cryptographic key length recommendation (March 2009), http://keylength.com
24. Goldwasser, S., Micali, S.: Probabilistic encryption. Journal of Computer and System Sciences 28(2), 270–299 (1984)
25. Goyal, V., Mohassel, P., Smith, A.: Efficient two party and multi party computation against covert adversaries. In: Smart, N.P. (ed.) EUROCRYPT 2008. LNCS, vol. 4965, pp. 289–306. Springer, Heidelberg (2008)
26. Ishai, Y., Kilian, J., Nissim, K., Petrank, E.: Extending oblivious transfers efficiently. In: Boneh, D. (ed.) CRYPTO 2003. LNCS, vol. 2729, pp. 145–161. Springer, Heidelberg (2003)
27. Jarecki, S., Shmatikov, V.: Efficient two-party secure computation on committed inputs. In: Naor, M. (ed.) EUROCRYPT 2007. LNCS, vol. 4515, pp. 97–114. Springer, Heidelberg (2007)
28. Jarrous, A., Pinkas, B.: Secure hamming distance based computation and its applications. In: Applied Cryptography and Network Security (ACNS 2009). LNCS, vol. 5536, pp. 107–124. Springer, Heidelberg (2009)
29. Karatsuba, A., Ofman, Y.: Multiplication of many-digital numbers by automatic computers. Proceedings of the SSSR Academy of Sciences 145, 293–294 (1962)
30. Kerschbaum, F.: Practical privacy-preserving benchmarking. In: International Information Security Conference (SEC 2008), vol. 278, pp. 17–31 (2008)
31. Kolesnikov, V.: Gate evaluation secret sharing and secure one-round two-party computation. In: Roy, B. (ed.) ASIACRYPT 2005. LNCS, vol. 3788, pp. 136–155. Springer, Heidelberg (2005)
32. Kolesnikov, V., Sadeghi, A.-R., Schneider, T.: Improved garbled circuit building blocks and applications to auctions and computing minima. Cryptology ePrint Archive, Report 2009/411 (2009)
33. Kolesnikov, V., Schneider, T.: Improved garbled circuit: Free XOR gates and applications. In: Aceto, L., Damgård, I., Goldberg, L.A., Halldórsson, M.M., Ingólfsdóttir, A., Walukiewicz, I. (eds.) ICALP 2008, Part II. LNCS, vol. 5126, pp. 486–498. Springer, Heidelberg (2008)
34. Lindell, Y., Pinkas, B.: A proof of Yao's protocol for secure two-party computation. ECCC Report TR04-063, Electronic Colloquium on Computational Complexity, ECCC (2004)
35. Lindell, Y., Pinkas, B.: An efficient protocol for secure two-party computation in the presence of malicious adversaries. In: Naor, M. (ed.) EUROCRYPT 2007. LNCS, vol. 4515, pp. 52–78. Springer, Heidelberg (2007)

36. Lipmaa, H.: Verifiable homomorphic oblivious transfer and private equality test. In: Laih, C.-S. (ed.) ASIACRYPT 2003. LNCS, vol. 2894, pp. 416–433. Springer, Heidelberg (2003)
37. Malkhi, D., Nisan, N., Pinkas, B., Sella, Y.: Fairplay — a secure two-party computation system. In: USENIX (2004)
38. Naor, M., Pinkas, B.: Efficient oblivious transfer protocols. In: ACM-SIAM Symposium On Discrete Algorithms (SODA 2001). Society for Industrial and Applied Mathematics, pp. 448–457 (2001)
39. Naor, M., Pinkas, B., Sumner, R.: Privacy preserving auctions and mechanism design. In: ACM Conference on Electronic Commerce, pp. 129–139 (1999)
40. Nielsen, J.B., Orlandi, C.: Lego for two-party secure computation. In: Reingold, O. (ed.) TCC 2009. LNCS, vol. 5444, pp. 474–495. Springer, Heidelberg (2009)
41. Paillier, P.: Public-key cryptosystems based on composite degree residuosity classes. In: Stern, J. (ed.) EUROCRYPT 1999. LNCS, vol. 1592, pp. 223–238. Springer, Heidelberg (1999)
42. Paus, A., Sadeghi, A.-R., Schneider, T.: Practical secure evaluation of semi-private functions. In: Applied Cryptography and Network Security (ACNS 2009). LNCS, vol. 5536, pp. 89–106. Springer, Heidelberg (2009)
43. Qi, Y., Atallah, M.J.: Efficient privacy-preserving k-nearest neighbor search. In: International Conference on Distributed Computing Systems (ICDCS 2008), pp. 311–319. IEEE, Los Alamitos (2008)
44. Shaneck, M., Kim, Y., Kumar, V.: Privacy preserving nearest neighbor search. In: International Conference on Data Mining - Workshops (ICDMW 2006), pp. 541–545. IEEE, Los Alamitos (2006)
45. Yao, A.C.: Protocols for secure computations. In: Symposium on Foundations of Computer Science (SFCS 1982), pp. 160–164. IEEE, Los Alamitos (1982)
46. Yao, A.C.: How to generate and exchange secrets. In: IEEE Symposium on Foundations of Computer Science (FOCS 1986), pp. 162–167. IEEE, Los Alamitos (1986)

A   Input/Output Conversion Protocols

A.1   Private Inputs and Outputs

Private S Input: Inputs privately known to the circuit constructor S are easiest to deal with. For each of these inputs i, S sends the garbled value $\widetilde{w}_i^{v_i}$ corresponding to the plain value v_i to evaluator C. As described in [42], in case of semi-honest constructor (i.e., with no cut-and-choose), the inputs of S can also be securely incorporated into the garbled circuit. This optimization avoids to transfer any additional data for S's private inputs and the size of the GC can be reduced as well. However, in many applications it is beneficial even in the semi-honest scenario to separate conversion of the inputs from creation of the garbled circuit, as this allows S to create the garbled circuit in an offline pre-computation phase already before its private inputs are known.

Private C Input: For private inputs w_i of the evaluator C, both parties execute an OT protocol for each input bit in which constructor S inputs the two garbled t'-bit values $\widetilde{w}_i^0, \widetilde{w}_i^1$ and C inputs its plain value v_i to obtain $\widetilde{w}_i^{v_i}$ as output. For ℓ input bits, the OTs can be executed in a parallel $OT_{t'}^\ell$ protocol which can efficiently be extended to OT_t^t as described in 2.3.

Private S Output: If the output of the functionality is a private output w_i of the evaluator $\mathcal{C}$, constructor $\mathcal{S}$ provides $\mathcal{C}$ with the output decryption table for w_i, i.e., the permutation bit π_i chosen when creating the garbled value $\widetilde{w}_i^0 = \langle k_i^0, \pi_i \rangle$.

Private C Output: For private outputs w_i of the constructor $\mathcal{S}$, evaluator $\mathcal{C}$ does not get an output decryption table but sends the obtained permutation bit π_i of the obtained garbled value $\widetilde{w}_i = \langle k_i, \pi_i \rangle$ back to $\mathcal{S}$ who can deduce the corresponding plain value from this. Clearly, this works only if $\mathcal{C}$ is semi-honest as otherwise he could easily flip the output bit. This can be prevented by requiring $\mathcal{C}$ to send the output key k_i instead.

A.2   Secret-Shared Inputs and Outputs

Secret-Shared Input: As proposed in [19], a bit b can be secret-shared between $\mathcal{C}$ holding share $b_{\mathcal{C}}$ and $\mathcal{S}$ holding share $b_{\mathcal{S}}$, with $b = b_{\mathcal{C}} \oplus b_{\mathcal{S}}$. A secret-shared input bit b can be converted into a garbled input $\widetilde{b}$ using an $\mathrm{OT}_{t'}^\ell$ protocol: $\mathcal{C}$ inputs $b_{\mathcal{C}}$ and $\mathcal{S}$ inputs the two corresponding garbled values in the usual order $\widetilde{b}^0, \widetilde{b}^1$ if $b_{\mathcal{S}} = 0$ or swaps them to $\widetilde{b}^1, \widetilde{b}^0$ otherwise. It is easy to verify that $\mathcal{C}$ obliviously obtains the correct garbled value $\widetilde{b}$ for the shared bit b.

Secret-Shared Output: A similar method can be used for a secret-shared output bit b. $\mathcal{S}$ chooses a random share $b_{\mathcal{S}}$ and provides $\mathcal{C}$ with an output decryption table (cf. private output to $\mathcal{C}$) in the correct order in case $b_{\mathcal{S}} = 0$ or with swapped entries otherwise. $\mathcal{C}$ decrypts the garbled output to $b_{\mathcal{C}}$ which satisfies $b = b_{\mathcal{C}} \oplus b_{\mathcal{S}}$.

A.3   Homomorphically Encrypted Inputs and Outputs

In the scenario of secure two-party computation based on homomorphic encryption, one party, say client $\mathcal{C}$, generates a key-pair of the homomorphic encryption scheme and sends the (verifiably correct) public key and its inputs encrypted under the public key to $\mathcal{S}$. Afterwards, $\mathcal{S}$ can perform operations on the ciphertexts which result in corresponding operations on the encrypted plaintext data (using the homomorphic property of the cryptosystem). In order to compute operations that are not compatible with the homomorphic property (e.g., multiplication of two ciphertexts encrypted with an additively homomorphic encryption scheme), additional communication rounds must be performed. In the following we show how computing on homomorphically encrypted data can be combined with a garbled circuit to efficiently evaluate non-linear functions, such as comparison, minimum search, or other functionalities in a constant number of rounds.

Homomorphically Encrypted Input: If $\mathcal{S}$ holds an ℓ-bit value $[\![\mathbf{x}^\ell]\!]$, additively homomorphically encrypted under $\mathcal{C}$'s public key, this value can be converted into a garbled value $\widetilde{\mathbf{x}}^\ell$ output to $\mathcal{C}$ as follows: $\mathcal{S}$ chooses a random value r from the plaintext space P and adds this to the encrypted value: $[\![y]\!] = [\![\mathbf{x}^\ell + r]\!]$. In order to avoid an overflow, this requires that $\ell + \kappa \leq |P|$ for a statistical correctness parameter κ (e.g., $\kappa = 40$). $\mathcal{S}$ sends $[\![y]\!]$ to $\mathcal{C}$ who decrypts into y. Afterwards, both parties evaluate a garbled circuit which takes off the additive blinding: the private input of $\mathcal{S}$ into this garbled circuit are the ℓ least significant bits of r, $\mathbf{r}^\ell = r \mod 2^\ell$, and $\mathcal{C}$ inputs the ℓ least significant bits of y, $\mathbf{y}^\ell = y \mod 2^\ell$. The garbled circuit is an ℓ-bit subtraction circuit (cf. §3.1) which recovers the plaintext value from the blinded value as $\widetilde{x}^\ell = \widetilde{y}^\ell - \widetilde{r}^\ell$. This conversion protocol from additively homomorphically encrypted values into garbled values was used in [12,28]. A detailed proof and efficiency improvements (by packing together multiple values and converting the packed value at once) is given in [4].

Homomorphically Encrypted Output: This is similar to an homomorphically encrypted input (add random $(\ell + \sigma)$-bit mask in GC and remove it under homomorphic encryption) and included in the full version of this paper [32].

Multi Party Distributed Private Matching, Set Disjointness and Cardinality of Set Intersection with Information Theoretic Security

G. Sathya Narayanan[1,*], T. Aishwarya[1,**], Anugrah Agrawal[2],
Arpita Patra[3,***], Ashish Choudhary[3,†], and C. Pandu Rangan[3,‡]

[1] National Institute of Technology, Trichy, India
{sathya.phoenix,aishwarya.t}@gmail.com
[2] Institute of Technology, BHU, India
anugrah.agrawal.apm06@itbhu.ac.in
[3] Department of Computer Science and Technology, Indian Institute of Technology
Madras, Chennai India
{ arpitapatra_10,partho_31}@yahoo.co.in, prangan55@yahoo.com

Abstract. In this paper, we focus on the specific problems of Private Matching, Set Disjointness and Cardinality of Set Intersection in *information theoretic* settings. Specifically, we give *perfectly secure protocols* for the above problems in n party settings, tolerating a *computationally unbounded semi-honest* adversary, who can passively corrupt at most $t < n/2$ parties. To the best of our knowledge, these are the first such information theoretically secure protocols in a multi-party setting for all the three problems. Previous solutions for Distributed Private Matching and Cardinality of Set Intersection were *cryptographically* secure and the previous Set Disjointness solution, though information theoretically secure, is in a two party setting. We also propose a new model for Distributed Private matching which is relevant in a multi-party setting.

Keywords: Privacy preserving Set operations, Multiparty Computation.

1 Introduction

Consider the following problem: Alice has a set A of values and there exists an element $a \in A$. Bob also has a set of values B. Alice wants to check if her element a belongs to Bob's set B or not; i.e., if $a \in B$ or not. Alice does not want to reveal her element a to Bob and nor does Bob want Alice to know about any of

* Work supported by IIT-M Summer Fellowship 2009, IIT Madras.
** Work supported by Project No. CSE/05-06/075/MICO/CPAN on Foundation Research in Cryptography, sponsored by Microsoft Research, India.
*** Financial Support from Microsoft Research India Acknowledged.
† Financial Support from Infosys Technology India Acknowledged.
‡ Work supported by Project No. CSE/05-06/076/DITX/CPAN on Protocols for Secure Communication and Computation, sponsored by Department of Information Technology, Govt. of India.

J.A. Garay, A. Miyaji, and A. Otsuka (Eds.): CANS 2009, LNCS 5888, pp. 21–40, 2009.
© Springer-Verlag Berlin Heidelberg 2009

the elements in his set. Alice should ultimately learn if her element belongs to Bob's set or not and nothing more. And Bob should not learn anything (neither about Alice's value nor about its presence in his set). This is the *private matching problem*. In the distributed private matching problem proposed by Ye et al. [16], Bob's dataset B is distributed across n servers such that t or less servers cannot come together and reconstruct his dataset.

Consider another problem: Alice and Bob have sets A and B respectively and Alice wishes to find out if $A \cap B = \phi$. Alice and Bob also do not want to reveal any other information about their datasets to either party. The only information that Alice should gain is whether $A \cap B = \phi$ or not and Bob learns no new information. This is known as the *private set disjointness* test [6].

Suppose Alice and Bob have the sets A and B respectively and Alice wants to find out the cardinality of the set $A \cap B$. The solution should only reveal $|A \cap B|$ to Alice and should not reveal any more information about Bob's dataset to Alice and at the end of the solution, Bob should not gain any extra information about Alice's dataset. This is the *Cardinality Set-Intersection Problem* [3].

Distributed Private Matching has a lot of motivating examples from real life. For example, assume Alice has a highly sensitive information and wants to know if Bob has any record of the same. Bob, concerned about the security of his data and in order to cater to needs across the globe, has distributed all the information he has, in a database over n servers. Bob is willing to help Alice, but at the same time is not ready to reveal any other information that might help Alice get his dataset. Also, Alice does not want to reveal her sensitive information to Bob. For example, Alice could be a credit card service provider and Bob could have the set of all credit faulters from a single service. Alice might want to check if her customer belongs to bad credit union before agreeing to provide services and Bob would not want Alice to gain information about anyone else on the list. The Distributed Private Matching protocol gives solution to such problems.

As another example, suppose that a community social services centre has a list of drug abusers in the age group 11-19. A school administration in the local- ity wants to find out if its school is 'clean' or not. Since this is highly sensitive information the centre would not like to reveal any information or names in the list and would be willing to reveal only whether there are report cases of drug abuse from the school. This is an example of private set disjointness test. Again, if the school wants to know the number of students on the list and the centre is willing to reveal the number but does not want to divulge any more information, it becomes an example of the cardinality of set intersection.

Existing Literature: Private matching was introduced as a private *two-party* matching problem by Freedman et al.[3], who solved the problem under *cryp- tographic* assumptions, using oblivious polynomial evaluation with a public-key homomorphic encryption system. In the protocol, Alice has an element a and Bob has a dataset $B = \{b_1, b_2, \cdots, b_m\}$. At the end of the protocol, Alice gets to know if his element a belongs to Bob's Dataset B. The protocol does not reveal Alice's element a to Bob and Alice knows nothing more than whether $a \in B$. Ye, Wang and Pieprzyk [16] extended this problem to a Distributed scenario.

In Distributed Private Matching, Alice has a value a and Bob has his dataset $B = \{b_1, b_2, \cdots, b_m\}$ distributed among n servers such that t or less servers cannot discover Bob's Original Dataset while $t + 1$ or more servers can reconstruct Bob's dataset. In [16], a protocol for distributed private matching problem under *cryptographic* assumptions is provided.

Quite a few protocols exist for the set disjointness problem. Freedman et al. [3] proposed a protocol for two-party set disjointness under *cryptographic* assumptions, based on the representation of datasets as roots of a polynomial and oblivious polynomial evaluation techniques. The protocol reveals information about the cardinality of set intersection. It is very efficient against honest-but-curious adversaries but invokes expensive sub-protocols to work against malicious adversaries. Hohenberger and Weis [5] had used similar construction as in [3] and proposed a protocol in cryptographic setting (the security proof relies on the hardness of the discrete logarithm problem). Their protocol assumes an honest Alice while Bob can be malicious and again reveals information about the cardinality of the set intersection. Kiayias and Mitrofanova [7] proposed three protocols for set disjointness. The first protocol works on a relatively smaller domain for set disjointness, the second uses a new primitive called superposed encryption based on Pedersen commitments [13], the third uses a multi variate polynomial to reduce the high round complexity of the second protocol. Both [5] and [7] work in the two party setting. The first information theoretic solution to the private set disjointness problem is provided in [17]. The authors in [17] presented two protocols for two-party set disjointness using Sylvester matrices technique with round complexity of $O(1)$. While the first protocol is secure against honest-but-curious adversaries, the second protocol is secure against malicious adversaries.

The Cardinality Set-Intersection problem was previously studied in [3] in the two-party setting. In [7], Kiayias et al. studied private set disjointness as mentioned above which can be looked at as a restricted version of Cardinality of Set Intersection . In [8], Kisner et al. studied the problem in the multi-party setting and proposed efficient solutions for both honest-but-curious and malicious adversary under *cryptographic* assumptions, using zero knowledge proofs. Vaidya and Clifton [15] presented a protocol for cardinality of set intersection that is scalable and efficient in cryptographic settings and hence suitable for data mining applications.

Also, multi-party set intersection problems in information theoretic settings have been studied in [9] and [12]. Though there exist protocols for set intersection and in general for Multi-Party Computations, using them to solve set disjointness or cardinality of set intersection will be an overkill and highly inefficient. Our goal is to design efficient customised protocols as opposed to using generic abstract protocols.

Our Motivation and Contribution: From the literature, we find that existing solutions for private matching are in cryptographic settings. Also, the Distributed Private Matching proposed in [16] is essentially between two parties, where the data set of the second party is distributed among n servers. In this paper, we propose the first information theoretically secure protocol for Distributed Private Matching in the model proposed in [16]. We then propose a new model for Distributed Private Matching in a multi-party setting. The distributed private matching in our new model can be looked at as a general n-party Private Matching, where each party has a dataset and Alice has a value a and wants to know if a belongs to any of the n datasets. Here, the parties distribute their datasets among themselves such that t or less parties cannot come together and gain information about any honest party's dataset. Thus the parties themselves act as the servers used in the 2-party setting. The n-party Distributed Private Matching is useful in many scenarios. For example, suppose there are n trading agencies who store information about the available resources in a region. This information is sensitive and to ensure its safety, they share it among each other, so that any set of t or less agencies cannot get the information of any other agency. Now assume that Alice is a trader interested in setting up a factory over this region but needs to know if she can get the necessary resources for her production from any of the trading agencies. But Alice does not want to reveal her requirements to the agencies till she can get a confirmation that they will be of help. In such a case, n-party Distributed Private Matching is helpful. We also propose an information theoretically secure protocol for n-party Distributed Private Matching, secure against a semi-honest adversary.

Set Disjointness has been handled in information theoretic setting previously [17], but only in a 2-party setting. We provide the first multi party information theoretically secure protocol for set disjointness, secure against a semi-honest adversary. In our model, there are n parties where each party's dataset is distributed among the n parties, such that $t + 1$ or more parties need to come together to reconstruct the entire dataset, similar to our proposed model for n-party Distributed Private Matching.

Privacy law is the area of law concerned with the protection and preservation of the privacy rights of individuals. The law of privacy regulates the type of information which may be collected and how this information may be used. Many privacy rules and regulations like HIPAA, GLBA and SOX [10] exist that restrict companies from sharing their data as it is to other parties. For example, there could be a hospital database and there could be a vendor who wants to check if the technology used by his mobile results in complaints such as ear ache, headache etc. Also there could be a vendor who wants to check for multiple ailments which could result from the use of his product. So, he would like to find out if a threshold number of customers have complained of these ailments by checking with the hospital database. Also, the hospital's database would be governed by privacy rules like the ones mentioned above. Hence, this problem is an example for multi-party cardinality of set intersection.

The existing solutions for Cardinality of Set Intersection problem (both in 2-party and n-party setting) are in *cryptographic* settings. We provide the first multi party information theoretically secure protocol for Cardinality of Set Intersection problem, secure against a semi-honest adversary.

Hence our contribution in this paper is to provide information theoretically secure protocols for Private Matching, Set Disjointness and Cardinality Set-Intersection in a multi-party setting against a semi-honest adversary. We also show how to adapt our protocols to work against an active adversary in the same model. To the best of our knowledge, this is the first work to address these problems in a multi-party scenario, in information theoretic settings.

2 Model Definitions and Preliminaries

In this paper, we will be considering two different models. The first model is adapted from [16], while the second model is proposed by us. We provide a perfectly secure protocol for Two party distributed private matching problem in the first model, while we propose perfectly secure protocols for n-party distributed private matching, Cardinality of Set Intersection and Set Disjointness in the second model. We now briefly discuss these models. We also give the details of various existing sub-protocols, used in this paper.

2.1 Model for 2-Party Distributed Private Matching [16]

Here Alice and Bob are two parties. Alice has a secret value $a \in \mathbb{F}$ and Bob has a private dataset $B = \{b^{(1)}, \ldots, b^{(m)}\}$, consisting of m elements from a finite prime field $\mathbb{F}$, where $|\mathbb{F}| > n$. The dataset of Bob is distributed among n servers in a manner as explained in section 2.5, where $n \geq 2t + 1$. There exists a *passive adversary* with *unbounded computing power*, who can control at most t servers out of the n servers. We assume that Alice does not interact with Bob directly. Instead Alice contacts the set of n servers to perform the private-matching operations. We assume also that no server colludes with Alice to cheat and only Alice learns the output of any operation. More precisely, the following conditions should hold [16]:

1. CORRECTNESS: *If Alice and the servers honestly follow the steps of the protocol, then protocol works and Alice learns the correct result of the operation specified in the protocol.*

2. ALICE'S SECURITY : *If Alice is honest, then at the end of the protocol, the adversary controlling t servers should not get any information whatsoever about a.*

3. BOB'S SECURITY : *Provided that no server colludes with Alice, the protocol ensures that Alice does not get any extra information other than the output of the operation. In addition, any t or less servers should not be able to find out any information about Bob's dataset.*

2.2 Model for n-Party Distributed Private Matching, Set Disjointness and Cardinality of Set Intersection

Here we consider a complete synchronous network of n parties, denoted as $\mathcal{P} = \{P_1, \ldots, P_n\}$, who are pairwise connected by a secure channel. There exists a centralized adversary, having *unbounded computing power*, who can passively control at most $t < n/2$ parties. This is a valid assumption as information theoretic MPC against a computationally unbounded t-active passive adversary is possible iff $n \geq 2t + 1$ [2]. By passive adversary, we mean that all the parties under the control of adversary follow the prescribed steps of the protocol, but may try to learn *something extra* from the messages seen during the execution of the protocol. Each party P_i has a private data set $B_i = \{b^{(i,1)}, \ldots, b^{(i,m)}\}$, consisting of m elements from a finite prime field $\mathbb{F}$ where $|\mathbb{F}| > n$ (the protocols presented in this paper will also work if the number of elements in each data set is different). All computation and communication in our protocols are done over $\mathbb{F}$. To ensure the secrecy and distributed nature of datasets, each party P_i distributes his dataset among all other parties, as shown in Section 2.5. We now state the security definition, associated with n-party distributed private matching, Cardinality of Set Intersection and Set Disjointness.

Security Definition for n-Party Distributed Private Matching: Here Alice has an element $a \in \mathbb{F}$, whose presence she wants to check for in any of the n datasets. For this, she interacts with the n parties. As in [16], we assume that no party colludes with Alice and only Alice learns the output of any operation. More precisely, the following should hold as in [16]:

1. CORRECTNESS: *If Alice and the parties honestly follow the steps of the protocol, then protocol works and Alice learns the correct result of the operation specified in the protocol.*

2. ALICE'S SECURITY : *If Alice is honest, then at the end of the protocol, the adversary controlling t parties should not get any information whatsoever about a.*

3. PARTY'S SECURITY : *Provided that no party colludes with Alice, the protocol ensures that Alice does not get any extra information other than the output of the operation. In addition, if P_i is honest, then his dataset B_i is secure against a passive adversary controlling at most t parties.*

Security Definition for n-Party Set Disjointness: Here the n parties want to know whether $(B_1 \cap B_2 \cap \cdots \cap B_n) = \phi$ or not and nothing more. More specifically, the following conditions should be satisfied at the end of the protocol, even if $t < n/2$ parties are passively corrupted by a computationally unbounded adversary:

1. CORRECTNESS: *If the parties honestly follow the steps of the protocol, then they learn if $(B_1 \cap B_2 \cap \cdots \cap B_n) = \phi$ or not.*

2. PARTY'S SECURITY : *The adversary should not get any extra information about the input and output of honest parties, other than what can be inferred by the input of t corrupted parties (i.e., the dataset of these parties) and the output of t corrupted parties (which is $(B_1 \cap B_2 \cap \cdots \cap B_n) \stackrel{?}{=} \phi$).*

Security Definition for n-Party Cardinality of Set Intersection : Here the parties want to know $|B_1 \cap B_2 \cap \cdots \cap B_n|$ and nothing more. More specifically, the following conditions should be satisfied at the end of the protocol, even if $t < n/2$ parties are passively corrupted by a computationally unbounded adversary:

1. CORRECTNESS: *If the parties honestly follow the steps of the protocol, then the parties learn* $|B_1 \cap B_2 \cap \cdots \cap B_n|$.

2. PARTY'S SECURITY : *The adversary should not get any extra information about the input and output of honest parties, other than what can be inferred by the input of t corrupted parties (i.e., the dataset of these parties) and the output of t corrupted parties (which is* $|B_1 \cap B_2 \cap \cdots \cap B_n|$).

2.3   Sharing a Value s among n Parties

Consider the following problem: there exists a *dealer* $D \in \mathcal{P}$. D has a secret $s \in \mathbb{F}$, which he wants to share among $P_1, \ldots, P_n$, such that if t or less parties pool their shares, then they will know nothing about s. On the other hand, if $t + 1$ or more parties pool their shares, then they can reconstruct s. This problem is called *secret sharing*. One of the methods to solve this problem is Shamir Secret Sharing [14], where to share s, D chooses a random polynomial $f(x)$ of degree t, such that $f(0) = s$. D then gives P_i his share $s_i = f(\alpha_i)$, where each α_i is a publicly known distinct element from $\mathbb{F}$. To reconstruct s, each party produces his share s_i. Once all the n shares are available, anyone can interpolate the t degree polynomial $f(x)$ passing through (α_i, s_i)'s and hence reconstruct $s = f(0)$. It is easy to see that if t parties pool their shares, then they will know nothing about s [14].

d-Sharing and its Properties [1]: We say a value $s \in \mathbb{F}$ is *d-shared* among the parties in $\mathcal{P}$, if every (honest) party $P_i \in \mathcal{P}$ is holding a share s_i of s, such that there exists a degree-d polynomial $p(\cdot)$ with $p(0) = s$ and $p(\alpha_i) = s_i$ for every $P_i \in \mathcal{P}$. The vector of shares $(s_1, \ldots, s_n)$ is called a *d-sharing* of s, and is denoted by $[s]_d$. *In the rest of the paper, whenever we say that the parties have $[s]_d$ for some $s \in \mathbb{F}$, we mean to say that each party is holding his share corresponding to d-sharing of s.*

Shamir sharing is a t-sharing scheme and generates t-sharing $[s]_t$ of secret s. Notice that Shamir sharing satisfies the following properties:

1. $[a]_t + [b]_t = [a + b]_t$, for any $a, b \in \mathbb{F}$.
2. $[a]_t[b]_t = [ab]_{2t}$, for any $a, b \in \mathbb{F}$.

Thus if the parties hold $[a]_t$ and $[b]_t$, then they can locally generate $[a + b]_t$, without doing any communication, by simply adding their respective shares of a and b. On the other hand, if the parties simply multiply their respective shares of a and b, then this will generate $[ab]_{2t}$. To generate $[ab]_t$ from $[a]_t$ and $[b]_t$, we need to use the multiplication protocol specified in section 2.4.

2.4   Multiplying Shared Values

Let a and b be two values, which are Shamir shared (i.e., t-shared) among $P_1, \ldots, P_n$ using degree-t polynomials $f(x)$ and $g(x)$ respectively. Thus party P_i has shares a_i and b_i of a and b respectively. Then the parties $P_1, \ldots, P_n$ can generate the Shamir shares of $c = ab$ by using the multiplication protocol of [4] as follows: Let $d_i = a_i b_i$. Each party P_i Shamir share d_i, say using degree-t polynomial $h_i(x)$. This results in party P_i holding the share-share $d_{1i}, \cdots, d_{ni}$ of $d_1, \ldots, d_n$ respectively. Then from Lagrange's interpolation, the degree-t polynomial $h(x) = \sum_{i=1}^{n} w_i h_i(x)$ is the polynomial that Shamir shares c, where

$$ w_i = \prod_{j=1, j \neq i}^{n} \alpha_j / (\alpha_j - \alpha_i) \qquad (1) $$

To get j^{th} share of c, party P_j computes $c_j = h(\alpha_j) = \sum_{i=1}^{n} w_i h_i(\alpha_j) = \sum_{i=1}^{n} w_i d_{ij}$. It is easy to see that during this process, an adversary passively controlling at most t parties does not get any information about a, b and c [4]. Also, the method works only if $n > 2t$ which holds in our case.

Lemma 1. *The above multiplication protocol communicates $\mathcal{O}(n^2)$ field elements and takes one round of communication.*

PROOF: In the protocol, each party Shamir shares a value, which involves a communication complexity of $O(n)$ field elements and one round of communication. Hence the lemma.   □

Lemma 2. *Suppose the parties have $[a^{(1)}]_t, \ldots, [a^{(\ell)}]_t$ and $[b^{(1)}]_t, \ldots, [b^{(\ell)}]_t$, where each $a^{(l)}$ and $b^{(l)}$ belongs to $\mathbb{F}$ and $\ell \geq 1$. Then the parties can generate $[a^{(l)} b^{(l)}]_t$, for $l = 1, \ldots, \ell$ in 1 communication round using the above multiplication protocol. On the other hand the parties can generate $[a^{(1)} \ldots a^{(\ell)}]_t$ in $\log_2 \ell$ communication rounds.*

PROOF: Computing $[a^{(l)} b^{(l)}]_t$ for $l = 1, \ldots, \ell$ will require 1 round because each $a^{(l)} b^{(l)}$ is independent of the other and we can generate these products in parallel. On the other hand, generating $[a^{(1)} \ldots a^{(\ell)}]_t$ requires $\log_2 \ell$ communication rounds because we can multiply two operands at a time, say a_i and $a_{i + \lfloor \frac{\ell}{2} \rfloor}$ for $i = 1, \cdots, \lfloor \frac{\ell}{2} \rfloor$, and find their products in the first round and then make pairs among the resulting products(after the first round) and multiply them in the next round and so on.   □

2.5   Dataset Distribution of Parties

In both the models, namely the one presented in section 2.1 and section 2.2, the parties distribute their dataset in a specific manner. We now give the details of how this is done.

Dataset Distribution for Two Party Distributed Private Matching:
Here Bob on having the data set $B = \{b^{(1)}, \ldots, b^{(m)}\}$ does the following: Bob forms a polynomial $F(x)$ such that the elements of his set B are roots of the polynomial; i.e., $F(x) = \prod_{i=1}^{m}(x - b^{(i)}) = \sum_{i=0}^{m} C_i x^i$, such that $C_m = 1$. Bob then Shamir shares each C_i among the n servers. It is easy to see that even if t or less servers combine their shares, they will have no information about B. On the other hand, B can be reconstructed by pooling the shares of any $t + 1$ or more servers.

Dataset Distribution for n-party Distributed Private Matching: Here each party P_i on having a dataset $B_i = \{b^{(i,1)}, \ldots, b^{(i,m)}\}$ distributes it in the following way: P_i forms a polynomial $F_i(x)$ such that the elements of his set B_i are roots of the polynomial; i.e., $F_i(x) = \prod_{j=1}^{m}(x - b^{(i,j)}) = \sum_{j=0}^{m} C^{(i,j)} x^j$, such that $C^{(i,m)} = 1$. Party P_i then Shamir shares each $C^{(i,j)}$ among the n parties. It is easy to see that even if t or less parties combine their shares, they will have no information about B_i. On the other hand, B_i can be reconstructed by pooling the shares of any $t + 1$ or more servers.

Dataset Distribution for n-party Set Disjointness and Cardinality of Set Intersection: Here each party P_i on having dataset $B_i = \{b^{(i,1)}, \ldots, b^{(i,m)}\}$, distributes it in the following way: for $j = 1, \ldots, m$, party P_i Shamir shares $b^{(i,j)}$ among the parties in $\mathcal{P}$. Since each element in the dataset is individually shared using a t-degree polynomial, it implies that if P_i is honest, then each element of his dataset B_i is secure against a passive adversary controlling at most t parties. Moreover, any set of $t + 1$ or more parties can reconstruct B_i by pooling their shares.

2.6   Checking If a Shared Value Is Zero

Nishide and Ohta [11] present an efficient and deterministic protocol to check if a shared value is zero or not. More specifically, the protocol takes $[s]_t$ as input, where s is shared using Shamir sharing and outputs the following:

1. If $s = 0$, then the protocol generates $[1]_t$.
2. If $s \neq 0$, then the protocol generates $[0]_t$.

The protocol performs $81l$ multiplications of shared values, where $l = \log(|\mathbb{F}|)$ and takes 8 rounds. In the rest of the paper, we use this protocol for testing if a shared value is zero. We shall henceforth refer to this protocol as TEST-IF-ZERO.

Remark 1. The TEST-IF-ZERO protocol of [11] is a deterministic protocol, without any error, which we use in the rest of this paper. A drawback of this protocol is that it performs very large number of multiplications. In the last section of this paper, we present a simple protocol for testing if a shared value is zero, involving significantly less number of multiplications. However, this protocol is probabilistic and gives the correct output, except with an error probability of $\frac{1}{|\mathbb{F}|}$.

3 Two-Party Distributed Private Matching Protocol

Recall that in the 2-party distributed private matching, Bob has a private data set B of m elements, which he has distributed among n servers, say $S_1, \ldots, S_n$, as explained in section 2.5. Alice has a secret element $a \in \mathbb{F}$, whose presence she wants to check in Bob's dataset. For this she interacts with the servers. We now present a perfectly secure protocol for this problem. Before proceeding further, we give the following trivial lemma:

Lemma 3. *The value a belongs to B iff $F(a) = 0$, where $F(x) = \prod_{i=1}^{m}(x - b^{(i)}) = \sum_{i=0}^{m} C_i x^i$.*

PROOF: The proof is obvious and it follows from the definition of $F(x)$.   □

The high level idea of the protocol is as follows: Alice first Shamir shares the values $a, \ldots, a^m$ among n servers. Hence all the servers, apart from having the shares of the coefficients of $F(x)$, now also have the shares of $a, \ldots, a^m$. The servers then compute the Shamir shares of $V_j = C_j a^j$ for $0 \le j \le m$, using the multiplication protocol (see section 2.4). Since $F(a) = \sum_{j=0}^{m} V_j$, by a linear combination of all the shares that a server has, each server gets his share of $F(a)$. Till this point, the servers have generated the Shamir shares of $F(a)$. Now if the servers give their shares of $F(a)$ to Alice, then Alice could reconstruct $F(a)$ and find whether a belongs to B. But directly revealing $F(a)$ to Alice will violate Bob's security, as Alice would come to know about one point on $F(x)$.

Since Alice wants to know only if $a \in B$ or not, all we need to find out is if $F(a) = 0$ or not. For this, all the servers run the TEST-IF-ZERO protocol on the shares of $F(a)$ and reconstruct the output towards Alice. If the output is one then Alice concludes that $F(a) = 0$, otherwise $F(a) \neq 0$. Accordingly, Alice concludes that a belongs (does not belong) to B. The protocol called 2-party DPMP is formally given in Table 1. Before proceeding further to prove the properties of 2-Party DPMP, we make the following claim.

Claim. In protocol 2-Party DPMP if Alice is honest, then a passive adversary controlling at most t servers does not get any information about a even after knowing t shares of $a, \ldots, a^m$.

PROOF: The proof follows easily from the properties of Shamir sharing and simple linear algebra. For a complete proof, see **APPENDIX A**.   □

Lemma 4. *Protocol 2-party DPMP satisfies the properties of 2-party distributed private matching.*

PROOF: The CORRECTNESS property is trivial. The secrecy of Bob's dataset against a passive adversary controlling at most t servers follows from the properties of Shamir sharing. The secrecy of Bob's data set against a passive Alice follows from the secrecy of TEST-IF-ZERO. Finally, secrecy of Alice's a follows from Claim 3.   □

Lemma 5. *Protocol 2-party DPMP communicates $\mathcal{O}(n^2 m)$ field elements and involves one invocation of TEST-IF-ZERO. The protocol takes two rounds.*

PROOF: In the setup phase, the parties communicates $\mathcal{O}(nm)$ field elements for data set distribution. In the computation phase, there are m multiplications and hence it communicates $\mathcal{O}(n^2m)$ field elements. Since all the multiplications are independent, by lemma 2 it can be done in parallel in one round. Moreover, setup phase takes one round.   □

Table 1. Protocol for 2-party DPMP

2-Party DPMP
Setup Phase:
Alice Shamir shares $a, \ldots, a^m$ among n servers. Bob distributes his dataset $B = \{b^{(1)}, \ldots, b^{(m)}\}$ among n servers as explained in section 2.5. Let $F(x) = \prod_{j=1}^{m}(x - b^{(j)}) = \sum_{j=0}^{m} C_j x^j$. Thus the parties have $[C_j]_t$ for $j = 0, \ldots, m$.
Computation (by each server):
1. The servers compute $[V_j]_t = [C_j a^j]_t$ for $0 \le j \le m$ using the multiplication protocol described in section 2.4.
2. The servers then compute $[F(a)]_t = [V_0]_t + \ldots + [V_m]_t$.
3. Finally the servers run the protocol TEST-IF-ZERO on $[F(a)]_t$ to generate $[v]_t$, where $v = 1(0)$, if $F(a) = 0(\ne 0)$.
Reconstruction Phase:
The servers give their shares of v to Alice. Alice reconstruct v and checks if $v = 1$. If $v = 1$, then $a \in B$ else $a \notin B$.

4   n-Party Distributed Private Matching Protocol

We now present a perfectly secure protocol called n-party DPMP for distributed private matching in n-party settings. For this, we use the model presented in section 2.2. Recall that in this model, there are n parties denoted as $\mathcal{P} = \{P_1, \ldots, P_n\}$, where each P_i has a private dataset $B_i = \{b^{(i,1)}, \ldots, b^{(i,m)}\}$ represented using $F_i(x) = \prod_{j=1}^{m}(x - b^{(i,j)}) = \sum_{j=0}^{m} C^{(i,j)} x^j$. Moreover, the dataset B_i is Shamir shared among the n parties; i.e., the parties hold $[C^{(i,j)}]_t$, for $i = 1, \ldots, n$ and $j = 0, \ldots, m$ (see Section 2.5). Alice has a secret element a. Alice wants to know if $a \in (B_1 \cup \ldots \cup B_n)$. Before proceeding further, we give the following lemma.

Lemma 6. $a \in (B_1 \cup B_2 \cup \cdots \cup B_n)$ iff atleast one of the $F_i(a) = 0$.

The high level idea of protocol n-party DPMP is as follows: Alice first Shamir shares the values $a, \ldots, a^m$ among n servers. Thus parties hold $[a^j]_t$ for $j = 1, \ldots, m$. The parties then compute $[V^{(i,j)}]_t = [C^{(i,j)} a^j]_t$ using the multiplication protocol. The parties then compute the Shamir shares of $[F_i(a)]_t = \sum_{j=0}^{m} [V^{(i,j)}]_t$. Since shamir sharing is linear, this can be done locally. The parties then compute $[F(a)]_t$ where

$$F(a) = \prod_{i=1}^{n} F_i(a) \tag{2}$$

After this step, the parties have $[F(a)]_t$. To ensure that no more information is revealed to Alice than what is necessary, the parties run the TEST-IF-ZERO protocol on $[F(a)]_t$ and reconstruct the output towards Alice, so that Alice gets to know only if $F(a)$ is zero or not and nothing more. Alice checks if the reconstructed value is 0 or not to find if $a \in (B_1 \cup B_2 \cup \cdots \cup B_n)$. Protocol n-party DPMP is formally given in the following table.

---

n-Party DPMP

Setup Phase:
Alice shamir shares $a, a^2, a^3, \cdots, a^m$ among the n parties and each party P_i for $1 \le i \le n$ distributes his dataset B_i using the polynomial $F_i(x) = \sum_{j=0}^{m} C^{(i,j)} x^j$ among n parties using Dataset Distribution scheme described in section 2.5. Thus the parties have $[a^j]_t$ for $1 \le j \le m$ and $[C^{(i,j)}]_t$ for $1 \le i \le n$ and $0 \le j \le m$.

Local Computation (by each party):

1. The parties compute $[V^{(i,j)}]_t = [C^{(i,j)} a^j]_t$ for $0 \le j \le m$ and $1 \le i \le n$ and compute $[F_i(a)]_t = \sum_{j=0}^{m} [V^{(i,j)}]_t$.

2. The parties compute $[F(a)]_t = \prod_{i=1}^{n} [F_i(a)]_t$ by running the multiplication protocol specified in section 2.4.

3. The parties now run the protocol TEST-IF-ZERO on $[F(a)]_t$ to generate $[v]_t$, where $v = 1(0)$, if $F(a) = 0 (\neq 0)$.

Reconstruction Phase: The parties give their shares of v to Alice. Alice reconstructs v and checks if $v = 1$. If $v = 1$, then $a \in B_1 \cup B_2 \cup \cdots \cup B_n$ else $a \notin B_1 \cup B_2 \cup \cdots \cup B_n$.

Lemma 7. *Protocol n-party DPMP satisfies the properties of n-party distributed private matching.*

PROOF: Follows directly from the protocol steps and properties of Shamir sharing and TEST-IF-ZERO protocol. □

Lemma 8. *Protocol n-party DPMP communicates $\mathcal{O}(n^3 m)$ field elements and executes one instance of* TEST-IF-ZERO. *The protocol involves $\mathcal{O}(\log(n))$ communication rounds.*

PROOF: The communication complexity is easy to analyze. In setup phase, $\mathcal{O}(nm)$ values are t shared and it communicates $\mathcal{O}(n^2 m)$ field elements. During computation phase, in step 1, we first do nm multiplications simultaneously which can be done in 1 round. In step 2, n shared values need to be multiplied. Since these values are dependent, the multiplications totally take $\mathcal{O}(\log(n))$ communication rounds and communicates $\mathcal{O}(n^3 m)$ field elements by lemma 1 and lemma 2. □

5 n-Party Set Disjointness Protocol

We now present a perfectly secure protocol called n-party Set Disjointness for n-party set disjointness problem. Recall that in this problem, there are n parties $\mathcal{P} = \{P_1, \ldots, P_n\}$, where each P_i has a private dataset $B_i = \{b^{(i,1)}, \ldots, b^{(i,m)}\}$. Moreover, each $b^{(i,j)}$ is Shamir shared among the n parties for $j = 1, \ldots, m$. Thus the parties hold $[b^{(i,j)}]_t$. The parties want to know if $(B_1 \cap \ldots \cap B_n) = \phi$ or not. We first present a protocol called Gen-E^l, which we will later use in solving the n-party set disjointness as well as cardinality of set intersection problem.

5.1   Protocol Gen-E^l

In this section we give a protocol that helps in solving the set cardinality and disjointness problems. We call this protocol as Gen-E^l, which generates shares of E^l (which we will define subsequently). We first observe that, $(B_1 \cap \cdots \cap B_n) \subseteq B_r$ for any $1 \leq r \leq n$. Hence we fix a party $P_r \in \mathcal{P}$ as reference and refer to the elements of his dataset as $\{a^1, \ldots, a^m\}$ for convenience. Now we check if some element of B_r is present in the the dataset of each party in $P \setminus P_r$. If there exists any such element $a^l \in \{a^1, \ldots, a^m\}$, then the sets B_i's are not disjoint. Protocol Gen-E^l checks for the presence of any such element a^l.

Gen-E^l

Setup Phase:

1. Each party P_i on having dataset $B_i = \{b^{(i,1)}, \ldots, b^{(i,m)}\}$ Shamir shares each $b^{(i,j)}$. Thus the parties hold $[b^{(i,j)}]_t$, for $1 \leq i \leq n$ and $1 \leq j \leq m$.

2. The parties fix one arbitrary party among them as the reference, say, the lowest index party, P_1. For the ease of presentation, let $\{a^1, \ldots, a^m\}$ denote the element of P_1's dataset.

Computation (by each party):

1. The parties first compute $[V^{(l,i,j)}]_t = [b^{(i,j)}]_t - [a^l]_t$ for $i = 2, \ldots, n, j = 1 \ldots, m$ and $l = 1 \ldots, m$.

2. The parties use the multiplication protocol to generate $[C^{(i,l)}]_t = \prod_{j=1}^{j=m} [V^{(l,i,j)}]_t$ for $i = 2, \ldots, n$ and $l = 1 \ldots, m$.

3. The parties then compute $[E^l]_t = \sum_{i=2}^{i=n} [C^{(i,l)}]_t$.

Lemma 9. *If $E^l = 0$ is zero, then the element $a_l \in B_1$ is also present in $(B_1 \cap \cdots \cap B_n)$.*

PROOF : The proof follows easily from the protocol steps.   □

Lemma 10. *Protocol Gen-E^l communicates $\mathcal{O}(n^3 m^2)$ field elements in setup phase. The protocol takes $\mathcal{O}(\log(nm^2))$ communication rounds.*

PROOF: The communication complexity and the number of multiplications is easy to analyze. Since there are nm^2 multiplications and all of them are dependent, from Lemma 2, the number of rounds needed to perform these multiplications is $\mathcal{O}(\log(nm^2))$.   □

5.2   Protocol for Multi Party Set Disjointness

To compute multi party set disjointness, the parties first execute protocol Gen-E^l to generate $[E^l]_t$ for $l = 1, \ldots, m$. The parties then compute $[E]_t = \prod_{l=1}^{l=m}(E^l)$ by using the multiplication protocol. The parties then execute TEST-IF-ZERO on $[E]_t$ to generate $[v]_t$. Finally v is reconstructed by each party, after which the parties conclude whether the sets are disjoint or not.

---

n-party Set Disjointness

1. The parties run protocol Gen-E^l to generate $[E^l]_t$ for $l = 1, \ldots, m$.
2. The parties now run multiplication protocol to compute $[E]_t = \prod_{l=1}^{l=m}([E^l]_t)$.
3. The parties then run the TEST-IF-ZERO protocol on $[E]_t$ to generate $[v]_t$.

Reconstruction Phase:

The parties produce their respective share of v. Once v is reconstructed, the parties check if $v = 0$ or 1. If $v = 0$, then $(B_1 \cap \cdots \cap B_n) = \phi$ else $(B_1 \cap \cdots \cap B_n) \neq \phi$.

Lemma 11. *Protocol n-party Set Disjointness satisfies the properties of n-party set disjointness.*

PROOF: The proof follows easily from the protocol steps, properties of protocol Gen-E^l and Shamir secret sharing.   □

Lemma 12. *Protocol n-party Set Disjointness communicates $\mathcal{O}(n^3m^2 + n^2m)$ field elements, and executes one instance of TEST-IF-ZERO. The protocol takes $\mathcal{O}(\log(nm^2) + \log(m))$ communication rounds.*

PROOF: Communication complexity is easy to analyze. In the protocol, Gen-E^l performs nm^2 multiplications. Computing $[E]_t$ require m multiplications. Hence the total number of multiplications done is $(nm^2 + m)$. Since these multiplications are dependent, we require $\mathcal{O}(\log(nm^2) + \log(m))$ rounds to perform them. It is easy to see that protocol executes one instance of TEST-IF-ZERO.   □

6 n-Party Cardinality of Set Intersection

We now present a perfectly secure protocol called n-party Cardinality of Set Intersection for n-party cardinality of set intersection. Recall that in this problem, there are n parties $\mathcal{P} = \{P_1, \ldots, P_n\}$, where each P_i has a private dataset $B_i = \{b^{(i,1)}, \ldots, b^{(i,m)}\}$. Moreover, each $b^{(i,j)}$ is Shamir shared among the n parties for $j = 1, \ldots, m$. Thus the parties hold $[b^{(i,j)}]_t$. The parties want to only know the value $|B_1 \cap \cdots \cap B_n|$ and nothing more. Protocol n-party Cardinality of Set Intersection is formally given in the following table.

---

n-party Cardinality of Set Intersection

1. The parties run the protocol Gen-E^l to generate the $[E^l]_t$ for $l = 1, \ldots, m$.

2. For $l = 1, \ldots, m$, the parties run TEST-IF-ZERO on $[E^l]_t$ to generate $[v^l]_t$.

3. The parties then compute $[v]_t = \sum_{l=1}^{l=m} [v^l]_t$.

Reconstruction Phase: All the parties produce their respective shares of v to reconstruct v. The value v is $|B_1 \cap \cdots \cap B_n|$.

Lemma 13. *Protocol n-party Cardinality of Set Intersection satisfies the properties of n-party cardinality of set intersection.*

PROOF: The security of honest parties' datasets is satisfied because the elements of the datasets of honest parties are Shamir shared and hence are safe against a passive adversary having control over t parties. Also the final outcome of the protocol is only the cardinality of $(B_1 \cap \cdots \cap B_n)$ and not anything more. The correctness property follows from Lemma 9 and protocol steps.   □

Lemma 14. *Protocol n-party Cardinality of Set Intersection communicates $\mathcal{O}(n^3 m^2)$ field elements and invokes m instances of TEST-IF-ZERO protocol. The protocol has a round complexity of $\mathcal{O}(\log(nm^2))$.*

PROOF: The protocol communicates $\mathcal{O}(n^2 m)$ field elements for sharing the datasets and does nm^2 multiplications. Since all the multiplications are dependent, from Lemma 2, it takes $\mathcal{O}(\log(nm^2))$ communication rounds to perform them. From protocol, we can clearly see that it involves m invocations of TEST-IF-ZERO protocol.   □

7   A Simple Protocol for Checking If a Shared Value Is Zero

In section 2.6, we gave the details of protocol TEST-IF-ZERO, which checks whether a shared value is zero or not. The protocol is used in all our protocols as a black-box. However, protocol TEST-IF-ZERO involves a large number of multiplications. We now present a protocol NEW-TEST-IF-ZERO, which takes $[a]_t$ as an input and produces $[V]_t$ as the output, where $a \in \mathbb{F}$ is a random value. The protocol has the following properties:

1. If $a = 0$, then $V = 0$.
2. If $V = 0$, then except with probability $\frac{1}{|\mathbb{F}|}$, $a = 0$.
3. If $a \neq 0$ then except with probability $\frac{1}{|\mathbb{F}|}$, V can be any random non-zero value.
4. Even if V is reconstructed by each party, a passive adversary controlling at most t parties/servers will have no information about a.
5. The protocol performs significantly less number of multiplications in comparison to protocol TEST-IF-ZERO.

The protocol is formally given in the following table.

NEW-TEST-IF-ZERO
Setup Phase:
1. Each party P_i chooses a random value $r^{(i)} \in \mathbb{F}$ and Shamir shares $r^{(i)}$. So now the parties have $[a]_t$ and $[r^{(i)}]_t$ for $1 \leq i \leq n$.
Computation (by each party):
1. The parties compute $[R]_t = \sum_{i=1}^{n} [r^{(i)}]_t$.
2. The parties compute $[V]_t = [R]_t [a]_t$ by using the multiplication protocol.
Reconstruction Phase:
The parties reconstruct the value V. If V is non-zero, then the parties conclude that a is also non-zero. If V is 0, then the parties conclude that a is 0 with very high probability.

Lemma 15. *Protocol* NEW-TEST-IF-ZERO *satisfies all the properties mentioned above.*

PROOF: Before proceeding further, we first note that $R = \sum_{i=1}^{n} r^{(i)}$ is completely random. This is because at least one $r^{(i)}$ in R is shared by an honest P_i and hence $r^{(i)}$ is random, implying that R is also random. It is easy to see that if $a = 0$ then $V = Ra$ will be also zero. Thus if $V = 0$, the probability that $a \neq 0$ is same as the probability $R = 0$, which is $\frac{1}{|\mathbb{F}|}$. Moreover, if $a \neq 0$, then except with probability $\frac{1}{|\mathbb{F}|}$, V can be any random value. This is because R can be any random value. If a passive adversary knows t shares of a, then even after knowing $V = Ra$, the value a remains information theoretically secure due to the random R. Finally it is easy to see that the protocol performs only one multiplication.   □

7.1 Application of Protocol NEW-TEST-IF-ZERO

By seeing the properties of protocol NEW-TEST-IF-ZERO, we find that it can be used as a substitute of TEST-IF-ZERO in any protocol, where we just want to know whether the shared value a is zero or not. In protocols 2-Party DPMP, n-party DPMP and n-party Set Disjointness, protocol TEST-IF-ZERO was used to just check whether a shared value a is zero or not. So we can replace TEST-IF-ZERO with our new protocol NEW-TEST-IF-ZERO in 2-Party DPMP, n-party DPMP and n-party Set Disjointness. Since NEW-TEST-IF-ZERO requires less number of multiplications than TEST-IF-ZERO, the resultant protocols for 2-Party DPMP, n-party DPMP and n-party Set Disjointness becomes more efficient. However, since protocol NEW-TEST-IF-ZERO involves a negligible error probability, the resultant protocols for 2-Party DPMP, n-party DPMP and n-party Set Disjointness will also now involve a negligible error probability in correctness.

Notice that we cannot use our new protocol NEW-TEST-IF-ZERO as a substitute of TEST-IF-ZERO in protocol n-party Cardinality of Set Intersection.

This is because if $a \neq 0$, then NEW-TEST-IF-ZERO outputs $[V]_t$, where except with probability $\frac{1}{|\mathbb{F}|}$, V can be any random non-zero value. On the other hand, protocol TEST-IF-ZERO would output $[0]_t$ in this case. Now in protocol n-party Cardinality of Set Intersection , the parties added the outcome of each instance of TEST-IF-ZERO to count the number of elements in the intersection of n sets. However, the parties cannot do so if TEST-IF-ZERO is replaced by NEW-TEST-IF-ZERO.

8   Some Optimization Issues

8.1   Further Reduction of Communication Complexities

In all our protocols, we have used the multiplication protocol of [4], which communicates $\mathcal{O}(n^2)$ field elements to generate the t-sharing of the product of two t-shared values, in the presence of a computationally unbounded passive adversary, controlling $t < n/2$ parties/servers. To further optimise the communication complexity of our protocols, we can use the multiplication protocol in [1]. In [1] the authors have presented a perfectly secure multiplication protocol tolerating a t-active *malicious* adversary. The MPC protocol of [1], when executed with $n = 2t + 1$ parties, in the presence of a *passive* adversary, controlling at most t parties, will communicate $\mathcal{O}(n)$ field elements to generate the t-sharing of the product of two t-shared values. Moreover, the protocol will take $\mathcal{O}(1)$ communication rounds.

8.2   Adapting Our Protocols to Work against Malicious Adversary

All our protocols can be extended to work against a t-active malicious adversary[1], having unbounded computing power, by doing the following steps:

1. Taking $n = 3t + 1$, instead of $n = 2t + 1$. This is required because from [2], secure computation tolerating an all powerful, t-active malicious adversary is possible iff $n \geq 3t + 1$.
2. Using *Verifiable Secret Sharing* (VSS) [2] instead of Shamir's secret sharing scheme for $n = 3t + 1$.
3. Using multiplication protocol of [1] secure against a malicious adversary.
4. Modifying the TEST-IF-ZERO protocol as suggested in [11] to work against a malicious adversary.

9   Conclusion and Open Problems

In this paper, we have given perfectly secure protocols for private matching, set disjointness and cardinality of set intersection problems in information theoretic settings, secure against a computationally unbounded passive adversary.

---

[1] A malicious adversary takes complete control of the parties under its control and can make them behave in any arbitrary fashion during the protocol execution.

Future work would be to come up with efficient protocols that can work against more powerful adversaries such as byzantine and mixed adversaries. The security model can be made stronger by allowing collusion between Alice and the servers; protocols that work in this model are interesting open problems. Also, improving the communication complexity and efficiency of the protocols presented in this paper are other interesting future directions.

References

1. Beerliová-Trubíniová, Z., Hirt, M.: Perfectly-secure mpc with linear communication complexity. In: TCC, pp. 213–230 (2008)
2. Ben-Or, M., Goldwasser, S., Wigderson, A.: Completeness theorems for non-cryptographic fault-tolerant distributed computation (extended abstract). In: STOC, pp. 1–10 (1988)
3. Freedman, M.J., Nissim, K., Pinkas, B.: Efficient private matching and set intersection. In: Cachin, C., Camenisch, J.L. (eds.) EUROCRYPT 2004. LNCS, vol. 3027, pp. 1–19. Springer, Heidelberg (2004)
4. Gennaro, R., Rabin, M.O., Rabin, T.: Simplified vss and fact-track multiparty computations with applications to threshold cryptography. In: PODC, pp. 101–111 (1998)
5. Hohenberger, S., Weis, S.A.: Honest-verifier private disjointness testing without random oracles. In: Danezis, G., Golle, P. (eds.) PET 2006. LNCS, vol. 4258, pp. 277–294. Springer, Heidelberg (2006)
6. Kiayias, A., Mitrofanova, A.: Testing disjointness of private datasets. In: S. Patrick, A., Yung, M. (eds.) FC 2005. LNCS, vol. 3570, pp. 109–124. Springer, Heidelberg (2005)
7. Kiayias, A., Mitrofanova, A.: Syntax-driven private evaluation of quantified membership queries. In: Zhou, J., Yung, M., Bao, F. (eds.) ACNS 2006. LNCS, vol. 3989, pp. 470–485. Springer, Heidelberg (2006)
8. Kissner, L., Song, D.X.: Privacy-preserving set operations. In: Shoup, V. (ed.) CRYPTO 2005. LNCS, vol. 3621, pp. 241–257. Springer, Heidelberg (2005)
9. Li, R., Wu, C.: An unconditionally secure protocol for multi-party set intersection. In: Katz, J., Yung, M. (eds.) ACNS 2007. LNCS, vol. 4521, pp. 226–236. Springer, Heidelberg (2007)
10. Natan, R.B.: Implementing Database Security and Auditing. Elsevier, Amsterdam (2005)
11. Nishide, T., Ohta, K.: Multiparty computation for interval, equality, and comparison without bit-decomposition protocol. In: Okamoto, T., Wang, X. (eds.) PKC 2007. LNCS, vol. 4450, pp. 343–360. Springer, Heidelberg (2007)
12. Patra, A., Choudhary, A., Rangan, C.P.: Information theoretically secure multi party set intersection re-visited. Cryptology ePrint Archive, Report 2009/116 (2009), http://eprint.iacr.org/
13. Pedersen, T.P.: Non-interactive and information-theoretic secure verifiable secret sharing. In: Feigenbaum, J. (ed.) CRYPTO 1991. LNCS, vol. 576, pp. 129–140. Springer, Heidelberg (1992)
14. Shamir, A.: How to share a secret. Communications of the ACM 22(11), 612–613 (1979)
15. Vaidya, J., Clifton, C.: Secure set intersection cardinality with application to association rule mining. Journal of Computer Security 13(4), 593–622 (2005)

16. Ye, Q., Wang, H., Pieprzyk, J.: Distributed private matching and set operations. In: Chen, L., Mu, Y., Susilo, W. (eds.) ISPEC 2008. LNCS, vol. 4991, pp. 347–360. Springer, Heidelberg (2008)
17. Ye, Q., Wang, H., Pieprzyk, J., Zhang, X.-M.: Efficient disjointness tests for private datasets. In: Mu, Y., Susilo, W., Seberry, J. (eds.) ACISP 2008. LNCS, vol. 5107, pp. 155–169. Springer, Heidelberg (2008)

APPENDIX A: Properties of Protocol 2-Party DPMP

Claim 3: *In protocol 2-Party DPMP if Alice is honest, then a passive adversary controlling at most t servers does not get any information about a even after knowing t shares of $a, \ldots, a^m$.*

PROOF: Without loss of generality, let the adversary passively control the servers $S_1, \ldots, S_t$. Thus the adversary will know the first t shares of $a, \ldots, a^m$. We first show that by knowing t shares of a and a^2, the adversary does not get any extra information than by just knowing the t shares of a. So let a and a^2 be Shamir shared using degree-t polynomials $f(x)$ and $g(x)$ respectively, where $f(0) = a$ and $g(0) = a^2$. Moreover, for $i = 1, \ldots, t$, we have $f(\alpha_i) = a_i$ and $g(\alpha_i) = a_i^2$. Here a_i and a_i^2 denotes i^{th} share of a and a^2 respectively. Moreover, $\alpha_1, \ldots, \alpha_t$ are publicly known distinct elements from $\mathbb{F}$.

From the shares of a, the adversary can form a $t-1$ degree polynomial $f_{int}(x)$ such that $f_{int}(x) = f(x)$, for $x = \alpha_1, \ldots, \alpha_t$. The polynomial $f(x)$ can thus be expressed in terms of $f_{int}(x)$ in the following way :

$$f(x) = f_{int}(x) + \gamma(x - \alpha_1) \ldots (x - \alpha_t) \tag{3}$$

The adversary knows $f_{int}(0)$ and $\alpha_1, \ldots, \alpha_t$. However, $f(0)$ is information theoretically secure because of the fact that γ is still unknown to the adversary. Thus the security of a lies on the inability of the adversary to gain information on γ.

Similarly, the adversary can form a $t-1$ degree polynomial $g_{int}(x)$, such that

$$g(x) = g_{int}(x) + \beta(x - \alpha_1) \ldots (x - \alpha_t) \tag{4}$$

If Alice would have only Shamir shared a^2, then a^2 would be information theoretically secure because adversary would have no information about β. However, from Eqn (3) and Eqn (4), the adversary can form the following system of equations:

$$f(x) = f_{int}(x) + \gamma(x - \alpha_1) \ldots (x - \alpha_t) \tag{5}$$
$$g(x) = g_{int}(x) + \beta(x - \alpha_1) \ldots (x - \alpha_t) \tag{6}$$

With the above two equations for $f(x)$ and $g(x)$, the adversary can obtain the following relation :

$$a = f_{int}(0) + \gamma(-\alpha_1) \ldots (-\alpha_t) \tag{7}$$
$$a^2 = g_{int}(0) + \beta(-\alpha_1) \ldots (-\alpha_t) \tag{8}$$

Using the relation between a and a^2, adversary can obtain the following relation between γ and β:

$$(f_{int}(0) + (-1)^t \gamma \times \prod_{i=1}^{t} \alpha_i)^2 = (g_{int}(0) + (-1)^t \beta \times \prod_{i=1}^{t} \alpha_i) \qquad (9)$$

As we see, we can only get β in terms of γ and vice versa. As long as γ is secure, the whole set of dependent information is secure and we can see that, because of the dependency, the security of a^2 is also in terms of γ. The argument can be further extended to other powers of a or to any set of values that are dependent on a showing that ultimately all their securities depend on γ and hence on the security of a as is the case when we share just a (and not its powers).   □

On Cryptographic Schemes Based on Discrete Logarithms and Factoring

Marc Joye

Thomson R&D, Security Competence Center
1 avenue de Belle Fontaine, 35576 Cesson-Sévigné Cedex, France
marc.joye@thomson.net
http://joye.site88.net/

Abstract. At CRYPTO 2003, Rubin and Silverberg introduced the concept of torus-based cryptography over a finite field. We extend their setting to the ring of integers modulo N. We so obtain compact representations for cryptographic systems that base their security on the discrete logarithm problem and the factoring problem. This results in smaller key sizes and substantial savings in memory and bandwidth. But unlike the case of finite fields, analogous trace-based compression methods cannot be adapted to accommodate our extended setting when the underlying systems require more than a mere exponentiation. As an application, we present an improved, torus-based implementation of the ACJT group signature scheme.

Keywords: Torus-based cryptography, ring $\mathbb{Z}_N$, discrete logarithm problem, factoring problem, compression, ACJT group signatures.

— To Isabelle Déchène

1 Introduction

Groups where the discrete logarithm problem is assumed to be intractable are central in the design of public-key cryptography. This was first pointed out by Diffie and Hellman in their seminal paper [7]. The security of the Diffie-Hellman key-distribution system relies on the intractability of the discrete logarithm problem in the multiplicative group of finite fields. Such groups also allow one to construct encryption schemes, digital signature schemes, and many other cryptographic primitives and protocols [19].

Several improvements were proposed to improve the efficiency of the so-obtained schemes. In [24], Schnorr suggests to work in a prime-order subgroup of $\mathbb{F}_p^\times$ rather than in the whole group $\mathbb{F}_p^\times$. Building on [26], Lenstra [15] extends this idea to the cyclotomic subgroup of $\mathbb{F}_{p^r}^\times$. He states that the underlying field is really $\mathbb{F}_{p^r}$ and not some intermediate subfield thereof. More recently, Rubin and Silverberg [21] (see also [22,23]) rephrased cyclotomic subgroups in terms of algebraic tori over $\mathbb{F}_p$. They also proposed the CEILIDH cryptosystem. The main advantage of their approach resides in the compact representation of the

J.A. Garay, A. Miyaji, and A. Otsuka (Eds.): CANS 2009, LNCS 5888, pp. 41–52, 2009.

elements. Previous prominent proposals featuring a compact representation include LUC [26] and XTR [16]. Several speedups and simplifications for XTR are described in [28]. Optimizations for CEILIDH and a comparison with XTR can be found in [11].

Variants of Diffie-Hellman key-distribution system in the multiplicative group $\mathbb{Z}_N^\times$ where N is the product of two primes are proposed in [17,25]. The goal is to combine the security of the original scheme with the difficulty of factoring large numbers. In [17], McCurley argues that it may be desirable to design cryptographic systems with the property that breaking them requires to solve two different computational problems.

In this paper, we introduce torus-based cryptography over the ring $\mathbb{Z}_N$. It finds applications in settings similar to those considered by McCurley. It also reveals useful to increase the performance of cryptographic schemes whose security requires *both* the integer factorization assumption and the discrete logarithm assumption, or related assumptions (e.g., [3,2,10]). Substantial savings both in memory and in transmission are achieved without security loss. The representation used in [27] offers the same savings as one-dimensional tori over $\mathbb{Z}_N$. Unfortunately, its usage is mostly restricted to exponentiation: [27] presents an analogue of RSA. Numerous applications however require full use of multiplication. Tori over $\mathbb{Z}_N$ embed a group structure and therefore suit a much wider range of applications. We consider this as the main feature of torus-based cryptography.

As an illustration, we consider the ACJT group signature scheme [1], used in the design of the protocol standardized by the Trusted Computing Group [29] to protect privacy of the device's user. Group signature schemes, as introduced by Chaum and van Heyst [5], allow a group member to sign anonymously on behalf of the group. However, the group manager is able to recover the signer's identity. The ACJT scheme makes use of arithmetic modulo N, where $N = pq$ is a strong RSA modulus. Each group member possesses a membership certificate $[A, e]$ satisfying $A^e = a^x\, a_0 \pmod{N}$ where $\{a, a_0, N\}$ are common public parameters and x denotes the member's private key. *As the group manager may know the factorization of N, the secrecy of private key x is only guaranteed modulo p and q. As remarked in [4], if we wish to disallow the group manager to frame group members, the length of modulus N should typically be doubled.* Based on current understanding, a torus-based implementation offers the same security level but without requiring to increase the length of N. For example, for an expected 80-bit security level, the size of the resulting signatures is about 11 kb (this is half the amount of the original scheme) and the generation of a signature is more than three times faster.

The rest of this paper is organized as follows. In the next section, we provide some background on algebraic tori. We present a compact representation of one-dimensional tori from the geometric interpretation of the group law on Pell conics. We also mention compact representations for higher-dimensional tori. In Section 3, we extend torus-based representations over rings. The main focus is put on the ring $\mathbb{Z}_N$ where N is an RSA modulus. We compare the so-obtained representations with Lucas-based representations and explain why the latter are not appropriate. Section 4 addresses applications of our torus-based compression. We present a detailed implementation of the ACJT group signature scheme using a torus-based representation and discuss the performance of the resulting scheme. Finally, we conclude in Section 5.

2   Torus-Based Cryptography

Let $\mathbb{F}_q$ denote the finite field with $q = p^r$ elements. The order of the multiplicative group $\mathbb{F}_{p^r}^\times = \mathbb{F}_{p^r} \setminus \{0\}$ is $p^r - 1$. Note that $p^r - 1 = \prod_{d|r} \Phi_d(p)$ where $\Phi_d(x)$ represents the r-th cyclotomic polynomial. We let $\mathsf{G}_{p,r} \subseteq \mathbb{F}_{p^r}^\times$ denote the cyclic subgroup of order $\Phi_r(p)$.

In [21], Rubin and Silverberg identify $\mathsf{G}_{p,r}$ with the $\mathbb{F}_p$-points of an algebraic torus $T_r(\mathbb{F}_p)$. Namely, they consider

$$T_r(\mathbb{F}_p) = \{\alpha \in \mathbb{F}_{p^r}^\times \mid N_{\mathbb{F}_{p^r}/F}(\alpha) = 1 \text{ whenever } \mathbb{F}_p \subseteq F \subsetneq \mathbb{F}_{p^r}\},$$

that is, the elements of $\mathbb{F}_{p^r}^\times$ whose norm is one down to every intermediate subfield F. The key observation is that $T_r(\mathbb{F}_p)$ forms a group whose elements can be represented with only $\varphi(r)$ elements of $\mathbb{F}_p$, where φ denotes Euler's totient function. The compression factor is thus of $r/\varphi(r)$ over the field representation [21,9].

2.1   Parametrization of $T_2(\mathbb{F}_p)$

We detail a compact representation of $T_r(\mathbb{F}_p)$ for the case $r = 2$. We have $|\mathbb{F}_{p^2}^\times| = p^2 - 1$, $\Phi_2(p) = p + 1$, and $\mathsf{G}_{p,2} = \{\alpha \in \mathbb{F}_{p^2}^\times \mid \alpha^{\Phi_2(p)} = 1\}$. We assume p odd and write $\mathbb{F}_{p^2} = \mathbb{F}_p(\sqrt{\Delta})$ for some non-square $\Delta \in \mathbb{F}_p^\times$. We have $\mathsf{G}_{p,2} = \{x + y\sqrt{\Delta} \mid x, y \in \mathbb{F}_p \text{ and } (x + y\sqrt{\Delta})^{p+1} = 1\}$. Since $(x + y\sqrt{\Delta})^p = x - y\sqrt{\Delta}$, it follows that $(x + y\sqrt{\Delta})^{p+1} = (x - y\sqrt{\Delta})(x + y\sqrt{\Delta}) = x^2 - \Delta y^2$.

So, the group $\mathsf{G}_{p,2}$ can be seen as the set of $\mathbb{F}_p$ points on the genus 0 curve C over $\mathbb{F}_p$ given by the Pell equation

$$C_{/\mathbb{F}_p} : x^2 - \Delta y^2 = 1. \tag{1}$$

We have $\mathsf{G}_{p,2} \cong T_2(\mathbb{F}_p) \cong C(\mathbb{F}_p)$ [21, Lemma 7] (see also [18, Theorem 4.5]). If we denote by $\oplus$ the group law on $C(\mathbb{F}_p)$, given two points $(x_1, y_1), (x_2, y_2) \in C(\mathbb{F}_p)$, we have

$$(x_1, y_1) \oplus (x_2, y_2) = (x_1 x_2 + \Delta y_1 y_2, x_1 y_2 + x_2 y_1).$$

The neutral element is $\mathcal{O} = (1, 0)$ and the inverse of (x, y) is $(x, -y)$.

As remarked in [6, Chapter 3], the geometric interpretation of the group law on $C(\mathbb{F}_p)$ gives rise to a compact representation. Let $P = (x_1, y_1)$ and $Q = (x_2, y_2)$ be two points of $C(\mathbb{F}_p)$. The group law on $C(\mathbb{F}_p)$ is given by the so-called 'chord-and-tangent' rule [20] (see also [14, § 1] for a detailed account). We denote by $\ell_{P,Q}$ the line passing through P and Q; $\ell_{P,Q}$ represents the tangent line at P if $P = Q$. The parallel line, say ℓ', to $\ell_{P,Q}$ that passes through

$\mathcal{O} = (1,0)$ intersects (counting multiplicity) the Pell conic C at precisely one other point (x_3, y_3), which is defined as $P \oplus Q$. If m denotes the slope of $\ell_{P,Q}$ then the equation of ℓ' is given by $y = m(x-1)$. Therefore, (x_3, y_3) satisfies $x_3{}^2 - \Delta y_3{}^2 = 1$ and $y_3 = m(x_3 - 1)$. We get $x_3{}^2 - \Delta m^2(x_3 - 1)^2 = 1 \iff (x_3 - 1)\big(x_3(1 - \Delta m^2) + \Delta m^2 + 1\big) = 0$. From $y_3 = m(x_3 - 1)$, we find

$$(x_3, y_3) = \left(\frac{\Delta m^2 + 1}{\Delta m^2 - 1}, \frac{2m}{\Delta m^2 - 1} \right) = \left(\frac{(\Delta m)^2 + \Delta}{(\Delta m)^2 - \Delta}, \frac{2(\Delta m)}{(\Delta m)^2 - \Delta} \right).$$

Let now $P = (x, y)$ be a point in $C(\mathbb{F}_p) \setminus \{\mathcal{O}\}$. Since $P = P + \mathcal{O}$, we have a map

$$\psi : \mathbb{F}_p \to C(\mathbb{F}_p) \setminus \{\mathcal{O}\}, \bar{m} \mapsto P = \left(\frac{\bar{m}^2 + \Delta}{\bar{m}^2 - \Delta}, \frac{2\bar{m}}{\bar{m}^2 - \Delta} \right) \tag{2}$$

where $\bar{m} = \Delta m$ and m is the slope of the line $\ell_{P,\mathcal{O}}$ passing through P and $\mathcal{O}$.[1] Note that $\bar{m}^2 - \Delta \neq 0$ for all $\bar{m} \in \mathbb{F}_p$ since Δ is a non-square in $\mathbb{F}_p$.

Proposition 1. *The set of solutions satisfying Eq.* (1) *is given by*

$$\{\psi(\bar{m}) \mid \bar{m} \in \mathbb{F}_p\} \cup \{\mathcal{O}\}.$$

Proof. It is easy to see that ψ is injective. Indeed, assuming $\psi(\bar{m}_1) = \psi(\bar{m}_2)$, we get

$$\begin{cases} (\bar{m}_1^2 + \Delta)(\bar{m}_2^2 - \Delta) = (\bar{m}_2^2 + \Delta)(\bar{m}_1^2 - \Delta) \\ 2\bar{m}_1(\bar{m}_2^2 - \Delta) = 2\bar{m}_2(\bar{m}_1^2 - \Delta) \end{cases}$$

$$\implies \begin{cases} \bar{m}_1^2 = \bar{m}_2^2 \\ 2\bar{m}_1(\bar{m}_2^2 - \Delta) = 2\bar{m}_2(\bar{m}_1^2 - \Delta) \end{cases}$$

$$\implies \bar{m}_1 = \bar{m}_2.$$

This concludes the proof by noting that there are $(p+1)$ solutions to Eq. (1). $\square$

The inverse map is given by

$$\psi^{-1} : C(\mathbb{F}_p) \setminus \{\mathcal{O}\} \to \mathbb{F}_p, (x, y) \mapsto \bar{m} = \frac{\Delta y}{x - 1}. \tag{3}$$

By augmenting $\mathbb{F}_p$ with ∞, maps ψ and ψ^{-1} yield an isomorphism $C(\mathbb{F}_p) \xrightarrow{\sim} \mathbb{F}_p \cup \{\infty\}$ by defining $\psi(\infty) = \mathcal{O}$ and $\psi^{-1}(\mathcal{O}) = \infty$.

We use this latter representation and define

$$\mathcal{T}_2(\mathbb{F}_p) = \{\bar{m} \mid \bar{m} = \psi^{-1}(x, y) \text{ with } (x, y) \in C(\mathbb{F}_p)\}. \tag{4}$$

The neutral element in $\mathcal{T}_2(\mathbb{F}_p)$ is ∞. The inverse of $\bar{m}$ is $-\bar{m}$. Let $\bar{m}_1, \bar{m}_2 \in \mathcal{T}_2(\mathbb{F}_p) \setminus \{\infty\}$ and write $\otimes$ for the group law in $\mathcal{T}_2(\mathbb{F}_p)$. If $\bar{m}_1 = -\bar{m}_2$ then $\bar{m}_1 \otimes \bar{m}_2 = \infty$. If $\bar{m}_1 \neq -\bar{m}_2$, we get

$$\bar{m}_1 \otimes \bar{m}_2 = \psi^{-1}\big(\psi(m_1) \oplus \psi(m_2)\big) = \frac{\bar{m}_1 \bar{m}_2 + \Delta}{\bar{m}_1 + \bar{m}_2}. \tag{5}$$

---

[1] We consider $\bar{m}$ rather than m to get slightly faster arithmetic. This corresponds to the map presented in [21, §5.2].

As a result, we can do cryptography in $\mathcal{T}_2(\mathbb{F}_p)$ by doing all arithmetic directly in $\mathbb{F}_p$.

2.2 Trace-Based Compression

The trace map is defined by

$$\mathrm{Tr} : \mathbb{F}_{p^2} \to \mathbb{F}_p : \alpha \mapsto \mathrm{Tr}(\alpha) = \alpha + \alpha^p.$$

Then $\alpha \in \mathsf{G}_{p,2}$ and its conjugate α^p are the roots of polynomial $(X - \alpha)(X - \alpha^p) = X^2 - \mathrm{Tr}(\alpha)X + 1$. Define $V_k = \mathrm{Tr}(\alpha^k)$. Since $V_k = \alpha^k + \alpha^{-k}$, it is easily verified that $V_{i+j} = V_i V_j - V_{i-j}$. In particular, we get $V_{2i} = V_i^2 - 2$ and $V_{2i+1} = V_{i+1}V_i - \mathrm{Tr}(\alpha)$. Therefore, if ℓ is the binary length of k, $\mathrm{Tr}(\alpha^k)$ can be quickly evaluated with only ℓ multiplications and ℓ squarings in $\mathbb{F}_p$ using the Montgomery ladder (e.g., see [13, Fig. 4]). As we see, trace-based representations are well suited for exponentiation.

Note that letting $P = \mathrm{Tr}(\alpha)$, $V_k = V_k(P, 1)$ corresponds to the kth item of Lucas sequence $\{V_k(P, Q)\}$ with parameter $Q = 1$. Moreover, since $\alpha \in \mathsf{G}_{p,2}$, it follows that $\alpha \neq \alpha^p$ and $\Delta := \mathrm{Tr}(\alpha)^2 - 4 = P^2 - 4$ is a non-square. Let $\{U_k(P, 1)\}$ denote the companion Lucas sequence where $U_k \in \mathbb{F}_p$ satisfies $V_k + U_k\sqrt{\Delta} = 2\alpha^k$. Noting that $\sqrt{\Delta} = \alpha - \alpha^{-1}$, we have $U_k = (\alpha^k - \alpha^{-k})/(\alpha - \alpha^{-1})$. We also have $V_k^2 - \Delta U_k^2 = (V_k + U_k\sqrt{\Delta})(V_k - U_k\sqrt{\Delta}) = (2\alpha^k)(2\alpha^{-k}) = 4$. Consequently, an element $\alpha = x + y\sqrt{\Delta} \in \mathsf{G}_{p,2}$ can be equivalently written as $\alpha = \frac{V_1}{2} + \frac{U_1}{2}\sqrt{\Delta}$ and $\alpha^k = \frac{V_k}{2} + \frac{U_k}{2}\sqrt{\Delta}$ such that $\left(\frac{V_k}{2}\right)^2 - \Delta\left(\frac{U_k}{2}\right)^2 = 1$. In other words, we have $C(\mathbb{F}_p) = \{\left(\frac{V_k}{2}, \frac{U_k}{2}\right) \mid 0 \leq k \leq p\}$.

Enhanced trace-based representation. Trace-based representations over $\mathbb{F}_p$ can be 'enhanced' to allow the multiplication of two compressed elements. For $y \in \mathbb{F}_p$, we define the parity bit of y as $\mathrm{par}(y) = y \bmod 2$. As prime p is odd, we obviously have $\mathrm{par}(-y) = 1 - \mathrm{par}(y)$ if $y \in \mathbb{F}_p \setminus \{0\}$. Hence, a point $P = (x, y) \in C(\mathbb{F}_p)$ is uniquely identified by the pair (x, β) where $\beta = \mathrm{par}(y)$. We call this the enhanced trace-based representation. Hence, being given (x_1, β_1) and (x_2, β_2) (corresponding to P_1 and $P_2 \in C(\mathbb{F}_p)$), the compressed value (x_3, β_3) (corresponding to $P_3 = P_1 \oplus P_2$) can be obtained as follows: evaluate square roots $\sqrt{(x_1^2 - 1)/\Delta}$ and $\sqrt{(x_2^2 - 1)/\Delta}$ over $\mathbb{F}_p$; recover $P_1 = (x_1, y_1)$ and $P_2 = (x_2, y_2)$ from β_1 and β_2; compute $(x_3, y_3) = (x_1, y_1) \oplus (x_2, y_2)$; and output (x_3, β_3) with $\beta_3 = y_3 \bmod 2$. Compared to torus-based representation, this is however at the expense of the computation of two square roots and of further memory requirements.

2.3 Parametrization of Higher Dimensional Tori

The next cases for which the ratio $r/\varphi(r)$ is large (and thus leading to optimal compression factors) are $r = 6$ and $r = 30$. An explicit compact representation of $\mathcal{T}_6(\mathbb{F}_p)$ is detailed in [21, Section 5.1]. For the case $r = 30$, we refer the reader to [8, Section 5].

3   Compact Representations over the Ring $\mathbb{Z}_N$

Let $N = pq$ be the product of two large primes. We let $\mathbb{Z}_N$ denote the ring of integers modulo N. The isomorphism $\mathbb{Z}_N \cong \mathbb{F}_p \times \mathbb{F}_q$ induces an isomorphism between $\mathcal{T}_r(\mathbb{Z}_N)$ and $\mathcal{T}_r(\mathbb{F}_p) \times \mathcal{T}_r(\mathbb{F}_q)$.

Current knowledge in cryptanalytic techniques implies that the hardness of factoring an RSA modulus N or computing discrete logarithms in a finite field of the size of N is broadly the same. Assuming that p and q are of equal size, the discrete logarithm problem in $\mathcal{T}_r(\mathbb{F}_p)$ and in $\mathcal{T}_r(\mathbb{F}_q)$ will thus not be easier than factoring N provided that $r \geq 2$ [12]. For efficiency reasons, a smaller value for r yields better performance. Henceforth, we focus on $\mathcal{T}_r(\mathbb{Z}_N)$ with $r = 2$.

3.1   Tori $\mathcal{T}_2(\mathbb{Z}_N)$ and $\widetilde{\mathcal{T}_2}(\mathbb{Z}_N)$

Consider the Pell equation over $\mathbb{Z}_N$,

$$C_{/\mathbb{Z}_N} : x^2 - \Delta y^2 = 1 \,,$$

where $\Delta \in \mathbb{Z}_N^\times$ is a non-square modulo p and modulo q. By Chinese remaindering, the set of points $(x, y) \in \mathbb{Z}_N \times \mathbb{Z}_N$ satisfying this equation form a group, $C(\mathbb{Z}_N) = C(\mathbb{F}_p) \times C(\mathbb{F}_q)$, under the 'chord-and-tangent' law (see § 2.1). The neutral element is $\mathcal{O} = (1, 0)$. For each point $P \in C(\mathbb{Z}_N)$, there exists a unique pair of points $P_p \in C(\mathbb{F}_p)$ and $P_q \in C(\mathbb{F}_q)$ such that $P \bmod p = P_p$ and $P \bmod q = P_q$. We denote this equivalence by $P = [P_p, P_q]$.

We can now extend the previous compression map (cf. Eq. (3)) to $\mathbb{Z}_N$. The only complication is that they are some points of the form $[P_p, \mathcal{O}_q]$ or $[\mathcal{O}_p, P_q]$. To deal more easily with these points, we consider a projective representation for the compressed result. We write $\bar{m}$ as a pair $(M : Z)$ and say that $\bar{m} = (M : Z)$ is equivalent to $\bar{m}' = (M' : Z')$ if there exists some $t \in \mathbb{Z}_N^\times$ such that $M' = t M$ and $Z' = t Z$. So we define

$$\psi^{-1} : C(\mathbb{Z}_N) \to \mathbb{P}^1(\mathbb{Z}_N), (x, y) \mapsto \bar{m} = (\Delta y : x - 1). \tag{6}$$

This in turn leads to the definition of $\mathcal{T}_2(\mathbb{Z}_N)$,

$$\mathcal{T}_2(\mathbb{Z}_N) = \left\{ \bar{m} \mid \bar{m} = \psi^{-1}(x, y) \text{ with } (x, y) \in C(\mathbb{Z}_N) \right\}. \tag{7}$$

Group law. We note $\otimes$ the group law on $\mathcal{T}_2(\mathbb{Z}_N)$. The neutral element is $(t : 0)$ for some $t \in \mathbb{Z}_N^\times$. The inverse of an element $\bar{m} = (M : Z)$ is $(-M : Z)$. From Eq. (5), given $\bar{m}_1 = (M_1 : Z_1)$ and $\bar{m}_2 = (M_2 : Z_2)$ in $\mathcal{T}_2(\mathbb{Z}_N)$, a simple calculation shows that

$$(M_1 : Z_1) \otimes (M_2 : Z_2) = (M_1 M_2 + \Delta Z_1 Z_2 : M_1 Z_2 + M_2 Z_1). \tag{8}$$

Observe that the group law is complete: it works for all inputs $\bar{m}_1, \bar{m}_2 \in \mathcal{T}_2(\mathbb{Z}_N)$.

Affine parametrization. The map given by Eq. (6) does not yield a compact representation for $T_2(\mathbb{Z}_N)$ since each element $\bar{m}$ then requires two elements of $\mathbb{Z}_N$. A possible workaround is to ignore input points of the form $[P_p, \mathcal{O}_q]$ or $[\mathcal{O}_p, P_q]$ and to restrict to subset $\widetilde{C}(\mathbb{Z}_N) = \{(x, y) \in C(\mathbb{Z}_N) \mid x - 1 \in \mathbb{Z}_N^{\times}\} \cup \{\mathcal{O}\}$. A point $P = (x, y) \in \widetilde{C}(\mathbb{Z}_N)$ corresponds to $\bar{m} = \left(\frac{\Delta y}{x - 1} : 1\right)$ if $P \neq \mathcal{O}$, and $\bar{m} = (1 : 0)$ otherwise. We define

$$\widetilde{T}_2(\mathbb{Z}_N) = \{\bar{m} \mid \bar{m} = \psi^{-1}(x, y) \text{ with } (x, y) \in \widetilde{C}(\mathbb{Z}_N)\}. \tag{9}$$

From the above observation, an element $\bar{m}$ in $\widetilde{T}_2(\mathbb{Z}_N)$ can be represented by an element of $\mathbb{Z}_N$ plus one bit: $\left(\frac{\Delta y}{x-1}, 1\right)$ or $(1, 0)$. Yet another possibility is to represent $\bar{m}$ as an element of $\mathbb{A}^1(\mathbb{Z}_N) \cup \{\infty\}$. Namely, if $P = (x, y) \in \widetilde{C}(\mathbb{Z}_N)$ then $\bar{m} = \frac{\Delta y}{x-1}$ if $P \neq \mathcal{O}$, and $\bar{m} = \infty$ otherwise. In both cases, we get a compact representation for $\widetilde{T}_2(\mathbb{Z}_N)$.

The group $T_2(\mathbb{Z}_N)$ consists of all the elements of $\widetilde{T}_2(\mathbb{Z}_N)$ together with a number of elements of the form $(M : Z)$ with $\gcd(Z, N) = p$ or q (corresponding to points $[P_p, \mathcal{O}_q]$ and $[\mathcal{O}_p, P_q]$ in $C(\mathbb{Z}_N)$). The 'chord-and-tangent' law on $\widetilde{C}(\mathbb{Z}_N)$, whenever it is defined, coincides with the group law on $C(\mathbb{Z}_N) = C(\mathbb{F}_p) \times C(\mathbb{F}_q)$. The same holds for $\widetilde{T}_2(\mathbb{Z}_N)$. In practice, for cryptographic applications, N is the product of two large primes. It is therefore extremely unlikely that operation $\otimes$ is not defined on $\widetilde{T}_2(\mathbb{Z}_N)$.

3.2   Torus-Based vs. Trace-Based Compression

Similarly to § 2.2, Lucas sequences can be defined over the ring $\mathbb{Z}_N$ by Chinese remaindering. Trace-based or equivalently Lucas-based compressions are well suited to exponentiation. For example, Smith and Lennon proposed in [27] an analogue to RSA using Lucas sequence $\{V_k(P, 1)\}$ over $\mathbb{Z}_N$.

When more than a mere exponentiation is required, trace-based representations are not applicable. Indeed, let $P_1 = (x_1, y_1), P_2 = (x_2, y_2) \in C(\mathbb{Z}_N)$. Computing $P_3 = P_1 \oplus P_2$ being given P_1 and P_2 is easy: we have $P_3 = (x_1 x_2 + \Delta y_1 y_2, x_1 y_2 + x_2 y_1)$. However, computing $x_3 = x_1 x_2 + \Delta y_1 y_2$ being only given x_1 and x_2 is not possible.

Even an enhanced trace-based representation (cf. § 2.2) does not seem helpful when working over $\mathbb{Z}_N$. Here is an example of such an enhanced compression for Blum integers N (i.e., $N = pq$ with primes $p, q \equiv 3 \pmod 4$). As before, for $y \in \mathbb{Z}_N$, we define $\mathrm{par}(y) = y \bmod 2$. We also define $\mathrm{chr}(y) = 0$ if $\left(\frac{y}{N}\right) = 1$ and $\mathrm{chr}(y) = 1$ otherwise, where $\left(\frac{y}{N}\right)$ denotes the Jacobi symbol of y modulo N. Since $p, q \equiv 3 \pmod 4$, we have $\left(\frac{-1}{p}\right) = \left(\frac{-1}{q}\right) = -1$. It is therefore easily verified that a point $P = (x, y) \in C(\mathbb{Z}_N)$ is uniquely identified by the tuple (x, β, χ) where $\beta = \mathrm{par}(y)$ and $\chi = \mathrm{chr}(y)$, that is, with one element of $\mathbb{Z}_N$ and two bits. Unfortunately, decompressing (x, β, χ) into $P = (x, y)$ requires the knowledge of p and q, which are, in most settings, private values. Unlike the finite field case, we do not know enhanced trace-based representation over $\mathbb{Z}_N$ allowing to multiply compressed elements. Only torus-based representation over $\mathbb{Z}_N$ is available in this case to get a compact representation.

3.3 Extensions and Generalizations

Because the problems of computing discrete logarithms and of factoring were assumed to be balanced for an RSA modulus $N = pq$, we focused on the case $\mathcal{T}_2(\mathbb{Z}_N)$. But the same methodology extends to higher-dimensional tori. It also generalizes to more general moduli; for example, to RSA moduli made of three prime factors. This allows for different trade-offs between the two computational problems.

4 Applications

Our compression technique reduces the parameter size (typically by a factor of two). This in turn reduces the requirements for storage and transmission. It saves a significant amount in applications where many group elements are evaluated.

As an example, we consider the ACJT group signature scheme. We pointed out in the introduction (see also [4]) that the group manager in the original scheme may know the factors of RSA modulus $N = pq$ and so can frame group members if the computation of discrete logarithms modulo the factors of N is feasible. As will be apparent, a torus-based implementation allows one to keep the security of the original ACJT scheme even when the group manager is not entirely trustworthy — without increasing the length of RSA modulus N.

To simplify the presentation, we omit the various security lengths ($\lambda_1, \lambda_2, \gamma_1, \gamma_2$) and corresponding ranges (Λ, Γ). We refer the reader to [1] for details. Slight modifications need to be brought. The original ACJT group signature scheme makes use of a strong RSA modulus, that is, $N = pq$ with $p = 2p' + 1$ and $q = 2q' + 1$ for primes p', q'. Since $\mathsf{G}_{p,2}$ (resp. $\mathsf{G}_{q,2}$) has order $p+1$ (resp. $q+1$), we choose an RSA modulus $N = pq$ with $p = 4p' - 1$ and $q = 4q' - 1$ for primes p', q'; note that doing so -1 is a non-square modulo p and modulo q (i.e., $p, q \equiv 3 \pmod{4}$), which yields faster arithmetic. We let $\mathbb{T}_2$ denote the subgroup of order $p'q'$ in $\mathcal{T}_2(\mathbb{Z}_N)$. Finally, we let $\widetilde{\mathbb{T}}_2 = \mathbb{T}_2 \cap \widetilde{\mathcal{T}}_2(\mathbb{Z}_N)$.

Being a group signature scheme, our modified scheme consists of five algorithms. We use the notation of [1].

Setup Select two random primes p', q' such that $p = 4p' - 1$ and $q = 4q' - 1$ are prime. Set the modulus $N = pq$. Choose random elements a, a_0, g, h in $\widetilde{\mathbb{T}}_2$. Choose a random element $x \in \mathbb{Z}_{p'q'}^{\times}$ and set $y = g^x \in \widetilde{\mathbb{T}}_2$.
The group public key is $\mathcal{Y} = (N, a, a_0, y, g, h)$. The corresponding secret key (known only to the group manager) is $\mathcal{S} = (p', q', x)$.

Join Each user U_i interactively constructs with the group manager a membership certificate $[A_i, e_i]$ satisfying $A_i^{e_i} = a^{x_i} \otimes a_0$ in $\mathbb{T}_2$ for some prime e_i. Parameter x_i is the private key of U_i (and is unknown to the group manager).

Sign Generate a random value w and compute in $\widetilde{\mathbb{T}}_2$

$$T_1 = A_i \otimes y^w, \quad T_2 = g^w, \quad T_3 = g^{e_i} \otimes h^w.$$

Randomly choose values r_1, r_2, r_3, r_4 and compute

1. $d_1 = T_1{}^{r_1} \otimes (a^{r_2} \otimes y^{r_3})^{-1}$, $d_2 = T_2{}^{r_1} \otimes (g^{r_3})^{-1}$, $d_3 = g^{r_4}$ and $d_4 = g^{r_1} \otimes h^{r_4}$
   (all in $\widetilde{\mathbb{T}}_2$);
2. $c = \mathcal{H}(\mathcal{Y}\|T_1\|T_2\|T_3\|d_1\|d_2\|d_3\|d_4\|m)$ where m is the message being signed;
3. $s_1 = r_1 - c(e_i - 2^{\gamma_1})$, $s_2 = r_2 - c(x_i - 2^{\lambda_1})$, $s_3 = r_3 - ce_i w$ and $s_4 = r_4 - cw$
   (all in $\mathbb{Z}$).

The signature on message m is $\sigma = (c, s_1, s_2, s_3, s_4, T_1, T_2, T_3)$.

Verify Compute in $\widetilde{\mathbb{T}}_2$

$$d_1' = a_0{}^c \otimes T_1{}^{s_1 - c2^{\gamma_1}} \otimes (a^{s_2 - c2^{\lambda_1}} \otimes y^{s_3})^{-1}, \quad d_2' = T_2{}^{s_1 - c2^{\gamma_1}} \otimes (g^{s_3})^{-1},$$

$$d_3' = T_2{}^c \otimes g^{s_4}, \quad d_4' = T_3{}^c \otimes g^{s_1 - c2^{\gamma_1}} \otimes h^{s_4}.$$

Accept the signature if and only if $c' := \mathcal{H}(\mathcal{Y}\|T_1\|T_2\|T_3\|d_1'\|d_2'\|d_3'\|d_4'\|m)$ is equal to c (and if the signature components belong to appropriate ranges).

Open Check the signature's validity. The group manager then recovers $A_i = T_1 \otimes (T_2{}^x)^{-1}$ (in $\widetilde{\mathbb{T}}_2$).

We now discuss the performance of our modified scheme and compare it with the original ACJT scheme.

Let ℓ_N denote the binary length of modulus N. The system secret key $\mathcal{S}$ requires $2\ell_N$ bits. As shown in § 3.1, an element in $\widetilde{\mathbb{T}}_2 \setminus \{\infty\}$ can be coded with ℓ_N bits using an affine parametrization. Hence, the common public key $\mathcal{Y}$ consisting of 6 elements of $\widetilde{\mathbb{T}}_2$ requires $6\ell_N$ bits. The size of exponent e_i in membership certificate $[A_i, e_i]$ and of corresponding private key x_i are about the size of N^2; therefore, a membership certificate requires roughly $3\ell_N$ bits and the user's private key roughly $2\ell_N$ bits. Since the size of s_j $(1 \le j \le 4)$ is about the size of N^2, a signature $\sigma = (c, s_1, s_2, s_3, s_4, T_1, T_2, T_3)$ requires approximatively $11\ell_N$ bits. Typically, for a 80-bit security level (i.e., 2048-bit modulus for the ACJT scheme and 1024-bit modulus for its torus-based implementation), we have:

Table 1. Performance comparison: Typical lengths

	ACJT scheme	Torus-based scheme
Common public key	12 kb	6 kb
System secret key	4 kb	2 kb
Membership certificate	6 kb	3 kb
User's private key	4 kb	2 kb
Signature (approx.)	22 kb	11 kb

Our torus-based signatures are not only shorter, as we will see, they are also faster to generate. We neglect the cost of additions and hash computations. For the sake of comparison, we assume that exponentiations are done with the basic square-and-multiply algorithm and that multi-exponentiations are evaluated with the simultaneous binary exponentiation algorithm (see e.g. [19, Algorithm 14.88]). A k-exponentiation with exponent of binary length ℓ then amounts to $(\ell-1)\cdot(\mathsf{S}+\frac{2^k-1}{2^k}\mathsf{M})$, on average, where S and M represent the cost of a squaring and of a multiplication in $\widetilde{\mathbb{T}}_2$. It also requires $(2^k - 2)\mathsf{M}$ for the precomputation. Since T_1, T_2 involve exponents of size about ℓ_N bits and T_3, d_1, d_2, d_3, d_4 involve exponents of size about $2\ell_N$ bits, the generation of a signature takes about

$$\ell_N(\mathsf{S} + \tfrac{1}{2}\mathsf{M}) + \ell_N(\mathsf{S} + \tfrac{1}{2}\mathsf{M}) + 2\ell_N(\mathsf{S} + \tfrac{3}{4}\mathsf{M}) + 2\ell_N(\mathsf{S} + \tfrac{7}{8}\mathsf{M})$$
$$+ 2\ell_N(\mathsf{S} + \tfrac{3}{4}\mathsf{M}) + 2\ell_N(\mathsf{S} + \tfrac{1}{2}\mathsf{M}) + 2\ell_N(\mathsf{S} + \tfrac{3}{4}\mathsf{M}) = 12\ell_N\mathsf{S} + 8.25\ell_N\mathsf{M}, \quad (10)$$

neglecting the precomputation.

In our scheme, we have $p, q \equiv 3 \pmod 4$. We can thus take $\Delta = -1$. In this case, using projective coordinates, the multiplication of two elements $\bar{m}_1 = (M_1 : Z_1)$ and $\bar{m}_2 = (M_2 : Z_2)$ in $\widetilde{\mathbb{T}}_2$, $\bar{m}_3 = \bar{m}_1 \otimes \bar{m}_2$, simplifies to $\bar{m}_3 = (M_3 : Z_3)$ with $M_3 = M_1 M_2 + Z_1 Z_2$ and $Z_3 = M_1 Z_2 + M_2 Z_1 = (M_1 + Z_1)(M_2 + Z_2) - M_3$. Let s and m denote the cost of a square and a multiplication in $\mathbb{Z}_N$. The multiplication of two elements of $\widetilde{\mathbb{T}}_2$ requires thus $3\mathsf{m}$. Note that for a mixed multiplication (i.e., when one of the two operands has its Z-coordinate equal to 1), the cost reduces to $2\mathsf{m}$. Squaring $\bar{m}_1 = (M_1 : Z_1)$ can be evaluated as $\bar{m}_3 = (M_3 : Z_3)$ with $Z_3 = 2M_1 Z_1$ and $M_3 = (M_1 + Z_1)^2 - Z_3$ and requires thus $1\mathsf{s} + 1\mathsf{m}$. If the precomputed values in the k-exponentiation are expressed in affine way (this can be done with a single inversion and a few multiplications in $\mathbb{Z}_N$ using the so-called Montgomery's trick), we have $\mathsf{M} = 2\mathsf{m}$ and $\mathsf{S} = 1\mathsf{s} + 1\mathsf{m}$. Therefore, neglecting the cost of this inversion in $\mathbb{Z}_N$ and assuming $\mathsf{s} = 0.8\mathsf{m}$, we obtain that the cost of a torus-based ACJT group signature is about $(12 \cdot 1.8 + 8.25 \cdot 2)\ell_N\mathsf{m} = 38.1\,\ell_N\mathsf{m}$.

Similarly, from Eq. (10), we obtain that the cost of a regular ACJT group signature is about $(12 \cdot 0.8 + 8.25)\ell_N\mathsf{m} = 17.85\,\ell_N\mathsf{m}$, assuming again $\mathsf{s} = 0.8\mathsf{m}$. But since the length of ℓ_N is twice smaller in $\widetilde{T}_2$, the expected speed-up factor amounts to

$$\frac{17.85\,\ell_N \cdot (\ell_N)^2}{38.1\,(\ell_N/2) \cdot (\ell_N/2)^2} \approx 3.75.$$

In practice, the expected speed-up factor is even more spectacular as the above value assumes that the same exponentiation algorithms are being used; however, for the same amount of memory, the torus-based implementation can be sped up using more pre-computed values and higher-order methods. Note also that the above analysis neglects the cost of inversion in $\mathbb{Z}_N^\times$ (in the evaluation of d_1 and d_2) for the regular ACJT signatures.

5   Conclusion

This paper extended the concept of torus-based cryptography over the ring of integers modulo N. Our extended setting finds applications in cryptographic schemes whose security is related to factoring and discrete logarithms. Typically, it results in twice shorter keys and outputs and offers faster computation. This was exemplified with a torus-based implementation of the ACJT group signature scheme.

References

1. Ateniese, G., Camenisch, J., Joye, M., Tsudik, G.: A practical and provably secure coalition-resistant group signature scheme. In: Bellare, M. (ed.) CRYPTO 2000. LNCS, vol. 1880, pp. 255–270. Springer, Heidelberg (2000)
2. Barić, N., Pfitzmann, B.: Collision-free accumulators and fail-stop signature schemes without trees. In: Fumy, W. (ed.) EUROCRYPT 1997. LNCS, vol. 1233, pp. 480–494. Springer, Heidelberg (1997)
3. Boneh, D.: The decision Diffie-Hellman problem. In: Buhler, J. (ed.) ANTS 1998. LNCS, vol. 1423, pp. 48–63. Springer, Heidelberg (1998)
4. Camenisch, J., Groth, J.: Group signatures: Better efficiency and new theoretical aspects. In: Blundo, C., Cimato, S. (eds.) SCN 2004. LNCS, vol. 3352, pp. 120–133. Springer, Heidelberg (2005)
5. Chaum, D., van Heyst, E.: Group signatures. In: Davies, D.W. (ed.) EUROCRYPT 1991. LNCS, vol. 547, pp. 257–265. Springer, Heidelberg (1991)
6. Déchène, I.: Generalized Jacobians in Cryptography. PhD thesis, McGill University, Montreal, Canada (2005)
7. Diffie, W., Hellman, M.E.: New directions in cryptography. IEEE Transactions on Information Theory 22(6), 644–654 (1976)
8. van Dijk, M., Granger, R., Page, D., Rubin, K., Silverberg, A., Stam, M., Woodruff, D.P.: Practical cryptography in high dimensional tori. In: Cramer, R. (ed.) EUROCRYPT 2005. LNCS, vol. 3494, pp. 234–250. Springer, Heidelberg (2005)
9. van Dijk, M., Woodruff, D.: Asymptotically optimal communication for torus-based cryptography. In: Franklin, M.K. (ed.) CRYPTO 2004. LNCS, vol. 3152, pp. 157–178. Springer, Heidelberg (2004)
10. Fujisaki, E., Okamoto, T.: Statistical zero-knowledge protocols to prove modular polynomial equations. In: Kaliski Jr., B.S. (ed.) CRYPTO 1997. LNCS, vol. 1294, pp. 16–30. Springer, Heidelberg (1997)
11. Granger, R., Page, D., Stam, M.: A comparison of CEILIDH and XTR. In: Buell, D.A. (ed.) ANTS 2004. LNCS, vol. 3076, pp. 235–249. Springer, Heidelberg (2004)
12. Granger, R., Vercauteren, F.: On the discrete logarithm problem on algebraic tori. In: Shoup, V. (ed.) CRYPTO 2005. LNCS, vol. 3621, pp. 66–85. Springer, Heidelberg (2005)
13. Joye, M., Yen, S.-M.: The Montgomery powering ladder. In: Kaliski Jr., B.S., Koç, Ç.K., Paar, C. (eds.) CHES 2002. LNCS, vol. 2523, pp. 291–302. Springer, Heidelberg (2003)
14. Lemmermeyer, F.: Higher descent on Pell conics, III (2003) (preprint)
15. Lenstra, A.K.: Using cyclotomic polynomials to construct efficient discrete logarithm cryptosystems over finite fields. In: Mu, Y., Pieprzyk, J., Varadharajan, V. (eds.) ACISP 1997. LNCS, vol. 1270, pp. 127–138. Springer, Heidelberg (1997)
16. Lenstra, A.K., Verheul, E.R.: The XTR public key system. In: Bellare, M. (ed.) CRYPTO 2000. LNCS, vol. 1880, p. 119. Springer, Heidelberg (2000)
17. McCurley, K.S.: A key distribution system equivalent to factoring. Journal of Cryptology 1(2), 95–105 (1988)

18. Menezes, A.J.: Elliptic Curve Public Key Cryptosystems. Kluwer Academic Publishers, Dordrecht (1993)
19. Menezes, A.J., van Oorchot, P.C., Vanstone, S.A.: Handbook of Applied Cryptography. CRC Press, Boca Raton (1997)
20. Niewenglowski, B.: Note sur les équations $x^2 - ay^2 = 1$ et $x^2 - ay^2 = -1$. Bulletin de la Société Mathématique de France 35, 126–131 (1907)
21. Rubin, K., Silverberg, A.: Torus-based cryptography. In: Boneh, D. (ed.) CRYPTO 2003. LNCS, vol. 2729, pp. 349–365. Springer, Heidelberg (2003)
22. Rubin, K., Silverberg, A.: Using primitive subgroups to do more with fewer bits. In: Buell, D.A. (ed.) ANTS 2004. LNCS, vol. 3076, pp. 18–41. Springer, Heidelberg (2004)
23. Rubin, K., Silverberg, A.: Compression in finite fields and torus-based cryptography. SIAM Journal on Computing 37(5), 1401–1428 (2008)
24. Schnorr, C.-P.: Efficient signature generation by smart cards. Journal of Cryptology 4(3), 161–174 (1991)
25. Shmuely, Z.: Composite Diffie-Hellman public key generating systems hard to break. Technical Report 356, Israel Institute of Technology, Computer Science Department, Technion (February 1985)
26. Smith, P., Skinner, C.: A public-key cryptosystem and a digital signature system based on the Lucas function analogue to discrete logarithms. In: Safavi-Naini, R., Pieprzyk, J. (eds.) ASIACRYPT 1994. LNCS, vol. 917, pp. 357–364. Springer, Heidelberg (1995)
27. Smith, P.J., Lennon, M.J.J.: LUC: A new public key system. In: Dougall, E.G. (ed.) 9th International Conference on Information Security (IFIP/Sec 1993). IFIP Transactions, vol. A-37, pp. 103–117. North-Holland, Amsterdam (1993)
28. Stam, M., Lenstra, A.K.: Speeding up XTR. In: Boyd, C. (ed.) ASIACRYPT 2001. LNCS, vol. 2248, pp. 125–143. Springer, Heidelberg (2001)
29. Trusted Computing Group. TCG TPM specification 1.2 (2003), http://www.trustedcomputinggroup.org/

Asymptotically Optimal and Private Statistical Estimation*
(Invited Talk)

Adam Smith

Pennsylvania State University, University Park, PA, USA

Abstract. *Differential privacy* is a definition of "privacy" for statistical databases. The definition is simple, yet it implies strong semantics even in the presence of an adversary with arbitrary auxiliary information about the database.

In this talk, we discuss recent work on measuring the *utility* of differentially private analyses via the traditional yardsticks of statistical inference. Specifically, we discuss two differentially private estimators that, given i.i.d. samples from a probability distribution, converge to the correct answer at the same rate as the optimal nonprivate estimator.

1 Differential Privacy

Differential privacy is a definition of "privacy" for statistical databases. Roughly, a statistical database is one which is used to provide aggregate, large-scale information about a population, without leaking information specific to individuals. Think, for example, of the data from government surveys (*e.g.* the decennial census or epidemiological studies), or data about a company's customers that it would like a consultant to analyze.

Differential privacy is a condition on the algorithm used by the server/agency to analyze and release information about the database. Informally, it ensures that no matter what an adversary knows ahead of time about the data set, the adversary learns the same thing about an individual whether or not the individual's data appears in the data set. For example, a differentially private release of epidemiological statistics might reveal that the H1N1 virus is prevalent among US university students, but would not allow an attacker to learn whether or not a particular student was infected. For formal statements of this guarantee, see Dwork [4] and Kasiviswanathan and Smith [11].

We model the server/agency as a randomized function $\mathcal{A}$ that takes the data set as input and outputs the released information (or the transcript, in the case of an interactive service). Due to a composition property of differential privacy [7], we do not need to distinguish between the interactive and non-interactive settings.

We model the database x as a set of rows, each containing one person's data. For example if each person's data is a vector of d real numbers, then

---

* Supported in part by US National Science Foundation (NSF) TF award #0747294 and NSF CAREER/PECASE award #0729171.

J.A. Garay, A. Miyaji, and A. Otsuka (Eds.): CANS 2009, LNCS 5888, pp. 53–57, 2009.
© Springer-Verlag Berlin Heidelberg 2009

$x \in (\{0,1\}^d)^n$, where n is the number of individuals in the database. We say databases x and x' *differ in at most one element* if one is a subset of the other and the larger database contains just one additional row.

Definition 1 ([7,4]). *A randomized function A gives ε-differential privacy if for all data sets x and x' differing on at most one element, and all $S \subseteq Range(A)$,*

$$\Pr[A(x) \in S] \leq \exp(\varepsilon) \times \Pr[A(x') \in S], \tag{1}$$

where the probability space in each case is over the coin flips of A.

Several different approaches have been used to design differentially private functions. For example, one can calculate the average value of a numerical variable, add random noise to the result, and release the noisy value [1,7]. More generally, one can sample from an appropriately constructed distribution on objects of interest [12], such as synthetic data sets [2,16]. See [5] for a survey.

2   Differential Privacy and Statistical Theory

Initially, work on differential privacy concentrated on "function approximation" tasks—given a function f, how accurately can we approximate $f(x)$ differnetially privately? In this talk, we discuss recent work on measuring the utility of differentially private analyses via the traditional yardsticks of statistical inference.

For simplicity, suppose our data set X consists of n i.i.d. observations $X_1, ..., X_n$ drawn according to probability distribution P. How accurately can we describe the properties of P privately? Even without the complications of privacy, this question is involved. However, over the past century, statisticians have developed a remarkable set of tools for answering it. In particular, there are many settings where asymptotically optimal estimators are known. Here we mention two simple settings where a similar result holds even with the added constraint of privacy.

Other Work on Differential Privacy and Statistical Theory. In addition to the results described here, a number of recent papers have investigated the connections between differential privacy and statistical theory. Dwork and Lei [6] adapt ideas from *robust statistics* to develop differentially private estimators for location, scale and linear regression problems. Wasserman and Zhou [16] (following Blum, Ligett and Roth [2]) discuss the generation of synthetic data with precise statistical guarantees, as well as nonparametric density estimators (see below). Kasiviswanathan *et al.* [10] consider differentially private algorithms with PAC learning guarantees. Recently, Chaudhuri and Monteleoni [3] and McSherry and Williams [13] investigated estimators for logistic regression. Dwork and Smith [9] survey some recent work and discuss open problems.

2.1   Differential Privacy and Maximum Likelihood Estimation

We recently showed that, for every "well behaved" parametric model, there exists a differentially private point estimator which behaves much like the maximum likelihood estimate (MLE) [15]. This result exhibits a large class of settings in which the perturbation added for differential privacy is provably negligible compared to the sampling error inherent in estimation (such a result had been previously proved only for specific settings [8]).

Specifically, one can combine the *sample-and-aggregate* technique of Nissim *et al.* [14] with the *bias-corrected* MLE from classical statistics to obtain an estimator that satisfies differential privacy and is *asymptotically efficient*, meaning that the averaged squared error of the estimator is $(1 + o(1))/(nI(\theta))$, where n is the number of samples in the input, $I(\theta)$ denotes the Fisher information of f at θ and $o(1)$ denotes a function that tends to zero as n tends to infinity.

This estimator's average error is optimal even among estimators with no confidentiality constraints. In a precise sense, then, differential privacy comes at no asymptotic cost to accuracy for parametric point estimates.

Consider a parameter estimation problem defined by a model $f(x; \theta)$ where θ is a real-valued vector in a bounded space $\Theta \subseteq \mathbb{R}^p$ of diameter Λ, and x takes values in a domain D (typically, either a real vector space or a finite, discrete set). The assumption of bounded diameter is made for convenience and to allow for cleaner final theorems.

Given i.i.d. random variables $X = (X_1, ..., X_n)$ drawn according to the distribution $f(\cdot; \theta)$, we would like to estimate θ using an estimator t that takes as input the data x as well an additional, independent source of randomness R (used, in our case, for perturbation):

$$\theta \to X \to t(X, R) = \mathcal{T}(X)$$
$$\uparrow$$
$$R$$

Theorem 1 ([15]). *Under appropriate regularity conditions on f, there exists a (randomized) estimator $\mathcal{T}^*$ which is asymptotically efficient and ε-differentially private, where $\lim_{n\to\infty} \varepsilon = 0$.*

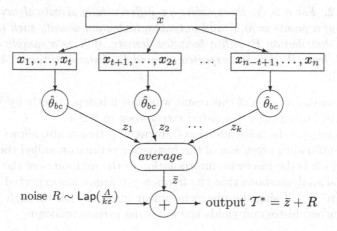

Fig. 1. The estimator $\mathcal{T}^*$. When the number of blocks k is $\omega(\sqrt{n})$ and $o(n^{2/3})$, and ε is not too small, $\mathcal{T}^*$ is asymptotically efficient.

The idea is to apply the "sample-and-aggregate" method of [14], similar in spirit to the parametric bootstrap. The procedure is quite general and can be instantiated in several variants (see Figure 1 for an example). See [15] for details.

3　Private Histogram Estimators

Consider now a different setting, where we are given samples $X = \{X_1, ..., X_n\}$ from an unknown probability distribution P on the interval $[0, 1]$. A histogram estimator approximates the density of P by a step function: given a set of intervals (called "bins") $B_1, ..., B_t$ which partition $[0, 1]$, the histogram estimator assigns mass $\frac{\#(X \cap B_i)}{n}$ to the interval B_i, spreading the mass uniformly in the interval. That is, $\hat{h}(y) = \frac{\#(X \cap B_i)}{n \cdot length(B_i)}$ when $y \in B_i$. In a fixed-width histogram, the bins have the form $B_i = [(i-1)w, iw]$, where $w = \frac{1}{t}$ is the width of each bin.

A classic result of nonparametric statistics states that if the distribution P is continuous with a bounded density, then there is a fixed-width histogram estimator that approximates P well in expectation for large n. Specifically, if the number of bins t is $\Theta(n^{2/3})$, then the expected squared L_2 distance (*integrated mean square error*, or IMSE) between P and $\hat{h}$ is $O(n^{-2/3})$. Moreover, this rate is optimal (sometimes called "*minimax*")—no fixed-width histogram estimator converges faster for all bounded-density distributions.

Here, we note that this result has a differentially private analogue. Dwork, McSherry, Nissim and Smith [7] showed that given a partition of an underlying domain D, one can release the proportion of points that lie in each piece of the partition with relatively little noise while satisfying ε-differential privacy. In the context of histograms, it suffices to add noise from a Laplace distribution with variance $\Theta(\frac{1}{n^2 \varepsilon^2})$ and expectation 0 to each of the observed proportions $\frac{\#(X \cap B_i)}{n}$ [7]. Constructing a histogram density estimator from the perturbed proportions results in the following theorem:

Theorem 2. *For $n > \frac{1}{\varepsilon^3}$, there exists a ε-differentially private algorithm $\mathcal{A}$ that takes a set of n points in $[0, 1]$ and outputs a histogram density such that for any continuous distribution P with a bounded density, if X is a sample of n points drawn i.i.d. from P, then the expected squared L_2 distance between $\mathcal{A}(X)$ and P is $O(n^{-2/3})$.*

(A slightly weaker version of this result was shown independently by Wasserman and Zhou [16], using the *exponential mechanism* of [12].)

An advantage of the noise addition framework is that it also allows for analyzing a straightforward extension of the histogram estimator, called the frequency polygon, which is the piecewise interpolation of the midpoints of the histogram. The classical analysis shows that the frequency polygon has expected squared L_2 error of $O(n^{-4/5})$. Using the piecewise linear interpolation through the centers of the perturbed histogram yields the following private analogue:

Theorem 3. *For $n > \frac{1}{\varepsilon^{5/2}}$, there exists a ε-differentially private algorithm $\mathcal{A}$ that takes a set of n points in $[0, 1]$ and outputs a piecewise linear density such* *that for any continuous distribution P with a bounded density, if X is a sample of n points drawn i.i.d. from P, then the expected squared L_2 distance between $\mathcal{A}(X)$ and P is $O(n^{-4/5})$.*

References

1. Blum, A., Dwork, C., McSherry, F., Nissim, K.: Practical Privacy: The SuLQ framework. In: Proceedings of the 24th ACM SIGMOD-SIGACT-SIGART Symposium on Principles of Database Systems (June 2005)
2. Blum, A., Ligett, K., Roth, A.: A Learning Theory Approach to Non-Interactive Database Privacy. In: Proceedings of the 40th ACM SIGACT Symposium on Thoery of Computing (2008)
3. Chaudhuri, K., Monteleoni, C.: Privacy-preserving logistic regression. In: Proceedings of the 22nd Annual Conference on Neural Information Processing Systems (NIPS) 2008, pp. 289–296 (2008)
4. Dwork, C.: Differential Privacy. In: Bugliesi, M., Preneel, B., Sassone, V., Wegener, I. (eds.) ICALP 2006. LNCS, vol. 4052, pp. 1–12. Springer, Heidelberg (2006)
5. Dwork, C.: Differential Privacy: A Survey of Results. In: Agrawal, M., Du, D.-Z., Duan, Z., Li, A. (eds.) TAMC 2008. LNCS, vol. 4978, pp. 1–19. Springer, Heidelberg (2008)
6. Dwork, C., Lei, J.: Differential Privacy and Robust Statistics. In: Proceedings of the 41st ACM Symposium on Theory of Computing (2009)
7. Dwork, C., McSherry, F., Nissim, K., Smith, A.: Calibrating Noise to Sensitivity in Private Data Analysis. In: Halevi, S., Rabin, T. (eds.) TCC 2006. LNCS, vol. 3876, pp. 265–284. Springer, Heidelberg (2006)
8. Dwork, C., Nissim, K.: Privacy-Preserving Datamining on Vertically Partitioned Databases. In: Franklin, M. (ed.) CRYPTO 2004. LNCS, vol. 3152, pp. 528–544. Springer, Heidelberg (2004)
9. Dwork, C., Smith, A.: Differential Privacy for Statistics: What we Know and What we Want to Learn. In: NCHS/CDC Data Confidentiality Workshop (2008)
10. Kasiviswanathan, S., Lee, H., Nissim, K., Raskhodnikova, S., Smith, S.: What Can We Learn Privately? In: Proc. 49th Annual IEEE Symposium on Foundations of Computer Science (2008)
11. Kasiviswanathan, S., Smith, S.: A Note on Differential Privacy: Defining Resistance to Arbitrary Side Information. Preprint arxiv:0803.3946 (2008), www.arxiv.org
12. McSherry, F., Talwar, K.: Mechanism Design via Differential Privacy. In: Proceedings of the 48th Annual Symposium on Foundations of Computer Science (2007)
13. McSherry, F., Williams, O.: Probabilistic Inference and Differential Privacy. Manuscript (2009)
14. Nissim, K., Raskhodnikova, S., Smith, A.: Smooth Sensitivity and Sampling in Private Data Analysis. In: Proceedings of the 39th ACM Symposium on Theory of Computing, pp. 75–84 (2007)
15. Smith, A.: Efficient, Differentially Private Point Estimators. Preprint arxiv:0809.4794 (2008), http://www.arxiv.org
16. Wasserman, L., Zhou, S.: A statistical framework for differential privacy. Preprint arXiv:0811.2501v1 (2008), http://www.arxiv.org

Linear (Hull) and Algebraic Cryptanalysis of the Block Cipher PRESENT

Jorge Nakahara Jr.[1], Pouyan Sepehrdad[1], Bingsheng Zhang[2,*],
and Meiqin Wang[3,**]

[1] EPFL, Lausanne, Switzerland
[2] Cybernetica AS, Estonia and University of Tartu, Estonia
[3] Key Laboratory of Cryptologic Technology and Information Security, Ministry of
Education, Shandong University, Jinan 250100, China
{jorge.nakahara,pouyan.sepehrdad}@epfl.ch, b.zhang2009@gmail.com,
mqwang@sdu.edu.cn

Abstract. The contributions of this paper include the first linear hull
and a revisit of the algebraic cryptanalysis of reduced-round variants
of the block cipher PRESENT, under known-plaintext and ciphertext-
only settings. We introduce a pure algebraic cryptanalysis of 5-round
PRESENT and in one of our attacks we recover half of the bits of the key
in less than three minutes using an ordinary desktop PC. The PRESENT
block cipher is a design by Bogdanov *et al.*, announced in CHES 2007
and aimed at RFID tags and sensor networks. For our linear attacks,
we can attack 25-round PRESENT with the whole code book, $2^{96.68}$ 25-
round PRESENT encryptions, 2^{40} blocks of memory and 0.61 success
rate. Further we can extend the linear attack to 26-round with small
success rate. As a further contribution of this paper we computed linear
hulls in practice for the original PRESENT cipher, which corroborated
and even improved on the predicted bias (and the corresponding attack
complexities) of conventional linear relations based on a single linear
trail.

Keywords: block ciphers, RFID, linear hulls, algebraic analysis, systems of sparse polynomial equations of low degree.

1 Introduction

This paper describes linear (hull) and algebraic cryptanalysis of reduced-round
versions of the PRESENT block cipher, a design by Bogdanov *et al.* aimed at
restricted environments such as RFID tags [3] and sensor networks. For the linear

---

* This author is supported by Estonian Science Foundation, grant #8058, the Eu-
ropean Regional Development Fund through the Estonian Center of Excellence in
Computer Science (EXCS), and the 6th Framework Programme project AEOLUS
(FP6-IST-15964).
** This author is supported by 973 Program of China (Grant No.2007CB807902) and
National Outstanding Young Scientist fund of China (Grant No. 60525201).

J.A. Garay, A. Miyaji, and A. Otsuka (Eds.): CANS 2009, LNCS 5888, pp. 58–75, 2009.

case, our analysis include linear hulls of reduced-round variants of PRESENT, which unveils the influence of the linear transformation in the clustering effect of linear trails. The computation of linear hulls also served to determine more accurately the overall bias of linear relations, and consequently, more precise complexity figures of the linear attacks.

Previous known analyses on (reduced-round) PRESENT, including the results in this paper, are summarized in Table 4 along with attack complexities.

Our efficient attacks reach 25-round PRESENT under a known-plaintext setting and 26-round with small success rate, and 15 rounds under a ciphertext-only setting. The algebraic attacks, on the other hand, can recover keys from up to 5-round PRESENT in a few minutes, with only five known plaintext-ciphertext pairs.

This paper is organized as follows: Sect. 2 briefly details the PRESENT block cipher; Sect. 3 presents our algebraic analysis on PRESENT; Sect. 4 describes our linear cryptanalysis of reduced-round PRESENT; Sect. 5 describes our linear hull analysis of PRESENT; Sect. 6 concludes the paper.

2   The PRESENT Block Cipher

PRESENT is an SPN-based block cipher aimed at constrained environments, such as RFID tags and sensor networks. It was designed to be particularly compact and competitive in hardware. PRESENT operates on 64-bit text blocks, iterates 31 rounds and uses keys of either 80 or 128 bits. This cipher was designed by Bogdanov *et al.* and announced at CHES 2007 [3]. Each (full) round of PRESENT contains three layers in the following order: a bitwise exclusive-or layer with the round subkey; an S-box layer, in which a fixed 4×4-bit S-box (Table 5) is applied sixteen times in parallel to the intermediate cipher state; a linear transformation, called pLayer, consisting of a fixed bit permutation. Only the xor layer with round subkeys is an involution. Thus, the decryption operation requires the inverse of the S-box (Table 5) and of the pLayer. After the 31st round there is an output transformation consisting of an exclusive-or with the last round subkey. One full round of PRESENT is depicted in Fig. 1. Our

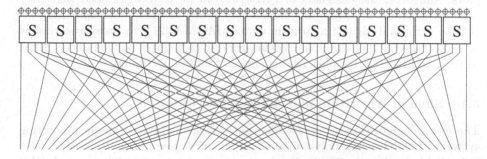

Fig. 1. One full round of PRESENT

attacks are independent of the key schedule algorithm. Further details about the key schedule, for each key size, can be found in [3].

3   Algebraic Analysis

Algebraic cryptanalysis is attributed to C.E. Shannon, who mentioned in [27] that breaking a good cipher should require "as much work as solving a system of simultaneous equations in a large number of unknowns of a complex type". Contrary to linear and differential attacks, that require a large number of chosen or known plaintexts (which makes it roughly impractical in reality), algebraic cryptanalysis requires a comparatively small number of text pairs. Any algebraic attack consists of two distinct stages: the adversary writes the cipher as a system of polynomial equations of low degree over $\mathbb{GF}(2)$ or $\mathbb{GF}(2^k)$ [7,23]. Then, it solves the corresponding system which turns out to be overdefined and sparse. The methods already proposed for solving polynomial system of equations are Gröbner basis including Buchberger algorithm [4], F4 [15], F5 [16] and algorithms like ElimLin [9], XL [6] and its family [7], and Raddum-Semaev algorithm [26]. Converting these equations to Boolean expressions in Conjunctive Normal Form (CNF) [9] and deploying various SAT-solver programs is another strategy [14]. Algebraic attacks since the controversial paper of [7] has gotten considerable attention, has been applied to several stream ciphers (see [8]) and is able to break some of them but it has not been successful in breaking real life block ciphers, except Keeloq [11,18].

In this paper we deploy ElimLin algorithm proposed by Courtois against DES [9] and F4 algorithm by Faugére [15] and we break up to 5-round PRESENT for both key sizes. Then we compare our results using these two approaches. Courtois and Debraize in [10] have already proposed a Guess-then-Algebraic attack on 5-round PRESENT only for the 80-bit key version. In fact, our main focus in this paper is a comparison between the efficiency of ElimLin algorithm which uses simple linear algebra and the recent so called efficient implementation of F4 algorithm under PolyBori framework. Although there exist other types of attacks for larger number of rounds, we believe this result is interesting, because we can recover many key bits with a relatively few known plaintext-ciphertext pairs. Moreover, the designers of PRESENT in [3] have mentioned that they were unsuccessful to obtain any satisfactory result in reasonable time using algebraic cryptanalysis (F4 algorithm under MAGMA [21]) to break two rounds of a smaller version of the cipher having only seven S-boxes per round compared to the real PRESENT cipher having sixteen S-boxes per round.

3.1   ElimLin Algorithm and Attack Description

The *ElimLin* algorithm stands for *Eliminate Linear* and is a technique for solving systems of multivariate polynomial equations of degree mostly 2, 3 or 4 over a finite field, specifically $\mathbb{GF}(2)$. Originally, it was proposed in [9] to attack DES and was reported to break 5-round DES.

ElimLin is composed of two distinct stages, namely: *Gaussian Elimination* and *Substitution*. All the linear equations in the linear span of initial equations are found. Subsequently, one of the variables is nominated in each linear equation and is substituted in the whole system. This process is repeated up to the time when no linear equation is obtained in the linear span of the system.

3.2   F4 Algorithm under PolyBori Framework

F4 is currently the most efficient algorithm for computing the Gröbner basis of an ideal. The most efficient implementation of F4 is available under PolyBori framework [2] running alone or under SAGE algebra system. PolyBori is a C++ library designed fundamentally to compute Gröbner basis applied to Boolean Polynomials. The ring of Boolean Polynomials is a quotient ring over $\mathbb{GF}(2)$, where the field equation for each variable is $x^2 = x$. A Python interface is used, surrounding the C++ core, announced by the designers to be used from the beginning to facilitate "parsing of complex polynomial systems" and "sophisticated and easy extendable strategies for Gröbner base computation" [2]. It uses zero-suppressed binary decision diagrams (ZDDs) [17] as a high level data structure for storing Boolean Polynomials which results in the monomials to be stored more efficiently with respect to the space they occupy in memory and making the computational speed faster compared with other computer algebra systems. We used polybori-0.4 in our attacks.

3.3   Algebraic Representation of PRESENT

It is a straightforward procedure to demonstrate that every 4×4-bit S-box has at least 21 quadratic equations. The larger the number of equations, the weaker the S-box. In fact, the S-box of PRESENT has exactly 21 equations. Writing the whole 80-bit key variant of PRESENT as a system of quadratic equations for 5 rounds, we obtained 740 variables and 2169 equations. In our attack, we fix some of the key bits and we recover the remaining unknown ones. In fact, we introduce an attack on both PRESENT with key sizes of 80 and 128 bits. Notice that for both key sizes one pair is not enough to recover the key uniquely and we need at least two pairs.

The summary of our results is in Table 1. All the timings were obtained under a 2Ghz CPU with 1Gb of RAM and we used an efficient implementation of ElimLin available in [12]. As it is depicted in Table 1, the timing results of ElimLin and PolyBori are comparable except the time in which PolyBori crashed[1] due to probably running out of memory. As our experiments revealed, in all cases ElimLin used much less memory compared to F4 under PolyBori which turns out to be due to the Gröbner basis approach of increasing the degree of polynomials in the intermediate stages.

---

[1] In Appendix, we give the intermediate results of ElimLin for one of the cases in which PolyBori crashes.

Table 1. Algebraic attack complexities on reduced-round PRESENT

# rounds	#key bits	#key bits fixed	full key (hours)	# plaintexts	notes
5	80	40	0.04	5 KP	ElimLin
5	80	40	0.07	5 KP	PolyBori
5	80	37	0.61	10 KP	ElimLin
5	80	37	0.52	10 KP	PolyBori
5	80	36	3.53	16 KP	ElimLin
5	80	36	Crashed!	16 KP	PolyBori
5	80	35	4.47	16 KP	ElimLin
5	80	35	Crashed!	16 KP	PolyBori
5	128	88	0.05	5 KP	ElimLin
5	128	88	0.07	5 KP	PolyBori

Since the time complexity of the experiment depends on the system instance, Table 1 represents the average time complexity. We had some instances revealing that the times it takes to recover 45 bits of the key is much less than that for 44 bits. This seems very surprising at the first glance, but it can be justified by considering that the running time of ElimLin implementation in [12] is highly dependable on the sparsity of equations. So, our intuition is that as we have picked distinct plaintext and key randomly in each experiment, by pure chance the former system of equations turns out to be sparser than the latter and it is also probable that more linear equations are generated due to some combination of plaintexts and keys randomly picked.

In [1], Albrecht and Cid compared their result with exhaustive key search over PRESENT assuming that checking an arbitrary key takes at least 2 CPU cycles which seems ambitious implying that recovering 45 bits of the key should take at least more than 9 hours, while we could recover the key in less than two hours using only five KP in our best attack.

We tried to break 6 rounds of PRESENT by ElimLin and F4, but ElimLin did not give us any satisfactory result and PolyBori crashed after a while due to probably running out of memory for 6-round PRESENT. In [10], the results are compared with F4 implementation under MAGMA which is specified not to yield any satisfactory results in reasonable time. Although PolyBori crashes in much fewer cases, we could not get anything better by using F4 under PolyBori compared to ElimLin in this specific case.

4   Linear Analysis

Linear cryptanalysis (LC) typically operates under a known-plaintext (KP) or a ciphertext-only (CO) setting, and its origin dates back to the works of Matsui on DES [22]. The main tool for this attack is the *linear relation*, which consists of a linear combination of text and key bits, holding with a relatively high parity deviation from the uniform parity distribution. The effectiveness of a linear relation is measured by a parameter called bias, denoted ϵ, which is the absolute value of the difference between the parity of the linear relation from $1/2$. The higher the bias, the more attractive the linear relations are, since they demand less plaintext-ciphertext pairs. These relations form the core of a linear distinguisher, namely, a tool that allows one to distinguish a given cipher from a random permutation, or to recover subkey bits.

Our linear analysis of the PRESENT cipher started with a study of the Linear Approximation Table (LAT) [22] (Table 6) of its 4×4 S-box (Table 5). In Table 6, the acronym IM stands for the Input Mask (leftmost column); OM for the Output Mask (top row); the entries for non-zero masks are either 0, 2, -2, 4 or -4, that is, the S-box is linearly 4-uniform. Thus, the largest entry for non-trivial (non-zero) bitmasks corresponds to a bias of $4/16 = 2^{-2}$. Thus, entries in the LAT correspond to $16 \cdot \epsilon$, except for the sign; a negative entry implies that the parity is closer to '0', while the absence of a sign means the parity is closer to '1'.

One-round and multiple-round linear approximations were derived by combining the LAT with the bit permutation pLayer that follows each S-box layer. Our analysis indicated that the most promising linear relations shall exploit

- one-bit-input-to-one-bit-output bitmasks in order to minimize the number of active S-boxes per round;
- the pLayer bit permutation following the S-box layer has order three, that is, if we denote this permutation by P, then $P(P(P(X))) = P^3(X) = X$, for all text blocks X; this fact motivated us to look for iterative relations across three rounds. Particular bit positions of pLayer, though, have much smaller order, such as the leftmost bit in a block which is unaffected by pLayer, that is, it is a fix-point. There are four such fix-points in the pLayer. Thus, the branch number [13] of pLayer is just two. This means that diffusion is quite poor in PRESENT. Due to the fix-points of pLayer, and the LAT profile of the S-box, iterative linear relations exist for any number of rounds.

Taking the order of pLayer into account, it is straightforward to find 3-round iterative linear relations with only three active S-boxes (there cannot be less than one active S-box per round due to the SPN structure). Nonetheless, the S-box design minimizes the bias of single-bit linear approximations. The bias for each such approximation is $2/16 = 2^{-3}$, which gives a maximum bias of $2^{2-3-3-3} = 2^{-7}$ for any 3-round linear relation.

Let us denote a 64-bit mask by

$$\Gamma = \gamma_0 \gamma_1 \gamma_2 \gamma_3 \gamma_4 \gamma_5 \gamma_6 \gamma_7 \gamma_8 \gamma_9 \gamma_{10} \gamma_{11} \gamma_{12} \gamma_{13} \gamma_{14} \gamma_{15}$$

where $\gamma_i \in \mathbb{Z}_2^4$, $0 \leq i \leq 15$, that is, a nibble (4 bits). An interesting example of 1-round linear relation for PRESENT is

$$8000000000000000_x \xrightarrow{1r} 8000000000000000_x \qquad (1)$$

where the linear approximation $8 \xrightarrow{S} 8$ for the S-box was used for the leftmost nibble (the leftmost S-box), with bias 2^{-3}, and $\xrightarrow{1r}$ means one round transition. Note that this bias is not the highest possible, but the non-zero bit position in the mask is a fix-point for the pLayer. Thus, (1) is an iterative linear relation with a single active S-box. If we denote a text block as $X = (x_0, x_1, \ldots, x_{63})$, then (1) can be expressed simply as

$$p_0 \oplus c_0 = k_0^1 \oplus k_0^2, \qquad (2)$$

where k_0^1 and k_0^2 are the leftmost (most significant) bits of the first two round sub-keys. For a distinguish-from-random attack, the complexity for a single round is $N = 8 \cdot (2^{-3})^{-2} = 2^9$ KP and equivalent parity computations (2). Notice, though, that if the plaintext is composed of ASCII text, then the most significant bit of every plaintext byte is always zero, and the attack actually requires 2^9 cipher-texts only (CO). Iterating (1) for up to 14 rounds requires $8 \cdot (2^{13-3*14})^{-2} = 2^{61}$ CO and an equivalent amount of parity computations. If we allow a lower success probability, we can attack up to 15-round PRESENT using $4 \cdot (2^{14-3*15})^{-2} = 2^{64}$ CO, and equivalent number of parity computations. But, since the codebook is exhausted, the KP or CO settings are the same.

Other two 1-round iterative linear relations with the bias 2^{-3}, also based on fix-points of pLayer are

$$0000000000200000_x \xrightarrow{1r} 0000000000200000_x, \qquad (3)$$

and

$$0000040000000000_x \xrightarrow{1r} 0000040000000000_x. \qquad (4)$$

A linear relation based on the fourth fix-point of pLayer is not effective since the LAT entry is zero.

An example of 2-round (non-iterative) linear relation for PRESENT with maximum bias is

$$1000000000000000_x \xrightarrow{1r} 0000800000008000_x \xrightarrow{1r} 0808080808080000_x, \qquad (5)$$

with bias $2^{2-2-2-2} = 2^{-4}$, and three active S-boxes. The local S-box approximations used were $1 \xrightarrow{S} 5$ and $8 \xrightarrow{S} 14$, both with bias 2^{-2}. Reducing the number of active S-boxes to only two across two rounds would decrease the bias to $2^{1-3-3} = 2^{-5}$. Thus, the trade-off of three active S-boxes versus the bias, across two rounds, is the best possible. The attack complexity is $N = 8 \cdot (2^{-4})^{-2} = 2^{11}$ KP and equivalent parity computations.

For three rounds, one of the simplest, most-biased and iterative linear relations we have found is

$$0800000000000000_x \xrightarrow{1r} 4000000000000000_x \xrightarrow{1r}$$
$$0000800000000000_x \xrightarrow{1r} 0800000000000000_x, \qquad (6)$$

where the S-box linear approximations were $8 \xrightarrow{S} 8$ and $4 \xrightarrow{S} 4$, both with bias 2^{-3}. The overall bias is $2^{2-3-3-3} = 2^{-7}$. Relation (6) is an example that demon-strates a trade-off between the number of active S-boxes per rounds versus the overall bias of linear relations involving single-bit-input-single-bit-output masks.

Relation (6) allows to mount a distinguish-from-random linear attack on 3-round PRESENT with $N = 8 \cdot (2^{-7})^{-2} = 2^{17}$ KP and equivalent number of parity computations (less than one encryption) and negligible memory. For six rounds, the attack complexity becomes $N = 8 \cdot (2^{1-2*7})^{-2} = 2^{29}$ KP and equivalent parity computations. For nine rounds, the complexity becomes $N = 8 \cdot (2^{2-3*7})^{-2} = 2^{41}$ KP and equivalent number of parity computations. For twelve rounds, the attack complexity becomes $N = 8 \cdot (2^{3-4*7})^{-2} = 2^{53}$ KP and equivalent parity computations. For fifteen rounds, if we allow a smaller success rate, then $N = 4 \cdot (2^{4-5*7})^{-2} = 2^{64}$ KP are required, and an equivalent number of parity computations (which is about the effort of one-round computation). Actually, in the 12-round case, the first and last round S-box approximations can be improved, leading to

$$
\begin{aligned}
&7700000000000000_x \xrightarrow{1r} 0000C00000000000_x \xrightarrow{1r} 0800000000000000_x \xrightarrow{12r} \\
&0800000000000000_x \xrightarrow{1r} 4000000000000000_x \xrightarrow{1r} 8000000080008000_x,
\end{aligned}
\tag{7}
$$

where the S-box approximations were $7 \xrightarrow{S} 4$ with bias 2^{-2} in the 1st round; $C \xrightarrow{S} 8$ with bias 2^{-2} in the 2nd round; $8 \xrightarrow{S} 8$ with bias 2^{-3} in the 15th round; $4 \xrightarrow{S} B$, with bias 2^{-2} in the last round. The notation $\xrightarrow{xr}$ means an x-round transition. The overall bias is $2^{-2-2-2-3\cdot12-3-2+16} = 2^{-31}$. A distinguish-from-random attack using the 16-round relation (7) costs $N = 4 \cdot (2^{-31})^{-2} = 2^{64}$ KP.

Additional 16-round linear approximations can be derived taking into account other fix-points of pLayer. For instance, using (3):

$$
\begin{aligned}
&00000000A0A00000_x \xrightarrow{1r} 0000000000A00000_x \xrightarrow{1r} 0000000000200000_x \xrightarrow{13r} \\
&0000000000200000_x \xrightarrow{1r} 0020000000200020_x,
\end{aligned}
\tag{8}
$$

where the S-box approximations were $A \xrightarrow{S} 2$ with bias 2^{-2} in the 1st and 2nd rounds; $2 \xrightarrow{S} B$ with bias 2^{-2} in the last round. The overall bias is $2^{-3\cdot(2+13)-2+16} = 2^{-31}$.

Further, using (1), we have

$$
\begin{aligned}
&CC00000000000000_x \xrightarrow{1r} C000000000000000_x \xrightarrow{1r} 8000000000000000_x \xrightarrow{13r} \\
&8000000000000000_x \xrightarrow{1r} 8000800080000000_x,
\end{aligned}
\tag{9}
$$

where the S-box approximations were $C \xrightarrow{S} 8$ with bias 2^{-2} in the 1st and 2nd rounds; $8 \xrightarrow{S} E$ with bias 2^{-2} in the last round. The overall bias is $2^{-3\cdot(13+2)-2+16} = 2^{-31}$.

A 1R key-recovery attack can be applied at the top end of (9) would require guessing the subkeys on top of four S-boxes, because $CC00000000000000_x$ has four active bits. It means a complexity of $2^{64+16}/4 = 2^{78}$ 1-round computations, or $2^{78}/17 \approx 2^{73.91}$ 17-round computations. The memory complexity is a 16-bit counter and the success rate [28] is about 0.37. Recovering subkeys at the bottom end requires guessing only twelve subkey bits since 8000800080000000_x has only three active bits. It means $2^{64+12} \cdot 3/(16 \cdot 17) \approx 2^{69.50}$ 17-round computations. The memory complexity is a 12-bit counter and the success rate is about 0.63. Applying this attack at both ends (2R attack) requires $2^{64+16+12} \cdot 7/(16 \cdot 18) \approx 2^{86.64}$ 18-round computations (applies only to 128-bit keys). The success rate is about 0.03.

For the remaining 100 key bits, we use (8), which has the same bias as (9). So, the effort to recover further 24+12 key bits is $2^{64+24+12} \cdot 8/(16 \cdot 18) \approx 2^{96.83}$ 18-round computations. The remaining 64 key bits can be found by exhaustive search.

5 Linear Hulls

The concept of linear hulls was first announced by Nyberg in [24]. A linear hull is the LC counterpart to differentials in differential cryptanalysis. Therefore, a linear hull stands for the collection of all linear relations (across a certain number of rounds) that have the same (fixed) input and output bitmasks, but involves different sets of round subkeys according to different linear trails. Consequently, the bias of a linear hull stands for the actual bias of a linear relation involving a given pair of input and output bitmasks. When there is only a single linear trail between a given pair of input and output bitmasks, the concepts of linear relation and linear hull match.

The linear hull effect accounts for a clustering of linear trails, with the consequence that the final bias may become significantly higher than that of any individual trail. Due to Nyberg [24], given the input and output masks a and b for a block cipher $Y = Y(X, K)$, the potential of the corresponding linear hull is denoted

$$ALH(a,b) = \sum_c (P(a \cdot X \oplus b \cdot Y \oplus c \cdot K = 0) - \frac{1}{2})^2 = \epsilon^2 \qquad (10)$$

where c is the mask for the subkey bits. Then, key-recovery attacks such as Algorithm 2 in [22] apply with

$$N = \frac{t}{ALH(a,b)} = \frac{t}{\epsilon^2}$$

known plaintexts, where t is a constant. An advantage to use linear hulls in key-recovery attacks, such as in Algorithm 2, is that the required number of known plaintexts can be decreased for a given success rate. Keliher et al. exploited this method to attack the Q cipher [19].

For PRESENT, in particular, it makes sense to choose input/output masks that affect only a few S-boxes, because it minimizes the number of key bits to guess in key-recovery attacks around the linear hull distinguisher. Moreover, minimizing the number of active S-boxes in the first round may also minimize the number of linear trails to look for, which speeds up our search program for all possible linear paths and the corresponding bias computation.

Our approach to determine linear hulls for PRESENT used a recursive algorithm (in ANSI C) employing a depth-first search strategy. It is a classical technique to find exhaustively all linear trails systematically and with low memory cost. Fixed input and output (I/O) bitmasks and a number of rounds were provided as parameters, and the algorithm computed all possible linear trails starting and ending in the given I/O masks, the corresponding biases and the number of active S-boxes. One objective of the linear hull search was to double check if the linear relations we derived in (1), (5) and (6) actually had the predicted biases (which have been confirmed).

An interesting phenomenon we have observed is the rate of decrease of the bias in linear hulls with some fixed input/output bitmasks of low Hamming Weight (HW), for increasing number of rounds. In particular, we have focused on a few cases, where the input and output masks are the same (iterative linear relations) and have low HW. We have studied all 64-bit input and output bit masks with $HW = 1$. Further, to optimize the search, we have focused only on the linear trails with the highest bias (single-bit trails), which we call the best trails (with a single active S-box per round). The best results we obtained concern the mask 0000000000200000_x (both at the input and at the output). Table 2 summarizes our experiments, where "computed bias" denotes the bias of the linear hull for the given number of rounds computed according to 10. For up to four rounds, all trails were found. For five rounds or more, only the trails with highest biases were accounted for. The values under the title "expected bias" indicate the bias as computed by the Piling-up lemma.

From Table 2, we observe that the bias for linear hulls in PRESENT does not decrease as fast, with increasing number of rounds, as in linear relations as dictated by the Piling-up lemma. Fig.2 compares the computed and the predicted bias values in Table 2. Our experimental results indicate that the linear hull effect is significant in PRESENT even for a small number of rounds. For five rounds or more, we could not determine all linear trails, but we looked for the ones with the highest bias values, so that their contribution to the overall ALH would be significant. We have searched for linear trails with r up to $r + 2$ active S-boxes across r rounds. Thus, the values for more than four rounds represent a lower bound on the overall bias of the linear hulls.

In Table 2, consider the linear hull across five rounds. We have found nine trails with bias 2^{-11} inside this linear hull. Repeating it three times, we arrive at 9^3 15-round linear trails. The ALH $(0000000000200000_x, 0000000000200000_x)$ for 15 rounds is $(2^{-31})^2 \cdot 9^3 = 2^{-62+9.51} = 2^{-52.49}$. We extend this 15-round linear hull to a 17-round linear hull with 9^3 17-round linear trails by choosing an additional 1-round relation at the top and at the bottom ends of it:

$$0000000000A00000_x \xrightarrow{1r} 0000000000200000_x \xrightarrow{15r}$$
$$0000000000200000_x \xrightarrow{1r} 0020000000200020_x, \tag{11}$$

where the ALH for the 17-round linear hull is $(2^{-33})^2 \cdot 9^3 = 2^{-66+9.51} = 2^{-56.49}$. This linear hull can be used to distinguish 17-round PRESENT from a random permutation with $2^{56.49} \cdot 8 = 2^{59.49}$ KP, and equivalent parity computations.

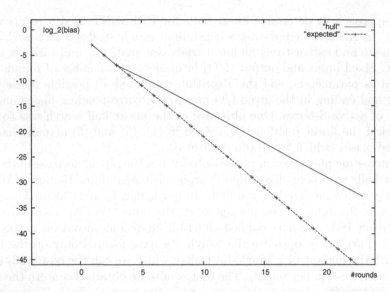

Fig. 2. Behaviour of linear hull bias (ALH) against expected bias (by Piling-up lemma) for increasing number of rounds of PRESENT (data from Table 2)

Table 2. Computed bias (cb) and expected bias (eb) of linear hulls in PRESENT for input/output mask 0000000000200000_x

# rounds	1	2	3	4	5	6	7	8
# trails	1	1	1	9	9	27	72	192
(cb)	2^{-3}	2^{-5}	2^{-7}	$2^{-8.20}$	$2^{-9.40}$	$2^{-10.61}$	$2^{-11.90}$	$2^{-13.19}$
(eb)	2^{-3}	2^{-5}	2^{-7}	2^{-9}	2^{-11}	2^{-13}	2^{-15}	2^{-17}
# rounds	9	10	11	12	13	14	15	16
# trails	512	1344	3528	9261	24255	63525	166375	435600
(cb)	$2^{-14.48}$	$2^{-15.78}$	$2^{-17.08}$	$2^{-18.38}$	$2^{-19.71}$	$2^{-21.02}$	$2^{-22.33}$	$2^{-23.63}$
(eb)	2^{-19}	2^{-21}	2^{-23}	2^{-25}	2^{-27}	2^{-29}	2^{-31}	2^{-33}
# rounds	17	18	19	20	21	22	23	
# trails	1140480	2985984	7817472	20466576	53582633	140281323	367261713	
(cb)	$2^{-24.94}$	$2^{-26.25}$	$2^{-27.55}$	$2^{-28.85}$	$2^{-30.16}$	$2^{-31.47}$	$2^{-32.77}$	
(eb)	2^{-35}	2^{-37}	2^{-39}	2^{-41}	2^{-43}	2^{-45}	2^{-47}	

Applying a key-recovery (1R attack) at the top end of (11) requires guessing only eight bits because there are only two active bits in $0000000000A00000_x$. The attack complexity becomes $2^{59.49+8} \cdot 2/(16 \cdot 18) = 2^{60.33}$ 18-round computations. The memory complexity is just an 8-bit counter and the success rate is about 0.997.

For six rounds, and still using mask 0000000000200000_x, we have found 27 trails, each with bias 2^{-13} inside this linear hull. Concatenating the linear hull three times, we arrive at 27^3 18-round trails. The ALH (0000000000200000_x, 0000000000200000_x) for 18 rounds is $(2^{-37})^2 \cdot 27^3 = 2^{-59.73}$. Extending it to 20 rounds by choosing carefully an additional relation on top and at the bottom of it results in

$$0000000000A00000_x \xrightarrow{1r} 0000000000200000_x \xrightarrow{18r}$$
$$0000000000200000_x \xrightarrow{1r} 0020000000200020_x. \tag{12}$$

The ALH of (12) is $(2^{-39})^2 \cdot 27^3 = 2^{-63.73}$. Thus, (12) can be used to distinguish 20-round PRESENT from a random permutation using the full codebook. A key-recovery (1R) attack at the top of (12) leads to a complexity of $2^{64+8}/(8 \cdot 21) \approx 2^{64.60}$ 21-round computations. The memory needed is an 8-bit counter and the success rate is about 0.246. For 80-bit keys, the remaining $80-12 = 68$ key bits can be found by exhaustive search.

For nine rounds and mask 0000000000200000_x, we have found 512 trails, each with bias 2^{-19} inside this linear hull. Concatenating the linear hull twice, we arrive at 512^2 18-round trails. The ALH $(0000000000200000_x, 0000000000200000_x)$ for 18 rounds is $(2^{-37})^2 \cdot 512^2 = 2^{-56}$. Extending it to 20 rounds, just like (12), leads to an ALH of $(2^{-39})^2 \cdot 512^2 = 2^{-60}$. A key-recovery (1R) attack at the top of this linear hull results in a complexity of $8 \cdot 2^{60+8}/(8 \cdot 21) \approx 2^{63.60}$ 21-round computations. The memory needed is an 8-bit counter and the success rate is about 0.997. For 80-bit keys, the remaining 72 key bits can be found by exhaustive search, leading to a complexity of 2^{72} encryptions.

For ten rounds and mask 0000000000200000_x, we have found 1344 trails, each with bias 2^{-21} inside this linear hull. Concatenating the linear hull twice, we arrive at 1344^2 20-round trails. The corresponding ALH $(0000000000200000_x, 0000000000200000_x)$ for 20 rounds is $(2^{-41})^2 \cdot 1344^2 = 2^{-61.22}$. Extending it to 21 rounds leads to

$$0000000000A00000_x \xrightarrow{1r} 0000000000200000_x \xrightarrow{20r}$$
$$0000000000200000_x, \tag{13}$$

with an ALH of $(2^{-42})^2 \cdot 1344^2 = 2^{-63.21}$. A key-recovery (2R) attack at both ends of this linear hull requires guessing 16 key bits. The effort becomes $2^{63.21+16}/(16 \cdot 23) \approx 2^{60.68}$ 23-round computations. The memory needed is a 16-bit counter. For 80-bit keys the remaining 64 key bits can be found by exhaustive search, leading to a final complexity of 2^{64} encryptions.

For the 21-round linear hull, with bitmask 0000000000200000_x, we have found 53582633 trails with bias 2^{-43} and the accumulated bias is $2^{-30.16}$. These trails always have one single active S-box per round. In order to improve the accumulated bias, we identify the second best trails across 21 rounds in which 23 active S-boxes are involved. Unlike the best trails, the second best ones have a '2-way branching' that is the trail splits from one to two S-boxes. This branching later merges back into a single S-box (Fig. 3) after three rounds. We developed another depth-first search program to find the 2nd-best trails for a variable number of rounds. The results are listed in Table 3. From the empirical results in Table 3, the number of 2nd best trails seems to be (# rounds-3) times more than the number of best trails. This means that the contribution of the second best trails to the overall bias of the 22-round hull is about $\sqrt{18 \cdot 53582633} \cdot 2^{-47}$ or

Table 3. Number of second best trails using bitmask 0000000000200000_x

# rounds	# best trails	# 2nd best trail	bias of 2nd best trails
5	9	18	$2^{-12.915}$
6	27	81	$2^{-13.830}$
7	72	288	$2^{-14.915}$
8	192	960	$2^{-16.046}$
9	512	3072	$2^{-17.207}$
10	1344	9536	$2^{-18.376}$
11	3528	28896	$2^{-19.565}$
12	9261	85995	$2^{-20.771}$
13	24255	252021	$2^{-21.990}$
14	63525	730235	$2^{-23.219}$

$2^{-32.077}$. Combining the biases of the 1st best trails and 2nd best trails results in $\sqrt{(2^{-30.16})^2 + (2^{-32.077})^2} \sim 2^{-30.11}$.

We now make the key recovery attack on 25-round by guessing 20 bits at both ends of the 21-round linear hull. This means $2^{64+16+16+4+4} \cdot 10/16 = 2^{103.33}$ 1-round computations, or $2^{103.33}/25 \approx 2^{98.68}$ 25-round computations, which only applies to 128-bit keys. For the remaining 88-bit subkey, we can search it exhaustively. The success rate is 0.61.

For the 22-round linear hull, with bitmask 0000000000200000_x, we have found 140281323 trails, each with bias 2^{-45} and $\sqrt{19 \cdot 140281323} \cdot 2^{-49}$ trails each with bias 2^{-49}. The corresponding ALH is $(2^{-45})^2 \cdot 140281323 + (19 \cdot 140281323) \cdot 2^{-49} \approx 2^{-62.83}$, which means an accumulated bias of $2^{-31.415}$. We use this 22-round linear hull to make a key recovery attack on 26-round PRESENT. This means $2^{64+16+16+4+4} \cdot 10/16 = 2^{103.33}$ 1-round computations, or $2^{103.33}/26 \approx 2^{98.62}$ 26-round computations, which only applies to 128-bit keys. The success rate is only 0.00002.

It is reasonable that the linear trails in a linear hull could not be independent. Kaliski et al., though, showed that the linear dependency of the linear approximations has no effect for the attack [20].

6   Conclusions

This paper described the first linear hull attacks and revisited algebraic attacks with a comparison between two distinct algorithms on reduced-round versions of the block cipher PRESENT. The analysis based on linear hulls were used to detect any significant variation in the bias, which would impact the linear attack complexities; and, to assess the linear hull effect in PRESENT and its resilience to LC. We have confirmed that the linear hull effect is significant even for a small number of rounds of PRESENT.

Table 4 lists the attack complexities for PRESENT for increasing number of rounds and in increasing order of time complexity.

Table 4. Attack complexities on reduced-round PRESENT block cipher

#Rounds	Time	Data	Memory	Key Size (bits)	Source	Comments
5	2.5 min	5 KP	—	80	Sect. 3	KR†, AC
5	2.5 min	5 KP	—	128	Sect. 3	KR†, AC
5	1.82 h	64 KP	—	80	[10]	KR, AC
6	2^{26}	2^{26} KP	—	all	eq. (6)	DR* + KR, LC
7	$2^{100.1}$	$2^{24.3}$ CP	2^{77}	128	[30]	IC
14	2^{61}	2^{61} CO	—	all	eq. (1)	DR* + KR, LC
15	$2^{35.6}$	$2^{35.6}$ CP	2^{32}	all	[5]	KR, SC
15	2^{64}	2^{64} KP	—	all	eq. (1)	DR*, LC
16	2^{62}	2^{62} CP	1Gb	all	[1]	KR, AC + DC
16	2^{64}	2^{64} CP	2^{32}	all	[29]	KR, DC
17	$2^{69.50}$	2^{64} KP	2^{12}	80	eq. (9)	KR, LC
17	$2^{73.91}$	2^{64} KP	2^{16}	80	eq. (9)	KR, LC
17	2^{104}	2^{63} CP	2^{53}	128	[25]	KR, RKR
17	2^{93}	2^{62} CP	1Gb	128	[1]	KR, AC + DC
18	2^{98}	2^{62} CP	1Gb	128	[1]	KR, AC + DC
19	2^{113}	2^{62} CP	1Gb	128	[1]	KR, AC + DC
24	2^{57}	2^{57} CP	2^{32}	all	[5]	KR, SSC
25	$2^{98.68}$	2^{64} KP	2^{40}	128	Table 2	KR, LH
26	$2^{98.62}$	2^{64} KP	2^{40}	128	Table 2	KR, LH

*: time complexity is number of parity computations; †: recover half of the user key;
DR: Distinguish-from-Random attack; KR: Key Recovery attack
LC: Linear Cryptanalysis; DC: Differential Cryptanalysis; AC: Algebraic Crypt.;
SSC: Statistical Saturation analysis; IC: Integral Cryptanalysis;
RKR: Related-Key Rectangle; LH: Linear Hull; ML: Multiple Linear;
CP: Chosen Plaintext; KP: Known Plaintext; CO: Ciphertext Only.

A topic for further research is to look for the 3rd and 4th best trails inside a linear hull. The issue is to find out their contribution to the overall bias of the linear hulls, that is, if they can further improve the bias as the 2nd best trails did.

Acknowledgements

We would like to thank anonymous reviewers for their very important comments.

References

1. Albrecht, M., Cid, C.: Algebraic Techniques in Differential Cryptanalysis. In: Dunkelman, O. (ed.) FSE 2009. LNCS, vol. 5565, pp. 193–208. Springer, Heidelberg (2009)
2. Brickenstein, M., Dreyer, A.: PolyBoRi: A framework for Gröbner basis computations with Boolean polynomials. Electronic Proceedings of MEGA (2007), http://www.ricam.oeaw.ac.at/mega2007/electronic/26.pdf

3. Bogdanov, A., Knudsen, L.R., Leander, G., Paar, C., Poschmann, A., Robshaw, M.J.B., Seurin, Y., Vikkelsoe, C.: PRESENT: An ultra-lightweight block cipher. In: Paillier, P., Verbauwhede, I. (eds.) CHES 2007. LNCS, vol. 4727, pp. 450–466. Springer, Heidelberg (2007)
4. Buchberger, B.: An Algorithm for Finding the Basis Elements of the Residue Class Ring of a Zero Dimensional Polynomial Ideal. Ph.D Dissertation (1965)
5. Collard, B., Standaert, F.X.: A Statistical Saturation Attack against the Block Cipher PRESENT. In: CT-RSA 2009. LNCS, vol. 5473, pp. 195–210. Springer, Heidelberg (2009)
6. Courtois, N., Shamir, A., Patarin, J., Klimov, A.: Efficient Algorithms for Solving Overdefined Systems of Multivariate Polynomial Equations. Adv. In: Preneel, B. (ed.) EUROCRYPT 2000. LNCS, vol. 1807, pp. 392–407. Springer, Heidelberg (2000)
7. Courtois, N., Pieprzyk, J.: Cryptanalysis of Block Ciphers with Overdefined Systems of Equations. In: Zheng, Y. (ed.) ASIACRYPT 2002. LNCS, vol. 2501, pp. 267–287. Springer, Heidelberg (2002)
8. Courtois, N., Meier, W.: Algebraic Attacks on Stream Ciphers with Linear Feedback. In: Biham, E. (ed.) EUROCRYPT 2003. LNCS, vol. 2656, pp. 345–359. Springer, Heidelberg (2003)
9. Courtois, N.T., Bard, G.V.: Algebraic cryptanalysis of the data encryption standard. In: Galbraith, S.D. (ed.) Cryptography and Coding 2007. LNCS, vol. 4887, pp. 152–169. Springer, Heidelberg (2007)
10. Courtois, N.T., Debraize, B.: Specific S-Box Criteria in Algebraic Attacks on Block Ciphers with Several Known Plaintexts. In: Lucks, S., Sadeghi, A.-R., Wolf, C. (eds.) WEWoRC 2007. LNCS, vol. 4945, pp. 100–113. Springer, Heidelberg (2008)
11. Courtois, N.T., Bard, G.V., Wagner, D.: Algebraic and slide attacks on keeLoq. In: Nyberg, K. (ed.) FSE 2008. LNCS, vol. 5086, pp. 97–115. Springer, Heidelberg (2008)
12. Courtois, N.T.: Tools for experimental algebraic cryptanalysis, http://www.cryptosystem.net/aes/tools.html
13. Daemen, J., Rijmen, V.: The Design of Rijndael: AES - The Advanced Encryption Standard. Springer, Heidelberg (2002)
14. Eén, N., Sörensson, N.: MiniSat 2.0. An open-source SAT solver package, http://www.cs.chalmers.se/Cs/Research/FormalMethods/MiniSat/
15. Faugére, J.: A new efficient algorithm for computing Gröbner bases (F4). Journal of Pure and Applied Algebra, 61–69 (1999)
16. Faugére, J.: A new efficient algorithm for computing Gröbner bases without reduction to zero (F5). In: Symbolic and Algebraic Computation - ISSAC, pp. 75–83 (2002)
17. Ghasemzadeh, M.: A New Algorithm for the Quantified Satisfiability Problem, Based on Zero-suppressed Binary Decision Diagrams and Memorization. Ph.D. thesis, Potsdam, Germany, University of Potsdam (2005), http://opus.kobv.de/ubp/volltexte/2006/637/
18. Indesteege, S., Keller, N., Dunkelman, O., Biham, E., Preneel, B.: A practical attack on keeLoq. In: Smart, N.P. (ed.) EUROCRYPT 2008. LNCS, vol. 4965, pp. 1–18. Springer, Heidelberg (2008)
19. Keliher, L., Meijer, H., Tavares, S.: High Probability Linear Hulls in Q. In: Second NESSIE Conference (2001)
20. Kaliski, B.S., Robshaw, M.J.B.: Linear Cryptanalysis Using Multiple Approximations and FEAL. In: Preneel, B. (ed.) FSE 1994. LNCS, vol. 1008, pp. 249–264. Springer, Heidelberg (1995)

21. Magma, software package, http://magma.maths.usyd.edu.au/magma/
22. Matsui, M.: Linear cryptanalysis method for DES cipher. In: Helleseth, T. (ed.) EUROCRYPT 1993. LNCS, vol. 765, pp. 386–397. Springer, Heidelberg (1994)
23. Murphy, S., Robshaw, M.J.B.: Essential Algebraic Structure within AES. Adv. In: Yung, M. (ed.) CRYPTO 2002. LNCS, vol. 2442, pp. 1–16. Springer, Heidelberg (2002)
24. Nyberg, K.: Linear approximation of block ciphers. Adv. In: De Santis, A. (ed.) EUROCRYPT 1994. LNCS, vol. 950, pp. 439–444. Springer, Heidelberg (1995)
25. Özen, O., Varici, K., Tezcan, C., Kocair, Ç.: Lightweight Block Ciphers Revisited: Cryptanalysis of Reduced Round PRESENT and HIGHT. In: ACISP 2009. LNCS, vol. 5594, pp. 90–107. Springer, Heidelberg (2009)
26. Raddum, H., Semaev, I.: New technique for solving sparse equation systems. Cryptology ePrint Archive, Report 2006/475 (2006), http://eprint.iacr.org/2006/475
27. Shannon, C.E.: Claude Elwood Shannon collected papers. Wiley-IEEE Press, Piscataway (1993)
28. Selçuk, A.A., Biçak, A.: On probability of success in linear and differential cryptanalysis. In: Cimato, S., Galdi, C., Persiano, G. (eds.) SCN 2002. LNCS, vol. 2576, pp. 174–185. Springer, Heidelberg (2003)
29. Wang, M.: Differential Cryptanalysis of reduced-round PRESENT. In: Vaudenay, S. (ed.) AFRICACRYPT 2008. LNCS, vol. 5023, pp. 40–49. Springer, Heidelberg (2008)
30. Z'aba, M.R., Raddum, H., Henricksen, M., Dawson, E.: Bit-Pattern Based Integral Attack. In: Nyberg, K. (ed.) FSE 2008. LNCS, vol. 5086, pp. 363–381. Springer, Heidelberg (2008)

A   ElimLin Intermediate Results

Table 7 depicts the intermediate ElimLin results for 5-round PRESENT-80 where 36 bits of the key are fixed and we try to recover the remaining key bits. In the third column, T represents the total number of monomials and Ave is the average number of monomials per equation.

Table 5. The 4×4-bit S-box of PRESENT and the inverse S-box

x	0	1	2	3	4	5	6	7	8	9	10	11	12	13	14	15
$S[x]$	12	5	6	11	9	0	10	13	3	14	15	8	4	7	1	2
$S^{-1}[x]$	5	14	15	8	12	1	2	13	11	4	6	3	0	7	9	10

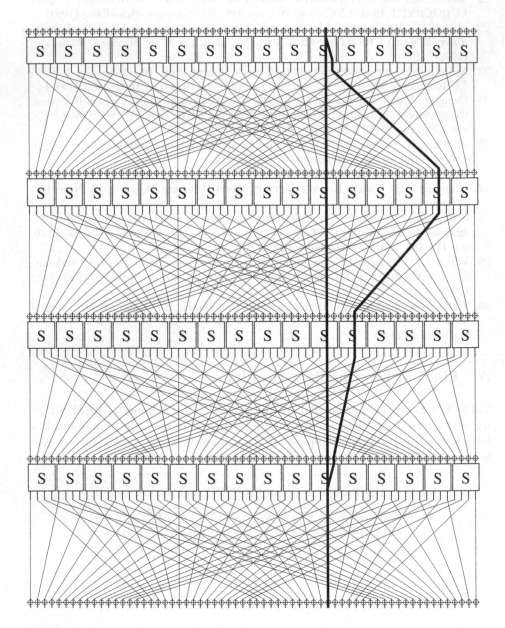

Fig. 3. Example of branching inside a trail, from single S-box to two S-boxes, and merging back to one S-box

Table 6. Linear Approximation Table (LAT) of the S-box of PRESENT

IM	OM															
	0	1	2	3	4	5	6	7	8	9	10	11	12	13	14	15
0	8	0	0	0	0	0	0	0	0	0	0	0	0	0	0	0
1	0	0	0	0	0	-4	0	-4	0	0	0	0	0	-4	0	4
2	0	0	2	2	-2	-2	0	0	2	-2	0	4	0	4	-2	2
3	0	0	2	2	2	-2	-4	0	-2	2	-4	0	0	0	-2	-2
4	0	0	-2	2	-2	-2	0	4	-2	-2	0	-4	0	0	-2	2
5	0	0	-2	2	-2	2	0	0	2	2	-4	0	4	0	2	2
6	0	0	0	-4	0	0	-4	0	0	-4	0	0	4	0	0	0
7	0	0	0	4	4	0	0	0	0	-4	0	0	0	0	4	0
8	0	0	2	-2	0	0	-2	2	-2	2	0	0	-2	2	4	4
9	0	4	-2	-2	0	0	2	-2	-2	-2	-4	0	-2	2	0	0
10	0	0	4	0	2	2	2	-2	0	0	0	-4	2	2	-2	2
11	0	-4	0	0	-2	-2	2	-2	-4	0	0	0	2	2	2	-2
12	0	0	0	0	-2	-2	-2	-2	4	0	0	-4	-2	2	2	-2
13	0	4	4	0	-2	-2	2	2	0	0	0	0	2	-2	2	-2
14	0	0	2	2	-4	4	-2	-2	-2	-2	0	0	-2	-2	0	0
15	0	4	-2	2	0	0	-2	-2	-2	2	4	0	2	2	0	0

Table 7. ElimLin result for 5-round PRESENT-80 when 36 bits of key are fixed

# Variables	# Equations	(Ave/ # Monomials)	# Linear Equations
10340	46980	7/ T= 46321	6180
4160	46980	8/ T= 48744	1623
2537	46980	9/ T= 40763	1069
1468	46980	14/ T= 43155	405
1063	46980	76/ T= 73969	165
898	46980	158/ T= 145404	201
697	46980	77/ T= 85470	584
113	46980	0/ T= 413	113

Saturation Attack on the Block Cipher HIGHT

Peng Zhang[1], Bing Sun[1], and Chao Li[1,2]

[1] Department of Mathematics and System Science, Science College of National
University of Defence Technology, Changsha, 410073, China
[2] State Key Laboratory of Information Security, Institute of Software of Chinese
Academy of Sciences, Beijing, 100190, China
cheetahzhp@163.com

Abstract. HIGHT is a block cipher with 64-bit block length and 128-bit key length, which was proposed by Hong et al. in CHES 2006 for extremely constrained environments such as RFID tags and sensor networks. In this paper, a new saturation attack on HIGHT is discussed. We first point out and correct an error in the 12-round saturation distinguishers given by the HIGHT proposers. And then two new 17-round saturation distinguishers are described. Finally, we present a 22-round saturation attack on HIGHT including full whitening keys with $2^{62.04}$ chosen plaintexts and $2^{118.71}$ 22-round encryptions.

Keywords: Block cipher, HIGHT, Distinguisher, Saturation attack.

1 Introduction

With the establishment of the AES[18], the need for new block ciphers has been greatly diminished. The AES is an excellent and preferred choice for almost all block cipher applications. However, despite recent implementation advances, the AES is not suitable for extremely constrained environments such as RFID(Radio Frequency Identification) tags and sensor networks which are low-cost with limited resources. So some block ciphers suitable for these environments have been designed, such as TEA[25], HIGHT[10], SEA[22], CGEN[20], mCrypton[13], PRESENT[5] and so on.

HIGHT[10] is a block cipher with 64-bit block length and 128-bit key length, which is suitable for low-cost, low-power and ultra-light implementations, such as RFID systems. It has a 32-round iterative structure which is a variant of generalized Feistel network. Due to the simple operations such as XOR, addition mod 2^8 and left bitwise rotation, HIGHT is especially efficient in hardware implementations.

Square attack, proposed by Daemen et al. in[6], is a dedicated attack on the block cipher SQUARE. And it has been applied to some other block ciphers based on the SPN structure. In order to apply square attack to the Feistel structure, Lucks introduced the saturation attack on the Twofish cipher in FSE 2001[16], which is a variation of square attack. Then in[4], Biryukov et al. proposed the multiset attack by which we can break a 4-round SPN cipher even

J.A. Garay, A. Miyaji, and A. Otsuka (Eds.): CANS 2009, LNCS 5888, pp. 76–86, 2009.

if the S-box is unknown. And in[12], a more generalized attack, named integral attack, was proposed by Knudsen et al.. Furthermore, Z'aba et al. presented the bit-pattern based integral attack against bit-based block ciphers in[28], which is a new type of integral attack. Consequently, some famous ciphers have been analyzed based on the idea of the attacks shown above, such as Rijndael[8], FOX[26], Camellia[7,9,27], SMS4[14], CLEFIA[21,23] and so on.

In[10], the HIGHT proposers had analyzed its security against some existing cryptanalytic attacks, they described a differential attack[3], a linear attack[17] and a boomerang attack[24] on 13-round HIGHT, a truncated differential attack[11] and a saturation attack[16] on 16-round HIGHT, an impossible differential attack[1] on 18-round HIGHT, and finally a related-key boomerang attack[2] on 19-round HIGHT. Subsequently, Lu gave an impossible differential attack on 25-round HIGHT, a related-key rectangle attack on 26-round HIGHT, and a related-key impossible differential attack on 28-round HIGHT[15]. Furthermore, Özen et al. presented an impossible differential attack on 26-round HIGHT and a related-key impossible differential attack on 31-round HIGHT[19].

Particularly in [10], the proposers gave some 12-round saturation distinguishers which were applied to the attack on 16-round HIGHT with 2^{42} chosen plaintexts and 2^{51} 16-round encryptions. In this paper, we first point out and correct an error in the 12-round saturation distinguishers shown in[10]. Then two new 17-round saturation distinguishers are described. Finally, we present a 22-round saturation attack on HIGHT including full whitening keys with $2^{62.04}$ chosen plaintexts and $2^{118.71}$ 22-round encryptions.

The paper is organized as follows. In Section 2, we briefly describe some notations and the HIGHT block cipher. Section 3 corrects an error in the 12-round distinguishers shown in[10] and gives two new 17-round saturation distinguishers. Then the attacks are discussed in section 4. Finally, we conclude this paper and summarize our findings in Section 5.

2 Preliminaries

2.1 Notations

In order to clearly illustrate the following encryption and attack, some notations and symbols are defined as follows:

$\oplus$: XOR (exclusive OR);

$\boxplus$: addtion modulo 2^8;

$A^{<<<s}$: s-bit left rotation of an 8-bit value A;

$P = (P_7, P_6, \cdots, P_0)$: the plaintext;

$C = (C_7, C_6, \cdots, C_0)$: the ciphertext;

$X_i = (X_{i,7}, X_{i,6}, \cdots, X_{i,0})$: the output of the i-th round($1 \leq i \leq 32$);

$X_{i-1} = (X_{i-1,7}, X_{i-1,6}, \cdots, X_{i-1,0})$: the input of the i-th round($1 \leq i \leq 32$);

$X_{i,j}^{(k)}(0 \leq k \leq 7)$: the k-th bit of $X_{i,j}$;

$MK = (MK_{15}, MK_{14}, \cdots, MK_0)$: the master key;

$MK_{i,j}(0 \leq j \leq 7)$: the j-th bit of MK_i;

$WK_i(0 \leq i \leq 7)$: the whitening keys;

$WK_{i,j}(0 \leq j \leq 7)$: the j-th bit of WK_i;

$SK_i(0 \leq i \leq 127)$: the round keys;

$SK_{i,j}(0 \leq j \leq 7)$: the j-th bit of SK_i.

2.2 The HIGHT Block Cipher

The encryption procedure can be described as follows:

Step1. Perform the Initial Transformation, which transforms a plaintext P to the input of the first round X_0 by using the four whitening keys, WK_0, WK_1, WK_2 and WK_3.

$$X_{0,0} = P_0 \boxplus WK_0; X_{0,1} = P_1; X_{0,2} = P_2 \oplus WK_1; X_{0,3} = P_3;$$
$$X_{0,4} = P_4 \boxplus WK_2; X_{0,5} = P_5; X_{0,6} = P_6 \oplus WK_3; X_{0,7} = P_7.$$

Step2. For $i = 1, 2, \cdots, 32$, Round Function transforms X_{i-1} to X_i as follows, which is shown in Fig.1.

$$X_{i,0} = X_{i-1,7} \oplus (F_0(X_{i-1,6}) \boxplus SK_{4i-1});$$
$$X_{i,1} = X_{i-1,0};$$
$$X_{i,2} = X_{i-1,1} \boxplus (F_1(X_{i-1,0}) \oplus SK_{4i-2});$$
$$X_{i,3} = X_{i-1,2};$$
$$X_{i,4} = X_{i-1,3} \oplus (F_0(X_{i-1,2}) \boxplus SK_{4i-3});$$
$$X_{i,5} = X_{i-1,4};$$
$$X_{i,6} = X_{i-1,5} \boxplus (F_1(X_{i-1,4}) \oplus SK_{4i-4});$$
$$X_{i,7} = X_{i-1,6}.$$

And the functions F_0 and F_1 are defined as:

$$F_0(x) = (x^{<<<1}) \oplus (x^{<<<2}) \oplus (x^{<<<7}),$$
$$F_1(x) = (x^{<<<3}) \oplus (x^{<<<4}) \oplus (x^{<<<6}).$$

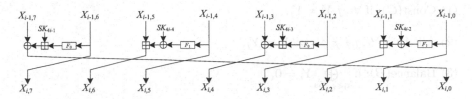

Fig. 1. The i-th Round Function of HIGHT for $i = 1, 2, \cdots, 32$

Step3. Perform the Final Transformation, which untwists the swap of the last round function and transforms X_{32} to the ciphertext C by using the four whitening keys, WK_4, WK_5, WK_6 and WK_7.

$$C_0 = X_{32,1} \boxplus WK_4; C_1 = X_{32,2}; C_2 = X_{32,3} \oplus WK_5; C_3 = X_{32,4};$$
$$C_4 = X_{32,5} \boxplus WK_6; C_5 = X_{32,6}; C_6 = X_{32,7} \oplus WK_7; C_7 = X_{32,0}.$$

The key schedule of HIGHT consists of two algorithms, which generate 8 whitening key bytes $WK_i (0 \le i \le 7)$ and 128 subkey bytes $SK_j (0 \le j \le 127)$. Firstly, the 128-bit master key is considered as a concatenation of 16 bytes and denoted by $MK = (MK_{15}, MK_{14}, \cdots, MK_0)$. Then the whitening key bytes are generated as follows:

$$WK_i = MK_{i+12}(i = 0, 1, 2, 3);$$
$$WK_i = MK_{i-4}(i = 4, 5, 6, 7).$$

And the subkey bytes are generated as follows:

$$SK_{16 \cdot i + j} = MK_{j-i \bmod 8} \boxplus \delta_{16 \cdot i + j}(0 \le i, j \le 7),$$
$$\text{or } SK_{16 \cdot i + j + 8} = MK_{(j-i \bmod 8)+8} \boxplus \delta_{16 \cdot i + j + 8}(0 \le i, j \le 7),$$

where $\delta_{16 \cdot i + j}$ and $\delta_{16 \cdot i + j + 8}$ are public constants.

3 Saturation Distinguishers

Now we will point out and correct an error in the saturation distinguishers shown in[10]. Furthermore, two new 17-round saturation distinguishers will be presented.

3.1 Distinguishers based on Byte Saturation

Let $S = \{Y_i | Y_i = (Y_{i,7}, Y_{i,6}, \cdots, Y_{i,0}) \in \{0, 1\}^8, 0 \le i < 2^8\}$ be a set of 8-bit values, where $Y_{i,k}(0 \le k \le 7)$ is the k-th bit of Y_i. Then we categorize the status of the set S into five groups depending on the conditions defined as follows:

(1) Const(C): if $\forall i, j, Y_i = Y_j$,

(2) All(A): if $\forall i, j, i \neq j \iff Y_i \neq Y_j$,

(3) Balance(B): if $\displaystyle\bigoplus_{0 \leq i < 2^8} Y_i = 0$,

(4) Balance$_k$(B_k): if $\displaystyle\bigoplus_{0 \leq i < 2^8} Y_{i,k} = 0$,

(5) Unknown(U): unknown.

Thus, using the conditions, we can get two 11-round saturation distinguishers based on byte saturation which are shown as follows. And the details of distinguisher(I) is shown in Fig.2.

$$\text{(I)} \quad (A, C, C, C, C, C, C, C) \xrightarrow{11r} (U, U, U, U, B_0, U, U, U),$$

$$\text{(II)} \quad (C, C, C, C, A, C, C, C) \xrightarrow{11r} (B_0, U, U, U, U, U, U, U).$$

In[10], the HIGHT proposers had described that distinguisher(I) is a 12-round one. From the details shown in Fig.2, we can see that it is only a 11-round one. In addition, we also implement the distinguishers through the computer simulation, and we find that there are no Balance or Balance$_k$ sets in the output of the 12-round encryptions. That is to say, the distinguishers based on byte saturation are only 11-round ones.

3.2   17-Round Distinguishers

With the two 11-round distinguishers shown above, we can deduce two 17-round ones, respectively. Hereinafter, we will give the details.

Firstly, we explain how to extend to 12-round distinguishers. Let $A_{(16)}$ be an All state of 16-bit values, and it is divided into two segments as $A_{(16)} = A_{1(16)} | A_{0(16)}$. Then we can get the 12-round distinguishers as follows:

$$\text{(I)} \quad (A_{1(16)}, A_{0(16)}, C, C, C, C, C, C) \xrightarrow{12r} (U, U, U, U, B_0, U, U, U),$$

$$\text{(II)} \quad (C, C, C, C, A_{1(16)}, A_{0(16)}, C, C) \xrightarrow{12r} (B_0, U, U, U, U, U, U, U).$$

These distinguishers can be explained in the following way. After the first round of distinguisher(I), $(A_{1(16)}, A_{0(16)}, C, C, C, C, C, C)$ becomes $(A_{0(16)}, C, C, C, C, C, C, A'_{1(16)})$, where the concatenated segment $A'_{1(16)} | A_{0(16)}$ is also an All sate. It can be regarded as that $(A_{0(16)}, C, C, C, C, C, C, A'_{1(16)})$ contains 2^8 structures of (A, C, C, C, C, C, C, C), where the first rightmost constant takes all possible 2^8 8-bit values. Therefore, the Balance$_0$ state is kept at the output, which is presented in the above 11-round distinguishers. In addition, distinguishers(II) can also be explained using the similar method.

The extensions to 13,14,15,16 and 17-round distinguishers can be obtained in the similar way. Let $A_{(56)}$ be an All state of 56-bit values, which is divided into

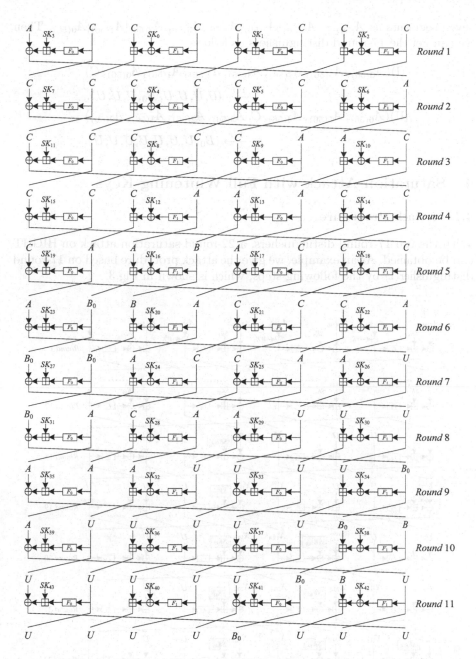

Fig. 2. 11-Round Distinguisher(I) Based on Byte Saturation

seven segments as $A_{(56)} = A_{6(56)}|A_{5(56)}|A_{4(56)}|A_{3(56)}|A_{2(56)}|A_{1(56)}|A_{0(56)}$. Then we can get the 17-round distinguishers as follows.

$$(\mathrm{I})\,(A_{6(56)}, A_{5(56)}, A_{4(56)}, A_{3(56)}, A_{2(56)}, A_{1(56)}, A_{0(56)}, C)$$
$$\xrightarrow{17r} (U, U, U, U, U, B_0, U, U, U),$$
$$(\mathrm{II})\,(A_{6(56)}, A_{5(56)}, A_{4(56)}, C, A_{3(56)}, A_{2(56)}, A_{1(56)}, A_{0(56)})$$
$$\xrightarrow{17r} (B_0, U, U, U, U, U, U, U).$$

4   Saturation Attack with Full Whitening Keys

4.1   Attack Procedure

With the two 17-round distinguishers, a 22-round saturation attack on HIGHT can be obtained. As an example, we give the attack procedure based on 17-round distinguisher(I) by the following steps, which is shown in Fig.3.

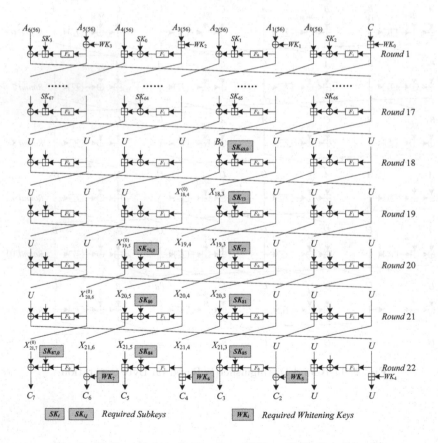

Fig. 3. 22-Round Attack Based on 17-Round Distinguisher(I)

Step1. Chose a set of 2^{56} plaintexts which has the following format:

$$S = (A_{6(56)}, A_{5(56)}, A_{4(56)}, A_{3(56)}, A_{2(56)}, A_{1(56)}, A_{0(56)}, C).$$

Ask for the corresponding ciphertexts C_7, C_6, C_5, C_4, C_3 and C_2 after the 22-round encryptions.

Step2. Guess the subkeys and the whitening keys as follows:

$$SK_{69,0}, SK_{73}, SK_{76,0}, SK_{77}, SK_{80}, SK_{81}, SK_{84}, SK_{85}, SK_{87,0},$$
$$WK_5, WK_6, WK_7.$$

Since the value $X_{17,3}$ which is a $Balance_0$ state can be calculated by the guessed keys and the ciphertexts obtained in Step1. Then the correct keys can be checked by the following equation.

$$\bigoplus_{P \in S} X_{17,3}^{(0)} = \bigoplus_{P \in S} F(C_i, SK_{j,0}, SK_k, WK_l) = 0, \tag{1}$$

where P is the plaintext, $i=7,6,5,4,3,2$, $j = 69, 76, 87$, $k = 73, 77, 80, 81, 84, 85$ and $l = 5, 6, 7$.

If the equation is satisfied, accept the guessed keys as the candidates for the correct keys. Otherwise, discard the keys.

Step3. For the key candidates kept in Step2, chose another set which has the format shown in Step1 and check whether Eq.1 is satisfied. If it is satisfied, keep the keys. Otherwise, discard the ones. Repeat this step, until all the correct keys are obtained exclusively.

Using the similar method, we can get another 22-round saturation attack based on 17-round distinguisher(II). And the following keys require to be guessed in the attack.

$$SK_{71,0}, SK_{75}, SK_{78,0}, SK_{79}, SK_{82}, SK_{83}, SK_{85,0}, SK_{86}, SK_{87},$$
$$WK_4, WK_5, WK_7.$$

4.2   Complexity Analysis

From the key schedule, it is clear that all the guessed keys in the attack are generated by the master key bytes $MK_i (0 \leq i \leq 15)$, respectively. And the generations are shown in Table 1 and Table 2.

In the tables, we can find that some guessed keys are generated by the same master key byte. Thus we need not guess all of them. In addition, the guessed keys in the attack based on distinguisher(II) which have been obtained in the attack based on distinguisher(I) are not required anymore.

Consequently, the guessed keys in the attack based on distinguisher(I) are listed as follows:

$$MK_0, MK_1, MK_2, MK_3, MK_4, MK_7, MK_{8,0}, MK_9, MK_{13}.$$

Table 1. Generation of Guessed Keys in the Attack Based on Distinguisher(I)

$SK_{69,0}$	SK_{73}	$SK_{76,0}$	SK_{77}	SK_{80}	SK_{81}	SK_{84}	SK_{85}	$SK_{87,0}$	WK_5	WK_6	WK_7
$MK_{1,0}$	MK_{13}	$MK_{8,0}$	MK_9	MK_3	MK_4	MK_7	MK_0	$MK_{2,0}$	MK_1	MK_2	MK_3

Table 2. Generation of Guessed Keys in the Attack Based on Distinguisher(II)

$SK_{71,0}$	SK_{75}	$SK_{78,0}$	SK_{79}	SK_{82}	SK_{83}	$SK_{85,0}$	SK_{86}	SK_{87}	WK_4	WK_5	WK_7
$MK_{3,0}$	MK_{15}	$MK_{10,0}$	MK_{11}	MK_5	MK_6	$MK_{0,0}$	MK_1	SK_2	MK_0	MK_1	MK_3

And we merely need guess the following keys in the attack based on distinguisher(II):

$$MK_5, MK_6, MK_{10,0}, MK_{11}, MK_{15}.$$

In the attack, the probability that a key candidate in the key space survives the discarding step is expected to be 2^{-1}. And we need to guess 65-bit and 33-bit key values in the attack based on distinguisher(I) and (II), respectively. Then if we suppose that N sets are required, it is satisfied that

$$2^{65} \cdot 2^{-N} < 1, \text{ and } 2^{33} \cdot 2^{-N} < 1.$$

Besides, the sets used in the attack based on distinguisher(I) can also be used in the attack based on distinguisher(II). Therefore, 66 sets of 2^{56} plaintexts are required to obtain the correct key values exclusively. That is to say, we need $66 \times 2^{56} \approx 2^{62.04}$ plaintexts in the attack. Additionally, we need $(2^{56} \times 2^{65} + 2^{56} \times 2^{64} + \cdots + 2^{56}) + (2^{56} \times 2^{33} + 2^{56} \times 2^{32} + \cdots + 2^{56}) \approx 2^{122}$ F-function computations to get the correct keys. Consequently, the time complexity is $2^{122} \times \frac{9}{22 \times 4} \approx 2^{118.71}$ 22-round encryptions.

5   Conclusion

In this paper, a new saturation attack on HIGHT is discussed. We first point out and correct an error in the 12-round saturation distinguishers shown by the HIGHT proposers. And then two new 17-round saturation distinguishers are described. Finally, we present a 22-round saturation attack on HIGHT including full whitening keys with $2^{62.04}$ chosen plaintexts and $2^{118.71}$ 22-round encryptions. The attack presented in this paper shows that the reduced versions of HIGHT are less secure than they should be.

Acknowledgments. The authors would like to thank the anonymous reviewers for many helpful comments and suggestions. The work in this paper is supported by the Natural Science Foundation of China (No:60803156) and the open research fund of State Key Laboratory of Information Security(No: 01-07).

References

1. Biham, E., Biryukov, A., Shamir, A.: Cryptanalysis of Skipjack reduced to 31 rounds using impossible differentials. In: Stern, J. (ed.) EUROCRYPT 1999. LNCS, vol. 1592, pp. 12–23. Springer, Heidelberg (1999)
2. Biham, E., Dunkelman, O., Keller, N.: Related-key boomerang and rectangling attacks. In: Cramer, R. (ed.) EUROCRYPT 2005. LNCS, vol. 3494, pp. 507–525. Springer, Heidelberg (2005)
3. Biham, E., Shamir, A.: Differential cryptanalysis of DES-like cryptosystems. In: Menezes, A.J., Vanstone, S.A. (eds.) CRYPTO 1990. LNCS, vol. 537, pp. 2–21. Springer, Heidelberg (1991)
4. Biryukov, A., Shamir, A.: Structural cryptanalysis of SASAS. In: Pfitzmann, B. (ed.) EUROCRYPT 2001. LNCS, vol. 2045, pp. 394–405. Springer, Heidelberg (2001)
5. Bogdanov, A., Knudsen, L.R., Leander, G., Paar, C., Poschmann, A., Robshaw, M.J.B., Seurin, Y., Vikkelsoe, C.: PRESENT: An Ultra-Lightweight Block Cipher. In: Paillier, P., Verbauwhede, I. (eds.) CHES 2007. LNCS, vol. 4727, pp. 450–466. Springer, Heidelberg (2007)
6. Daemen, J., Knudsen, L.R., Rijmen, V.: The Block Cipher SQUARE. In: Biham, E. (ed.) FSE 1997. LNCS, vol. 1267, pp. 149–165. Springer, Heidelberg (1997)
7. Duo, L., Li, C., Feng, K.: New observation on Camellia. In: Preneel, B., Tavares, S. (eds.) SAC 2005. LNCS, vol. 3897, pp. 51–64. Springer, Heidelberg (2006)
8. Ferguson, N., Kelsey, J., Lucks, S., Schneier, B., Stay, M., Wagner, D., Whiting, D.: Improved Cryptanalisis of Rijndael. In: Schneier, B. (ed.) FSE 2000. LNCS, vol. 1978, pp. 213–230. Springer, Heidelberg (2001)
9. He, Y., Qing, S.: Square Attack on Reduced Camellia Cipher. In: Qing, S., Okamoto, T., Zhou, J. (eds.) ICICS 2001. LNCS, vol. 2229, pp. 238–245. Springer, Heidelberg (2001)
10. Hong, D., Sung, J., Hong, S., Kim, J., Lee, S., Koo, B.-S., Lee, C., Chang, D., Lee, J., Jeong, K., Kim, H., Kim, J., Chee, S.: HIGHT: A New Block Cipher Suitable for Low-Resource Device. In: Goubin, L., Matsui, M. (eds.) CHES 2006. LNCS, vol. 4249, pp. 46–59. Springer, Heidelberg (2006)
11. Knudsen, L.R.: Trucated and higher order differentials. In: Preneel, B. (ed.) FSE 1994. LNCS, vol. 1008, pp. 196–211. Springer, Heidelberg (1995)
12. Knudsen, L.R., Wagner, D.: Integral cryptanalysis. In: Daemen, J., Rijmen, V. (eds.) FSE 2002. LNCS, vol. 2365, pp. 112–127. Springer, Heidelberg (2002)
13. Lim, C., Korkishko, T.: mCrypton — A Lightweight Block Cipher for Security of Low-cost RFID Tags and Sensors. In: Song, J.-S., Kwon, T., Yung, M. (eds.) WISA 2005. LNCS, vol. 3786, pp. 243–258. Springer, Heidelberg (2006)
14. Liu, F., Ji, W., Hu, L., Ding, J., Lv, S., Pyshkin, A., Weinmann, R.-P.: Analysis of the SMS4 block cipher. In: Pieprzyk, J., Ghodosi, H., Dawson, E. (eds.) ACISP 2007. LNCS, vol. 4586, pp. 158–170. Springer, Heidelberg (2007)
15. Lu, J.: Cryptanalysis of Reduced Versions of the HIGHT BLOCK Cipher from CHES 2006. In: Nam, K.-H., Rhee, G. (eds.) ICISC 2007. LNCS, vol. 4817, pp. 11–26. Springer, Heidelberg (2007)
16. Lucks, S.: The Saturation Attack—A Bait for Twofish. In: Matsui, M. (ed.) FSE 2001. LNCS, vol. 2355, pp. 1–15. Springer, Heidelberg (2002)
17. Matsui, M.: Linear cryptanalysis method for DES cipher. In: Helleseth, T. (ed.) EUROCRYPT 1993. LNCS, vol. 765, pp. 386–397. Springer, Heidelberg (1994)

18. National Institute of Standards and Technology: FIPS 197: Advanced Encryption Standard (2001), http://csrc.nist.gov
19. Özen, O., Vaici, K., Tezcan, C., Kocair, C.: Lightweight Block Cipher Revisited: Cryptanalysis of Reduced Round PRESENT and HIGHT. In: Boyd, C., González Nieto, J. (eds.) ACISP 2009. LNCS, vol. 5594, pp. 90–107. Springer, Heidelberg (2009)
20. Robshaw, M.J.B.: Searching for compact algorithms: CGEN. In: Nguyên, P.Q. (ed.) VIETCRYPT 2006. LNCS, vol. 4341, pp. 37–49. Springer, Heidelberg (2006)
21. Sony Corporation: The 128-bit Blockcipher CLEFIA: Security and Performance Evaluation. Revision 1.0 (2007)
22. Standaert, F.-X., Piret, G., Gershenfeld, N., Quisquater, J.-J.: SEA: A Scalable Encryption Algorithm for Small Embedded Applications. In: Domingo-Ferrer, J., Posegga, J., Schreckling, D. (eds.) CARDIS 2006. LNCS, vol. 3928, pp. 222–236. Springer, Heidelberg (2006); IFIP International Federation for Information Processing (2006)
23. Wang, W., Wang, X.: Saturation cryptanalysis of CLEFIA(in Chinese). Journal on Communications 29(10), 88–92 (2008)
24. Wangner, D.: The boomerang attack. In: Knudsen, L.R. (ed.) FSE 1999. LNCS, vol. 1636, pp. 156–170. Springer, Heidelberg (1999)
25. Wheeler, D., Needham, R.: TEA, a Tiny Encryption Algorithm. In: Preneel, B. (ed.) FSE 1994. LNCS, vol. 1008, pp. 363–366. Springer, Heidelberg (1995)
26. Wu, W., Zhang, W., Feng, D.: Integral Cryptanalysis of Reduced FOX Block Cipher. In: Won, D.H., Kim, S. (eds.) ICISC 2005. LNCS, vol. 3935, pp. 229–241. Springer, Heidelberg (2006)
27. Yeom, Y., Park, S., Kim, I.: On the security of Camellia against the Square attack. In: Daemen, J., Rijmen, V. (eds.) FSE 2002. LNCS, vol. 2365, pp. 89–99. Springer, Heidelberg (2002)
28. Z'aba, M.R., Raddum, H., Henricksen, M., Dawson, E.: Bit-Pattern Based Integral Attack. In: Nyberg, K. (ed.) FSE 2008. LNCS, vol. 5086, pp. 363–381. Springer, Heidelberg (2008)

Extensions of the Cube Attack Based on Low Degree Annihilators

Aileen Zhang[1], Chu-Wee Lim[1], Khoongming Khoo[1], Lei Wei[2],
and Josef Pieprzyk[3]

[1] DSO National Laboratories
20 Science Park Drive, Singapore 118230
{zyinghui,lchuwee,kkhoongm}@dso.org.sg
[2] School of Physical and Mathematical Sciences
Nanyang Technological University, Singapore
wl@pmail.ntu.edu.sg
[3] Department of Computing, Macquarie University, Australia
josef@ics.mq.edu.au

Abstract. At Crypto 2008, Shamir introduced a new algebraic attack
called the cube attack, which allows us to solve black-box polynomials if
we are able to tweak the inputs by varying an initialization vector. In a
stream cipher setting where the filter function is known, we can extend
it to the cube attack with annihilators: By applying the cube attack to
Boolean functions for which we can find low-degree multiples (equivalently annihilators), the attack complexity can be improved. When the
size of the filter function is smaller than the LFSR, we can improve
the attack complexity further by considering a sliding window version of
the cube attack with annihilators. Finally, we extend the cube attack to
vectorial Boolean functions by finding implicit relations with low-degree
polynomials.

Keywords: Cube Attack, Algebraic Attack, Low-Degree Annihilators.

1 Introduction

In the history of cryptography, algebraic cryptanalysis is a rather recent trend.
The underlying idea behind this attack is rather simple: in trying to attack a
cryptosystem, write the problem as a set of polynomial equations with coefficients and unknowns in some common finite field K, most probably of characteristic 2. One then employs whatever means at one's disposal to solve this system
of polynomial equations.

It has been long known that the general problem of solving such a system
is NP-complete, even if the system comprises of only quadratic equations over
$\mathbb{F}_2$ (see [13]). Nevertheless, many cryptographic systems appear susceptible to
attacks via this approach. Indeed, a large arsenal of attacks have been designed
with the algebraic approach in mind, including (but not restricted to) linearization, relinearization [9], eXtended Linearization [4], Gröbner basis [7,8] and the
fast algebraic attack [5].

J.A. Garay, A. Miyaji, and A. Otsuka (Eds.): CANS 2009, LNCS 5888, pp. 87–102, 2009.

In Aug 2008, during the Crypto conference, Adi Shamir [12] presented a new approach to algebraic attacks in an invited lecture. Termed *cube attack*, his method requires the attacker to launch an active attack (e.g. chosen-IV or chosen-PT) in order to extract useful information from the bits obtained. Roughly speaking, by skillfully choosing the bits in a systematic manner, the attacker may lower the degree of the polynomial quickly.

In Section 2, we shall give a description of Shamir's cube attack. Then, we offer several variations to the basic cube attack. In Section 3, we extend the cube attack to polynomials f for which we can find a low degree g such that fg is also of low degree, and we apply this to the Toyocrypt cipher as an example. We call this the cube attack with low degree annihilator and show that it has better attack complexity than the basic cube attack and algebraic attack.

In Section 3.2, we refine the cube attack with low degree annihilators to the special case where the size of the filter function is smaller than the LFSR. We call this refinement the sliding window cube attack. We demonstrate several scenarios where the sliding window cube attack has better attack complexity than the cube attack with annihilators. We also compare our attack with the re-synchronization attack of Daemon et. al. [10] since it is also applicable in this case, and conclude that our attack gives better attack complexity under suitable conditions.

In Section 4, we consider the cube attack when applied to vectorial filter functions. We show that there always exist equations describing the vectorial functions, which has lower degree than low degree multiples of single-bit output Boolean functions. Thus this shows that theoretically, we have better attack complexity when we apply the vectorial cube attack rather than attacking single bit output of the S-boxes by the cube attack with low degree annihilators. Finally in Section 5, we summarize our findings and propose some further research directions.

2 Preliminaries: Cube Attack

First let us give a brief overview of the cube attack [12]. Throughout this article, all polynomials have coefficients in $\mathbb{F}_2$, and $\mathbf{x}$ (*resp.* $\mathbf{v}$) denotes the vector $(x_0, x_1, \ldots x_{n-1})$ (*resp.* $(v_0, v_1, \ldots, v_{m-1})$).

The primary idea behind this attack lies in the following theorem:

Theorem 1. *Let $f(\mathbf{x})$ be a polynomial in n variables of degree d. Suppose $0 < k \leq d$ and t is the monomial $x_0 x_1 \ldots x_{k-1}$. Write f in the form:*

$$f(\mathbf{x}) = t \cdot P_t(\mathbf{x}) + Q_t(\mathbf{x}),$$

where none of the terms in $Q_t(\mathbf{x})$ is divisible by t. Note that $\deg(P_t) \leq d - k$.

Then the sum of f over all $(x_0, \ldots, x_{k-1}) \in \mathbb{F}_2^k$, considered as a polynomial in $x_k, x_{k+1}, \ldots$, equals

$$P_t(\overbrace{1, \ldots, 1}^{k}, x_k, x_{k+1}, \ldots, x_{n-1})$$

and hence is a polynomial of degree at most $d - k$.

Proof. Consider the equality $f = t \cdot P_t + Q_t$. Split the sum into $\sum_{(x_0,\ldots,x_{k-1})} t \cdot P_t$ and $\sum_{(x_0,\ldots,x_{k-1})} Q_t$. In the first sum, $t = 0$ unless $x_0 = x_1 = \cdots = x_{k-1} = 1$ in which case

$$\sum_{(x_0,\ldots,x_{k-1})\in\mathbb{F}_2^k} t \cdot P_t = P_t(\overbrace{1,\ldots,1}^{k}, x_k, x_{k+1}, \ldots, x_{n-1}).$$

On the other hand Q_t is a sum of monomials, each of which is not divisible by t. Let m be any one of these monomials. Since m is not divisible by t, it excludes x_i for some $0 \leq i \leq k - 1$. If it excludes (say) x_0, then the sum across all $(x_0, \ldots, x_{k-1}) \in \mathbb{F}_2^k$ can be further split into two sums: the sum for $x_0 = 0$ and for $x_0 = 1$. These two sums are equal since x_0 does not appear in m. Hence

$$\sum_{(x_0,\ldots,x_{k-1})\in\mathbb{F}_2^k} m = 0 \quad \Longrightarrow \quad \sum_{(x_0,\ldots,x_{k-1})\in\mathbb{F}_2^k} Q_t = 0.$$

This completes our proof of the theorem.

Let us apply this theorem to cryptanalyze a stream cipher. Write the cipher in the form:

$$z = f(\mathbf{x}, \mathbf{v}),$$

which takes in an n-bit key $\mathbf{x}$ and an m-bit IV $\mathbf{v}$, and outputs the first bit of the keystream. Suppose $d = \deg f \leq m$. We describe the cube attack for the term $t = v_0 v_1 \cdots v_{d-2}$.

Fix the IV bits $v_{d-1}, v_d, v_{d+1}, \cdots \in \mathbb{F}_2$ and write C for the set of $\mathbf{v}$ with these values of $v_{d-1}, v_d, \ldots$. Thus $|C| = 2^{d-1}$. Sum $f(\mathbf{x}, \mathbf{v})$ over $\mathbf{v} \in C$. By applying Theorem 1 to t, this sum is linear in $\mathbf{x}$:

$$\sum_{\mathbf{v}\in C} f(\mathbf{x}, \mathbf{v}) = L(\mathbf{x}). \tag{1}$$

If $L(\mathbf{x}) \neq 0$, we call t a **maxterm** in accordance with [12], and obtain one linear relation in the key bits. To obtain $n - 1$ more such relations, we can do the following.

- Use the same f, but use a different maxterm t.
- Use a different f, e.g. by using the second bit of the keystream.

With n linearly independent relations of the key bits, we can easily find them via Gaussian elimination.

Hence, given access to such a function f, *Cube Attack* proceeds to find the unknown vector $\mathbf{x}$ according to the following stages:

1. *First: the preprocessing stage.* This involves finding the coefficients of L for n such L. Each L has $n + 1$ coefficients including the constant term; to find them, we need to compute the sum (1) for $n + 1$ keys:

$$\mathbf{x} = 0, \mathbf{e}_0, \mathbf{e}_1, \ldots, \mathbf{e}_{n-1},$$

where $\mathbf{e}_i$ is the vector where the i-th component is 1 and the rest are 0. The amount of work required is $n(n+1)2^{d-1}$ evaluations of f.

We also compute the inverse of the matrix of linear relations. This requires n^3 operations at most so the amount of work is upper-bounded by

$$n(n+1)2^{d-1} + n^3.$$

2. *Second: the online stage.* Now we apply a chosen-IV attack on the cipher. Compute the sum (1) for n linear relations L. Each sum requires 2^{d-1} evaluations of f, so we need $n2^{d-1}$ evaluations of f in all. Since we already have the inverse of the L-matrix, we only need to perform matrix multiplication which takes n^2 operations. Hence, the amount of work is upper-bounded by

$$n2^{d-1} + n^2.$$

Notice that the attack only assumes $\deg f \leq d$, and that we can evaluate f. No knowledge of the coefficients of f is required.

Remark 1. For a given maxterm t, in the case where $\deg(t) < n - 1$, we may be able to derive multiple equations, since each maxterm gives an equation that may have monomials containing terms in the IV as well as in the key. Hence, substituting in different values for the terms in the IV that are not in the maxterm may produce different equations.

3   Cube Attack with Annihilators

In 2003, Courtois and Meier[5] observed that for some polynomials f, we can find a low degree g such that $h := fg$ is also of low degree. We shall apply this observation to derive an enhanced version of the cube attack.

As before, let $z = f(\mathbf{x}, \mathbf{v})$ represent the first output bit, where $\mathbf{x}$ is the key and $\mathbf{v}$ is the IV. Let $g(\mathbf{x}, \mathbf{v})$ be a polynomial such that:

- $g(\mathbf{x}, \mathbf{v})$ is of low degree e;
- $h(\mathbf{x}, \mathbf{v}) := f(\mathbf{x}, \mathbf{v})g(\mathbf{x}, \mathbf{v})$ is of degree $d \leq \deg(f)$ and $d > e$.

Our attack works as follows: suppose we pick the maxterm $v_0 v_1 \cdots v_{d-e-1}$. Fix the IV-bits $v_{d-e}, v_{d-e+1}, \cdots \in \mathbb{F}_2$ and let C be the set of $\mathbf{v}$ which has these values of $v_{d-e}, v_{d-e+1} \ldots$. Consider the sum:

$$\sum_{\mathbf{v} \in C} h(\mathbf{x}, \mathbf{v}) = \sum_{\mathbf{v} \in C} f(\mathbf{x}, \mathbf{v})g(\mathbf{x}, \mathbf{v}).$$

By Theorem 1, on the left, we get a polynomial in $\mathbf{x}$ of degree at most $d - (d-e) = e$. On the right, note that $f(\mathbf{x}, \mathbf{v})$ is known since it is a keystream bit, so we get a polynomial of degree $\leq e$. Now we can solve for the secret bits by applying a range of techniques, such as linearization [4] or Gröbner basis techniques [7,8].

We shall term this method *Cube Attack with Annihilators*[1]. Given the filter function f, assume we have a low degree multiple g of degree e such that $h = fg$ has low degree d. There are many efficient algorithms in literature to find f and g. See [1] and [11] for example. The attack proceeds as follows:

1. *First, the preprocessing stage.* We need to compute the polynomial

$$P(\mathbf{x}) := \sum_{\mathbf{v} \in C} h(\mathbf{x}, \mathbf{v})$$

   which is of degree $\leq e$. Since this is a linear combination of $\sum_{i=0}^{e} \binom{n}{i}$ monomials, we need to evaluate this sum $\sum_{i=0}^{e} \binom{n}{i}$ times (by pumping in different $\mathbf{x}$'s) to find the coefficients. This requires $2^{d-e} \sum_{i=0}^{e} \binom{n}{i}$ evaluations of h to compute the coefficients of a single P. For linearization to work, we need $\sum_{i=0}^{e} \binom{n}{i}$ such polynomials, so the total amount of operations is:

$$2^{d-e} \left(\sum_{i=0}^{e} \binom{n}{i} \right)^2$$

   evaluations of h.
2. *Second: the online phase.* For each of the $\sum_{i=0}^{e} \binom{n}{i}$ maxterms, we must compute $\sum_{\mathbf{v} \in C} f(\mathbf{x}, \mathbf{v}) g(\mathbf{x}, \mathbf{v})$. The polynomial $g(\mathbf{x}, \mathbf{v})$, for a fixed $\mathbf{v} \in C$ has typically $\sum_{i=0}^{e} \binom{n}{i}$ terms. Hence, the computation of the term $f(\mathbf{x}, \mathbf{v}) g(\mathbf{x}, \mathbf{v})$ requires $2^{d-e} \sum_{i=0}^{e} \binom{n}{i}$ computations. For $\sum_{i=0}^{e} \binom{n}{i}$ such maxterms, we require $2^{d-e} \left(\sum_{i=0}^{e} \binom{n}{i} \right)^2$ computations. Finally linearization of

$$\sum_{\mathbf{v} \in C} f(\mathbf{x}, \mathbf{v}) g(\mathbf{x}, \mathbf{v}) = \sum_{\mathbf{v} \in C} h(\mathbf{x}, \mathbf{v}) = P(\mathbf{x})$$

   gives a system of $\sum_{i=0}^{e} \binom{n}{i}$ linear equations which requires $\left(\sum_{i=0}^{e} \binom{n}{i} \right)^3$ operations to solve. Hence, the total amount of operations is about:

$$2^{d-e} \left(\sum_{i=0}^{e} \binom{n}{i} \right)^2 + \left(\sum_{i=0}^{e} \binom{n}{i} \right)^3.$$

We note that both the basic cube attack and the cube attack with annihilators are chosen IV resynchronization attacks. However, some of their differences are as follows:

1. This variation of the cube attack requires us to compute $h = fg$ for an appropriate polynomial g. To find such a g, we most likely need to express f in algebraic normal form.
2. Here, we cannot perform the matrix inversion during the preprocessing stage, because the entries of the matrix depends on the keystream output.

---

[1] We have used the term annihilators in naming our attack because from [1], the existence of low degree multiples is equivalent to the existence of low degree annihilators.

3. Each polynomial evaluation (of g or h) requires $\sum_{i=0}^{e} \binom{n}{i}$ computations if we express the polynomials in algebraic normal form.

In the next subsection, we shall provide a concrete example of this variant of cube attack.

3.1   Application to the Toyocrypt Cipher with Re-synchronization

The main Toyocrypt cipher [15] comprises of a 128-bit MLFSR (modular linear feedback shift register), filtered through a nonlinear function f of degree 63. This f is given by:

$$f(s_0, \ldots, s_{127}) = s_{127} + \sum_{i=0}^{62} s_i s_{\alpha_i} + s_{10} s_{23} s_{32} s_{42} +$$

$$s_1 s_2 s_9 s_{12} s_{18} s_{20} s_{23} s_{25} s_{26} s_{28} s_{33} s_{38} s_{41} s_{42} s_{51} s_{53} s_{59} + \prod_{i=0}^{62} s_i,$$

where α_i, $0 \leq i \leq 62$, is a permutation of the set $\{63, \ldots, 125\}$. The output of the filter function gives a keystream bit. Upon the next clocking, the MLFSR clocks once and passes through the filter function to give the next keystream bit. For simplicity of explanation, we can treat the MLFSR as a LFSR because as shown in [15], there is a one-to-one linear transformation between the states of the MLFSR and an LFSR.

In [3], Courtois described an algebraic attack on Toyocrypt. He observed that f can be approximated by a degree-4 polynomial g by ignoring the two terms of degree 17 and 63 respectively. The error rate in this approximation is given by 2^{-17} which is good enough for practical purposes. Later, in [5], Courtois and Meier found an even better attack by noting that the polynomials

$$f \cdot (s_{23} + 1) \quad \text{and} \quad f \cdot (s_{42} + 1)$$

are cubic since the variables s_{23} and s_{42} occur in all terms of f of degree at least 4.

The above observations will come in handy when we apply the two variants of cube attack on Toyocrypt. *We shall assume a simplified variant, where during initialization, an n-bit key and m-bit IV are linearly mixed to fill up the LFSR.*

Let us replace f with a quartic polynomial g as mentioned above. We may then write the first bit of the keystream as a quartic polynomial in the key (x_i) and the IV (v_j). In applying cube attack, we require a preprocessing work factor of $n^3 + 8n(n+1)$ and an online work factor of $n^2 + 8n$.

The attack fails if $f \neq g$ for one of the evaluations. We may safely assume that this does not occur during preprocessing (since checks can easily circumvent that); hence, the probability of success is

$$(1 - 2^{-17})^{8n}.$$

Even in the extreme case of $n = 128$, this is greater than 99%.

Cube Attack with Annihilators: We can find a degree-1 polynomial g such that $fg = h$ is cubic. Hence the cube attack with annihilators requires only $2^{3-1}n^2 = 4n^2$ evaluations of a linear function during the preprocessing stage. During the online phase, the amount of work is $8n + n^3$.

A Comparison: In Table 1, we compare the above variants of the cube attack (quartic approximation, low degree annihilators) with the basic cube attack on the filter function of degree 63 and the algebraic attack of [6,14] using cubic equations on the Toyocrypt cipher with $n = 128$. We see that our cube attack variant has lower complexities and requires much fewer keystream bits.

Table 1. Comparison of Improved Variants of Cube Attack with the Basic Cube Attack and Algebraic Attack on Toyocrypt with 128-bit State Function Linearly Initialized by 128-bit Key and IV

	Fast Algebraic Attack [6, 14]	Basic Cube Attack [12]	Basic Cube Attack with Quartic Approximation	Cube Attack with Annihilators (new)
Keystream Bits	2^{18}	$2^{62} \times 128$	$2^3 \times 128$	$2^2 \times 128$
Pre-Computation	2^{30}	2^{76}	2^{21}	2^{16}
Online Complexity	2^{20}	2^{69}	2^{14}	2^{21}

Note that the keystream bits for algebraic attack can be obtained from one keystream while those of the other attacks have to be obtained across different keystreams from re-synchronizations. E.g., in the 4th column, we need $2^2 \times 128 = 2^9$ keystream bits from 4 re-synchronizations, each having 128 bits.

Implementation: We implemented both variants of the cube attack (quartic approximation and low-degree annihilators). In both versions, the Toyocrypt cipher can be broken in a few milliseconds on an ordinary PC. Although both variants seem to have comparable pre-computation + online attack time from Table 1, the cube attack with annihilators runs about twice as fast as the basic cube attack using quartic approximation. It also has the slight advantage of being 100% reliable and uses fewer re-synchronizations.

In a Nutshell: The example of Toyocrypt is used to illustrate our cube attack variant. It demonstrates its effectiveness against ciphers in which multiplying f with a low-degree polynomial g dramatically lowers its degree.

3.2 Sliding Window Cube Attack on Filter Function Taking Few Inputs

Consider the case where our key is of size N and our filter function f takes only $n < N$ inputs from the state. Suppose we have the following two conditions:

- Linear initialization
- Linear feedback

There is a known re-synchronization attack on such a cipher with complexity $\lceil N/n \rceil \times 2^n$, see [10, Section 3]. We shall describe an extension of the cube attack with annihilators on this cipher where the complexity is generally better than the re-synchronization attack of [10].

Because of the linear initialization, we can write the inputs from the state at time t as $l_t(\mathbf{x}, \mathbf{v})$, where l_t is linear, and consider the filter function as a function of the inputs (as opposed to the entire state). Now its output at time t is

$$z_t = f(l_t(\mathbf{x}, \mathbf{v})) = f(\mathbf{y}_t + l_t(\mathbf{0}, \mathbf{v}))$$

where $\mathbf{y}_t = l_t(\mathbf{x}, \mathbf{0})$.

As before, suppose we have g and h of low degrees e and d respectively such that $fg = h$ and $e < d \le \deg(f)$. Let us write $f_t(\mathbf{y}_t, \mathbf{v}) = f(\mathbf{y}_t + l_t(\mathbf{0}, \mathbf{v}))$, and define g_t and h_t similarly. We can apply the cube attack with annihilators to $f_t g_t = h_t$ to find $\mathbf{y}_t$ for any t. We choose $\lceil N/n \rceil$ values of t such that the corresponding $\mathbf{y}_t$ give us N linearly independent equations in the (x_i), and solve for the $\mathbf{y}_t$. We can then solve the N linear equations in (x_i) by Gaussian elimination.

Suppose we have found low degree g and h such that $h = fg$. The attack works as follows:

1. *First: The preprocessing stage.* We pick $\lceil N/n \rceil$ values of t such that the $\mathbf{y}_t$ give us N linearly independent equations in the (x_i). For each value of t, we pick $\sum_{i=0}^{e} \binom{n}{i}$ maxterms. For a given maxterm, we denote C to be the cube of 2^{d-e} vectors which have all possible combinations of values for the terms in the maxterm, and have all other terms fixed in some configuration. For each maxterm, we compute $\sum_C h_t(\mathbf{y}_t, \mathbf{v})$ by finding the coefficient of every $\mathbf{y}_t$-monomial, of which there are $\sum_{i=0}^{e} \binom{n}{i}$, so h_t gets evaluated $2^{d-e} \sum_{i=0}^{e} \binom{n}{i}$ times.
   Hence the total complexity of this stage is

$$\lceil N/n \rceil 2^{d-e} \sum_{i=0}^{e} \binom{n}{i}$$

2. *Second: The online phase.* Each value of t has $\sum_{i=0}^{e} \binom{n}{i}$ corresponding maxterms, and for each maxterm we can compute $\sum f_t(\mathbf{y}_t, \mathbf{v}) g_t(\mathbf{y}_t, \mathbf{v})$ since we know the keystream bits $f_t(\mathbf{y}_t, \mathbf{v})$. This is a polynomial in $\mathbf{y}_t$, and we find the coefficient of every $\mathbf{y}_t$-monomial as before. This has complexity $2^{d-e} \sum_{i=0}^{e} \binom{n}{i}$. We then equate it to $\sum h_t(\mathbf{y}_t, \mathbf{v})$ to obtain an equation in $\mathbf{y}_t$ of degree at most e. Since there are $\sum_{i=0}^{e} \binom{n}{i}$ maxterms for each t, we get $\sum_{i=0}^{e} \binom{n}{i}$ equations in $\mathbf{y}_t$ of degree at most e, and we can solve for $\mathbf{y}_t$ by linearization. This has complexity $(\sum_{i=0}^{e} \binom{n}{i})^3$.
   After solving for all $\lceil N/n \rceil$ of the $\mathbf{y}_t$, which are linear combinations of (x_i), we get N linear equations in (x_i), and can then solve for $\mathbf{x}$ using Gaussian elimination with complexity N^3.
   The total complexity of this stage is

$$\lceil N/n \rceil \left(2^{d-e} \left(\sum_{i=0}^{e} \binom{n}{i} \right)^2 + \left(\sum_{i=0}^{e} \binom{n}{i} \right)^3 \right) + N^3$$

Remark 2. This argument also applies in the more general case where the filter function can be written as a function of $\alpha_t(\mathbf{x})$ and $\beta_t(\mathbf{v})$ where α_t, β_t are not necessarily linear, and the α_t are of degree at most c for some small c. In this case, we need to solve for $\binom{n}{c}$ of the $\alpha_t(\mathbf{x})$, and then solve for $\mathbf{x}$ by linearization.

3.3 Applications of the Sliding Window Cube Attack

In this section, we give two examples of the sliding window cube attack on a filter function generator where the filter function has size $n = 128$ and the key has size $N = 256$ and 10000.

Example 1. Consider a filter function generator where a 256-bit key is linearly mixed with a 256-bit IV to fill up a 256 bit LFSR. Let the filter function be a Toyocrypt-like function defined by:

$$f(s_0, \ldots, s_{127}) = s_{127} + \sum_{i=0}^{62} s_i s_{\alpha_i} + s_0 s_1 s_2 \ldots s_{31} + s_{32} s_{33} s_{34} \ldots s_{62},$$

where α_i, $0 \le i \le 62$, is a permutation of the set $\{63, \ldots, 125\}$. This function is balanced, has algebraic degree 32 and like the Toyocrypt filter function, near optimal nonlinearity $2^{127} - 2^{64}$ for protection against correlation attack. However, it is easy to see that we can multiply it by $(s_0 + 1)(s_{32} + 1)$ to get a degree 4 equation. Thus $e = 2$ and $d = 4$.

1. The complexity of the sliding window cube attack from Section 3.2 on this cipher is:

$$(256/128) \times \left(\left(\sum_{i=0}^{2} \binom{128}{i} \right)^2 \times 2^{4-2} + \left(\sum_{i=0}^{2} \binom{128}{i} \right)^3 \right) + 256^3 \approx 2^{40.03}$$

and it needs $\left(\sum_{i=0}^{2} \binom{128}{i} \right) \times (256/128) \approx 2^{14.01}$ keystream bits from each of $2^{4-2} = 4$ re-synchronizations.

2. The complexity of the cube attack with annihilator from Section 3 on this cipher is:

$$\left(\sum_{i=0}^{2} \binom{256}{i} \right)^2 \times 2^{4-2} + \left(\sum_{i=0}^{2} \binom{256}{i} \right)^3 \approx 2^{45.02}$$

and it needs $\left(\sum_{i=0}^{2} \binom{256}{i} \right) \approx 2^{15.00}$ keystream bits from each of $2^{4-2} = 4$ re-synchronizations.

3. The complexity of the direct cube attack [12] on this cipher by taking the filter function degree as 32 is:

$$256^2 \times 2^{32-1} + 256^3 \approx 2^{47.01}$$

and it needs 256 keystream bits from each of $2^{32-1} = 2^{31}$ re-synchronizations.

4. The complexity of the re-synchronization attack [10] on this cipher by taking the filter function input size as 128 bits is

$$(256/128) \times 2^{128} = 2^{129}$$

and it needs $(256/128) = 2$ keystream bits from each of 128 re-synchronizations.

5. The complexity of the fast algebraic attack [6] on this cipher (where we combine the pre-processing and online complexity) is:

$$\Big(\sum_{i=0}^{2} \binom{256}{i}\Big)\Big(\sum_{i=0}^{4} \binom{256}{i}\Big) + \Big(\sum_{i=0}^{2} \binom{256}{i}\Big)^3$$

$$+ \Big(\sum_{i=0}^{4} \binom{256}{i}\Big) \log \Big(\sum_{i=0}^{4} \binom{256}{i}\Big) \approx 2^{45.23}$$

and it needs $\sum_{i=0}^{2} \binom{256}{i} \approx 2^{15.00}$ bits from one keystream.

Thus we see that the sliding window cube attack has better attack complexity than the other attacks when applied to on this filter function generator. □

When we increase the size of the LFSR as in the following example, we can see that the cube attack with annihilator may even perform worse than the direct cube attack [12] but the sliding window cube attack will still have better attack complexities than the other attacks.

Example 2. Consider a filter function generator where we use the same filter function as that in Example 1 but increase the size of the key and IV to 10000 bits and used that to initialize a 10000-bit LFSR. Then by replacing $N = 256$ with $N = 10000$ in Example 1, we have the following complexity for the various attacks:

1. Sliding Window Cube Attack: Attack Complexity $= 2^{45.37}$ and it needs $2^{19.32}$ keystream bits from each of 4 re-synchronizations.
2. Cube Attack with Annihilators: Attack Complexity $= 2^{76.73}$ and it needs $2^{25.58}$ keystream bits from each of 4 re-synchronizations.
3. Direct Cube Attack. Attack Complexity$=2^{57.58}$ and it needs $2^{13.29}$ keystream bits from each of 2^{31} re-synchronizations.
4. Resynchronization Attack. Attack Complexity $= 2^{134.30}$ and it needs $2^{6.30}$ keystream bits from each of 128 re-synchronizations.
5. Fast Algebraic Attack. Attack Complexity $= 2^{76.95}$ and it needs $2^{25.58}$ bits from one keystream.

Again, we see that the sliding window cube attack has better attack complexity than the other attacks when applied to on this filter function generator. □

4   Cube Attack on Vectorial Filter Function with Low (x, v)-Degree

4.1   Applying the Cube Attack to Vectorial Filter Functions

We now consider the case where the state function (e.g. LFSR) is filtered by a vectorial Boolean function $F : \mathbb{F}_2^n \to \mathbb{F}_2^r$, $r > 1$. In 2005, Canteaut [2] introduced a method for finding implicit equations of the form $G(\mathbf{x}, \mathbf{v}, \mathbf{z}) = 0$ where $\mathbf{z} = F(\mathbf{x}, \mathbf{v})$ and $G(\mathbf{x}, \mathbf{v}, \mathbf{z})$ has low $(\mathbf{x}, \mathbf{v})$-degree and is of unrestricted degree in the output variable $\mathbf{z}$. Then this low degree equation can be solved by XL or linearization methods to recover the secret key.

In a similar way, we can extend the cube attack with annihilators to vectorial filter functions. Let a vectorial filter function be denoted by

$$\mathbf{z} = F(\mathbf{x}, \mathbf{v})$$

where $\mathbf{x}$ is the key of size n, $\mathbf{v}$ is the IV of size m, and $\mathbf{z}$ is a vector of multiple output bits. We can find $G(\mathbf{x}, \mathbf{v}, \mathbf{z})$ of low $(\mathbf{x}, \mathbf{v})$-degree e such that $H(\mathbf{x}, \mathbf{v}) := G(\mathbf{x}, \mathbf{v}, F(\mathbf{x}, \mathbf{v}))$ also has low $(\mathbf{x}, \mathbf{v})$-degree d, with $e < d \le \deg(F)$. Proposition 1 in Section 4.2 ensures that we can always find such functions $G(\mathbf{x}, \mathbf{v}, \mathbf{z})$ and $H(\mathbf{x}, \mathbf{v})$.

We can apply an adaptation of the attack on 1-bit filter functions to G and H. For a given maxterm, we denote C to be the cube of 2^{d-e} vectors which have all possible combinations of values for the terms in the maxterm, and have all other terms fixed in some configuration. For each $\mathbf{v} \in C$, $\mathbf{z} = F(\mathbf{x}, \mathbf{v})$ is known as it is a keystream bit, so by substituting these keystream bits into $\sum_C G(\mathbf{x}, \mathbf{v}, \mathbf{z}) = \sum_C H(\mathbf{x}, \mathbf{v})$, we get a polynomial of degree at most e. We do this for $\sum_{i=0}^{e} \binom{n}{i}$ maxterms to find $\sum_{i=0}^{e} \binom{n}{i}$ polynomials of degree at most e, and then solve for $\mathbf{x}$ by linearization.

The attack proceeds as follows:

1. *First: the preprocessing stage.* First, we pick $\sum_{i=0}^{e} \binom{n}{i}$ maxterms. For each maxterm, we compute $\sum_C H(\mathbf{x}, \mathbf{v})$ by finding the coefficient of every $\mathbf{x}$-monomial, of which there are $\sum_{i=0}^{e} \binom{n}{i}$, so H gets evaluated $2^{d-e} \sum_{i=0}^{e} \binom{n}{i}$ times.

   The total complexity of this stage is

   $$2^{d-e} \left(\sum_{i=0}^{e} \binom{n}{i} \right)^2$$

2. *Second: the online phase.* For each maxterm we can compute $\sum_C G(\mathbf{x}, \mathbf{v}, \mathbf{z})$ as a polynomial of $\mathbf{x}$, since we have the keystream bits $\mathbf{z}$. This has complexity $2^{d-e} \sum_{i=0}^{e} \binom{n}{i}$. We equate this to $\sum H(\mathbf{x}, \mathbf{v})$ to obtain an equation in $\mathbf{x}$ of degree at most e. Since there are $\sum_{i=0}^{e} \binom{n}{i}$ maxterms, we get $\sum_{i=0}^{e} \binom{n}{i}$ equations in $\mathbf{x}$ of degree at most e, and we can solve for $\mathbf{x}$ by linearization. This has complexity $(\sum_{i=0}^{e} \binom{n}{i})^3$.

   The total complexity of this stage is

   $$2^{d-e} \left(\sum_{i=0}^{e} \binom{n}{i} \right)^2 + \left(\sum_{i=0}^{e} \binom{n}{i} \right)^3$$

Remark 3. Given a stream cipher filtered by the vectorial function

$$\mathbf{z} = (z_1, \ldots, z_r) = F(\mathbf{x}, \mathbf{v}).$$

A straightforward attack would be to apply the cube attack on a linear combination of output bits, which we denote by $z = \sum_{i \in I} z_i = f(\mathbf{x}, \mathbf{v})$. If the attacker is able to find a multiple $f(\mathbf{x}, \mathbf{v})g(\mathbf{x}, \mathbf{v})$ of low degree d where $zg(\mathbf{x}, \mathbf{v})$ has low degree e, the attack complexity can be much reduced as in the attack on Toyocrypt in Section 3.1.

However, it is easy to see that low-degree equations $f(\mathbf{x}, \mathbf{v})g(\mathbf{x}, \mathbf{v})$ and $zg(\mathbf{x}, \mathbf{v})$ are special cases of the equation $G(\mathbf{x}, \mathbf{v}, F(\mathbf{x}, \mathbf{v}))$ of low $(\mathbf{x}, \mathbf{v})$-degree d and $G(\mathbf{x}, \mathbf{v}, \mathbf{z})$ of low $(\mathbf{x}, \mathbf{v})$-degree e, considered in Section 4. Therefore we expect the vectorial cube attack in Section 4 to utilize lower degree equations than the single-bit cube attack. This will translate into lower attack complexity when we linearize and solve the resulting system of equations for the secret keys.

4.2   Existence of Low Degree Equations for Vectorial Cube Attack

In contrast with Canteaut's method [2], we need not to have $H(\mathbf{x}, \mathbf{v}) = 0$ for all $\mathbf{x}, \mathbf{v}$, so the condition that $2^r \sum_{i=0}^e \binom{n+m}{i} > 2^{n+m}$ is not necessary. Instead, we require a weaker condition stated as an existence result in Proposition 1 below. The proposition also implies that finding low degree annihilators for vectorial Boolean function case $(r > 1)$ is no harder than the single output bit case $(r = 1)$.

Proposition 1. *(Existence of Low Degree Equations) Let a vectorial Boolean function $F : \{0,1\}^n \times \{0,1\}^m \to \{0,1\}^r$ be denoted by $\mathbf{z} = F(\mathbf{x}, \mathbf{v})$ where $\mathbf{x}$ is the key of size n, $\mathbf{v}$ is the IV of size m, and $\mathbf{z}$ is a vector of multiple output bits. For $0 < e < d \leq \deg(F)$, if*

$$(2^r - 1) \sum_{i=0}^e \binom{n+m}{i} + \sum_{i=0}^d \binom{n+m}{i} > 2^{n+m},$$

then there exists $G(\mathbf{x}, \mathbf{v}, \mathbf{z})$ of $(\mathbf{x}, \mathbf{v})$-degree e such that $H(\mathbf{x}, \mathbf{v}) := G(\mathbf{x}, \mathbf{v}, F(\mathbf{x}, \mathbf{v}))$ is of degree d.

Proof. Construct a matrix M with each row indexed by a value of $(\mathbf{x}, \mathbf{v})$, there are 2^{n+m} rows. Let the columns range over all $(\mathbf{x}, \mathbf{v}, \mathbf{z})$-monomials with $(\mathbf{x}, \mathbf{v})$-degree at most e, $\mathbf{z}$-degree unrestricted except that $\mathbf{z} = \mathbf{0}$, as well as all the $(\mathbf{x}, \mathbf{v})$-monomials with degree at most d. The number of columns $n_C := (2^r - 1) \sum_{i=0}^e \binom{n+m}{i} + \sum_{i=0}^d \binom{n+m}{i}$.

Define the (i,j)-th entry of M to be the value of the monomial corresponding to the j-th column evaluated with the value of $(\mathbf{x}, \mathbf{v})$ at the i-th row, with $\mathbf{z} = \mathbf{F}(\mathbf{x}, \mathbf{v})$.

If $n_C > 2^{n+m}$, we can find a column vector $\mathbf{y} \in \mathcal{F}_2^{n_C}$ by Gaussian elimination such that $M\mathbf{y} = \mathbf{0}$. Then for every value of $(\mathbf{x}, \mathbf{v})$, corresponding to row i, we have

$$\sum_j y_j M_{ij} = 0.$$

Namely, each solution for $\mathbf{y}$ corresponds to a linear combination of some of the column index monomials, the sum of which evaluates to 0 for every value of

$(\mathbf{x}, \mathbf{v})$. Let G be the sum of all $(\mathbf{x}, \mathbf{v}, \mathbf{z})$-monomials and H be the sum of all $(\mathbf{x}, \mathbf{v})$-monomials. The proof is done.   □

Remark 4. For a general single-bit output Boolean function, it may have algebraic immunity $n/2$, in which case, the best we can do is $d = e = n/2$ for the annihilator cube attack. But for the vectorial case, as shown above, we always get low degree equations when the existence condition holds and in many cases, d and e are lower than $n/2$. Thus theoretically the vectorial cube attack is better than the single-bit output cube attack with annihilator.

4.3   Results from Searching Implicit Low Degree Equations for Vectorial Boolean Functions

We have implemented the algorithm in the proof of Proposition 1 for a few well-known vectorial Boolean functions. Experimental results obtained seem to be even better than the above stated theoretical bound, namely, even if the condition for existence does not hold, as long as the number of columns exceeds the number of non-zero rows in the Reduced Row Echelon Form (RREF) of the matrix, we are still able to find low degree annihilators for $F(\mathbf{x}, \mathbf{v})$. The results are presented in Appendix A.

The algorithm is of at least exponential space complexity to $n + m$. However, it may provide a good motivation and starting point to find efficient algorithms to search for such low degree implicit equations for vectorial cube attack.

5   Conclusion

We have proposed several variants of the cube attack, which makes use of low degree equations. First, the cube attack with annihilators combines the low degree multiples used in algebraic attack with cube attack. The complexity of this combined attack is better than just a direct application of the cube attack or the algebraic attack by itself. This is demonstrated in the attack on Toyocrypt where the attack complexities are lower and the keystream needed is greatly reduced as shown in Table 1.

Second, when the size of the filter function is smaller than the LFSR, we proposed the sliding window cube attack with annihilators. As shown in Examples 1 and 2, it has better complexity than the cube attack with annihilators and the related resynchronization attack of Daemen et. al. [10].

Finally, the vectorial cube attack works on multi-output stream ciphers and it combines the cube attack with a new form of low degree implicit equations. The existence of such equations can be ensured by the rank computation of certain "monomial" matrices. Because the upper bound of the degree of the implicit equations in the vectorial cube attack is less than that of the degree of annihilators in the single-bit case, we see that theoretically, the vectorial cube attack has better attack complexity than applying the cube attack with annihilators to single-bit output of the vectorial function. We also did some experiments to find low degree implicit equations for the vectorial cube attack and got some results which are better than that expected by theory. These findings may serve as a motivation to find an efficient algorithm to find the low degree vectorial equations $G(\mathbf{x}, \mathbf{v}, F(\mathbf{x}, \mathbf{v}))$ and $G(\mathbf{x}, \mathbf{v}, \mathbf{z})$ for the vectorial cube attack of Section 4.

Acknowledgements

The research of Wei Lei is in part supported by the Singapore Ministry of Education under Research Grant T206B2204.

References

1. Armknecht, F., Carlet, C., Gaborit, P., Knzli, S., Meier, W., Ruatta, O.: Efficient Computation of Algebraic Immunity for Algebraic and Fast Algebraic Attacks. In: Vaudenay, S. (ed.) EUROCRYPT 2006. LNCS, vol. 4004, pp. 147–164. Springer, Heidelberg (2006)
2. Canteaut, A.: Open Problems Related to Algebraic Attacks on Stream Ciphers. In: Ytrehus, Ø. (ed.) WCC 2005. LNCS, vol. 3969, pp. 120–134. Springer, Heidelberg (2006)
3. Courtois, N.: Higher Order Correlation Attacks, XL algorithm and Cryptanalysis of Toyocrypt. In: Lee, P.J., Lim, C.H. (eds.) ICISC 2002. LNCS, vol. 2587, pp. 182–199. Springer, Heidelberg (2003)
4. Courtois, N., Klimov, A., Patarin, J., Shamir, A.: Efficient Algorithms for Solving Overdefined Systems of Multivariate Polynomial Equations. In: Preneel, B. (ed.) EUROCRYPT 2000. LNCS, vol. 1807, pp. 392–407. Springer, Heidelberg (2000)
5. Courtois, N., Meier, W.: Algebraic Attacks on Stream Ciphers with Linear Feedback. In: Biham, E. (ed.) EUROCRYPT 2003. LNCS, vol. 2656, pp. 345–359. Springer, Heidelberg (2003)
6. Courtois, N., Meier, W.: Fast Algebraic Attacks on Stream Ciphers with Linear Feedback. In: Boneh, D. (ed.) CRYPTO 2003. LNCS, vol. 2729, pp. 176–194. Springer, Heidelberg (2003)
7. Faugère, J.-C.: A New Efficient Algorithm for Computing Gröbner Bases (F_4). Journal of Pure and Applied Algebra 139(1), 61–88 (1999)
8. Faugère, J.-C.: A New Efficient Algorithm for Computing Gröbner Bases Without Reduction to Zero (F_5). In: Proceedings of the 2002 international symposium on Symbolic and algebraic computation (ISSAC), pp. 75–83. ACM Press, New York (2002)
9. Kipnis, A., Shamir, A.: Cryptanalysis of the HFE Public Key Cryptosystem by Relinearization. In: Wiener, M. (ed.) CRYPTO 1999. LNCS, vol. 1666, p. 19. Springer, Heidelberg (1999)
10. Daemen, J., Govaerts, R., Vandewalle, J.: Resynchronization Weaknesses in Synchronous Stream Ciphers. In: Helleseth, T. (ed.) EUROCRYPT 1993. LNCS, vol. 765, pp. 159–167. Springer, Heidelberg (1994)
11. Didier, F., Tillich, J.-P.: Computing the Algebraic Immunity Efficiently. In: Robshaw, M.J.B. (ed.) FSE 2006. LNCS, vol. 4047, pp. 359–374. Springer, Heidelberg (2006)

12. Dinur, I., Shamir, A.: Cube Attacks on Tweakable Black Box Polynomials. In: Ghilardi, S. (ed.) Eurocrypt 2009, vol. 5479, pp. 278–299. Springer, Heidelberg (2009)
13. Garey, M.R., Johnson, D.S.: Computers and Intractability: A Guide to the Theory of NP-Completeness. Freeman and Company, New York (1979)
14. Hawkes, P., Rose, G.: Rewriting Variables: The Complexity of Fast Algebraic Attacks on Stream Ciphers. In: Franklin, M. (ed.) CRYPTO 2004. LNCS, vol. 3152, pp. 390–406. Springer, Heidelberg (2004)
15. Mihaljevic, M., Imai, H.: Cryptanalysis of Toyocrypt-HS1 Stream Cipher. IEICE Transactions on Fundamentals E85-A, 66-73 (2002)

Appendix

A Implicit Low Degree Equations for Vectorial Boolean Functions

Let a configuration be $(n + m, r, e, d)$, we restrict the output of $F(\mathbf{x}, \mathbf{v})$ to the first r bits. Let n_R be the number of rows in the matrix M, n_C the number of columns, n_{RREF} be the row rank of M, namely, the number of non-zero rows when M is reduced to the reduced row echelon form (RREF). Let n_S be the number of low degree equations obtained. When working in $GF(2^n)$, the irreducible polynomial is denoted $m(x)$.

Table 2. $F : \{0, 1\}^8 \rightarrow \{0, 1\}^8$, F is the S-Box of AES

$n+m$	r	e	d	n_R	n_C	n_{RREF}	n_S
8	2 1 2	2^8	64	64	0
8	2 1 3	2^8	120	120	0
8	2 1 4	2^8	190	190	0
8	2 2 3	2^8	204	203	1
8	2 2 4	2^8	274	248	26
8	2 3 4	2^8	442	256	186
8	3 1 2	2^8	100	100	0
8	3 1 3	2^8	156	156	0
8	3 1 4	2^8	226	224	2
8	3 2 3	2^8	352	256	96
8	3 2 4	2^8	422	256	166
8	3 2 4	2^8	422	256	166
8	3 3 4	2^8	814	256	558
8	4 1 2	2^8	172	172	0
8	4 1 3	2^8	228	225	3
8	4 1 4	2^8	298	256	42
8	4 2 3	2^8	648	256	392
8	4 2 4	2^8	718	256	462
8	4 3 4	2^8	1558	256	1295

Table 3. $F : GF(2^9) \to GF(2^9)$, the inverse function, $m(x) = x^9 + x^4 + 1$

$n+m$	r	e	d	n_R	n_C	n_{RREF}	n_S
9	2 1 2	2^9	76	76	0
9	2 1 3	2^9	160	160	0
9	2 1 4	2^9	286	286	0
9	2 2 3	2^9	268	268	0
9	2 2 4	2^9	394	384	10
9	2 3 4	2^9	646	511	135
9	3 1 2	2^9	116	116	0
9	3 1 3	2^9	200	200	0
9	3 1 4	2^9	326	326	0
9	3 2 3	2^9	452	441	11
9	3 2 4	2^9	578	510	68
9	3 3 4	2^9	1166	512	654
9	4 1 2	2^9	196	196	0
9	4 1 3	2^9	280	280	0
9	4 1 4	2^9	406	403	3
9	4 2 3	2^9	820	512	308
9	4 2 4	2^9	946	512	434
9	4 3 4	2^9	2206	512	1694

Table 4. $F : GF(2^{10}) \to GF(2^{10})$, the inverse function, $m(x) = x^{10} + x^3 + 1$

$n+m$	r	e	d	n_R	n_C	n_{RREF}	n_S
10	2 1 2	2^{10}	89	89	0
10	2 1 3	2^{10}	209	209	0
10	2 1 4	2^{10}	419	419	0
10	2 2 3	2^{10}	344	344	0
10	2 2 4	2^{10}	554	554	0
10	2 3 4	2^{10}	914	873	41
10	3 1 2	2^{10}	133	133	0
10	3 1 3	2^{10}	253	253	0
10	3 1 4	2^{10}	463	463	0
10	3 2 3	2^{10}	568	568	0
10	3 2 4	2^{10}	778	751	27
10	3 3 4	2^{10}	1618	1024	594
10	4 1 2	2^{10}	221	221	0
10	4 1 3	2^{10}	341	341	0
10	4 1 4	2^{10}	551	551	0
10	4 2 3	2^{10}	1016	993	23
10	4 2 4	2^{10}	1226	1024	202
10	4 3 4	2^{10}	3026	1024	2002

An Analysis of the Compact XSL Attack on BES and Embedded SMS4

Jiali Choy, Huihui Yap, and Khoongming Khoo

DSO National Laboratories, 20 Science Park Drive, S118230, Singapore
{cjiali,yhuihui,kkhoongm}@dso.org.sg

Abstract. The XSL attack when applied on BES-128 has been shown to have an attack complexity of 2^{100}, which is faster than exhaustive search. However at FSE 2007, Lim and Khoo analyzed the eprint XSL attack on BES and showed that the attack complexity should be 2^{401}. Later at IEEE-YCS 2008, Qu and Liu counter-proposed that the compact XSL attack on BES-128 works and has complexity 2^{97}. In this paper, we point out some errors in the attack of Qu and Liu. We also show that the complexity of the compact XSL attack on BES-128 is at least $2^{209.15}$. At Indocrypt 2007, Ji and Hu claimed that the eprint XSL attack on ESMS4 has complexity 2^{77}. By the same method we used to analyze BES, we also show that the complexity of compact XSL attack on ESMS4 is at least $2^{216.58}$. Our analysis adapts the approach of Lim and Khoo to the compact XSL attack, and improves on it by considering the T' method that grows the number of equations.

Keywords: Compact XSL Attack, BES, ESMS4.

1 Introduction

The eXtended Sparse Linearization method [3], introduced in 2002 by Courtois and Pieprzyk, is a refinement of the XL algorithm and is supposed to work on special types of ciphers such as AES. One main improvement is to take advantage of the sparseness of the multivariate quadratic system of the cipher: the equations should only be multiplied by "carefully selected monomials" so as to generate a new system where there are more equations than monomials. This will allow the new equations to be solved via linearization [3].

There are different versions of the XSL algorithm. The eprint version was described in [3] and a compact version was introduced in [4]. For both versions, there are two types of attacks. In the first XSL attack, the key schedule equations are not considered but more plaintext-ciphertext pairs are required. The second XSL attack includes the key schedule equations and uses less plaintext-ciphertext pairs, but is more specific as the key schedule needs to have a similar structure to the encryption. We shall focus on the second XSL attack as its complexity is much better than the first XSL attack.

Although the complexities of both the eprint and compact XSL attacks on AES (the Advanced Encryption Standard by NIST) are worse than or close

J.A. Garay, A. Miyaji, and A. Otsuka (Eds.): CANS 2009, LNCS 5888, pp. 103–118, 2009.

to that of exhaustive search, the breakthrough of the XSL attack is that the complexity is polynomial in the number of rounds, unlike traditional block cipher attacks whose complexities are exponential in the number of rounds.

At Crypto 2002, Murphy and Robshaw [7] showed that we can embed AES into a big encryption system (BES) where all operations are done over $GF(2^8)$. The advantage of this embedding is that the multivariate quadratic system becomes even more sparse. This results in a tremendous improvement in the attack complexity which makes the XSL attack not just a theoretically interesting attack, but also a plausible one. When the eprint XSL attack is applied on BES, the complexity is only 2^{100} (or 2^{87} if we take the complexity of solving N linear equations to be $N^{2.376}$ [3]). However, it was shown by Lim and Khoo [6], through an alternative analysis of the equations of the XSL attack, that the actual complexity of the eprint XSL attack on BES should be 2^{401}. Therefore, they conclude that BES (and thus AES) is still secure against the eprint XSL attack.

In a later paper, Qu and Liu claimed that the BES cipher can be efficiently cryptanalyzed by the compact XSL attack with complexity 2^{97} [9]. They tried to improve on the compact XSL attack by introducing a new concept of using S'-boxes where they only consider the conjugacy relations for most of the S-boxes in the key schedule. However, we show in Section 3.3 that this process will disregard many inversion relations and introduce too many free variables for the system of equations to be solvable. We also explain that some of the observations on [6] made by the authors of [9] are erroneous.

In Section 3.4, we shall give an analysis of the compact XSL attack on BES based on the corrected attack of [9], where S'-boxes are disallowed. It can be seen that this corrected attack reduces to an application of the compact XSL attack [4] on BES. By adapting the analysis of the eprint XSL attack on BES in [6] to the compact XSL attack, we show that the actual complexity of compact XSL attack on BES is at least $2^{209.15}$. Moreover, we also improved the approach of [6] by taking into account the T' method. Thus BES-128 is still secure against the compact XSL attack. Furthermore, we also show that the actual complexities of the compact XSL attack on BES-192 and BES-256 are $2^{231.50}$ and $2^{267.21}$ respectively.

In [5], Ji and Hu used the method of [7] to embed the SMS4 cipher (the Chinese National Encryption Standard) in $GF(2^8)$. They called the embedded cipher the ESMS4 cipher. They applied the eprint second XSL attack on ESMS4 and found the attack complexity to be 2^{77}. We shall analyze the compact XSL attack on ESMS4 with a method similar to our analysis of BES and show that the attack complexity is at least $2^{216.58}$. Therefore ESMS4 is also secure against the compact XSL attack.

2   Preliminaries

2.1   Embedding Block Ciphers in $GF(2^8)$

One difficulty in the cryptanalysis of AES is the tension between the operations in two different fields, $GF(2)$ and $GF(2^8)$. The S-box inversion is defined over $GF(2^8)$ whereas diffusion is defined over $GF(2)$. To address this issue, Robshaw and Murphy introduced the Big Encryption System (BES) embedding for the AES cipher [7] at Crypto 2002. In this embedding, all operations are performed over the finite field $GF(2^8)$ and this allows for easier analysis of structural features of the cipher. Since the algebraic structure of SMS4 is similar to that of AES, we can also embed SMS4 in a larger cipher, ESMS4 which is an Extension of SMS4 [5], to achieve the same aim.

Let $\mathbf{F}$ denote the field $GF(2^8)$. The main idea is to represent each byte $a \in \mathbf{F}$ by a set of 8-byte vector conjugates in $\mathbf{F}^8$ with the use of a vector conjugate mapping ϕ from $\mathbf{F}$ to a subset of $\mathbf{F}^8$ as follows:

$$\phi(a) = (a^{2^0}, a^{2^1}, a^{2^2}, a^{2^3}, a^{2^4}, a^{2^5}, a^{2^6}, a^{2^7}).$$

This definition extends naturally to a vector conjugate mapping ϕ from $\mathbf{F}^n$ to a subset of $\mathbf{F}^{8n}$ where $\mathbf{a} = (a_0, \ldots, a_{n-1}) \in \mathbf{F}^n$ is mapped to

$$\tilde{\mathbf{a}} = \phi(\mathbf{a}) = (\phi(a_0), \ldots, \phi(a_{n-1})).$$

Both the block cipher and its embedded system use a state vector of bytes, which is transformed by basic operations within a round. In both cases, the plaintext is the input state vector and the ciphertext is the output state vector. Denote the state spaces of the block cipher and its embedded system by $\mathbf{A}$ and $\mathbf{B}$ respectively. For AES, $\mathbf{A} = \mathbf{F}^{16}$ and $\mathbf{B} = \mathbf{F}^{128}$ while for SMS4, $\mathbf{A} = \mathbf{F}^4$ and $\mathbf{B} = \mathbf{F}^{32}$. Since ϕ is additive and preserves inversion, any state vector in the vector space $\mathbf{A}$ can be embedded in the vector space $\mathbf{B}$ with the vector conjugate map ϕ.

As described in detail in [7] and [5], all AES and SMS4 operations can be replicated by simple operations on the conjugates. In particular, all $GF(2)$-linear transformations can be extended to $GF(2^8)$ operations by virtue of the following result which is a direct consequence of dimension counting over $GF(2)$.

Lemma 1. *Consider the finite field $K = GF(2^8)$. Then any $GF(2)$-affine map $K \to K$ can be written in the form:*

$$f(x) = c + a_0 x + a_1 x^2 + \cdots + a_{n-1} x^{2^{n-1}},$$

for some constants $c, a_0, a_1, \ldots, a_{n-1} \in K$.

In essence, the encryption of AES and SMS4 can now be described exclusively in terms of $GF(2^8)$ operations. The advantage of this rewriting lies in the simplicity of the S-box equation: by first conveniently ignoring the case when $x = y = 0$, we have $xy = 1$ immediately instead of 8 quadratic equations in the input and output bits. Furthermore, in introducing the conjugates to the S-boxes, we have to express their relationship as $x_{i+1} = x_i^2$, where the subscript is taken from $\mathbb{Z}/8\mathbb{Z}$ (the integers modulo 8). If we denote the input and output variables by $x_0, x_1, \ldots, x_7 \in \mathbf{F}$ and $y_0, y_1, \ldots, y_7 \in \mathbf{F}$ respectively, then

$$x_0 y_0 = 1, \ x_1 y_1 = 1, \ x_2 y_2 = 1, \ \ldots, x_7 y_7 = 1,$$
$$x_0^2 = x_1, \ x_1^2 = x_2, \ x_2^2 = x_3, \ \ldots, x_7^2 = x_0,$$
$$y_0^2 = y_1, \ y_1^2 = y_2, \ y_2^2 = y_3, \ \ldots, y_7^2 = y_0.$$

Hence, this gives 24 equations for each S-box. Observe that the number of monomials is substantially reduced to 41 for both BES and ESMS4. These monomials are $1, x_i, y_i, x_i^2, y_i^2, x_i y_i$ for $0 \leq i \leq 7$.

To conclude this section, we make the following definition which will be used in the subsequent sections.

Definition 1. *[6] Let x_i be an input variable of an S-box and y_i be the corresponding output variable such that $x_i y_i = 1$. We shall say x_i and y_i are dual to each other.*

3   Analysis of the Compact XSL Attack on BES from [9]

This section is structured as follows:

(1) In Section 3.1, we start by describing the compact XSL attack on BES from [9].
(2) In Section 3.2, we describe the analysis of the eprint XSL attack on BES from [6]. Later, the techniques of [6] shall be modified and adapted to our analysis of the attack of [9].
(3) In Section 3.3, we point out and correct some errorneous assumptions made by the authors of [9] in their compact XSL attack on BES. These corrected assumptions will be used in our anaysis of the attack of [9]. We also correct some errorneous observations made by the authors of [9] on the analysis of [6].
(4) Finally in Section 3.4, we present our analysis of the attack from [9].

3.1   Description of the Attack of [9]

In [9], the authors presented an analysis of the *compact* XSL attack applied to the BES with the key schedule involved. They distinguished between two different types of S-boxes: one used both inversion and conjugacy relations (**regular** S-boxes); the other used purely conjugacy relations (**S'**-boxes). With their approach, they claimed to obtain a complexity estimate of 2^{97} for BES-128. In this section, we describe their strategy. We shall adopt their notation:

N_r: number of encryption rounds;

N_a, N_b: number of rows and columns in the encryption state;

N_k: number of columns in the cipher key state;

w_{nijl}, x_{nijl}: input and output of the S-box inversion in the encryption state in round n for

component in column i, row j and conjugate l, with $1 \leq n \leq N_r$, $0 \leq i < N_b$, $0 \leq j < N_a$,

$0 \leq l < 8$;

k_{nijl}: round key used in round n for component in column i, row j, and conjugate l, with

$1 \leq n \leq N_r, 0 \leq i < N_b, 0 \leq j < N_a, 0 \leq l < 8$;

s_{n3jl}: S-box inversion of the subkey k_{n3jl} which goes through an S-box during the key schedule;

S: number of *regular* S-boxes;

s: number of bits on an S-box;

r: number of equations in a *regular* S-box;

t: number of monomials in the *regular* S-box equations;

t': number of terms in the *basis* for one *regular* S-box that can be multiplied by some fixed variable

in the same S-box and still belong to the *basis*;

S': number of S'-boxes in the key schedule;

r_k: number of equations in an S'-box;

t_k: number of monomials in the S'-box equations;

T: total number of monomials in the system;

T': total number of monomials in the system which can be multiplied by a fixed variable in some

regular S-box and still remain in the same set of T system monomials.

R: total number of equations in the system;

R': number of linear equations in the BES cipher.

S-box Inversion Equations. All variables $w_{nijl}, x_{nijl}, k_{nijl}$, and s_{n3jl} are elements in $GF(2^8)$. By introducing the vector conjugate mapping to the S-boxes, we have $r = 24$ equations for each inverse S-box:

$$w_{nijl}x_{nijl} = 1, w_{nijl}^2 = w_{nij(l+1)}, x_{nijl}^2 = x_{nij(l+1)}$$

with $t = 41$ terms:

$$\{1, w_{nijl}x_{nijl}, w_{nijl}^2, x_{nijl}^2, w_{nijl}, x_{nijl}\}.$$

Subscript addition is computed modulo 8 and the case where $w = 0$ is disregarded. A natural basis of $t - r = 17$ elements $\{1, w_{nijl}, x_{nijl}\}$ of each S-box is chosen and we get $t' = 5^1$.

Key Schedule S and S'-boxes Equations. Next, consider the key schedule equations. Using the 1280 key schedule linear equations, the authors express all round key variables k_{nijl} as a linear combination of *key basis elements* defined as follows:

Key Basis Elements =

{Initial cipher key variables k_{0ijl}} ∪ {key schedule S-box output bits s_{n3jl}}.

Based on this approach, they reduced the number of S-box equations and monomials used, thereby reducing the size of the expanded equation system in the compact XSL attack. The details involving the key schedule quadratic equations used in their attack are outlined below:

---

[1] In [9], the value of t' was stated to be 4. However, we have verified that this is incorrect and that in fact, $t' = 5$. For example, for the variable w_{nij0}, the terms involved in t' are $\{1, w_{nij0}, x_{nij0}, x_{nij1}, x_{nij7}\}$.

(1) Out of the 40 S-boxes in the key schedule, the 4 S-boxes corresponding to the last column of the cipher key should be included in the regular S-box set since both their input k_{03jl} and output s_{03jl} are in the key basis;

(2) The rest of the cipher key variables $k_{00jl}, k_{01jl}, k_{02jl}$ do not pass through any S-box. Furthermore, none of the the remaining key basis variables s_{n3jl}'s input counterparts are included in the basis. Therefore, the authors ignored the inversion equations in the remaining 36 S-boxes of the key schedule. Then, the only relevant equations are:

$$k_{0ijl}^2 = k_{0ij(l+1)}$$

and

$$s_{n3jl}^2 = s_{n3j(l+1)}, \qquad 1 \le n \le N_r - 1.$$

In the above equations, the authors define each conjugate set (consisting of 8 conjugate equations) as an S'-box, and treat them in a similar way as S-boxes. There are 48 conjugate sets arising from the remaining $40 - 4 = 36$ S-boxes and $16 - 4 = 12$ bytes of cipher key (the authors exclude the 8 bytes k_{03jl} and s_{03jl} from the conjugate sets because they are the input/output of a regular S-box).

In light of the preceeding argument, they used $r_k = 8$, $t_k = 16$ for each S'-box, and the sets $\{k_{0ijl}\}$ and $\{s_{n3jl}\}$ ($1 \le n \le N_r - 1$) are chosen as the basis of the S'-boxes. In total, there will be 164 S-boxes and 48 S'-boxes that will be utilized in the attack.

Cipher Linear Layer Equations. The linear diffusion layer consists of matrix multiplication by a 128×128 matrix in $GF(2^8)$, giving 128 linear equations in each round of the form

$$w_{(n+1)ijl} + \sum_{ijl} d_{nijl} \cdot x_{nijl} + [k_{nijl}], 1 \le n \le N_r.$$

Adding in 128 more linear equations $w_{1ijl} + p_{ijl} + k_{0ijl}$ generated by the initial AddRoundKey (where p_{ijl} is the plaintext), there is a total of $R' = 1408$ cipher linear equations. Here, each $[k_{nijl}]$ is k_{nijl} expressed as a linear combination of terms that are in the basis of the relevant key schedule S and S'-boxes.

Multiplying Linear Layer Equations. After deriving all the S, S'-boxes and cipher linear equations, choose a parameter P, where P needs to satisfy certain conditions to be explained later. Multiply each of the cipher linear equations by the product of $(P - 1)$ monomials from the bases of different S and S'-boxes. Use the quadratic relations such as $x_{nijl}^2 = x_{nij(l+1)}$ to simplify the resulting equations where possible. Following this step, the authors got

$$R \approx R' \sum_{i=0}^{P-1} (t - r)^i \binom{S}{i} (t_k - r_k)^{P-1-i} \binom{S'}{P-1-i}$$

(not all linearly independent) equations. The total number of terms used in the attack is approximately

$$T \approx \sum_{i=0}^{P}(t-r)^{i}\binom{S}{i}(t_k - r_k)^{P-i}\binom{S'}{P-i}.$$

The T' Method. If the number of equations R is more than the number of terms T, then it would be possible to solve for the key via linearization, i.e. treat each monomial as an independent variable and solve by Gaussian elimination. If not, use the T' method described in [3,4] to obtain new linearly independent equations without creating any new monomials. In their attack, T' is about

$$T' \approx t' \sum_{i=0}^{P-1}(t-r)^{i}\binom{S-1}{i}(t_k - r_k)^{P-1-i}\binom{S'}{P-1-i}.$$

In order to implement the T' method, P needs to be chosen such that $R > T - T'$, i.e. so that there are more equations than terms to apply the linearization attack.

Summary. The compact XSL attack presented in [9] is as follows:

(1) Obtain the set of S and S'-box equations.
(2) Define the basis for each S and S'-box.
(3) Express the terms in each cipher linear layer equation as a linear combination of terms from the bases of S and S'-boxes.
(4) For each linear equation, multiply it by a product of $(P-1)$ monomials from the bases of different S and S'-boxes.
(5) Use the S and S'-box equations to simplify the monomials in the resulting system.
(6) Apply the T' method and solve for the key via linearization attack.

Results for BES-128. For BES-128, the condition that $R > T - T'$ is satisfied when $P \geq 3$. Then taking the Gaussian reduction exponent to be 3, Qu and Liu [9] obtained the result that the complexity of the XSL attack on BES-128 is $WF \approx 2^{97}$.

3.2 Description of the Analysis on the ePrint XSL Attack on BES from [6]

Lim and Khoo argued in [6] that for the smallest value of P such that the system of equations can be solved, the complexity of the attack is 2^{401} for BES-128. Their attack was based on *non-compact* eprint XSL attack published in [3] using key schedule relations. We highlight here that all the S-boxes considered in [6] are regular S-boxes having 16 input and output variables. We employ the following notation:

S_{total}: number of *regular* S-boxes in the cipher and key schedule;

L: number of linear equations in the cipher and key schedule.

The (eprint non-compact) XSL attack can be summarized as follows:

(1) Pick P distinct S-boxes and choose one of them to be **active**. The rest are called **passive** S-boxes. From the active S-box, choose one equation and multiply it by a product of monomials from each of the $P-1$ passive S-boxes. The resulting equation is called an **extended S-box equation**. The set of all extended S-box equations is denoted by Σ_S. Each monomial occuring in the equations of Σ_S is called an **extended S-box monomial**.

(2) Pick a linear equation and choose $P-1$ distinct S-boxes. Multiply the linear equation by a product of monomials from each of the $P-1$ S-boxes. The resulting equation is called an **extended linear equation** and the set of all extended linear equations is denoted by Σ_L.

(3) Solve the system of equations $\Sigma_S \cup \Sigma_L$ via linearisation.

(4) If there are insufficient equations for a complete solution, apply the T' method to grow the number of linearly independent equations.

Definition 2. *Let $\alpha = \alpha_1 \alpha_2 \ldots \alpha_Q$ be an extended S-box monomial, where each α_i is a variable belonging to some S-box. Then α is defined to be* reduced *if no two variables belong to the same S-box. The set of reduced S-box monomials of degree Q is denoted by Φ_Q.*

Due to the peculiarity of each S-box equation which is always an equation of two monomials, we can get the following result:

Fact 1. *[6, Theorem 1] Every extended S-box monomial α is equivalent to a unique reduced monomial β.*

Therefore, upon solving the extended S-box equations, the linearly independent terms are exactly $\Phi_0 \cup \Phi_1 \cup \ldots \Phi_P$. The upshot of the preceding argument is that the XSL method can be summarized as follows:

(1) Obtain the set Σ_S of extended S-box equations.

(2) For each linear equation, multiply it by a *reduced monomial* from $\Phi_0 \cup \Phi_1 \cup \ldots \cup \Phi_{P-1}$ and obtain the set Σ_L' of extended linear equations.

(3) Solve $\Sigma_S \cup \Sigma_L'$ together via linearisation.

The authors then considered the hypothetical situation where we only do steps 2 and 3 (i.e. $\Sigma_S = \emptyset$). This is equivalent to removing the S-boxes, giving $8S_{total}$ free variables taken to be the input variables of the S-boxes. Therefore, the number of linearly independent terms is at least the number of reduced monomials formed by these free variables. This is given by

$$D_1 = \sum_{i=0}^{P} \binom{S_{total}}{i} 8^i.$$

We are then faced with the question of whether step 1 provides sufficiently many equations to remove this number of linearly independent terms. The sole purpose of the set Σ_S is to substitute each monomial in Σ'_L with a corresponding reduced one. Hence, they obtained the following result:

Fact 2. *[6, Theorem 2] When solving Σ'_L with the extended S-box equations, the only useful extended S-box equations are of the form*

$$(v)(m_1) = m_2,$$

where:

(1) $m_1, m_2 \in \Phi_0 \cup \Phi_1 \dots \Phi_{P-1}$;
(2) v is an S-box variable such that it or its dual occurs in m_1;
(3) the remaining variables in m_1 are among the $8S_{total}$ free (input) variables.

An extended S-box equation is **relevant** if it is of the form stated in Fact 2. The number of such equations is given by

$$D_2 = 24S_{total} \times \sum_{i=0}^{P-2} \binom{S_{total} - 1}{i} 8^i.$$

So in order to solve the system of equations $\Sigma_S \cup \Sigma'_L$, we must have $D_2 \geq D_1$ and this imposes a condition on P. Note that in [6], the authors left the T' method out of their discussion on the basis that it can only be applied if the original number of equations is already very close to being sufficient.

3.3   Corrections on Some Claims in [9]

Before embarking on our analysis of the compact XSL attack on BES, we first clarify several major erroneous claims made in [9].

Qu and Liu eliminated many S-box inversion equations in the key schedule and replaced them with S'-boxes using only conjugacy relations. This was based on the reasoning that all the subkey bits can be expressed as a linear combination of a reduced set of subkey bits. However, this step will introduce many free variables s_{n3jl} ($1 \leq n \leq N_r - 1$) and therefore, should be disallowed. This means that we need to take all the S-box inversion and conjugacy relations into account in both the cipher and key schedule, as in the traditional approach. Thus, in the next section, we shall leave out the discussion of the S'-boxes. Consequently, all the cipher and key schedule linear equations should be explicitly involved in the multiplication by monomials.

In [9], it was stated that the authors of [6] excluded the key schedule from their analysis. However, this is incorrect since in Section 2.1 of [6], variables such as S_{total} and L are specifically defined to take the key schedule into account. S_{total} and L are defined to be the number of S-boxes and linear equations in the cipher *and key schedule*. Consequently, all subsequent calculations take key schedule variables and equations into consideration.

It was also claimed that Lim and Khoo presented their analysis of the attack over $GF(2)$, which is untrue since their paper focuses on XSL attack on BES which is over $GF(2^8)$.

Furthermore, it was claimed in [9] that the compact XSL attack is more efficient than the eprint XSL attack (both involving the key schedule). However contrary to their claim, the complexity of the second XSL attack on AES-128 was estimated to be 2^{298} in Courtois and Pieprzyk's compact XSL paper [4, Section 7.1], while the complexity in their eprint XSL attack paper [3, Section 8.1] is only 2^{230}.

3.4   An Analysis of the Compact XSL Attack in [9]

As explained in the previous section, we shall leave out the discussion of S'-boxes, that is, all the S-boxes involved are regular ones. Comparing the descriptions in Sections 3.1 and 3.2, we can easily spot the parallel between the two approaches due to the simple form of the S-box equations in BES:

(1) Multiplying Linear Layer Equations:
   - In the analysis of [6], each of the linear equations are multiplied by a reduced monomial comprising of a product of variables from $P - 1$ distinct S-boxes.
   - In compact XSL attack, each of the linear equations are multiplied by a product of $(P - 1)$ monomials from the basis of different S boxes in both the cipher and key schedule. For each S-box, the set with 17 elements, comprising 1, the input and output components of the S-box inversion, is chosen as the basis.
   - From this, we can see that the product of monomials in compact XSL attack is, in actual fact, a reduced monomial.
(2) Solving the System:
   - In the analysis of [6], the set Σ_S of extended S-box equations are used to substitute each monomial in Σ'_L with a corresponding reduced one.
   - In compact XSL attack, the S-box quadratic equations are used to simplify the resulting post-multiplication equations.
   - Since each S-box is simply an equality of one monomial with another, these two steps are equivalent.

Based on this correspondence, we can easily apply the arguments in Section 3.2 to the compact XSL attack on BES. However, we also note the following difference between the two attacks in Sections 3.1 and 3.2:

(1) T' Method:
   - In the analysis of [6], the T' method was not considered.
   - In compact XSL attack of [9], the authors used the T' method.
   - To make our analysis more complete, we shall include the T' method as applied to the compact XSL attack in the discussion that follows.

As in Section 3.2,

$$D_1 = \sum_{i=0}^{P} \binom{S_{total}}{i} 8^i$$

and

$$D_2 = 24 S_{total} \times \sum_{i=0}^{P-2} \binom{S_{total} - 1}{i} 8^i.$$

(Recall that S_{total} is the number of *regular* S-boxes in the cipher and key schedule.)

The T' Method in our Analysis. The T' method in [3,4] works as follows: Suppose we have R equations in T monomials of degree at most d. Choose a subset of monomials $\mathbf{T}'$ such that when we multiply them by the variable x_i, the resulting monomial still belongs to the original set of T monomials. By a single Gaussian elimination, we express each monomial not in $\mathbf{T}'$ in terms of the monomials in $\mathbf{T}'$. If $R > T - T'$ (where T' is the size of $\mathbf{T}'$), there will be some equations which only consists of the monomials of the set $\mathbf{T}'$. When we multiply these equations by the variable x_i, we obtain *extra* equations without growing the number of monomials. We apply the T' method to the same system of equations several times with respect to distinct variables $x_1, x_2, \dots$. Then we express the *extra* equations of one system in terms of the T' monomials of another to obtain even more equations. In this way, we can *grow* the number of equations from R to T whenever $R > T - T'$. An illustrative example of the T' method can be found in [3, Appendix E] or [4, Appendix B].

Now suppose an adversary has found a P such that $R > T - T'$ in the compact XSL attack on BES. Then he can *grow* another $T - R$ equations by the T' method to make the total number of equations T. If we partially solve the original R equations, we have shown that there would be at least $D_1 - D_2$ unsolved monomials, i.e. $D_1 - D_2$ degrees of freedom. Assuming each of the $T - R$ extra equations were able to help solve for another monomial, i.e. decrease a degree of freedom, we would need $D_1 - D_2 \le T - R$. Since $T - R < T'$, we see that $D_1 - D_2 < T'$ is a necessary condition for the system of equations to be solved by linearization.

As pointed out in Section 3.3, the formulas of Section 3.1 obtained from the concept of S'-boxes are errorneous. The corrected formulas should be:

$$R \approx R'(t - r)^{P-1} \binom{S_{total}}{P - 1},$$

$$T \approx (t - r)^{P} \binom{S_{total}}{P},$$

$$T' \approx t'(t - r)^{P-1} \binom{S_{total} - 1}{P - 1}.$$

Note that now, unlike in Section 3.1, these expressions for R, T and T' no longer consider S'-boxes.

Table 1. D_1, D_2 and T' values corresponding to various P for compact XSL attack on BES

P		BES-128	BES-192	BES-256
2	D_1	1.288×10^6	5.554×10^6	8.020×10^6
	D_2	4.824×10^3	1.001×10^4	1.202×10^4
	$D_1 - D_2$	1.283×10^6	5.544×10^6	8.008×10^6
	T'	1.700×10^4	3.536×10^4	4.250×10^4
3	D_1	6.839×10^8	6.149×10^9	1.067×10^{10}
	D_2	7.723×10^6	3.332×10^7	4.811×10^7
	$D_1 - D_2$	6.762×10^8	6.115×10^9	1.063×10^{10}
	T'	2.876×10^7	1.247×10^8	1.803×10^8
$\vdots$	$\vdots$	$\vdots$	$\vdots$	$\vdots$
6	D_1	2.235×10^{16}	1.850×10^{18}	5.596×10^{18}
	D_2	1.281×10^{15}	5.048×10^{16}	1.269×10^{17}
	$D_1 - D_2$	2.107×10^{16}	1.799×10^{18}	5.469×10^{18}
	T'	1.800×10^{16}	7.195×10^{17}	1.812×10^{18}
7	D_1	4.985×10^{18}	8.691×10^{20}	3.166×10^{21}
	D_2	4.021×10^{17}	3.329×10^{19}	1.007×10^{20}
	$D_1 - D_2$	4.583×10^{18}	8.359×10^{20}	3.066×10^{21}
	T'	9.946×10^{18}	8.378×10^{20}	2.541×10^{21}
8	D_1			1.565×10^{24}
	D_2			6.647×10^{22}
	$D_1 - D_2$			1.498×10^{24}
	T'			3.049×10^{24}

We see in Table 1 the minimum values of P required to satisfy the condition $D_1 - D_2 < T'$ are $P = 7$ for BES-128, BES-192 and $P = 8$ for BES-256. This will translate to an attack complexity of at least:

$$WF = T^3 \approx (9.710 \times 10^{20})^3 \approx 2^{209.15} \text{ for BES-128;}$$

$$WF = T^3 \approx (1.697 \times 10^{23})^3 \approx 2^{231.50} \text{ for BES-192;}$$

$$WF = T^3 \approx (6.492 \times 10^{26})^3 \approx 2^{267.21} \text{ for BES-256.}$$

This shows that the compact XSL attack is worse than exhaustive search for BES-128, 192 and 256. Furthermore, we have to keep in mind that this is just a lower bound on the attack complexity because we have assumed that each 'extra' equation generated by the T' method can solve for a remaining unsolved monomial of the original R equations. Therefore, the complexity estimate in [9] is far too optimistic.

It is easy to see that the corrected formulas used in our analysis are exactly those of the compact XSL attack in [4]. Thus we have essentially shown that the compact XSL attack of [4] when applied to BES-128, 192 and 256 does not work.

Remark 1. We used T in the calculation of the complexity of the attack, unlike in [6] where D_1 was used instead. Since D_1 is a lower bound for the number of linearly independent terms, either term would be applicable to our analysis of the attack.

Remark 2. Note that we have assumed that the complexity to solve N linear equations by Gaussian elimination is N^3 so that it is a fair comparison with the complexity 2^{97} claimed in [9]. If we assume the Gaussian elimination complexity to be $N^{2.376}$, then the attack complexity for BES-128 will be 2^{143} which is still worse than exhaustive search.

4 Analysis of the Compact XSL Attack on ESMS4

Following the notation used in [5], let $(X_0, X_1, X_2, X_3) \in (\mathbf{F}^{32})^4$ be the input plaintext block of ESMS4 and the corresponding ciphertext be $(X_{32}, X_{33}, X_{34}, X_{35}) \in (\mathbf{F}^{32})^4$. For rounds $i = 1, \cdots, 32$,

$W_{i,(j,m)}$: the $(8j + m)$-th component of the input of the affine transformation in the S-box;

$V_{i,(j,m)}$: the $(8j + m)$-th component of the input variables of the inversion transformation;

$Y_{i,(j,m)}$: the $(8j + m)$-th component of the output variables of the inversion transformation;

$Z_{i,(j,m)}$: the $(8j + m)$-th component of the output variables of the linear diffusion transformation;

$K_{i,(j,m)}$: the $(8j + m)$-th component of the round key,

where $j = 0, \cdots, 3$ and $m = 0, \cdots, 7$.

According to [5], the probability of 0-inverse occurring is only $O(2^{-10})$. Based on this assumption, the encryption of ESMS4 can be fully described by the following system:

$$0 = X_{i,(j,m)} + X_{i+1,(j,m)} + X_{i+2,(j,m)} + K_{i,(j,m)} + W_{i,(j,m)}$$

$$0 = C_{(j,m)} + V_{i,(j,m)} + \sum_{(j',m')} \alpha_{(j,m),(j',m')} W_{i,(j',m')}$$

$$0 = D_{(j,m)} + Z_{i,(j,m)} + \sum_{(j',m')} \beta_{(j,m),(j',m')} Y_{i,(j',m')}$$

$$0 = X_{i-1,(j,m)} + Z_{i,(j,m)} + X_{i+3,(j,m)}$$

$$0 = V_{i,(j,m)} Y_{i,(j,m)} + 1$$

$$0 = V_{i,(j,m)}^2 + V_{i,(j,m+1)}$$

$$0 = Y_{i,(j,m)}^2 + Y_{i,(j,m+1)},$$

where $C_{(j,m)}$, $D_{(j,m)}$, $\alpha_{(j,m),(j',m')}$ and $\beta_{(j,m),(j',m')}$ are some constants.

Since the key scheduling algorithm is similar to its encryption, we can also describe the key schedule with a multivariate quadratic system over $\mathbf{F}$. Hence, for ESMS4, we have $N_r = 32$, $S_{total} = 256$, $t = 41$, $r = 24$ and $s = 8$. It was calculated in [5] that there are 8192 linear equations, 6144 quadratic equations and

17536 terms for the whole system in total. As the above multivariate quadratic system is very sparse, ESMS4 seems to be a potential candidate for XSL attack to be carried out.

The XSL technique used in [5] is based on the eprint second XSL attack where the key schedule is taken into account. Different from the compact version, all the t monomials from each S-box are used to multiply the S-box and diffusion layer equations, as opposed to the use of a basis of $t - r$ monomials for each S-box. By taking the Gaussian reduction exponent to be 3, Ji and Hu [5] calculated the complexity of eprint XSL second attack to be approximately 2^{77}. However, as our paper is devoted to the study of compact XSL attack, we shall limit our analysis of the XSL attack on ESMS4 to the compact version with the key schedule equations being taken into account.

We now proceed to evaluate the effectiveness of the compact XSL attack against ESMS4 by using a similar approach as BES to estimate the complexity of the attack. For each S-box in the cipher, choose $\{1, V_{i,(j,m)}, Y_{i,(j,m)}\}$ with 17 elements as the basis.

For $i = 1, \cdots, 32$, $j = 0, \cdots, 3$ and $m = 0, \cdots, 7$, denote the $(8j + m)$-th component of the input and output of the S-box inversion in the key schedule by $G_{i,(j,m)}$ and $H_{i,(j,m)}$ respectively. Similarly, for each S-box in the key schedule, we choose $\{1, G_{i,(j,m)}, H_{i,(j,m)}\}$ with 17 elements as the basis. Hence, as before, we get $t' = 5$.

With the following formula,

$$D_1 = \sum_{i=0}^{P} \binom{256}{i} 8^i,$$

Table 2. D_1, D_2 and T' values corresponding to various P for compact XSL on ESMS4

$P = 2$	D_1	2.091×10^6
	D_2	6.144×10^3
	$D_1 - D_2$	2.085×10^6
	T'	2.168×10^4
$P = 3$	D_1	1.417×10^9
	D_2	1.254×10^7
	$D_1 - D_2$	1.404×10^9
	T'	4.680×10^7
$\vdots$	$\vdots$	$\vdots$
$P = 6$	D_1	9.690×10^{16}
	D_2	4.339×10^{15}
	$D_1 - D_2$	9.256×10^{16}
	T'	6.132×10^{16}
$P = 7$	D_1	2.770×10^{19}
	D_2	1.743×10^{18}
	$D_1 - D_2$	2.596×10^{19}
	T'	4.343×10^{19}

$$D_2 = 24 \times 256 \times \sum_{i=0}^{P-2} \binom{255}{i} 8^i,$$

$$T \approx 17^P \times \binom{256}{P},$$

$$T' \approx 5 \times 17^{P-1} \binom{255}{P-1},$$

we obtain the values for D_1, D_2 and T' values corresponding to various P and hence estimate the complexity of the attack. The results are summarized in Table 2.

From Table 2, the smallest P such that $D_1 - D_2 < T'$ is $P = 7$. Hence the complexity for compact XSL second attack on ESMS4 is

$$WF = T^3 = (5.401 \times 10^{21})^3 \approx 2^{216.58},$$

which shows that the compact XSL attack on ESMS4 is worse than exhaustive search.

Remark 3. Here, we assume that the complexity to solve N linear equations by Gaussian elimination is N^3 so that it is a fair comparison with the complexity 2^{77} claimed in [5]. If we assume the Gaussian elimination complexity to be $N^{2.376}$, then the attack complexity for ESMS4 will be $2^{171.53}$ which is still worse than exhaustive search.

5   Conclusion

We have adapted the analysis on the eprint XSL attack on BES in [6] to the compact XSL attack on BES and ESMS4 in this paper. Moreover, we have considered the T' method in our analysis to make it more accurate. We showed that the compact XSL attack on both BES (for all 3 key sizes) and ESMS4 are worse than exhaustive search. Further research in this direction could be undertaken to include the T' method in the analysis of the eprint XSL attack on both BES and ESMS4.

Acknowledgement

We would like to thank Chu-Wee Lim for many interesting discussions and providing insightful comments on the paper.

References

1. Cid, C., Leurent, G.: An Analysis of the XSL Algorithm. In: Roy, B. (ed.) ASIACRYPT 2005. LNCS, vol. 3788, pp. 333–352. Springer, Heidelberg (2005)
2. Courtois, N., Klimov, A., Patarin, J., Shamir, A.: Efficient Algorithms for Solving Systems of Multivariate Polynomial Equations. In: Preneel, B. (ed.) EUROCRYPT 2000. LNCS, vol. 1807, pp. 392–407. Springer, Heidelberg (2000)

3. Courtois, N., Pieprzyk, J.: Cryptanalysis of Block Ciphers with Overdefined Systems of Equations, IACR eprint server 2002/044 (March 2002),
   http://www.iacr.org
4. Courtois, N., Pieprzyk, J.: Cryptanalysis of Block Ciphers with Overdefined Systems of Equations. In: Zheng, Y. (ed.) ASIACRYPT 2002. LNCS, vol. 2501, pp. 267–287. Springer, Heidelberg (2002)
5. Ji, W., Hu, L.: New Description of SMS4 by an Embedding over $GF(2^8)$. In: Srinathan, K., Rangan, C.P., Yung, M. (eds.) INDOCRYPT 2007. LNCS, vol. 4859, pp. 238–251. Springer, Heidelberg (2007)
6. Lim, C.-W., Khoo, K.: An Analysis of XSL Applied to BES. In: Biryukov, A. (ed.) FSE 2007. LNCS, vol. 4593, pp. 242–253. Springer, Heidelberg (2007)
7. Murphy, S., Robshaw, M.: Essential Algebraic Structure Within the AES. In: Yung, M. (ed.) CRYPTO 2002. LNCS, vol. 2442, pp. 1–16. Springer, Heidelberg (2002)
8. Murphy, S., Robshaw, M.: Comments on the Security of the AES and the XSL Technique. Electronic Letters 39, 26–38 (2003)
9. Qu, B., Liu, L.: An XSL Analysis on BES. In: Proceedings of the 9th International Conference for Young Computer Scientist, pp. 1418–1423. IEEE Press, Los Alamitos

RFID Distance Bounding Protocol with Mixed Challenges to Prevent Relay Attacks

Chong Hee Kim and Gildas Avoine

Université catholique de Louvain, B-1348 Louvain-la-Neuve, Belgium
{chong-hee.kim,gildas.avoine}@uclouvain.be

Abstract. RFID systems suffer from different location-based attacks such as distance fraud, mafia fraud and terrorist fraud attacks. Among them mafia fraud attack is the most serious since this attack can be mounted without the notice of both the reader and the tag. An adversary performs a kind of man-in-the-middle attack between the reader and the tag. It is very difficult to prevent this attack since the adversary does not change any data between the reader and the tag. Recently distance bounding protocols measuring the round-trip time between the reader and the tag have been researched to prevent this attack.

All the existing distance bounding protocols based on binary challenges, without final signature, provide an adversary success probability equal to $(3/4)^n$ where n is the number of rounds in the protocol. In this paper, we introduce a new protocol based on binary mixed challenges that converges toward the expected and optimal $(1/2)^n$ bound. We prove its security in case of both noisy and non-noisy channels.

Keywords: RFID, authentication, distance bounding protocol, relay attack.

1 Introduction

RFID (radio frequency identification) tags or contactless smart cards are often used for proximity authentication. For example, Texas Instrument (TI) manufactured an RFID device called a Digital Signature Transponder (DST). The DST serves as a theft-deterrent in millions of automobiles. Present as a tiny, concealed chip in the ignition key of the driver, the DST authenticates the key to a reader near the key slot as a precondition for starting the engine. The DST is also present in SpeedPass^{TM} wireless payment devices, used by millions of customers primarily at ExxonMobil petrol stations in North America.

RFID tags and contactless smart cards are normally passive; they operate without any internal battery and receive the power from the reader. This offers long lifetime but results in short read ranges and limited processing power. They are also vulnerable to different attacks related to the location: distance fraud and relay attacks. Relay attacks occur when a valid reader is tricked by an adversary into believing that it is communicating with a valid tag and vice versa. That is, the adversary performs a kind of man-in-the-middle attack between the reader

J.A. Garay, A. Miyaji, and A. Otsuka (Eds.): CANS 2009, LNCS 5888, pp. 119–133, 2009.

and the tag. It is difficult to prevent these attacks since the adversary does not change any data between the reader and the tag. Therefore relay attacks cannot be prevented by cryptographic protocols that operates at the application layer.

Although one could verify location through the use of GPS coordinates, RFID tags do not lend themselves to such applications. *Distance bounding protocols* is a good solution to prevent such distance fraud and relay attacks. These protocols measure the signal strength or the round-trip time between the reader and the tag. However the proof based on measuring signal strength is not secure as an adversary can easily amplify signal strength as desired or use stronger signals to read from afar.

1.1   Distance Fraud and Relay Attacks

There are three types of attacks related with distance between the reader and the tag. The dishonest tag may claim to be closer than he really is. This attack is called **distance fraud attack**. There are two types of relay attacks: mafia fraud and terrorist fraud attacks.

Mafia fraud attack was first described by Desmedt [2]. In this attack scenario, both the reader (R) and the tag (T) are honest, but a malicious adversary is performing a man-in-the-middle attack between the reader and the tag by putting fraudulent tag ($\overline{T}$) and receiver ($\overline{R}$). The fraudulent tag $\overline{T}$ interacts with the honest reader R and the fraudulent reader $\overline{R}$ interacts with the honest tag T. $\overline{T}$ and $\overline{R}$ cooperate together. It enables $\overline{T}$ to convince R of a statement related to the secret information of an honest tag T, without actually needing to know anything about the secret information.

Terrorist fraud attack is an extension of the mafia fraud attack. The tag T is not honest and collaborate with fraudulent tag $\overline{T}$ in this attack. The dishonest tag T uses $\overline{T}$ to convince the reader that he is close, while in fact he is not. $\overline{T}$ does not know the long-term private or secret key of T.

Among these attacks, mafia fraud attack is the most serious since this attack can be mounted without the notice of neither the reader nor the tag. Many works are devoted to prevent this attack [1,3,4,5,6,7].

1.2   Distance Bounding Protocols

In 1993, Brands and Chaum presented their distance bounding protocol [1]. It consists of a *fast bit exchanges* phase where the reader sends out one bit and starts a timer. Then the tag responds to the reader with one bit that stops the timer. The reader uses the round trip time to extract the propagation time. After series of n rounds (n is a security parameter), the reader decides whether the tag is within the limitation of the distance. In order to extract the propagation time, the processing time of the tag must be as short and invariant as possible. The communication method used for these exchanges is different from the used one for the ordinary communication. An ultra wide band (UWB) channel is used to achieve a resolution of 10 cm. It does not contain any error detection or correction mechanism in order not to make additional variable cycles of processing.

Reader		**Tag**
(secret K)		(secret K)

Pick a random N_a Pick a random N_b

$$\xrightarrow{\quad N_a \quad}$$
$$\xleftarrow{\quad N_b \quad}$$

$$\{H\}^{2n} = h(K, N_a, N_b)$$
$$\{v^0\} = H_1 || H_2 || \dots || H_n$$
$$\{v^1\} = H_{n+1} || H_{n+2} || \dots || H_{2n}$$

Start of fast bit exchange
for $i = 1$ to n

Pick $C_i \in \{0, 1\}$

Start Clock $\qquad\qquad\qquad\xrightarrow{\quad C_i \quad}$

$$R_i = \begin{cases} v_i^0, & \text{if } C_i = 0 \\ v_i^1, & \text{if } C_i = 1 \end{cases}$$

Stop Clock $\qquad\qquad\xleftarrow{\quad R_i \quad}$

Check correctness of
R_i's and $\Delta t_i \leq t_{max}$

End of fast bit exchange

Fig. 1. Hancke and Kuhn's protocol

Hancke and Kuhn proposed a distance bounding protocol (HKP) [3] that has been chosen as a reference-point because it is the most popular distance bounding protocol in the RFID framework. As depicted in Fig. 1, the protocol is carried out as follows. After exchanges of random nonces (N_a and N_b), the reader and the tag compute two n-bit sequences, v^0 and v^1, using a pseudorandom function (typically a MAC algorithm, a hash function, etc.). Then the reader sends a random bit for n times. Upon receiving a bit, the tag sends back a bit R_i from v^0 if the received bit C_i equals 0. If C_i equals 1, then it sends back a bit from v^1. After n iterations, the reader checks the correctness of R_i's and the propagation time.

In each round, the probability that the adversary sends a correct response is *a priori* $\frac{1}{2}$. However the adversary can query the tag in advance with some arbitrary C_i's, between the nonces are sent and the rapid bit exchange starts. Doing so, the adversary obtains n bits of the registers. For example, if the adversary queries the with some zeroes only, he will entirely get v^0. In half of all cases, the adversary will have the correct guesses, that is $C_i' = C_i$, and therefore will have obtained in advance the correct value R_i that is needed to satisfy the reader. In the other half of all cases, the adversary can reply with a guessed bit, which will be correct in half of all cases. Therefore, the adversary has $\frac{3}{4}$ probability of replying correctly.

One of the solutions to reduce the probability less than $(\frac{3}{4})^n$ is to include signed messages [1,6,7]. However signed messages could not be sent with UWB

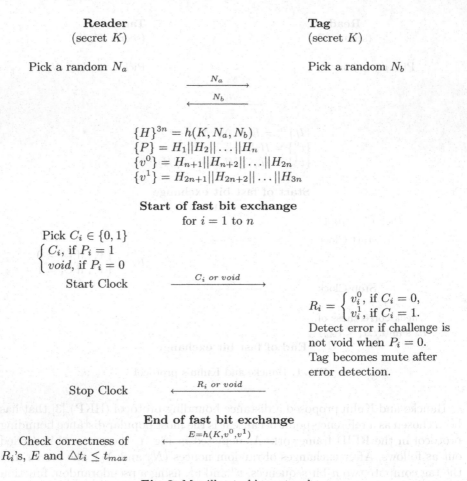

Reader
(secret K)

Pick a random N_a

Tag
(secret K)

Pick a random N_b

$$\{H\}^{3n} = h(K, N_a, N_b)$$
$$\{P\} = H_1||H_2||\dots||H_n$$
$$\{v^0\} = H_{n+1}||H_{n+2}||\dots||H_{2n}$$
$$\{v^1\} = H_{2n+1}||H_{2n+2}||\dots||H_{3n}$$

Start of fast bit exchange
for $i = 1$ to n

Pick $C_i \in \{0, 1\}$
$$\begin{cases} C_i, & \text{if } P_i = 1 \\ void, & \text{if } P_i = 0 \end{cases}$$

Start Clock

$$R_i = \begin{cases} v_i^0, & \text{if } C_i = 0, \\ v_i^1, & \text{if } C_i = 1. \end{cases}$$
Detect error if challenge is
not void when $P_i = 0$.
Tag becomes mute after
error detection.

Stop Clock

End of fast bit exchange

Check correctness of
R_i's, E and $\triangle t_i \leq t_{max}$

Fig. 2. Munilla et al.'s protocol

as it is very sensitive to the background noise. It should be sent by normal communication method with error detection or correction technique. Therefore this approach would put an overload on computation of a tag as well as communication, which causes the protocol slower.

Munilla, Ortiz, and Peinado [4,5] modified the Hancke and Kuhn's protocol by applying "void challenges" in order to reduce the success probability of the adversary. As shown in Fig. 2, the challenges from the reader are divided into two categories, *full challenge* and *void challenge*. After exchanges of random nonces (N_a and N_b), the reader and the tag compute $3n$-bit sequence, $P||v^0||v^1$, using a pseudorandom function. The string P indicates the void-challenges; that is, if $P_i = 1$ reader sends a random challenge and if $P_i = 0$ it does not. These void-challenges allow the tag to detect if an adversary is trying to get the responses in advance. When the tag detects an adversary it stops sending responses. The protocol ends with a message to verify that no adversary has been detected.

The adversary can choose between two main attack strategies: asking in advance to the tag, taking the risk that the tag uncovers him, and without taking in advance and trying to guess the responses to the challenges when they occur. The adversary's success probability depends on p_f, the probability of the occurrence of full challenge, and can be calculated:

$$p_{MP} = \begin{cases} (1 - \frac{p_f}{2})^n, \text{if } p_f \leq \frac{4}{5} \text{ (without asking in advance)}, \\ (p_f \cdot \frac{3}{4})^n, \text{if } p_f > \frac{4}{5} \text{ (asking in advance)}. \end{cases} \quad (1)$$

The adversary's success probability is the lowest when $p_f = 4/5$, but it is not easy to generate a bit string P with such value of p_f. However, the value $p_f = 3/4$ is close to $4/5$ and it is much easier to generate. By generating a random $2n$-bit P and letting '00', '01', or '10' as $P_i = 1$ and '11' as $P_i = 0$, we can get $p_f = 3/4$. If the responses of the tag are taken out from one edge of the bit-string (LSB, the least significant bit) or from the other one (MSB, the most significant bit) depending on the challenge, $n + 1$ bits are enough to generate $v^0 || v^1$. Therefore total $3n + 1$ bits ($2n$ bits for P, $n + 1$ bits for responses) are required to store. The success probability of the adversary is $(\frac{5}{8})^n$ if the string P is random [5], which is less than $(\frac{3}{4})^n$.

Note that the final confirmation message $h(K, v^0, v^1)$ does not take any challenges C_i as an input. So it can be pre-computed before the start of the fast bit exchange. On the other side, the disadvantage of their solution is that it requires three (physical) states: 0, 1, and *void*, which may be difficult to implement. Furthermore the success probability of the adversary is higher than $(\frac{1}{2})^n$.

2 Distance Bounding Protocol Using Mixed Challenges

2.1 Description

To overcome the disadvantage of Munilla et al.'s protocol, we present a modification using *mixed challenges*: the challenges from the reader to the tag in the fast bit exchanges are divided into two categories, *random challenges* and *predefined challenges*. The earlier are random bits from the reader and the latter are predefined bits known to both the reader and the tag in advance.

As shown in Fig. 3, the reader and the tag compute $4n$-bit sequence for $T || D || v^0 || v^1$, after exchange of random nonces (N_a and N_b). The string T indicates random-challenges: if $T_i = 1$ the reader sends a random bit $S_i \in \{0, 1\}$ and if $T_i = 0$ it sends a predefined bit D_i to the tag. From the point of the adversary's view, all C_i's from the reader look like random. Therefore he cannot distinguish random challenges from predefined challenges. However, thanks to these predefined challenges, the tag is able to detect an adversary sending random challenges in order to get responses in advance. Upon reception of a challenge C_i from the reader, either $T_i = 1$ (random challenge), in this case the tag sends out the bit $v_i^{C_i}$; or $T_i = 0$ (predefined challenge), in this case the tag sends out the bit v_i^0 if $C_i = D_i$ and a random bit if $C_i \neq D_i$ (it detects an error).

Reader
(secret K)

Tag
(secret K)

Pick a random N_a

Pick a random N_b

$$\xrightarrow{\quad N_a \quad}$$
$$\xleftarrow{\quad N_b \quad}$$

$$\{H\}^{4n} = h(K, N_a, N_b)$$
$$\{T\} = H_1 \| H_2 \| \ldots \| H_n$$
$$\{D\} = H_{n+1} \| H_{n+2} \| \ldots \| H_{2n}$$
$$\{v^0\} = H_{2n+1} \| H_{2n+2} \| \ldots \| H_{3n}$$
$$\{v^1\} = H_{3n+1} \| H_{3n+2} \| \ldots \| H_{4n}$$

Start of fast bit exchange
for $i = 1$ to n

Pick $S_i \in \{0, 1\}$
$$C_i = \begin{cases} S_i, & \text{if } T_i = 1 \\ D_i, & \text{if } T_i = 0 \end{cases}$$

Start Clock

$$\xrightarrow{\quad C_i \quad}$$

If $T_i = 1$, then
$$R_i = \begin{cases} v_i^0, & \text{if } C_i = 0 \\ v_i^1, & \text{if } C_i = 1 \end{cases}$$
If $T_i = 0$, then
$$R_i = \begin{cases} v_i^0, & \text{if } C_i = D_i \\ \text{random}, & \text{if } C_i \neq D_i \end{cases}$$
(error detected)

* After error detection,
sends a random bit
for all subsequent iterations.

Stop Clock

$$\xleftarrow{\quad R_i \quad}$$

Check: $\Delta t_i \leq t_{max}$
Check: correctness of R_i

End of fast bit exchange

Fig. 3. Distance bounding protocol using mixed challenges

From the moment the tag detects an error, it replies a random value to all the subsequent challenges. By doing this, both the reader and the tag fight the adversary.

We note that we do not use any confirmation message after the end of fast bit exchanges, which improves efficiency in terms of computation and communication compared to Munilla et al. [5].

2.2   Discussion about the Tag's Behavior after an Error Is Detected

In our protocol, the tag always replies with a random bit after detection of an error. This is a conservative behavior but some other ones are also possible.

Interrupt the protocol. One may think that the tag can simply interrupt the protocol when an error is detected. However, the reader will simply conclude in such a case that the protocol failed, while in practice it could be interesting for the reader to be able to distinguish a failure from an attack; it could so react accordingly.

Complementary bits. Another variant consists for the tag in sending the complementary bits of the right answers once a received predefined challenge is wrong. In this way, the tag helps the reader to detect early that an attack occurs. Indeed, if the strategy of the adversary is to exactly send to the reader what he previously received from the tag, then his probability of success is 0 once he sent a wrong challenge. However, with such a variant, a better strategy for the adversary is to expect an early wrong challenge, and then to flip all the subsequent responses from the tag. His probability of success becomes 1 once he sent a wrong challenge.

Half-time complementary bits. To thwart an attack based on the "flip strategy", one way consists in flipping only half of the responses. Thus, after an error is detected, the tag sends the right response when $T_i = 0$ but sends the complementary one when $T_i = 1$. As the adversary cannot distinguish between $T_i = 0$ and $T_i = 1$, he cannot decide when he delivers the response as it is or not. Consequently, after an error, the probability for the adversary to get the right response is 1 if $T_i = 0$, and 0 if $T_i = 1$.

Use the obsolete D_is. Instead of using the complementary approach or generating new random bits as in the basic protocol, the tag may reply with the remaining D_is after an attack is detected. Indeed, after a wrong challenge is received, the D_is become useless. This approach has two advantages: (a) to avoid generating new random values; (b) to help the reader to detect earlier an attack (the reader detects that the answers match the D_is). Here, because the D_is are still used for the reader's challenges when $T_i = 0$, this variant gives to the adversary the ability to observe that he has been detected by the tag. He may so interrupt the protocol, expecting the reader to conclude that a failure occurs instead of an attack.

Use the obsolete T_is. As in the previous variant, after an error is detected, the tag uses an already generated random register that is no longer in use; however, T is used instead of D because T does not reveal that the adversary is detected. This variant presents the same two advantages than the previous one without revealing the attack detection.

A detailed analysis of the success probability of the adversary follows in the next section. We consider in this analysis the basic version of the protocol, where the tag replies with some random bits after an error is detected.

3 Analysis

We define p_d as the probability that a challenge is a *predefined challenge*. Similarly p_r is defined as the probability that a challenge is a *random challenge*.

Therefore we have $p_d + p_r = 1$. We start the analysis from the noise-free case then analyze in the noisy case.

3.1 Analysis in the Noise-Free Case

The adversary can choose between two main attack strategies. First he can guess the responses to the challenges without asking in advance to tag. Secondly he can ask in advance to the tag, taking the risk that the tag uncovers him. Let us denote the adversary's probability of success without asking as P_{no-ask} and that with asking as P_{ask}. The adversary's probability of success of Munilla et al.'s protocol depends on the probability of the occurrence of full or void challenges as shown in Eq. 1. Therefore P_{no-ask} and P_{ask} are different according to the probability of the occurrence of full challenges. With the recommended $p_f = 3/4$, the adversary's probability of the success without asking, P_{no-ask}, is $(5/8)^n$, which is higher than P_{ask}. Therefore it is better for the adversary to choose the first attack approach in Munilla et al.'s protocol.

In our proposed protocol the adversary's probability of the success without asking in advance, P_{no-ask}, is always $(1/2)^n$. Therefore we compute P_{ask} and compare it with P_{no-ask} in the next section.

Adversary's probability of success of not being detected by reader. To compute P_{ask}, we assume that an adversary asks in advance to tag, taking the risk that the tag uncovers him. If the challenge, C_i^*, that the adversary asks to the tag in advance is the same than the challenge, C_i, that the reader sends to the tag, he sends the response received from the tag to the reader. If $C_i^* \neq C_i$, then he sends a random response to the reader.

Although the adversary's attack is detected by the tag there is still a chance of not being detected by the reader. To analyze this probability, we define some events as follows:

- A_i: the event that the adversary's attack is detected by the reader in the ith round,
- $\overline{A}_i$: the event that the adversary's attack is *not* detected by the reader in the ith round,
- B_i: the event that the adversary's attack is detected by the tag in the ith round,
- $\overline{B}_i$: the event that the adversary's attack is *not* detected by the tag in the ith round.

The event of not being detected by the reader in the ith round, $\overline{A}_i$, depends on the event of being detected by the tag in the $(i-1)^{th}$ round. Because tag gives random values once it detects an error and the adversary does not know that it is an original or a random value.

The probability of not being detected by the reader in the ith round provided that it is not detected by the tag in the $(i-1)^{th}$ round, $P(\overline{A}_i|\overline{B}_{i-1})$, is

$$P(\overline{A}_i|\overline{B}_{i-1}) = P_{\text{random ch. \& not detected}} + P_{\text{predefined ch. \& not detected}}$$

$$= p_r \cdot \left(P_{C_i^*=C_i \& \text{ not det. by Reader}} + P_{C_i^* \neq C_i \& \text{ not det. by Reader}}\right)$$

$$+ p_d \cdot \left(P_{C_i^*=C_i \& \text{ not det. by Reader}} + P_{C_i^* \neq C_i \& \text{ not det. by Reader}}\right)$$

$$= p_r\left(\frac{1}{2} \cdot 1 + \frac{1}{2} \cdot \frac{1}{2}\right) + p_d\left(\frac{1}{2} \cdot 1 + \frac{1}{2} \cdot \frac{1}{2}\right)$$

$$= \frac{3}{4}p_r + \frac{3}{4}p_d$$

$$= \frac{3}{4}. \tag{2}$$

If the challenge is random and $C_i^* = C_i$, then the adversary can correctly answer the response. If the challenge is random and $C_i^* \neq C_i$, then the adversary have a chance of $\frac{1}{2}$ of giving a correct response. If the challenge is predefined and $C_i^* = C_i$, he can correctly answer the response. If the challenge is predefined and $C_i^* \neq C_i$, he has a chance of $\frac{1}{2}$ of giving a correct response to the reader although he is always detected by the tag.

The probability of not being detected by the reader in the ith round provided that it is detected by the tag in the $(i-1)^{th}$ round, $P(\overline{A}_i|B_{i-1})$, is

$$P(\overline{A}_i|B_{i-1}) = P_{\text{random ch. \& not detected}} + P_{\text{predefined ch. \& not detected}}$$

$$= p_r \cdot \left(P_{C_i^*=C_i \& \text{ not det. by Reader}} + P_{C_i^* \neq C_i \& \text{ not det. by Reader}}\right)$$

$$+ p_d \cdot \left(P_{C_i^*=C_i \& \text{ not det. by Reader}} + P_{C_i^* \neq C_i \& \text{ not det. by Reader}}\right)$$

$$= p_r\left(\frac{1}{2} \cdot \frac{1}{2} + \frac{1}{2} \cdot \frac{1}{2}\right) + p_d\left(\frac{1}{2} \cdot \frac{1}{2} + \frac{1}{2} \cdot \frac{1}{2}\right)$$

$$= \frac{1}{2}p_r + \frac{1}{2}p_d$$

$$= \frac{1}{2}. \tag{3}$$

If the challenge is random and $C_i^* = C_i$, then the adversary sends the same response from the tag (not knowing that he was detected in the previous round). This response from the tag is a random value as the tag detected an error in the previous round. So he has a chance of $\frac{1}{2}$ of not detected by the reader. If the challenge is random and $C_i^* \neq C_i$, then the adversary have a chance of $\frac{1}{2}$ of giving correct response. Because he chooses a random response by himself. If the challenge is predefined and $C_i^* = C_i$, he again sends the same response from the tag. Therefore he has a chance of $\frac{1}{2}$ of not being detected by the reader. If the challenge is predefined and $C_i^* \neq C_i$, he has a chance of $\frac{1}{2}$ of giving a correct response to the reader as he chooses a random response by himself.

The probability of not being detected by the reader in the ith round is computed by

$$P(\overline{A}_i) = P(\overline{A}_i|\overline{B}_{i-1})P(\overline{B}_{i-1}) + P(\overline{A}_i|B_{i-1})P(B_{i-1})$$

$$= \frac{3}{4}P(\overline{B}_{i-1}) + \frac{1}{2}P(B_{i-1}), \tag{4}$$

where, $P(\overline{A}_1) = \frac{3}{4}$ and $i = 2, 3, \dots$.

The probability of being detected by the tag in the ith round depends on the probability of being detected by the tag in the $(i-1)^{th}$ round. Therefore we have

$$P(B_i) = P(B_i|B_{i-1})P(B_{i-1}) + P(B_i|\overline{B}_{i-1})P(\overline{B}_{i-1}). \tag{5}$$

The $P(B_i|B_{i-1})$ is always 1 since the tag already detected an error in the $(i-1)^{th}$ round. The $P(B_i|\overline{B}_{i-1})$ is $\frac{1}{2}p_d$ since the tag can detect an error when the challenge is a predefined one and $C_i^* \neq C_i$. Therefore we can rewrite Eq 5 as follows:

$$P(B_i) = P(B_{i-1}) + \frac{1}{2}p_d P(\overline{B}_{i-1}) \tag{6}$$

$$= P(B_{i-1}) + \frac{1}{2}(1 - p_r)P(\overline{B}_{i-1}), \tag{7}$$

where, $P(B_0) = 0$. From Eq. 4 and Eq. 7, we compute:

$$P(\overline{A}_i) = \frac{1}{2} + \frac{1}{4}\left(\frac{1}{2} + \frac{1}{2}p_r\right)^{i-1}.$$

When $p_r = \frac{1}{2}$, we obtain:

$$P(\overline{A}_i) = \frac{1}{2} + \frac{1}{4}\left(\frac{3}{4}\right)^{i-1}.$$

For example, when $p_r = \frac{1}{2}$, $P(\overline{A}_1) = \frac{3}{4}$ $P(\overline{A}_2) = \frac{11}{16}$, $P(\overline{A}_3) = \frac{41}{64}$, $P(\overline{A}_4) = \frac{155}{256}$, etc.

We depict the probabilities of not being detected by the reader by varying p_r and n in Fig. 4. For the comparison we also show the success probabilities of the adversary of HKP and MP. The adversary's probability of success of our protocol is smaller than those of HKP and MP. And the adversary's probability of success with asking, P_{ask}, is higher than that with no asking, $P_{no-ask} = (\frac{1}{2})^n$. So it is better for the adversary to choose the strategy of asking advance. However as the number of iterations increases, the success probability of the adversary approaches $(\frac{1}{2})^n$.

Distance fraud attack. Until now, we suppose that the tag is honest and the adversary tries to perform a mafia fraud attack. In this section, we consider the case of a dishonest tag. That is, we analyze the distance fraud attack on our protocol.

The tag knows the predefined challenges before the start of the fast bit exchanges as he knows T. Therefore he may try to deceive the reader with distance fraud attack. The probability of the success of the distance fraud attack by the dishonest tag for a round is

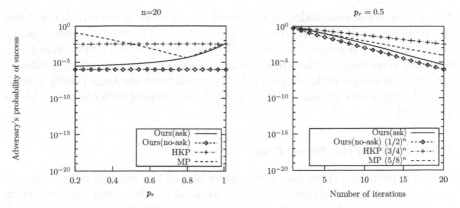

Fig. 4. Probability of not being detected by the reader

$$P_{\text{distance fraud attack}} = P_{\text{random challenge and deceive}} + P_{\text{predefined challenge and deceive}}$$
$$= p_r \cdot (P_{v_i^0 = v_i^1 \text{ and deceive}} + P_{v_i^0 \neq v_i^1 \text{ and deceive}}) + p_d$$
$$= p_r \cdot (\frac{1}{2} \cdot 1 + \frac{1}{2} \cdot \frac{1}{2}) + p_d$$
$$= \frac{3}{4} p_r + p_d$$
$$= 1 - \frac{1}{4} p_r.$$

If the challenge is a random challenge ($T_i = 1$) and i^{th} bits of v^0 and v^1 are equal, then the tag can send its response early. If i^{th} bits of v^0 and v^1 are not equal when $T_i = 1$, then the tag chooses the response randomly. Finally if the challenge is a predefined challenge ($T_i = 0$), then he can send its response early.

We depict the success probabilities according to the variation of p_r and n in Fig. 5. If $p_r = \frac{1}{2}$, which is the average case when we use a pseudorandom function to generate a bit string, then we have $(\frac{7}{8})^n$.

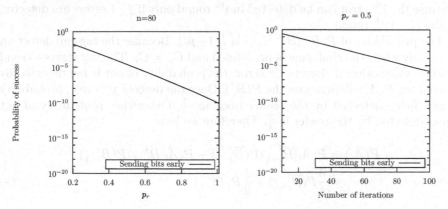

Fig. 5. Probability of distance fraud attack

We note that the probability of success of the Distance fraud attack decreases as p_r is closer to 1. However that of the Mafia fraud attack increases as p_r becomes higher. Therefore The trade-off between these two attacks should be considered according to the applications. In an environment where Distance fraud attack does not make sense (that is, you can trust the tag), taking a small p_r, possibly 0, clearly increases the security of the protocol with respect to Mafia fraud attack.

3.2   Analysis in the Noisy Case

In a real application, there may exist errors due to the channel noise although the adversary does not attack. Therefore, we need to allow some channel errors for practical reasons. However the adversary may get benefit from this allowance of channels errors. We analyze the success probability of the adversary in the noisy case.

We assume that the maximum $(j-1)$ errors are allowed in the tag. That is, the tag starts to send random values after j^{th} error is detected. We define some events as follows:

- A_i: the event that the error is detected by the reader in the ith round,
- $\overline{A}_i$: the event that the error is *not* detected by the reader in the ith round,
- B_i^j: the event that total j errors are detected by the tag until the ith round,
- $\overline{B}_i^j$: the event that less than j errors are detected by the tag until the ith round.

Then we have

$$P(B^j{}_i) = P(B^j{}_i|B^j{}_{i-1})P(B^j{}_{i-1}) + P(B^j{}_i|\overline{B}^j{}_{i-1})P(\overline{B}^j{}_{i-1})$$
$$= P(B^j{}_i|B^j{}_{i-1})P(B^j{}_{i-1}) + P(B^j{}_i|B^{j-1}{}_{i-1})P(B^{j-1}{}_{i-1})P(\overline{B}^j{}_{i-1}).$$

The probability of $P(B^j{}_i|B^j{}_{i-1})$ is 1 as j errors are already detected in $(i-1)^{th}$ round. The probability of $P(B^j{}_i|\overline{B}^j{}_{i-1})$ is equal to $P(B^j{}_i|B^{j-1}{}_{i-1})$ $P(B^{j-1}{}_{i-1})$. Because the jth error can be detected in i^{th} round only if $j-1$ errors are detected in the $(i-1)^{th}$ round.

The probability of $P(B^j{}_i|B^{j-1}{}_{i-1})$ is $\frac{1}{2}(1-p_r)$. Because the tag can detect an error only when the challenge is predefined and $C_i^* \neq C_i$. The tag starts to send random values after it detects j^{th} error, the probabilty of not being detected by the reader, $P(\overline{A}_i)$, depends on the $P(B_i^j)$. Once tag detects j^{th} error, probability of not being detected by the reader becomes $\frac{1}{2}$. Otherwise, probability of not being detected by the reader is $\frac{3}{4}$. Therefore we have

$$P(\overline{A}_i) = P(\overline{A}_i|\overline{B}^j_{i-1})P(\overline{B}^j_{i-1}) + P(\overline{A}_i|B^j_{i-1})P(B^j_{i-1})$$
$$= \frac{3}{4}P(\overline{B}^j_{i-1}) + \frac{1}{2}P(B^j_{i-1}), \tag{8}$$

where, $P(\overline{A}_1) = \frac{3}{4}$ and $i = 2, 3, ...$

For example, suppose that $j = 2$ and $p_r = 0.5$. It means that only one error is allowed in the tag and the tag sends random values after it detects two errors. Then, we have

$$P(B^2{}_i) = P(B^2{}_i|B^2{}_{i-1})P(B^2{}_{i-1}) + P(B^2{}_i|B^1{}_{i-1})P(B^1{}_{i-1})P(\overline{B}^2{}_{i-1}),$$

$$= P(B^2{}_{i-1}) + \frac{1}{4}P(B^1{}_{i-1})P(\overline{B}^2{}_{i-1}). \tag{9}$$

To compute $P(B^2{}_i)$, we need to compute $P(B^1{}_i)$ that is the same with $P(B_i)$ in the previous section. Therefore, we have $P(B_1^1) = \frac{1}{4}$, $P(B_2^1) = \frac{7}{16}$, $P(B_3^1) = \frac{37}{64}$, etc. The $P(B_1^2) = 0$ as two errors can not be detected in the first round. From Eq. 9, we have $P(B_2^2) = \frac{1}{16}$, $P(B_3^2) = \frac{169}{1024}$, etc. Finally we have $P(\overline{A}_1) = \frac{3}{4}$ $P(\overline{A}_2) = \frac{3}{4}$, $P(\overline{A}_3) = \frac{46}{64}$, $P(\overline{A}_4) = \frac{2734}{4096}$, etc. from Eq. 8. The probability of not being detected by reader until i^{th} round, $P_{ask}[i] = \prod_{k=1}^{i} P(\overline{A}_i)$. We depict it in Fig. 6.

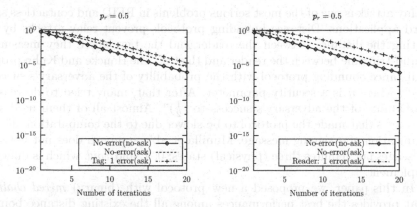

Fig. 6. Probability of not being detected by the reader in noisy channel: left) when tag allows errors, right) when reader allows errors

Now we assume that the maximum $(j - 1)$ errors are allowed in the reader. That is, the reader decides that the attack occurs after j^{th} error is detected. Contrary to the error detection by the tag, the detection of the error by the reader does not change $P(A_i)$ nor $P(B_i)$. Therefore the probability of not being detected by reader until i^{th} round, $P_{ask}[i]$ is

$$P_{ask}[i] = \begin{cases} \sum_{\mathbb{C}} \prod_{k=1}^{i}\{a_k P(A_k) + \overline{a}_k P(\overline{A}_k)\}, & i \geq j, \\ 1, & i < j, \end{cases}$$

Where, $a_k \in \{0, 1\}$ and $\mathbb{C} = \{(a_1, ..., a_i) | \sum a_k < j\}$. That is, $P_{ask}[i]$ is the probability that less than j errors are detected by the reader until i^{th} round. For example, if $j = 2$, then we have

$$P_{ask}[1] = 1,$$
$$P_{ask}[2] = P(\overline{A}_1)P(\overline{A}_2) + P(A_1)P(\overline{A}_2) + P(\overline{A}_1)P(A_2),$$
$$P_{ask}[3] = P(\overline{A}_1)P(\overline{A}_2)P(\overline{A}_3) + P(A_1)P(\overline{A}_2)P(\overline{A}_3) + P(\overline{A}_1)P(A_2)P(\overline{A}_3) + P(\overline{A}_1)P(\overline{A}_2)P(A_3),$$
$$P_{ask}[4] = P(\overline{A}_1)P(\overline{A}_2)P(\overline{A}_3)P(\overline{A}_4) + P(A_1)P(\overline{A}_2)P(\overline{A}_3)P(\overline{A}_4) + P(\overline{A}_1)P(A_2)P(\overline{A}_3)P(\overline{A}_4) + P(\overline{A}_1)P(\overline{A}_2)P(A_3)P(\overline{A}_4) + P(\overline{A}_1)P(\overline{A}_2)P(\overline{A}_3)P(A_4),$$
$$P_{ask}[5] = \dots$$

If $p_r = 0.5$, we have $P_{ask}[i] = \{1, 0.922, 0.776, 0.600, 0.432, \dots\}$. We depict the result in Fig. 6.

4   Conclusion

Relay attack is one of the most serious problems in RFID and contactless smart-card applications. Distance bounding protocols prevent relay attacks by computing the distance between the reader and the tag, where they measure the round-trip time between the reader and the tag. The Hancke and Kuhn proposed a distance bounding protocol with the probability of the adversary's success of $(\frac{3}{4})^n$, where n is a security parameter. After that, many tried to decrease the probability of the adversary's success to $(\frac{1}{2})^n$. Almost all of them used signed messages that made the protocol to be slower due to the computation and communication of a signing message. Munilla et al.'s approach does not use signed messages but requires three (physical) states: 0, 1, and *void*, which is difficult to implement.

In this paper, we proposed a new protocol with (binary) *mixed challenges* that provides the best performances among all the existing distance bounding protocols with binary challenges that do no use final signature. Indeed, while all these protocols provide a probability of the adversary success equal to $(3/4)^n$, the probability quickly converges toward $(1/2)^n$ in our case.

References

1. Brands, S., Chaum, D.: Distance-Bounding Protocols. In: Helleseth, T. (ed.) EUROCRYPT 1993. LNCS, vol. 765, pp. 344–359. Springer, Heidelberg (1994)
2. Desmedt, Y.: Major security problems with the Unforgeable Feige)-Fiat-Shamir proofs of identiy and how to overcome them. In: SecuriCom 1988, pp. 15–17 (1988)
3. Hancke, G., Kuhn, M.: An RFID distance bounding protocol. In: The 1st International Conference on Security and Privacy for Emergin Areas in Communications Networks (SECURECOMM 2005), pp. 67–73. IEEE Computer Society, Los Alamitos (2005)
4. Munilla, J., Ortiz, A., Peinado, A.: Distance bounding protocols with void-challenges for RFID. In: Workshop on RFID Security - RFIDSec 2006 (2006)

5. Munilla, J., Peinado, A.: Distance bounding protocols for RFID enhanced by using void-challenges and analysis in noisy channels. Wireless communications and mobile computing. Published online, an earlier version appears in [4], January 17 (2008)
6. Singelée, D., Preneel, B.: Distance bounding in noisy environments. In: Stajano, F., Meadows, C., Capkun, S., Moore, T. (eds.) ESAS 2007. LNCS, vol. 4572, pp. 101–115. Springer, Heidelberg (2007)
7. Tu, Y.-J., Piramuthu, S.: RFID distance bounding protocols. In: The 1st International EURASIP Workshop in RFID Technology, Vienna, Austria (2007)

Anonymizer-Enabled Security and Privacy
for RFID

Ahmad-Reza Sadeghi[1], Ivan Visconti[2], and Christian Wachsmann[1]

[1] Horst Görtz Institute for IT-Security, Ruhr-University Bochum, Germany
{ahmad.sadeghi,christian.wachsmann}@trust.rub.de
[2] Dipartimento di Informatica ed Applicazioni, University of Salerno, Italy
visconti@dia.unisa.it

Abstract. RFID-based systems are becoming a widely deployed pervasive technology that is more and more used in applications where privacy-
sensitive information is entrusted to RFID tags. Thus, a careful analysis
in appropriate security and privacy models is needed before deployment
to practice.

Recently, Vaudenay presented a comprehensive security and privacy
model for RFID that captures most previously proposed privacy models.
The strongest achievable notion of privacy in this model (*narrow-strong
privacy*) requires public-key cryptography, which in general exceeds the
computational capabilities of current cost-efficient RFIDs. Other privacy
notions achievable without public-key cryptography heavily restrict the
power of the adversary and thus are not suitable to realistically model
the real world.

In this paper, we extend and improve the current state-of-the art for
privacy-protecting RFID by introducing a security and privacy model for
anonymizer-enabled RFID systems. Our model builds on top of
Vaudenay's model and supports anonymizers, which are separate devices specifically designated to ensure the privacy of tags. We present
a privacy-preserving RFID protocol that uses anonymizers and achieves
narrow-strong privacy without requiring tags to perform expensive
public-key operations (i.e., modular exponentiations), thus providing a
satisfying notion of privacy for cost-efficient tags.

1   Introduction

Radio frequency identification (RFID) is a technology that enables RFID *readers* to perform fully automatic wireless identification of objects that are labeled
with RFID *tags*. Initially, this technology was mainly used for electronic labeling
of pallets, cartons and products to enable seamless supervision of supply chains.
Today, RFID technology is widely deployed and studied for its applications, including animal identification [1], library management [2], access control [1,3,4,5],
electronic tickets [4,3,5,6] and passports [7] and even human implantation [8].

As pointed out in previous publications (see, e.g., [9,8]), this prevalence of
RFID technology introduces various risks, in particular concerning the privacy
of its users and holders. The most deterrent privacy risk concerns the tracking of

J.A. Garay, A. Miyaji, and A. Otsuka (Eds.): CANS 2009, LNCS 5888, pp. 134–153, 2009.
© Springer-Verlag Berlin Heidelberg 2009

users. Thus, an RFID system should provide *anonymity* (i.e., confidentiality of the tag identity) as well as *untraceability* (i.e., unlinkability of the communication of a tag) even in case the state of (i.e., the information stored on) the tag has been disclosed.[1] RFID applications in practice must also achieve various security and functional goals. The security goals include *authentication*, which prevents an adversary from impersonating and forging tags and *availability*, which means the resilience to remote tampering that allows denial-of-service attacks. The functional goals include *efficiency* (e.g., fast verification of cost-efficient tags) and *scalability* (i.e., support of a huge number of tags).

Most currently used RFID systems do not offer privacy at all (e.g., [11,12,4,5]). This is mainly because current cost-efficient tags do not provide the necessary computational resources to run privacy-preserving protocols [1,4], which heavily rely on public-key cryptography. Moreover, as pointed out in Section 2, privacy-preserving solutions without public-key cryptography do not fulfill important security or functional requirements and thus, are inapplicable to real-world applications.

As elaborated in related work (see Section 2), a promising approach towards solving these problems and our focus in this paper, are *anonymizers*. These are special devices that take off the computational workload (i.e., the public-key operations) from the tags and enable privacy-preserving protocols with cost-efficient tags. Note that an anonymizer-based RFID system is not equivalent to a straight forward extension of a resource constrained RFID system to one with higher capabilities (such as public-key cryptography). The anonymizer-enabled approach in general requires an additional protocol between tags and anonymizers that opens attack surfaces and thus, must be carefully considered. Indeed, an anonymizer shall not be able to impersonate or to copy the tags it anonymizes since this would violate authentication. Moreover, to ensure availability, the protocol between a tag and the anonymizer must be secure against attacks where an adversary aims to manipulate the tag.

Anonymizers can be incorporated into the standard RFID system model in different ways. One approach is to provide public anonymizers that can be controlled by the operator of the RFID system or by one of several independent anonymizing service providers the user may choose from. Alternatively, each user may have his/her own personal anonymizer that could be implemented as a software running on the user's mobile phone or PDA[2], allowing a very cost-efficient implementation of anonymizers. The main advantage of anonymizer-enabled protocols is that they allow operators of RFID systems to enable privacy for the concerned users (who may buy his/her own personal anonymizer) with no or only minor extra costs.

---

[1] To distinguish tracing in past or future protocol-runs, the notions of *forward untraceability* (i.e., unlinkability of the communication of the tag that has been recorded *before* disclosure) and *backward untraceability* (i.e., unlinkability of the communication of the tag that takes place *after* disclosure) are defined in [10].

[2] An increasing number of mobile phones and PDAs support the Near Field Communication (NFC) standard, which enables them to communicate to RFID devices.

However, as pointed out in Section 2, current anonymizer-enabled solutions are vulnerable to impersonation attacks. Hence, the design of a secure and privacy-preserving RFID system requires an appropriate security and privacy model to enable a careful analysis of the underlying schemes. On the other hand, existing security and privacy models for RFID (e.g., [13,14,15,16]) suffer from various shortcomings. As discussed in Section 2, these models do not consider important aspects like adversaries with access to *auxiliary information* (i.e., whether the identification of a tag was successful or not) or the privacy of *corrupted* tags (i.e., whose state has been disclosed). Both are essential to ensure anonymity and untraceability in practice. Another drawback is that most of these models are incomparable, which leads to the problem that a protocol can be proven secure in one privacy model while being insecure in another model.[3] Therefore, it is crucial to develop a widely accepted security and privacy model for RFID.

Recently, a comprehensive security and privacy model that generalizes and improves many previous works has been proposed in [18] and refined in [19,20]. The strongest *achievable* privacy notion in this model (*narrow-strong privacy*) allows the adversary to arbitrarily corrupt tags but does not capture the availability of auxiliary information. If auxiliary information is of concern, the weaker notions of *destructive* and *forward privacy* must be considered while *weak privacy* does not adequately model the capabilities of real-world adversaries since weak privacy does not allow tag corruption. However, narrow-strong privacy requires the use of public-key cryptography [18], which in general clearly exceeds the capabilities of current cost-efficient RFIDs [1,4]. Moreover, it has been shown that forward privacy can be achieved but at the cost of using public-key cryptography [18] (which in general is too expensive).

We observe that the model of [18] does not include anonymizers, which play a critical role for going beyond the barrier of simultaneously achieving a strong privacy notion with protocols that are suitable for cost-efficient tags. Therefore, we investigate the use of anonymizers in the model of [18] and show an anonymizer-enabled scheme that provides important security and privacy properties while fulfilling the functional requirements of real-world applications.

Contribution. We introduce a formal framework for privacy-preserving RFID systems, which extends the security and privacy model of [18] to support anonymizers and at the same time is backwards-compatible to it. Given the granularity of the different security and privacy notions of [18], our anonymizer-based model is the first universal security and privacy model for anonymizer-enabled RFID systems. Moreover, we propose a privacy-preserving RFID protocol that can be proven secure and private in the anonymizer-enabled model (with random oracles). The protocol that we propose enjoys several appealing features that were not simultaneously achieved by any previous proposal. Indeed, our protocol is very efficient for all involved entities, in particular for tags that only have

---

[3] For instance, the OSK protocol [17] can be proven secure in the model of [13] although a tracing attack can be shown in the model of [14].

to perform minimal computations. Further, the protocol enjoys the strongest achievable[4] privacy notion defined in [18], which is narrow-strong privacy. Our protocol also provides forward privacy, which restricts the adversary's capability to corrupt tags but instead allows him to access auxiliary information. We finally stress that our protocol is provably secure against impersonation attacks and forgeries even if the adversary can corrupt the anonymizers. Therefore, we require the existence of (honest) anonymizers in the system only to guarantee privacy (anonymity and untraceability) of the tags. This assumption gracefully matches the realistic scenario where many anonymizers are spread in the system and an adversary can be successful in corrupting many of them with the purpose of violating the security of the system. At the same time, privacy is guaranteed as long as tags are frequently anonymized by an uncorrupted anonymizer.

2   Related Work

Privacy-Preserving RFID Protocols. A general problem with privacy-preserving authentication of low-cost tags that are incapable of public-key operations is how to inform the reader which key should be used for the authentication.[5] Essentially there are two approaches that address this problem. The first approach is that the reader performs an exhaustive search for the secret key that is used by the authenticating tag [9]. Solutions to optimize this approach (see, e.g., [2,30]) suffer from inefficiency since tag verification depends on the total number of tags in the system. Clearly, this violates the efficiency and scalability requirements of most practical RFID systems. In the second approach, a tag updates its identity after each interaction such that the new identity is unlinkable and only known to the tag and the authorized readers, which allows readers to identify tags in constant time (see, e.g., [31,32,33,10,34]). This approach requires each tag to be always synchronized with all readers in the system. However, in general, it is easy to mount denial-of-service attacks that desynchronize the tag and the readers (see e.g., [31,33]). For a broad overview about privacy issues in RFID systems, see also [35].

Anonymizer-Enabled RFID Protocols. A promising approach to enhance privacy of RFID without lifting the computational requirements on tags are anonymizer-enabled protocols, where external devices (*anonymizers*) are in charge of providing anonymity of tags. Anonymizer-enabled RFID protocols are very

---

[4] Note that the impossibility of achieving strong privacy [18] trivially holds in our anonymizer-enabled model since any protocol in the anonymizer-enabled model also works in the model of [18] by simply requiring that the anonymization protocol (i.e., the protocol run between tags and anonymizers) is played locally inside tags.

[5] A prominent family of lightweight authentication protocols proposed in the context of RFID are the HB protocols (see e.g., [21,22,23,24]). However, these protocols are subject to man-in-the-middle attacks [25,26,27,28], require the reader to perform an exhaustive search for the (shared) authentication secret of the authenticating tag and have a high communication complexity (many rounds of interaction) [29]. Moreover, tag corruption is usually not considered in the security evaluation of the HB protocols.

suitable for many practical scenarios with privacy needs that use cost-efficient tags. The main concept of existing anonymizer-enabled protocols [36,37,38,39] is that each tag stores a ciphertext that encrypts the information carried by the tag (e.g., the tag identifier) under the public key of the reader. This ciphertext is sent to the reader each time the tag authenticates. Since this ciphertext is static data and can be used to track and to identify the tag, it must be frequently changed to provide anonymity and unlinkability. However, current RFIDs [1,4] are not capable of updating their ciphertext on their own and thus, privacy in these protocols relies on anonymizers that frequently refresh the ciphertexts stored on the tags. The first proposal to use anonymizers [36] considers a plan by the European Central Bank to embed RFID tags into Euro banknotes to aggravate forgeries [40]. It proposes to store a ciphertext of the serial number of a banknote on the RFID tag that is attached to the banknote. Each time the banknote is spent, anonymizers in shops or banks re-encrypt the ciphertext stored on the tag. The drawback of this scheme is that the serial number of a banknote must be optically scanned before its ciphertext can be re-encrypted. In [37], the authors introduce a primitive called *universal re-encryption*, which is an extension of the El Gamal encryption scheme where re-encryption is possible without knowledge of the corresponding (private and public) keys. In this approach, an adversary can "mark" tags such that he can recognize them even after they have been re-encrypted. This issue has been addressed in [38] that shows tracing attacks and proposes solutions. In [39], the authors improve the ideas of [37] and [38] by introducing the notion of *insubvertible encryption*, which adds a signature on the blinded public key of the reader that is linked to the ciphertexts stored on the tags. Re-randomization involves this signature in a way that prevents the adversary from marking tags.

All known anonymizer-enabled schemes are subject to impersonation attacks since authentication is only based on the ciphertext that the tag sends to the reader. Moreover, existing security models do not capture RFID systems that use anonymizers.

Privacy Models for RFID. One of the first privacy models for RFID [17] defines anonymity and backward untraceability based on a security game where an adversary must distinguish a random value from the output of a tag. It does not consider forward untraceability. A privacy model specific for RFIDs that cannot perform any cryptographic operations [41] is based on assumptions on the number of queries an adversary can make to a tag but does not capture adversaries who can corrupt tags. Thus, it does not cover backward and forward untraceability, which is required to realistically model adversaries against cost-efficient tags in practice. Another privacy model [13,42] provides various flexible definitions for different levels of privacy based on a security experiment where an adversary must distinguish two known tags. This model is extended in [14] by the notion of auxiliary information. In [15], a *completeness* and *soundness* requirement is added to the definition of [14], which means that a reader must accept *all* but *only* valid tags. The definition of [14] has been further improved in [43] to cover backwards untraceability. Another privacy model [16] is based on the universal composability (UC) framework and claims to be the first model that considers availability. However, it does not allow the adversary to corrupt tags and thus does not capture backwards untraceability.

Recently, [18] presented a privacy model that generalizes and classifies previous RFID privacy models by defining eight levels of privacy that correspond to real-world adversaries of different strength. The strongest privacy notion of [18] captures anonymity, backward and forward untraceability and adversaries with access to auxiliary information. Moreover, it provides a security definition equivalent to [15] that covers authentication. The model of [18] has been extended in [19] to consider reader authentication whereas [20] aims at reducing the mentioned eight privacy classes to three privacy classes. Recently in [44,45] other privacy notions have been considered along with denial of service attacks. Since [18] classifies the most significant RFID privacy notions, we focus on this security model and extend it to support anonymizers.

3 Our Anonymizer-Enabled RFID System

3.1 Trust Relations and Assumptions

Before presenting our anonymizer-enabled RFID system, we first give an informal description of the underlying trust relations that are formalized in Section 4.1.

Roles and Trust Relations. An anonymizer-enabled RFID system consists of readers R, anonymizers A and tags T. The readers R set up tags that can later be identified by all the readers R in he system. A tag T that has been set up by an honest R is called *legitimate*. The task of the anonymizers A is to enforce the privacy goals of legitimate tags.

As most RFID privacy models, we assume the readers R to be trusted. This means that the readers R will behave as intended, which means that they do nothing that violates the security and privacy goals of legitimate tags. Tags are considered to be untrusted since an adversary can obtain full control of the tags and the data stored on them. Similar to tags, we consider anonymizers to be untrusted and an adversary can get full control over many anonymizers and their secrets.

Assumptions. Following the majority of existing RFID models, we make the following assumptions.

Reader. We assume that all readers R are connected to the same backend system (e.g., a database d). Thus, all honest readers R have access to the same information and thus can be subsumed as *one single* reader entity R. Moreover, the reader R can perform public-key cryptography and can handle multiple instances of the identification protocol with different tags in parallel.

Tags. The tags considered in this paper are passive devices, which means that they do not have own power supply but are powered by the electromagnetic field of the reader R. Thus, tags cannot initiate communication, have a narrow communication range (e.g., a few centimeters to meters) and are constrained in

their computational and storage capabilities, which limits them to basic cryptographic functions like hashing, random number generation and symmetric-key
encryption [1,4].

Anonymizer. Anonymizers can perform public-key cryptography and can handle
multiple parallel instances of the anonymization protocol with different tags.
Since a tag T does not posses the required computational resources to update
its state, it can always be tracked between two anonymizations. Therefore, to
provide anonymity and unlinkability, it must be guaranteed that each tag T
is frequently anonymized by an honest anonymizer (e.g., every few minutes).
In practice, this is achieved by a dense network of public anonymizers or a
personal anonymizer. At this point we stress that in order to eavesdrop on every
interaction of a tag with a reader or an anonymizer, an adversary must always
be within reading range of the tag. Due to the limited communication range
of RFID this implies that the adversary is following the user of the tag, which
obviously violates the tag user's privacy even if he would not carry an RFID tag.
Thus, a privacy-preserving RFID system can at most offer privacy guarantees
against adversaries that do not have permanent access to the tags. Moreover,
an adversary in practice can at most corrupt a limited number of anonymizers,
which ensures that there is at least one honest anonymizer in the system.

3.2 Notation and Preliminaries

General Notation. For a finite set S, $|S|$ denotes the size of set S whereas for
an integer n the term $|n|$ means the bit-length of n. The term $s \in_R S$ means
the assignment of a uniformly chosen element from S to variable s. Let A be
a probabilistic algorithm. Then $y \leftarrow A(x)$ means that on input x, algorithm A
assigns its output to variable y. $A_K(x)$ means that the output of A depends on x
and some additional parameter K (e.g., a secret key). Probability $\epsilon(l)$ is called
negligible if for all polynomials $f(\cdot)$ it holds that $\epsilon(l) \leq 1/f(l)$ for all sufficiently
large l. Moreover, probability $1 - \epsilon(l)$ is called *overwhelming* if $\epsilon(l)$ is negligible.

Encryption Schemes. An encryption scheme ES is a tuple of algorithms
(Genkey, Enc, Dec) where Genkey is the key generation, Enc is the encryption
and Dec is the decryption algorithm. ES is called *homomorphic* if there are
two operations $(\circ, \bullet)$ such that for every pair of ciphertexts $c_1 = \text{Enc}(m_1)$ and
$c_2 = \text{Enc}(m_2)$ it holds that $c_1 \bullet c_2 = \text{Enc}(m_1 \circ m_2)$ (see, e.g., [46,37,47]). We
indicate homomorphic encryption schemes by $\text{ES}^\flat = (\text{Genkey}^\flat, \text{Enc}^\flat, \text{Dec}^\flat)$. A
public-key encryption scheme is said to be *CPA-secure* [48,49] if every proba-
bilistic polynomial time (p.p.t.) adversary $\mathcal{A}$ has at most negligible advantage
of winning the following security experiment. An algorithm $\mathcal{S}^{\text{cpa}}$ (called *CPA-
challenger*), generates an encryption key pair $(sk, pk) \leftarrow \text{Genkey}(1^l)$ and gives
the public encryption key pk to $\mathcal{A}$. Now, $\mathcal{A}$ must respond with two messages m_0
and m_1. $\mathcal{S}^{\text{cpa}}$ then randomly chooses a bit $b \in_R \{0,1\}$, encrypts $c_b \leftarrow \text{Enc}_{pk}(m_b)$
and returns the resulting ciphertext c_b to $\mathcal{A}$, who now must return a bit b' that
indicates whether c_b encrypts m_0 or m_1. $\mathcal{A}$ wins if $b' = b$.

Random Oracles. A random oracle RO [50] is an oracle that responds with a random output for each given input. More precisely, RO starts with an empty look-up table τ. When queried with an input m, RO first checks if it already knows a value $\tau[m]$. If this is not the case, RO chooses a random value r and updates τ such that $\tau[m] = r$. Finally, RO returns $\tau[m]$.

3.3 Protocol Description and Specification

Our Goals. Our scheme combines and extends some of the schemes proposed in [18] and employs anonymizers, which brings several improvements that are important for practical applications. Our protocol achieves both narrow-strong and forward privacy, allows tags to be verified in constant time and provides basic protection against denial-of-service attacks. Therefore, our protocol achieves the most important security, privacy and functional requirements of practical RFID systems for both adversaries with and without access to auxiliary information.

We stress that our scheme only considers anonymity and untraceability of the communication between tags and the reader that takes place when a tag is used to access some service. Therefore, our protocol does not consider privacy of the communication between tags and anonymizers. Notice that all tags access anonymizers and thus from a rerandomization there is no special information given to the adversary about the use of a given tag obtaining access to a given service (i.e., when the tag communicates with a reader). Moreover, the use of services can be selective, since only some tags can have access to some services and thus privacy is critical in this phase. Finally note that the crucial issue is that an adversary must not be able to obtain any information about which tag accessed any service and about whether the same tag has obtained access to some services.

Our protocol provides basic availability, which means that an adversary cannot manipulate (i.e., invalidate) legitimate tags without physically attacking an anonymizer (and thus criminalizing himself). However, this is sufficient for most practical scenarios since a stolen or damaged public anonymizer can be detected and thus such attacks are unlikely to happen just to violate privacy. Further, public anonymizers can be physically secured (e.g., by a robust housing as it is used for surveillance cameras). Moreover, in the scenario of personal anonymizers, the damage that can be done by a corrupted anonymizer is limited only to the tags of one single user (since only the key of this single user's anonymizer is revealed). Obviously, a potential success in a security violation (i.e., in impersonating a legitimate tag) could motivate an adversary since he would obtain unauthorized access to services, which in turn means that he would get some economic advantages. However, our protocols turn out to be secure against impersonation attacks even against adversaries that corrupt anonymizers.

We do not consider unclonability of tags since this seems to be infeasible to achieve without hardware assumptions for the tags (which would significantly increase the costs of the tags). Further, we do not consider tracing or identification attacks based on the physical characteristics of tags, which in practice seems to be a problem that cannot be prevented by protocols on the logical layer [51].

One of the main features of our scheme is that we give a generic structure that allows one to instantiate our scheme using various cryptographic primitives (i.e., any CPA-secure homomorphic encryption scheme) based on different number-theoretic assumptions with different performance properties. Our protocol does not require tags to perform public-key cryptography (beyond the homomorphic operation that usually does not resort to a modular exponentiation) and thus, is not limited to the use of special lightweight public-key encryption schemes. This opens the possibility to employ optimized schemes, e.g., with short keys (in particular when using a prime as modulus) and ciphertexts to reduce the memory requirements to tags and to decrease the size of the protocol messages.

Protocol Overview. Our RFID scheme consists of two protocols: The tag identification and the tag anonymization protocol. The former protocol is executed by the reader R and a tag T and allows R to check if T is legitimate. The latter protocol ensures anonymity and untraceability of T in the identification protocol by updating the authentication secrets of T.

System Setup. The reader R and the anonymizers A are initialized as follows.

Reader Setup. Given a security parameter $l_R = (l_\mathfrak{h}, l_s)$, the reader R generates a key pair $(sk_R, pk_R) \leftarrow \mathsf{Genkey}^\mathfrak{h}(1^{l_\mathfrak{h}})$ for a CPA-secure homomorphic public-key encryption scheme. Moreover, R initializes a secret database $d \leftarrow \{\}$ that later stores the identities and authentication secrets of all legitimate tags. The secret key of R is sk_R whereas the corresponding public key is $(l_\mathfrak{h}, l_s, pk_R)$. For brevity, we write pk_R to mean the complete tuple.

Anonymizer Setup. Given a security parameter $l_A = (l_a, l_s)$, the anonymizer A generates a key pair $(sk_A, pk_A) \leftarrow \mathsf{Genkey}(1^{l_a})$ for the CPA-secure public-key encryption scheme. The secret key of A is sk_A whereas the corresponding public key is the tuple (l_a, l_s, pk_A). We write pk_A to mean the complete tuple.[6]

Tag Creation. A tag T with identifier ID is initialized by the reader R as follows: first, R generates a random long-term secret K and an ephemeral secret T, that are used later in the authentication protocol to authenticate T to R. Moreover, R generates a symmetric encryption key $A \leftarrow \mathsf{Genkey}(1^{l_s})$, which is used later by T to encrypt the communication of the anonymization protocol. Moreover, R computes three public-key encryptions $E \leftarrow \mathsf{Enc}_{pk_A}(A)$, $F \leftarrow \mathsf{Enc}^\mathfrak{h}_{pk_R}(T)$ and $G \leftarrow \mathsf{Enc}^\mathfrak{h}_{pk_R}(\mathrm{ID})$. The ciphertext E is used to transport the symmetric key A from T to A in the anonymization protocol whereas F and G are used to transport the ephemeral secret T and the identifier ID from T to R in the identification protocol. Finally, R updates its database $d \leftarrow d \cup \{(\mathrm{ID}, K)\}$ and initializes T with the state $S \leftarrow (A, T, E, F, G, \mathrm{ID}, K)$.

---

[6] Note that personal anonymizers (i.e., those running on the users' mobile phone or PDA) can have different user-specific keys. However, this requires the user of a personal anonymizer to indicate to the tag issuing entity which anonymizer shall be used later to anonymize the newly created tag.

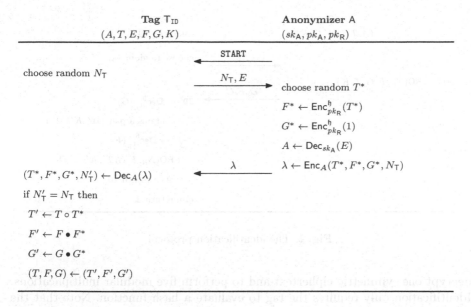

Fig. 1. The anonymization protocol

Anonymization Protocol. The anonymization protocol is illustrated in Figure 1. It is a protocol between a tag T with identifier ID and an anonymizer A and aims at updating the state S of T. First, T randomly chooses N_T and sends (N_T, E) to A. Then, A chooses a new ephemeral tag secret T^* and encrypts it to $F^* \leftarrow \mathsf{Enc}^{\mathfrak{h}}_{pk_R}(T^*)$. Moreover, A encrypts $G^* \leftarrow \mathsf{Enc}^{\mathfrak{h}}_{pk_R}(1)$ of the identity w.r.t. to the homomorphic operation $\circ$ of the public-key encryption scheme. Finally, A decrypts $A \leftarrow \mathsf{Dec}_{sk_A}(E)$, encrypts $\lambda \leftarrow \mathsf{Enc}_A(T^*, F^*, G^*, N_T)$ and sends λ to T. Then, T decrypts $(T^*, F^*, G^*, N'_T) \leftarrow \mathsf{Dec}_A(\lambda)$ and checks if $N'_T = N_T$. If this is the case, T computes a new ephemeral authentication secret $T' \leftarrow T \circ T^*$, the (homomorphic) public-key encryption $F' \leftarrow F \bullet F^*$ of the new ephemeral key T' and a new (re-randomized) encryption $G' \leftarrow G \bullet G^*$ of the tag identifier ID. Finally, T updates its state $(T, F, G) \leftarrow (T', F', G')$. If $N'_T \neq N_T$, T aborts the anonymization protocol without updating its state.

Identification Protocol. Figure 2 illustrates the identification protocol, which takes place between a tag T with identifier ID and the reader R with the goal to identify T on the reader side. R sends a random N_R to T, which then computes $D \leftarrow \mathsf{RO}(N_R, F, G, T, K)$ and responds with (D, F, G). Then, R decrypts ID' $\leftarrow \mathsf{Dec}^{\mathfrak{h}}_{sk_R}(G)$ and checks if its secret database d contains a tuple (ID', K'). If this is the case, R decrypts $T' \leftarrow \mathsf{Dec}^{\mathfrak{h}}_{sk_R}(F)$ and accepts T by returning ID' if $D = \mathsf{RO}(N_R, F, G, T', K')$. Otherwise, R rejects T and returns $\bot$.

Technical Feasibility. Using the (homomorphic) El Gamal public-key encryption scheme, our protocol requires tags to provide about 0.6 KBytes of non-volatile memory. Anonymization requires the tag to generate a random number,

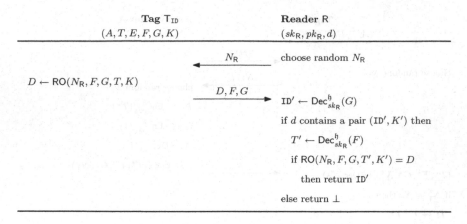

Fig. 2. The identification protocol

decrypt one symmetric ciphertext and to perform five modular multiplications. Identification only requires the tag to evaluate a hash function. Note that the anonymization protocol is completely transparent to the user whereas identification usually requires the user to wait (e.g., at a door) until the authentication protocol completes. Thus, in contrast to the anonymization protocol, most practical applications have strict time constraints on the identification protocol. Our scheme should be implementable with widely available RFID tags.

4   The Anonymizer-Enabled RFID Model

To prove the security and privacy properties claimed in Section 3.3, an appropriate security and privacy model is needed. Since existing RFID security and privacy models do not capture anonymizer-enabled protocols (see Section 2), we extend the model of [18] to the first universal security and privacy model for anonymizer-enabled RFID systems.

4.1   System Model

To form the anonymizer-enabled model, the original system model of [18] must additionally consider the anonymizers A and the corresponding protocols. This means that there must be a procedure to set up A and an interactive protocol where A updates the state of the tags. Following [18], we now define an anonymizer-enabled RFID system.

SetupReader(1^{l_R}) $\rightarrow$ (sk_R, pk_R, d) on input of a security parameter l_R, this function initializes the reader R by creating some public parameters pk_R that are known to all entities and some secret parameters sk_R that are only known to R. This function also creates a secret database d that can only be accessed by R and that stores the identities and authentication secrets of all legitimate tags.

SetupAnon($1^{l_A}, pk_R$) $\rightarrow$ (sk_A, pk_A) on input of a security parameter l_A and the public key pk_R of R, this function initializes the anonymizer A by creating some public parameters pk_A that are known to all entities and some secret parameters sk_A that are only known to A.

SetupTag$_{pk_R}$(ID, pk_A) $\rightarrow$ (K, S) generates a tag-specific secret K and uses the public key pk_R of R to create an initial state S for the tag T with identifier ID. T is initialized with S and (ID, K) is stored in the secret database d of R. Since T must identify the anonymizer A in the anonymization protocol, this procedure involves pk_A.

AnonTag[$T_{ID}(S) \leftrightarrow$ A(sk_A, pk_A, pk_R)] $\rightarrow S'$ is an interactive protocol that is (frequently) run between the tag T with identifier ID and the anonymizer A. The goal of this protocol is to update the state S of T to a new indistinguishable state S'.

IdentTag[$T_{ID}(S) \leftrightarrow$ R(sk_R, pk_R, d)] $\rightarrow out$ is an interactive protocol between the tag T with identifier ID and the reader R. The goal of this protocol is to identify T and to verify whether T is legitimate. With overwhelming probability, R returns $out = $ ID if T is legitimate and $out = \perp$ otherwise.[7]

4.2   Adversary Model

The adversary model of [18] defines the privacy and security objectives as a security experiment, where a polynomially bounded adversary can interact with a set of oracles that model the capabilities of the adversary.

In the anonymizer-enabled model, an adversary may obtain information from the anonymization protocol. This ability is modeled by allowing the adversary to launch new anonymization protocol sessions and to interact with the anonymizer. To consider the case where the adversary controls a set of anonymizers, we allow the adversary to obtain the secrets of the anonymizers by corrupting them. However, as discussed in Section 3.1 and stated in Assumption 1, we assume that there is at least one honest anonymizer in the system whose communication cannot be eavesdropped or manipulated by the adversary. In the anonymizer-enabled model, the adversary has access to the oracles described below.

CreateTag^b(ID, pk_A) This oracle allows the adversary to set up a tag with identifier ID. This oracle internally calls SetupTag$_{pk_R}$(ID, pk_A) to create (K, S) for tag ID. If input $b = 1$, the adversary chooses the tag to be legitimate, which means that (ID, K) is added to the secret database d of the reader R. For input $b = 0$, the adversary can create illegitimate tags where (ID, K) is not added to d. This models the fact that an adversary can obtain (e.g., buy) legitimate tags and create forgeries.

DrawTag(Δ) $\rightarrow$ ($vtag_1, b_1, \ldots, vtag_n, b_n$) Initially, the adversary cannot interact with any tag but must query the DrawTag oracle to get access to a set of tags that has been chosen according to a given tag distribution Δ. This

---

[7] A *false negative* occurs when T is legitimate but $out = \perp$, a *false positive* happens if T is not legitimate and $out \neq \perp$. An *incorrect identification* occurs if the tag T with identifier ID is legitimate but $out \notin \{$ID$, \perp\}$.

models the fact that the adversary can only interact with the tags within his reading range. The adversary usually only knows the tags he can interact with by some temporary identifiers $vtag_1, \ldots, vtag_n$ (e.g., in our protocol the tuple (F, G) can be seen as virtual identifier). The DrawTag oracle manages a secret look-up table $\mathcal{T}$ that keeps track of the real identifier ID_i that is associated with each temporary identifier $vtag_i$, i.e., $\mathcal{T}(vtag_i) = ID_i$. Moreover, the DrawTag oracle also provides the adversary with information on whether the corresponding tags are legitimate ($b_i = 1$) or not ($b_i = 0$). This models the availability of auxiliary information to the adversary.[8]

FreeTag(vtag) Contrary to the DrawTag oracle, the FreeTag oracle makes a tag vtag inaccessible to the adversary, which means that the adversary cannot interact with vtag any longer until it is made accessible again (under a new temporary identifier $vtag'$) by another DrawTag query. This models the fact that a tag can get out of the reading range of the adversary.

LaunchIdent( ) $\rightarrow \pi_R$ makes the reader to start a new instance π_R of the IdentTag protocol, which allows the adversary to start different parallel IdentTag protocol instances with the reader R.

LaunchAnon( ) $\rightarrow \pi_A$ makes the anonymizer to start a new instance π_A of the AnonTag protocol, which allows the adversary to start different parallel AnonTag protocol instances with an honest anonymizer.

SendTag(m, vtag) $\rightarrow m'$ sends a message m to the tag T that is known as vtag to the adversary. The tag T responds with message m'. This allows the adversary to perform active attacks against both the AnonTag and the IdentTag protocol.

SendReader(m, π_R) $\rightarrow m'$ sends a message m to the instance π_R of the IdentTag protocol that is executed by the reader R, which responds with message m'. This allows the adversary to perform active attacks against the IdentTag protocol.

SendAnon(m, π_A) $\rightarrow m'$ sends a message m to the instance π_A of the AnonTag protocol that is executed by an honest anonymizer A, which responds with message m'. This allows the adversary to perform active attacks against the AnonTag protocol.

Result(π_R) returns 1 if the instance π_R of the IdentTag protocol has been completed but the tag T that participates in the protocol has not been accepted by the reader R. In case R identified a legitimate tag, Result returns 0. This allows the adversary to obtain auxiliary information on whether the authentication of T was successful or not.

CorruptTag(vtag) $\rightarrow S$ returns the current state S of the tag T that is known as vtag to the adversary. This models (physical) attacks on tags that disclose the current tag state.

CorruptAnon(A) $\rightarrow (sk_A)$ returns the secret parameter sk_A of anonymizer A. This models (physical) attacks against honest anonymizers that disclose the secret sk_A of anonymizer A.

---

[8] For instance, in an access control scenario, the adversary may notice that a tag $vtag_i$ is legitimate by observing its communication with a reader at a locked door and then watching whether the door opens or not.

As discussed in Section 3.1, we make the following assumption:

Assumption 1. *A tag* T *with identifier* ID *always runs* AnonTag$[\mathsf{T}_{\mathrm{ID}} \leftrightarrow \mathsf{A}]$ *with an honest anonymizer* A *at least once before each execution of* IdentTag$[\mathsf{T}_{\mathrm{ID}} \leftrightarrow \mathsf{R}^*]$ *with a (potentially malicious) reader* R^* *and before each* CorruptTag(vtag) *query where* $\mathcal{T}(\text{vtag}) = \text{ID}$.

Adversary Classes. The original model of [18] distinguishes the following four major adversary classes that represent adversaries of different strength:

- *Weak adversaries* cannot corrupt tags and are limited to active attacks on the protocols. This assumes that corruption of tags is infeasible (e.g., due to tamper-resistant hardware), which is clearly not the case for low-cost RFIDs.
- *Forward adversaries* cannot interact with the RFID system (i.e., all the oracles described above) any longer after corrupting any of the tags for the first time but they can still make CorruptTag queries to all other tags. This models the case where the secrets of the tags become known when the life of the system is over.
- *Destructive adversaries* can never use a tag again after it has been corrupted but can still query all oracles for any of the remaining non-corrupted tags. This assumes that tags are destroyed when they are corrupted (e.g., due to tamper-evident hardware).
- *Strong adversaries* have full access to all of the oracles at any time.

Moreover, [18] defines *narrow* variants of the four adversary classes described above. A *narrow adversary* cannot obtain auxiliary information (i.e., on whether a tag is legitimate or not). This may be the case in application scenarios where the result of the identification protocol cannot be observed by the adversary. Therefore, a narrow adversary cannot query the Result oracle and is not given the values $(b_1, \ldots, b_n)$ from the DrawTag oracle, which both are the only sources of auxiliary information.

4.3   Security Definition

The security definition of [18] considers attacks where the adversary aims to impersonate or to forge a legitimate tag. More precisely, the definition is based on a security experiment $\mathbf{Exp}^{\mathrm{sec}}_{\mathcal{A}_{\mathrm{sec}}}$ where a *strong adversary* must create an instance LaunchIdent( ) $\rightarrow \pi_{\mathsf{R}}$ of the IdentTag protocol with the reader R and finish this protocol instance π_{R} with a query SendReader(m, π_{R}). Note that $\mathcal{A}_{\mathrm{sec}}$ can arbitrarily interact with all of the oracles defined in Section 4.2 at any time during the experiment. The adversary $\mathcal{A}_{\mathrm{sec}}$ wins if (i) R identifies a legitimate tag ID in the instance π_{R} of the IdentTag protocol, (ii) tag ID has not been corrupted and (iii) tag ID and R have not run any instance π_{R}' of the IdentTag protocol that generated the same messages as instance π_{R} (i.e., π_{R} is not a *replay* of an old transcript π_{R}'). Let $\mathbf{Exp}^{\mathrm{sec}}_{\mathcal{A}_{\mathrm{sec}}} = 1$ denote the case where the adversary $\mathcal{A}_{\mathrm{sec}}$ wins this security experiment.

Definition 1 (Security [18]). *An RFID system (as defined in Section 4.1) is secure if for any strong adversary A_{sec} the probability $\Pr[\mathbf{Exp}^{sec}_{A_{sec}} = 1]$ is negligible.*

Definition 1 can be used in the anonymizer-enabled model as it is. Note that the adversary A_{sec} is allowed to corrupt all the anonymizers when playing the security experiment described above. This models the fact that anonymizers should not be able to clone or to forge tags.

4.4   Privacy Definition

The privacy definition of [18] is very flexible and, dependent on the class of adversaries considered (in Section 4.2), it covers different notions of privacy. For strong adversaries the definition considers anonymity, backward and forward untraceability.

The privacy definition requires the communication of a tag T to not reveal any information that helps an adversary A_{prv} to trace or to identify T. It is based on the existence of a simulator B that can simulate the communication of T to A_{prv} without using any of the secrets of the RFID system. B must answer all queries of A_{prv} by only using the inputs and outputs of the oracle queries that A_{prv} previously made (i.e., B "sees" what A_{prv} "sees"). In case the success probability of A_{prv} does not change significantly when interacting with B instead of the real RFID system, the communication of T does not help A_{prv} to break the privacy properties of the RFID scheme. In [18], B is called *blinder* and an adversary A_{prv}^{B} who interacts with B is called *blinded adversary*.

More formally, the privacy definition considers a security game $\mathbf{Exp}^{prv}_{A_{prv}}$ where an adversary A_{prv} must distinguish whether he interacts with the real RFID system or a blinder B. Therefore, A_{prv} first performs an attack phase that is followed by an analysis phase. In the attack phase, A_{prv} is allowed to interact with the oracles described in Section 4.2 in an arbitrary way. In the analysis phase, A_{prv} cannot access the oracles any more but is given access to the secret table T of the DrawTag oracle, which allows A_{prv} to link the temporary identifiers vtag of all the tags he interacted with to their corresponding real identities ID. Finally, A_{prv} must return a bit b to indicate whether he interacted with a blinder B ($b = 1$) or the real RFID system ($b = 0$). This leads to the privacy definition described below.

Definition 2 (Privacy [18]). *Let P be one of the adversary classes defined in Section 4.2. An RFID system (as defined in Section 4.1) is P-private if for any adversary A_{prv} of class P there exists a blinder B such that $|\Pr[\mathbf{Exp}^{prv}_{A_{prv}} = 1] - \Pr[\mathbf{Exp}^{prv}_{A^{B}_{prv}} = 1]|$ is negligible.*

The communication of a tag is modeled by the LaunchIdent, SendReader, SendTag and the Result oracle. Thus, a blinder B must simulate these oracles. In the anonymizer-enabled model, we additionally have the LaunchAnon and the SendAnon oracles that model the interaction of a tag with the anonymizer. However, as discussed in Section 3.3, we are not concerned of the privacy of the communication between tags and the anonymizer. Thus, the LaunchAnon and the SendAnon oracle need not to be simulated by $\mathcal{B}$. Note that the CorruptTag query is not simulated by $\mathcal{B}$ because Definition 2 only captures the privacy loss of the wireless communication of tags.

5   Security Analysis

Theorem 1. *The RFID system presented in Section 3.3 is correct, secure in the random oracle model, narrow-strong and forward private in the random oracle model under Assumption 1 if the homomorphic public-key encryption scheme is CPA-secure.*

Note that Assumption 1 is *only* required to ensure the privacy properties of our scheme. Security (against impersonation attacks) also holds if there is no (honest) anonymizer in the system.

Due to space restrictions, we only give proof sketches and provide full proofs in the full version of the paper [52].

Correctness. No false negative can be produced since each legitimate tag T will always be accepted by the reader R. A false positive cannot be produced since the decryption of G outputs a unique ID and, if ID is not in the database d, R immediately rejects the identification.   □

Security. The idea of the security proof is as follows: by contradiction, we assume that there is a narrow-strong adversary $\mathcal{A}_{sec}$ (as defined in Section 4.2), who wins the security game of Definition 1. Given $\mathcal{A}_{sec}$, one can construct a p.p.t. algorithm that finds a collision to the random oracle with non-negligible probability. However, by the pseudorandomness of the random oracle, this can happen with at most negligible probability.   □

Narrow-Strong Privacy. The idea of the privacy proof is as follows: by contradiction, we assume that there is a narrow-strong adversary $\mathcal{A}_{prv}$ (as defined in Section 4.2), who wins with non-negligible probability the game of Definition 2. Given such an adversary $\mathcal{A}_{prv}$, one can construct a p.p.t. algorithm that breaks the CPA-security of the homomorphic public key encryption scheme with non-negligible probability. However, since the encryption scheme is assumed to be CPA-secure, this can happen with at most negligible probability, which is a contradiction.   □

Forward-Privacy. To prove forward-privacy, we can use the following lemma from [18]:

Lemma 1. *For every secure RFID scheme that has the property that, whenever a legitimate tag T and the reader R have executed a complete run of the IdentTag protocol in a secure environment (i.e., where no adversary can manipulate the protocol-run), the output out of R is never ⊥ (i.e., R does never reject a legitimate tag), it holds that narrow-forward privacy implies forward-privacy.*

Our scheme is narrow-strong private, which implies narrow-forward privacy [18]. Moreover, it is correct and secure, which means that it fulfills all requirements to apply Lemma 1. Since the original proof of Lemma 1 is also valid in the anonymizer-enabled model, we can apply Lemma 1 to prove that our scheme achieves forward privacy.   □

Acknowledgments

The second author wishes to thank Paolo D'Arco and Alessandra Scafuro for several useful discussions about RFID privacy notions.

The work of the authors has been supported in part by the European Commission through the EU ICT program under Contract ICT-2007-216646 ECRYPT II. The work of the second author has also been supported in part by the European Commission through the FP7 Information Communication Technologies programme, under Contract FET-215270 FRONTS.

References

1. Atmel Corporation: Innovative IDIC solutions (2007),
   http://www.atmel.com/dyn/resources/prod_documents/doc4602.pdf
2. Molnar, D., Wagner, D.: Privacy and security in library RFID: Issues, practices, and architectures. In: Proceedings of the 11th ACM Conference on Computer and Communications Security, pp. 210–219. ACM Press, New York (2004)
3. Calypso Networks Association: Web site of Calypso Networks Association. (May 2007), http://www.calypsonet-asso.org/
4. NXP Semiconductors: MIFARE smartcard ICs (September 2008),
   http://www.mifare.net/products/smartcardics/
5. Sony Global: Web site of Sony FeliCa (June 2008),
   http://www.sony.net/Products/felica/
6. Sadeghi, A.R., Visconti, I., Wachsmann, C.: User privacy in transport systems based on RFID e-tickets. In: International Workshop on Privacy in Location-Based Applications (PiLBA), Malaga, Spain, October 9 (2008)
7. I.C.A. Organization: Machine Readable Travel Documents, Doc 9303, Part 1 Machine Readable Passports, Fifth Edition (2003)
8. Juels, A.: RFID security and privacy: A research survey. Journal of Selected Areas in Communication 24(2), 381–395 (2006)
9. Weis, S.A., Sarma, S.E., Rivest, R.L., Engels, D.W.: Security and privacy aspects of low-cost radio frequency identification systems. In: Hutter, D., Müller, G., Stephan, W., Ullmann, M. (eds.) Security in Pervasive Computing. LNCS, vol. 2802, pp. 50–59. Springer, Heidelberg (2004)
10. Lim, C.H., Kwon, T.: Strong and robust RFID authentication enabling perfect ownership transfer. In: Ning, P., Qing, S., Li, N. (eds.) ICICS 2006. LNCS, vol. 4307, pp. 1–20. Springer, Heidelberg (2006)
11. Spirtech: CALYPSO functional specification: Card application, version 1.3. (October 2005), http://calypso.spirtech.net/

12. Octopus Holdings: Web site of Octopus Holdings (June 2008), http://www.octopus.com.hk/en/
13. Avoine, G.: Adversarial model for radio frequency identification. Cryptology ePrint Archive, Report 2005/049 (2005)
14. Juels, A., Weis, S.A.: Defining strong privacy for RFID. Cryptology ePrint Archive, Report 2006/137 (2006)
15. Damgård, I., Østergaard, M.: RFID security: Tradeoffs between security and efficiency. Cryptology ePrint Archive, Report 2006/234 (2006)
16. Burmester, M., van Le, T., de Medeiros, B.: Provably secure ubiquitous systems: Universally composable RFID authentication protocols. In: Proceedings of Second International Conference on Security and Privacy in Communication Networks (SecureComm), pp. 1–9. IEEE Computer Society, Los Alamitos (2006)
17. Ohkubo, M., Suzuki, K., Kinoshita, S.: Cryptographic approach to privacy-friendly tags (November 2003)
18. Vaudenay, S.: On privacy models for RFID. In: Kurosawa, K. (ed.) ASIACRYPT 2007. LNCS, vol. 4833, pp. 68–87. Springer, Heidelberg (2007)
19. Paise, R.I., Vaudenay, S.: Mutual authentication in RFID: Security and privacy. In: ASIACCS 2008: Proceedings of the 2008 ACM Symposium on Information, Computer and Communications Security, pp. 292–299. ACM Press, New York (2008)
20. Ng, C.Y., Susilo, W., Mu, Y., Safavi-Naini, R.: RFID privacy models revisited. In: [53], pp. 251–256
21. Juels, A., Weis, S.A.: Authenticating pervasive devices with human protocols. In: Shoup, V. (ed.) CRYPTO 2005. LNCS, vol. 3621, pp. 293–308. Springer, Heidelberg (2005)
22. Katz, J., Shin, J.S.: Parallel and Concurrent Security of the HB and HB+ Protocols. In: Vaudenay, S. (ed.) EUROCRYPT 2006. LNCS, vol. 4004, pp. 73–87. Springer, Heidelberg (2006)
23. Katz, J., Smith, A.: Analyzing the HB and HB+ protocols in the large error case. Cryptology ePrint Archive, Report 2006/326 (2006)
24. Katz, J.: Efficient Cryptographic Protocols Based on the Hardness of Learning Parity with Noise. In: Galbraith, S.D. (ed.) Cryptography and Coding 2007. LNCS, vol. 4887, pp. 1–15. Springer, Heidelberg (2007)
25. Gilbert, H., Robshaw, M., Silbert, H.: An active attack against HB+ — A provable secure leightweight authentication protocol. Cryptology ePrint Archive, Report 2007/237 (2007)
26. Gilbert, H., Robshaw, M.J.B., Seurin, Y.: Good Variants of HB+ Are Hard to Find. In: Tsudik, G. (ed.) FC 2008. LNCS, vol. 5143, pp. 156–170. Springer, Heidelberg (2008)
27. Ouafi, K., Overbeck, R., Vaudenay, S.: On the Security of HB# against a Man-in-the-Middle Attack. In: Pieprzyk, J. (ed.) ASIACRYPT 2008. LNCS, vol. 5350, pp. 108–124. Springer, Heidelberg (2008)
28. Frumkin, D., Shamir, A.: Un-Trusted-HB: Security Vulnerabilities of Trusted-HB. Cryptology ePrint Archive, Report 2009/044 (2009)
29. Levieil, E., Fouque, P.A.: An Improved LPN Algorithm. In: De Prisco, R., Yung, M. (eds.) SCN 2006. LNCS, vol. 4116, pp. 348–359. Springer, Heidelberg (2006)
30. Tsudik, G.: YA-TRAP: Yet Another Trivial RFID Authentication Protocol. In: Security in Pervasive Computing. LNCS, vol. 2802, pp. 640–643. IEEE Computer Society, Los Alamitos (2006)

31. Henrici, D., Müller, P.: Hash-based enhancement of location privacy for radio-frequency identification devices using varying identifiers. In: Proceedings of the Second IEEE Annual Conference on Pervasive Computing and Communications Workshops, pp. 149–153. IEEE Computer Society, Los Alamitos (2004)

32. Ohkubo, M., Suzuki, K., Kinoshita, S.: Efficient hash-chain based RFID privacy protection scheme. In: International Conference on Ubiquitous Computing (UbiComp), Workshop Privacy: Current Status and Future Directions (September 2004)

33. Dimitriou, T.: A lightweight RFID protocol to protect against traceability and cloning attacks. In: Proceedings of the First International Conference on Security and Privacy for Emerging Areas in Communications Networks (SecureComm), pp. 59–66. IEEE Computer Society, Los Alamitos (2005)

34. Song, B., Mitchell, C.J.: RFID authentication protocol for low-cost tags. In: Proceedings of the First ACM Conference on Wireless Network Security, pp. 140–147. ACM Press, New York (2008)

35. Sadeghi, A.R., Visconti, I., Wachsmann, C.: Location privacy in RFID applications. In: Bettini, C., et al. (eds.) Privacy in Location-Based Applications: Research Issues and Emerging Trends. LNCS, vol. 5599, pp. 127–150. Springer, Heidelberg (2009)

36. Juels, A., Pappu, R.: Squealing Euros: Privacy protection in RFID-enabled banknotes. In: Wright, R.N. (ed.) FC 2003. LNCS, vol. 2742, pp. 103–121. Springer, Heidelberg (2003)

37. Golle, P., Jakobsson, M., Juels, A., Syverson, P.: Universal re-encryption for mixnets. In: Okamoto, T. (ed.) CT-RSA 2004. LNCS, vol. 2964, pp. 163–178. Springer, Heidelberg (2004)

38. Saito, J., Ryou, J.C., Sakurai, K.: Enhancing privacy of universal re-encryption scheme for RFID tags. In: Yang, L.T., Guo, M., Gao, G.R., Jha, N.K. (eds.) EUC 2004. LNCS, vol. 3207, pp. 879–890. Springer, Heidelberg (2004)

39. Ateniese, G., Camenisch, J., de Medeiros, B.: Untraceable RFID tags via insubvertible encryption. In: Proceedings of the 12th ACM Conference on Computer and Communications Security, pp. 92–101. ACM Press, New York (2005)

40. Economist: Security technology: Where's the smart money? The Economist, 69–70 (February 2002)

41. Juels, A.: Minimalist cryptography for low-cost RFID tags (extended abstract). In: Blundo, C., Cimato, S. (eds.) SCN 2004. LNCS, vol. 3352, pp. 149–164. Springer, Heidelberg (2005)

42. Avoine, G., Dysli, E., Oechslin, P.: Reducing time complexity in RFID systems. In: Preneel, B., Tavares, S. (eds.) SAC 2005. LNCS, vol. 3897, pp. 291–306. Springer, Heidelberg (2006)

43. Ha, J.H., Moon, S.J., Zhou, J., Ha, J.C.: A new formal proof model for RFID location privacy. In: [53], pp. 267–281.

44. D'Arco, P., Scafuro, A., Visconti, I.: Semi-Destructive Privacy in DoS-Enabled RFID systems. In: Proceedings of RFIDSec 2009 (July 2009)

45. D'Arco, P., Scafuro, A., Visconti, I.: Revisiting DoS attacks and privacy in rfid-enabled networks. In: Dolev, S. (ed.) ALGOSENSORS 2009. LNCS, vol. 5804, p. 263. Springer, Heidelberg (2009)

46. Paillier, P.: Public-key cryptosystems based on composite degree residuosity classes. In: Stern, J. (ed.) EUROCRYPT 1999. LNCS, vol. 1592, pp. 223–238. Springer, Heidelberg (1999)

47. Prabhakaran, M., Rosulek, M.: Homomorphic encryption with CCA security. Cryptology ePrint Archive, Report 2005/079 (2008)

48. Goldwasser, S., Micali, S.: Probabilistic encryption. Journal of Computer and System Sciences 28, 270–299 (1984)
49. Bellare, M., Desai, A., Pointcheval, D., Rogaway, P.: Relations among notions of security for public-key encryption schemes. In: Krawczyk, H. (ed.) CRYPTO 1998. LNCS, vol. 1462, pp. 26–45. Springer, Heidelberg (1998)
50. Bellare, M., Rogaway, P.: Random oracles are practical: A paradigm for designing-efficient protocols. In: Proceedings of the Annual Conference on Computer and Communications Security (CCS) (1994)
51. Danev, B., Heydt-Benjamin, T.S., Capkun, S.: Physical-layer Identification of RFID Devices. In: 18th USENIX Security Symposium, Montreal, Canada, August 10-14, pp. 199–214 (2009)
52. Sadeghi, A.R., Visconti, I., Wachsmann, C.: Efficient RFID security and privacy with anonymizers. In: Proceedings of RFIDSec 2009 (July 2009)
53. Jajodia, S., Lopez, J. (eds.): ESORICS 2008. LNCS, vol. 5283, p. 602. Springer, Heidelberg (2008)

Blink 'Em All: Scalable, User-Friendly and Secure Initialization of Wireless Sensor Nodes

Nitesh Saxena and Md. Borhan Uddin

Computer Science and Engineering
Polytechnic Institute of New York University
nsaxena@poly.edu, borhan@cis.poly.edu

Abstract. Wireless sensor networks have several useful applications in commercial and defense settings, as well as user-centric personal area networks. To establish secure (point-to-point and/or broadcast) communication channels among the nodes of a wireless sensor network is a fundamental security task. To this end, a plethora of so-called *key pre-distribution* schemes have been proposed in the past, e.g., [25][9][19][8][5]. All these schemes, however, rely on shared secret(s), which are assumed to be pre-loaded onto the sensor nodes, e.g., during the manufacturing process.

In this paper, we consider the problem of user-assisted secure initialization of sensor network necessary to bootstrap key pre-distribution. This is a challenging problem due to the level of user burden involved in initializing multiple (often large number of) sensor nodes and lack of input and output user-interfaces on sensor motes. We propose a novel method for secure sensor node initialization based on a visual out-of-band channel that utilizes minimal output interface in the form of LED(s) already available on most off-the-shelf sensor motes. The proposed method requires only a little extra cost, is efficient and reasonably scalable. Moreover, based on a usability study that we conducted, the method turns out to be quite user-friendly and easy to administer by everyday computer users.

Keywords: Wireless Sensor Networks, Authentication, Key Distribution.

1 Introduction

Wireless sensor networks (WSN) have several useful applications in monitoring diverse aspects of the environment. Ready examples include monitoring of: structural/seismic activity, wildlife habitat, air pollution, border crossings, nuclear emission and water quality. In addition to commercial and defense settings, WSNs appeal to a variety of user-centric applications in personal area networks [11,36,22]. In some applications, sensor nodes operate in a potentially hostile environments and security measures are needed to inhibit or detect node compromise and/or tampering with inter-node or node-to-sink communication. A large body of literature has been accumulated in the last decade dealing with many aspects of sensor network security, e.g., key management, secure routing and DoS detection [24,14,9,7].

In a WSN environment, the nodes might need to communicate sensitive data among themselves and with the base station (also referred to as "sink"). The communication

J.A. Garay, A. Miyaji, and A. Otsuka (Eds.): CANS 2009, LNCS 5888, pp. 154–173, 2009.
© Springer-Verlag Berlin Heidelberg 2009

among the nodes might be point-to-point and/or broadcast, depending upon the application. These communication channels are easy to eavesdrop on and to manipulate, raising the very real threat of the so-called *Man-in-the-Middle* (MiTM) attacker. A fundamental task, therefore, is to secure these communication channels.

Key Pre-Distribution and the Underlying Assumption: A number of so-called "key pre-distribution" techniques to bootstrap secure communication in a WSN have been proposed, e.g., [25,9,19,8,5]. However, all of them assume that, before deployment, sensor nodes are somehow pre-installed with secret(s) shared with other sensor nodes and/or the sink. The TinySec architecture [15] also assumes that the nodes are loaded with shared keys prior to deployment. This might be a reasonable assumption in some, but certainly not all, cases. Consider, for example, a user-centric application of WSN: an individual user (Bob) wants to install a sensor network to monitor the perimeter of his property; he purchases a set of commodity noise-and-vibration sensor nodes at some retailer and wants to deploy the sensor nodes with his home computer acting as the sink. Being off-the-shelf, these sensor nodes are not sold with any built-in secrets. Some types of sensor nodes might have a USB (or similar) connector that allows Bob to plug each sensor node into his computer to perform secure initialization. This would be immune to both eavesdropping and MiTM attacks. However, sensor nodes might not have any interface other than wireless, since having a special "initialization" interface influences the complexity and the cost of the sensor node. Also, note that Bob would have to perform security initialization manually and separately for each sensor node. To initialize N motes, Bob will have to perform $O(N)$ amount of work. This undermines the scalability of the approach since potentially a large number of sensor nodes might be involved.

Furthermore, we argue that keys can not always be pre-loaded during the manufacturing phase because eventual customers might not trust the manufacturer. Moreover, an application might involve motes produced by multiple manufacturers. A PKI-based solution might be infeasible as it would require a global infrastructure involving many diverse manufacturers.[1]

Secure Initialization Approach: The best possible strategy would be for the network administrator or user of WSN to himself/herself perform the key distribution on-site. Due to lack of hardware interfaces (such as USB interfaces) on sensor nodes and for usability reasons, this key distribution should be performed wirelessly. Prior key pre-distribution schemes assume the existence of some pre-installed secret (such as a point on a bivariate polynomial $f(x, y)$ in [8]) using which the shared keys can be derived. Therefore, the task of key distribution is reduced to establishing a secure channel between the administrator's computer (the sink node) and each sensor node. The resulting secure channels can in turn be used to securely transfer, from the sink to each node, the shared secrets necessary to bootstrap key pre-distribution. Since the administrator might need to initialize a large number of sensor nodes, the process needs to be repeated

---

[1] The problem that we consider in this paper is very similar to the problem of "wireless device pairing," the premise of which is also based on the fact that the devices wanting to communicate with each other do not share any pre-shared secrets or a common PKI with each other [2].

in batches. The larger the number of sensor nodes in each batch, the more *scalable* is the secure initialization method.

Out-of-Band Channels: In quest of a scalable sensor node initialization method, we consider out-of-band (OOB) channels. The OOB (audio, visual or tactile) channels have recently been utilized in the context of secure device pairing application [2,21,12,29], used to establish shared keys between two previously un-associated devices (we review these methods and their applicability to sensor node initialization in the following section). Unlike the wireless communication channel, the OOB channels are both perceivable and manageable by the human user(s) operating the devices, and thus can be used to authenticate information exchanged over the wireless channel. Unlike the wireless channel, the attacker can not remain undetected if it interferes with the OOB channel, although it can still eavesdrop upon it.

Our Contributions: We develop a scalable sensor node initialization method based on a visual OOB channel. Our system builds on an existing protocol of Saxena et al. [29]. However, we make two important extensions to realize the proposed system. First, we develop a new visual channel consisting of simultaneously blinking LEDs[2] as transmitters on sensor nodes and a video camera on the administrator's computer (the sink). This enables efficient transmission of OOB data from sensor motes to the sink with little involvement from the administrator. Second, we design a very intuitive yet effective interface on the sink that allows the administrator to easily discard any potential "attacked" sensor nodes.

Our experiments show that with an inexpensive web cam connected to a laptop or desktop computer, we are efficiently able to use the above visual channel to securely initialize 16 sensor nodes per batch. To evaluate our proposal at the "usability layer," we pursue a thorough and systematic usability study. The results of our study show that our system is both user-friendly as well as robust to errors (human or otherwise).

Organization: In the following section, we review the prior related work. Next, we describe the security model and summarize the relevant protocol we use. This is followed by the description of the design and implementation of our scheme. Finally the results of our experimental testing are presented and discussed.

2   Related Work

The problem of secure sensor node initialization has been considered only recently. Most closely related to our proposal is the sensor network initialization method, called "Message-In-a-Bottle" (MiB) by Kuo et al. [6]. In MiB, the key distribution takes place inside a Faraday Cage, which is used to shield communication from eavesdropping and outside interference. MiB can support key distribution onto multiple sensor nodes[3] and from the administrator's perspective, it is quite user-friendly. However, it has some

---

[2] Most commercially available sensor motes possess multiple (typically three) LEDs. (For example, refer to Mica2 specifications [1].)

[3] Although it is not clear from the experiments presented in [6], at most how many motes can be initialized per batch.

drawbacks. The first problem is the need to obtain and carry around a specialized piece of equipment – a Faraday Cage. As illustrated in [6], building a truly secure Faraday Cage might be a challenge. The cost and the physical size of the Cage can also be problematic. In other words, only a very few sensor motes could be supported in each batch with a reasonably priced and reasonably sized cage. The second drawback with MiB is that if the initialization process fails for only one sensor node or if there is an error (e.g., if the cage was not properly closed), the entire batch of sensor nodes needs to be re-initialized and re-keyed from scratch. Third, a batch of sensor motes must consist of homogeneous sensor motes with similar weights (the weight is used to calculate the number of motes inside the Cage [6]). Fourth, two additional motes (called "keying device" and "keying beacon") that possess physical interfaces, such as USB connectors, are needed along with the base station and un-initialized nodes. These increase both the cost and the complexity of the system.

The method we propose in this paper can be viewed as an alternative to MiB; the former provides (authenticated key exchange) protocol level security whereas the latter offers physical layer security. Our method also addresses aforementioned drawbacks with MiB (we will discuss this in the final section of the paper). As opposed to MiB [6], our proposal is based on public-key cryptography. We note, however, that most commercial sensor motes are efficiently able to perform public key cryptography [20]. Elliptic-Curve Cryptography has particularly been shown to be very promising on sensor motes [18].

Prior to the MiB method of [6], following schemes were proposed. However, these schemes were aimed at associating only two sensor nodes at a time and not multiple nodes, which is the focus of our paper. The "Shake-them-up" [4] scheme suggests a simple manual technique for pairing two sensor motes that involves shaking and twirling them in very close proximity to each other, in order to prevent eavesdropping. While being shaken, two sensor motes exchange packets and agree on a key one bit at a time, relying on the adversary's inability to determine the sending sensor node. However, it turns out that the sender can be identified using radio fingerprinting [27] and the security of this scheme is uncertain.

Other two related schemes are: "Smart-Its Friends" [13] and "Are You with Me?" [17]. Both use human-controlled movement to establish a secret key between two devices. In addition to having the same drawbacks as "Shake-Them-Up", these schemes would require an accelerometer on each sensor mote to measure movement. Most sensor motes are not equipped with accelerometers, however.

The initialization method that we propose in this paper is similar to the device pairing schemes that use an OOB channel. Thus, we also review most relevant device pairing methods and argue whether or not they can be extended for the application of scalable sensor node initialization. In their seminal work, Stajano and Anderson [35] proposed to establish a shared secret between two devices using a link created through a physical contact (such as an electric cable). As pointed out previously, this approach requires interfaces not available on most sensor motes. Moreover, the approach would be unscalable.

Balfanz, et al. [2] extended the above approach through the use of infrared as an OOB channel – the devices exchange their public keys over the wireless channel followed by exchanging (at least 80-bits long) hashes of their respective public keys over infrared. Most sensor motes do not possess infrared transmitters. Moreover, infrared is not easily perceptible by humans.

Based on the protocol of Balfanz et al. [2], McCune et al. proposed the "Seeing-is-Believing" (SiB) scheme [21]. SiB involves establishing two unidirectional visual OOB channels – one device encodes the data into a two-dimensional barcode and the other device reads it using a photo camera. To apply SiB for sensor node initialization, one would need to affix a static barcode (during the manufacturing phase) on each mote, which can be captured by a camera on the sink node. However, this will only provide unidirectional authentication, since the sensor motes can not afford to have a camera each. Note that it will also not be possible to manually input on each sensor mote the hash of the public key of the sink, since most motes do not possess keypads and even if they do, this will not scale.

Saxena et al. [29] proposed a new scheme based on visual OOB channel. The scheme uses one of the protocols based on Short Authenticated Strings (SAS) [23], [16], and is aimed at pairing two devices (such as a cell phone and an access point), only one of which has a relevant receiver (such as a camera). The protocol is depicted in Figure 1 and as we will see in the next section, this is the protocol that we utilize in our proposal. In this paper, we extend the above scheme to a "many-to-one" setting applicable to key distribution in sensor networks. Basically, the novel OOB channel that we build consists of multiple devices blinking their SAS data simultaneously, which is captured using a camera connected to the sink. In addition, we design an intuitive user interface on the sink that facilitates human users to clearly discard any potential "attacked" sensor nodes.

Recently, Soriente et al. [34] consider the problem of pairing two devices based on an audio channel. Their scheme can be based on the protocol of [29], with the unidirectional SAS channel consisting of one device encoding its SAS data into audio, and the other device capturing it using a microphone. Extending this scheme to initialize multiple sensor nodes in a scalable manner seems hard as it will be hard to decode simultaneously "beeping" sensor nodes.

There are a variety of other pairing schemes, based on manual comparison/transfer of OOB data: [12,37] can not be used on motes as they require displays; [33,26] are applicable on sensor motes but would not scale well due to their manual nature.

3   Communication and Security Model, and Protocol

Model: The protocol that we utilize in our initialization method is based upon the following communication and adversarial model [38]. The devices being paired are connected via two types of channels: (1) a short-range, high-bandwidth bidirectional wireless channel, and (2) auxiliary low-bandwidth physical OOB channel(s). Based on device types, the OOB channel(s) can be device-to-device (d2d), device-to-human (d2h) and/or human-to-device (h2d). An adversary attacking the pairing protocol is assumed to have full control on the wireless channel, namely, it can eavesdrop, delay, drop, replay and modify messages. On the OOB channel, the adversary can eavesdrop on but can not modify messages. In other words, the OOB channel is assumed to be an authenticated channel. The security notion for a pairing protocol in this setting is

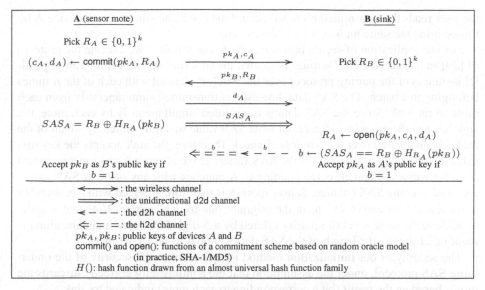

A (sensor mote)

Pick $R_A \in \{0,1\}^k$

$(c_A, d_A) \leftarrow \mathsf{commit}(pk_A, R_A)$ $\xrightarrow{\quad pk_A, c_A \quad}$

$\xleftarrow{\quad pk_B, R_B \quad}$

$\xrightarrow{\quad d_A \quad}$

$SAS_A = R_B \oplus H_{R_A}(pk_B)$ $\xRightarrow{\quad SAS_A \quad}$

Accept pk_B as B's public key if
$b = 1$

B (sink)

Pick $R_B \in \{0,1\}^k$

$R_A \leftarrow \mathsf{open}(pk_A, c_A, d_A)$

$\xLeftarrow{\quad b \quad} b \leftarrow (SAS_A == R_B \oplus H_{R_A}(pk_B))$

Accept pk_A as A's public key if
$b = 1$

$\xrightarrow{\qquad}$: the wireless channel
$\xRightarrow{\qquad}$: the unidirectional d2d channel
$\xleftarrow{\ -\ -\ -}$: the d2h channel
$\xLeftarrow{\ =\ =\ =}$: the h2d channel
pk_A, pk_B: public keys of devices A and B
commit() and open(): functions of a commitment scheme based on random oracle model
(in practice, SHA-1/MD5)
$H()$: hash function drawn from an almost universal hash function family

Fig. 1. The protocol by Saxena et al. based on the SAS protocol of Pasini-Vaudenay

adopted from the model of authenticated key agreement due to Canneti and Krawczyk [3]. In this model, a multi-party setting is considered wherein a number of parties simultaneously run multiple/parallel instances of pairing protocols. In practice, however, it is reasonable to assume only two-parties running only a few serial/parallel instances of the pairing protocol. For example, during authentication for an ATM transaction, there are only two parties, namely the ATM machine and a user, restricted to only three authentication attempts. The security model does not consider denial-of-service (DoS) attacks. Note that on wireless channels, explicit attempts to prevent DoS attacks might not be useful because an adversary can simply launch an attack by jamming the wireless signal.

In a communication setting involving two users restricted to running three instances of the protocol, the SAS protocol of [29] [30] need to transmit only k ($= 15$) bits of data over the OOB channels. As long as the cryptographic primitives used in the protocols are secure, an adversary attacking these protocols can not win with a probability significantly higher than 2^{-k} ($= 2^{-15}$). This gives us security equivalent to the security provided by 5-digit PIN-based ATM authentication.

Protocol: The protocol that we utilize [29][30] is depicted in Figure 1 (we base the protocol upon the SAS protocol of [23], although it can similarly work with other SAS protocol [16] as well). The protocol works as follows. Over the wireless channel, devices A (sensor mote) and B (sink) follow the underlying SAS protocol (due to lack of space, we omit describing the protocol steps over the wireless channel and refer the reader to [29]). Then a unidirectional OOB channel is established by device A transmitting the SAS data, over the d2d channel. This is followed by device B comparing the received data with its own copy of the SAS data, and transmitting the resulting bit b of comparison over the 1-bit d2h OOB channel (say, displayed on its screen). Finally, the user reads the transmitted bit b and accordingly indicates the result to device A by transmitting the same bit b over an h2d input channel.

For our application of secure initialization of sensor nodes, we execute the protocol of [29] in a "many-to-one" setting. Basically, the sink runs serial or (preferably) parallel instances of the pairing protocol over the wireless channel with each of the n motes belonging to a batch. The SAS data, however, is transmitted simultaneously from each mote to the sink. Since the SAS data is transmitted simultaneously by each mote, the sink has no efficient way to figure out what SAS value was transmitted by which of the motes it discovered over the wireless channel. Therefore, the sink accepts the key distribution on a particular mote A if the SAS value (derived from information transmitted over the wireless channel) corresponding to A matches with any of the n SAS values received over the SAS channel. Sensor mote A is therefore accepted with a probability at most $n2^{-k}$ instead of 2^{-k} as in the original "one-to-one" setting. Note that in order to achieve the same level of security offered by a 5-digit PIN-based authentication (as mentioned above), the length of the SAS data should now be $15 + log_2(n)$.

The security of our initialization method is equivalent to the security of the underlying SAS protocol, under the assumption that the administrator correctly discards the motes based on the result (bit b corresponding to each mote) indicated by sink.

4   Our Proposal: Secure Initialization Using a Visual Channel

In this section, we describe the design and implementation of an efficient, scalable, user friendly and commercially viable method of secure initialization for sensor nodes. The core of our solution relies on the protocol of [29] executed in a many-to-one setting, as mentioned in the previous section. For transmitting the SAS data of all motes simultaneously over the visual channel, the LEDs of sensor motes are used for ON-OFF encoding, and for receiving the data, video frame based image processing is used on the receiver side.

4.1   Set-Up of the Mechanism

In our setup (Figure 3), the administrator's computer (the sink) is connected (using a USB interface) with a sensor node having the functionality of a base station. The sink is also connected with a video camera (a web cam). The motes and the sink communicate over the wireless channel. Sensor motes have their on board displays implemented using two types of LEDs – one Sync LED (used for synchronizing the data transmission between the mote and the sink) and at least one Data LED (used for transmitting SAS data). The Data LEDs can be of any color (same or different), but their color(s) should be different from the color of the Sync LED. The blinking LEDs on motes are used to transmit the SAS data, which is captured using the camera on the sink. The sink matches the received SAS data with its own copy of the acquired SAS data for each mote and based on this, learns whether a particular sensor mote "passed" or "failed" during the process. The sink also displays on its screen the result corresponding to each sensor mote. Based on the result indicated, the administrator must remove or turn off the failed motes. In case the sink is also connected with a printer, the screen indicating the result can also be printed, to better assist the administrator.

4.2 Role of the Administrator

The administrator needs to follow the steps shown in Figure 2. On completion of Step 5, the sink makes use of the resulting secure channels between itself and each sensor mote to bootstrap any of the key pre-distribution schemes, e.g., [8].

Step 1. The administrator turns on the sensor motes and places them on a table, one by one.

Step 2. The administrator presses the "Start" button on the sink. This triggers the sink to sense the nearby motes and signal them over the wireless channel to start an instance each of the protocol of Figure 1. Once done with their SAS data computation, the motes show a "Ready" signal to the administrator by lighting up their red LEDs, and the sink shows the message "Focus the Camera on Ready Motes and Press OK".

Step 3. The administrator adjusts the camera accordingly to capture the LEDs of the ready motes and presses the "OK" button on the sink. The sink sends a "Start Transmission" signal over wireless channel to all sensor motes simultaneously to transmit their SAS data. All the motes transmit their SAS data simultaneously and the camera on the sink captures and decodes the data, and shows the result on the screen and/or prints it out.

Step 4. The administrator turns off the failed motes based on the on-screen or printed output. The turning off of a mote is to be implemented in such a manner that it is equivalent to the mote *rejecting* the protocol instance it executed with the sink. If the administrator does not turn off a particular mote, within an (experimentally determined) time period Δ, by default, the protocol instance will be accepted by the mote. (The default acceptance mechanism is adopted in order to improve the usability of our method. Under normal circumstances, i.e., when no attacks or errors occur, the administrator does not need to turn off any mote.)

Step 5. Steps 1-4 are repeated, batch by batch, until all motes are initialized successfully.

Fig. 2. The Administrator's Role

4.3 Design and Implementation

Our sensor node initialization method requires three phases: (1) the device discovery phase, whereby the sink discovers each mote (over the wireless channel)[4], (2) protocol execution phase, whereby the first three rounds of the SAS protocol of Figure 1 are executed between the sink and each mote, and (3) the SAS data transmission, whereby the sensor motes simultaneously transmit their SAS data, the sink captures them, matches each of them with the local copies and accordingly indicates to the administrator to discard any failed motes.

---

[4] The sink as well as the motes need to know the actual number n of motes being initialized in one batch, since the length k of random nonces R_A and R_B and of SAS_A in the protocol of Figure 1, should ideally be equal to $15 + log_2(n)$ (as discussed in Section 3). However, an adversary might influence the value of n the sink and the motes determine by sensing over the wireless channel. Therefore, one can hard-code the value of k on the motes and on the sink, based on the expected maximum number of motes to be initialized in a batch. For example, one can safely set k to be 20, if it is expected that at most only 32 motes will be initialized in a batch.

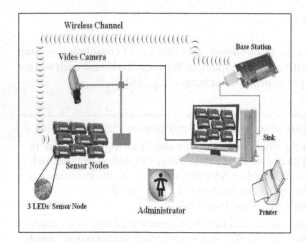

Fig. 3. The Overall Set-up of Mechanism

Fig. 4. Experimental setup: Receiver is Web camera, Transmitters are LEDs on Breadboard

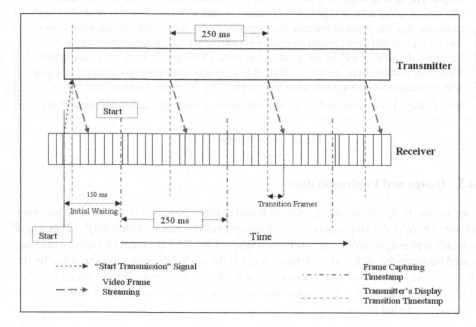

Fig. 5. Synchronization of Transmission (using LEDs) and Reception (on sink) of Data

We were most interested in the third phase as it is an essential element of our proposal. To this end, we have developed an application in Microsoft Visual C# that simulates our sensor node initialization process. The application has two parts – the transmitter simulating the sensor nodes and the receiver simulating the sink; running on two different computers. The transmitter encodes and transmits the SAS data using the display consisting of three blinking LEDs per sensor mote. All motes show the ith bit of their respective SAS data simultaneously. The sink captures the transmitted data as a video stream using its camera, extracts the SAS data for each mote, compares it with its own local copy for the corresponding mote and displays the result on screen and/or prints it out through a printer connected to it. Instead of dealing with real motes[5], we simulated the display of motes using LEDs on a breadboard, integrated with the transmitter through the parallel port of the transmitting computer. It is important to note that our simulated set-up very closely resembles a real system as viewed from usability perspective, which is the primary focus of our evaluation.

Encoding using LEDs: In our simulation, each mote is equipped with one Sync LED (red color LED) for synchronization at the beginning and end of SAS data transmission and two Data LEDs(green color LEDs) for transmitting the SAS data. We simulated the display of a total of 16 sensor motes on a breadboard(Figure 6) each having three LEDs as most commercially available motes; however, our implementation supports an arbitrary number of LEDs (with an arbitrary physical topology) and two distinct but not fixed color LEDs(for differentiating Sync and Data LEDs).

The sync LED (kept "ON" at the beginning and end of SAS data transmission; "OFF", otherwise) is used to indicate the beginning and end of the SAS data transmission and to detect any synchronization delays, adversarial or otherwise, between the motes and the sink.

The data LEDs are used for SAS data transmission by indicating different bits ('0'/'1') using different states (OFF/ON) of LEDs. If N is the number of Data LEDs, the transmitter can display N bits of SAS data at a time. The states of the sync and data LEDs are kept unchanged for a certain time period (named "hold time"; experimentally determined as 250ms); so that, a stable state (named "BitFrame") can be easily captured in the video stream of the receiver video camera. After every 250 ms, next N bits of the SAS data are simultaneously shown by each mote in the next frame. This process continues until all bits of SAS data are transmitted. If the last frame does not have N number of SAS bits to show, the beginning required LEDs show the data bits and the remaining are kept OFF.

For discovering the location, color, dimension of LEDs for each mote at the receiver side, two extra frames are needed at the beginning of data transmission – an "All-ON" frame having all LEDs in ON state and an "All-OFF" frame having all LEDs in OFF state. In addition to All-ON and All-OFF frames, another frame is required at the end of SAS data transmission, to detect synchronization delays having the Sync LED in ON state and the data LEDs in OFF state. Therefore, overall a total of three extra frames are required. Thus, for 20-bit SAS data transmission(recall that $[15 + log_2(16)]$-bit long SAS is required for 16 motes) the total number of frames to be transmitted is $\lceil \frac{20}{N} \rceil + 3$, which yields a total transmission time of $(\lceil \frac{20}{N} \rceil + 3) \times 250$ ms. For transmitting 20-bit SAS data using N=2 data LEDs, there is requirement of a total of 13 frames and thus a total of 3.25 seconds of transmission and capturing time.

---

[5] Since we wanted to deal with a number of motes, a testbed consisting of real motes was not affordable, nor was it necessary.

Decoding using a Video Camera: For successfully decoding the data transmitted using the LEDs of motes, the receiver video camera must have a frame rate higher than the transmission rate. If frames are not carefully captured from the video stream, there is a likelihood of obtaining the counterfeit frames, which contain the transition state of LEDs.

Resolving the Timing Issue of Frame Capturing: We assume that the transmission delay of "Start Transmission" (ST) signal from the receiver to the transmitter is negligible (5-6 ms) compared to the "hold time" (HT) (of 250 ms) and the receiver video camera also has a delay (about 30-40 ms, since most common cameras have a rate of 30-40 frames per second) of capturing the frame from video stream. Bases on this assumption, the receiver captures the first frame from the video stream after a time, equal to $0.6\times$ HT (i.e. after 150 ms), termed as "initial waiting" (IW), after sending the signal. The sink pre-calculates capturing (saving frames into memory from video stream buffer) timestamps for all frames by adding the IW + (HT (250ms) ×"frame_index"), with the timestamp of sending of the ST signal. The frames are captured into memory at the corresponding timestamps. Figure 5 depicts the synchronization of transmission and reception of SAS data. In this figure each small rectangle on the receiving window denotes a video frame of video stream and brown arrow marked with "Video Frame Streaming" denotes the propagation of transmitted signal to streamed frame in the video stream, which implies that there is some propagation delay of an input transition from transmitter's side to the receiver's video stream.

Detection of LEDs and Retrieval of SAS data: The frames are processed after the completion of capturing of all required frames. Our LED location and dimension detection algorithm is simple yet fast, robust and efficient, unlike existing object/face detection algorithms [28,31,39]. The algorithm detects the position and dimension of LEDs deterministically. It is able to detect any shape/geometry of LEDs unlike [39] and does not require any prior training unlike [28,31]. The algorithm uses the color threshold adjustment technique like [40] to detect the position and dimension of LEDs.

The maximal differences of RGB values, $max(dR, dG, dB)$ (denoted as μ), of each pixel of All-OFF and All-ON frames are measured and kept in memory. Using a threshold value for μ, bit-strings are built for each row of pixels. For example, if μ exceeds a certain threshold, the corresponding bit in the string becomes '1', otherwise it becomes a '0'.

Each bit-string is matched against a regular expression for consecutive 1s. For each matching bit-string, its center is calculated and its safeness and centeredness as an LED center is checked by matching against the already explored LEDs and exploring only the nearby pixels of this center in the frame. If its safeness and centeredness is proved, it is accepted as an LED and its coordinates are included in the explored list of LEDs. This process continues up to a number of times by adjusting the threshold value of μ and constructing the new bit strings until all LEDs are detected. In Figure 7, we show an example of detection of LEDs from the bit-string.

After successful discovery of LEDs, the length, width, average RGB values of ON and OFF states of LED area, for each LED are stored in memory for detecting the ON/OFF state of LEDs in subsequent BitFrames. Successfully discovered LEDs are

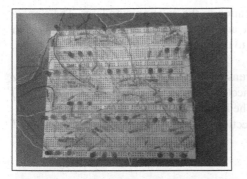

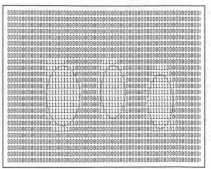

Fig. 6. Transmitter: Breadboard with 48 LEDs Simulating Displays of 16 Motes

Fig. 7. Detected LEDs from BitString

clustered according to a threshold value of proximity among themselves, for identifying the displays of different sensor motes.

After successful detection of all motes, the data LEDs of each mote are sorted according to the left-to-right and top-to-bottom ordering of coordinates. Now SAS data for each mote is extracted from the BitFrames by comparing the average RGB values of LEDs with previously saved (from All-OFF and All-ON frames) OFF and ON state RGB values of LEDs. For each extracted SAS, the sink matches it with its own computed list of "free" SAS values. If there is a match, the sink marks the corresponding computed SAS as "used" and the mote as "SAS Matched". If extracted SAS of a mote does not match with any free SAS values, the corresponding mote and all motes having the same SAS are marked as "SAS Mismatched". Each BitFrame is then examined: the Sync LEDs of all motes should be in the OFF state, except for the last frame, where the Sync LED should be in the ON state and all data LEDs of all motes should be in the OFF state. If this is not the case, it implies that a synchronization error occurred.

If for a mote, both "SAS Matched" and "Sync Matched" are true, the sink accepts the mote as a "passed"; otherwise, it rejects the mote as a "failed" due to mismatch of SAS and/or synchronization errors. The LEDs of a passed mote are marked with a rectangle of green color; and the LEDs of a failed mote are crossed out with red color (Figure 8). Additionally, an automatic printing of the result-screen is done by the printer connected to the sink. By observing the graphical result on screen of the sink and/or the printed result, the administrator discards the failed motes.

5 Experiments and Results

5.1 Experimental Setup

To test our simulator implementing the sensor node initialization method, we used the following set-up. The sink is running on a DELL Vostro 1500 Laptop (1.6 GHz CPU, 2GB RAM, WinXP Pro SP2) connected with a USB Web Camera (Microsoft LifeCam VX6000, up to 30 frames/sec, live video streaming of resolution $640 X 480$ pixels) and a wireless printer. The webcam can be replaced with any similar camera with a frame rate 30 fps or higher, without any modification to the existing simulator. The camera is set in NON-STOP video capturing mode and frames are taken setting the camera in preview mode. Camera controller is added to the simulator to allow adjusting the focus, tilt and pan of camera as needed.

The transmitting side of the simulator runs on a DELL desktop computer (1.8 GHz CPU, 1 GB RAM, WinXP Pro SP2) connected with LEDs on breadboard (Figure 6) through parallel port (DB25 Connector). The laptop and the desktop computer are connected with our university's wireless connection (54 Mbps). Figure 4 has a snapshot of our set-up.

5.2 Usability Testing

In order to test how our method fares with non-expert users, and especially to figure out if the users are easily and correctly able to discard the failed sensor motes based on the result screen (and/or print-out), we performed a usability study.

Testing Framework: For creating an automated testing framework, we extended the transmitter application running on the desktop computer by implementing the usability testing and user feedback collection functionality on it. The sink application running on the laptop was configured to send the result (indicating passed or failed motes) to the desktop application, as soon as it was determined. As there is no interface on breadboard using which the users can turn off the failed mote(s), we simulated the "turning off" mechanism in the desktop application. As soon as the desktop application receives the result from the laptop application, it shows the layout of the mote field (i.e., the breadboard) on screen, associating each sensor mote with a transparent button with the layout of the mote in the background. The users are instructed to transfer the result from the laptop screen to the desktop screen by clicking on the buttons (on the desktop screen) corresponding to the failed motes shown on the laptop screen. After test completion, the desktop application has the functionality of showing the questionnaires to obtain user feedback and logging the data. In our current tests, we did not make use of the printed output.

Test Cases: We created five categories of test cases to evaluate our method against different types of possible attacks and errors. These included (1) matching SAS and no synchronization errors (to simulate normal execution scenarios, where no attacks or faults occur), (2) (single- and multiple-bit) SAS mismatch on a varying number of motes; (3) missing, pre-mature and delayed turning on of the Sync LED (to simulate synchronization errors), (4) both SAS mismatch and synchronization errors, and (5) variable distance (from 0.5 to 2 feet) between the camera and the transmitters. Ten test cases for each category were created. Each user executed a total of five test cases, one each selected randomly from each of the five categories.

A (portion of the) screenshot of the result of execution of one of the test cases is shown in Figure 8.

Test Participants: We recruited 21 subjects[6] for our usability testing. Subjects were chosen on a first-come first-serve basis from respondents to recruiting posters and email

---

[6] It is well-known that a usability study performed by 20 participants captures over 98% of usability related problems [10].

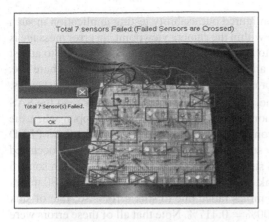

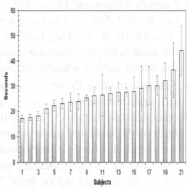

Fig. 8. Result Screen: 7 Failed Motes (marked by "red cross"), 9 Passed Motes (marked by "green rectangle")

Fig. 9. Average time (per test case execution) taken by 21 subjects with standard error. Subjects are sorted by average time.

ads. At the end of the tests, the participants were asked to fill out an on screen questionnaire through which we obtained user demographics and their feedback on the method tested.

Recruited subjects were mostly university students, both graduate and undergraduate, with CS and non-CS backgrounds. This resulted in a fairly young (ages between 22-31 [*mean*=25.48, *se*=0.5417]), well-educated participant group. All participants were regular computer users. 19 out of 21 participants reported they have previously used a PC camera (for internet chat). None of the study participants reported any physical impairments that could have interfered with their ability to complete given task. The gender split was: 17 males and 4 females.

Testing Process: Our study was conducted in a graduate student laboratory of our university. Each participant was given a brief overview of our study goals and our experimental set-up. Each participating user was then asked to follow on-screen instructions on the laptop and desktop computer. No training of any sort was given. Basically, the participants played the role of the administrator in the sensor node initialization method, as depicted in Figure 2. Sink output, user interactions throughout the tests and timings were logged automatically by the testing framework.

After completing the deputed test cases in the above manner, the participants were asked to give some qualitative feedback on how easy or hard they found to focus the camera on all LEDs, to read the result of the output screen and about the overall ease/difficulty of the method. Participants demographic information such as age, gender, educational qualification, visual disability, computer and camera experience is also collected through this questionnaire. All user data and feedback was logged by the testing framework for future analysis.

Test Results: Each of our 21 subjects executed 5 test cases, leading to a total of 105 test cases. Most of the test cases executed successfully giving expected results. In some cases, however, we observed a few errors, which we categorize and describe below.

- *Camera Adjustment Error:* We configured our usability testing application in such manner that if all the LEDs are not within the camera viewpoint, an error message is shown to the user asking him/her to re-execute. In our tests, 2 users failed to adjust the camera on one occasion each and thus they had to repeat the tests. Therefore, the rate of camera adjustment error equals $\frac{2}{(105+2)} \times 100\% = 1.87\%$ of test cases.

- *Sink Mis-reading Error:* Sometimes the sink is not able to correctly read the SAS string(s) transmitted by one or more sensor nodes. This could happen when the camera is too distant (> 2 feet) from the sensor motes or due to reflection of LED light on the table and other nearby surfaces. In our tests, this type of error occurred for a total of 7 motes, where SAS strings of 1 or 2 motes were mis-read in some 5 test cases. In 105 testcases, the sink dealt with a total of $(105 \times 16) = 1680$ motes on breadboard and out of them 7 motes failed due to sink errors. So, rate of sink mis-reading error equals $\frac{7}{1680} \times 100\% = 0.417\%$. Note that all of these errors were only false positives, i.e., the mistakenly marked a passed mote as a failed one.

- *User Error:* A user error occurs when the user is not able to correctly transfer the result, from the laptop screen to the desktop screen (simulating switching off the failed mote). In our tests, 3 users accidentally clicked, on one occasion each, a passed mote on the desktop screen (this implies that a passed mote was turned off). However, it is important to note that on no occasions did a user miss clicking on a failed mote. In other words, we did get a few false positives but no false negatives whatsoever. Thus, rate of user errors from our tests turned out to be equal to $\frac{3}{1680} \times 100\% = 0.18\%$.

The average time taken by each user (over the 5 test cases), to complete Steps 2 to 4 of Figure 2, is depicted in Figure 9. As we see, the time taken by all of our users to perform a test is less than a minute [*mean*=26.5 seconds, *se*=1.37]. Note that these numbers arise when we assume a fairly conservative setting, one where both normal scenarios and attacks or faults occur with equal likelihood. However, in practice, attacks or faults are less likely. Therefore, considering only the normal test case, we find that that on an average a user only takes 19.18 seconds [*se*=1.11] to complete the whole process.

The results we obtained through the user feedback questionnaire are shown in Table 1. Clearly, most users found the method robust and quite easy to work with. We did not find any notable correlation of the subjects' age, gender and technical expertise with the results obtained for the method, however.

Table 1. User Feedback (numbers denote the number of users)

Easiness	Very Easy	Easy	Medium Difficult	Difficult	Very Difficult	Impossible
Camera Adjustment over LEDs	5	13	3	0	0	0
Detection of Failed Motes	11	10	0	0	0	0
Easiness of Mechanism	7	10	4	0	0	0

6   Discussion

We proposed a novel method for secure initialization of sensor nodes. Based on the results of our testing with the proposed method, we discuss the following properties.

6.1   Efficiency

Using N Data LEDs and one Sync LED per sensor node, the transmission requires $[\lceil \frac{20}{N} \rceil +$ $3] \times 250$ ms. This is equal to 3.25 sec for N=2 and 20-bit SAS data. Extraction of SAS data from captured frames and displaying the result on screen require less than 3-4 seconds. So, execution time of the method is 7-8 seconds. Overall, as our experiment results show, most users took less than a minute to perform the whole process. Also, as shown in [20], most existing commercial sensor motes (e.g. Mica2) can efficiently execute (within a minute) the public key operations (private and public key generation, and one exponentiation). Note that these operations constitute the dominant costs in the SAS protocol (of Figure 1) that a sensor node executes with the sink. The sink, on the other hand, is assumed to be a computer with a fairly strong computational power and therefore can efficiently execute n parallel protocol instances with each of the sensor nodes.

Based on the above numbers, we recommend setting $\Delta = 2$ minutes, as the time period (to complete Steps 2 to 4 of Figure 2) by which the key initialization will be accepted by each sensor node, by default. As our experiments show, within 2 minutes, a human user can safely complete the initialization process, turning off any (failed) sensor nodes, if necessary.

6.2   Robustness

Our method is quite robust to varying distances between the transmitter and receiver. The distance between the camera and sensor motes on breadboard can be up to 2 feet. The method also works quite well in varying lighting and brightness conditions as it deterministically learns the environment using the first two, All-OFF and All-ON, frames in each session. The method could fail in presence of background noise during transmission and reception of SAS data. Huge variations of lighting conditions during transmission of SAS data which exceed color threshold of LEDs or shaking or displacement of all sensor motes/camera while transmission of SAS data exceeding the dimension threshold of LEDs will also cause failure of the method. However, these will only lead to false positives and not to an attack. Except for the camera adjustment errors (as discussed previous section), all errors occurring with our method are localized i.e, if a single sensor mote fails due to some reason, only that particular sensor node needs to be re-initialized. Note that this is unlike the MiB scheme of [6], where any errors lead to the re-initialization/re-keying of the whole batch of sensor motes. Even when camera adjustment occur in our method, only the SAS data transmission needs be repeated, not the whole initialization process. On the other hand, MiB is less prone to user errors than our method. However, our results indicate that our user errors only lead to false positives and are negligible nevertheless. In our future work, we plan to explore how default rejection (as opposed to our current default acceptance mechanism) would impact the efficiency, usability and scalability of our method. It will clearly improve security.

6.3 Scalability

Our method can be used to initialize multiple sensor nodes per batch. We tested the method with 16 sensor nodes having three LEDs each. Compared to prior work, which only allows for initialization of two motes, this is a significant improvement.[7] By using good quality wide-angle cameras (which will somewhat increase the overall cost of the system), this number can be further improved, we believe. We are currently exploring ways to make our method more scalable. Note that increase in the number of sensor nodes will come at only a slight cost of increase in the length of SAS data. For example, to support 128 sensor nodes, we would need to transmit 22 SAS bits. However, this will make the task of detecting failed motes much harder for the administrator in case the system is under attack by a man-in-the-middle attacker.

6.4 Usability

Via a systematic usability study, we find that our method is quite user friendly. It does not require any expertise or prior training. Little or no acquaintance with the method is enough to administer the process. It is easy to work with and enables safe detection of failed sensor nodes by observing the result on the screen of the sink. Unlike the MiB scheme of [6], the administrator does not have to deal with a specialized and often cumbersome Faraday Cage. Of course, the administrator has to deal with a camera in our method; however, most users are getting more and more familiar with cameras as they become ubiquitous and our usability study (Table 1) shows that most users found the task "camera adjustment to LEDs" to be easy . Moreover, a camera can be used for purposes other than key distribution and is thus not truly specialized. Also note that the sensor motes per batch do not need to be homogeneous. They can have different number, color of LEDs, in any topology/orientation whatsoever (the only requirement being they all possess one RED colored LED to act as the Sync LED). Recall that this is unlike MiB [6], which can only support homogeneous sensor motes with very similar weights. We consider this as an important issue with respect to usability – an administrator might need to initialize a diverse pool of sensor motes and should not need to group them up.

6.5 Power Requirements

From [20], we know that most available commercial motes can do public key crypto operations using only a small amount of power. Now, we show that the SAS data transmission through blinking LEDs also incurs a minimal overhead on motes in terms of power. For 20-bit SAS data transmission, the three LEDs on each mote light-up 13 times (for a period of 250ms), i.e., for a duration of $13 \times 250 = 3.25$ seconds. Each LED has a drop voltage, $V = 2.9$ Volts (typical range 1.7-3.3 Volts); Current Rating, $I = 2.2$ mA (typical range 2-3 mA). Therefore, the maximum energy consumption per mote (3 LEDs), $E = 3 \times (V \times I \times t) = 3 \times (2.9 \times 2.2 \times 10^{-3} \times 3.25)$ Volt-A-seconds $= 0.062205$ Joules.

---

[7] Although the MiB [6] method considers multiple sensor nodes, the maximum number of nodes that can be securely initialized per batch is not clear from the experiments and results presented in [6]. We believe this number would be limited by the size of the Faraday Cage used.

As stated in [20], the Energizer No. E91, two AA batteries used in Mica2 motes, have a total energy of $2 \times (1.5 \times 2.850 \times 3600)=30780$ Joules. So, our SAS data transmission requires $\frac{0.062205}{30780} \times 100\% = 0.0002\%$ of battery life of Mica2. As shown in [20], public key generation requires 0.816 Joules of energy. Thus, our SAS data transmission is more that 13.11 times better than the public key generation in terms of power consumption

6.6 Simplicity and Economic Viability

The sink needs only a camera and each sensor node requires at least two LEDs (one Sync and one Data) which are very cheap and commonly available. In fact, most existing commercial sensor motes have three LEDs. Our method is quite economic, as opposed to MiB [6] which requires a specialized Faraday Cage and two additional motes having USB interfaces (called "keying device" and "keying beacon") along with a base station and un-initialized nodes.

6.7 Resistance to Malicious Sensor Nodes

Our method offers a natural protection against corrupted or malicious sensor nodes[8]. Our method is based on an authenticated key exchange protocol following the security model of [3]. This model guarantees that an adversary who learns session key(s) corresponding to some corrupted session(s), does not learn any information about the keys corresponding to other uncorrupted sessions. This is unlike MiB [6], where a single corrupted sensor node can compromise keys corresponding to all other sensor nodes[9].

7 Conclusion and Future Work

In this paper, we presented a novel scalable method of secure sensor node initialization. The proposed method offers (authenticated key exchange) protocol level security for key pre-distribution process using visual OOB channel. This is a promising alternative to MiB [6], the only prior work in this area, which offers physical layer security by attenuating and jamming the wireless signals. We believe that achieving physical layer security on insecure wireless channel might be a tough task and require specialized and expensive equipment. Via a thorough and systematic usability study, we showed that our method has several advantages over MiB in terms of scalability, usability, simplicity and economic viability. Our future work includes usability study of the default rejection mechanism as discussed in previous section and improvement of the scalability of the mechanism by using slightly better quality cameras.

Acknowledgements

We would like to thank Jonathan Voris for his comments on an earlier version of this paper and CANS'09 anonymous reviewers for their feedback.

---

[8] A manufacturer could possibly sneak in malicious sensor node(s) along with normal sensor nodes shipped to a customer, as pointed out in [6].

[9] [6] suggests using a software-based attestation technique [32] to prevent this attack.

References

1. Mica2 specifications, http://www.xbow.com/Products/Product_pdf_files/Wireless_pdf/MICA2_Datasheet.pdf
2. Balfanz, D., Smetters, D., Stewart, P., Wong, H.C.: Talking to strangers: Authentication in ad-hoc wireless networks. In: Network & Distributed System Security (NDSS) (2002)
3. Canetti, R., Krawczyk, H.: Analysis of key-exchange protocols and their use for building secure channels. In: Pfitzmann, B. (ed.) EUROCRYPT 2001. LNCS, vol. 2045, p. 453. Springer, Heidelberg (2001)
4. Castelluccia, C., Mutaf, P.: Shake them up!: a movement-based pairing protocol for cpu-constrained devices. In: International Conference on Mobile Systems, Applications, and Services (MobiSys) (2005)
5. Chan, H., Perrig, A., Song, D.X.: Random key predistribution schemes for sensor networks. IEEE Security & Privacy (2003)
6. Cynthia, K., Luk, M., Negi, R., Perrig, A.: Message-in-a-bottle: User-friendly and secure key deployment for sensor nodes. In: ACM Conference on Embedded Networked Sensor Systems (SenSys) (2007)
7. Du, W., Deng, J., Han, Y., Chen, S., Varshney, P.: A key management scheme for wireless sensor networks using deployment knowledge. In: IEEE INFOCOM 2004 (March 2004)
8. Du, W., Deng, J., Han, Y.S., Varshney, P.K.: A pairwise key pre-distribution scheme for wireless sensor networks. In: ACM Computer and Communications Security, CCS (2003)
9. Eschenauer, L., Gligor, V.D.: A key-management scheme for distributed sensor networks. In: ACM Computer and Communications Security (CCS) (2002)
10. Faulkner, L.: Beyond the five-user assumption: Benefits of increased sample sizes in usability testing. Behavior Research Methods, Instruments, & Computers 35(3), 379–383 (2003)
11. Giorgetti, G., Manes, G., Lewis, J.H., Mastroianni, S.T., Gupta, S.K.S.: The personal sensor network: a user-centric monitoring solution. In: BodyNets 2007: Proceedings of the ICST 2nd international conference on Body area networks (2007)
12. Goodrich, M.T., Sirivianos, M., Solis, J., Tsudik, G., Uzun, E.: Loud and Clear: Human-Verifiable Authentication Based on Audio. In: International Conference on Distributed Computing Systems (ICDCS) (2006)
13. Holmquist, L.E., Mattern, F., Schiele, B., Alahuhta, P., Beigl, M., Gellersen, H.-W.: Smart-Its Friends: A Technique for Users to Easily Establish Connections between Smart Artefacts. In: Abowd, G.D., Brumitt, B., Shafer, S. (eds.) UbiComp 2001. LNCS, vol. 2201, p. 116. Springer, Heidelberg (2001)
14. Hu, F., Sharma, N.: Security considerations in ad hoc sensor networks. Ad Hoc Networks 3 (2005)
15. Karlof, C., Sastry, N., Wagner, D.: Tinysec: a link layer security architecture for wireless sensor networks. In: ACM Conference on Embedded Networked Sensor Systems (SenSys) (2004)
16. Laur, S., Asokan, N., Nyberg, K.: Efficient mutual data authentication based on short authenticated strings. In: Pointcheval, D., Mu, Y., Chen, K. (eds.) CANS 2006. LNCS, vol. 4301, pp. 90–107. Springer, Heidelberg (2006)
17. Lester, J., Hannaford, B., Borriello, G.: Are You with Me? - Using Accelerometers to Determine If Two Devices Are Carried by the Same Person. In: International Conference on Pervasive Computing (Pervasive) (2004)
18. Liu, A., Ning, P.: Tinyecc: A configurable library for elliptic curve cryptography in wireless sensor networks. In: Information Processing in Sensor Networks (IPSN) (2008)
19. Liu, D., Ning, P.: Establishing pairwise keys in distributed sensor networks. In: ACM Computer and Communications Security (CCS) (2003)
20. Malan, D.J., Welsh, M., Smith, M.D.: A public-key infrastructure for key distribution in tinyos based on elliptic curve cryptography. In: IEEE Communications Society Conference on Sensor, Mesh and Ad Hoc Communications and Networks (SECON) (2004)

21. McCune, J.M., Perrig, A., Reiter, M.K.: Seeing-is-believing: Using camera phones for human-verifiable authentication. IEEE Security & Privacy (2005)
22. Milenkovic, A., Otto, C., Jovanov, E.: Wireless sensor networks for personal health monitoring: Issues and an implementation. Computer Communications 29(13-14), 2521–2533 (2006)
23. Pasini, S., Vaudenay, S.: SAS-based authenticated key agreement. In: Yung, M., Dodis, Y., Kiayias, A., Malkin, T.G. (eds.) PKC 2006. LNCS, vol. 3958, pp. 395–409. Springer, Heidelberg (2006)
24. Perrig, A., Stankovic, J., Wagner, D.: Security in wireless sensor networks. Communications of the ACM 47, 53–57 (2004)
25. Perrig, A., Szewczyk, R., Wen, V., Culler, D.E., Tygar, J.D.: Spins: security protocols for sensor netowrks. In: ACM Annual International Conference on Mobile Computing and Networking (MOBICOM) (2001)
26. Prasad, R., Saxena, N.: Efficient device pairing using human-comparable audiovisual patterns. In: Bellovin, S.M., Gennaro, R., Keromytis, A.D., Yung, M. (eds.) ACNS 2008. LNCS, vol. 5037, pp. 328–345. Springer, Heidelberg (2008)
27. Rasmussen, K.B., Capkun, S.: Implications of radio fingerprinting on the security of sensor networks. In: International Conference on Security and Privacy in Communication Networks (SecureComm) (2007)
28. Rowley, H.A., Baluja, S., Kanade, T.: Neural network-based face detection. In: Pattern Analysis and Machine Intelligence (PAMI) (1998)
29. Saxena, N., Ekberg, J.-E., Kostiainen, K., Asokan, N.: Secure device pairing based on a visual channel. IEEE Security & Privacy, short paper (2006)
30. Saxena, N., Uddin, B.: Automated device pairing for asymmetric pairing scenarios. In: Chen, L., Ryan, M.D., Wang, G. (eds.) ICICS 2008. LNCS, vol. 5308, pp. 311–327. Springer, Heidelberg (2008)
31. Schneiderman, H., Kanade, T.: A statistical method for 3d object detection applied to faces and cars. In: IEEE Conference on Computer Vision and Pattern Recognition (June 2000)
32. Seshadri, A., Perrig, A., van Doorn, L., Khosla, P.K.: Swatt: Software-based attestation for embedded devices. IEEE Security & Privacy (2004)
33. Soriente, C., Tsudik, G., Uzun, E.: BEDA: Button-Enabled Device Association. In: International Workshop on Security for Spontaneous Interaction (IWSSI) (2007)
34. Soriente, C., Tsudik, G., Uzun, E.: HAPADEP: Human-assisted pure audio device pairing. In: Wu, T.-C., Lei, C.-L., Rijmen, V., Lee, D.-T. (eds.) ISC 2008. LNCS, vol. 5222, pp. 385–400. Springer, Heidelberg (2008)
35. Stajano, F., Anderson, R.J.: The resurrecting duckling: Security issues for ad-hoc wireless networks. In: Security Protocols Workshop (1999)
36. Tatbul, N., Buller, M., Hoyt, R., Mullen, S., Zdonik, S.: Confidence-based data management for personal area sensor networks. In: DMSN 2004:1st international workshop on Data management for sensor networks. ACM, New York (2004)
37. Uzun, E., Karvonen, K., Asokan, N.: Usability analysis of secure pairing methods. In: Dietrich, S., Dhamija, R. (eds.) FC 2007 and USEC 2007. LNCS, vol. 4886, pp. 307–324. Springer, Heidelberg (2007)
38. Vaudenay, S.: Secure communications over insecure channels based on short authenticated strings. In: Shoup, V. (ed.) CRYPTO 2005. LNCS, vol. 3621, pp. 309–326. Springer, Heidelberg (2005)
39. Viola, P., Jones, M.: Rapid object detection using a boosted cascade of simple features. In: IEEE Conference on Computer Vision and Pattern Recognition (2001)
40. Weszka, J.: A survey of threshold selection techniques. Computer Graphics and Image Processing 7 (1978)

DepenDNS: Dependable Mechanism against DNS Cache Poisoning

Hung-Min Sun, Wen-Hsuan Chang, Shih-Ying Chang, and Yue-Hsun Lin

Information Security Laboratory, Department of Computer Science,
National Tsing Hua University, Taiwan R.O.C
hmsun@cs.nthu.edu.tw,
{eifie,godspeed,tenma}@is.cs.nthu.edu.tw

Abstract. DNS cache poisoning attacks have been proposed for a long time. In 2008, Kaminsky enhanced the attacks to be powerful based on nonce query method. By leveraging Kaminsky's attack, phishing becomes large-scale since victims are hard to detect attacks. Hence, DNS cache poisoning is a serious threat in the current DNS infrastructure. In this paper, we propose a countermeasure, DepenDNS, to prevent from cache poisoning attacks. DepenDNS queries multiple resolvers concurrently to verify an trustworthy answer while users perform payment transactions, e.g., auction, banking. Without modifying any resolver or authority server, DepenDNS is conveniently deployed on client side. In the end of paper, we conduct several experiments on DepenDNS to show its efficiency. We believe DepenDNS is a comprehensive solution against cache poisoning attacks.

1 Introduction

Domain Name System (DNS) provides name resolutions between memorable domain names and machine-friendly IP addresses. Almost all network applications have to resolve given domain names to the corresponding IP addresses, such as http, ftp, email, etc. However, design of DNS is prone to suffering attacks [1,2]. The most critical one is DNS cache poisoning attacks. Main concept of DNS cache poisoning is to alter cache records of a DNS resolver. While user machines query the DNS resolver, they would obtain poisoned mapping information and connect to the forged IP addresses.

Nowadays, phishing is a huge threat on the Internet [3]. Traditionally, attackers often exploit the similar domain name to cheat the victims to access the faked websites. For instance, attackers can impersonate legal banks to despatch e-mails. While victims receive the mails, the mails ask them to provide private information on a phished website. When the victims login to conduct transaction, their bank accounts or authorized codes of credit cards are going to be stolen. The above attack is called *URL obfuscation*. However, *URL obfuscation* is hard to accomplish since those mails may be filtered by the spam-mail engine. Or users may observe this attack since they pay more attention. Compared to *URL obfuscation*, phishing becomes more practical and large-scale by leveraging DNS

J.A. Garay, A. Miyaji, and A. Otsuka (Eds.): CANS 2009, LNCS 5888, pp. 174–188, 2009.

cache poisoning attacks. An attacker does not need to deliver faked url links within spam mails. Instead, he only alters mapped IP addresses of banks cached in DNS resolvers. Moreover, the scale of phishing becomes larger since users who query to the poisoned DNS resolvers would be phished.

Numerous researchers have proposed solutions to prevent DNS cache poisoning. Popular countermeasures are based on cryptography, e.g., DNSSEC [4], TKEY [5], TSIG [6], or based on SSL [7]. They provide integrity of DNS message through cryptographic functions. However, these approaches require a major overhaul to the current DNS infrastructure. For instance, DNSSEC uses public key cryptography to authenticate communications. It requires a public key infrastructure (PKI) to distribute public keys. Another approaches use traffic analysis or construct a model to detect DNS cache poisoning attacks [8,9,10,11]. These approaches do not work well due to high false positive.

On the other hand, several solutions against cache poisoning attacks were implemented on DNS resolvers without cryptography. DoX [12] adopts the peer-to-peer system(P2P) to enhance DNS security, the cache updating is based on the trust between resolvers. In 2008, Dagon et al. proposed DNS-0x20 [13] to raise the low bound of DNS cache poisoning attacks. They utilized case encoding of the queries to increase the attacked complexity. However, these two approaches still need to modify DNS resolvers.

There is a proposed approach implemented on DNS clients, ConfiDNS [14]. ConfiDNS that utilizes cooperative DNS resolver systems, CoDNS [15], provides more security against DNS cache poisoning attacks. CoDNS groups mutually-trusted nodes agreement to resolve each other's queries while the local infrastructure is failed, but its security is weak since any corruption or misbehavior of a single resolver can easily propagate throughout the system. Although ConfiDNS improves the security of CoDNS, their mechanisms are not strong against Kaminsky's DNS cache poisoning attacks.

In this paper, we introduce the proposed security mechanism, *DepenDNS*, against DNS cache poisoning attacks. DepenDNS utilizes multi-DNS resolvers lookup mechanism, e.g., obtaining a dependable answer via querying multiple resolvers. In brief, DepenDNS has the following advantages.

Practical. DepenDNS is a client program based on built-in DNS lookup utility. Without modifying any DNS resolver or back-end authority server, DepenDNS is convenient to be deployed on user machines.

Efficient. DepenDNS is a lightweight approach. The average lookup time is only 241.8ms when DepenDNS queries 20 DNS resolvers. And the storage for history usage is also tiny, 0.0739KB for a single domain name with one IP address.

Secure. Comparing with existing DNS security mechanisms, DepenDNS provides sufficient security. Since poisoning multiple DNS resolvers at the same time is difficult and infeasible, DepenDNS utilizes the IP addresses through querying multiple resolvers concurrently. Moreover, the IP addresses returned by multiple resolvers are verified by our algorithm π. Instead of returning queried IP addresses only, π chooses the trustworthy and dependable IP addresses as output.

The rest of this paper is organized as follows. Section 2 presents some related background about DNS architecture and DNS cache poisoning attacks. Section 3 depicts the details of DepenDNS. Section 4 analyzes the security of DepenDNS. Section 5 shows our experiments about availability and overhead. Finally, discussion and conclusion are described in Section 6 and Section 7.

2   Backgrounds

2.1   DNS Architecture Overview

DNS is commonly regarded as a distributed architecture in the form of an inverted hierarchical tree. Any DNS server containing a complete copy of the domain's zone file is authoritative for that domain. These DNS servers are called *authority servers*. In DNS architecture, authority servers can delegate a sub-domain to another authority server and only maintain a referral to that server. Besides, two other components typically existed in DNS architecture, DNS clients and DNS recursive resolvers. Generally, user machines are capable of DNS client programs. When a DNS client wants to resolve a domain name, the client queries the DNS resolver via lookup procedure, instead of interacting with authority servers. Another component is DNS recursive resolver, also called resolver for simple. When a client sends queries to the resolver, the resolver will execute lookups processes from root authority server to leaf authority server. Once the resolver gets the response, the resolver sends the answer to the client and also maintains this answer in its cache. Here we use a brief instance to depict how DNS works. Resolving a domain name, e.g.,*www.example.com.*, requires the following steps:

1. DNS client D_c sends the query to the DNS resolver D_r. This query is usually triggered by user applications, e.g. web browser.
2. D_r searches the corresponding domain name in its cache records. If a record is matched, the IP addresses would be returned to D_c. Otherwise, go to the next step.
3. D_r starts to hierarchically traverse down the DNS authority servers by querying sub-domain recursively until the original query is returned by a DNS authority server which is responsible for authoritatively answer. At the beginning, the root authority server (dot server, ".") points out a downward delegation of the *com.* to other DNS authority servers. Similarly, the *com.* authority server delegates to *example.com.* authority server. Then D_r obtains a response for *www.example.com.* which is returned by *example.com.* authority server.
4. The answer is returned to D_c, and cached by D_r to assist in further resolutions. Note that each answer of DNS authority servers would be cached until its TTL values expires. In the example, D_r caches the following IP addresses.
   (a) IP addresses of DNS authority servers which are responsible for *com.*
   (b) IP addresses of DNS authority servers which are responsible for *example.com.*
   (c) IP addresses of *www.example.com.*

2.2   DNS Cache Poisoning Attacks

DNS cache poisoning attacks are tampering with the cache records stored in resolvers. In Section 2.1, the resolver caches IP addresses to facilitate further queries. Once other clients query the same domain names, the resolver would directly return the IP addresses from cache records. If attackers could tamper with the cached records of resolvers, clients would receive malicious IP addresses and be phished to the malicious websites.

The poisoning attacks could be achieved as follows. In Step 3 of the lookup procedure in Section 2.1, resolvers starts recursively querying with the backward authority servers when there is no matching records. Then authority servers would return appropriate answer to the resolver. Concurrently, the attacker could generates counterfeit packets and forwards them to the resolver before the legal packets of authority server reaches.

To defend against such attacks, researchers proposed two well-known counter-measures, transaction ID authentication and port randomization. In the first approach, transaction ID is to authenticate connections between authority servers and resolvers. The attacker should forge legal answer packets with matched 16-bit ID. However, the probability of guessing the matched transaction ID could be much higher based on weaknesses in the random number generators and birthday attacks [13,12]; even if there are $2^{16} = 65536$ possibilities, it is feasible to guess the ID value successfully. The second approach is port randomization which also increases the computation overhead on DNS cache poisoning. In port randomization, DNS resolvers would randomly choose a 16-bit source port number in the query packet. Once it receives the answer packet from authority server, this field is also adopted for authentication. However, several source ports could not be utilized, such as the well-known ports, e.g., port numbers less than 1024. Leveraging these two approaches concurrently, an attacker only guess about 2^{30} to 2^{32} combinations. Nevertheless, not all DNS resolvers are capable of port randomization, especially the DNS resolvers in embedded systems [13]. Thus, existing mechanisms are not secure enough for current DNS architecture.

2.3   Dan Kaminsky's DNS Cache Poisoning

The primitive DNS cache poisoning attack spends a large amount of time for waiting for TTL value to expire. The attacker will have to wait for TTL value to expire when the targeted DNS resolvers have already the specific records (the records are for some domain names which were targeted by the attacker).

In 2008, Dan Kaminsky substantially reduces the attack time of DNS cache poisoning attacks. The skill is called nonce query. The attacker queries a series of nonce queries to a DNS resolver. Each query is with a different random prefix and contains additional records with genuine owner domain names but malicious IP addresses. If the DNS poisoning attack fails to match the correct transaction ID, a new nonce query with a distinct prefix is generated. Because each round of the attacks has the different prefix of domain name, the DNS resolver will consult with the backward DNS authority servers each time, no need to wait for

Query Packet:
    Query Section
    - Domain Name: <nonce>.example.com
Spoof Response Packet:
    Query Section
    - Domain Name: <nonce>.example.com
    Answer Section
    - Resource Data: arbitrary IP addresses
    Authority Section
    - Resource Data: example.com
    Additional Section
    - Additional Data: 123.456.78.90

Fig. 1. Example of Dan Kaminsky's DNS Cache Poisoning

TTL value expiration. Here we use a brief instance to depict how Kaminsky's DNS cache poisoning works (See Fig. 1).

At the beginning, an attacker sends a nonce query of victim domain name (e.g., *example.com*) to the target resolver. The prefix of a nonce query is randomly generated, for domain *www.example.com*, could be *abc*. Hence, the nonce query is *abc.example.com*. Because the records of this nonce query are always not exist in target resolver, the resolver will consult with the backward DNS authority servers every time. Concurrently, the attacker sends a large number of spoofed response packets to the target resolver. These spoofed response packets include malicious informations in authority section and addition section. The functionality of authority section and addition section are used to update corresponding records in the resolvers. However, the attacker utilizes these two sections to tamper with the records in resolvers. Kaminsky substantially reduces the attack time from weeks to seconds. An attacker could succeed to launch attacks within about 6 seconds on most networks [13].

2.4   Attack Model

We assume that attackers can generate any packet forged and forward the packet to any resolver targeted. The goal of the attackers is to tamper with the cache records in resolvers and direct clients to malicious websites.

The patterns of such attacks are similar even in Dan Kaminsky's DNS cache poisoning attacks. The attackers exhaustively generate and forward the forged answer packets to the resolver, until a answer packet is same as that one generated from the authority server. Therefore, several countermeasures against such attacks is to increase complexity of guessing correct answer packets, e.g., random transaction ID and port randomization. In other words, if the complexity gets higher, the countermeasure is stronger against cache poisoning attacks.

Besides that, we assume that the caches of the resolvers are poisoned with independent success probability. This assumption is reasonable because the most countermeasures are randomized algorithms. The attackers hardly take advantage of the previous success attacks since the countermeasures in each query are usually with different randomness.

3 DepenDNS

3.1 Multi-DNS Resolvers Lookup

The multi-DNS resolvers lookup mechanism protects users against DNS cache poisoning attacks. Traditionally, when a user visits a site *www.example.com*, the browser sends the query to one default resolver provided by ISP. Instead of querying single resolver, DepenDNS duplicates the query and sends them to multiple resolvers according to the predefined DNS resolver list. Once DepenDNS gathers multiple responses from the resolvers, it will choose some trustworthy IP addresses according to the proposed matching algorithm π. (See Section 3.2)

DepenDNS executes the following actions when a user wants to visit a website through network applications, e.g., web browser. These actions are shown in Fig. 2.

1. The browser triggers DepenDNS to handle the domain name x of website.
2. DepenDNS sends the duplicated queries of x to multiple resolvers from the resolver list. Without loss of generality, we assume the number of resolvers is t.
3. DepenDNS looks up the history data H_x for x. If a matching record is found, this means there was the same query in the past.
4. Let $R = R_1 \cup R_2 \cup ... \cup R_t$ where R_i is the set of IP addresses returned from i^{th} resolver. Given R and H_x as inputs, run algorithm π.
5. According to algorithm π, DepenDNS chooses a set of trustworthy IP addresses $A \subseteq R$.
6. Based on A, the browser will connect to the trustworthy IP addresses.
7. Store A into history data for further utilization.

The successful probability of poisoning multiple resolvers is quite lower than that of poisoning one resolver in a time period. The more resolvers we query, the higher accuracy of queried IP addresses we obtain. The detail analysis is presented in Section 4.

3.2 Matching Algorithm π

If we only consider the countermeasure against DNS poisoning attacks, choosing the most appeared IP addresses is intuitive. However, this approach may destroy the load-balance mechanisms since users all connect to the most appeared IP address [16,17,18]. Instead, algorithm π picks up a set of IP addresses A which are dependable and trustworthy.

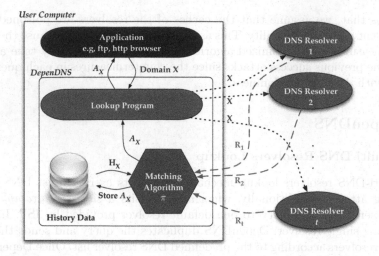

Fig. 2. The Proposed Multi-DNS Resolvers Lookup Scheme

To design algorithm π, we observe some properties of IP addresses returned from multi-DNS resolvers according to the experiments (See Section 5). We exploit these properties to select IP addresses. First, we define n_j^i, n^i, and c^k as follows.

$$n_j^i = \begin{cases} 1, & \text{if IP}_i \in R_j. \\ 0, & \text{otherwise.} \end{cases}, \text{ and } n^i = \sum_{j=1}^{t} n_j^i.$$

$$c^k = \sum n^i \text{ , where IP}_i \text{ exists in the } k^{th} \text{ class B IP address.}$$

According to the above definitions, n^i counts the number of IP$_i$ appearing in the responses from all resolvers, and c^k counts the number of IP addresses in k^{th} class B. Furthermore, let $n^{max} = Max(n^1, n^2, ..., n^{|R|})$, where $Max(.)$ outputs the max value of a set and $|R|$ denotes the size of set R. Next, we define the following three policies according to the properties. Each unique IP address is individually verified, and the following shows the verification of IP$_i$.

- α: Because each domain name has different number of IP addresses, and the associated authority servers decide how many number of IP addresses is returned by the resolvers in each query. The authority servers not always return all IP addresses but instead they return a subset of all IP addresses according to their load-balance mechanisms. This means that the responses from multiple DNS resolvers will be dispersed to each IP$_i$. According to our observation, when we query to multiple resolvers, each n^i will be quite closed if the load-balance mechanism is to averagely allocate traffic to each IP$_i$ (such as Round-Robin DNS [17]). The following is the evaluation of α, where 20% is the tolerance of n^i. We can decrease the tolerance to increase security.

$$\alpha = \begin{cases} 1, \text{ if } n^i \geq n^{max}(1 - 20\%). \\ 0, \text{ otherwise.} \end{cases}$$

- β: According to our experiments, almost all domain names did not change their IP addresses at all in one week. It implies that the IP_i is trustworthy if this IP_i exists in history data at the same domain name. The following is the evaluation of β:

$$\beta = \begin{cases} 1, \text{ if } IP_i \text{ exists in history data at the same domain name.} \\ 0, \text{ otherwise.} \end{cases}$$

- γ: A domain name may have several IP addresses but these IP addresses usually belong to some same class B, which means the first 2-octet prefix of the IP addresses are the same. Although the returned IP addresses will not be all the same in each query, the proportion of specific class B IP addresses will be almost the same in each query (See the example in Table 2). The following is the evaluation of γ, where 10% is the tolerance of c^k. We can decrease the tolerance to increase security. $c^k_{current}$ and $c^k_{history}$ represent the c^k of current query and history data, respectively.

$$\gamma = \begin{cases} 1, \text{ if } IP_i \text{ belongs to } k^{th} \text{ class B and} \\ \quad -10\% \leq c^k_{current} - c^k_{history} \leq 10\% \\ 0, \text{ otherwise.} \end{cases}$$

Next, we define the *regions* of IP addresses, denoted by N, as follows.

- N: As we mentioned above, the responses from multiple DNS resolvers will be dispersed to each IP address, and we use N to represent the dispersion strength. For example, if a domain name totally has three IP addresses (IP_1, IP_2, IP_3) and returns only one IP address to each resolver. We say that this domain name has three regions, $N = 3$. The following is the calculating of N:

$$N = \frac{|R_\alpha \cup R_\beta \cup R_\gamma|}{Mode(|R_1|, |R_2|, ..., |R_t|)}$$

In the denominator, $Mode(.)$ outputs the value that occurs the most frequently in the set. We use $Mode(.)$ to reduce the effect of that the attacker uses some poisoned R_i to affect the value of N, namely, reducing the effect of outliers.

According to the above definitions, N means the *regions* of IP addresses, and each n^i will be smaller if N is bigger. However, attacker is easier to pass the verification of α if N is bigger (n^{max} is smaller). Therefore, we decrease the weight of α and instead increase the weight of β and γ when N increases. In this way, besides passing verification of α, the attacker should pass β, γ or both to pass the verification of π.

Next, we calculate a grade G for each IP_i according to the three policies. The grade represents the strength of reliability about IP_i. G_α and $G_{\beta\gamma}$ represent the weights of α and $\beta\gamma$ where $G_\alpha + G_{\beta\gamma} = 100$. We set G_α and $G_{\beta\gamma}$ to 60 and 40 respectively in our experiment.

$$G = \alpha * (G_\alpha - (N-1) * 10)\% + \tfrac{1}{2} * (\beta + \gamma) * (G_{\beta\gamma} + (N-1) * 10)\%$$

Finally, we select the IP addresses whose grades are greater than or equal to 60, store these IP addresses in set A and return A to browser. We set $G_\alpha = 60$ because we think a IP address is undoubtedly trustworthy if most resolvers return this IP address. More specifically, when $N = 1$, passing verification of α is enough. According to our experiments (See Section 5), the grades of more than 94.67% IP addresses are over 60 in normal cases. Although some IP addresses of a website fail the verification, the experiments show that at least one IP address would pass the verification. In other words, the IP addresses which fail the verification would not result in unreachable websites. Furthermore, it is extremely difficult to obtain a high grade of malicious IP address in DNS cache poisoning attacks (See Section 4).

3.3 History Data Records

Because DepenDNS utilizes the history data in algorithm π, the reliability of history data is an important issue. If a name-to-IP mapping has been the same for an extended period of time, the history data may be reliable. On the contrary, a frequently changed name-to-IP mapping may indicate that a server frequently migrated to other IP addresses and this causes the history data unreliable. In previous study [19,14], more than 85% of domain names did not change their IP addresses in one month and only 2% of domain names change IP addresses more than once per week. Therefore, it is very likely that the attack occurs if some IP addresses which does not exist in 7-day history data are returned.

4 Security Analysis

In this section, we analyze the probability of passing α, β and γ verification according to the attack model.

The probability of passing α verification (P^α). We assume that there are t selected DNS resolvers $d_1, d_2, ..., d_t$, whose probabilities of poisoning their caches are $p_1, p_2, ..., p_t$, respectively and $\bar{p}_i = 1 - p_i$. Let $J(d_1 d_2 \bar{d}_3 ... \bar{d}_t)$ denotes that the probability of poisoning exact d_1 and d_2 in a time period. Let E_i denotes the probability of poisoning any i of t resolvers, e.g., $E_0 = J(\bar{d}_1 \bar{d}_2 \bar{d}_3 ... \bar{d}_t)$ and $E_1 = J(d_1 \bar{d}_2 \bar{d}_3 ... \bar{d}_t) + J(\bar{d}_1 d_2 \bar{d}_3 ... \bar{d}_t) + ... + J(\bar{d}_1 \bar{d}_2 ... d_{t-1} d_t)$. Let P_i denotes the probability of poisoning at least i of t resolvers.

$$P_i = 1 - \sum_{j=0}^{i-1} E_j. \tag{1}$$

Recall that for a IP_i, α is set to 1 while n^i is greater than $n^{max}(1-20\%)$. The best case, the least number of resolvers the attacker should poison, occurs while $n^{max} = n^j$ and $n^i = 1, i \neq j$. In this best case, the attacker should poison at least x resolvers to pass α where $x \geq (n^{max} - x)(1 - 20\%)$. Consequently, we can infer $P^\alpha \leq P_x$ where $x \geq \frac{4}{9}n^{max}$.

To simplifying estimation, we assume that these probabilities $p_1, p_2, ..., p_t$ are equivalent and independent. Namely, we assume $p_1 = p_2 = ... = p_t = p$ and $J(d_1 d_2...d_{t-1}\bar{d}_t) = p_1 p_2...p_{t-1}\bar{p}_t = p^{t-1}\bar{p}$. (1) is simplified to (2).

$$P_i = 1 - \sum_{j=0}^{i-1} \binom{t}{j} p^j \bar{p}^{t-j}. \tag{2}$$

In most cases, N is fixed and determines n^{max}. If we assume that $t = 20$ and $N = 1, 2, 3, 4$, n^{max} is about $20, 10, 7, 5$, respectively and the numbers of resolvers the attacker should poison are $9, 5, 4, 3$ in the best cases, respectively. If we assume that $p = 2^{-16}$, the success probabilities are $2^{-126.64}$, $2^{-66.080}$, $2^{-51.758}$, $2^{-37.845}$, respectively. For $p = 2^{-32}$, the success probabilities are $2^{-270.64}$, $2^{-146.08}$, $2^{-115.76}$, $2^{-85.844}$, respectively. Since all probabilities are quite smaller than that of poisoning single resolver, α in DepenDNS is significant for security if n^{max} is in a reasonable range.

The probability of affecting N. N determines the weights of α, β and γ in grade G. It threatens the security of the proposed scheme that attackers can easily adjust the value of N. More specifically, if the attackers set bigger N, they have higher probability to pass α verification. Therefore, N should be hardly affected. According to the calculation of N, the probability of affecting numerator is $P^{\alpha \vee \beta \vee \gamma}$ and that of affecting denominator is usually close to $P_{\frac{t}{2}}$ because the values of $|R_i|, i = 1, 2, ..., t$ are usually all the same according to our experiments. In short, affecting N is infeasible.

The probability of passing β verification (P^β). P^β is equal to P^π since all IP addresses in history data should have ever passed the verification of DepenDNS.

The probability of passing γ verification (P^γ). P^γ refers to the hardness of applying a specific class B IP address. In practice, to apply such a IP address is not trivial because the application for specific class B IP address needs some licenses. In addition, even if the attacker gets the same class B IP address with legal owner's class B IP address, she still needs to let the value of $c^k_{current}$ close to $c^k_{history}$. Therefore, the attacker is difficult to pass the verification of γ.

The probability of passing DepenDNS (P^π). For an attacker, P^π is equal to the probability of obtaining a grade which is equal or greater than 60. Since the malicious IP address would not be in history data before and N, which is hard to affect, is less than 6, the IP address should pass α and γ. Therefore, $P^\pi = P^\alpha P^\gamma$. According to the above analysis, P^π would be quite smaller than that of poisoning a single resolver. This demonstrates that DepenDNS is more secure than other conventional mechanisms whose security depends on a single resolver.

5   Experiment Conduction

The history data was collected in 7 days - from 26th Mar, 2009 to 2nd Apr, 2009, and we query 20 resolvers for different domain names once per half an hour in this week, so each site has 336 lookup records in history data. After that, we use DepenDNS to query the same sites 200 times.

5.1   Availability

We assume that $\bar{\alpha}$ denotes the cases of failing the verification of α, $\alpha \wedge \beta$ denotes the cases of passing both verifications of α and β, and $\beta|\alpha$ denotes the cases of passing the verification of β in the condition of passing the verification α already.

Table 1 shows the percentage of IP addresses matching our policies α, β and γ in the varied conditions. In the result, most of the IP addresses matches the policies so the policies are consistent with the way that the authority servers handle the domain name queries. More specifically, α refers to that the responses from multiple resolvers are dispersed to each IP_i. β refers to that the authority servers rarely change the IP addresses in one week. γ refers to that the authority servers usually use the IP addresses in some same class B and the proportions of different class B IP addresses are usually fixed each time.

In addition, none of α, β and γ is removable because each of them has its importance. We explain some special cases as follows. $\beta \wedge \gamma|\bar{\alpha}$ represents that even if some legal IP_i with small n^i fail the verification of α, they still probably pass the verification of DepenDNS. As shown in the example in Table 2, when N is big ($N = 6$), some legal IP addresses are hard to pass the verification of α but these addresses can get sufficient grades from the verification of β and γ. $\alpha \wedge \gamma|\bar{\beta}$ represents the cases of new IP addresses (maybe the domain name adopts dynamic IP addresses). We notice that over 80% new IP addresses are accepted according to the successful verification of α and γ. Furthermore, the

Table 1. The percentage of IP addresses passing DepenDNS verification ($G \geq 60$). " − " denotes non-existed condition and "x" denotes the cases of $G < 60$.

Policy	$N = 1$	$1 < N \leq 2$	$2 < N \leq 3$	$3 < N$
α	100%	82.98%	72.22%	84.62%
β	98.78%	100%	100%	96.24%
γ	98.63%	95.74%	88.89%	82.69%
$\alpha|\bar{\beta} \wedge \bar{\gamma}$	100%	x	x	x
$\beta \wedge \gamma|\bar{\alpha}$	-	87.5%	80.0%	50%
$\alpha \wedge \gamma|\bar{\beta}$	92.33%	-	-	83.35%
$\alpha \wedge \beta|\bar{\gamma}$	98.91%	50%	50%	55.56%
$\alpha \wedge \beta \wedge \gamma$	97.63%	80.85%	66.67%	78.85%

Table 2. Example: Part of results about *www.live.com*. This domain name totally has more than 10 IP addresses and each resolver returns just 2 IP addresses in each query, so $|R| \geq 10$ and $Mode\ (|R_1|, |R_2|, ..., |R_t|) = 2$.

History Data		Current Data				
IP address	Percentage	Times	Percentage	α	β	γ
1th class B: 203.133.x.x	23.0%	4	23.4%	-	-	-
IP_1: 203.133.9.8	11.5%	2	11.7%	0	1	1
IP_2: 203.133.9.17	11.5%	2	11.7%	0	1	1
2th class B: 61.30.x.x	23.1%	4	23.4%	-	-	-
IP_3: 61.30.236.135	11.5%	2	11.7%	0	1	1
IP_4: 61.30.236.136	6.7%	2	11.7%	0	1	1
IP_5: 61.30.236.134	4.9%	-	-	-	-	-
The *region* is 6, $N = 6$ and $t = 20$.						

percentages of some cases are 50% ∼ 70%, but the number of IP addresses in these cases is rare (less than 5% IP addresses).

Table 3 shows average grades. The grades of 94.67% IP addresses are greater than or equal to 60. This means that only 5.33% IP addresses are not selected, so we almost retain the load-balance mechanism. Note that the unselected 5.33% IP addresses would not make some websites unreachable since at least one IP address verified would be returned by DepenDNS according to our experiments.

5.2 Overhead Analysis

To analyze overhead, we examined the multi-DNS lookup time. As shown in Fig. 3, the lookup time slightly increases when the number of resolvers increases. The average lookup time is 146.4ms for one resolver and 241.8ms for 20 resolvers. Obviously, the time overhead increases slightly. Moreover, for history data, the average storage for a domain name with one IP address is 0.0739KB, and is only

Table 3. The grades versus different regions. The proportion of IP address in $N \leq 1$, $1 < N \leq 2$, $2 < N \leq 3$ and $N > 3$ are 50.0%, 13.91%, 5.33% and 30.76%, respectively.

Region	Resolver Number = 20		
Region N	Grade (The value of G)		
	100	60 ∼ 99	0 ∼ 59
$N \leq 1$	97.63%	2.37%	0%
$1 < N \leq 2$	80.85%	8.51%	10.64%
$2 < N \leq 3$	66.67%	27.78%	5.56%
$N > 3$	78.85%	9.62%	11.53%
Total	87.87%*	6.80%	5.33%

*87.5% = 97.63 * 50.0% + 80.85 * 13.91% + 66.67 * 5.33% + 78.85 * 30.76%.

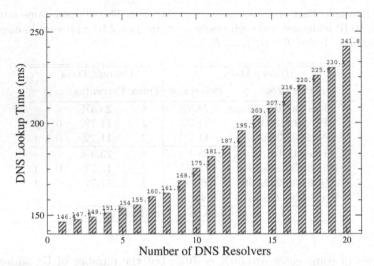

Fig. 3. DNS lookup time versus number of DNS resolvers

0.739KB for a domain name with ten IP addresses. The storage consumption is quite negligible nowadays.

6   Discussion

Cooperation with Existing Solutions. DepenDNS which runs on client side is independent of DNS resolvers and authority servers. Therefore, DepenDNS can cooperate with existing solutions implemented on DNS resolvers, e.g, DNS-0x20 [13] and DoX [12]. More specifically, we assign a weight to each DNS resolver according to the security strength of the solutions adopted. In this way, we can provide a more robust mechanism to verify IP addresses through weighting DNS resolvers.

Survivability from DoS Attacks. Since DepenDNS utilizes multi-DNS re-solvers lookup mechanism, it can keep operations even if some DNS resolvers suffered from DoS attacks. Thus, DepenDNS can decrease the impact from DoS attacks.

Accommodation to the Queries with No History Data. Because the policies β and γ are determined by history data, they can not work without history data about specific domain name. To accommodate to the initial query, we utilize a centralized server database to offer users to look up when they query a new domain name, or adjust the value of G_α and $G_{\beta\gamma}$ in the first time.

Adoption of Other Policies. Two policies we have considered but did not adopt by DepenDNS, Time to Live (TTL) and authoritative answer. First, TTL, the time of IP addresses to live in a cache, refers to how freshness the cache is. But TTL value can be altered by attacker since she could counterfeits response packets. Second, authoritative answer represents this response is directly returned from authority server. The authoritative answer is more trustworthy; however, it is quite rare from resolvers. Therefore, we consider adopting TTL and authoritative answer in DepenDNS is not significant.

Security of SSL Certificate. SSH [20] and HTTPS [21] with self-signed certificates utilize the popularity of "Trust-on-first-use" (Tofu) authentication to improve the completely insecure protocols (e.g., Telnet, HTTP). However, even if the certificates are bad or out-of-date, most users decide to ignore the warnings and continue to connect. Due to user carelessness, accepting all certificates on the initial connections threatens the security of SSL connection [22]. Based on this, "Man-in-the-Middle" (MitM) attacks would be achieved in Tofu authentication. In other words, SSL connection cannot totally protect users against phishing.

7 Conclusion and Future Work

Combining with DNS cache poisoning attacks, phishing becomes serious and large-scale for defrauding transactions. Although there were many proposed solutions for combating poisoning attacks, they are barely adopted by all resolvers for several reasons, such as expensive upgrade cost, compatibility problem and so on. Therefore, we propose a novel approach based on querying multi-DNS resolvers, called DepenDNS. DepenDNS is a lightweight client-end program without modification to the current infrastructure. The analysis and experiment results show DepenDNS has sufficient security and feasibility.

In the near future, we plan to design an extension to evaluate the confident levels of querying DNS resolvers. If we can weight the responses from different resolvers, this feature would makes DepenDNS more robust and accurate. For convenient use, we plan to implement DepenDNS as a patch for current DNS lookup capability, e.g., nslookup. We believe DepenDNS would protect users from suffering large-scale phishing.

References

1. Ariyapperuma, S., Mitchell, C.J.: Security vulnerabilities in dns and dnssec. In: ARES 2007: Proceedings of The Second International Conference on Availability, Reliability and Security, pp. 335–342. IEEE Computer Society, Los Alamitos (2007)
2. Chatzis, N.: Motivation for behaviour-based dns security: A taxonomy of dns-related internet threats. In: International Conference on Emerging Security Information, Systems, and Technologies. SecureWare 2007, pp. 36–41 (October 2007)
3. Ollmann, G.: The phishing guide. Next Generation Security Software Ltd. (2004)
4. Friedlander, A., Mankin, A., Maughan, W., Crocker, S.: DNSSEC: a protocol toward securing the internet infrastructure. Communications of the ACM 50(6), 44–50 (2007)
5. Eastlake, D.: Secret key establishment for DNS (TKEY RR). RFC 2930 (September 2000)
6. Vixie, P., Gudmundsson, O., Eastlake, D., Wellington, B.: Secret key transaction authentication for DNS (TSIG). RFC 2845 (May 2000)

7. Oppliger, R., Hauser, R., Basin, D.: SSL/TLS session-aware user authentication–Or how to effectively thwart the man-in-the-middle. Computer Communications 29(12), 2238–2246 (2006)
8. Ju, Y.W., Song, K.H., Lee, E.J., Shin, Y.T.: Cache poisoning detection method for improving security of recursive DNS. In: The 9th International Conference on Advanced Communication Technology, vol. 3, pp. 1961–1965 (2007)
9. Ren, P., Kristoff, J., Gooch, B.: Visualizing DNS traffic. In: VizSEC 2006: Proceedings of the 3rd international workshop on Visualization for computer security, pp. 23–30. ACM, New York (2006)
10. Zdrnja, B.: Security Monitoring of DNS traffic. CompSci780 project, University of Auckland (May 2006)
11. Roolvink, S.: Detecting attacks involving dns servers: a netflow data based approach (December 2008), http://essay.utwente.nl/58497/
12. Yuan, L., Kant, K., Mohapatra, P., Chuah, C.N.: Dox: A peer-to-peer antidote for DNS cache poisoning attacks. In: ICC 2006: Proceedings of the International Conference on Communications, vol. 5, pp. 2345–2350 (2006)
13. Dagon, D., Antonakakis, M., Vixie, P., Jinmei, T., Lee, W.: Increased dns forgery resistance through 0x20-bit encoding: security via leet queries. In: CCS 2008: Proceedings of the 15th ACM conference on Computer and communications security, pp. 211–222. ACM, New York (2008)
14. Poole, L., Pai, V.S.: Confidns: leveraging scale and history to improve dns security. In: WORLDS 2006: Proceedings of the 3rd conference on USENIX Workshop on Real, Large Distributed Systems, p. 3. USENIX Association, Berkeley (2006)
15. Park, K., Pai, V., Peterson, L., Wang, Z.: CoDNS: improving DNS performance and reliability via cooperative lookups. In: OSDI: Proceedings of the 6th USENIX Symposium on Operating Systems Design and Implementation, USENIX Association Berkeley, CA, USA, pp. 14–14 (2004)
16. Brisco, T.: DNS support for load balancing. RFC 1794 (April 1995)
17. Cardellini, V., Colajanni, M., Yu, P.: Dynamic load balancing on web-server systems. Internet Computing, IEEE 3(3), 28–39 (1999)
18. Hong, Y., No, J., Kim, S.: Dns-based load balancing in distributed web-server systems. In: SEUS 2006/WCCIA 2006: Proceedings of the Fourth International Workshop on Software Technologies for Future Embedded and Ubiquitous Systems and the 2006 Second International Workshop on Collaborative Computing, Integration, and Assurance, vol. 4 (April 2006)
19. Ballani, H., Francis, P.: Mitigating DNS dos attacks. In: CCS 2008: Proceedings of the 15th ACM conference on Computer and communications security, pp. 189–198. ACM, New York (2008)
20. Ylonen, T., Lonvick, C.: Rfc 4251: The secure shell (ssh) protocol architecture (January 2006), http://www.ietf.org/rfc/rfc4251.txt
21. Rescorla, E.: Http over TLS (May 2000)
22. Wendlandt, D., Andersen, D., Perrig, A.: Perspectives: improving SSH-style host authentication with multi-path probing. In: USENIX 2008 Annual Technical Conference on Annual Technical Conference table of contents, USENIX Association Berkeley, CA, USA, pp. 321–334 (2008)

Privacy-Preserving Relationship Path Discovery in Social Networks

Ghita Mezzour[1], Adrian Perrig[1], Virgil Gligor[1], and Panos Papadimitratos[2]

[1] Carnegie Mellon University, Pittsburgh, PA 15213
{mezzour,perrig,gligor}@cmu.edu
[2] EPFL, Lausanne, Switzerland
panos.papadimitratos@epfl.ch

Abstract. As social networks sites continue to proliferate and are being used for an increasing variety of purposes, the privacy risks raised by the full access of social networking sites over user data become uncomfortable. A decentralized social network would help alleviate this problem, but offering the functionalities of social networking sites is a distributed manner is a challenging problem. In this paper, we provide techniques to instantiate one of the core functionalities of social networks: discovery of paths between individuals. Our algorithm preserves the privacy of relationship information, and can operate offline during the path discovery phase. We simulate our algorithm on real social network topologies.

1 Introduction

Social Networks have proliferated over the past few years, offering numerous services that attract millions of subscribers. In fact, Social Networking Sites are among the most frequently visited Internet sites (e.g., Myspace, Facebook). Their novel functionality and the ways of personal interaction they offer fuel their tremendous success.

A fundamental feature of social networks is the *relationship graph* that connects users. This graph enables two individuals to find the relationship paths that connect them. These paths are useful to express trustworthy users: nearby people (with short relationship paths connecting them) often deserve a higher level of trust. The path discovery mechanism can be used as a building block for many social networking applications: (1) discovering a relationship path to a recruiter may boost the chances of a job applicant to get the position; vice-versa, discovering a relationship path to an applicant could help the recruiter get a more trusted judgment on the applicant; (2) relationship path discovery can provide a basis for access control mechanisms suitable for Social Networks, where users determine the authorized users based on their distance to themselves in the social network; (3) a path to a person submitting an online review can boost confidence in the review; and (4) ensuring the receiver of an email that the sender is nearby in her social network can help avoid falsely flagging the email as spam.

Although the relationship graph is at the core of the usefulness of social networks, personal relationships represent sensitive, private information that can also be misused. A primary concern is the unwelcome linkage among users. For example, two professionals employed by rival companies that have a connection may trigger suspicion. Or, connections of innovators and venture capitalists could alert the competition by giving leads to upcoming technological developments. Or, simply, a social relationship can

J.A. Garay, A. Miyaji, and A. Otsuka (Eds.): CANS 2009, LNCS 5888, pp. 189–208, 2009.

correspond to a sensitive personal real-world relationship. Of greater concern is the discovery of entire relationship paths and, in the end, of the entire graph. A significant negative consequence of this discovery is the large-scale targeting, tracking, and monitoring of multiple individuals in real life based on discovered relationship paths. Other privacy concerns can arise from graph operations; e.g., user de-anonymization through merging of relationship graphs [15].

Problem Scope. Protecting the privacy of relationship paths in a social network is a separate concern from that of protecting the privacy of (1) other content of individual users, (2) pairwise inter-user relationships, or (3) inter-user relationships based on a single intermediary contact. The stored content of a social network site can be protected using cryptographic primitives [4,10,13]. Pairwise-private relationship protection arise in practical settings such as instant messaging [12].

The protection of a user's pairwise private relationships has received extensive coverage outside social networks; e.g., "trust-negotiation" between a client and a server. In trust negotiations, a client's relationships-revealing credentials could not be disclosed to a server, and the server's relationship-revealing access policies could not be disclosed to the client. In other settings, private set-intersection protocols (viz., Section 3) can maintain a limited type relationship privacy, namely help discover one intermediate contact between two users privately.

In all these examples, enforcement of privacy protection is a matter of *local* user policies; e.g., client-server policies, pairwise-private policies. In contrast, protecting relationship paths among users in social networks requires (uniform) enforcement of *global* privacy policies because all users are affected by unauthorized privacy breaches. This is a significantly more challenging problem, whose solutions, nevertheless, present opportunities for novel use in a variety of other applications.

Decentralized Access. Any effective solution to relationship-path privacy would be based on decentralized access control mechanisms and policies. Decentralized access control is required by both privacy and robustness concerns. First, mobile users should be able to discover other mobile users with whom they have a relationship without having to connect to a centrally administered social network site, which could track their movement or which would be unreachable without network access. For example, two nearby users should still be able to discover their private relationship paths even in the absence of Internet connectivity.

Second, perhaps more important, a single point of privacy failure should be avoided. The centralized, full access that Social Networking Sites have over relationship data pose serious privacy threats. Some Social Networking Sites have permissive privacy policies that enable them to use such data for commercial purposes. Moreover, even sites with strict privacy policies can be subject to disclose private information disclosure. For example, sites can be subpoenaed to disclose user information indiscriminately, which would be ineffective whenever such information would be decentralized; i.e., subpoenas would have to be network-node selective, need-to-know based and would yield limited user information. Another risk are malicious insiders with access to private information that can disclose private information. Finally, administrator errors or security vulnerabilities can also lead to privacy disclosures, as we have frequently witnessed recently in the case of leaked databases containing credit card and social security numbers.

Design Challenges. Implementing decentralized access control to protect relationship-path privacy poses significant challenges, and obvious approaches do not address the problem. An approach where people make their relationships list visible to their neighbors is not viable because that would disclose sensitive information. Another approach would be to flood relationship path discovery messages throughout the social network, but that would incur high overhead and latency, especially if some users are offline; they would delay path discovery until they are online. Furthermore, these challenges are exacerbated during relationship-path discovery. For example, if two users want to discover private relationship paths to each other, the path-discovery protocol could not rely on all intermediate users sites to be on-line and help find relationship paths. In other words, two users should be able to discover private relationship paths to each other based only on their local communication.

Contributions. In this paper we propose a system that enables users to benefit from social networking applications without the privacy exposure associated with having all their data stored at a central site. The key contribution of this paper is thus our privacy-preserving multi-hop relationship path discovery mechanism. Our design enables new access control policies based on social relationships, new trust establishment policies, and new ways to manage communication applications (e.g., e-mail spam).

2   System Model and Problem Statement

A *relationship path* between two users u and v is a sequence of users whose pairwise relationships (i.e., friendship) connect the two users. We call the two users the *end users* of the relationship path.

The *distance* (depth) d between two users on a relationship path is the number of edges (aka hops) from one user to the other on that relationship path.

A *bridge contact* user u_i, of user u to a user v on some relationship path is the direct relationship (i.e., edge) of user u to user v on that relationship path.

A *private relationship path* from user u to user v is a tuple (u, u_i, d, v) encoding a relationship path $(u, u_i, ..., v_j, v)$ of distance d. By symmetry, (v, v_i, d, u) is the private relationship path from user v to user u of distance d. Whenever a relationship path $(u, u_i, ..., v_j, v)$ is private, neither user (u, v) can discover any intermediate users on that path to the other user beyond its respective bridge contact (u_i, v_j). Of course, there may be multiple private relationship paths from u to v with the same bridge contact and different distances, and an user may have multiple bridge contacts each on a different private relationship path to the other user.

A bridge contact u_i has different roles: (1) it can facilitate the introduction of u to v. This can be relevant, for example, in a job search scenario. (2) It helps u assign a trust value to v, which can be useful when the private relationship paths are used to enforce access control. (3) It helps u track/blacklist misbehaving users. Assume that in the email whitelisting scenario, many of u_i's friends send spam messages to u. u may no longer want to whitelist a user v to which u has a private relationship path (u, u_i, d, v).

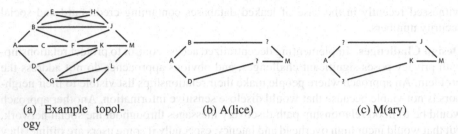

(a) Example topology   (b) A(lice)   (c) M(ary)

Fig. 1. Example topology and private relationship paths of distance $d \leq 3$ discovered by A(lice) and M(ary)

Whenever there are multiple private relationship paths (u, u_i, d_i, v), then u would assign trust value v based on the shortest private relationship path.

An *obfuscated private relationship path* $(u, u_i, d, ?)$ is a private relationship path, where the identity of one end user is unknown.

The example topology of Figure 1(a) is provided to illustrate the use of our system entities defined above. In that figure, users A and M discover private relationship paths to each other at distances $d \leq 3$. A discovers the two private relationship paths $(A, B, 3, M)$ and $(A, D, 3, M)$, where B and D are the bridge contacts and 3 is the distance of the private relationship paths. These private relationship paths captured by graph edges $A - B - ? - M$ and $A - D - ? - M$, depicted in Figure 1(b), where "?" means that user A does not know the identity of the corresponding user. User M discovers the two private relationship paths $(M, J, 3, A)$ and $(M, K, 3, A)$, corresponding to $M - J - ? - A$ and $M - K - ? - A$ depicted in Figure 1(c).

2.1 Assumptions

We assume that each user v is free to choose the users with whom she wishes to establish a relationship, but that relationships can only be established based on the consent of both users. We assume the existence of a secure and authenticated channel between each pair of friends. We assume that friends can communicate anonymously, for example using Tor. We assume that v has a public-private homomorphic key pair. We also assume that two users who want to discover private relationship paths can authenticate each other's homomorphic public keys.

In this work, we also assume the typical case where relationship paths decline in value with increasing length. Hence, users do not gain any benefits from increasing the length of their own relationship paths. This assumption is based on the observation that for all social network applications in use today, shorter paths are more desirable and take precedence over longer paths.

2.2 Desired Properties

Privacy. The central goal of our protocol is to enable two parties to discover private relationship paths to each other in an efficient manner while minimizing the additional information about relationships that users can learn. We will compare our approaches to an *ideal scheme*, which returns private relationship paths without disclosing any additional information. We say that our approach provides privacy if it does not leak any more information than the ideal scheme.

Path Integrity. The protocol needs to provide *path integrity* for the discovered paths, i.e., an adversary cannot alter discovered paths. An adversary may attempt to violate path integrity in two ways: alter the bridge contact, or shorten the discovered path.

Completeness. Our protocol should also be *complete* in the sense that each party discovers all the discoverable private relationship paths. On a discoverable private relationship path, all users on the corresponding relationship path do not object the discovery.

Offline discovery. Our protocol should enable two users to discover private relationship paths offline, i.e., based only on their physical interaction and without the help of intermediate users. Not requiring intermediate users to be online during the discovery phase is very appealing for most situations. The ability of two users to discover private relationship paths based exclusively on their local interaction, without being connected to the Internet, is particularly relevant for mobile users.

Low overhead. We aim for low communication and computation overhead.

2.3 Adversary Model

We distinguish between two kinds of adversaries: internal and network adversaries. An internal adversary sets up an account and creates relationships with users they can e.g., her friends in real life. The network adversary controls the communication channel between users, can eavesdrop, stop or inject messages. This adversary is, however, not part of the social network, and does not participate in our protocol.

The internal adversary is free to arbitrarily deviate from our protocols. The adversary may wish to alter the topology of the social network to bring more value to themselves. For example, an adversary may want to make other users see her nearby (i.e., at a small distance from them). The adversary M may also want to alter the bridge contact in a private relationship path discovered by some user u to M. v may trust one bridge contact more than other. Similarly, the adversary may try to deny value to honest users. The adversary may also want to break the privacy of the social network by discovering the relationships of other users. A misbehavior is not considered an attack if it only results in an outcome permitted in social networks as described in Section 2.1. For example, a user may not want an extremely sensitive relationship to be leaked. This user suppresses all private relationship paths corresponding to a relationship path containing that relationship.

The network adversary can observe the network traffic of users in order to learn about their relationships. It should be noted that people who use regular e-mails or encrypted e-mails are already vulnerable to having this kind of adversary learn their relationships. The network adversary can observe the source and destination of their e-mails, as well as the frequency at which they send and receive e-mails from each person. This combined information enables the network adversary to infer the friends of a user as the people with whom the user has many e-mail exchanges. People concerned about this kind of adversary usually use some form of anonymous communication. We do not discuss this kind of adversary further in the paper.

3 Background on Privacy-Preserving Cryptographic Techniques

In this section, we present background on privacy-preserving cryptographic techniques that we will build on. Private set intersection protocols [3, 8, 11] enable two or more parties that each hold a set of inputs drawn from a large domain to jointly calculate the intersection of their inputs, without leaking additional information. The private set intersection proposed by Freedman et al. [8][1] is a two-party protocol between a client C and a server S. C's input is a set of size k_C, drawn from some domain of size N; S's input is a set of size k_S drawn from the same domain. At the conclusion of the protocol, C learns which specific inputs are shared by both C and S. That is, if C inputs $X = x_1,\ldots,x_{k_C}$ and S inputs $Y = y_1,\ldots,y_{k_S}$, C learns $X \cap Y : \{x_i | \exists j, x_i = y_j\}$. The protocol is based on the presentation of sets as roots of a polynomial and on the use of homomorphic cryptosystems. The protocol follows the basic structure. C defines a polynomial P whose roots are her inputs: $P(y) = (x_1 - y)(x_2 - y)\cdots(x_{k_C} - y) = \sum_{w=0}^{k_C} \alpha_w y^w$. C sends to S homomorphic encryptions of the coefficients of this polynomial. S uses the homomorphic properties of the encryption system to evaluate the polynomial at each of her inputs. She then multiplies each result by a fresh random number n to get an intermediate result, and she adds to it an encryption of the value of her input. That is, S computes $Enc(n.P(y_j) + y_j)$ for each of her inputs $y_j \in Y$. S randomly permutes this set and returns it to C. C decrypts each element of this set. For each element in the intersection of the two parties' inputs, the result of this computation is the value of the corresponding element, whereas for all others the result is random. The computation overhead for C mainly consists of k_C homomorphic encryptions and k_S homomorphic decryptions. The k_C homomorphic encryptions only need to be computed once per input set. The computation overhead for S is mainly due to the evaluation of each of her inputs on a degree-k_C polynomial. The use of multiple-low degree polynomials and Horner's Rule make the asymptotic computation overhead $O(k_C + k_S \ln \ln k_C)$ exponentiations.

A slight variation of the above protocol is provably secure in the random oracle model against malicious adversaries, where a malicious adversary may behave arbitrarily. The security definition of private set intersection protocols is based on a comparison between an ideal and a real implementation. In an ideal implementation, a trusted third party receives the inputs of the two parties and outputs the result of the intersection, whereas in the real implementation there is no trusted third party. The security model requires that in the real implementation of the protocol, each party does not learn more information than the ideal implementation. Such a security model does not deal with attacks that apply to the ideal model. For example, a party may lie about her input set.

4 Protocol Overview

The goal of our protocol is to enable two users to discover private relationship paths to each other offline, without disclosing superfluous relationships. This requires users to store topology information that enables the discovery, without revealing relationships.

---

[1] The paper refers to the protocol both by private set intersection and private matching. We only refer to the protocol as private set intersection in order to avoid confusion with matchmaking protocols.

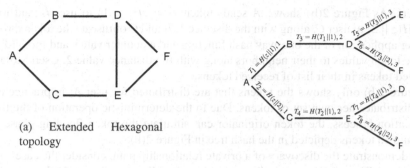

(a) Extended Hexagonal topology

(b) Tokens propagated

Fig. 2. Basic scheme. Extended hexagonal topology and obfuscated tokens propagated for originator A.

Our protocol operates in two phases: a *token flooding phase* and a *path discovery phase*. During the infrequent token flooding phase, we assume that users are online so they can exchange obfuscated topology information. The path discovery phase runs when two users want to discover private relationship paths to each other, offline based on the collected obfuscated topology information; it returns to the two users private relationship paths.

During the token flooding phase, users disseminate cryptographic tokens that become associated with obfuscated relationship paths. Each user issues tokens to her neighbors to explore relationship paths starting from herself. Other users will cryptographically obfuscate these tokens and continue the flooding process up to a pre-determined number of hops. The obfuscation operation is deterministic and independent of the identity of the user. This enables the originator to compute the value of all possible tokens at any distance. The originator will in fact simulate the flooding and compute the obfuscated token values at all distances up to the maximum flooding depth. The originator associates each computed token with an obfuscated relationship path. The path discovery phase runs when two users want to discover private relationship paths to each other. The two users perform a private set intersection protocol: one user plays the role of a client and enters the tokens she has computed as originator, and the other plays the role of a server and enters the tokens received. The first user learns the existing private relationship paths based on the common tokens. A second run of the private set intersection protocol with the roles of the two users inverted enables the second user to learn existing private relationship paths. The second private set intersection protocol does not run if v is not interested in discovering private relationship paths.

We now explain some more details based on the example in Figure 2(a). We consider the token flooding phase where node A is the originator and floods her tokens through the network. To limit the flood up to a maximum distance, each token is accompanied with the distance from the originator. Figure 2(b) depicts the hash tree that is created during the flooding process for A's tokens. To initiate the token flooding process, A picks a random number z. To prevent nodes from inferring relationship topology information during the flooding process, each token is obfuscated through a one-way hash function H and a counter value – through this approach each token represents a unique sequence of users. As Figure 2(b) shows, A sends token $T_1 = H(z \parallel 1)$ to user B, and token $T_2 = H(z \parallel 2)$ to user C, along with the distance 1. B and C obfuscate the token through another application of the one-way hash function and a counter value, and forwards the unique token values to their neighbors along with the distance value 2. Users store the received tokens in their list of received tokens.

Figure 2(b) only shows the tokens that are distributed for user A; in practice each user distributes her own set of tokens. Due to the deterministic operation of the token obfuscation process, the token originator can simulate the token flooding phase and compute all tokens depicted in the hash tree in Figure 2(b).

To demonstrate the discovery of a private relationship path, consider the case where users A and F meet. One user will use the computed list of obfuscated tokens, and the other user will use the list of all received tokens. In our example in Figure 2(b), A will compute tokens $T_1, T_2, \ldots, T_8$, and F will use all received tokens, which include tokens T_6 and T_8. After performing private set intersection of the two lists, they find a match with tokens T_6 and T_8. Since A knows that token T_6 was derived from token T_1 which was handed to user B, B is the bridge contact for that private relationship path. A further knows that T_6 is at distance 3 hops. Analogously, A knows that C is the bridge contact for the 3-hop long path represented by T_8.

5   Protocol Description

We describe next our protocol in more detail, assuming all users follow the protocol. Sections 6.1, and 6.2 discuss protocol deviations. Table 1 describes our notation.

Table 1. Notation.

o	originator
r	relay node
v, u	users in the social network
x_i	i^{th} friend of x
$\|x\|$	number of friends of user x
d	distance
d_{max}	maximum distance of private relationship paths
deg_{max}	maximum number of neighbors of any node

5.1   Basic Scheme

In this section, we first describe a basic version of our protocol and extend it in Section 5.2.

Token Flooding Phase. As we describe in Section 4, the originator floods tokens that represent obfuscated paths and that can later be used to discover private relationship paths. Algorithm 1 is used by the originator o to construct and send tokens to her friends. To start, the originator o generates a random number z and uses it as a seed to compute nodes at depth 1 of a hash tree. o computes $T_i = H(z\|i), i = 1, \ldots, |o|$ and sends $(d = 1, T_i)$ to each friend o_i.

Algorithm 1. Basic scheme. Token propagation. Originator o.

Generate a random number z
Send $(d = 1, T_i = H(z||i))$ to $o_i, i = 1, \ldots, |o|$

Algorithm 2. Basic scheme. Token propagation. Relay r.

for reception of (d, T) from r_i **do**
    Insert (r_i, d, T) into list of received tokens

    if $d < d_{max}$ **then**
        Send $(d + 1, T_j = H(T||j))$ to $r_j, j = 1, \ldots, |r|, j \neq i$
    end if
end for

Algorithm 3. Basic scheme. Hash tree construction. Originator o.

Insert$(o_i, 1, T_i = H(z||i)), i = 1, \ldots, |o|$ into hash tree
while Unmarked (o_i, d, T) **do**
    Mark (o_i, d, T)

    if $d < d_{max}$ **then**
        Insert $(o_i, d + 1, T_j = H(T||j)), j = 1, \ldots, deg_{max}$ in hash tree
    end if
end while

Algorithm 2 is used by the friends of the originator to construct and send tokens to their friends. This propagation continues hop-by-hop up to distance $d = d_{max}$ from the originator. A user performing this propagation is termed a *relay* r. When r receives (d, T) from her friend r_i, she inserts T into her list of received tokens. The relay r_i computes $T_j = H(T||j), j \in \{1, \ldots, |r|\}, j \neq i$, and sends $(d + 1, T_j)$ to each of her friends $r_j \neq r_i$. Figure 2(b) presents the hash tree spanned by A's token propagation, where A plays the role of an originator.

The originator o can reconstruct the hash tree created during the token flooding phase by users at distance $d < d_{max}$. This way, the originator has a common token with each user at distance $d \leq d_{max}$. Each of these tokens is associated with an obfuscated relationship path. The tokens that are common between the originator and users at distance $d \leq d_{max}$ constitute the basis for discovering private relationship paths during the path discovery phase (explained in detail below). Algorithm 3 is used by the originator to reconstruct all tokens from the hash tree. The originator o constructs a hash tree based on the seed z, the maximum degree deg_{max} and depth d_{max}. Each token of the hash tree is associated with an obfuscated private relationship path. A token T_i at depth 1 is a token associated with the trivial obfuscated private relationship paths $(o, o_i, 1, ?)$, as the token T_i was sent to the friend o_i. A node T at depth d with ancestor T_i is associated with the obfuscated private relationship path $(o, o_i, d, ?)$.

Path Discovery Phase. The path discovery phase runs when two users u and v want to discover private relationship paths. Users u and v run a private set intersection protocol (as described in Section 3) where u plays the role of the client entering the tokens in her hash tree and v plays the role of the server entering the tokens in her list of received tokens. The private set intersection outputs to u the tokens T_k in u's hash tree corresponding to the entries (u_k, d_k, T_k) in u's hash tree. This enables u to discover private relationship paths (u, u_k, d_k, v). A second private set intersection protocol with the roles of u and v inverted. That is, v plays the role of the client and enters the tokens in her hash tree as input, and u plays the role of the server and enters the tokens in her list of received tokens as input. This second private set intersection enables v to discover private relationship paths to u.

We consider the extended hexagonal topology, with users A and F running the path discovery phase. A and F run a private set intersection protocol where A enters the tokens in her hash tree as input, while F enters the tokens in her list of received tokens as input. A's hash tree contains tokens $T_1, \cdots, T_8$ depicted in figure 2(b). During the token flooding phase, F received tokens T_6, T_8. The private set intersection protocol outputs T_6, T_8 for A. T_6 enables A to learn $(A, B, 3, F)$. A knows that B is the bridge contact since A knows that T_6 was derived from T_1 that was handed to B. A further knows that F is at distance 3 since T_6 was derived from T_1 by applying three times a hash function to the seed z. Analogously, A discovers the private relationship path $(A, C, 3, F)$ from T_8. A second private set intersection protocol is run with the roles of A and F inverted. F learns the private relationship paths $(F, D, 3, A)$ and $(F, E, 3, A)$.

5.2   Extended Scheme

The basic scheme, unfortunately, suffers from a privacy leak: it is possible to learn whether discovered private relationship paths intersect at specific intermediate users. Assume that in the extended hexagonal topology, A runs the path discovery phase with F and subsequently with D. A discovers the private relationship paths $(A, B, 3, F)$, $(A, C, 3, F)$ based on T_6, T_8 and later $(A, B, 2, D)$, $(A, C, 3, D)$ based on T_3, T_7. Since A knows that T_3 is the ancestor of T_6 in the hash tree, A learns that D is the user at distance 2 in the private relationship path $(A, B, 3, F)$. Similarly, since T_7 and T_8 have the same ancestor, A learns that $(A, C, 3, D)$ and $(A, C, 3, F)$ have the some intermediate user at distance 2. The basic scheme also has a very large overhead since an originator computes a hash tree of degree deg_{max}. Since most users have fewer than deg_{max} friends, the basic scheme computes and stores many unnecessary tokens in the hash tree.

We present an extended scheme to address these shortcomings. The extended scheme utilizes a randomization technique to seal the privacy leak and an optimization to reduce the overhead associated with the computation of unnecessary tokens. The goal of the randomization technique is to prevent an originator from learning information about intermediate users on a private relationship path beyond what can be directly inferred from discovered private relationship paths. This holds even after the originator runs a path discovery phase with multiple users. In the extended scheme, we separate the computation of the token of a user v from the computation of tokens of intermediate users between v and the bridge contact. More specifically, for a given bridge contact, we separate the computation of tokens based on the distance to the originator. A user at distance $2 \leq d \leq d_{max}$ receives a token randomly chosen from the set of tokens computed for d.

Our basic observation towards designing the extended scheme is that since the bridge contact is disclosed with the private relationship path, the bridge contact can directly

---

Algorithm 4. Extended scheme. Token propagation. Originator o.

Generate a random number z

Send $(d = 1, T_i^1 = H(z||'1'||i))$ to o_i, $i = 1, \ldots, |o|$

---

create tokens for all subsequent users at $d \leq d_{max}$. Since in practice $d_{max} = 3$, the bridge contacts creates tokens for distances 2 and 3. The bridge contact obtains the number of tokens for distance 3 by asking her friends about their degrees and summing these degrees. Tokens for distance 3 are computed separately from the ones for distance 2. The bridge contact sends to each of her friends one token for distance 2 and a number of tokens for distance 3 proportional to the friend's degree. The tokens transmitted are randomly chosen from the computed tokens. The friend stores the token for distance 2 and to each of her friends one of the received tokens for distance 3. This approach addresses the inefficiency of the basic scheme, because each bridge contact can inform the originator of how many tokens were distributed.

We now describe the extended scheme in more detail for the case of $d_{max} = 3$, which we consider to be the largest value for d_{max} that is viable in practice. Algorithm 4 presents the algorithm used by the originator o to construct and forward tokens to her friends. o generates a random number z and uses it to create tokens for her friends as $T_i^1 = H(z||'1'||i), i = 1, \cdots, |o|$ and sends $(1, T_i^1)$ to each of her friends. These friends constitute the bridge contacts. Algorithm 5 presents the algorithm used by a bridge contact $b = o_i$ to construct and send tokens. The algorithm assumes that b previously received the value of $|b_i|$ from each of her friends. When b receives a token T from o, b computes tokens for all subsequent users at distances 2 and 3. Tokens for distance 2 are computed as $T_i^2 = H(T||'2'||i), i = 1, \cdots, p = |b| - 1$. Tokens for distance 3 are computed as $T_j^3 = H(T||'3'||j), j = 1, \cdots, q = \sum_{i=1, b_i \neq o}^{|b|}(|b_i| - 1)$. Similarly to the basic scheme, each token encodes the distance to the originator. b sends to each of her friends b_i a token for distance 2 and $|b_i| - 1$ tokens for distance 3. These tokens are randomly chosen from tokens computed for distances 2 and 3 respectively. At the end, b informs o about (p, q) the number of tokens computed for distances 2 and 3. A subsequent user on the relationship path is termed as a *relay* r. Algorithm 6 presents the algorithm used by r to forward tokens to her friends. When r receives from a bridge contact $((2, T^2), T_i^3, i = 1, \cdots, |r| - 1)$, r stores T^2. r sends to each of her friends $(3, T^3)$, where T^3 is randomly chosen from the received tokens T_i^3. Figure 5.2 presents the hash tree spanned by A's tokens propagation.

Similarly to the basic scheme, the originator computes the hash tree of tokens received in the network, where each token is associated with an obfuscated private relationship path. Algorithm 7 presents the algorithm used by the originator. The algorithm takes as input (p_i, q_i), the number of tokens computed for distances 2 and 3 by each bridge contact o_i. For each bridge contact o_i, the originator constructs a token for distance 1 as $T_i^1 = H(z||'1'||i)$, where z is the previously generated seed. The originator constructs tokens for distance 2 as $T_j^2 = H(T^1||'2'||j), j = 1, \cdots, p_i$ and associates them with obfuscated private relationship paths $(o, o_i, 2, ?)$. Similarly, tokens for distance 3 are constructed as $T_j^3 = H(T^1||'3'||j), j = 1, \cdots, q_i$ and associated with obfuscated private relationship paths $(o, o_i, 3, ?)$. The hash tree constructed by this algorithm has a different structure than the one for the basic scheme. The depth of the tree is 2. The

---

Algorithm 5. Extended scheme. Token propagation. Bridge contact b.

Require: $|b_i|$, $i = 1, \cdots, |b|$
  for reception of $(1, T)$ from o **do**
    Insert T in list of received tokens

    $p = |b| - 1$
    $T_i^2 = H(T||'2'||i)$, $i = 1, \cdots, p$

    $q = \sum_{i=1, b_i \neq o}^{|b|} (|b_i| - 1)$
    $T_j^3 = H(T||'3'||j)$, $j = 1, \cdots, q$

    Shuffle T_i^2
    Shuffle T_j^3
    Send $(2, T_i^2)$ and $(|r_i| - 1)$ values from T_j^3 to r_i
    Send (p, q) to o
  end for

Algorithm 6. Extended scheme. Token propagation. Relay r.

  for reception of $((2, T), T_i^3, i = 1, \ldots, |r| - 1)$ from bridge contact b **do**
    Insert T in list of received tokens
    Shuffle T_i^3
    Send $(3, T_i^3)$ to r_i, $i = 1, \ldots, |r|$, $r_i \neq b$
  end for

  for reception of $(3, T)$ from r_i **do**
    Insert T in list of received tokens
  end for

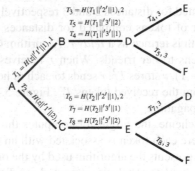

Fig. 3. Extended scheme. Obfuscated tokens propagated for originator A in the extended hexagonal topology.

root z has degree $|o|$, whereas a token for distance 1 handed to o_i has degree $p_i + q_i$. For example, T_1 has 3 children: T_3, T_4 and T_5.

The path discovery phase runs similarly to the basic scheme. A private set intersection between u and v enables u to discover common tokens and learn private relationship paths.

Algorithm 7. Extended scheme. Hash tree construction. Originator o.

Require: (p_i, q_i) number of tokens for distances 2 and 3 computed by bridge contacts o_i, $i = 1, \cdots, |o|$

    for $i = 1, \cdots, |o|$ **do**

        Insert $(o_i, 1, T^1 = H(z||'1'||i))$ in hash tree

        Insert $(o_i, 2, T_j^2 = H(T^1||'2'||j))$, $j = 1, \cdots, p_i$ in hash tree

        Insert $(o_i, 3, T_j^3 = H(T^1||'3'||j))$, $j = 1, \cdots, q_i$ in hash tree

    end for

In practice, a user does not inform others about her exact degree. Instead, the user adds to it a positive noise in order to further hide the network topology. To be consistent with the noise added, the user creates and sends dummy tokens as if she had additional friends. The amount of the noise to be added in order to appropriately hide the network topology is beyond the scope of this paper. It should be noted that the extended scheme does not require additional trustworthiness from the bridge contact. The bridge contact is free to deviate from the protocol in order to suppress private relationship paths containing her relationships, as any other intermediate user.

In our discussion, we considered $d_{max} = 3$. However, the scheme can be easily extended to a larger d_{max}.

6 Evaluation

In this section, we evaluate the extended scheme. Sections 6.1 and 6.2 provide a security and a privacy analysis. Section 6.3 analyzes the scheme overhead. Simulations based on real social network topologies are presented in Section 6.4. We consider $d_{max} = 3$.

6.1 Security Analysis

Completeness. Assume there exists a relationship path $(u, u_i, \cdots, v_j, v)$ of distance $d \leq d_{max}$. Completeness requires that when u and v run a path discovery phase, u discovers (u, u_i, d, v) and v discovers (v, v_j, d, u) in case all users on the relationship path consent to the discovery. In our protocol, a user is provided with the flexibility to suppress all private relationship paths corresponding to paths containing one of her relationships. This relationship may be extremely sensitive.

We focus on relationship paths with distance 3. The discussion can be easily adapted to relationship paths with distance 2. We consider the discovery by u, the one by v being very similar. We list sufficient conditions for u's discovery: (1) during the token flooding phase, v receives a well constructed token T originating from u and u computes T as part of her hash tree and associates it with the obfuscated private relationship path $(u, u_i, d, ?)$; and (2) during the path discovery phase, both u and v enter T to the private set intersection protocol instance where u plays the role of a client and v plays the role of a server. The second condition holds if u

The first condition holds if u sends a token associated with $d = 1$ to u_i, u_i receives an upper bound on $|v_j|$ from v_j, constructs and forwards as many tokens for $d = 3$ (one of these tokens being T) to v_j, and v_j forwards T to v. The second condition holds if u

computes all tokens encoding $d = 3$ forwarded by u_i to v_j, and associates them with the obfuscated private relationship path $(u, u_i, 3, ?)$. u can perform such computation if u_i informs her about the total number of constructed tokens encoding $d = 3$. It can be seen therefore that the first condition holds if all users on the relationship path consent to the discovery by performing the required actions. Similarly, the second condition only depends on the consent of u and v.

In order to further clarify why completeness holds in our protocol, we consider one specific protocol deviation. Assume an intermediate user on the relationship path $(u, u_i, \cdots, v_j, v)$ does not forward tokens to subsequent users. u and v will not be able to discover the corresponding private relationship paths. This, however, does not break the completeness property. In our protocol, a user is provided with the flexibility of suppressing private relationship paths corresponding to a relationship path containing one of her relationships.

Path Integrity. Path integrity requires than an adversary M cannot alter the bridge contact or shorten the discovered path. In our protocol, u discovers a private relationship path (u, u_i, d, M) when running a path discovery phase with M if: (1) M owns a token that u associates with the obfuscated private relationship path $(u, u_i, d, ?)$ and (2) both u and M enter that token to the private set intersection protocol.

When M is at distance $d > d_{max}$ from u, M does not receive any token originating from u. M is only left with the option of generating random tokens. This attack is considered of limited scope. We now consider the existence of one relationship path $(u, u_i, \cdots, M)$ of distance $d \leq d_{max}$. M receives a token T corresponding to the private relationship path (u, u_i, d, M). Additionally, M has access to tokens corresponding to longer private relationship paths with the same bridge contact because of the hop-by-hop token propagation. T cannot be used to compute tokens corresponding to shorter obfuscated paths with or with a different bridge contact, thanks to the preimage resistance property of a cryptographic hash function. When there are multiple relationship paths of distance $d \leq d_{max}$ from u to M, M receives a token originating from u for each of these relationship paths. Here also, the preimage resistance property of a cryptographic hash function prevents these tokens from being used to compute a token corresponding to a shorter obfuscated private relationship path or one with a different bridge contact.

In our protocol, M can make a user v appear at a shorter distance to another user u. Consider a relationship path (u, u_i, M, v). M can forward its own token to v, making v appear at distance 2 to u. This misbehavior causes disturbance to the system, but does not directly benefit M.

6.2   Privacy of the Network Topology

We compare our protocol to an ideal scheme, which returns private relationship paths without disclosing any additional information. we first consider an honest but curious adversary that follows our protocol, but performs all possible computation on available data. Later, we consider a malicious adversary that deviates from our protocol.

Honest but Curious Adversary. During the token flooding phase, users learn no information about the network topology in an ideal scheme. In our protocol, we distinguish between the three roles played by each user v: an originator, a bridge contact and a relay. As an originator, v learns a noisy version of the number of users at distances 2 and 3, for each of her bridge contacts. We clarify that a user having two distinct relationship paths of distance 2 or 3 to v is counted twice in the number learnt by v. v receives this information directly from the bridge contacts as described in Algorithm 5. As a bridge contact, v does not learn additional information, compared to what v already learnt as an originator. As a relay, v can infer the number of users at distances 2 and 3 through a simple count of the received tokens. v already learnt this information as an originator. v does not learn whether two received tokens have the same originator, when v did not originate any of the two tokens. This is achieved thanks to the way tokens are constructed. However, v can recognize that she is the originator of a received token. v receives one of her tokens when v is on a cycle of distance $d \leq d_{max} = 3$. Cycles of distances 1 and 2 are trivial. The ones of distance 3 are the only interesting case. These cycles enable v to learn about the existence of a friendship relationship between any of her two friends. Although, in the ideal scheme, v does not learn about the existence of these relationships during the token flooding phase, v can easily learn about them by running a path discovery phase with her friends. Similarly, v does not know whether the same intermediate user (beyond direct friends) was involved in the forwarding of two received tokens, if v was not involved in the forwarding of any of the two tokens. v can be involved in the forwarding of a token either as a relay, bridge contact or originator. We consider again the example topology in Figure 1(a). Figure 6.2 presents A(lice)'s perception of the social network topology by the end of the token flooding phase. The figure does not depict the noise added by users to their degrees.

During the path discovery phase, the two parties u and v exclusively learn existing private relationship paths, in an ideal scheme. In our protocol, u learns the common tokens between her hash tree and v's list of received tokens. The construction of these tokens and the randomization technique prevent these tokens from leaking information beyond existing private relationship paths to u. However, the perception of the social network topology gained by u by the end of the token flooding phase may help her gain some additional knowledge. Such knowledge is limited in case users have a large number of friends. It is further reduced by the noise added by users to their degrees. We consider the example topology in Figure 1(a). Assume that A(lice) runs a path discovery phase with M(ary) and subsequently with G(ary). A discovers $(A,B,3,M)$, $(A,D,3,M)$ and subsequently $(A,D,2,G)$. The tokens of M and G were randomly chosen from the total set of tokens computed by D for distances 2 and 3 respectively. This prevents A from learning whether G is D's friend present in $(A,D,3,M)$. However, from Figure 6.2, A knows that $|D| = 3$. A gains more confidence up to whether G is D's friend present in $(A,D,3,F)$, compared to the ideal scheme, where A does not know $|D|$. This confidence is limited in typical social network topologies where $|D|$ would be much larger. It is further reduced by the noise added by D to $|D|$.

Malicious Adversary. We examine relationship information that can be leaked to a malicious adversary, but not to a honest but curious one. More specifically, we are interested in misbehaviors that aim at breaking relationship privacy. In our protocol, relationship information can be learnt in two ways:(1) during the token flooding phase, through analysis of data received from friends; and (2) by running a path discovery phase with

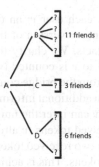

Fig. 4. A's perception of the topology by the end of the token flooding phase for the topology from Figure 1(a)

other users. During the token flooding phase, a malicious adversary M learns additional relationship information compared to an honest but curious one if it can influence the received data in a way that leaks private information. However, the only data that M can both influence and receive is the one propagated through cycles of distance $d \leq d_{max} = 3$ containing M. Even if M associates the smallest possible distance 1 with a transmitted token, other users will stop the propagation after d_{max} hops from M. Such misbehavior can help M discover friendship relationships between any of her friends. However, an honest but curious adversary already knows about these relationships, as was already discussed. During the path discovery phase with a user v, a malicious adversary can enter to the private set intersection protocol more or less tokens than prescribed in our protocol. When M plays the role of a server, this kind of misbehavior is not beneficial as the private set intersection protocol does not output anything to the server. As a client, M does not gain anything from entering less tokens to the private set intersection protocol. When M enters additional tokens, M learns whether these tokens are in v's list of received tokens. Because of the large space of tokens, we assume that M has a token T that is in v's list of received tokens only when M was involved in the construction or the propagation of T. We first consider when M does not deviate during the token flooding phase. Beyond tokens in its hash tree, M has access to the tokens it received and to the ones that can be constructed from them. Entering these tokens to the private set intersection protocol makes M discover private relationship paths to v of distance $d \leq d_{max} - 1$. These private relationship paths are also discovered in the normal run of the protocol. By deviating during the token flooding phase, M can only receive maliciously constructed tokens through cycles of distance $d \leq d_{max}$, as was already explained. Tokens propagated through these cycles do not help M learn additional information.

6.3   Overhead Analysis

We distinguish the overhead of the token flooding phase and that of the path discovery phase. The token flooding phase needs to run very infrequently. The path discovery phase runs when two users u and v need to discover private relationship paths *for the first time*. We introduce a new variable for the purpose of simplifying the notation for this section. F_v^i is the fan-out of a user v at depth i. That is F_v^i is the number of relationship paths starting at v, of distance $d \leq i$. In the extended hexagonal topology, $F_A^1 = 2, F_A^2 = 4$ and $F_A^3 = 8$.

Token Flooding Phase. As an originator, v computes F_v^3 tokens in constructing the hash tree and transmits F_v^1 tokens. As a bridge contact, v computes $O(F_v^1 . F_v^2)$ tokens and transmits a similar number of them. v receives F_v^1 tokens. As a relay, v does not perform any computation. v receives $O(F_v^3 + F_v^1 . F_v^2)$ and transmits $O(F_v^1 . F_v^2)$ tokens. Therefore, during the token flooding phase, the overhead is $O(F_v^3 + 2F_v^1 . F_v^2)$ hash computation and exchange.

Path Discovery Phase. We evaluate the overhead when u discovers private relationship paths to v. The overhead originates from the private set intersection protocol run where u plays the role of the client and enters the F_u^3 tokens in her hash tree and v plays the role of the server and enters the F_v^3 tokens in her list of received tokens. From Section 3, the computation overhead of u consists of F_u^3 homomorphic encryptions and F_v^3 homomorphic decryptions. The F_u^3 homomorphic encryptions only need to be computed once per input set. The computation overhead of v consists of $O(F_u^3 + F_v^3 \ln \ln F_u^3)$ exponentiations. The communication overhead of this step consists of $O(F_u^3 + F_v^3)$ exchange of homomorphic ciphertexts. The overhead for v's discovery can be obtained through a similar analysis.

6.4 Simulations

We carried out our overhead analysis based on graphs of major social networking sites: Flickr, LiveJournal, Orkut, YouTube. The graphs were crawled by Mislove et al. [14] in late 2006. Table 2 presents statistics about the social network topologies used.

Token Flooding Phase. Figure 5 presents the computation and communication overhead during the token flooding phase to an individual user. It presents the cdf of the number of tokens computed and exchanged in logarithmic scale. For Flickr, LiveJournal and YouTube, about 90% of users exchange less than 10^5 hash values, which is equivalent to 2 MB, given that a hash value consists of 20 B. For Orkut, more than 75% of users exchange less than 10^6 hash values equivalent to 20 MB and more than 90% of users exchange less than 10^7 tokens, equivalent to 200 MB. Similar trend applies to the computation overhead.

Path Discovery Phase. We consider $F_u^3 = F_v^3$. Figure 6 presents the computation overhead when user u discovers private relationship paths to user v. The overhead follows a similar trend compared to the token flooding phase. The main difference is that u performs homomorphic decryptions and v performs exponentiations. These operations are more expensive than hash computations. The communication overhead is not depicted. It follows a similar trend compared to the token flooding phase. The difference is that the items transmitted are homomorphic ciphertexts and not hash values. It should be noted, however, that the path discovery phase only needs to run once between two particular users u and v. After the first run, u and v can mark the common tokens with the identity of the other party. u and v can also establish a shared symmetric key for future use.

Table 2. Statistics about the social network topologies used

	Flickr	LiveJournal	Orkut	YouTube
Number of users	1,846,198	5,284,457	3,072,441	1,157,827
Estimated fraction of user population crawled	26.9 %	95.4 %	11.3 %	unknown
Number of friend links	22,613,981	77,402,652	223,534,301	4,945,382

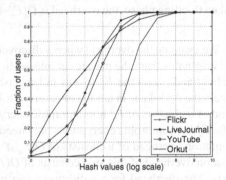

Fig. 5. Token flooding phase. Overhead per user in the number of hash values computed and exchanged.

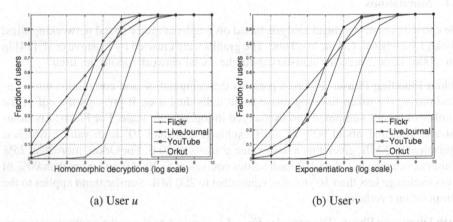

(a) User u (b) User v

Fig. 6. Path Discovery phase. Computation overhead when user u discovers private relationship paths to user v.

7   Related Work

In this section, we discuss decentralized social networks that were proposed. We then discuss previous schemes to discover relationship paths between users.

Several centralized social networking sites enable users to find relationship paths. For instance, LinkedIn, a professional social networking site, enables users to find private relationship paths to others. Unfortunately, these centralized sites know the entire topology and offer no privacy.

Decentralized social networks were proposed [1,2] to circumvent the lack of interoperability among current social networking sites, where users store their profiles locally and directly communicate with their friends. We did not find any support for a discovery of private relationship paths in a fully decentralized manner in any of these works. A decentralized social network was also proposed by Popescu et al. [16] to resist government monitoring. The social network provided search capabilities for sensitive items, but unfortunately, discovery of private relationship paths is not supported.

The most closely related work is by Freedman and Nicolosi [7], and their preceding paper [9]. Techniques to verify social proximity between users are presented as a mechanism to whitelist emails from the social network of the recipient. The paper mainly focuses on verifying friend of friend relationships, i.e., a relationship path of distance 2, while only disclosing common friends to one party. The paper suggests an extension to check for longer relationship paths, but unfortunately, their extension discloses all the relationships on the relationship path at the time of verifying the social proximity.

Carminati et al. [5] propose techniques to discover relationship paths between users in a decentralized social network. The paper assumes an untrusted central node, and discloses to one party all the relationships on the discovered relationship paths. Domingo-Ferrer [6] propose a mechanism to discover private relationship paths in a decentralized social network. When u wishes to discover paths to v, u floods her social network at that time. A major issue of this approach is that the discovery to be arbitrarily delayed if intermediate users are offline. Moreover, after some time has elapsed, u cannot know whether there does not exist a relationship path, or simply that some intermediate user did not happen to be online.

8 Conclusion

Social networks are increasing in importance. The majority of current social networking sites rely on a centralized server, which unfortunately offer no privacy for users' sensitive data. Given the highly privacy-sensitive nature of social networking topology (i.e., friendship relationships), a challenge is how to construct privacy-preserving social networks that provide the ability to find relationship paths without disclosing superfluous relationships. We take this problem one step further and consider decentralized social networks, where users can discover relationship paths *offline* (in a privacy-preserving manner) with people they meet. Our proposed approach provides the property to users who casually meet to discover relationship paths without disclosing their private relationships. More efficient schemes are the subject of our future research.

Acknowledgment

This research was supported in part by CyLab at Carnegie Mellon under grants DAAD19-02-1-0389 and MURI W 911 NF 0710287 from the Army Research Office, and by the iCAST project, National Science Council, Taiwan under the Grant NSC96-3114-P-001-002-Y. The views and conclusions contained here are those of the authors and should not be interpreted as necessarily representing the official policies or endorsements, either express or implied, of ARO, CMU, or the U.S. Government or any of its agencies.

We would like to express our gratitude to Ahren Studer and Bo-Yin Yang for useful feedback and interesting discussions. We would also like to thank Hazel Diana Mary for her help with the simulations.

References

1. foaf-o-matic, http://www.foaf.com/
2. Ackermann, M., Hymon, K., Ludwig, B., Wilhelm, K.: Helloworld: An open source, distributed and secure social network. In: W3C Workshop on the Future of Social Networking (January 2009)
3. Agrawal, R., Evfimievski, A., Srikant, R.: Information sharing across private databases. In: Proceedings of ACM SIGMOD international conference on Management of data (June 2003)
4. Baden, R., Bender, A., Spring, N., Bhattacharjee, B., Starin, D.: Persona: An online social network with user-defined privacy. In: Proceedings of ACM SIGCOMM (August 2009)
5. Carminati, B., Ferrari, E., Perego, A.: Private relationships in social networks. In: International Conference on Data Engineering Workshops (2007)
6. Domingo-Ferrer, J.: A public-key protocol for social networks with private relationships. In: Torra, V., Narukawa, Y., Yoshida, Y. (eds.) MDAI 2007. LNCS (LNAI), vol. 4617, pp. 373–379. Springer, Heidelberg (2007)
7. Freedman, M.J., Nicolosi, A.: Efficient private techniques for verifying social proximity. In: Proceedings of International Workshop on Peer-to-Peer Systems (2007)
8. Freedman, M.J., Nissim, K., Pinkas, B.: Efficient private matching and set intersection. In: Cachin, C., Camenisch, J.L. (eds.) EUROCRYPT 2004. LNCS, vol. 3027, pp. 1–19. Springer, Heidelberg (2004)
9. Garriss, S., Kaminsky, M., Freedman, M.J., Karp, B., Mazieres, D., Yu, H.: RE: Reliable email. In: Symposium on Networked Systems Design and Implementation (NSDI) (May 2006)
10. Katz, J., Sahai, A., Waters, B.: Predicate encryption supporting disjunctions, polynomial equations, and inner products. In: Smart, N.P. (ed.) EUROCRYPT 2008. LNCS, vol. 4965, pp. 146–162. Springer, Heidelberg (2008)
11. Kissner, L., Song, D.: Privacy preserving set operations. In: Shoup, V. (ed.) CRYPTO 2005. LNCS, vol. 3621, pp. 241–257. Springer, Heidelberg (2005)
12. Laurie, B.: Apres: A system for anonymous presence (2004), http://www.apache-ssl.org/apres.pdf/
13. Lucas, M., Borisov, N.: Flybynight: mitigating the privacy risks of social networking. In: Proceedings of ACM workshop on Privacy in the Electronic Society (WPES) (October 2008)
14. Mislove, A., Marcon, M., Gummadi, K.P., Druschel, P., Bhattacharjee, B.: Measurement and analysis of online social networks. In: Proceedings of the ACM SIGCOMM conference on Internet measurement (IMC) (2007)
15. Narayanan, A., Shmatikov, V.: De-anonymizing social networks. In: IEEE Symposium on Security and Privacy (May 2009)
16. Popescu, B.C., Crispo, B., Tanenbaum, A.S.: Safe and private data sharing with turtle: Friends team-up and beat the system. In: Christianson, B., Crispo, B., Malcolm, J.A., Roe, M. (eds.) Security Protocols 2004. LNCS, vol. 3957, pp. 221–230. Springer, Heidelberg (2006)

Verifying Anonymous Credential Systems in Applied Pi Calculus

Xiangxi Li[1], Yu Zhang[2], and Yuxin Deng[1],*

[1] Department of Computer Science and Engineering, Shanghai Jiao Tong University
Shanghai, China
[2] Laboratory for Computer Science, Institute of Software, Chinese Academy of Sciences
Beijing, China

Abstract. Anonymous credentials are widely used to certify properties of a credential owner or to support the owner to demand valuable services, while hiding the user's identity at the same time. A credential system (a.k.a. pseudonym system) usually consists of multiple interactive procedures between users and organizations, including generating pseudonyms, issuing credentials and verifying credentials, which are required to meet various security properties. We propose a general symbolic model (based on the applied pi calculus) for anonymous credential systems and give formal definitions of a few important security properties, including pseudonym and credential unforgeability, credential safety, pseudonym untraceability. We specialize the general formalization and apply it to the verification of a concrete anonymous credential system proposed by Camenisch and Lysyanskaya. The analysis is done automatically with the tool ProVerif and several security properties have been verified.

1 Introduction

The use of anonymous credential systems (sometimes called pseudonym systems) [11] is by far the best known idea to protect personal information in communications. These systems use pseudonyms generated by special random processes instead of users' private information to identify the user in order to guarantee the anonymity of users. A credential can be issued to a pseudonym, and the corresponding user can show her possession of the credential, without revealing any information beyond the bare fact that she owns such a credential. For a credential system to be useful, some basic properties must be satisfied. For example, a user should not be able to make transactions with others by using a credential not issued by some valid organization (*credential unforgeability*), and transactions carried out by the same user cannot be linked (*unlinkability* or *untraceability*). In some applications it might be desirable that a credential can only be used once (*one-show credential*) or a user cannot lend her credentials to others (*non-transferability*).

It is widely known that designing good security protocols is an error-prone task. There were protocols which had been used in practical applications for many years but

---

* The first and the third authors are supported by the National Natural Science Foundation of China (Grant No. 60703033).

J.A. Garay, A. Miyaji, and A. Otsuka (Eds.): CANS 2009, LNCS 5888, pp. 209–225, 2009.
© Springer-Verlag Berlin Heidelberg 2009

later on were found to be flawed. Examples include Needham-Schroeder [14], SSL [20] and PKCS [19]. Formal methods were introduced as a promising technique to analyze security protocols and in many cases the analysis can be done with automatic tools. As a case study in this respect, we formalize the anonymous credential system proposed by Camenisch and Lysyanskaya [10] (which we shall refer to as the CL system in the sequel) in the applied pi calculus [3] and we employ the tool ProVerif [8] to analyze several security properties of the system.

To our knowledge, this is the first formal, automated verification (at the symbolic level) of this type of security systems, although many credential systems, as a main research concern of cryptography, have been verified using traditional, semi-formal approaches in cryptography. The main part of our work is devoted to the mechanized analysis of the CL system, which is probably the most complex pseudonym system targeting many security requirements but remaining efficient. We have checked that the system satisfies two very basic properties: unforgeability (of both pseudonyms and credentials) and user privacy (pseudonym untraceability). A less arresting property, which we call *credential safety* and aims at preventing unauthorized use (or stealing) of credentials, is not met by the system — the traditional replay attack breaks safety.

However, we regard our contribution as more than just a case study using ProVerif. As credential systems become more and more widely used in large-scale security applications and many protocols have been proposed, we have been very careful in formalizing the system to make our model scalable. In particular, we provide a general modeling framework in the applied pi calculus and we believe that the formalization of most credential systems falls into it. In fact, we are currently studying other systems with significantly different implementation from the CL system.

The rest of the paper is structured as follows: In Section 2 we briefly introduce the applied pi calculus, as well as the formalization of zero-knowledge proofs by Backes *et al*. Section 3 gives a general description of credential systems and an overall modeling structure in applied pi. The next section formalizes four most important security properties: pseudonym unforgeability, credential unforgeability, credential safety and pseudonym untraceability, and summarizes the verification results in ProVerif of the basic credential system of Camenisch and Lysyanskaya. Section 5 discusses related work and Section 6 concludes the paper.

2   The Applied pi Calculus

2.1   Syntax and Semantics

We briefly recall the syntax and operational semantics of the applied pi calculus; more details can be found in [3].

A *signature* Σ is a finite set of function symbols. Given a signature Σ, an infinite set of *terms* is defined by the following grammar:

$$
\begin{array}{lll}
M, N := a, b, c, \ldots, k, \ldots & & \text{names} \\
\quad \mid \ x, y, z & & \text{variables} \\
\quad \mid \ f(M_1, \ldots, M_l) & & \text{function applications}
\end{array}
$$

where f ranges over the functions in Σ and l matches the arity of f. Terms are equipped with an *equational theory* E which consists of a set of equations over terms. We write $\Sigma \vdash M = N$ when the equation $M = N$ is in the theory associated with Σ, and $\Sigma \nvdash M = N$ for the opposite.

The grammar for *plain process* is similar to the one in the pi calculus [16], except that here messages can contain terms rather than names.

$$
\begin{array}{llll}
P, Q, R & := & 0 & \text{null processes} \\
& | & P|Q & \text{parallel composition} \\
& | & !P & \text{replication} \\
& | & \nu n \,.\, P & \text{name restriction} \\
& | & \text{if } M = N \text{ then } P \text{ else } Q & \text{conditional} \\
& | & u(x) \,.\, P & \text{message input} \\
& | & \overline{u}(N) \,.\, P & \text{message output}
\end{array}
$$

The null process 0 does nothing and is usually omitted from process specifications. The process $P|Q$ executes P and Q in parallel, and $!P$ stands for an infinite copies of P running in parallel. The process $\nu n.P$ generates a fresh name n and behaves as P. The process if $M = N$ then P else Q behaves as P if $\Sigma \vdash M = N$, and as Q otherwise. The input process $u(x).P$ can receive a message N from channel a and behaves as $P\{N/x\}$. We often take the abbreviation $\nu(\widetilde{u}).P$ for $\nu u_1.\cdots.\nu u_n.P$ and $u(=M).P$ for

$$u(x).\,\text{if } x = M \text{ then } P \text{ else } 0.$$

The output process $\overline{u}\langle N\rangle.P$ sends message N on channel a and behaves as P.

Extended processes are defined with *active substitutions*:

$$
\begin{array}{llll}
A, B, C & ::= & P & \text{plain process} \\
& | & A|B & \text{parallel composition} \\
& | & \nu x.A & \text{variable restriction} \\
& | & \{M/x\} & \text{active substitution} \\
& | & \text{event}(x_1, \ldots, x_n) & \text{events}
\end{array}
$$

where $\{M/x\}$ is the substitution that replaces the variable x with the term M. The process $\nu x.(\{M/x\}|P)$ restricts the scope of substitution in P and is often written as *let* $x = M$ *in* P. As usual, names and variables have scopes, which are delimited by restrictions and inputs. We write $fv(A)$ and $bv(A)$ (resp. $fn(A)$ and $bn(A)$) for the sets of free and bound variables (resp. names) of A. An extended process is *closed* when every variable is either bound or defined by an active substitution. Events are supported by ProVerif and are used to define traces of processes.

Every extended process can be mapped to a *frame* $\varphi(A)$ by replacing every plain process embedded in A with 0. Thus, a frame is built up from 0 and active substitutions by parallel composition and restriction. The frame $\varphi(A)$ can be viewed as the static knowledge exposed by A to the environment, but not as $A's$ dynamic behavior. The domain $dom(\varphi)$ of a frame φ is the set of variables that φ exports.

An *evaluation context* is a context (a process with a hole) whose hole is not under a replication, a conditional, an input, or an output. A context $C[\_]$ closes A when $C[A]$ is closed.

The semantics of the applied pi calculus are defined by structural equivalence and internal reduction. *Structural equivalence* $\equiv$ is the smallest equivalence relation on extended processes that satisfies the following rules and that is closed under α-renaming of names and variables and under application of evaluation contexts.

$A \equiv A\|0$	PAR-0	$\nu n.0 \equiv 0$	NEW-0
$A\|(B\|C) \equiv (A\|B)\|C$	PAR-A	$\nu m.\nu n.A \equiv \nu n.\nu m.A$	NEW-C
$A\|B \equiv B\|A$	PAR-C	$\nu x.\{M/x\} \equiv 0$	ALIAS
$!P \equiv P\|!P$	REPL	$\{M/x\} \equiv \{N/x\},$	REWRITE
$\{M/x\}\|A \equiv A\{M/x\}$	SUBST	if $\Sigma \vdash M = N$	

Internal reduction $\rightarrow$ is the smallest relation on extended processes closed by structural equivalence and application of evaluation contexts such that:

$$\overline{u}\langle x \rangle.P \mid u(x).Q \ \rightarrow \ P \mid Q \qquad \text{COMM}$$
$$\text{if } M = M \text{ then } P \text{ else } Q \rightarrow P \qquad \text{THEN}$$
$$\text{if } M = N \text{ then } P \text{ else } Q \rightarrow Q \qquad \text{ELSE}$$
$$\text{for all ground terms } M, N \text{ s.t. } \Sigma \nvdash M = N$$

Observational equivalence is an important relation for the applied pi calculus. Intuitively, two processes are observationally equivalent if no evaluation context can distinguish them; evaluation contexts are often used to model attackers. We write $A \Downarrow a$ when A can send a message on a, i.e $A \rightarrow^* C[\overline{a}\langle M \rangle.P]$ for some evaluation context C that does not bind a.

Definition 1. *Observational equivalence ($\approx$) is the largest symmetric relation $\mathcal{R}$ between closed extended processes with the same domain such that $A \mathcal{R} B$ implies:*

1. *if $A \Downarrow a$, then $B \Downarrow a$;*
2. *if $A \rightarrow^* A'$, then $B \rightarrow^* B'$ and $A' \mathcal{R} B'$ for some B'.*
3. *$C[A] \mathcal{R} C[B]$ for all closing evaluation contexts $C[\_]$.*

Trace properties are also important in process calculi. *Correspondence* was introduced to capture these properties [8].

Definition 2 (Correspondence). *The closed process P satisfies the* correspondence*:*

$$P \Vdash \text{event}(f(x_1, \ldots, x_i)) \rightsquigarrow \text{event}(f'(y_1, \ldots, y_j))$$

means that if the event $f(x_1, \ldots, x_i)$ is executed, then event $f'(y_1, \ldots, y_j)$ must have been executed. The closed process P satisfies the injective correspondence*:*

$$P \Vdash \text{event}(f(x_1, \ldots, x_i)) \rightsquigarrow \text{event}(f'(y_1, \ldots, y_j))$$

means that for each event $f(x_1, \ldots, x_i)$ being executed, there is a unique event $f'(y_1, \ldots, y_j)$ which has been previously executed.

We refer the reader to [8] for the technical definition of correspondences.

2.2 Representing Zero-Knowledge Proofs in Applied pi

Zero-knowledge proofs become a widely used technique in constructing modern cryptographic protocols [18], including many credential systems. Loosely speaking, a zero-knowledge proof consists of a message or a sequence of messages that constitute a proof of a statement, which yields nothing but the validity of the statement. The applied pi calculus does not natively support the verification of security protocols involving zero-knowledge proofs, but Backes *et al.* have extended the tool ProVerif to enable modeling and analyzing non-interactive zero-knowledge proofs [5].

Let Σ_{base} be a base signature including logic and arithmetic operations as well as basic cryptographic primitives such as encryption, decryption, digital signature, etc., and E_{base} be an equational theory for Σ_{base}. For representing zero-knowledge, we need to extend the equational theory, based on an extended signature:

$$\Sigma_{ZK} = \Sigma_{base} \cup \{\text{ZK}_{i,j}, \text{Ver}_{i,j}, \text{Public}_i, \text{Formula}, \text{true} \mid i, j \in \mathbb{N}\}$$

A non-interactive zero-knowledge proof is formalized as a term $\text{ZK}_{i,j}(\widetilde{M}, \widetilde{N}, F)$, where $\widetilde{M}$ denotes the term sequence $M_1, \ldots, M_i$ which represent the *private components* of the statement that are not revealed to the verifier and the adversary, and $\widetilde{N}$ denotes the term sequence $N_1, \ldots, N_j$ which represent the *public components* of the statement, and F constitutes a formula over these terms. In particular, we fix a distinguished set of variables $ZV = \{\alpha_1, \alpha_2, \ldots, \beta_1, \beta_2, \ldots\}$ which are only used to construct zero-knowledge formulas. Intuitively, α variables can be substituted by private components and β variables by public components. We call a term F an (i, j)-*formula* if it contains no names and $fv(F) \subseteq \{\alpha_1, \ldots, \alpha_i, \beta_1, \ldots, \beta_j\}$. $\text{Public}_i, \text{Formula}$ are operations for retrieving, respectively, the i-th public element and the formula from a proof, and $\text{Ver}_{i,j}$ is the function which verifies a proof against a formula. We shall often omit the arities of $\text{ZK}_{i,j}$ and $\text{Ver}_{i,j}$ when they are clear from the context.

The equational theory E_{ZK} for representing zero-knowledge is the smallest theory satisfying all equations in E_{base} and the following equations defined over all terms $\widetilde{M}, \widetilde{N}, F$:

$$\text{Public}_l(\text{ZK}_{i,j}(\widetilde{M}, \widetilde{N}, F)) = N_l, \qquad 1 \leq l \leq j,$$
$$\text{Formula}(\text{ZK}_{i,j}(\widetilde{M}, \widetilde{N}, F)) = F,$$
$$\text{Ver}_{i,j}(F, \text{ZK}_{i,j}(\widetilde{M}, \widetilde{N}, F)) = \text{true} \quad \text{iff } E_{ZK} \vdash F\{\widetilde{M}/\widetilde{\alpha}\}\{\widetilde{N}/\widetilde{\beta}\} = \text{true}$$
$$\text{and } F \text{ is an } (i, j)\text{-formula.}$$

Backes *et al.* also supply several techniques for dealing with infinite equational theories, so as to enforce the termination of the verification in ProVerif. We refer the reader to [18] for details.

3 A General Description of Credential Systems in Applied pi

In general, a credential system consists of two types of agents: *users* who wish to anonymously prove part of their personal information or use valuable services through credentials, and *organizations* who issue credentials to users and verify the validity of credentials shown by users. In *anonymous* credential systems, a user needs first to interact with an organization to establish a pseudonym before demanding a credential. The user is then known at the organization by the pseudonym, which is usually based on some information known by the organization about the user (e.g. an account in a bank). A real individual can have a pseudonym at each organization, or even multiple pseudonyms at one organization. However, organizations should not be able to link two different pseudonyms belonging to the same user. Credentials are issued by organizations to pseudonyms instead of real identities: when demanding a credential, the user interacts with the organization in the name of the pseudonym that he has established, and obtains a credential which can be shown by the user to another organization in the verifying procedure.

3.1   Modeling Credential Systems

We give a general framework of modeling credential systems in applied pi, by defining the overall structure of processes without concrete definitions. In later sections we shall see how a real credential system can be modeled following the structure. The model of a credential system in applied pi generally consists of two types of processes: the user processes and the organization processes.

When a user enters the system, she must first demand a pseudonym at some organization, and then use this pseudonym to demand and show credentials, hence a user process can be generally defined as

$$UP \stackrel{\text{def}}{=} v(\widetilde{u}) \, . \, !\text{ckey}(j, pk_j) \, . \, UN(j, pk_j) \, . \, (!UC(j, pk_j, nym_j) \mid !UV(j, pk_j, nym_j))$$

$\widetilde{u}$ is a set of secret channels inside the user process which are basically used to transmit secret data like keys, randoms, and so on. The process $UN(j, pk_j)$ models the user's behavior of establishing a pseudonym at the organization O_j and the user must receive the correct public key pk_j properly (e.g., from a secret channel ckey shared between users and organizations). The process $UC(j, pk_j, nym_j)$ models the user's behavior of demanding a credential from the organization O_j, in name of the pseudonym nym_j that has been established in $UN(j, pk_j)$. In the end of the process, the generated credential must be recorded together with the ID of the issuing organization. The process $UV(j, pk_j, nym_j)$ models the user's behavior of showing a credential using the pseudonym nym_j. There are in general two manners of showing a credential:

$$UV(j, pk_j, nym_j) \stackrel{\text{def}}{=} !u_i(= j, cred_j) \, . \, UV^1(j, pk_j, cred_j)$$
$$\mid !u_i(l, cred_l) \, . \, UV^2(j, pk_j, nym_j, l, pk_l, cred_l),$$

where $UV^1(j, pk_j, cred_j)$ models the behavior of showing a single credential $cred_j$ issued by the organization O_j and $UV^2(j, pk_j, nym_j, l, pk_l, cred_l)$ models the behavior of showing a credential $cred_l$ issued by the organization O_l, using the pseudonym nym_j (known at O_j. Note that in the first procedure, the credential can be essentially shown to any valid organization, while in the latter it can only be shown to O_j, i.e., the organization who knows the pseudonym nym_j. If the system guarantees unlinkability of pseudonyms, this does not break the anonymity.

Correspondingly, an organization process consists of generating a pseudonym, issuing a credential and verifying a credential:

$$OP \stackrel{\text{def}}{=} \nu(\bar{o}) . !ON . (!OC(nym) \mid !OV^1(l, pk_l, cred) \mid !OV^2(l, pk_l, nym, cred)),$$

where the process ON models the organization's behavior of establishing a pseudonym nym, $OC(nym)$ models the behavior of issuing a credential to nym, $OV^1(l, pk_l, cred)$ models the behavior of verifying the credential $cred$ and $OV^2(l, pk_l, nym, cred)$ models the behavior of verifying the credential $cred$, plus the statement that its owner has the pseudonym nym. Note that nym in UV^2 must be the pseudonym established by the organization, but $cred$ in UV^1 and UV^2 can be an arbitrary credential issued by a valid organization (presumably the organization O_l). It is possible that some organizations only do the verification and never issue credentials, and in modeling a concrete system, one can safely remove the corresponding processes.

The whole credential system is then modeled as a set of user processes and organization processes running in parallel.

3.2 Events

As we shall see in Section 4, many security properties are defined using the notion of *correspondence* between events, which must be added at right places when we define processes. We summarize here a set of events which can be commonly defined in many credential systems and are sufficient for defining and verifying their security properties.

- **NymGenerated**(U, n): The user U executes this event when she establishes a pseudonym n with some organization.
- **NymApproved**(O, n): The organization O executes this event when he approves that the pseudonym n is correctly formed.
- **CredIssued**(O, c, n): The organization O executes this event after she issues the credential c to the pseudonym n.
- **UserShow**(U, c): A user executes this event when he starts a session of showing a credential c with a verifying organization.
- **CredVerified**(O, c, O'): Verifier O executes this event after the credential c has been shown to her and she is convinced that c has been issued by organization O'.
- **CredNymVerified**(O, n, c, O'): The verifying organization O executes this event after the credential c has been shown to her with the pseudonym n and she is convinced that c has been issued by O' to the user. Note that n is not the pseudonym to which c has been issued to, but the one that is known by the verifier O. In principle, the user must possess another pseudonym n' (known by O')and has used n' to get the credential c before she shows it, but this pseudonym is irrelevant in the verifying procedure. in short, n hides the user's identity at the verifying organization and n' protects her at the issuing organization.

We remark that events are not necessary for modeling protocols, but rather for specifying security properties based on traces, so only processes representing honest agents will execute these events — adversaries never execute events. When defining security properties using these events, we often omit some parameters (replaced by _) when they are irrelevant.

4 Security Analysis of an Anonymous Credential System

In this section we apply the general modeling of the previous section to the anonymous credential system proposed by Camenisch and Lysyanskaya [10], and do a verification using ProVerif. The basic system consists of four protocols for, respectively, pseudonym generation, credential generation, showing a single credential and showing a credential w.r.t. a pseudonym. Due to the page limit, we only present here an abstract definition of the whole system. Detailed description of protocols and corresponding process specifications in applied pi can be found in Appendix A.

Definition 3 (Basic credential system). *The* basic credential system *of the CL system is a process in applied pi:*

$$BCS \overset{\text{def}}{=} \nu(\mathsf{ckey}) . (UP_1 \mid \ldots \mid UP_n \mid OP_1 \mid \ldots \mid OP_m),$$

where ckey *is a secret channel for transferring organizations' public keys between honest agents,* $UP_i \overset{\text{def}}{=} \nu(\mathsf{cu}_i) . !(\mathsf{ckey}(l, pk_l) . UN_i(l, pk_l))$, *modeling each user agent, and*

$$OP_j \overset{\text{def}}{=} \nu(seed_j) . \underline{\mathsf{let}} \ pk_j = \mathsf{pkey}(seed_j), \ sk_j = \mathsf{skey}(seed_j) \ \mathsf{in}$$
$$!\overline{\mathsf{ckey}}(j, pk_j) \mid !\overline{\mathsf{c}}(j, pk_j) \mid !ON_j \mid !VP_j$$

modeling each organization agent, where c *is a public channel allowing agents including adversaries to communicate with each other.*

Definitions of the processes UN_i, ON_j, VP_j can be found in Appendix A.

The rest of the section is devoted to the formalization and verification of basic security properties: unforgeability of credentials and pseudonyms, safety of credentials and user privacy (pseudonym untraceability), which are supposed to be met by the CL basic system.

Let *CS* be a model of a credential system defined in applied pi, such as *BCS* in Definition 3. We write $\mathcal{U}(CS)$ for the set $\{U_1, \ldots, U_n\}$ where each U_i is a process representing an honest user agent, and $O(CS)$ for the set $\{O_1, \ldots O_m\}$ where each O_j is a process representing an honest organization agent.

4.1 Unforgeability

Unforgeability of pseudonyms and credentials is the very basic security requirement of anonymous credential systems, which in principle prevents adversaries from forging fake credentials. Fake pseudonyms must be prevented too, since credentials are issued to pseudonyms, never to real identities.

Definition 4 (Pseudonym unforgeability). *A credential system CS respects* pseudonym unforgeability *if whenever an organization* $O' \in O(CS)$ *issues a credential c to a pseudonym* n', *she must have established this pseudonym with a user:*

$$CS \Vdash \mathbf{CredIssued}(O', c, n') \rightsquigarrow \mathbf{NymApproved}(O', n'), \tag{1}$$

and whenever an organization $O \in O(CS)$ verifies a credential c that is shown to her w.r.t. a pseudonym n and is claimed to be issued by O', the verifier must have established the pseudonym with the user:

$$CS \Vdash \texttt{CredNymVerified}(O, n, c, O') \rightsquigarrow \texttt{NymApproved}(O, n), \qquad (2)$$

Definition 5 (Credential unforgeability). *A credential system CS respects credential unforgeability if every successful showing (either single or with a pseudonym n at organization O) of a credential c, being claimed to be issued by an organization $O' \in O(CS)$, implies that O' has previously issued c to some pseudonym n':*

$$CS \Vdash \texttt{CredVerified}(\_, c, O') \rightsquigarrow \texttt{CredIssued}(O', c, n'). \qquad (3)$$

$$CS \Vdash \texttt{CredNymVerified}(O, n, c, O') \rightsquigarrow \texttt{CredIssued}(O', c, n'). \qquad (4)$$

If a credential system respects both pseudonym unforgeability and credential unforgeability, we say that it is an *unforgeable credential system*.

The definition of credential unforgeability does not exclude the case where the adversary can forge a *valid* credential that has indeed been generated by an honest organization, but not to the adversary. We shall consider it as the safety of credentials.

Theorem 1. *The CL basic credential system (Definition 3) is an unforgeable credential system, i.e., it respects the unforgeability of both pseudonyms and credentials.*

Proof. We check the four correspondences (1), (2), (3) and (4) in ProVerif.   □

4.2   Credential Safety

Credential *safety* aims at preventing adversaries from *stealing* or using unauthorizedly a valid credential. In other words, no one other than the honest user, to whom a valid credential has been issued to, can successfully show the credential to a verifier. However, there is a subtle situation where safety can be confused with unforgeability: if an adversary can forge a credential which has been generated by an honest organization to an honest user, we shall consider it as an attack to safety instead of unforgeability. In fact, when we talk about credential safety, we actually mean safety of *unforgeable credentials*, or safety in *unforgeable credential systems*.

Definition 6 (Credential safety). *A credential c in an unforgeable credential system CS is safe if there is an injective correspondence between the event indicating that c (being issued by $O \in Org(CS)$) is successfully verified, and the event indicating that some user $U \in \mathcal{U}(CS)$, who must be the owner of the credential, starts to show c, i.e, for all pseudonym n',*

$$CS \Vdash \texttt{CredVerified}(\_, c, O') \rightsquigarrow \texttt{CredIssued}(O', c, n')$$
$$\Rightarrow CS \Vdash \texttt{CredVerified}(\_, c, O') \leftrightsquigarrow \texttt{UserShow}(U, c) \qquad (5)$$
$$\wedge \texttt{NymApproved}(O', n') \rightsquigarrow \texttt{NymGenerated}(U, n').$$

If c is shown with respect to a pseudonym n at $O \in O(CS)$, then the user must be the owner of n, i.e., there exists another pseudonym n',

$$CS \Vdash \text{CredNymVerified}(O, n, c, O') \rightsquigarrow \text{CredIssued}(O', c, n')$$
$$\Rightarrow CS \Vdash \text{CredNymVerified}(O, n, c, O') \leftrightsquigarrow \text{UserShow}(U, c) \tag{6}$$
$$\wedge \text{NymApproved}(O', n') \rightsquigarrow \text{NymGenerated}(U, n')$$
$$\wedge \text{NymApproved}(O, n) \rightsquigarrow \text{NymGenerated}(U, n)$$

If an unforgeable credential system respects credential safety, we say it is a *safe credential system.*

Unfortunately, the CL basic system is not safe if we assume the channels between users and organizations is insecure as in normal networking environment. There is a replay attack to safety, which works as follows: in the case of showing a single credential, the adversary can record A, B in Protocol 3 and all messages in the zero-knowledge proof, then sends them repeatedly to a verifier. The reused zero-knowledge proof simply passes. ProVerif actually shows that the injective correspondence in (5) fails. The same attack simply applies in the case of showing a credential w.r.t. a pseudonym, but the adversary can only show the credential to the organization whom the user has shown it to, since she cannot change the pseudonym that is used in the verification.

Note that an adversary can steal a pseudonym too and even use it to demand a new credential, but she cannot show it as in that case she cannot forge a proof for showing the credential. In fact, in credential systems, what we care about is what people can do with credentials, not what they can do with pseudonyms, so we do not define a property like pseudonym safety.

4.3 Pseudonym Untraceability

Anonymous credential systems are designed essentially for providing user privacy. Credentials are issued to pseudonyms, so user privacy in credential systems indeed depends on what we can deduce based on pseudonyms. In particular, organizations should not be able to collectively distinguish pseudonyms that belong to different users. We call this property *pseudonym untraceability*, and in the applied pi calculus, this is formalized by the popular notion of *observational equivalence.*

Definition 7 (Pseudonym-untraceability). *A credential system CS respects pseudonym untraceability if for arbitrary users $U_i, U_j \in \mathcal{U}(CS)$ and a well formed public key pk_0,*

$$UN_i(\_, pk_0) \approx UN_j(\_, pk_0),$$

where UN_i (resp. UN_j), as defined in Definition 3, models the procedure of establishing a pseudonym by the user U_i (resp. U_j) and all her behavior involving the pseudonym in the system.

Theorem 2. *The basic credential system respects pseudonym untraceability, i.e. $UN_i(l, pk_l) \approx UN_j(l, pk_l)$.*

Proof. ProVerif supports proving observational equivalence of two processes which differ only in the choice of some terms. In our definition of pseudonym untraceability, we are actually proving: for all $U_i, U_j \in \mathcal{U}(BCS)$, $UN_i(0, pk_0) \approx UN_i(0, pk_0)[x_j/x_i]$. ProVerif shows that the above observational equivalence holds.   □

5   Related Work

A security model for anonymous credential systems was proposed by Pashalidis and Mitchell [17]. It follows the idea of Bellare and Rogaway [6] based on complexity theoretic arguments, which potentially leads to information theoretic anonymity metrics. The model does not specify how the credential system achieves its goals but defines what the goals are. Some basic properties such as credential unforgeability, non-transferability, pseudonym unlinkability, and pseudonym owner protection are formally defined and the relationships between them are explored. Compared with our definitions, their definitions are based on a computational model while ours are based on the applied pi calculus and allow an automatic verification.

Abadi and Fournet introduced the applied pi calculus [3] as a language for reasoning about security protocols. The calculus inherits communication and concurrency for the pure pi calculus [16], and introduces functions and equations to reason about complex messages transmitted in security protocols. ProVerif [7] is an automatic cryptographic protocol verifier for the analysis of trace-based security properties and observational equivalence. It accepts applied pi processes as inputs and translates them into Horn clauses. Using this tool, Blanchet *et al.* verified a protocol for certified emails [1], a protocol for secure file sharing on untrusted storage [9], as well as the JFK protocol [2]. Luo *et al.* [15] analyzed an electronic cash protocol, and Kremer and Ryan [12] verified an electronic voting protocol. Backes *et al.* [5] introduced an implementation of zero-knowledge in equational theories acceptable by ProVerif and applied it to the analysis of a remote electronic voting protocol [4].

6   Conclusion

In this paper we have presented a general formalization of credential systems and some important security properties. We apply them to the concrete credential system proposed by Camenisch and Lysyanskaya and have verified that the basic system satisfies unforgeability for both pseudonyms and credentials, and pseudonym untraceability. We also reveal an attack to the system which allow adversaries to steal and unauthorizedly use a credential.

We argue that the model that we propose in the paper is faithful enough. However, as the original protocol itself is not written in a formal language, there is no way to formally prove that our model faithfully specify the original protocol. Nevertheless, if we assume that the specification is correct, then the soundness of ProtoVerif already guarantees the correctness of the verification output.

One novelty of the CL system, compared with other credential systems, is the implementation of non-transferable credentials which prevent users from lending their credentials. As part of the future work, we shall investigate how to formalize non-transferability, as well as other interesting, advanced properties like non-reshowability. We are also trying to transplant our model to other credential systems, in order to establish a scalable model of analyzing credential systems with the applied pi calculus. Another interesting work would be to focus on improving the efficiency for verifying complex systems using zero-knowledge proofs heavily. How to optimize equational theories to speed up termination is still a challenging problem.

References

1. Abadi, M., Blanchet, B.: Computer-Assisted Verification of a Protocol for Certified Email. In: Cousot, R. (ed.) SAS 2003. LNCS, vol. 2694, pp. 316–335. Springer, Heidelberg (2003)
2. Abadi, M., Blanchet, B., Fournet, C.: Just fast keying in the pi calculus. In: Schmidt, D. (ed.) ESOP 2004. LNCS, vol. 2986, pp. 340–354. Springer, Heidelberg (2004)
3. Abadi, M., Fournet, C.: Mobile values, new names, and secure communication. In: Proceedings of the 28th ACM SIGPLAN-SIGACT Symposium on Principles of Programming Languages, vol. 36, pp. 104–115. ACM, New York (2001)
4. Backes, M., Hritcu, C., Maffei, M.: Automated verification of remote electronic voting protocols in the applied pi-calculus. In: Proceedings of the 21st IEEE Computer Security Foundations Symposium, pp. 195–209. IEEE Computer Society, Los Alamitos (2008)
5. Backes, M., Maffei, M., Unruh, D.: Zero-knowledge in the applied pi-calculus and automated verification of the direct anonymous attestation protocol. In: Proceedings of the IEEE Symposium on Security and Privacy, pp. 202–215. IEEE Computer Society, Los Alamitos (2008)
6. Bellare, M., Rogaway, P.: Entity authentication and key distribution. In: Stinson, D.R. (ed.) CRYPTO 1993. LNCS, vol. 773, pp. 232–249. Springer, Heidelberg (1994)
7. Blanchet, B.: Proverif: Cryptographic protocol verifier in the formal model, http://www.proverif.ens.fr/
8. Blanchet, B.: Automatic verification of correspondences for security protocols. Journal of Computer Security (2008) (to appear), http://arxiv.org/abs/0802.3444
9. Blanchet, B., Chaudhuri, A.: Automated formal analysis of a protocol for secure file sharing on untrusted storage. In: Proceedings of the IEEE Symposium on Security and Privacy, pp. 417–431. IEEE Computer Society, Los Alamitos (2008)
10. Camenisch, J., Lysyanskaya, A.: An efficient system for non-transferable anonymous credentials with optional anonymity revocation. In: Pfitzmann, B. (ed.) EUROCRYPT 2001. LNCS, vol. 2045, pp. 93–118. Springer, Heidelberg (2001)
11. Chaum, D.: Security without identification: Transaction systems to make big brother obsolete. Communications of the ACM 28(10), 1030–1044 (1985)
12. Kremer, S., Ryan, M.: Analysis of an electronic voting protocol in the applied pi calculus. In: Sagiv, M. (ed.) ESOP 2005. LNCS, vol. 3444, pp. 186–200. Springer, Heidelberg (2005)
13. Li, X., Zhang, Y., Deng, Y.: ProVerif scripts for verifying a non-transferable anonymous credential system, http://basics.sjtu.edu.cn/~xiangxi/credentialsys.rar
14. Lowe, G.: An attack on the needham-schroeder public-key authentication protocol. Information Processing Letters 56, 131–133 (1995)
15. Luo, Z., Cai, X., Pang, J., Deng, Y.: Analyzing an electronic cash protocol using applied pi calculus. In: Katz, J., Yung, M. (eds.) ACNS 2007. LNCS, vol. 4521, pp. 87–103. Springer, Heidelberg (2007)
16. Milner, R.: Communicating and Mobile Systems: the π-Calculus. Cambridge University Press, Cambridge (1999)
17. Pashalidis, A., Mitchell, C.J.: A security model for anonymous credential systems. In: Proceedings of the 19th International Workshop on Information Security, pp. 183–198. Kluwer, Dordrecht (2004)
18. Goldwasser, S., Micali, S., Rackoff, C.: The knowledge complexity of interactive proof-systems. SIAM Journal on Computing 18(1), 186–207 (1989)
19. Coron, J.-S., Joye, M., Naccache, D., Paillier, P.: New attacks on PKCS#1 v1.5 encryption. In: Preneel, B. (ed.) EUROCRYPT 2000. LNCS, vol. 1807, pp. 369–379. Springer, Heidelberg (2000)
20. Seifried, K.: The end of ssl and ssh? http://seifried.org/security/cryptography/20011108-end-of-ssl-ssh.html

A   Description and Modeling of the CL Basic System

A.1   Setup of the CL System

The CL system uses the asymmetric cryptography (typically RSA) to implement anonymous pseudonyms and credentials. Each organization O_j will have a public key PK_j consisted of an RSA modulus n_j, and five elements of QR_{n_j} : $(a_j, b_j, d_j, g_j, h_j)$, the corresponding secret key that contains the factorization of n_j. Each user U_i has a master secret key x_i.

A pseudonym N_{ij} — a name for U_i being known at O_j — consists of a user-generated part N_1 and organization-generated part N_2. Every pseudonym N_{ij} will be tagged with a *validating tag* P_{ij}. A credential issued by O_j to a pseudonym N_{ij} is pair (e, c), where e is a sufficiently long prime chosen by O_j, and $c = P_{ij}d_j^{1/e}$ (mod n_j). Under the strong RSA assumption, such tuples cannot be existentially forged for correctly formed tags even by an adaptive attack, since no one can generate c from e without knowing the factorization of n_j.

Zero knowledge is applied in the system to protect users' privacy. A proof of possession of a credential is realized by a proof of knowledge of a correctly formed tag P_{ij} and a credential on it. This is done by publishing statistically secure commitments to both the validating tag and the credential, and proving relationships between these commitments. It can also include a proof that the underlying secret key is the same in both the committed validating tag (corresponding to the pseudonym formed with the issuing organization) and the validating tag with the verifying organization.

The base signature for analyzing the CL system consists basically of two sorts of functions: basic crytpographic primitives (e.g., enc, dec for encryption and decryption and pkey, skey for generating asymmetric key pairs) and cyclic group arithmetic operations (e.g., add, mult, exp and inv for group addition, multiplication, exponentiation and inverse operation). The equational theory for this signature contains standard equations for cryptography and RSA arithmetic. Detailed definition can be found in [13].

A.2   Basic Credential System and Its Model in Applied pi

The CL basic system consists of four protocols for, respectively, pseudonym generation, credential generation, showing a single credential and showing a credential w.r.t. a pseudonym. We briefly describe these protocols and give the user and organization processes corresponding their behavior in each protocol.

Protocol 1 (Pseudonym generation). User U_i follows the protocol below to establish a pseudonym at organization O_j:

1. U_i chooses values N_1, r_1, r_2, r_3, sets $C_1 = g_j^{r_1} h_j^{r_2}, C_2 = g_j^{x_i} h_j^{r_3}$ and sends N_1, C_1, C_2 to O_j. U_i proves that C_1 and C_2 are formed correctly in

$$PK\{(\alpha_1, \alpha_2, \alpha_3, \alpha_4) : C_1 = g_j^{\alpha_1} h_j^{\alpha_2} \wedge C_2 = g_j^{\alpha_3} h_j^{\alpha_4}\}.$$

2. O_j generates two randoms r, N_2 and sends them to U_i.

3. U_i computes $s_{ij} = r_1 + r$, sets the pseudonym $N_{ij} = (N_1, N_2)$ and the validating tag $P_{ij} = a_j^{x_i} b_j^{s_{ij}}$, then sends P_{ij} to O_j. U_i proves that P_{ij} is formed correctly in

$$PK\{(\alpha, \beta, \gamma, \delta, \varepsilon) : C_1 = g_j^\alpha h_j^\beta \wedge C_2 = g_j^\gamma h_j^\delta \wedge P_{ij} = a_j^\gamma b_j^\varepsilon\}$$

In this protocol, every generated pseudonym N_{ij} corresponds to a validating tag P_{ij}, and they are stored in pair by both U_i and O_j. Here P_{ij} is used to distinguish different pseudonyms, and the representation of P_{ij} with respect to g_j and h_j is an essential part when showing a credential.

The user process and the organization process for the protocol are defined as:

$UN_i(j, pk_j) \overset{\text{def}}{=} \nu(N_1, r_1, r_2, r_3)$.

let $C_1 = \exp((g_j, h_j), (r_1, r_2))$, $C_2 = \exp((g_j, h_j), (x_i, r_3))$ in

let $zp_1 = ZK(r_1, r_2, x_i, r_3; C_1, C_2, g_j, h_j; F_1)$ in

$\overline{co}_j((N_1, C_1, C_2), zp_1) . cu(r, N_2)$.

let $N_{ij} = (N_1, N_2)$, $s_{ij} = r_1 + r$, $P_{ij} = \exp((a_j, b_j), (x_i, s_{ij}))$ in

let $zp_2 = ZK(r_1, r_2, x_i, r_3, s_{ij}; C_1, C_2, g_j, h_j, a_j, b_j, P_{ij}; F_2)$ in

NymGenerated$(U_i, N_{ij}) . \overline{co}_j(P_{ij}, zp_2)$.

$(!UC_i(j, N_{ij}, P_{ij}) \mid !(cu_i(l, cred) . UV_i^2(j, l, cred, N_{ij})))$,

$ON_j \overset{\text{def}}{=} co_j(N_1, C_1, C_2, z_1)$.

if $Ver(F_1, z_1) = \text{true}$ then

$\nu(r, N_2) . \overline{cu}(\langle r, N_2 \rangle) . co_j(P, z_2)$.

if $Ver(F_2, z_2) = \text{true}$ then

let $N = (N_1, N_2)$ in

NymApproved$(O_j, N) . (!OC_j(N, P) \mid !OV_j(N, P))$

where

$$F_1 \overset{\text{def}}{=} (\beta_1 = \exp((\beta_3, \beta_4), (\alpha_1, \alpha_2)) \wedge \beta_2 = \exp((\beta_3, \beta_4), (\alpha_3, \alpha_4))),$$
$$F_2 \overset{\text{def}}{=} (\beta_1 = \exp((\beta_3, \beta_4), (\alpha_1, \alpha_2)) \wedge \beta_2 = \exp((\beta_3, \beta_4), (\alpha_3, \alpha_4))$$
$$\wedge \beta_7 = \exp((\beta_5, \beta_6), (\alpha_3, \alpha_5))),$$

and co_j and cu are public channels, cu_i is a secret channel inside the process of U_i. $UC_i(j, N_{ij}, P_{ij})$ is the user process of demanding a credential from O_j, using the pseudonym N_{ij} (together with the validating tag P_{ij}). $UV_i^2(j, l, cred, N_{ij})$ is the user process of showing, to organization O_j, the credential $cred$ issued by organization O_l, using the pseudonym N_{ij}. $OC_j(N, P)$ and $OV_j(N, P)$ represent, respectively, the organization processes of issuing a credential to the pseudonym N and of verifying a credential sent by a user using the pseudonym N. Note that the credential issue and verification can be done separately at the organization side (by different processes in parallel), but they must be done sequentially in the user process.

The events **NymGenerated** and **NymApproved** are executed in this protocol: **NymGenerated** is executed by the user right after he receives the organization part of the pseudonym, and **NymApproved** is executed by the organization at the end of the protocol, when he receives the validating tag and verifies the validity of its form.

Note that the user is assumed to communicate with the organization via an *anonymous* channel, which is modeled using a global public channel c, and we rely on the scheduler to determine the right destination of messages transmitted on the channel. In particular, there is no successful trace where the response of an organization is sent to a wrong user, as in that case the organization will fail in checking the the second zero-knowledge proof.

Protocol 2 (Credential generation). User U_i follows the protocol below to demand a credential from organization O_j:

1. U_i sends (N_{ij}, P_{ij}) to O_j and proves the ownership in

$$PK\{(\alpha, \beta) : P_{ij} = a_j^\alpha b_j^\beta\}.$$

2. O_j checks that (N_{ij}, P_{ij}) is in its database, chooses a random prime e, computes $c = (P_{ij}d_j)^{1/e} \mod n_j$, sends c and e to U_i and stores (c, e) in its record for N_{ij}.
3. U_i checks if $c^e = P_{ij}d_j \mod n_j$; if so, she stores (c, e) in its record with organization O_j. The tuple (c, e) is called a *credential record*.

The cryptographic assumption ensures that an adversary who does not know the factorization of n_j should not be able to generate c from e.

The user process and the organization process of credential generation are:

$$UC_i(j, N_{ij}, P_{ij}) \overset{\text{def}}{=} \texttt{let } zp_3 = \texttt{ZK}(x_i, s_{ij}; P_{ij}, a_j, b_j; F_3) \texttt{ in}$$
$$\overline{\text{co}}_j(N_{ij}, P_{ij}, zp_3) . \text{cu}(c_{ij}, e_{ij}) .$$
$$\texttt{if } \texttt{exp}(c_{ij}, e_{ij}) = \texttt{mult}(P_{i,j}, d_j) \texttt{ then}$$
$$\texttt{let } cred_{ij} = (c_{ij}, e_{ij}) \texttt{ in}$$
$$!\overline{\text{cu}}_i(j, cred_{ij}) \mid !UV_i^1(j, cred_{ij})$$

$$OC_j(N, P) \overset{\text{def}}{=} \text{co}_j(N', P', z_3) .$$
$$\texttt{if } \texttt{Ver}(F_3, z_3) = \texttt{true then}$$
$$v(e) . \texttt{let } c = \texttt{exp}(\texttt{mult}(P, d_j), \texttt{inv}(e)), \ cred = (c, e) \texttt{ in}$$
$$\textbf{CredIssued}(O_i, cred, N) . \overline{\text{cu}}(c, e),$$

where

$$F_3 \overset{\text{def}}{=} \beta_1 = \texttt{exp}((\beta_2, \beta_3), (\alpha_1, \alpha_2)).$$

$UV_i^1(j, cred_{ij})$ is the user processes of showing the credential $cred_{ij}$ to an arbitrary organization (verifier).

When U_i receives a credential, she broadcasts it via the internal secret channel cu_i to all other sub-processes; when U_i wants to show a credential with respect to a pseudonym N_{ij}, she invokes the procedure UV_i^2 by sending the credential to the procedure via cu_i.

The event **CredIssued** is executed by the organization in this protocol after the credential is generated.

Protocol 3 (Showing a single credential). User U_i follows the protocol below to show a credential, issued by O_j, to a verifier V (without revealing the combined pseudonym):

1. U_i chooses r'_1, r'_2, computes $A = c_{ij}h_j^{r'_1}$, $B = h_j^{r'_1}g_j^{r'_2}$, and sends A, B with the credential to V.
2. U proves the validity of the credential in

$$PK\{(\alpha_1, \alpha_2, \alpha_3, \alpha_4, \alpha_5, \alpha_6, \alpha_7):$$

$$d_j = A^{\alpha_1}(\frac{1}{a_j})^{\alpha_2}(\frac{1}{b_j})^{\alpha_3}(\frac{1}{h_j})^{\alpha_4} \wedge B = h_j^{\alpha_5}g_j^{\alpha_6} \wedge 1 = B^{\alpha_1}(\frac{1}{h_j})^{\alpha_4}(\frac{1}{g_j})^{\alpha_7}\}$$

Those who can successfully show a single credential are assumed to know the representation of P_{ij} with respect to g_j, h_j as well as the credential pair (c_{ij}, e_{ij}), so the protocol should offer sufficiently security even when the transmission of a credential is unsafe.

The two processes engaged in this protocol are:

$$UV_i^1(j, cred_{ij}) \overset{\text{def}}{=} \nu(r'_1, r'_2).$$

$$\texttt{let } A = \exp((c_{ij}, h_j), (1, r'_1)), \ B = \exp((h_j, g_j), (r'_1, r'_2)) \texttt{ in}$$

$$\texttt{let } zp4 = \texttt{ZK}(e_{ij}, x_i, s_{ij}, \texttt{mult}(r'_1, e_{ij}), r'_1, r'_2, \texttt{mult}(r'_2, e_{ij});$$

$$A, B, a_j, b_j, d_j, g_j, h_j; F_4) \texttt{ in}$$

$$\textbf{UserShow}(U_i, cred_{ij}) \cdot \overline{\texttt{cv}}(j, c_{ij}, e_{ij}, A, B, zp4),$$

$$VP_j \overset{\text{def}}{=} \texttt{cv}(= j, c_{ij}, e_{ij}, A, B, z_4).$$

$$\texttt{if } \texttt{Ver}(F_4, z_4) = \texttt{true then } \textbf{CredVerified}(\_, cred, O_j),$$

where

$$F_4 \overset{\text{def}}{=} \beta_5 = \exp((\beta_1, \texttt{inv}(\beta_3), \texttt{inv}(\beta_4), \texttt{inv}(\beta_7)), (\alpha_1, \alpha_2, \alpha_3, \alpha_4))$$

$$\wedge \beta_2 = \exp((\beta_7, \beta_6), (\alpha_5, \alpha_6)) \wedge 1 = \exp((\beta_2, \texttt{inv}(\beta_7), \texttt{inv}(\beta_6)), (\alpha_1, \alpha_4, \alpha_7))$$

The events **UserShow** and **CredVerified** are executed in this protocol: **UserShow** is executed by the user right before he starts to show a credential and **CredVerified** is executed by the verifier after verifying the validity of the credential (the first parameter is omitted since in this protocol the identity of the verifier is irrelevant). In the process model, we explicitly let the user transmit the credential to the verifier, which is not included in the original protocol. This is only for the verifier process to be able to execute the event **CredVerified**. It is no harm of sending the credential over a public channel since any valid credential has been sent over a public channel when it is first generated.

Protocol 4 (Showing a credential w.r.t. a pseudonym). User U_i follows the protocol below to show a credential issued by organization O_l, to another organization O_j, using a pseudonym N_{ij} that she has established with O_j:

1. U_i chooses r'_1, r'_2, computes $A = c_{il} h_l^{r'_1}$ and $B = h_l^{r'_1}g_l^{r'_2}$, and sends N_{ij}, A, B to O_j.

2. U proves the validity of the credential and the ownership of N_{ij} in

$$PK\{(\alpha_1, \alpha_2, \alpha_3, \alpha_4, \alpha_5, \alpha_6, \alpha_7, \alpha_8) :$$

$$d_l = A^{\alpha_1}(\frac{1}{a_l})^{\alpha_2}(\frac{1}{b_l})^{\alpha_3}(\frac{1}{h_l})^{\alpha_4} \wedge B = h_l^{\alpha_5} g_l^{\alpha_6} \wedge 1 = B^{\alpha_1}(\frac{1}{h_l})^{\alpha_4}(\frac{1}{g_l})^{\alpha_7} \wedge P_{ij} = a_j^{\alpha_2} b_j^{\alpha_8}\}$$

The above proof has a fourth equation which proves that the same master secret key that is used in constructing the credential c_{il} (issued by O_l), is also used in P_{ij}, the attached validating tag of the pseudonym N_{ij} which is established with O_j.

The two processes engaged in this protocol are:

$$UV_i^2(j, l, cred, N_{ij}) \stackrel{\text{def}}{=} \nu(r_1', r_2').$$

$$\text{let } A = \exp((c_{ij}, h_j), (1, r_1')), \ B = \exp((h_j, g_j), (r_1', r_2')) \text{ in}$$

$$\text{let } zp_5 = \text{ZK}(e_{il}, x_i, s_{il}, \text{mult}(r_1', e_{il}), r_1', r_2', \text{mult}(r_2', e_{il}), s_{ij};$$

$$A, B, a_l, b_l, d_l, g_l, h_l, P_{ij}, a_j, b_j; F_5) \text{ in}$$

$$\textbf{UserShow}(U_i, cred).\overline{\text{co}}_j(k, cred, A, B, N_{ij}, zp_5)$$

$$OV_j(N, P) \stackrel{\text{def}}{=} \text{cv}(l, cred, A, B, = N, z_5).\text{ckey}(= l, pk_l).$$

$$\text{if } \text{Ver}(F_5, z_5) = \text{true then } \textbf{CredNymVerified}(O_j, N, cred, O_l)$$

where

$$F_4 \stackrel{\text{def}}{=} \beta_5 = \exp((\beta_1, \text{inv}(\beta_3), \text{inv}(\beta_4), \text{inv}(\beta_7)), (\alpha_1, \alpha_2, \alpha_3, \alpha_4))$$

$$\wedge \beta_2 = \exp((\beta_7, \beta_6), (\alpha_5, \alpha_6)) \wedge 1 = \exp((\beta_2, \text{inv}(\beta_7)), \text{inv}(\beta_6)), (\alpha_1, \alpha_4, \alpha_7))$$

$$\wedge \beta_8 = \exp((\beta_9, \beta_{10}), (\alpha_2, \alpha_8))$$

The two processes are similar as those for Protocol 3, except that the user needs to send a pseudonym to the verifier and it involves in the zero-knowledge proof.

Similar as **CredVerified**, the event **CredNymVerified** is executed in this protocol by the verifier after the verification, but it contains more information than **CredVerified**.

Transferable Constant-Size Fair E-Cash

Georg Fuchsbauer, David Pointcheval, and Damien Vergnaud

École normale supérieure, LIENS-CNRS-INRIA, Paris, France
http://www.di.ens.fr/{~fuchsbau,~pointche,~vergnaud}

Abstract. We propose a new blind certification protocol that provides interesting properties while remaining efficient. It falls in the Groth-Sahai framework for witness-indistinguishable proofs, thus extended to a certified signature it immediately yields non-frameable group signatures. We then use it to build an efficient (offline) e-cash system that guarantees user anonymity and transferability of coins without increasing their size. As required for fair e-cash, in case of fraud, anonymity can be revoked by an authority, which is also crucial to deter from double spending.

1 Introduction

1.1 Motivation

The issue of anonymity in electronic transactions was introduced for e-cash and e-mail in the early 1980's by Chaum, with the famous primitive of blind signatures [Cha83, Cha84]: a signer accepts to sign a message, without knowing the message itself, and without being able to later link a message-signature pair to the transaction it originated from. In e-cash systems, the message is a serial number to make a coin unique. The main security property is resistance to "one-more forgeries" [PS00], which guarantees the signer that after t transactions a user cannot have more than t valid signatures.

Blind signatures have thereafter been widely used for many variants of e-cash systems; in particular *fair* blind signatures [SPC95], which allow to provide revocable anonymity. They deter from abuse since in such a case the signer can ask an authority to reveal the identity of the defrauder. In order to allow the signer to control some part of the message to be signed, *partially* blind signatures [AO00] have been proposed.

Another primitive providing anonymity are group signatures [Cv91], enabling a user to sign as a member of a group without leaking any more information about his identity. The strong security model in [BSZ05] considers dynamic groups in which the group manager is not fully trusted: one thus requires that the latter cannot frame honest users.

For e-cash systems, the classical scenario is between a bank, a user and a merchant/shop: the user withdraws money from the bank and can then spend it in a shop. The latter deposits it at the bank to get its account credited. Literature tries to improve the withdrawal and the spending processes, e.g. with divisible e-cash [EO94, CG07]. However, for many applications, such as e-tickets

J.A. Garay, A. Miyaji, and A. Otsuka (Eds.): CANS 2009, LNCS 5888, pp. 226–247, 2009.

or coupons [NHS99], transferability [OO90, OO92, CG08] is a more desirable property. It is known that the size of coins grows linearly in the number of transfers [CP92]—a drawback we will avoid in our construction by modifying the model (cf. Sect. 1.3).

Classical e-cash requires that as long as a user does not spend a coin twice (double spending), she remains anonymous. Von Solms and Naccache [vSN92] pointed out that perfect anonymity enables perfect crimes, and thus suggested *fair* e-cash, where an authority can trace coins that were acquired illegally. Necessity to fight money laundering also encourages the design of fair e-cash systems enabling a trusted party to revoke the anonymity of users, whenever needed.

1.2 Contributions

Our first result is the definition and efficient pairing-based instantiation of a new primitive, which we call *partially-blind certification*. A protocol allows an issuer to interactively issue a *certificate* to a user, of which parts are then only known to the user and cannot be associated to a particular protocol execution by the issuer. The certificates are unforgeable in that from q runs of the protocol with the issuer cannot be derived more than q valid certificates. We then give two applications of the primitive:

- In order to achieve anonymity and unlinkability in group signatures, a common approach is the following: Using a signing key provided by the group manager, a user produces a signature, encrypts it and adds a proof of its validity. For this method to work efficiently in the standard model, these signing keys have to be constructed carefully. In [BW07] for example, it is the group manager that constructs the entire signing key—which means that he can impersonate (*frame*) users.
  Groth [Gro07] achieves *non-frameability* by using *certified signatures* (defined in [BFPW07]): The user chooses a verification key which is signed by the issuer. A signature produced with the corresponding signing key together with the verification key and the issuer's signature on it can then be verified under the issuer's key. Security of Groth's instantiation however relies on an unnatural assumption.
  We avoid this by observing the following: it is not necessary that the user choose the verification key, as long as she can be sure that the private key contains enough entropy. Since the blind component of our instantiation of our primitive can serve as signing key, our construction applies immediately to build non-frameable group signatures (see Sect. 4).
- Second, in e-cash, the serial number of a coin needs to contain enough entropy to avoid collisions, but again the user need not control it entirely. Partially-blind certificates are applicable here too.

1.3 Transferable Anonymous Constant-Size Fair E-Cash

The instantiation we give of our new primitive allows it to be combined with the results of Groth and Sahai [GS08], which is crucial to our main contribution:

an efficient standard-model anonymous fair e-cash system in the classical three-party scenario with the following novel features:

First, coins are transferable while remaining constant in size. We circumvent the impossibility results by introducing a new method to trace double spenders: the users keep *receipts* when receiving coins instead of storing all information about transfers inside the coin. The amount of data a user has to deal with is thus proportional to the number of coins he received, rather than the path a coin took until reaching him.

Second, partial blindness of our certificates provides the strongest possible notion of anonymity: a user remains anonymous even w.r.t. an entity issuing coins and able to *detect* double spendings.[1] Moreover, coins are unlinkable to anyone except the authority and the double-spending detector. We give an overview of our model before getting back to its security properties.

- The participants of the system are the following: the system manager (that registers users within the system), the bank (issuing coins), users (that withdraw, transfer or spend coins), merchants to which coins are spent, the double-spending detector, and a trusted authority, called tracer, that can trace coins, revoke anonymity and identify double-spenders.
- In order to get a coin, a user runs a withdrawal protocol with the bank, after which he holds a coin and a receipt to be kept even after transferring or spending the coin (to defend himself against wrongful accusation of double-spending).[2]
- Another protocol enables users to transfer coins to other users who, besides the coin, also get a receipt, which they keep too.
- To spend the coin, the user interacts with a merchant. The latter will deposit the coin at the bank who invokes the double-spending detector to check if it has already been spent. If it is the case, the tracer is invoked to reveal the double spender. He does so by tracing back the two instances of the coin by asking the receipts from the users that transferred the coins until identifying the double spender.

Note that the tracing authority identifies *innocent* users that merely transferred a coin that has been used fraudulently before. However, this does not weaken anonymity, which does not hold against the tracer anyway and since identities are not revealed to anyone else. Moreover, this can be proved to be unavoidable in order to achieve constant-size transferable coins. An inevitable shortcoming of our model is that a user who *loses* a receipt can be accused of double spending, since he cannot prove legal acquisition of the coin if he transferred it. The system satisfies the following security notions:

---

[1] In fair e-cash, there exists an authority that can trace users (user-tracing) and coins (coin-tracing) under a judge decision, in case of fraud suspicion (not necessarily double spending). We separate the notions of *detection* of double spendings, which is done on a regular basis when a coin is deposited, from that of *tracing*, which is performed by a trusted authority only when a fraud was committed.

[2] If one assumes a validity period for coins (after which the issuing key is changed), it suffices to keep a receipt only as long as the respective coin is valid.

- Any user who spends a coin twice is detected.
- As long as a user keeps all his receipts, he cannot be wrongfully accused of double spending, even if everyone else colludes against him.
- A user is anonymous even against collusions of the manager, the bank, the double-spending detector, merchants, and other users.
- Transfers of coins are *unlinkably* anonymous to collusions possibly comprising the manager, the bank, merchants, and other users. (The double-spending detector must necessarily be able to link two spendings of the same coin.)

Our construction is secure in the standard security model (i.e., without relying on the random oracle idealization [BR93])[3] and its security is based on a new (though natural) assumption that holds in the generic group model [Sho97].

1.4 Organization of the Paper

In the next section, we state the employed assumptions. In Sect. 3, we describe our new *Partially-Blind Certification* primitive, and apply it to group signatures in Sect. 4. In Sect. 5, we extend some techniques of Groth-Sahai, recapitulating re-randomization of commitments and introducing proofs for relations of values committed under *different* keys. In Sect. 6, we combine everything to construct our e-cash system.

2 Assumptions

We present the assumptions on bilinear groups on which our security results build. A *bilinear group* is a tuple $(p, \mathbb{G}, \mathbb{G}_T, e, G)$ where $(\mathbb{G}, +)$ and $(\mathbb{G}_T, \cdot)$ are two cyclic groups of order p, G is a generator of $\mathbb{G}$, and $e: \mathbb{G} \times \mathbb{G} \to \mathbb{G}_T$ is a non-degenerate bilinear map, i.e., $\forall U, V \in \mathbb{G} \; \forall a, b \in \mathbb{Z}: e(aU, bV) = e(U, V)^{ab}$, and $e(G, G)$ is a generator of $\mathbb{G}_T$.

The first two of the following assumptions are classical [DH76, BBS04]. The third is a simple extension of the Hidden Strong Diffie-Hellman Problem proposed by Boyen and Waters in [BW07].

Definition 1. *The* Computational Diffie-Hellman (CDH) Assumption *states that the following problem is intractable[4]: given $(G, \alpha G, \beta G) \in \mathbb{G}^3$, for $\alpha, \beta \in \mathbb{Z}_p$, output $\alpha\beta G$.*

---

[3] Note that in our context, due to re-randomization of encryptions (cf. Sect. 6.2 for details), it seems even impossible to replace the Groth-Sahai techniques with the Fiat-Shamir heuristic [FS87] to improve efficiency at the expense of relying on the random oracle model.

[4] We say that a computational problem is *intractable* if no probabilistic polynomial-time (p.p.t.) adversary can solve it with non-negligible probability. A decisional problem is *intractable* if no p.p.t. adversary can decide it with probability of non-negligibly more than $1/2$.

Definition 2. *The* Decisional Linear (DLIN) Assumption *states that the following problem is intractable: given* $(U, V, G, \alpha U, \beta V, \gamma G) \in \mathbb{G}^6$, *decide whether* $\gamma = \alpha + \beta$ *or not.*

Definition 3. *The* q-Double Hidden Strong Diffie-Hellman (DHSDH) Assumption *states that the following problem is intractable: given* $(G, H, K, \Gamma = \gamma G) \in \mathbb{G}^4$ *and* $q - 1$ *tuples*

$$\left(X_i = x_i G, \ X_i' = x_i H, \ Y_i = y_i G, \ Y_i' = y_i H, \ A_i = \frac{1}{\gamma + x_i}(K + y_i G)\right)$$

with $x_i, y_i \leftarrow \mathbb{Z}_p^*$ $(1 \leq i \leq q - 1)$, *output a new tuple* $\left(X = xG, \ X' = xH, \ Y = yG, \ Y' = yH, \ A = \frac{1}{\gamma + x}(K + yG)\right)$.

Note that a tuple (X, X', Y, Y', A) has the above format if and only if it satisfies

$$e(X, H) = e(G, X') \quad e(Y, H) = e(G, Y') \quad e(A, \Gamma + X) = e(K + Y, G)$$

Remark 4. Boneh and Boyen [BB04] introduced the Strong Diffie-Hellman (SDH) assumption stating that given a $(q + 1)$-tuple $(G, \gamma G, \gamma^2 G, \ldots, \gamma^q G) \in \mathbb{G}^{q+1}$ for a random $\gamma \leftarrow \mathbb{Z}_p^*$, it is infeasible to output a pair $(x, \frac{1}{\gamma + x} G) \in \mathbb{Z}_p \times \mathbb{G}$. Hardness of SDH implies hardness of the following two problems (the first implication is proven in [BB04], the second in the full version [FPV09]):

(I) Given $G, \gamma G \in \mathbb{G}$ and $q - 1$ distinct pairs $(x_i, \frac{1}{\gamma + x_i} G) \in \mathbb{Z}_p \times \mathbb{G}$, output a *new* pair $(x, \frac{1}{\gamma + x} G) \in \mathbb{Z}_p \times \mathbb{G}$.

(II) Given $G, K, \gamma G \in \mathbb{G}$ and $q - 1$ distinct triples $(x_i, y_i, \frac{1}{\gamma + x_i}(K + y_i G)) \in \mathbb{Z}_p^2 \times \mathbb{G}$, output a *new* triple $(x, y, \frac{1}{\gamma + x}(K + yG)) \in \mathbb{Z}_p^2 \times \mathbb{G}$.

The Hidden SDH problem defined in [BW07] is a variant of Problem (I), where instead of giving the x_i's explicitly, they are given as $(x_i G, x_i H)$. Similarly, the goal is to output a new triple $(xG, xH, \frac{1}{\gamma + x} G)$. Now the Double Hidden SDH assumption (Definition 3) transforms Problem (II) the same way: instead of being given explicitly, x_i and y_i are given as $(x_i G, x_i H, y_i G, y_i H)$. In the full version [FPV09] we discuss assumptions derived from SDH and their relations.

3   Partially-Blind Certification

3.1   Model

Definition 5. *A* partially-blind certification scheme (Setup, Sign, User, Verif) *is a 4-tuple of (interactive) probabilistic polynomial-time Turing machines (PPTs) such that:*

- Setup *is a PPT that takes as input an integer* k *and outputs a pair* (pk, sk) *of public (resp. secret) key. We call* k *the* security parameter.

Experiment $\mathsf{Exp}_{\mathcal{A}}^{\mathsf{blindness}-b}(k)$

$(pk, state) \leftarrow \mathcal{A}(\mathrm{FIND}, k)$

$\tau_0 \leftarrow \mathcal{T}$

$(\sigma_1, \tau_1) \ (\neq \perp) \leftarrow \mathsf{User}^{\mathcal{A}(state)}(pk)$

$b' \leftarrow \mathcal{A}(\mathrm{GUESS}, \tau_b)$

RETURN b'

Experiment $\mathsf{Exp}_{\mathcal{A}}^{\mathsf{forge}}(k)$

$(pk, sk) \leftarrow \mathsf{Setup}(k)$

$((\sigma_1, \tau_1), \ldots, (\sigma_\ell, \tau_\ell)) \leftarrow \mathcal{A}^{\mathsf{Sign}(sk, \cdot)}(pk)$

IF $\forall i \in [1, \ell], \mathsf{Verif}(pk, (\sigma_i, \tau_i)) = \mathtt{accept}$

AND $\forall(i, j) \in [1, \ell]^2, i \neq j \colon (\sigma_i, \tau_i) \neq (\sigma_j, \tau_j)$

AND $\ell > m$ RETURN 1

where m is the number of executions of the certificate issuing protocol where Sign outputs completed.

(1) Partial Blindness   (2) Unforgeability

Fig. 1. Security experiments for partially-blind certificates

- Sign *and* User *are interactive PPTs such that* User *takes as inputs a public key pk and* Sign *takes as input the matching secret key sk.* Sign *and* User *engage in the certificate-issuing protocol and when they stop,* Sign *outputs* completed *or* not-completed *while* User *outputs a pair of bit strings* (σ, τ) *or* $\perp$.
- Verif *is a deterministic polynomial-time Turing machine that on input a public key pk and a pair of bit strings* (σ, τ) *outputs either* accept *or* reject.

For all $k \in \mathbb{N}$, *all pairs* (pk, sk) *output by* Setup(k), *if* Sign *and* User *follow the certificate issuing protocol with input sk and pk respectively, then* Sign *outputs* completed *and* User *outputs a pair* (σ, τ) *that satisfies* Verif$(pk, (\sigma, \tau)) = $ accept. *A pair* (σ, τ) *is termed* valid *with regard to pk if on input* $(pk, (\sigma, \tau))$ Verif *outputs* accept, *in which case, we say that* (σ, τ) *is a* certificate *for pk and* τ *is termed the* blind component *of the certificate. We denote* $\mathcal{T} \subset \{0, 1\}^*$ *the set of bit-strings which are blind component of some certificate.*

Partial Blindness. To define partial blindness, we consider the *real-or-random* game (i.e., random experiment) among an adversarial signer $\mathcal{A}$ and a challenger presented in Fig. 1 (1).

- We define the advantage of $\mathcal{A}$ in breaking partial blindness by its advantage in distinguishing the two above experiments (with $b = 0$ or $b = 1$):

$$\mathsf{Adv}_{\mathcal{A}}^{\mathsf{blindness}}(k) := \Pr[\mathsf{Exp}_{\mathcal{A}}^{\mathsf{blindness}-1}(k) = 1] - \Pr[\mathsf{Exp}_{\mathcal{A}}^{\mathsf{blindness}-0}(k) = 1] \ ,$$

where the probability is taken over the coin tosses made by the challenger and $\mathcal{A}$.
- The scheme (Setup, Sign, User, Verif) is said to be *partially blind* if no adversary $\mathcal{A}$ running in probabilistic polynomial time has a non-negligible advantage $\mathbf{Adv}_{\mathcal{A}}^{\mathsf{blindness}}$.

Unforgeability. To define unforgeability, we introduce the game among an adversarial user $\mathcal{A}$ and an honest signer Sign depicted in Fig. 1 (2).

---

(1) User Choose $r, y_1 \leftarrow \mathbb{Z}_p$, compute and send: $R_1 := r(K + y_1 G)$, $T := rG$
and zero-knowledge proofs of knowledge of r and y_1 satisfying the relations (cf. Remark 7).

(2) Sign Choose $x, y_2 \leftarrow \mathbb{Z}_p$ and compute $R := R_1 + y_2 T$ (note that $R = r(K + yG)$ with $y := y_1 + y_2$.)

Send $(S_1 := \frac{1}{\omega + x} R,\ S_2 := xG,\ S_3 := xH,\ S_4 := y_2 G,\ S_5 := y_2 H)$

(3) User Check whether $(S_1, S_2, S_3, S_4, S_5)$ is correctly formed:

$$e(S_2, H) \stackrel{?}{=} e(G, S_3) \qquad e(S_4, H) \stackrel{?}{=} e(G, S_5) \qquad e(S_1, \Omega + S_2) \stackrel{?}{=} e(R, G)$$

If so, compute a certificate

$$(A := \tfrac{1}{r} S_1,\ X := S_2,\ X' := S_3,\ Y := y_1 G + S_4 = yG,\ Y' := y_1 H + S_5 = yH)$$

Fig. 2. Partially-blind certificate-issuing protocol

- We define the *success* of $\mathcal{A}$ in this game by

$$\mathbf{Succ}_{\mathcal{A}}^{\mathrm{unforge}}(k) := \Pr[\mathrm{Exp}_{\mathcal{A}}^{\mathrm{forge}}(k) = 1]\ ,$$

where the probability is taken over the coin tosses made by $\mathcal{A}$, Setup and Sign.
- The scheme (Setup, Sign, User, Verif) is said to be *unforgeable* if no adversary $\mathcal{A}$ running in probabilistic polynomial time has a non-negligible success $\mathbf{Succ}_{\mathcal{A}}^{\mathrm{unforge}}$.

Remark 6. In the experiment $\mathrm{Exp}_{\mathcal{A}}^{\mathrm{forge}}$, depending on the security model, the executions of the certificate issuing protocol are run sequentially or in a concurrent and interleaving way.

3.2 Instantiation

Let $(p, \mathbb{G}, \mathbb{G}_T, e, G)$ be a bilinear group and $G, H, K \in \mathbb{G}$ be public parameters; define the signer's key pair as $sk := \omega \leftarrow \mathbb{Z}_p$ and $pk = \Omega := \omega G$. A certificate is defined as

$$\mathsf{Crt}(\omega; x, y) := \left\{ A = \frac{1}{\omega + x}(K + yG) \quad \begin{array}{ll} X = xG & Y = yG \\ X' = xH & Y' = yH \end{array} \right.$$

for $x, y \leftarrow \mathbb{Z}_p$, with $\sigma := (A, X, X', Y)$ and the blind component $\tau := Y' \in \mathbb{G}$. It satisfies:

$$e(X, H) = e(G, X') \quad e(Y, H) = e(G, Y') \quad e(A, \Omega + X) = e(K + Y, G) \quad (1)$$

Fig. 2 depicts an efficient protocol to interactively generate such a certificate between the signer (issuer) that controls x and the user that partially controls

y: at the end, the signer has no information about y, except that it is uniformly distributed.

Remark 7. In the first round of the User protocol, one can use interactive Schnorr-like zero-knowledge proofs of knowledge (ZKPoK) [Sch90]. Extraction is then only possible for constant-depth concurrency [Oka06]. To achieve *full* concurrency, and at the same time reduce interactivity to 2 moves, one can use the following technique: Make linear commitments [GOS06] (cf. Sect. 5.1) to the bits of r and y_1 (which are extractable) and use the proof techniques from [FP09, Appendix A.3 of the full version]. The drawbacks of this approach are that security holds in the common reference string (CRS) model and we incur a loss of efficiency.

3.3 Security Results

Theorem 8. *Under DHSDH, the above certificates are unforgeable.*

Proof. Let $\mathcal{A}$ be an adversary impersonating corrupt users running the issuing protocol up to $q-1$ times and then outputting q different valid certificates. We build $\mathcal{B}$ solving q-DHSDH with the same probability by simulating the signer: $\mathcal{B}$ gets a q-DHSDH-instance $\left(G, H, K, \Omega, (A_i, X_i, X_i', Y_i, Y_i')_{i=1}^{q-1}\right)$. If the ZKPoK are non-interactive, it sets the CRS so that it can extract r and y_1 used in R_1 and T—if they are interactive, $\mathcal{B}$ rewinds $\mathcal{A}$ to extract. In each issuing, $\mathcal{A}$ first sends $(R_{1,i}, T_i)$ and proofs of knowledge. If the proofs are correct, $\mathcal{B}$ extracts r_i, $y_{1,i}$ from them and sends $(S_{1,i} := r_i A_i,\ S_{2,i} := X_i,\ S_{3,i} := X_i',\ S_{4,i} := Y_i - y_{1,i}G,\ S_{5,i} := Y_i' - y_{2,i}H)$. Finally, $\mathcal{B}$ checks the q certificates and forwards one different from those in the DHSDH-instance to its own challenger.   □

Theorem 9. *Under DLIN, the above certificates are partially blind.*

Proof. Consider $\mathcal{A}$, which after an execution of the blind issuing protocol can decide whether the blind component $\tau = Y'$ is real or random in $\mathbb{G}$. We build $\mathcal{B}$ deciding DLIN with the success probability of $\mathcal{A}$. The algorithm $\mathcal{B}$ gets a DLIN-instance (H, G, T, Z, K, R_1), i.e., it has to decide whether

$$R_1 \stackrel{?}{=} \left(\log_H Z + \log_G K\right) T \tag{2}$$

It gives $\mathcal{A}$ the triple (G, H, K) as public parameters (and a simulating CRS in case we use non-interactive ZKPoK) and gets Ω, the issuer's public key from $\mathcal{A}$. $\mathcal{B}$ runs the protocol User with $\mathcal{A}$, starting by sending R_1, and T from its DLIN instance and simulating the PoK.

After getting back $(S_1, \ldots, S_5)$, $\mathcal{B}$ checks its correctness and gives $\mathcal{A}$ the following: $Y' := Z + S_5$, with Z from its DLIN instance. ($\mathcal{B}$ can verify correctness of S without knowledge of y_1 and r by checking $e(S_2, H) = e(G, S_3)$, $(S_4, H) = e(G, S_5)$ and $e(S_1, \Omega + S_2) = e(R, G)$. Also note that $\mathcal{B}$ only needs to produce the last (blind) component of the certificate.) Finally $\mathcal{A}$ outputs a guess b', which $\mathcal{B}$ forwards to its DLIN challenger.

- If the DLIN instance is not a linear tuple then Z and therefore Y' is independently random.
- If (H, G, T, Z, K, R_1) is linear, then with $y_1 := \log_H Z$, $\kappa := \log_G K$, and $r := \log_G T$, we have $R_1 = (y_1 + \kappa)T$ by (2). Furthermore, for public parameters (G, H, K), we have

$$T = rG \qquad R_1 = (y_1 + \kappa)T = (y_1 + \kappa)rG = r(K + y_1 G) \qquad Z = y_1 H$$

which means that $Y' = Z + S_5$ is the blind component of a correctly produced certificate.

If $\mathcal{B}$ outputs the bit returned by $\mathcal{A}$, its success probability is equal to $\mathbf{Adv}_{\mathcal{A}}^{\text{blindness}}$.

□

4 A Fully-Secure Group Signature from Partially-Blind Certificates

As a first application of the certification protocol from Sect. 3.2, we construct *fully-secure* dynamic group signatures (in the sense of [BSZ05], in particular satisfying non-frameability and CCA-anonymity) without random oracles. We construct a *certified-signature* scheme, to which can then be applied Groth's [Gro07] methodology of transforming certified signatures that respect a certain structure into group signatures using Groth-Sahai NIZK proofs [GS08] and Kiltz' tag-based encryption [Kil06], both relying exclusively on the DLIN assumption.

The resulting scheme is less efficient than that from [Gro07]; however, it is based on a more natural assumption, while at the same time being of the same order of magnitude—especially compared to the first instantiations of fully-secure signatures in the standard model (e.g., [Gro06]). We think of the scheme as somehow being the "natural" extension of [BW07] in order to satisfy non-frameability.

Certified Signatures. A certified-signature scheme consists of a setup algorithm, a key-generation algorithm for the certification authority, an interactive protocol between the authority ("issuer") and a user letting the latter obtain a triple $(cert, vk, sk)$, where vk is a verification key for a signature scheme, sk is the corresponding signing key (unknown to the issuer) and $cert$ is a certificate on vk.

Besides correctness, Groth [Gro07] gives two security criteria that a certified signature must satisfy to be transformable into a secure group signature scheme: *Unfakeability* requires that no user can create a certificate for and a signature under a verification key that was not certified by the issuer. *Unforgeability* means that even a corrupt authority issuing a tuple $(cert, vk, sk)$ cannot forge a signature under vk.

Our Instantiation. Our certified signature is constructed from a certificate (A, X, X', Y, Y') by using (Y, Y') as a pair of public and secret key for Waters' signature scheme [Wat05]. A certified signature consists thus of the first four components of the certificate prepended to a Waters signature. Note that what is called *cert* above corresponds to (A, X, X') here, and (vk, sk) would be

Let $(U_i)_{i=0}^n \in \mathbb{G}^{n+1}$ be part of the public parameters; let Ω be the issuer's public key.

Certificate Generation. Run the certificate-creation protocol in Fig. 2, except that the issuer running Sign sends an extractable commitment of $S_4 = y_2 G$ before phase (1) and opens it in phase (2).

Signing. For a message $M = (m_1, \ldots, m_n) \in \{0,1\}^n$, denote $\mathcal{F}(M) := U_0 + \sum_{i=1}^n m_i U_i$. Given a certificate $C = (A, X, X', Y, Y')$, a signature on M using randomness $s \in \mathbb{Z}_p$ is defined as

$$\mathsf{Sig}(C, M\,;\,s) := (A, X, X', Y, Y' + s\mathcal{F}(M), -sG) .$$

Verification. A certified signature (A, X, X', Y, Z, Z') on message M is verified by checking

$$e(X, H) = e(G, X') \quad e(Y, H) = e(G, Z)\, e(Z', \mathcal{F}(M)) \quad e(A, \Omega + X) = e(K + Y, G)$$

Fig. 3. Chosen-message secure certified signature

(Y, Y'). The scheme is given in Fig. 3. Our construction satisfies the security requirements given by Groth:

Theorem 10. *The certified-signature scheme in Fig. 3 is perfectly correct, unfakeable under DHSDH, and existentially unforgeable under chosen-message attack under CDH.*

Proof. Correctness follows by inspection. The two other properties are proven similarly to Theorems 8 and 9, we thus highlight the differences.

(1) Unfakeability means that after running the issuing protocol multiple times with the issuer, no user is able to produce a valid tuple (A, X, X', Y, Z, Z') with Y different from those in the obtained certificates. The proof works similarly to that of Theorem 8 with the following modifications: For $0 \le i \le n$, $\mathcal{B}$ chooses $\mu_i \leftarrow \mathbb{Z}_p$ and sets the public parameters $U_i := \mu_i G$. In the issuing protocol, $\mathcal{B}$ simulates the additional commitment at the beginning. From a valid (A, X, X', Y, Z, Z') returned by $\mathcal{A}$, $\mathcal{B}$ can then extract a new certificate by setting $Y' := Z + (\mu_0 + \sum m_i \mu_i) Z'$.

(2) Existential unforgeability under chosen-message attack (EUF-CMA) follows from partial blindness of certificates and security of Waters signatures, which is implied by CDH (Def. 1): Let $\mathcal{A}$ be an adversary impersonating the issuer and mounting a chosen-message attack. We construct $\mathcal{B}$ against EUF-CMA of Waters signatures. $\mathcal{B}$ is given a Waters public key $V \in \mathbb{G}$ and a signing oracle.

$\mathcal{B}$ runs the certificate-generation protocol playing the role of User with $\mathcal{A}$. When $\mathcal{A}$ sends a commitment to S_4 in the first phase of the protocol, $\mathcal{B}$ extracts S_4 from it. It then chooses r, sends $R_1 := r(K + V - S_4)$ and $T := rG$ and simulates the zero-knowledge proofs. (Note that this implicitly sets $V = (y_1 + y_2)G$.) If $\mathcal{A}$ returns a valid tuple $(S_1, S_2, S_3, S_4, S_5)$, $\mathcal{B}$ can compute an (incomplete) certificate $(A := \frac{1}{r}S_1, X := S_2, X' := S_3, Y := V)$ which suffices to answer $\mathcal{A}$'s signing queries, as $\mathcal{B}$ can get the last two components by querying its own oracle. When $\mathcal{A}$ returns a successful forgery, $\mathcal{B}$ returns the last two components, i.e., a Waters signature under public key V. □

5 New Techniques for Groth-Sahai Proof Systems

5.1 Preliminaries

We briefly review the results of [GS08] relevant to our paper: witness-indistinguishable (WI) proofs that elements in $\mathbb{G}$ that were committed to via *linear commitments* satisfy *pairing-product equations*. We refer to the original work for more details and proofs.

Let $P \in \mathbb{G}$ be a generator. We define a key for *linear commitments*. Choose $\alpha, \beta, r_1, r_2 \in \mathbb{Z}_p$ and define $U = \alpha P$, $V = \beta P$, and

$$\mathbf{u}_1 := (U, 0, P) \qquad \mathbf{u}_2 := (0, V, P) \qquad \mathbf{u}_3 := (W_1, W_2, W_3) \qquad (3)$$

where $W_1 := r_1 U$, $W_2 := r_2 V$, and W_3 is either

- soundness setting: $W_3 := (r_1 + r_2)P$ (which makes $\vec{\mathbf{u}}$ a binding key)
- WI setting: $W_3 := (r_1 + r_2 - 1)P$ (which makes $\vec{\mathbf{u}}$ a hiding key)

Under key $ck = (U, V, W_1, W_2, W_3)$, a commitment to a group element $X \in \mathbb{G}$ using randomness $(s_1, s_2, s_3) \leftarrow \mathbb{Z}_p^3$ is defined as (with $\iota(X) := (0, 0, X)$)

$$\mathsf{Com}(ck, X; (s_1, s_2, s_3)) := \iota(X) + \sum_{i=1}^{3} s_i \mathbf{u}_i$$
$$= (s_1 U + s_3 W_1, \; s_2 V + s_3 W_2, \; X + s_1 P + s_2 P + s_3 W_3) \; .$$

Note that in the soundness setting, given the *extraction key* $ek := (\alpha, \beta)$, the committed value can be extracted from a commitment $\mathbf{c} = (c_1, c_2, c_3)$:

$$\mathsf{Extr}((\alpha, \beta), \mathbf{c}) := c_3 - \tfrac{1}{\alpha} c_1 - \tfrac{1}{\beta} c_2$$
$$= X + (s_1 + s_2 + s_3(r_1 + r_2))P - \tfrac{1}{\alpha}(s_1 + s_3 r_1)U - \tfrac{1}{\beta}(s_2 + s_3 r_2)V = X \; ,$$

since $\tfrac{1}{\alpha} U = P$ and $\tfrac{1}{\beta} V = P$. On the other hand, in the WI setting we have (with $s_1' := s_1 + s_3 r_1$ and $s_2' = s_2 + s_3 r_2$): $\mathbf{c} = (s_1' U, s_2' V, X + (s_1' + s_2' - s_3)P)$, which is equally distributed for every X. The two settings are indistinguishable by DLIN since for soundness (W_1, W_2, W_3) is linear w.r.t. (U, V, P), whereas in the WI setting it is not.

For the sake of readability and consistency with the work of [GS08], we stick to their abstract notation, which we sketch briefly:

- For a vector $\vec{\mathcal{X}} = (\mathcal{X}_1, \dots, \mathcal{X}_n)^\top \in \mathbb{G}^n$, let $\vec{\mathcal{X}} \cdot \vec{\mathcal{Y}} := \prod_{i=1}^{n} e(\mathcal{X}_i, \mathcal{Y}_i)$.
- Bold letters denote triples, e.g., $\mathbf{d} = (d_1, d_2, d_3) \in \mathbb{G}^{1 \times 3}$, $\vec{\mathbf{d}}$ denotes a column vector of triples, thus a matrix in $\mathbb{G}^{n \times 3}$. Furthermore, define $\widetilde{F}(\mathbf{c}, \mathbf{d}) := [e(c_i, d_j)]_{i,j=1,3} \in \mathbb{G}_T^{3 \times 3}$. In $\mathbb{G}_T^{3 \times 3}$, "+" denotes entry-wise multiplication of matrix elements. Define $\mathbf{c} \bullet \mathbf{d} := \sum_{i=1}^{n} \left(\frac{1}{2} \widetilde{F}(c_i, d_i) + \frac{1}{2} \widetilde{F}(d_i, c_i) \right)$.

A *pairing-product equation* is an equation for variables $\mathcal{Y}_1, \ldots, \mathcal{Y}_n \in \mathbb{G}$ of the form

$$\prod_{i=1}^{n} e(\mathcal{A}_i, \mathcal{Y}_i) \prod_{i=1}^{n} \prod_{j=1}^{n} e(\mathcal{Y}_i, \mathcal{Y}_j)^{\gamma_{i,j}} = t_T ,$$

with $\mathcal{A}_i \in \mathbb{G}$, $\gamma_{i,j} \in \mathbb{Z}_p$ and $t_T \in \mathbb{G}_T$. Setting $\Gamma := [\gamma_{i,j}]_{i,j=1,\ldots,n} \in \mathbb{Z}_p^{n \times n}$, this can be written as

$$(\vec{\mathcal{A}} \cdot \vec{\mathcal{Y}})(\vec{\mathcal{Y}} \cdot \Gamma \vec{\mathcal{Y}}) = t_T . \tag{4}$$

Set $H_1 := \begin{bmatrix} 0 & 1 & 0 \\ -1 & 0 & 0 \\ 0 & 0 & 0 \end{bmatrix}, H_2 := \begin{bmatrix} 0 & 0 & 1 \\ 0 & 0 & 0 \\ -1 & 0 & 0 \end{bmatrix}, H_3 := \begin{bmatrix} 0 & 0 & 0 \\ 0 & 0 & 1 \\ 0 & -1 & 0 \end{bmatrix}$, and $\iota_T(t_T) := \begin{bmatrix} 1 & 1 & 1 \\ 1 & 1 & 1 \\ 1 & 1 & t_T \end{bmatrix}$

for $t_T \in \mathbb{G}_T$.

Let $\vec{\mathbf{d}}$ be a vector of commitments to $\vec{\mathcal{Y}}$, i.e., $\vec{\mathbf{d}} := \iota(\vec{\mathcal{Y}}) + S\vec{\mathbf{u}}$ with $S \leftarrow \mathbb{Z}_p^{n \times 3}$ and $\iota(\vec{\mathcal{Y}}) := [\iota(\mathcal{Y}_i)]_{i=1,\ldots,n}$. The proof that the values committed in $\vec{\mathbf{d}}$ satisfy (4) is defined as

$$\Phi := S^{\top}\iota(\vec{\mathcal{A}}) + S^{\top}\Gamma\iota(\vec{\mathcal{Y}}) + S^{\top}\Gamma^{\top}\iota(\vec{\mathcal{Y}}) + S^{\top}\Gamma S\vec{\mathbf{u}} + \sum_{i=1}^{3} r_i H_i \vec{\mathbf{u}} , \tag{5}$$

with $r_1, r_2, r_3 \leftarrow \mathbb{Z}_p$, and is verified by

$$\iota(\vec{\mathcal{A}}) \bullet \vec{\mathbf{d}} + \vec{\mathbf{d}} \bullet \Gamma \vec{\mathbf{d}} = \iota_T(t_T) + \vec{\mathbf{u}} \bullet \Phi . \tag{6}$$

Soundness and WI of the proofs. In the soundness setting, if $\vec{\mathbf{d}}$ satisfies (6) for some Φ, then Extr extracts $\vec{\mathcal{Y}}$ satisfying (4). In the WI setting, let $\vec{c}$ and $\vec{\mathbf{d}}$ be commitments to $\vec{\mathcal{X}}$ and $\vec{\mathcal{Y}}$, resp., which both satisfy (4). Then Φ and Φ' constructed as in (5) for $\vec{c}$ and $\vec{\mathbf{d}}$, resp., are equally distributed.

5.2   Commitment Re-randomization and Proof Updating

As observed by [FP09] and [BCC+09], commitments of this form can be *re-randomized* and the corresponding proofs adapted without knowledge of the committed values nor the used randomness: Given a commitment $\vec{\mathbf{d}}$, set $\vec{c} := \vec{\mathbf{d}} + \widetilde{S}\vec{\mathbf{u}}$, for $\widetilde{S} \leftarrow \mathbb{Z}_p^{n \times 3}$, and *update* the proof Φ for $\vec{\mathbf{d}}$ to $\widetilde{\Phi}$ for $\vec{c}$:

$$\widetilde{\Phi} := \Phi + \widetilde{S}^{\top}\iota(\vec{\mathcal{A}}) + \widetilde{S}^{\top}\Gamma\vec{\mathbf{d}} + \widetilde{S}^{\top}\Gamma^{\top}\vec{\mathbf{d}} + \widetilde{S}^{\top}\Gamma\widetilde{S}\vec{\mathbf{u}} + \sum_{i=1}^{3}\widetilde{r}_i H_i \vec{\mathbf{u}} \tag{7}$$

with $\widetilde{r}_i \leftarrow \mathbb{Z}_p$. The pair $(\vec{c}, \widetilde{\Phi})$ satisfies (6) and some calculation shows that $\widetilde{\Phi}$ is constructed as in (5) for $\vec{c}$ being a commitment to $\vec{\mathcal{Y}}$ using randomness $S + \widetilde{S}$. (In particular (7) yields the same $\widetilde{\Phi}$ as (5) if in the latter the randomness used for the proof is $(r_i + \alpha_i + \widetilde{r}_i)_{i=1}^{3}$, where (r_1, r_2, r_3) is the randomness of Φ and $\alpha_1, \alpha_2, \alpha_3$ are such that $A := \widetilde{S}^{\top}\Gamma^{\top}S - S^{\top}\Gamma\widetilde{S} = \sum_{i=1}^{3}\alpha_i H_i$; such α_i exist since

A satisfies $\vec{u} \bullet A\vec{u} = 0$ and the H_i's form a basis for the matrices of this form; cf. [GS08, Chapter 4].)

5.3   Linear Equations and Different Commitment Keys

Consider two commitments c, d of Y, Z under *different* commitment keys $\vec{u}$ and $\vec{u}'$, respectively. We construct a re-randomizable WI proof that the committed values satisfy

$$e(H, Y) = e(G, Z) . \tag{8}$$

Let c be a commitment to Y w.r.t. key $\vec{u}$: $c := (s_{Y1}U + s_{Y3}W_1, s_{Y2}V + s_{Y3}W_2, Y + s_{Y1}P + s_{Y2}P + s_{Y3}W_3)$. The proof that the committed value Y satisfies (8) (in which Z is considered as a constant) is[5] $\pi := (s_{Y1}H, s_{Y2}H, s_{Y3}H)$, which is verified by

$$e(\pi_1, U)\, e(\pi_3, W_1) = e(H, c_1) \tag{9a}$$

$$e(\pi_2, V)\, e(\pi_3, W_2) = e(H, c_2) \tag{9b}$$

$$e(Z, G)\, e(\pi_1, P)\, e(\pi_2, P)\, e(\pi_3, W_3) = e(H, c_3) \tag{9c}$$

Regarding (9) as a set of equations over variables $c_1, c_2, c_3, Z, \pi_1, \pi_2, \pi_3$, we could just use the Groth-Sahai proof system a second time by committing to the new variables under key $\vec{u}'$ and making proofs for the equations in (9). However, this can be optimized, since we need not commit to c_1, c_2 and c_3. Correctness and soundness follow from a simple hybrid argument.

Let us consider witness indistinguishability. We show that every pair (Y, Z) satisfying (8) generates the same distribution of proofs once both keys $\vec{u}$ and $\vec{u}'$ are replaced by hiding keys. Let (Y, Z) satisfying (8) be arbitrarily fixed. Since u is perfectly hiding, for any given c there exist (s_1, s_2, s_3) s.t. $c = \iota(Y) + \sum_{i=1}^{3} s_i u_i$. Now WI under key $\vec{u}'$ (of the second layer of commitments/proofs) ensures that every (Z, π_1, π_2, π_3) satisfying (9) (with the c_i's fixed!) generates identically distributed proofs. Thus for $Z := (\log_G Y)H$, $\pi_i := s_i H$, the proof does not leak anything. We present the details:

We make commitments to Z, $\pi_1 = s_{Y1}H$, $\pi_2 = s_{Y2}H$, $\pi_3 = s_{Y3}H$ w.r.t. $\vec{u}'$:

$$d := \begin{bmatrix} s_{Z1}U' + s_{Z3}W_1' \\ s_{Z2}V' + s_{Z3}W_2' \\ Z + s_{Z1}P' + s_{Z2}P' + s_{Z3}W_3' \end{bmatrix} \quad p_i := \begin{bmatrix} t_{i,1}U' + t_{i,3}W_1' \\ t_{i,2}V' + t_{i,3}W_2' \\ s_{Yi}H + t_{i,1}P' + t_{i,2}P' + t_{i,3}W_3' \end{bmatrix} \tag{10}$$

for $1 \leq i \leq 3$. The proof ψ_i for the i-th equation in (9) is defined as follows:

$$\psi_{1,j} := t_{1,j}U + t_{3,j}W_1 \qquad \psi_{2,j} := t_{2,j}V + t_{3,j}W_2$$
$$\psi_{3,j} := s_{Zj}G + t_{1,j}P + t_{2,j}P + t_{3,j}W_3 \qquad \text{for } 1 \leq j \leq 3 \tag{11}$$

---

[5] Groth-Sahai proofs for linear equations reduce to 3 group elements; see Sect. 6.1 of the full version of [GS08].

The final verification relations are the following:

For (9a): $e(p_{1,1}, U)\, e(p_{3,1}, W_1) \;=\; e(\psi_{1,1}, U')\, e(\psi_{1,3}, W_1')$

$e(p_{1,2}, U)\, e(p_{3,2}, W_1) \;=\; e(\psi_{1,2}, V')\, e(\psi_{1,3}, W_2')$

$e(p_{1,3}, U)\, e(p_{3,3}, W_1) \;=\; e(H, c_1)\, e(\psi_{1,1}, P')\, e(\psi_{1,2}, P')\, e(\psi_{1,3}, W_3')$

For (9b): $e(p_{2,1}, V)\, e(p_{3,1}, W_2) \;=\; e(\psi_{2,1}, U')\, e(\psi_{2,3}, W_1')$

$e(p_{2,2}, V)\, e(p_{3,2}, W_2) \;=\; e(\psi_{2,2}, V')\, e(\psi_{2,3}, W_2')$

$e(p_{2,3}, V)\, e(p_{3,3}, W_2) \;=\; e(H, c_2)\, e(\psi_{2,1}, P')\, e(\psi_{2,2}, P')\, e(\psi_{2,3}, W_3')$

For (9c): $e(d_1, G)\, e(p_{1,1}, P)\, e(p_{2,1}, P)\, e(p_{3,1}, W_3) \;=\; e(\psi_{3,1}, U')\, e(\psi_{3,3}, W_1')$

$e(d_2, G)\, e(p_{1,2}, P)\, e(p_{2,2}, P)\, e(p_{3,2}, W_3) \;=\; e(\psi_{3,2}, V')\, e(\psi_{3,3}, W_2')$

$e(d_3, G)\, e(p_{1,3}, P)\, e(p_{2,3}, P)\, e(p_{3,3}, W_3) \;=$

$$e(H, c_3)\, e(\psi_{3,1}, P')\, e(\psi_{3,2}, P')\, e(\psi_{3,3}, W_3')$$

Re-randomization. Given commitments $\mathbf{c}, \mathbf{d}, \mathbf{p}_1, \mathbf{p}_2, \mathbf{p}_3$ and proofs ψ_1, ψ_2, ψ_3, we can re-randomize the commitments by choosing $s'_{Yi}, s'_{Zi}, t'_{i,j} \leftarrow \mathbb{Z}_p$ for $1 \leq i, j \leq 3$ and setting (cf. Sect. 5.2)

$$\widetilde{\mathbf{c}} := \begin{bmatrix} c_1 + s'_{Y3}U' + s'_{Y3}W_1' \\ c_2 + s'_{Y2}V' + s'_{Y3}W_2' \\ c_3 + s'_{Y3}P' + s'_{Y2}P' + s'_{Y3}W_3' \end{bmatrix} \qquad \widetilde{\mathbf{d}} := \begin{bmatrix} d_1 + s'_{Z1}U' + s'_{Z3}W_1' \\ d_2 + s'_{Z2}V' + s'_{Z3}W_2' \\ d_3 + s'_{Z1}P' + s'_{Z2}P' + s'_{Z3}W_3' \end{bmatrix}$$

$$\widetilde{\mathbf{p}}_i := \begin{bmatrix} p_{i,1} + t'_{i,1}U' + t'_{i,3}W_1' \\ p_{i,2} + t'_{i,2}V' + t'_{i,3}W_2' \\ p_{i,3} + s'_{Yi}H + t'_{i,1}P' + t'_{i,2}P' + t'_{i,3}W_3' \end{bmatrix} \qquad \text{for } 1 \leq i \leq 3$$

Note that $\widetilde{\mathbf{p}}_i$ not only re-randomizes $\mathbf{p}_i$ but at the same time updates the committed proofs π_i to the new randomness for the commitments to Y. The proofs ψ_i are updated as follows:

$$\widetilde{\psi}_{1,j} := \psi_{1,j} + t'_{1,j}U + t'_{3,j}W_1$$

$$\widetilde{\psi}_{2,j} := \psi_{2,j} + t'_{2,j}V + t'_{3,j}W_2 \qquad \text{for } 1 \leq j \leq 3$$

$$\widetilde{\psi}_{3,j} := \psi_{3,j} + s'_{Zj}G + t'_{1,j}P + t'_{2,j}P + t'_{3,j}W_3$$

5.4   Proofs That Commitments Open to the Same Value

Given the extraction key, one can prove that two commitments open to the same value without knowledge of the randomness used when committed. We start by showing how to prove that a commitment (c_1, c_2, c_3) opens to zero: given the extraction key $ek = (\alpha, \beta)$ define the proof as $(\pi_1 := \frac{1}{\alpha}c_1, \pi_2 := \frac{1}{\beta}c_2)$. It satisfies the following relations: $e(\pi_1, U) = e(c_1, P)$, $e(\pi_2, V) = e(c_2, P)$, $c_3 = \pi_1 + \pi_2$.

It is easily seen that the proofs are perfectly correct and perfectly sound. In addition, they do not leak information about the opener's secret key, since they can be produced without knowledge of ek, given the randomness used to commit and the "trapdoor" (r_1, r_2) for the W_i's: $c_1 = s_1 U + s_3 W_1 = \alpha(s_1 + s_3 r_1)P$, thus $\pi_1 = (s_1 + s_3 r_1)P$, and similarly $\pi_2 = (s_2 + s_3 r_2)P$. Now to show that $\mathbf{c}$ and $\mathbf{d}$ are two commitments to the same value, it suffices to prove that $\mathbf{c} - \mathbf{d}$ opens to 0.

6   Transferable Anonymous Constant-Size Fair E-Cash from Certificates

6.1   Formal Model

In our model for e-cash, there are the following protagonists: *users* $\mathcal{U}_i$ that—after registering—can withdraw, transfer and spend coins; the *system manager* $\mathcal{S}$, authorizing users to join the system; the *bank* $\mathcal{B}$, able to issue coins; *merchants* $\mathcal{M}_i$ who deposit the coins at the bank; the *double-spending detector* $\mathcal{D}$, that can detect if a coin was spent twice; and the *tracing authority* $\mathcal{T}$, able to trace users that misbehave in some way (e.g., tracing of a double spender or prosecution of criminal activities). The system comprises the following protocols and algorithms:

Setup   A protocol between $\mathcal{S}$ (who gets the manager key mk), $\mathcal{B}$ (who gets the issuing key ik), $\mathcal{D}$ (who gets dk), and $\mathcal{T}$ (who gets tk). The protocol also outputs the public parameters pp.

Join   A protocol between a user and $\mathcal{S}$ that registers the user in the system and gives him usk.

Withdraw   A protocol permitting a user to withdraw a coin from $\mathcal{B}$.

Transfer   A protocol between two users $\mathcal{U}_i$ and $\mathcal{U}_j$, where $\mathcal{U}_j$ gets a coin and a receipt from $\mathcal{U}_i$.

Spend   A protocol between a user and a merchant to spend a coin.

Detect   An algorithm enabling $\mathcal{D}$ to check for double spendings (without identifying the defrauder).

Trace$_{DS}$ A protocol conducted by $\mathcal{T}$ in order to trace a double spender.

Trace$_C$ An algorithm enabling $\mathcal{T}$ to match a withdrawal and a spending transcript of the same coin.

Trace$_S$ An algorithm that lets $\mathcal{T}$ reveal the identity of a spender from a spending transcript.

Besides correctness, which requires that honestly issued coins are accepted when transferred or spent by honestly registered users, and that the tracing algorithms work correctly, we define the following security notions for our model: *Anonymity of withdrawal* means that not even the bank colluding with the (double-spending) detector can tell to which withdrawal a coin corresponds. *Anonymity of transfer (or spending)* ensures that when transferring/spending a coin a user remains anonymous even with respect to the bank and malicious users the coin was transferred by.

Traceability of double spenders states that for each time a user spends a coin more than once he will be accused, whereas *Detectability of double spending*

Exp$_{\mathcal{A}}^{\text{anon-with}}(k)$

- Experiment plays: honest users $\mathcal{U}_0$ and $\mathcal{U}_1$
- $\mathcal{A}$ impersonates: $\mathcal{S}$, $\mathcal{B}$, $\mathcal{D}$, users
- $\mathcal{U}_0$, $\mathcal{U}_1$ run Join and Withdraw with $\mathcal{A}$ impersonating $\mathcal{S}$ and $\mathcal{B}$, resp.
- $b \leftarrow \{0, 1\}$; $\mathcal{A}$ receives the coin of $\mathcal{U}_b$
- $\mathcal{A}$ wins if it guesses b correctly

Exp$_{\mathcal{A}}^{\text{trace-DS}}(k)$

- Experiment plays: honest $\mathcal{S}$, $\mathcal{B}$
- $\mathcal{A}$ impersonates: users
- $\mathcal{A}$ gets keys: tk, dk (thus $\mathcal{T}$, $\mathcal{D}$ semi-honest)
- $\mathcal{A}$ gets oracles Join, Withdraw, Spend to communicate with $\mathcal{S}$, $\mathcal{B}$ and $\mathcal{D}$, resp.
- The experiment runs Detect and Trace on the spent coins
- Let q and d be the number of Withdraw and Spend queries, resp.; let a be the number of accusations by Trace. Then $\mathcal{A}$ wins if $a < d - q$

Exp$_{\mathcal{A}}^{\text{detect-DS}}(k)$

- Experiment plays: honest $\mathcal{B}$
- $\mathcal{A}$ impersonates: users, $\mathcal{S}$, $\mathcal{T}$
- $\mathcal{A}$ gets keys: dk (thus $\mathcal{D}$ semi-honest)
- $\mathcal{A}$ gets oracles Withdraw, Spend to communicate with $\mathcal{B}$ and $\mathcal{D}$, resp.
- The experiment runs Detect on the spent coins
- $\mathcal{A}$ wins if there where more accepted Spend than Withdraw calls and $\mathcal{D}$ does not detect double spending.

Exp$_{\mathcal{A}}^{\text{anon-trans}}(k)$

- Experiment plays: honest users $\mathcal{U}_0$ and $\mathcal{U}_1$
- $\mathcal{A}$ impersonates: $\mathcal{S}$, $\mathcal{B}$, users
- $\mathcal{U}_0$ and $\mathcal{U}_1$ run Join with $\mathcal{A}$ impersonating $\mathcal{S}$
- $\mathcal{A}$ can ask withdrawals, transfers and spendings of $\mathcal{U}_0$ and $\mathcal{U}_1$.
- $b \leftarrow \{0, 1\}$, $\mathcal{U}_b$ runs Transfer with $\mathcal{A}$ playing a user.
- $\mathcal{A}$ wins if it guesses b correctly.

Exp$_{\mathcal{A}}^{\text{trace-C/S}}(k)$

- Experiment plays: honest $\mathcal{S}$, $\mathcal{B}$
- $\mathcal{A}$ impersonates: users, $\mathcal{D}$
- $\mathcal{A}$ gets keys: tk (thus $\mathcal{T}$ semi-honest)
- Oracles for $\mathcal{A}$: Join, Withdraw
- $\mathcal{A}$ spends a coin and wins if
  - the spending cannot be matched to a withdrawal (traceability of coins); or
  - Trace$_{\mathcal{S}}$ returns $\perp$ (spender traceability)

Exp$_{\mathcal{A}}^{\text{non-fram}}(k)$

- Experiment plays an honest user $\mathcal{U}^*$
- $\mathcal{A}$ can impersonate: $\mathcal{S}$, $\mathcal{B}$, $\mathcal{D}$, $\mathcal{T}$, users
- $\mathcal{U}^*$ runs Join with $\mathcal{A}$ impersonating $\mathcal{S}$
- $\mathcal{A}$ can ask the user to withdraw coins, transfer and receive them and spend coins
- $\mathcal{A}$ wins if
  - it outputs a proof accusing $\mathcal{U}^*$ of double spending, which $\mathcal{U}^*$ cannot contest.
  - $\mathcal{U}^*$ is accused of a spending it did not perform

Fig. 4. Security experiments for constant-size e-cash

means that Detect will determine if a coin was spent multiple times. *Non-frameability* guarantees that even if everyone else colludes against an honest user, he cannot be wrongfully accused of a spending he did not perform, nor of double spending. See Fig. 4 for the details of the experiments. As for the BSZ-model of group signatures, we call protagonists *semi-honest* if $\mathcal{A}$ impersonates them but however follows protocols as prescribed. Note that in the experiment for non-frameability, $\mathcal{U}^*$ behaves honestly, so if he is asked to spend more coins than he withdrew he refuses; moreover, a malicious tracer can always *accuse* an honest user of not having a receipt, which the latter counters by showing it.

We say an e-cash system is traceable, non-frameable, etc., if no p.p.t. adversary can win the respective game with non-negligible probability (non-negligibly more than $1/2$ for the anonymity notions).

6.2   Instantiation

Overview.   The core of a coin in our system is a certificate from Sect. 3.2. Defining withdrawal as partially blind issuing guarantees that the bank does not know the last component C_5. Certificates were designed to consist of elements of $\mathbb{G}$ so that their verification relations are paring-product equations; the user can thus encrypt (in Groth-Sahai terminology: commit to) the coin and prove validity. Moreover, each time the coin is transferred, the receiver can re-randomize the encryption (cf. Sect. 5.2), which guarantees unlinkable anonymity.

To check for double spendings, the detector will get the decryption key to compare encrypted certificates. However, this straight-forward approach would not guarantee user anonymity when bank and detector cooperate. The blind component C_5 is thus encrypted under a *different* key than the rest (in Sect. 5.3 we showed how to construct the corresponding proofs). The detector gets only the key to decrypt C_5, which suffices to detect double spending. Since the the first 4 components remain hidden from the detector, *partial* blindness of certificates suffices. The other decryption key is given to the tracer, which enables tracing of a coin by comparing C_3 which is known to the bank.

The receipts, given when transferring and spending coins, are group signatures on them, the signing keys for which the users get when joining the system. This guarantees user traceability, while preserving anonymity (only the tracer, holding the group-signature opening key, can reveal users' identities). To identify a double spender, the tracer follows backwards the paths the certificate took before reaching the spender, by opening the receipts. A user that spent or transferred a coin twice is then unable to show two receipts. To guarantee soundness of tracing, we must ensure that each signature corresponds to at most one transfer. We achieve this by having the receiver choose a nonce which is added to the message the sender must sign.

Details.   Let $\mathcal{GS} = (\mathsf{Setup}_{\mathcal{GS}}, \mathsf{Join}_{\mathcal{GS}}, \mathsf{GSign}_{\mathcal{GS}}, \mathsf{GVer}_{\mathcal{GS}})$ be a dynamic non-frameable group-signature scheme.[6] Let $\mathcal{H}\colon \mathbb{G}^* \to \{0,1\}^n$ be a collision-resistant hash function.

Setup.   – Set up a group signature scheme $\mathcal{GS}$ such that $\mathcal{S}$ is the group's issuer (group manager) and $\mathcal{T}$ gets opening key ok. The group verification key gvk is added to pp.
   – Produce two keys for linear commitments $ck_{\mathcal{T}}$ and $ck_{\mathcal{D}}$. The corresponding extraction keys $ek_{\mathcal{T}}$ and $ek_{\mathcal{D}}$ are given to $\mathcal{T}$ (thus $tk = (ek_{\mathcal{T}}, ek_{\mathcal{D}}, ok)$). $\mathcal{D}$ receives $dk := ek_{\mathcal{D}}$.

---

[6] Encrypting the certified signatures from Sect. 4 and proving validity by adding a Groth-Sahai proof yields a (CPA-anonymous) non-frameable group signature scheme that does not require any further assumptions.

– Set up the CRS (if any) for the blind certificate-issuing scheme from Sect. 3.2. $\mathcal{B}$ picks issuing key $ik := \omega \leftarrow \mathbb{Z}_p$, adds $\Omega := \omega G$ to pp, and gets a group signing key $gsk_{\mathcal{B}}$ by joining $\mathcal{GS}$.

Join. A user $\mathcal{U}_i$ joins the system by running $\mathsf{Join}_{\mathcal{GS}}$ with $\mathcal{S}$ to obtain her group signing key gsk_i.

Withdraw. $\mathcal{U}_i$ runs the issuing protocol (Fig. 2) with $\mathcal{B}$ to get $(C_1, \ldots C_5) \in \mathbb{G}^5$ satisfying

$$e(C_1, \Omega + C_2) = e(K + C_4, G) \qquad \begin{aligned} e(C_2, H) &= e(G, C_3) \\ e(C_4, H) &= e(G, C_5) \end{aligned} \qquad (12)$$

$\mathcal{B}$ also gives the user a "receipt" $R_{\mathcal{B}} \leftarrow \mathsf{GSign}_{\mathcal{GS}}(gsk_{\mathcal{B}}, \mathcal{H}(C_1, C_2, C_3, \mathcal{U}_i)).$[7] $\mathcal{U}_i$ verifies the certificate and $R_{\mathcal{B}}$ and makes the following commitments:

$$\mathbf{c}_i := \mathsf{Com}(ck_{\mathcal{T}}, C_i), \quad \text{for } 1 \le i \le 4 \qquad \mathbf{c}_5 := \mathsf{Com}(ck_{\mathcal{D}}, C_5)$$

and proofs Φ_1, Φ_2, Φ_3 for the committed values satisfying each of the equations in (12). Φ_1 and Φ_2 are regular Groth-Sahai proofs; for the last equation on commitments under different keys, see Sect. 5.3. We call $(\vec{c}, \vec{\Phi})$ a *coin*, and refer to the full version [FPV09] for its concrete construction.

Transfer / Spend. When user $\mathcal{U}_i$ transfers a coin $(\vec{c}, \vec{\Phi})$ to user $\mathcal{U}_j$, she sends $R \leftarrow \mathsf{GSig}_{\mathcal{GS}}(gsk_{\mathcal{U}_i}, \mathcal{H}(\vec{c}, \mathcal{U}_j, N))$ as well, where N is a nonce set by $\mathcal{U}_j$. The receiver $\mathcal{U}_j$ checks correctness of $(\vec{c}, \vec{\Phi})$ and R, re-randomizes $\vec{c}$ and updates $\vec{\Phi}$ (cf. Sects. 5.2 and 5.3). Spending is defined as transferring.

Detect. After receiving new a coin, $\mathcal{D}$ uses extraction key $ek_{\mathcal{D}}$ to open $\mathbf{c}_5$: $C_5 := \mathsf{Extr}(ek_{\mathcal{D}}, \mathbf{c}_5)$ (cf. Sect. 5.1). He compares the tag C_5 with that of previously received coins to see if a coin was spent twice, in which case he charges $\mathcal{T}$ to trace the double spender.

Tracing of DS

– If multiple spendings $(\vec{c}^{(i)}, \vec{\Phi}^{(i)}, R^{(i)})$ with $\mathsf{Extr}(ek_{\mathcal{D}}, \mathbf{c}_5^{(i)}) = C_5^*$ for all i were detected, the tracer uses the key ok to open the signatures $R^{(i)}$ in order to reveal users $\mathcal{U}_0^{(i)}$.

– Each $\mathcal{U}_0^{(i)}$ has to *prove legal acquisition* of his coin, which a user $\mathcal{U}$ does as follows:

  • If the coin was obtained from the bank, show $C = (C_1, \ldots, C_5)$ and the receipt $R_{\mathcal{B}}$.
  $\mathcal{T}$ accepts if C is valid, $\mathsf{GVer}_{\mathcal{GS}}(gvk, \mathcal{H}(C_1, C_2, C_3, \mathcal{U}), R_{\mathcal{B}}) = 1$ and $C_5 = C_5^*$.

---

[7] Abusing notation slightly, we let $\mathcal{U}_i$ be a unique encoding of the user's identity in $\mathbb{G}$. Note that for the receipts from the issuer, no nonce is required, since the user contributes to the randomness of the certificate.

- If the coin was received from a user, show the receipt R received with it, and show $(\vec{c}', \vec{\Phi}')$, the received coin (i.e., before re-randomizing it), and the nonce N.
  $\mathcal{T}$ accepts if $(\vec{c}', \vec{\Phi}')$ is valid, $\mathsf{GVer}_{\mathcal{GS}}(gvk, \mathcal{H}(\vec{c}', \mathcal{U}, N), R) = 1$ and $\mathsf{Extr}(ek_{\mathcal{D}}, c'_5) = C_5^*$.

- In the second case (receipt produced by a user), $\mathcal{T}$ opens R to $\mathcal{U}_1^{(i)}$, who in turn has to prove legal acquisition of the coin. Moreover, the tracer only accepts a receipt if it has not been given to him before.
- Continuing this process for every i produces a chain of users $\mathcal{U}_0^{(i)}, \mathcal{U}_1^{(i)}, \ldots$ which either ends with the bank, or with a user failing to prove legal acquisition—in which case that user is accused.
- Correctness of tracing is proven by proving correctness of opening of group signatures and proving that two commitments contain the same certificate using the techniques from Sect. 5.4.

Tracing of coins and users. Given $ek_{\mathcal{T}}$, the tracer can recover C_3 from a coin and thus match withdrawn coins to spent coins. Spender anonymity is revoked by opening the group signature.

6.3   Security Results

We briefly argue why our instantiation satisfies the security definitions from Sect. 6.1. Each property follows by a straight-forward reduction to the security of the underlying building blocks.

Detectability and traceability of double spenders. (I) Assuming an honest bank, every certificate is only issued once with all but negligible probability; (II) by unforgeability of certificates (Theorem 8) and soundness of the WI proofs, opening all d spent coins leads to at most q different certificates, where q is the number of Withdraw queries. This proves detectability.

For every i let $s^{(i)}$ be the number of times certificate $C^{(i)}$ was spent. Then the tracing algorithm produces $s^{(i)}$ lists of users, beginning with the spenders and linked by their certificates. Unforgeability of group signatures and (I) guarantees that only one such list ends with the bank. Since $s^{(i)} - 1$ users are thus accused and by (II), we have $a = \sum_{i=1}^{q}(s^{(i)} - 1) = d - q$, which proves traceability.

Non-frameability. If $\mathcal{U}^*$ uses a random nonce each time then by collision resistance of $\mathcal{H}$, the probability of receiving the same valid receipt twice is negligible. The user can only be provably accused if he spent/transferred a coin of which he cannot justify acquisition. Non-frameability of group signatures guarantees that $\mathcal{U}^*$ only has to justify coins he actually transferred—and for each such coin he possesses a valid receipt. Note that if a malicious user transfers the same coin (possibly as two different randomizations) twice to $\mathcal{U}^*$ then $\mathcal{U}^*$ has two different signatures (due to the nonce) and can thus justify both coins.

Anonymity. Anonymity of withdrawal follows from partial blindness of issuing (indistinguishability of C_5) and witness indistinguishability of the commitments

$(\mathbf{c}_1, \ldots, \mathbf{c}_4)$ under key ck_T. Anonymity of transfer follows from WI of commitments under ck_T and ck_D and anonymity of group signatures.

Traceability. Traceability of coins follows from soundness of the WI proofs and unforgeability of certificates; traceability of spenders follows from traceability of group signatures.

Acknowledgments

The authors would like to thank the members of the PACE research project for the fruitful discussions that led to the new primitive discussed in this paper. This work was supported by the French ANR 07-TCOM-013-04 PACE Project, the European Commission through the IST Program under Contract ICT-2007-216646 ECRYPT II, and EADS.

References

[AO00]   Abe, M., Okamoto, T.: Provably secure partially blind signatures. In: Bellare, M. (ed.) CRYPTO 2000. LNCS, vol. 1880, pp. 271–286. Springer, Heidelberg (2000)

[BB04]   Boneh, D., Boyen, X.: Short signatures without random oracles. In: Cachin, C., Camenisch, J.L. (eds.) EUROCRYPT 2004. LNCS, vol. 3027, pp. 56–73. Springer, Heidelberg (2004)

[BBS04]   Boneh, D., Boyen, X., Shacham, H.: Short group signatures. In: Franklin, M. (ed.) CRYPTO 2004. LNCS, vol. 3152, pp. 41–55. Springer, Heidelberg (2004)

[BCC+09]   Belenkiy, M., Camenisch, J., Chase, M., Kohlweiss, M., Lysyanskaya, A., Shacham, H.: Randomizable proofs and delegatable anonymous credentials. In: Halevi, S. (ed.) CRYPTO 2009. LNCS, vol. 5677, pp. 108–125. Springer, Heidelberg (2009)

[BFPW07]   Boldyreva, A., Fischlin, M., Palacio, A., Warinschi, B.: A closer look at PKI: Security and efficiency. In: Okamoto, T., Wang, X. (eds.) PKC 2007. LNCS, vol. 4450, pp. 458–475. Springer, Heidelberg (2007)

[BR93]   Bellare, M., Rogaway, P.: Random oracles are practical: A paradigm for designing efficient protocols. In: Ashby, V. (ed.) ACM CCS 1993, pp. 62–73. ACM Press, New York (1993)

[BSZ05]   Bellare, M., Shi, H., Zhang, C.: Foundations of group signatures: The case of dynamic groups. In: Menezes, A. (ed.) CT-RSA 2005. LNCS, vol. 3376, pp. 136–153. Springer, Heidelberg (2005)

[BW07]   Boyen, X., Waters, B.: Full-domain subgroup hiding and constantsize group signatures. In: Okamoto, T., Wang, X. (eds.) PKC 2007. LNCS, vol. 4450, pp. 1–15. Springer, Heidelberg (2007)

[CG07]   Canard, S., Gouget, A.: Divisible E-cash systems can be truly anonymous. In: Naor, M. (ed.) EUROCRYPT 2007. LNCS, vol. 4515, pp. 482–497. Springer, Heidelberg (2007)

[CG08]   Canard, S., Gouget, A.: Anonymity in transferable E-cash. In: Bellovin, S.M., Gennaro, R., Keromytis, A.D., Yung, M. (eds.) ACNS 2008. LNCS, vol. 5037, pp. 207–223. Springer, Heidelberg (2008)

[Cha83]   Chaum, D.: Blind signatures for untraceable payments. In: Chaum, D., Rivest, R.L., Sherman, A.T. (eds.) CRYPTO 1982, pp. 199–203. Plenum Press, New York (1983)

[Cha84]   Chaum, D.: Blind signature system. In: Chaum, D. (ed.) CRYPTO 1983, p. 153. Plenum Press, New York (1984)

[CP92]   Chaum, D., Pedersen, T.P.: Transferred cash grows in size. In: Rueppel, R.A. (ed.) EUROCRYPT 1992. LNCS, vol. 658, pp. 390–407. Springer, Heidelberg (1993)

[Cv91]   Chaum, D., van Heyst, E.: Group signatures. In: Davies, D.W. (ed.) EUROCRYPT 1991. LNCS, vol. 547, pp. 257–265. Springer, Heidelberg (1991)

[DH76]   Diffie, W., Hellman, M.E.: New directions in cryptography. IEEE Transactions on Information Theory 22(6), 644–654 (1976)

[EO94]   Eng, T., Okamoto, T.: Single-term divisible electronic coins. In: De Santis, A. (ed.) EUROCRYPT 1994. LNCS, vol. 950, pp. 306–319. Springer, Heidelberg (1995)

[FP09]   Fuchsbauer, G., Pointcheval, D.: Proofs on encrypted values in bilinear groups and an application to anonymity of signatures. In: Shacham, H., Waters, B. (eds.) Pairing 2009. LNCS, vol. 5671, pp. 132–149. Springer, Heidelberg (2009), http://eprint.iacr.org/2008/528

[FPV09]   Fuchsbauer, G., Pointcheval, D., Vergnaud, D.: Transferable anonymous constant-size fair e-cash. Cryptology ePrint Archive, Report 2009/146 (2009), http://eprint.iacr.org/

[FS87]   Fiat, A., Shamir, A.: How to prove yourself: Practical solutions to identification and signature problems. In: Odlyzko, A.M. (ed.) CRYPTO 1986. LNCS, vol. 263, pp. 186–194. Springer, Heidelberg (1987)

[GOS06]   Groth, J., Ostrovsky, R., Sahai, A.: Non-interactive zaps and new techniques for NIZK. In: Dwork, C. (ed.) CRYPTO 2006. LNCS, vol. 4117, pp. 97–111. Springer, Heidelberg (2006)

[Gro06]   Groth, J.: Simulation-sound NIZK proofs for a practical language and constant size group signatures. In: Lai, X., Chen, K. (eds.) ASIACRYPT 2006. LNCS, vol. 4284, pp. 444–459. Springer, Heidelberg (2006)

[Gro07]   Groth, J.: Fully anonymous group signatures without random oracles. In: Kurosawa, K. (ed.) ASIACRYPT 2007. LNCS, vol. 4833, pp. 164–180. Springer, Heidelberg (2007)

[GS08]   Groth, J., Sahai, A.: Efficient non-interactive proof systems for bilinear groups. In: Smart, N.P. (ed.) EUROCRYPT 2008. LNCS, vol. 4965, pp. 415–432. Springer, Heidelberg (2008)

[Kil06]   Kiltz, E.: Chosen-ciphertext security from tag-based encryption. In: Halevi, S., Rabin, T. (eds.) TCC 2006. LNCS, vol. 3876, pp. 581–600. Springer, Heidelberg (2006)

[NHS99]   Nakanishi, T., Haruna, N., Sugiyama, Y.: Unlinkable electronic coupon protocol with anonymity control. In: Zheng, Y., Mambo, M. (eds.) ISW 1999. LNCS, vol. 1729, pp. 37–46. Springer, Heidelberg (1999)

[Oka06]   Okamoto, T.: Efficient blind and partially blind signatures without random oracles. In: Halevi, S., Rabin, T. (eds.) TCC 2006. LNCS, vol. 3876, pp. 80–99. Springer, Heidelberg (2006)

[OO90]   Okamoto, T., Ohta, K.: Disposable zero-knowledge authentications and their applications to untraceable electronic cash. In: Brassard, G. (ed.) CRYPTO 1989. LNCS, vol. 435, pp. 481–496. Springer, Heidelberg (1990)

[OO92]   Okamoto, T., Ohta, K.: Universal electronic cash. In: Feigenbaum, J. (ed.)
            CRYPTO 1991. LNCS, vol. 576, pp. 324–337. Springer, Heidelberg (1992)
[PS00]   Pointcheval, D., Stern, J.: Security arguments for digital signatures and
            blind signatures. Journal of Cryptology 13(3), 361–396 (2000)
[Sch90]   Schnorr, C.-P.: Efficient identification and signatures for smart cards. In:
            Brassard, G. (ed.) CRYPTO 1989. LNCS, vol. 435, pp. 239–252. Springer,
            Heidelberg (1990)
[Sho97]   Shoup, V.: Lower bounds for discrete logarithms and related problems.
            In: Fumy, W. (ed.) EUROCRYPT 1997. LNCS, vol. 1233, pp. 256–266.
            Springer, Heidelberg (1997)
[SPC95]   Stadler, M., Piveteau, J.-M., Camenisch, J.: Fair blind signatures. In: Guil-
            lou, L.C., Quisquater, J.-J. (eds.) EUROCRYPT 1995. LNCS, vol. 921,
            pp. 209–219. Springer, Heidelberg (1995)
[vSN92]   von Solms, S.H., Naccache, D.: On blind signatures and perfect crimes.
            Computers & Security 11(6), 581–583 (1992)
[Wat05]   Waters, B.R.: Efficient identity-based encryption without random oracles.
            In: Cramer, R. (ed.) EUROCRYPT 2005. LNCS, vol. 3494, pp. 114–127.
            Springer, Heidelberg (2005)

A Secure Channel Free Public Key Encryption with Keyword Search Scheme without Random Oracle

Liming Fang[1], Willy Susilo[2], Chunpeng Ge[1], and Jiandong Wang[1]

[1] College of Information Science and Technology
Nanjing University of Aeronautics and Astronautics, Nanjing, China
{fangliming,gecp}@nuaa.edu.cn
[2] Centre for Computer and Information Security Research
School of Computer Science and Software Engineering
University of Wollongong, Australia
wsusilo@uow.edu.au

Abstract. The public key encryption with keyword Search (PEKS) scheme, proposed by Boneh, Di Crescenzo, Ostrovsky and Persiano, enables one to search for encrypted keywords without compromising the security of the original data. Baek *et al.* noticed that the original notion of PEKS requires the existence of a secure channel, and they further extended this notion by proposing an efficient secure channel free public key encryption scheme with keyword search in the random oracle model. In this paper, we take one step forward by adopting Baek *et al.*'s model and propose a new and efficient scheme that does not require any secure channels, and furthermore, its security does not use random oracles.

Keywords: public key encryption with keyword search, searchable encryption, without random oracle.

1 Introduction

The public key encryption with keyword search (PEKS) scheme, proposed by Boneh *et al.* [4], enables one to search for encrypted keywords without revealing the security of the original data. PEKS schemes can be widely used and deployed in many practical applications. An interesting application of PEKS is intelligent email routing. Suppose Bob sends an encrypted email to Alice using Alice's public key. The contents of the email comprises a header information, a body of the email and a list of keywords that are encrypted. In this case, the mail gateway cannot observe the header information (as well as the body and the keywords attached) and hence, cannot make routing decisions. Alice uses different electronic devices to read her email, such as an iPhone or a PDA and a desktop computer. Alice may prefer emails to be routed to her devices depending on the associated keywords. For example, she may like to receive emails with the keyword "urgent" on her iPhone, meanwhile all other emails can be sent to her desktop computer instead. In particular, when Alice is away on holiday, she only wants to read emails with the keyword "urgent" that will require her urgent attention, instead of reading all of her work emails.

In short, PEKS provides a mechanism that enables Alice to provide the gateway with the ability to test whether "urgent" is a keyword in the original email, but additionally,

J.A. Garay, A. Miyaji, and A. Otsuka (Eds.): CANS 2009, LNCS 5888, pp. 248–258, 2009.

the gateway should learn nothing else about the email itself. More generally, the mail gateway can search the keywords required, but learn nothing else.

Waters *et al.* [16] introduced another interesting application of PEKS schemes, which is to let an untrusted logging device to maintain an encrypted audit log of privacy-sensitive data that is efficiently searchable by authorized auditors only. The entries in the audit log are encrypted under the public key of a PEKS scheme, of which the corresponding secret key is unknown to the logging device. If the device is ever confiscated, or if the logbook leaks, privacy of users and their actions is maintained. The secret key is known only to a trusted audit escrow agent, who provides (less trusted) authorized investigators with trapdoors for the keywords they want to search for.

Related Work

Despite a large number of research work related to the privacy of database data, Boneh *et al.* [4] noted that PEKS is different from the previously known solutions. Unlike the private-key setting, data collected by the mail server is from third parties, and can not be "organized" by the user in any convenient way. Furthermore, unlike the publicly available database, the data is not public, and hence the Public Information Retrieval (PIR) solutions will not be applicable. Shortly after Boneh *et al.*'s pioneering work, Waters *et al.* [16] showed that the PEKS scheme based on the bilinear pairing can be applied to build encrypted and searchable audit logs. Golle *et al.* [12] proposed schemes that allow for conjunctive keyword queries on encrypted data. Boneh and Waters [5] extended PEKS to support conjunctive, subset, and range comparisons over the keywords. Furthermore, the subsequent papers [9,18] investigated the secure combination of public key encryption with keyword search (PEKS) with public key data encryption (PKE). Since the fact that keywords are chosen from much smaller space than passwords and users usually use well-known keywords for search, the work in [6,13,15,17] studied the off-line keyword guessing attacks on PEKS.

The drawback of the Boneh *et al.*'s scheme [4] is that it uses a *secure channel* between Alice and the email server, which is usually costly [2]. Baek *et al.* [2] proposed an alternative solution to eliminate the need for a secure channel, by proposing public key encryption with keyword search scheme with a designated server. Throughout this paper, we refer this model as SCF-PEKS, which refers to *Secure Channel-Free PEKS*.

In 2007, Gu *et al.* [11] proposed an interesting construction of PEKS scheme based on pairings, in which there is no pairing operation involved in the encryption procedure. Then, they provided further discussion on removing the secure channel from PEKS, and presented an efficient secure channel free PEKS scheme. Rhee *et al.* [14] enhanced the Baek *et al.*'s security model [2] for SCF-PEKS in which an attacker is also allowed to obtain the relation between non-challenge ciphertexts and a trapdoor. They presented an SCF-PEKS scheme secure in the enhanced security model. The limitation of the existing schemes in the literature is that their security can only be guaranteed in the random oracle model. Unfortunately, a proof in the random oracle model has been shown to possibly lead to insecure schemes when the random oracles are implemented in the standard model [7]. Therefore, it is desirable to have a solution that does not require the random oracle model.

Our Contributions

In this paper, we present an efficient and secure channel free public key encryption with keyword search scheme. Based on the DBDH assumption and the truncated q-ABDHE assumption, we prove its indistinguishability of secure channel free PEKS against chosen keyword attack (IND-SCF-CKA) security without random oracle.

To summarize the existing knowledge, the three constructions of SCF-PEKS schemes are due to Baek *et al.* [2], Gu *et al.* [11] and Rhee *et al.*[14] that require random oracle model. Our work fills the gap in the literature by proposing an *efficient* secure channel free PEKS (SCF-PEKS) scheme that does not incorporate random oracle, as outlined in Table 1.

Table 1. Comparison Among Various SCF-PEKS Schemes

Properties	Baek *et al.* [2]	Gu *et al.* [11]	Rhee *et al.*[14]	This paper
Without ROM	×	×	×	✓
Assumption	BDH	BDH, q-BDHI	BDH, q-BDHI	DBDH, q-ABDHE

Paper Organization

The rest of this paper is organized as follows. In the next section, we will present some definitions and notations that will be used throughout this paper. In Section 3, we present our new and efficient scheme and analyze its security. Finally, Section 4 concludes the paper.

2 Definitions

In this section, we firstly review the complexity assumptions required in our schemes, and then provide the definition and security of a public key encryption with keyword search scheme.

2.1 Negligible Function

A function $\varepsilon(n) : N \to R$ is negligible in n if $1/\varepsilon(n)$ is a non-polynomially-bounded quantity in n.

2.1 Bilinear Maps

Let $\mathbb{G}_1$ and $\mathbb{G}_2$ be multiplicative cyclic groups of prime order p, and g be a generator of $\mathbb{G}_1$. (By $\mathbb{G}_1^*$ and $\mathbb{Z}_p^*$, we denote $\mathbb{G}_1 \backslash \{1\}$ where 1 is the identity element of $\mathbb{G}_1$, and $\mathbb{Z}_p \backslash \{0\}$ respectively). We say $e : \mathbb{G}_1 \times \mathbb{G}_1 \to \mathbb{G}_2$ is a bilinear map [3] if the following conditions hold.

1. $e(g_1{}^a, g_2{}^b) = e(g_1, g_2)^{ab}$ for all $a, b \in \mathbb{Z}_p$ and $g_1, g_2 \in \mathbb{G}_1$.
2. $e(g, g) \neq 1$.
3. There is an efficient algorithm to compute $e(g_1, g_2)$ for all $g_1, g_2 \in \mathbb{G}_1$.

2.2 The DBDH Assumption

Let $e : \mathbb{G}_1 \times \mathbb{G}_1 \to \mathbb{G}_2$ is a bilinear map. We define the advantage function

$$Adv_{\mathbb{G}_1, \mathcal{B}}^{DBDH}(\lambda)$$

of an adversary $\mathcal{B}$ as

$$|Pr[\mathcal{B}(g, g^a, g^b, g^c, e(g,g)^{abc}) = 1] - Pr[\mathcal{B}(g, g^a, g^b, g^c, e(g,g)^r) = 1]|$$

where $a, b, c, r \in \mathbb{Z}_p$ are randomly chosen. We say that the decisional bilinear Diffie Hellman assumption [3] relative to generator $\mathbb{G}_1$ holds if $Adv_{\mathbb{G}_1, \mathcal{B}}^{DBDH}(\lambda)$ is negligible for all PPT $\mathcal{B}$.

2.3 The Truncated (Decisional) q-ABDHE Assumption

Let $e : \mathbb{G}_1 \times \mathbb{G}_1 \to \mathbb{G}_2$ is a bilinear map. We define the advantage function

$$Adv_{\mathbb{G}_1, \mathcal{B}}^{q-ABDHE}(\lambda)$$

of an adversary $\mathcal{B}$ as

$$|Pr[\mathcal{B}(g, g^x, \cdots, g^{x^q}, g^z, g^{zx^{q+2}}, e(g,g)^{zx^{q+1}}) = 1] -$$
$$Pr[\mathcal{B}(g, g^x, \cdots, g^{x^q}, g^z, g^{zx^{q+2}}, e(g,g)^r) = 1]|$$

where $x, z, r \in \mathbb{Z}_p$ are randomly chosen. We say that the truncated decisional augmented bilinear Diffie-Hellman exponent assumption [10] relative to generator $\mathbb{G}_1$ holds if $Adv_{\mathbb{G}_1, \mathcal{B}}^{q-ABDHE}(\lambda)$ is negligible for all PPT $\mathcal{B}$.

2.4 Definition of SCF-PEKS

In the following, we will provide the definition of a SCF-PEKS scheme [2] and the game-based security definition model.

Definition 1 (SCF-PEKS). *A secure channel free public key encryption with keyword search scheme comprises the following algorithms:*

- *GlobalSetup(λ): Takes a security parameter λ generates a global parameter $\mathcal{GP}$.*
- *KeyGen$_{Server}$($\mathcal{GP}$): Takes as input the common parameters $\mathcal{GP}$. Output the public/secret pair (pk_S, sk_S) of server S.*
- *KeyGen$_{Receiver}$($\mathcal{GP}$): Takes as input $\mathcal{GP}$, generates public/secret pair $(pk_{\mathcal{R}}, sk_{\mathcal{R}})$ of receiver $\mathcal{R}$.*
- *SCF $-$ PEKS($\mathcal{GP}, pk_S, pk_{\mathcal{R}}, w$): Takes as input $\mathcal{GP}$, a receiver's public key $pk_{\mathcal{R}}$, a server's public key pk_S, and a keyword w. Return a PEKS ciphertext C under w.*
- *Trapdoor($\mathcal{GP}, sk_{\mathcal{R}}, w$): Takes as input $\mathcal{GP}$, a receiver's secret key $sk_{\mathcal{R}}$ and a keyword w. Generate a trapdoor T_w.*

- $Test(\mathcal{GP}, T_w, sk_S, C)$: *Takes as input a common parameter* $\mathcal{GP}$, *a trapdoor* T_w, *a server's secret key* sk_S *and a PEKS ciphertext* $C = SCF - PEKS(\mathcal{GP}, pk_S, pk_R, w')$. *Output a symbol "Correct" if* $w = w'$ *or "Incorrect" otherwise.*

We define the notion of consistency in a SCF-PEKS scheme, which is similar to the notion of consistency in a PEKS scheme from [1].

Definition 2 (Consistency). *Suppose there exits an adversary* A *that wants to make consistency fail. The consistency is formally defined as follows:*

Experiment $Exp_A{}^{cons}(\lambda)$

$(pk_R, sk_R) \leftarrow KeyGen_{Receiver}(\lambda); (pk_S, sk_S) \leftarrow KeyGen_{Server}(\lambda)$

$(w, w') \leftarrow A(pk_R, pk_S)$

$C \leftarrow SCF - PEKS(\mathcal{GP}, pk_S, pk_R, w); T_{w'} \leftarrow Trapdoor(\mathcal{GP}, sk_R, w')$

if $w \neq w'$ *and* $Test(\mathcal{GP}, T_{w'}, sk_S, C) = $ *"Correct",*

　　· *then return 1,*

　　　else return 0.

We define the advantage of A *as:*

$$Adv_A{}^{cons}(\lambda) = Pr[Exp_A{}^{cons}(\lambda) = 1]$$

The scheme is said to be computationally consistent if it is negligible for polynomial time adversaries A *to win the above experiment.*

In the following, we introduce the game-based security definition of SCF-PEKS, which we call indistinguishability of secure channel free PEKS against chosen keyword attack (IND-SCF-CKA). Informally, IND-SCF-CKA guarantees that the server that has not obtained the trapdoors for given keywords cannot tell which PEKS ciphertext encrypts which keyword, and the outside attacker that has not obtained the server's private key cannot make any decisions about the PEKS ciphertexts even though the attacker gets all the trapdoors for the keywords that it holds. Our definition is adopted from the definition by Baek *et al.* in [2]. Note that the attack models for these two types of attackers are described as Game 1 and Game 2, respectively, in the following definition.

Definition 3 (IND-SCF-CKA game). *Let* λ *be the security parameter and* A *be the adversary. We consider the following two games between* A *and the simulator* B.
Game 1: A *is assumed to be a server.*

1. *Setup: The common parameter generation algorithm GlobalSetup(λ), the two key generation algorithms* $KeyGen_{Server}(\mathcal{GP})$ *and* $KeyGen_{Receiver}(\mathcal{GP})$ *are executed. A common parameter* $\mathcal{GP}$, *private and public key pairs of the receiver and the server, which we denote by* (pk_R, sk_R) *and* (pk_S, sk_S) *respectively, are set. Then,* B *sends* (pk_S, sk_S) *and* pk_R *to* A.
2. *Query phase 1.* A *makes the queries a number of keywords, each of which is denoted by* w:
   - *Trapdoor query* $\langle w \rangle$: A *can adaptively asks* B *for the trapdoor* T_w *for any keyword* $w \in KS_w$ *of his choice.* B *responds the trapdoor* $T_w = Trapdoor(\mathcal{GP}, sk_R, w)$ *to* A.

3. *Challenge. Once $\mathcal{A}$ decides that Phase 1 is over, it outputs a target key pair (w_0, w_1). (Notice that none of w_0 nor w_1 has been queried for obtaining a corresponding trapdoor in Phase 1). Upon receiving this, $\mathcal{B}$ responds by choosing a random $b \in \{0, 1\}$, and creates a target PEKS ciphertext $C^* = SCF - PEKS(\mathcal{GP}, pk_S, pk_\mathcal{R}, w_b)$ and sends it to $\mathcal{A}$.*
4. *Query phase 2. $\mathcal{A}$ issues a number of trapdoor queries as in Phase 1. The restriction here is that w_0 and w_1 are not allowed to be queried as trapdoor queries.*
5. *Guess. $\mathcal{A}$ outputs the guess b'. The adversary wins if $b' = b$.*

We define $\mathcal{A}$'s advantage in Game 1 by $Adv_\mathcal{A}{}^{Game_1}(\lambda) = |Pr[b = b'] - 1/2|$.

Game 2: $\mathcal{A}$ is assumed to be an outside attacker (including the receiver).

1. *Setup: The common parameter generation algorithm $GlobalSetup(\lambda)$, the two key generation algorithms $KeyGen_{Server}(\mathcal{GP})$ and $KeyGen_{Receiver}(\mathcal{GP})$ are executed. A common parameter $\mathcal{GP}$, private and public key pairs of the receiver and the server, which we denote by $(pk_\mathcal{R}, sk_\mathcal{R}))$ and (pk_S, sk_S) respectively, are set. Then, $\mathcal{B}$ sends $(pk_\mathcal{R}, sk_\mathcal{R})$ and pk_S to $\mathcal{A}$.*
2. *Challenge. $\mathcal{A}$ outputs a target keyword pair (w_0, w_1). Upon receiving this, $\mathcal{B}$ responds by choosing a random $b \in \{0, 1\}$, and creates a target PEKS ciphertext $C^* = SCF - PEKS(\mathcal{GP}, pk_S, pk_\mathcal{R}, w_b)$ and sends it to $\mathcal{A}$.*
3. *Guess. $\mathcal{A}$ outputs the guess b'. The adversary wins if $b' = b$.*

We define $\mathcal{A}$'s advantage in Game 2 by $Adv_\mathcal{A}{}^{Game_2}(\lambda) = |Pr[b = b'] - 1/2|$. The SCF-PEKS scheme is said to be IND-SCF-CKA secure if $Adv_\mathcal{A}{}^{Game_i}(\lambda)$, where i is either 1 or 2, is negligible.

3 New SCF-PEKS Scheme

In this section, we will present our efficient construction of public key encryption with keyword search scheme without random oracle. Our scheme is based on Gentry's IBE in the standard model.

3.1 Our Construction

Our public key encryption with keyword search scheme is described as follows.

- *GlobalSetup(λ):* Let λ be the security parameter and $(p, g, \mathbb{G}_1, \mathbb{G}_2, e)$ be the bilinear map parameters. Select a one-way hash function $H : \{0, 1\}^* \to \mathbb{Z}_p{}^*$. The keyword space $KS_w = \mathbb{Z}_p{}^*$. The global parameters are $\mathcal{GP} = (p, g, \mathbb{G}_1, \mathbb{G}_2, e, H, KS_w)$.
- *KeyGen$_{Server}$($\mathcal{GP}$):* Choose $x \in \mathbb{Z}_p{}^*$ uniformly at random and compute $X = g^x$. Choose $Q \in \mathbb{G}_1{}^*$ uniformly at random. Return $pk_S = (\mathcal{GP}, Q, X)$ and $sk_S = (pk_S, x)$ as the server's public and private key, respectively.
- *KeyGen$_{Receiver}$($\mathcal{GP}$):* Choose $y \in \mathbb{Z}_p{}^*$ uniformly at random and compute $Y = g^y$. Choose $h \in \mathbb{G}_1{}^*$ uniformly at random. Return $pk_\mathcal{R} = (pk_S, Y, h)$ and $sk_\mathcal{R} = (pk_\mathcal{R}, y)$ as the receiver's public and private key, respectively.

- $SCF - PEKS(\mathcal{GP}, pk_S, pk_R, w)$: Choose $s, r \in \mathbb{Z}_p^*$ and compute $C_1 = g^s$, $t = H(e(X,Q)^s)$, $C_2 = (Yg^{-w})^{r/t}$, $C_3 = e(g,g)^r$, $C_4 = e(g,h)^r$. The PEKS ciphertext is $C = (C_1, C_2, C_3, C_4)$. Return C.
- $Trapdoor(\mathcal{GP}, sk_R, w)$: Choose $s_w \in \mathbb{Z}_p^*$ and compute $d_w = (hg^{-s_w})^{1/(y-w)}$. Let the trapdoor $T_w = (d_w, s_w)$. Return T_w.
- $Test(\mathcal{GP}, T_w, sk_S, C)$: Compute $t = H(e(C_1, Q)^x)$, check if $e(C_2^t, d_w)C_3^{s_w} = C_4$. If the equation holds return "Correct". Otherwise, return "Incorrect".

Correctness. In the following, we show that a correctly generated PEKS ciphertext can be correctly tested by the server who has the correct trapdoor. In the following, let a PEKS ciphertext $C = (C_1, C_2, C_3, C_4)$ associated with keyword w under the public key pk_S, pk_R. Let the trapdoor $T_w = (d_w, s_w)$. We have

$$t = H(e(C_1, Q)^x)$$
$$= H(e(g^s, Q)^x)$$
$$= H(e(g^x, Q)^s)$$
$$= H(e(X, Q)^s).$$

$$e(C_2^t, d_w)C_3^{s_w} = e(((Yg^{-w})^{r/t})^t, (hg^{-s_w})^{1/(y-w)})(e(g,g)^r)^{s_w}$$
$$= e(g^{(y-w)r}, (hg^{-s_w})^{1/(y-w)})e(g,g)^{rs_w}$$
$$= e(g,h)^r e(g^r, g^{-s_w})e(g,g)^{rs_w}$$
$$= C_4. \qquad \blacksquare$$

3.2   Consistency of Our SCF-PEKS

In this subsection, we prove the computational consistency of our scheme.

Theorem 1. *Our SCF-PEKS scheme is computationally consistent.*

Proof. Suppose there exists a polynomial-time adversary, $\mathcal{A}$, that can attack computational consistency of our scheme. Let (w, w') denote the pair of keywords that $\mathcal{A}$ returns in the consistency experiment, and assume without loss of generality that $w \neq w'$.

Let $s, r \in \mathbb{Z}_p^*$ denote the value chosen at random by $SCF-PEKS(\mathcal{GP}, pk_S, pk_R, w)$. Let $h = g^z$, $C_1 = g^s$, $t = H(e(X,Q)^s)$, $C_2 = (Yg^{-w})^{r/t}$, $C_3 = e(g,g)^r$, $C_4 = e(g,h)^r$.

Let $T_w = (d_{w'}, s_{w'})$ where $d_{w'} = (hg^{-s_{w'}})^{1/(y-w')} = g^{(z-s_{w'})/(y-w')}$ be the trapdoor of w'.

Note that $\mathcal{A}$ wins exactly when $w \neq w'$ and $e(C_2^t, d_{w'})C_3^{s_{w'}} = C_4$.

$e(C_2^t, d_{w'})C_3^{s_{w'}} = C_4$
$\iff e(((Yg^{-w})^{r/t})^t, g^{(z-s_{w'})/(y-w')})e(g,g)^{rs_{w'}} = e(g,g)^{zr}$
$\iff e(g^{(y-w)r}, g^{(z-s_{w'})/(y-w')})e(g,g)^{rs_{w'}} = e(g,g)^{zr}$
$\iff e(g,g)^{((y-w)/(y-w'))(z-s_{w'})r}e(g,g)^{rs_{w'}} = e(g,g)^{zr}$
$\iff e(g,g)^{((y-w)/(y-w'))zr}e(g,g)^{-((y-w)/(y-w'))s_{w'}r}e(g,g)^{rs_{w'}} = e(g,g)^{zr}$
$\iff ((y-w)/(y-w'))zr - ((y-w)/(y-w'))s_{w'}r + rs_{w'} = zr$
$\iff ((y-w)/(y-w') - 1)zr - ((y-w)/(y-w') - 1)s_{w'}r = 0$
$\iff ((w'-w)/(y-w'))zr - ((w'-w)/(y-w'))s_{w'}r = 0$
$\iff ((w'-w)/(y-w'))(z-s_{w'})r = 0.$

Since y, z is receiver's unknown secret key in $\mathbb{Z}_p^*$. Therefore, $Pr[s_{w'} = z] = 1/(p-1)$ and $Pr[w' = y] = 1/(p-1)$ where $p-1$ is the total element number in $\mathbb{Z}_p^*$.

As described above, under the condition $w \neq w'$ and $Test(\mathcal{GP}, T_{w'}, sk_S, C) =$ "Correct"

$$Adv_A{}^{cons}(\lambda) = Pr[Exp_A{}^{cons}(\lambda) = 1] = Pr[(s_{w'} = z) \vee (w' = y)] \leq 2/(p-1).$$

3.3 Security of Our SCF-PEKS

In this subsection, we analyze the security of our SCF-PEKS scheme without requiring any random oracle. The analysis of Game 1 and Game 2 is as follows.

Theorem 2. *The above scheme is IND-SCF-CKA secure without random oracle model assuming that the DBDH problem and q-ABDHE problem are intractable.*

Lemma 1. Let $q \geq q_k + 1$, where q_k is the number of trapdoor queries. Our scheme is semantically secure against a chosen keyword attack in Game 1 without random oracle model assuming q-ABDHE problem is intractable.

Proof. Suppose there exists a polynomial-time adversary, $\mathcal{A}$, in Game 1 that can attack our scheme in the standard model. Let q_k is the number of trapdoor queries. We build a simulator $\mathcal{B}$ that can play a q-ABDHE game. The simulation proceeds as follows:

We first let the challenger set the groups $\mathbb{G}_1$ and $\mathbb{G}_2$ with an efficient bilinear map e and a generator g of $\mathbb{G}_1$. Simulator $\mathcal{B}$ inputs a q-ABDHE instance $(g, g^x, g^{x^2}, \cdots, g^{x^q}, g^z, g^{zx^{q+2}}, T)$, and has to distinguish $T = e(g, g)^{zx^{q+1}}$ from a random element in $\mathbb{G}_2$.

1. Setup: Let λ be the security parameter and $(p, g, \mathbb{G}_1, \mathbb{G}_2, e)$ be the bilinear map parameters. Specify a one-way hash function $H : \{0,1\}^* \to \mathbb{Z}_p^*$. Let the keyword space be $KS_w = \mathbb{Z}_p^*$. The global parameters are $\mathcal{GP} = (p, g, \mathbb{G}_1, \mathbb{G}_2, e, H, KS_w)$. Choose $a \in \mathbb{Z}_p^*$ uniformly at random and compute $X = g^a$. Choose $Q \in \mathbb{G}_1^*$ uniformly at random. Let $pk_S = (\mathcal{GP}, Q, X)$ and $sk_S = (\mathcal{GP}, a)$ as the server's public and private key, respectively.
   Pick a random degree q polynomials $f(X)$ and define $Y = g^x, h = g^{f(x)}$. The receiver's public is $pk_{\mathcal{R}} = (pk_S, Y, h)$. Then, send $(pk_{\mathcal{R}}, pk_S, sk_S)$ to $\mathcal{A}$.

2. Query phase 1. $\mathcal{A}$ makes trapdoor query :
   – Trapdoor query $\langle w \rangle$: If $\mathcal{A}$ queries w to the trapdoor generation oracle, then $\mathcal{B}$ sets $s_w = f(w)$, computes $d_w = g^{(f(x)-f(w))/(x-w)}$, sends the trapdoor $T_w = \{d_w, s_w\}$ to $\mathcal{A}$. When $q \geq q_k + 1$, $s_w = f(w)$ is random value from $\mathcal{A}$'s view, since $f(X)$ is a random degree q polynomial.

3. Challenge. Once $\mathcal{A}$ decides that Phase 1 is over, it outputs a keyword pair (w_0, w_1). $\mathcal{B}$ responds by choosing a random $b \in \{0,1\}$, let $w^* = w_b$ and sets $\{s_{w^*} = f_k(w^*)\}$, then $\mathcal{B}$ computes $d_{w^*} = g^{(f(x)-f(w^*))/(x-w^*)}$. $\mathcal{B}$ randomly chooses $s^* \in \mathbb{Z}_p^*$ and computes $C_1^* = g^{s^*}, t^* = H(e(X, Q)^{s^*})$. Defines the degree $q + 1$ polynomial $F^*(X) = (X^{q+2} - (w^*)^{q+2})/(X - w^*) = \sum_{i=0}^{q+1} (F_i^* X^i)$. Computes

$$C_2^* = (g^{zx^{q+2}}(g^z)^{-(w^*)^{q+2}})^{1/t^*}$$

$$C_3^* = T^{F_{q+1}^*} e(g^z, \prod_{i=0}^{q} (g^{x^i})^{F_i^*})$$

$$C_4^* = e((C_2^*)^{t^*}, d_{w^*})(C_3^*)^{s_{w^*}}.$$

Sends the target PEKS ciphertext $C^* = (C_1{}^*, C_2{}^*, C_3{}^*, C_4{}^*)$ to $\mathcal{A}$.

Let $r^* = zF^*(x)$, if $T = e(g,g)^{zx^{q+1}}$, then

$C_2{}^* = (g^{zx^{q+2}}(g^z)^{-(w^*)^{q+2}})1/t^* = g^{(x-w^*)(z(x^{q+2}-(w^*)^{q+2})/(x-w^*))1/t^*} = g^{(x-w^*)r^*/t^*} = (Yg^{-w^*})^{r^*/t^*}$

$C_3{}^* = T^{F_{q+1}{}^*} e(g^z, \prod_{i=0}^{q}(g^{x^i})^{F_i{}^*}) = e(g,g)^{r^*}$

$C_4{}^* = e(g,h)^{r^*}$.

4. Query phase 2. $\mathcal{A}$ continues making queries as in the Query phase 1.
5. Guess. $\mathcal{A}$ outputs the guess b', if $b' = b$, then output 1 meaning $T = e(g,g)^{zx^{q+1}}$; else output 0 meaning $T = e(g,g)^r$.

Probability Analysis: If $T = e(g,g)^{zx^{q+1}}$, then the simulation is perfect, and $\mathcal{A}$ will guess the bit b correctly with probability $1/2 + \varepsilon$. Else, T is uniformly random, and thus $(C_2{}^*, C_3{}^*)$ is a uniformly random and independent element. In this case, the inequality $C_3{}^* \neq e((C_2{}^*)^{t^*}, g)^{1/(x-w^*)}$ holds with probability $1 - 1/p$. When the inequality holds, the value of

$$C_4{}^* = e((C_2{}^*)^{t^*}, d_{w^*})(C_3{}^*)^{s_{w^*}}$$
$$= e((C_2{}^*)^{t^*}, (h)^{1/(x-w^*)})((C_3{}^*)/(e((C_2{}^*)^{t^*}, g)^{1/(x-w^*)}))^{s_{w^*}}$$

is uniformly random and independent from $\mathcal{A}$'s view (except for the value $C_4{}^*$), since s_{w^*} is uniformly random (when $q \geq q_k + 1$, $s_{w^*} = f(w^*)$ are random values from $\mathcal{A}$'s view) and independent from $\mathcal{A}$'s view (except for the value $C_3{}^*$). Thus, $C_4{}^*$ is uniformly random and independent. Since $s^* \in \mathbb{Z}_p{}^*$ is randomly chosen, $C_1{}^* = g^{s^*}$ is uniformly random and independent from $(C_2{}^*, C_3{}^*, C_4{}^*)$ and $(C_1{}^*, C_2{}^*, C_3{}^*, C_4{}^*)$ can reveal no information regarding the bit b. This completes the proof of Game 1. ∎

Lemma 2. Our scheme is semantically secure against a chosen keyword attack in Game 2 without random oracle model assuming DBDH problem is intractable.

Proof. Suppose there exists a polynomial-time adversary, $\mathcal{A}$, in Game 2 that can attack our scheme in the standard model. We build a simulator $\mathcal{B}$ that can play a DBDH game. The simulation proceeds as follows:

We first let the challenger set the groups $\mathbb{G}_1$ and $\mathbb{G}_2$ with an efficient bilinear map e and a generator g of $\mathbb{G}_1$. $\mathcal{B}$ inputs a DBDH instance (g, g^a, g^b, g^c, T), and has to distinguish $T = e(g,g)^{abc}$ from a random element in $\mathbb{G}_2$.

1. Setup: Let λ be the security parameter and $(p, g, \mathbb{G}_1, \mathbb{G}_2, e)$ be the bilinear map parameters. Specify a one-way hash function $H : \{0,1\}^* \rightarrow \mathbb{Z}_p{}^*$. The global parameters are $\mathcal{GP} = (p, g, \mathbb{G}_1, \mathbb{G}_2, e, H, KS_w)$ where KS_w denotes a description of a keyword space.
   Let $X = g^a$ and $Q = g^b$, the server's public key is $pk_S = (\mathcal{GP}, Q, X)$. Choose $y \in \mathbb{Z}_p{}^*$ uniformly at random and compute $Y = g^y$. Choose $h \in \mathbb{G}_1{}^*$ uniformly at random. $pk_R = (pk_S, Y, h)$ and $sk_R = (pk_S, y)$ denote the receiver's public and private key respectively. Then, send (pk_R, sk_R, pk_S) to $\mathcal{A}$.
2. Challenge. $\mathcal{A}$ outputs a key pair (w_0, w_1). $\mathcal{B}$ responds by choosing a random $b \in \{0,1\}$, let the target keyword $w^* = w_b$, $C_1{}^* = g^c$ and $t^* = H(T)$, chooses $r \in \mathbb{Z}_p{}^*$, computes $C_2{}^* = (Yg^{-w^*})^{r/t^*}$, $C_3{}^* = e(g,g)^r$, $C_4{}^* = e(g,h)^r$. The PEKS ciphertext is $C^* = (C_1{}^*, C_2{}^*, C_3{}^*, C_4{}^*)$. Sends C^* to $\mathcal{A}$.

3. Guess. $\mathcal{A}$ outputs the guess b', if $b' = b$, then output 1 meaning $T = e(g, g)^{abc}$; else output 0 meaning $T = e(g, g)^r$.

Probability Analysis: Suppose there exists a polynomial-time adversary, $\mathcal{A}$, in Game 2 that can attack our scheme in the standard model with an advantage ε. Now we provide the probability of the simulator $\mathcal{B}$:

When $T = e(g, g)^{abc}$ then $\mathcal{A}$ must satisfy $|Pr[b = b'] - 1/2| \geq \varepsilon$. When T is uniform in $\mathbb{G}_2{}^*$ then $Pr[b = b'] = 1/2$. Therefore, when a, b, c are uniform in $\mathbb{Z}_p{}^*$ and T is uniform in $\mathbb{G}_2{}^*$. We have that

$$|Pr[\mathcal{B}(g, g^a, g^b, g^c, e(g, g)^{abc}) = 1] - Pr[\mathcal{B}(g, g^a, g^b, g^c, e(g, g)^r) = 1]| \geq |(1/2 \pm \varepsilon) - 1/2| = \varepsilon$$ as required. This completes the proof of Game 2. ∎

4 Conclusion and Future Work

In this paper, we constructed an efficient and secure channel free public key encryption with keyword search scheme without random oracle. This construction fills the gap in the literature that an efficient and secure channel free public key encryption with keyword search can be built without requiring any random oracle model.

References

1. Abdalla, M., Bellare, M., Catalano, D., Kiltz, E., Kohno, T., Lange, T., Malone-Lee, J., Neven, G., Paillier, P., Shi, H.: Searchable encryption revisited: Consistency properties, relation to anonymous IBE, and extensions. In: Shoup, V. (ed.) CRYPTO 2005. LNCS, vol. 3621, pp. 205–222. Springer, Heidelberg (2005)
2. Baek, J., Safavi-Naini, R., Susilo, W.: Public key encryption with keyword search revisited. In: Gervasi, O., Murgante, B., Laganà, A., Taniar, D., Mun, Y., Gavrilova, M.L. (eds.) ICCSA 2008, Part I. LNCS, vol. 5072, pp. 1249–1259. Springer, Heidelberg (2008)
3. Boneh, D., Boyen, X.: Efficient selective-ID Identity based encryption without random oracles. In: Cachin, C., Camenisch, J.L. (eds.) EUROCRYPT 2004. LNCS, vol. 3027, pp. 223–238. Springer, Heidelberg (2004)
4. Boneh, D., Di Crescenzo, G., Ostrovsky, R., Persiano, G.: Public key encryption with keyword search. In: Cachin, C., Camenisch, J.L. (eds.) EUROCRYPT 2004. LNCS, vol. 3027, pp. 506–522. Springer, Heidelberg (2004)
5. Boneh, D., Waters, B.: Conjunctive, subset, and range queries on encrypted data. In: Vadhan, S.P. (ed.) TCC 2007. LNCS, vol. 4392, pp. 535–554. Springer, Heidelberg (2007)
6. Byun, J.W., Rhee, H.S., Park, H.-A., Lee, D.-H.: Off-line keyword guessing attacks on recent keyword search schemes over encrypted data. In: Jonker, W., Petković, M. (eds.) SDM 2006. LNCS, vol. 4165, pp. 75–83. Springer, Heidelberg (2006)
7. Canetti, R., Goldreich, O., Halevi, S.: The random oracle methodology, revisited. In: Proc. of 30th ACM STOC, pp. 209–218. ACM Press, New York (1998)
8. Canetti, R., Halevi, S., Katz, J.: Chosen-ciphertext security from identity-based encryption. In: Cachin, C., Camenisch, J.L. (eds.) EUROCRYPT 2004. LNCS, vol. 3027, pp. 207–222. Springer, Heidelberg (2004)
9. Fuhr, T., Paillier, P.: Decryptable searchable encryption. In: Susilo, W., Liu, J.K., Mu, Y. (eds.) ProvSec 2007. LNCS, vol. 4784, pp. 228–236. Springer, Heidelberg (2007)

10. Gentry, C.: Practical identity-based encryption without random oracles. In: Vaudenay, S. (ed.) EUROCRYPT 2006. LNCS, vol. 4004, pp. 445–464. Springer, Heidelberg (2006)
11. Gu, C., Zhu, Y., Pan, H.: Efficient public key encryption with keyword search schemes from pairings. In: Pei, D., Yung, M., Lin, D., Wu, C. (eds.) Inscrypt 2007. LNCS, vol. 4990, pp. 372–383. Springer, Heidelberg (2008)
12. Golle, P., Staddon, J., Waters, B.: Secure Conjunctive Search over Encrypted Data. In: Jakobsson, M., Yung, M., Zhou, J. (eds.) ACNS 2004. LNCS, vol. 3089, pp. 31–45. Springer, Heidelberg (2004)
13. Jeong, I.R., Kwon, J.O., Hong, D., Lee, D.H.: Constructing PEKS schemes secure against keyword guessing attacks is possible? Computer Communications. Express 32(2), 394–396 (2009)
14. Rhee, H.S., Park, J.H., Susilo, W., Lee, D.H.: Improved searchable public key encryption with designated tester. In: Proc. of the 4th international Symposium on information, Computer, and Communications Security (ASIACCS 2009), pp. 376–379. ACM, New York (2009)
15. Rhee, H.S., Susilo, W., Kim, H.-J.: Secure searchable public key encryption scheme against keyword guessing attacks. IEICE Electron. Express 6(5), 237–243 (2009)
16. Waters, B., Balfanz, D., Durfee, G., Smetters, D.: Building an Encrypted and Searchable Audit Log. In: Proc. of Network and Distributed System Security Symposium (NDSS 2004) (2004)
17. Yau, W.-C., Heng, S.-H., Goi, B.-M.: Off-line keyword guessing attacks on recent public key encryption with keyword search schemes. In: Rong, C., Jaatun, M.G., Sandnes, F.E., Yang, L.T., Ma, J. (eds.) ATC 2008. LNCS, vol. 5060, pp. 100–105. Springer, Heidelberg (2008)
18. Zhang, R., Imai, H.: Generic combination of public key encryption with keyword search and public key encryption. In: Bao, F., Ling, S., Okamoto, T., Wang, H., Xing, C. (eds.) CANS 2007. LNCS, vol. 4856, pp. 159–174. Springer, Heidelberg (2007)

Private-Key Hidden Vector Encryption with Key Confidentiality

Carlo Blundo, Vincenzo Iovino, and Giuseppe Persiano

Dipartimento di Informatica ed Applicazioni,
Università di Salerno, 84084 Fisciano (SA), Italy
{carblu,iovino,giuper}@dia.unisa.it

Abstract. Predicate encryption is an important cryptographic primitive that has been recently studied [BDOP04, BW07, GPSW06, KSW08] and that has found wide applications. Roughly speaking, in a predicate encryption scheme the owner of the master secret key K can derive secret key $\tilde{K}$, for any *pattern* vector $\boldsymbol{k}$. In encrypting a message M, the sender can specify an *attribute* vector $\boldsymbol{x}$ and the resulting ciphertext $\tilde{X}$ can be decrypted only by using keys $\tilde{K}$ such that $P(\boldsymbol{x}, \boldsymbol{k}) = 1$, for a fixed predicate P. A predicate encryption scheme thus gives the owner of the master secret key fine-grained control on which ciphertexts can be decrypted and this allows him to delegate the decryption of different types of messages (as specified by the attribute vector) to different entities.

In this paper, we give a construction for *hidden vector encryption* which is a special case of predicate encryption schemes introduced by [BW07]. Here the ciphertext attributes are vectors $\boldsymbol{x} = \langle x_1, \ldots, x_\ell \rangle$ over alphabet Σ, key patterns are vectors $\boldsymbol{k} = \langle k_1, \ldots, k_\ell \rangle$ over alphabet $\Sigma \cup \{\star\}$ and we consider the $\mathsf{Match}(\boldsymbol{x}, \boldsymbol{k})$ predicate which is true if and only if $k_i \neq \star$ implies $x_i = k_i$. Besides guaranteeing the security of the attributes of a ciphertext, our construction also gives security guarantees for the key patterns. We stress that security guarantees for key patterns only make sense in a private-key setting and have been recently considered by [SSW09] which gave a construction in the symmetric bilinear setting with groups of composite (product of four primes) order. In contrast, our construction uses *asymmetric* bilinear groups of *prime* order and the length of the key is equal to the weight of the pattern, thus resulting in an increased efficiency. We remark that our construction is based on falsifiable (in the sense of [BW06, Nao03]) complexity assumptions for the asymmetric bilinear setting and are proved secure in the standard model (that is, without random oracles).

Keywords: private-key predicate encryption, key confidentiality.

1 Introduction

Predicate encryption is an important cryptographic primitive that has been recently studied [BDOP04, BW07, GPSW06, KSW08] and that has found wide applications. Roughly speaking, in a predicate encryption scheme the owner of

J.A. Garay, A. Miyaji, and A. Otsuka (Eds.): CANS 2009, LNCS 5888, pp. 259–277, 2009.

the master secret key SK, can derive secret key $\tilde{K}$, for any pattern vectors $\boldsymbol{k}$. Similarly, in encrypting a message M, the sender can specify an attribute vector $\boldsymbol{x}$ and the resulting ciphertext $\tilde{X}$ can be decrypted only by using keys $\tilde{K}$ such that $P(\boldsymbol{x}, \boldsymbol{k}) = 1$, for a fixed predicate P.

In this paper, we consider *hidden vector encryption* that is a special class of predicate encryptions first studied in [BW07]. In a hidden vector encryption scheme, ciphertexts are associated with *attribute vectors* $\boldsymbol{x}$ of length ℓ over an alphabet Σ and keys are associated with *pattern vectors* $\boldsymbol{k}$ of length ℓ over the alphabet $\Sigma \cup \{\star\}$. The predicate we are interested in is the Match predicate defined as follows: Match$(\boldsymbol{x}, \boldsymbol{k}) = 1$ if and only if for $i = 1, \ldots, \ell$ either $k_i = \star$ or $k_i = x_i$. Constructions for hidden vector encryption have been given in [BW07] (based on hardness assumptions in groups of composite order) and in [IP08] (based on hardness assumptions in groups of prime order).

Until now research has concentrated on guaranteeing the security of the ciphertext with respect to the cleartext and to the attribute vector and not much attention has been devoted to the security of the key. Specifically, one would like a key not to reveal the associated pattern. This is particularly important in some applications in which a user generates the key for a certain pattern and gives it to a third party to perform some operations. Knowledge of the pattern associated with the key might reveal some information about the operation being performed. Obviously, this is impossible to achieve in a public-key setting. Indeed an adversary $\mathcal{A}$ holding a key $\tilde{K}$ associated to a secret pattern $\boldsymbol{k}$ can simply produce a ciphertext $\tilde{X}$ with attribute $\boldsymbol{x}$ and then try to decrypt $\tilde{X}$ using $\tilde{K}$. If $\mathcal{A}$ succeeds in decrypting $\tilde{K}$ then $\mathcal{A}$ knows that $P(\boldsymbol{x}, \boldsymbol{k}) = 1$. This attack does not hold in the private key setting as $\mathcal{A}$ cannot produce ciphertext $\tilde{X}$. Simply keeping the public key secret from the adversary does not seem to work for previous predicate encryption schemes (see, for example [BW07, KSW08]) and the problem seems to call for a new construction. The scheme of [SSW09] is constructed modifying the previous scheme of [KSW08], likewise, we build our scheme from the scheme of [IP08].

Prior work and our contribution. Shen, Shi and Waters [SSW09] were the first to consider key confidentiality in the context of predicate encryption and they provided a construction for the inner-product predicate (that is, a key can decrypt a ciphertext if and only if the pattern vector of the key is orthogonal to the attribute vector of the ciphertext). In this paper we present a construction for an *hidden vector encryption scheme* which, besides guaranteeing privacy of the attribute vector of ciphertext, guarantees that keys do not leak any information on the associated pattern, *besides the location of the $\star$'s*. We stress that the construction of [SSW09] for the inner-product predicate implies (with a small loss of efficiency) a construction also for hidden vector encryption scheme. The security of the construction of [SSW09] is based on bilinear assumptions on groups of order product of *four* primes, and thus, it is less efficient. In our construction we show that, by slightly relaxing the notion of key confidentiality, we can obtain construction using asymmetric bilinear groups of prime order (which results in much more efficient constructions). We remark that our construction is based on falsifiable (in the sense of [BW06, Nao03]) complexity assumptions for the *asymmetric* bilinear setting for groups of prime order and are proved secure in the standard model (that is, without random oracles).

Moving from composite order groups to prime order groups, besides giving very efficient constructions, is also important since assumptions based on prime order groups are considered weaker than the corresponding assumptions that intertwine and compound potential vulnerabilities from factoring and pairings (see the discussion in [Boy08]).

Finally, we stress that the only previous construction of hidden vector encryption schemes based on prime order groups of [IP08] does not give any security guarantee for the key.

2 Hidden Vector Encryption Schemes

In this paper we consider a special type of predicate encryption schemes called *Hidden Vector Encryption Scheme*, (an HVE scheme, in short). We present the definition and the construction for $\Sigma = \{0, 1\}$. In Section 8 we briefly explain how the constructions can be extended to larger alphabets.

An HVE scheme consists of four algorithms:

1. MasterKeyGen($1^n, 1^\ell$): Given security parameter n, and number of attributes $\ell = \text{poly}(n)$, procedure MasterKeyGen outputs the private key SK.

2. Enc(SK, x): Given attribute vector $x \in \{0, 1\}^\ell$ and secret key SK, procedure Enc outputs an encrypted attribute vector $\tilde{X}$.

3. KeyGen(SK, k): Given private key SK, a pattern vector k of length ℓ over the alphabet $\{0, 1, \star\}$, procedure KeyGen outputs a *key* $\tilde{K}$ for the k.

4. Test($\tilde{X}, \tilde{K}$): given encrypted attribute vector $\tilde{X}$ and key $\tilde{K}$ corresponding to pattern k, procedure Test returns Match(x, k) except with negligible probability.

We state security in the selective attribute model using the following experiments.

2.1 Semantic Security

The first experiment considers an adversary that tries to learn information from an encryption. We model this using an indistinguishability experiment in which the adversary $\mathcal{A}$ selects two challenge attribute vectors z_0 and z_1 and receives an encrypted attribute vector corresponding to a randomly chosen challenge attribute vector. We allow the adversary to issue key queries for patterns y that match neither of z_0 and z_1 and to see encryption of attribute vectors of his choice (see Section 7 for a stronger notion). Following is the description of experiment SemanticExp$_{\mathcal{A}}$.

SemanticExp$_{\mathcal{A}}(1^n, 1^\ell)$

1. Initialization Phase. The adversary $\mathcal{A}$ announces two challenge attribute vectors $z_0, z_1 \in \{0, 1\}^\ell$.

2. Key-Generation Phase. The secret key SK is generated by the MasterKeyGen procedure.

3. Query Phase I. $\mathcal{A}$ can make any number of key and encryption query.
   A key query for pattern k is answered as follows. If $\mathsf{Match}(z_0, k) = 0$ and $\mathsf{Match}(z_1, k) = 0$ then $\mathcal{A}$ receives the output of $\mathsf{KeyGen}(\mathsf{SK}, k)$. Otherwise, $\mathcal{A}$ receives $\perp$. An encryption query for attribute vectors x is answered by returning $\mathsf{Enc}(\mathsf{SK}, x)$.

4. Challenge construction. η is chosen at random from $\{0, 1\}$ and $\mathcal{A}$ is given $\mathsf{Enc}(\mathsf{SK}, z_\eta)$.

5. Query Phase II. Identical to Query Phase I.

6. Output Phase. $\mathcal{A}$ returns η'.
   If $\eta = \eta'$ then the experiments returns 1 else 0.

Definition 1. *An* HVE *scheme* (MasterKeyGen, Enc, KeyGen, Test) *is semantically secure, if for all probabilistic poly-time adversaries* $\mathcal{A}$

$$\left| \mathrm{Prob}[\mathsf{SemanticExp}_{\mathcal{A}}(1^n, 1^\ell) = 1] - 1/2 \right|$$

is negligible in n *for all* $\ell = \mathsf{poly}(n)$.

2.2   Key Confidentiality

In this section we present our definition for key confidentiality. We model this property by using an indistinguishability experiment in which the adversary $\mathcal{A}$ outputs two challenge patterns k_0 and k_1 of his choice. $\mathcal{A}$ is then allowed to issue encryption queries for vectors x that match neither of k_0 and k_1 and key queries for patterns k of his choice. At the end $\mathcal{A}$ is presented with the key associated with a randomly chosen challenge pattern. In our notion of key confidentiality, the adversary is limited to challenges on patterns in which the "don't care" entries (that is, $\star$) are in the same positions.

$\mathsf{KeyExp}_{\mathcal{A}}(1^n, 1^\ell)$

1. Initialization Phase. The adversary $\mathcal{A}$ announces two challenge patterns $k_0, k_1 \in \{0, 1, \star\}^\ell$. If the set of positions for which k_0 and k_1 have $\star$ differ then the experiment returns 0.

2. Key-Generation Phase. The secret key SK is generated by the MasterKeyGen procedure.

3. Query Phase I. $\mathcal{A}$ can make any number of key and encryption query.
   A key query for pattern k is answered by returning $\mathsf{KeyGen}(\mathsf{SK}, k)$.
   An encryption query for attribute vector x is answered as follows.
   If $\mathsf{Match}(x, k_0) = \mathsf{Match}(x, k_1) = 0$ then $\mathcal{A}$ receives $\mathsf{Enc}(\mathsf{SK}, x)$. Otherwise, $\mathcal{A}$ receives $\perp$.

4. Challenge construction. η is chosen at random from $\{0, 1\}$ and receives $\mathsf{KeyGen}(\mathsf{SK}, k_\eta)$.

5. Query Phase II. Identical to Query Phase I.

6. Output Phase. $\mathcal{A}$ returns η'.

If $\eta = \eta'$ then the experiments returns 1 else 0.

Definition 2. *A predicate encryption scheme* (MasterKeyGen, Enc, KeyGen, Test) *is* key secure *if for all probabilistic poly-time adversaries* $\mathcal{A}$,

$$\left| \text{Prob}[\text{KeyExp}_{\mathcal{A}}(1^n, 1^\ell) = 1] - 1/2 \right|$$

is negligible in n *for all* $\ell = \text{poly}(n)$.

2.3 Secure HVE

Finally we have,

Definition 3. *An* HVE*scheme* (MasterKeyGen, Enc, KeyGen, Test) *is* secure *if it is both semantically secure and key secure.*

Remark on the notion of key confidentiality. In our notion of key confidentiality the key might reveal the position of the $\star$'s in the associated pattern, since no requirement is made for adversary choosing challenge patterns with $\star$'s in different positions. In some applications, this might not be a drawback. For example, predicate encryption can be used for performing searches on encrypted data. For example, a user interested in selecting ciphertexts for which *Name=Alex* and *Sex=M* gets a key corresponding to a pattern that has $\star$ in all positions other than *Name* and *Sex*. An eavesdropper learns that the user is searching the fields *Name* and *Sex* but no information is given on the name the user is searching for and whether the user is searching for a male or a female. We remark that the construction of [SSW09] hides all information of the key, but their construction is less efficient than ours since it uses groups of composite order of *four* primes. Roughly speaking, by slightly relaxing the security notion, we manage to build a more efficient scheme.

3   Complexity Assumptions

We work in asymmetric prime order bilinear groups of 'Type 3' (see [Boy08]). Specifically, we have cyclic multiplicative groups $\mathbb{G}_1, \mathbb{G}_2$ and $\mathbb{G}_T$ of order p such that there exists no efficiently computable morphism from $\mathbb{G}_1$ to $\mathbb{G}_2$ or from $\mathbb{G}_2$ to $\mathbb{G}_1$. In addition we have a non-degenerate pairing function $\mathsf{e} : \mathbb{G}_1 \times \mathbb{G}_2 \to \mathbb{G}_T$; that is, for all $x \in \mathbb{G}_1, y \in \mathbb{G}_2$, $x \neq 1$ or $y \neq 1$, we have $\mathsf{e}(x, y) \neq 1$ and for all $a, b \in \mathbb{Z}_p$ we have $\mathsf{e}(x^a, y^b) = \mathsf{e}(x, y)^{ab}$. We denote by g_1, g_2, and $\mathsf{e}(g_1, g_2)$ generators of $\mathbb{G}_1, \mathbb{G}_2$, and $\mathbb{G}_T$, respectively.

We call a tuple $\mathcal{I} = [p, \mathbb{G}_1, \mathbb{G}_2, \mathbb{G}_T, g_1, g_2, \mathsf{e}]$ an *asymmetric bilinear* instance and assume that there exists an efficient generation procedure $\mathcal{G}$ that, on input security parameter 1^n, outputs an instance with $|p| = \Theta(n)$.

We now present a new assumption, which we call the (d, m)-$\mathcal{Q}$ Assumption, on which we base the proof of key security of our construction. Semantic security is based instead on the Decision Linear Assumption and on the Bilinear Decision Diffie-Hellman Assumption which we review in Section 3. We present the assumption in the form of a game between a challenger Ch and a distinguisher $\mathcal{D}$ on input the security parameter n.

Game (d, m)-$Q(1^n)$

1. The challenger Ch picks a random asymmetric bilinear instance $\mathcal{I} = [p, \mathbb{G}_1, \mathbb{G}_2, \mathbb{G}_T, g_1, g_2, e]$ by running generator $\mathcal{G}$ on input security parameter 1^n and sets ChOutput $= \emptyset$.

2. For $i = 1, \ldots, d$ and $b = 0, 1$, Ch chooses random $\hat{t}_{i,b}, \hat{v}_{i,b} \in \mathbb{Z}_p$.

3. For $i = 1, \ldots, d$, Ch chooses random $\hat{a}_i \in \mathbb{Z}_p$ such that their sum is equal to 0.

4. Define set of pairs $JH = \{(j, h) | 1 \leq j \leq m, \ 1 \leq h \leq m, \ j \neq h$ or $j = h, \ m + 1 \leq j \leq d\}$.

   For $(j, h) \in JH$, Ch chooses a random $\hat{s}_{(j,h)} \in \mathbb{Z}_p$ and computes matrices $A_{j,h}$ and $B_{j,h}$ as follows, where $\times$ denotes a missing entry in the matrices:[1]

$$A_{j,h} = \begin{cases} \begin{bmatrix} g_1^{\hat{s}_{j,h}\hat{t}_{1,0}}, \ldots, & \times, & \ldots, g_1^{\hat{s}_{j,h}\hat{t}_{h,0}}, & \ldots, g_1^{\hat{s}_{j,h}\hat{t}_{d,0}} \\ g_1^{\hat{s}_{j,h}\hat{t}_{1,1}}, \ldots, g_1^{\hat{s}_{j,h}\hat{t}_{j,1}}, \ldots, & \times & \ldots, g_1^{\hat{s}_{j,h}\hat{t}_{d,1}} \end{bmatrix} & \text{if } j \neq h \text{ and } j, h \leq m \\[2em] \begin{bmatrix} g_1^{\hat{s}_{j,h}\hat{t}_{1,0}}, \ldots, & \times, & \ldots, g_1^{\hat{s}_{j,h}\hat{t}_{d,0}} \\ g_1^{\hat{s}_{j,h}\hat{t}_{1,1}}, \ldots, g_1^{\hat{s}_{j,h}\hat{t}_{j,1}}, \ldots, g_1^{\hat{s}_{j,h}\hat{t}_{d,1}} \end{bmatrix} & \text{if } j = h \text{ and } j > m \end{cases}$$

   and $B_{j,h} = \begin{cases} \begin{bmatrix} g_1^{\hat{s}_{j,h}\hat{v}_{1,0}}, \ldots, & \times, & \ldots, g_1^{\hat{s}_{j,h}\hat{v}_{h,0}}, \ldots, g_1^{\hat{s}_{j,h}\hat{v}_{d,0}} \\ g_1^{\hat{s}_{j,h}\hat{v}_{1,1}}, \ldots, g_1^{\hat{s}_{j,h}\hat{v}_{j,1}}, \ldots, & \times & \ldots, g_1^{\hat{s}_{j,h}\hat{v}_{d,1}} \end{bmatrix} & \text{if } j \neq zh \text{ and } j, h \leq m \\[2em] \begin{bmatrix} g_1^{\hat{s}_{j,h}\hat{v}_{1,0}}, \ldots, & \times, & \ldots g_1^{\hat{s}_{j,h}\hat{v}_{d,0}} \\ g_1^{\hat{s}_{j,h}\hat{v}_{1,1}}, \ldots, g_1^{\hat{s}_{j,h}\hat{v}_{j,1}}, \ldots, g_1^{\hat{s}_{j,h}\hat{v}_{d,1}} \end{bmatrix} & \text{if } j = h \text{ and } j > m \end{cases}$

   Ch appends the above matrices to ChOutput.

5. For $i = 1, \ldots, d$ and $b = 0, 1$, Ch computes and appends to ChOutput

$$C_{i,b} = g_2^{1/\hat{t}_{i,b}} \quad \text{and} \quad D_{i,b} = g_2^{1/\hat{v}_{i,b}}.$$

6. Ch chooses random $\eta \in \{0, 1\}$ and let $z = \langle z_1, \ldots, z_d \rangle = \eta^m \cdot 0^{d-m}$. For $i = 1, \ldots, d$, Ch computes

$$E_i = C_{i,z_i}^{\hat{a}_i} \quad \text{and} \quad F_i = D_{i,z_i}^{\hat{a}_i}$$

   and appends the values E_i and F_i to ChOutput.

---

[1] For the sake of simplicity of exposition, in the definition we have implicitly assumed that $j \leq h$.

7. Challenger Ch runs $\mathcal{D}$ on input sequence ChOutput and receives output η'.

We define the advantage $\mathsf{Adv}_{\mathcal{D}}(n, d, m)$ of distinguisher $\mathcal{D}$ in the Game (d, m)-$Q(1^n)$ as

$$\mathsf{Adv}_{\mathcal{D}}(n, d, m) = \left| \mathrm{Prob}[\eta = \eta'] - \frac{1}{2} \right|.$$

We are now ready to formally state Assumption (d, m)-Q.

Assumption 1 (Assumption (d, m)-Q). *For all probabilistic poly-time distinguishers $\mathcal{D}$, we have that $\mathsf{Adv}_{\mathcal{D}}(n, d, m)$ is negligible in n, for $d = \mathsf{poly}(n)$, and $1 \leq m \leq d$.*

The (d, m)-Q Assumption can be justified by extending the framework of the Uber-Assumption [BBG05, Boy08] to rational functions along the lines of [Boy08]. In the rest of this section we review other hardness assumptions used in the paper.

Bilinear Decision Diffie-Hellman. Given a tuple $[g_1, g_2, g_1^a, g_1^b, g_2^a, g_2^b, g_1^c, Z]$ for random exponents $a, b, c \in \mathbb{Z}_p$ it is hard to distinguish between $Z = \mathrm{e}(g_1, g_2)^{abc}$ and a random Z from $\mathbb{G}_T$. More specifically, for an algorithm $\mathcal{A}$ we define experiment $\mathsf{BDDHExp}_{\mathcal{A}}$ as follows.

$\mathsf{BDDHExp}_{\mathcal{A}}(1^n)$
1. Choose instance $\mathcal{I} = [p, \mathbb{G}_1, \mathbb{G}_2, \mathbb{G}_T, g_1, g_2, \mathrm{e}]$ with security parameter 1^n.
2. Choose $a, b, c \in \mathbb{Z}_p$ at random.
3. Choose $\eta \in \{0, 1\}$ at random.
4. If $\eta = 1$ then choose $z \in \mathbb{Z}_p$ at random; else, set $z = abc$.
5. Set $A = g_1^a, B = g_1^b, \hat{A} = g_2^a, \hat{B} = g_2^b, C = g_1^c$ and $Z = \mathrm{e}(g_1, g_2)^z$.
6. Let $\eta' = \mathcal{A}(\mathcal{I}, A, B, \hat{A}, \hat{B}, C, Z)$.
7. If $\eta = \eta'$ then return 1 else return 0.

Assumption 2 (Bilinear Decisional Diffie-Hellman (BDDH)). *For all probabilistic poly-time algorithms $\mathcal{A}$, $|\mathrm{Prob}[\mathsf{BDDHExp}_{\mathcal{A}}(1^n) = 1] - 1/2|$ is negligible in n.*

Decision Linear. Given a tuple $[g_1, g_2, g_1^{z_1}, g_1^{z_2}, g_2^{z_1}, g_2^{z_2}, g_1^{z_1 z_3}, g_1^s, Z]$ for random exponents $z_1, z_2, z_3, s \in \mathbb{Z}_p$ it is hard to distinguish between $Z = g_1^{z_2(s - z_3)}$ and a random Z from $\mathbb{G}_1$. More specifically, for an algorithm $\mathcal{A}$ we define experiment $\mathsf{DLExp}_{\mathcal{A}}$ as follows.

$\mathsf{DLExp}_{\mathcal{A}}(1^n)$
1. Choose instance $\mathcal{I} = [p, \mathbb{G}_1, \mathbb{G}_2, \mathbb{G}_T, g_1, g_2, \mathrm{e}]$ with security parameter 1^n.
2. Choose $u_1, u_2, u_3, u \in \mathbb{Z}_p$ at random.
3. Choose $\eta \in \{0, 1\}$ at random.
4. If $\eta = 1$ then choose $z \in \mathbb{Z}_p$ at random; else, set $z = u_2(u - u_3)$.
5. Set $U_1 = g_1^{u_1}, U_2 = g_1^{u_2}, \hat{U}_1 = g_2^{u_1}, \hat{U}_2 = g_2^{u_2}, U_{13} = g_1^{u_1 u_3}, U = g_1^u$, and $Z = g_1^z$.
6. Let $\eta' = \mathcal{A}(\mathcal{I}, U_1, U_2, \hat{U}_1, \hat{U}_2, U_{13}, U, Z)$.
7. If $\eta = \eta'$ then return 1 else return 0.

Assumption 3 (Decision Linear (DLinear)). *For all probabilistic poly-time algorithms* $\mathcal{A}$, $|\mathrm{Prob}[\mathsf{DLExp}_{\mathcal{A}}(1^n) = 1] - 1/2|$ *is negligible in* n.

Note that Decision Linear implies Decision BDDH and the Decision Linear assumption has been used in [BW06].

4   The Basic Scheme

In this section, we describe our proposal for a secure HVE.

The MasterKeyGen *procedure.* On input security parameter 1^n and the number of attributes $\ell = \mathrm{poly}(n)$, MasterKeyGen proceeds as follows.

1. Select an asymmetric bilinear instance $\mathcal{I} = [p, q, \mathbb{G}_1, \mathbb{G}_2, \mathbb{G}_T, g_1, g_2, \mathsf{e}]$ with $|N| = \Theta(n)$ by running $\mathcal{G}$.
2. Pick y at random in $\mathbb{Z}_p$ and set $Y = \mathsf{e}(g_1, g_2)^y$.
   For $i = 1, \ldots, \ell$,
   Choose $t_{i,0}, t_{i,1}, v_{i,0}, v_{i,1}$ at random from $\mathbb{Z}_p$.
   Set
   $$T_{i,0} = g_1^{t_{i,0}}, \quad T_{i,1} = g_1^{t_{i,1}}, \quad V_{i,0} = g_1^{v_{i,0}}, \quad V_{i,1} = g_1^{v_{i,1}},$$
   $$\bar{T}_{i,0} = g_2^{1/t_{i,0}}, \quad \bar{T}_{i,1} = g_2^{1/t_{i,1}}, \quad \bar{V}_{i,0} = g_2^{1/v_{i,0}}, \quad \bar{V}_{i,1} = g_2^{1/v_{i,1}}.$$

   Set $\mathsf{SK}_i = (T_{i,0}, T_{i,1}, V_{i,0}, V_{i,1}, \bar{T}_{i,0}, \bar{T}_{i,1} \bar{V}_{i,0}, \bar{V}_{i,1},)$.
3. Return $\mathsf{SK} = (\mathcal{I}, Y, y, \mathsf{SK}_1, \ldots, \mathsf{SK}_\ell)$.

The Enc *procedure.* On input secret key SK and attribute vector $\boldsymbol{x}$ of length ℓ, Enc proceeds as follows.

1. Pick s at random from $\mathbb{Z}_p$ and set $\Omega = Y^{-s}$.
2. For $i = 1, \ldots, \ell$,
   pick s_i at random from $\mathbb{Z}_p$.
   set $X_i = T_{i,x_i}^{s-s_i}$ and $Z_i = V_{i,x_i}^{s_i}$.
3. Return encrypted attribute vector $\tilde{X} = (\Omega, (X_i, Z_i)_{i=1}^\ell)$.

In the following sometimes will use the writing $\mathsf{Enc}(\mathsf{SK}, \boldsymbol{x}; s, s_1, \ldots, s_\ell)$ to denote the encrypted attribute vector $\tilde{X}$ output by Enc on input SK and $\boldsymbol{x}$ when using $s, s_1, \ldots, s_\ell$ as random elements.

The KeyGen *procedure.* On input secret key SK and pattern vector $\boldsymbol{k}$, KeyGen proceeds as follows.

1. Let $S_{\boldsymbol{k}}$ be the set of positions in which $k_i \neq \star$.
2. Choose $(a_i)_{i \in S_{\boldsymbol{k}}}$ at random in $\mathbb{Z}_p$ under the constraint that their sum is y.
3. For $i \in S_{\boldsymbol{k}}$, set $R_i = \bar{T}_{i,k_i}^{a_i}$ and $W_i = \bar{V}_{i,k_i}^{a_i}$.
4. Return $\tilde{K} = (i, R_i, W_i)_{i \in S_{\boldsymbol{k}}}$.

In the following sometimes will use the writing $\mathsf{KeyGen}(\mathsf{SK}, \boldsymbol{k}; (a_i)_{i \in S_{\boldsymbol{k}}})$ to denote the key $\tilde{K}$ computed by KeyGen on input SK and $\boldsymbol{k}$ and using $(a_i)_{i \in S_{\boldsymbol{k}}}$ as random elements.

The Test *procedure.* On input an encrypted attribute vector $\tilde{X} = (\Omega, (X_i, Z_i)_{i=1}^{\ell})$ and a key $\tilde{K} = ((i_1, R_{i_1}, W_{i_1}), \ldots, (i_m, R_{i_m}, W_{i_m}))$, Test proceeds as follows.

1. Compute $a = \Omega \cdot \prod_{j=1}^{m} \mathsf{e}(X_{i_j}, R_{i_j})\mathsf{e}(Z_{i_j}, W_{i_j})$.
2. If $a = 1$ then return TRUE else return FALSE.

We next prove that the quadruple is indeed a predicate encryption scheme.

Theorem 1. *The quadruple of algorithms* (MasterKeyGen, Enc, KeyGen, Test) *specified above is a predicate encryption scheme.*

Proof. It is sufficient to verify that the procedure Test returns 1 when $\mathsf{Match}(x, k)$ = 1. Let $\tilde{X} = (\Omega, (X_i, Z_i)_{i=1}^{\ell})$ be the output of $\mathsf{Enc}(\mathsf{SK}, x; s, s_1, \ldots, s_{\ell})$ and let $\tilde{K} = (i, R_i, W_i)_{i \in S_k}$ be the output of $\mathsf{KeyGen}(\mathsf{SK}, k; (a_i)_{i \in S_k})$. Then we have

$\mathsf{Test}(\tilde{X}, \tilde{K})$

$$= \Omega \cdot \prod_{i \in S_k} \mathsf{e}(X_i, R_i) \cdot \mathsf{e}(Z_i, W_i)$$

$$= \mathsf{e}(g_1, g_2)^{-ys} \cdot \prod_{i \in S_k} \mathsf{e}(T_{i,x_i}^{s-s_i}, \bar{T}_{i,k_i}^{a_i}) \cdot \mathsf{e}(V_{i,x_i}^{s_i}, \bar{V}_{i,k_i}^{a_i}) \text{ (since } x_i = k_i \text{ for } i \in S_k)$$

$$= \mathsf{e}(g_1, g_2)^{-ys} \cdot \prod_{i \in S_k} \mathsf{e}(T_{i,k_i}^{s-s_i}, \bar{T}_{i,k_i}^{a_i}) \cdot \mathsf{e}(V_{i,k_i}^{s_i}, \bar{V}_{i,k_i}^{a_i})$$

$$\text{(since } \mathsf{e}(T_{i,k_i}, \bar{T}_{i,k_i}) = \mathsf{e}(V_{i,k_i}, \bar{V}_{i,k_i}) = \mathsf{e}(g_1, g_2) \in \mathbb{G}_T)$$

$$= \mathsf{e}(g_1, g_2)^{-ys} \cdot \prod_{i \in S_k} \mathsf{e}(g_1, g_2)^{(s-s_i)a_i} \cdot \mathsf{e}(g_1, g_2)^{s_i a_i}$$

$$= \mathsf{e}(g_1, g_2)^{-ys} \cdot \prod_{i \in S_k} \mathsf{e}(g_1, g_2)^{s a_i} \text{ (since } \sum_{i \in S_k} a_i = y)$$

$$= \mathsf{e}(g_1, g_2)^{-ys} \cdot \mathsf{e}(g_1, g_2)^{ys} = 1. \qquad \square$$

5 Proof of Semantic Security

In this section, we prove that the scheme presented in Section 4 is semantically secure. Consider the following experiments, for $j = 0, \cdots, \ell$.

$\mathsf{SemanticExp}_{\mathcal{A}}(1^n, 1^{\ell}, z, j)$

1. Key-generation Phase. Compute $\mathsf{SK} = (\mathcal{I}, y, \mathsf{SK}_1, \cdots, \mathsf{SK}_{\ell})$ by executing $\mathsf{MasterKeyGen}(1^n, 1^{\ell})$.

2. Query Phase I. Answer Enc queries for attribute vectors x by using secret key SK.
   Answer KeyGen queries for pattern vectors k such that $\mathsf{Match}(z, k) = 0$ using secret key SK.

3. Challenge Construction.
   1. If $j = 0$ set $\Omega = \mathsf{e}(g_1, g_2)^{-ys}$.
   2. If $j \geq 1$ choose Ω uniformly at random from $\mathbb{G}_T$.

3. For $i = 1, \ldots, j - 1$,
   choose X_i and Z_i uniformly at random in $\mathbb{G}_1$.

4. If $j = 0$ set $\alpha = 1$ else set $\alpha = j$.

5. For $i = \alpha, \ldots, \ell$,
   choose s_i uniformly at random in $\mathbb{Z}_p$ and set $X_i = g_1^{t_{i,z_i}(s-s_i)}$ and $Z_i = g_1^{s_i v_{i,z_i}}$.

6. Set $\tilde{X} = (\Omega, (X_i, Z_i)_{i=1}^{\ell})$.

7. Query Phase II. Identical to Query Phase I.

8. **return:** $\mathcal{A}(\tilde{X})$.

We will use the writing $\mathsf{SemanticExp}_\mathcal{A}(1^n, 1^\ell, z, j; s, s_\alpha, \ldots, s_\ell)$ to denote the tuple $\tilde{X}$ computed by $\mathsf{SemanticExp}_\mathcal{A}(1^n, 1^\ell, z, j)$ using $s, s_\alpha, \ldots, s_\ell$ as random values, where $\alpha = 1$ for $j = 0$ and $\alpha = j$ for $j > 0$.

We will denote by $p_j^\mathcal{A}(z)$ the probability that experiment $\mathsf{SemanticExp}_\mathcal{A}(1^n, 1^\ell, z, j)$ returns 1. Notice that in $\mathsf{SemanticExp}_\mathcal{A}(1^n, 1^\ell, z, 0)$ adversary $\mathcal{A}$ receives a valid encrypted attribute vector $\tilde{X}$ for attribute vector z and secret key SK whereas in $\mathsf{SemanticExp}_\mathcal{A}(1^n, 1^\ell, z, \ell)$ adversary $\mathcal{A}$ receives $\tilde{X}$ consisting of one random element of $\mathbb{G}_T$ and 2ℓ random elements of $\mathbb{G}_1$. Next we prove that, under the Decision Linear assumption, for all attribute vectors z, the difference $|p_0^\mathcal{A}(z) - p_\ell^\mathcal{A}(z)|$ is negligible. This implies the semantic security of the scheme.

Due to space limitation we omit the proof of the next lemmata. Similar proofs can be found in [IP08].

Lemma 1. *Assume BDDH holds. Then for any attribute string z and for any adversary $\mathcal{A}$,*

$$|p_0^\mathcal{A}(z) - p_1^\mathcal{A}(z)|$$

is non-negligible.

Lemma 2. *Assume DLinear holds. Then, for any attribute string z, for any adversary $\mathcal{A}$, and for $1 \le j \le \ell - 1$*

$$|p_j^\mathcal{A}(z) - p_{j+1}^\mathcal{A}(z)|$$

is negligible.

Combining Lemma 1 and Lemma 2 and by noticing that DLinear implies BDDH, we have the following lemma.

Lemma 3. *Assume DLinear. Then predicate encryption* (MasterKeyGen, Enc, KeyGen, Test) *is semantically secure.*

6   Proof of Key Confidentiality

In this section, we prove the construction of Section 4 is key secure, under Assumption Q. We use the following experiments for $\eta \in \{0, 1\}$.

$\mathsf{KeyExp}_\mathcal{A}(1^n, 1^\ell, z_0, z_1, \eta)$

1. Key-Generation Phase. The secret key SK is generated by the MasterKeyGen procedure.

2. Query Phase I. $\mathcal{A}$ can make any number of key and encryption query.
   A key query for pattern $\boldsymbol{k}$ is answered by returning KeyGen(SK, $\boldsymbol{k}$).
   An encryption query for attribute vector $\boldsymbol{x}$ is answered as follows.
   If Match($\boldsymbol{x}, \boldsymbol{z}_0$) = Match($\boldsymbol{x}, \boldsymbol{z}_1$) = 0 then $\mathcal{A}$ receives Enc(SK, $\boldsymbol{x}$). Otherwise, $\mathcal{A}$ receives $\perp$.

3. Challenge construction.
   $\mathcal{A}$ receives KeyGen(SK, $\boldsymbol{z}_\eta$).

4. Query Phase II. Identical to Query Phase I.

5. Output Phase. $\mathcal{A}$ returns η'.

We denote by $p_\mathcal{A}(\boldsymbol{z}_0, \boldsymbol{z}_1, \eta)$ the probability that $\mathsf{KeyExp}_\mathcal{A}(1^n, 1^\ell, \boldsymbol{z}_0, \boldsymbol{z}_1, \eta)$ returns η. In the next lemma, we prove that, if $\boldsymbol{z}_0$ and $\boldsymbol{z}_1$ have no $\star$-entry and they differ in exactly m positions then the (ℓ, m)-Q assumption implies that

$$|p_\mathcal{A}(\boldsymbol{z}_0, \boldsymbol{z}_1, 0) - p_\mathcal{A}(\boldsymbol{z}_0, \boldsymbol{z}_1, 1)|$$

is negligible for all probabilistic poly-time adversaries. A similar (omitted) proof shows that, if $\boldsymbol{z}_0$ and $\boldsymbol{z}_1$ contain k $\star$'s in the same positions and differ in exactly m positions then the $(\ell - k, m)$-Q assumption implies that

$$|p_\mathcal{A}(\boldsymbol{z}_0, \boldsymbol{z}_1, 0) - p_\mathcal{A}(\boldsymbol{z}_0, \boldsymbol{z}_1, 1)|$$

is negligible.

Lemma 4. *Assume Assumption (ℓ, m)-Q holds. Then, for all probabilistic poly-time adversaries $\mathcal{A}$ and for all vectors $\boldsymbol{z}_0, \boldsymbol{z}_1 \in \{0, 1\}^\ell$ which differ in exactly m positions, we have that*

$$|p_\mathcal{A}(\boldsymbol{z}_0, \boldsymbol{z}_1, 0) - p_\mathcal{A}(\boldsymbol{z}_0, \boldsymbol{z}_1, 1)|$$

is negligible.

Proof. Write $\boldsymbol{z}_0 = \langle z_{0,1}, \ldots, z_{0,\ell} \rangle$ and $\boldsymbol{z}_1 = \langle z_{1,1}, \ldots, z_{1,\ell} \rangle$ and assume, without loss of generality, that $\boldsymbol{z}_0$ and $\boldsymbol{z}_1$ differ in exactly the first m positions and that $\boldsymbol{z}_0 = 0^m \cdot 0^{\ell-m}$ and $\boldsymbol{z}_1 = 1^m \cdot 0^{\ell-m}$.

We proceed by contradiction. We assume that the lemma does not hold for some probabilistic poly-time adversary $\mathcal{A}$, and prove that there exists a probabilistic poly-time distinguisher $\mathcal{B}$ that has a non-negligible advantage for Assumption (ℓ, m)-Q.

We now describe $\mathcal{B}$. $\mathcal{B}$ takes as input a challenge ChOutput for Assumption (ℓ, m)-Q, simulates $\mathsf{KeyExp}_\mathcal{A}$ with parameters $(1^n, 1^\ell, \boldsymbol{z}_0, \boldsymbol{z}_1, \eta)$ for $\mathcal{A}$ and uses $\mathcal{A}$'s output to obtain non-negligible advantage in the game of Assumption (ℓ, m)-Q.

Initialization Phase. $\mathcal{B}$ starts by choosing random $y \in \mathbb{Z}_p$ and by setting $Y = \mathsf{e}(g_1, g_2)^y$. Define $JH = \{(j, h) | 1 \leq j \leq m, 1 \leq h \leq m, j \neq h \text{ or } j = h, m+1 \leq j \leq d\}$. For $(j, h) \in JH$, $\mathcal{B}$ sets[2]

---

[2] Hereafter, we assume that $\mathsf{A}_{j,h}$'s ($\mathsf{B}_{j,h}$'s) rows are indexed by 0 and 1.

$$G_{j,h} = e(A_{j,h}[1,j], C_{j,1}).$$

Throughout the simulation we will consider secret key $\mathsf{SK} = (\mathcal{I}, Y, y, \mathsf{SK}_1, \ldots, \mathsf{SK}_\ell)$ *implicitly* defined by ChOutput, with $\mathsf{SK}_i = (T_{i,0}, T_{i,1}, V_{i,0}, V_{i,1}, \bar{T}_{i,0}, \bar{T}_{i,1}, \bar{V}_{i,0}, \bar{V}_{i,1})$, for $i = 1, \ldots, \ell$, where, for $i = 1, \ldots, \ell$ and $b = 0, 1$,

$$T_{i,b} = g_1^{\hat{t}_{i,b}}, \; V_{i,b} = g_1^{\hat{v}_{i,b}},$$
$$\bar{T}_{i,b} = C_{i,b}, \; \bar{T}_{i,1} = D_{i,1}.$$

This implies that, for $i = 1, \ldots, \ell$ and $b = 0, 1$,

$$t_{i,b} = \hat{t}_{i,b} \text{ and } v_{i,b} = \hat{v}_{i,b}.$$

Since, for $i = 1, \ldots, \ell$, and for $b = 0, 1$ the values $\hat{t}_{i,b}, \hat{v}_{i,b}$ are random from $\mathbb{Z}_p$, the key SK is uniformly distributed as the output of MasterKeyGen. We stress that $\mathcal{B}$ only has indirect access to SK through ChOutput and in what follows we show that this is sufficient for simulating KeyExp.

Answering encryption queries. To answer queries to the Enc oracle for attribute vectors $\boldsymbol{x} = \langle x_1, \ldots, x_\ell \rangle$, we distinguish two cases.

Case 1. The vector $\boldsymbol{x}$ is such that there exists and index $j \geq m + 1$ such that $x_j = 1$. $\mathcal{B}$ chooses $s', s'_1, \ldots, s'_\ell$ at random in $\mathbb{Z}_p$, sets $\Omega = G_{j,j}^{-ys'}$ and, for $i = 1, \ldots, \ell$, sets

$$X_i = (A_{j,j}[x_i, i])^{s' - s'_i} \quad \text{and} \quad Z_i = (B_{j,j}[x_i, i])^{s'_i}.$$

$\mathcal{B}$ returns $\tilde{X} = (\Omega, (X_i, Z_i)_{i=1}^\ell)$ as output of the query.

Case 2. The vector $\boldsymbol{x}$ is such that $x_j = 0$ for $m+1 \leq j \leq \ell$. Since $\mathsf{Match}(\boldsymbol{x}, \boldsymbol{z}_0) = \mathsf{Match}(\boldsymbol{x}, \boldsymbol{z}_1)$, then there exist two indices j and h such that $x_j = 1$ and $x_h = 0$. $\mathcal{B}$ chooses $s', s'_1, \ldots, s'_\ell$ at random in $\mathbb{Z}_p$, sets $\Omega = G_{j,h}^{-ys'}$ and, for $i = 1, \ldots, \ell$, sets

$$X_i = (A_{j,h}[x_i, i])^{s' - s'_i} \quad \text{and} \quad Z_i = (B_{j,h}[x_i, i])^{s'_i}.$$

$\mathcal{B}$ returns $\tilde{X} = (\Omega, (X_i, Z_i)_{i=1}^\ell)$ as output of the query.

We notice that, in both above described cases, $\mathcal{B}$ can perform the computation as it has access to the needed values from ChOutput and from the initialization phase. Let us now argue that the output returned by $\mathcal{B}$ has the same distribution as in KeyExp. By setting, in **Case 1**, $s = s' \hat{s}_{(j,j)}$ and $s_i = s'_i \hat{s}_{(j,j)}$, for $i = 1, \ldots, \ell$; and, in **Case 2**, $s = s' \hat{s}_{(j,h)}$ and $s_i = s'_i \hat{s}_{(j,h)}$, for $i = 1, \ldots, \ell$, we have that $X_i = T_{i,x_i}^{s - s_i}$ and $Z_i = V_{i,x_i}^{s_i}$. Thus, $\tilde{X} = \mathsf{Enc}(\mathsf{SK}, \boldsymbol{x}; s, s_1, \ldots, s_\ell)$. Moreover, since s and the s_i's are random and independently chosen from $\mathbb{Z}_p$ we can conclude that $\tilde{X}$ has the same distribution as the answers obtained by $\mathcal{A}$ in $\mathsf{KeyExp}_\mathcal{A}$.

Answering key queries. To answer to the queries to the KeyGen oracle for attribute vector $\boldsymbol{k} = \langle k_1, \ldots, k_\ell \rangle$, $\mathcal{B}$, for $i \in S_{\boldsymbol{k}}$, chooses random $a_i \in \mathbb{Z}_p$ such that their sum is y and sets

$$R_i = C_{i,k_i}^{a_i} \quad \text{and} \quad W_i = D_{i,k_i}^{a_i}.$$

$\mathcal{B}$ returns $\tilde{K} = (R_i, W_i)_{i \in S_k}$. Notice that, for $i = 1, \ldots, \ell$, we have $C_{i,k_i} = \bar{T}_{i,k_i}$ and $D_{i,k_i} = \bar{V}_{i,k_i}$. Therefore, we can conclude that $\tilde{K} = \mathsf{KeyGen}(\mathsf{SK}, k; (a_i)_{i \in S_k})$. Since the a_i are random in $\mathbb{Z}_p$ under the constraint that their sum is y, we can conclude that that $\tilde{K}$ has the same distribution as the answers obtained by $\mathcal{A}$ in $\mathsf{KeyExp}_{\mathcal{A}}$.

Challenge construction. We describe how $\mathcal{B}$ prepares the challenge for $\mathcal{A}$. $\mathcal{B}$ chooses, for $i = m + 1, \ldots, \ell$, random $b_i' \in \mathbb{Z}_p$ under the constraint that their sum is y and returns $\tilde{K} = ((R_1, W_1), \ldots, (R_\ell, W_\ell))$ computed as follows. For $i = 1, \ldots, m$, $\mathcal{B}$ sets

$$R_i = E_i \quad \text{and} \quad W_i = F_i;$$

while, for $i = m + 1, \ldots, \ell$, $\mathcal{B}$ sets

$$R_i = E_i \cdot C_{i,0}^{b_i'} \quad \text{and} \quad W_i = F_i \cdot D_{i,0}^{b_i'}.$$

Notice that, for $i = m + 1, \ldots, \ell$, we have $R_i = \bar{T}_{i,0}^{a_i}$ and $W_i = \bar{V}_{i,0}^{a_i}$ where $a_i = \hat{a}_i + b_i'$. In addition, for $i = 1, \ldots, m$, we have $R_i = \bar{T}_{i,z_{\eta_i}}^{a_i}$ and $W_i = \bar{V}_{i,z_{\eta_i}}^{a_i}$ where $a_i = \hat{a}_i$. Therefore, we can conclude that $\tilde{K} = \mathsf{KeyGen}(\mathsf{SK}, z_\eta, (a_1, \ldots, a_\ell))$. Finally, we observe that the a_i's are random in $\mathbb{Z}_p$ under the constraint that their sum is y. Thus, $\tilde{K}$ is distributed as in $\mathsf{KeyExp}_{\mathcal{A}}(1^n, 1^\ell, z_0, z_1, \eta)$.

Finally, when $\mathcal{A}$ halts and returns η', $\mathcal{B}$ halts and returns η'.

Since the simulation provided by $\mathcal{B}$ is perfect, by our assumption on $\mathcal{A}$'s advantage, we can conclude that the advantage of $\mathcal{B}$ is also non-negligible thus contradicting Assumption (d, m)-Q. □

We thus have the following lemma.

Lemma 5. *Under Assumptions (d, m)-Q predicate encryption scheme (MasterKeyGen,Enc,KeyGen,Test) is key secure.*

Combining Lemma 3 and Lemma 5 we have the main result of this paper.

Theorem 2. *Under Assumptions (d, m)-Q and Decision Linear predicate encryption scheme (MasterKeyGen,Enc,KeyGen,Test) is secure HVE.*

7 Match Concealing

In this section, we show that, under a given assumption, the scheme presented in Section 4 actually enjoys a stronger notion of semantic security in which the adversary $\mathcal{A}$ is allowed to make queries for keys associated to any pattern k provided only that $\mathsf{Match}(z_0, k) = \mathsf{Match}(z_1, k)$. We call this notion *match concealing*. In the notion presented in the main body of the paper, $\mathcal{A}$ is restricted to queries for patterns k such that $\mathsf{Match}(z_0, k) = \mathsf{Match}(z_1, k) = 0$. This latter notion is called *match revealing* (see [SBC+07]).

We now present the Double Decision Linear Assumption by means of the following experiment $\mathsf{DDLExp}_{\mathcal{A}}$.

$\mathsf{DDLExp}_{\mathcal{A}}(1^n)$

01. Choose instance $\mathcal{I} = [p, \mathbb{G}_1, \mathbb{G}_2, \mathbb{G}_T, g_1, g_2, \mathsf{e}]$ with security parameter 1^n.

02. Choose $u_1, u_2, u_3, u_4, u_5, u \in \mathbb{Z}_p$ at random.

03. Choose $\eta \in \{0, 1\}$ at random.

04. If $\eta = 1$, then

05.   set $Z = g_1^{u_2(u-u_3)}$ and $Z_0 = g_1^{u_1 u_3}$;

06.   else, set $Z = g_1^{u_5(u-u_3)}$ and $Z_0 = g_1^{u_4 u_3}$.

07. Set $U_1 = g_1^{u_1}, \hat{U}_1 = g_2^{u_1}, U_2 = g_1^{u_2}, U_4 = g_1^{u_4}, U_5 = g_1^{u_5}, U_{245} = g_2^{u_2 u_4 u_5}$.

08. Set $U_{145} = g_2^{u_1 u_4 u_5}, U_{125} = g_2^{u_1 u_2 u_5}, U_{124} = g_2^{u_1 u_2 u_4}, U = g_1^{u}$.

09. Let $\eta' = \mathcal{A}(\mathcal{I}, U_1, \hat{U}_1, U_2, U_4, U_5, U_{245}, U_{145}, U_{125}, U_{124}, U, Z, Z_0)$.

10. If $\eta = \eta'$ then return 1 else return 0,

Assumption 4 (Double Decision Linear (DDLinear)). *For all probabilistic poly-time algorithms $\mathcal{A}$, $|\mathrm{Prob}[\mathsf{DDLExp}_{\mathcal{A}}(1^n) = 1] - 1/2|$ is negligible in n.*

Suppose that z_0, z_1 are two attribute vectors in $\{0, 1\}^\ell$ which differ only in position j. Consider the following experiments.

$\mathsf{SemanticExp}_{\mathcal{A}}(1^n, 1^\ell, z_0, z_1, \eta)$

1. Key-generation Phase. Compute $\mathsf{SK} = (\mathcal{I}, y, \mathsf{SK}_1, \cdots, \mathsf{SK}_\ell)$ by executing $\mathsf{MasterKeyGen}(1^n, 1^\ell)$.

2. Query Phase I. Answer Enc queries for attribute vectors x by using secret key SK. Answer KeyGen queries for pattern vectors k such that $\mathsf{Match}(z_0, k) = \mathsf{Match}(z_1, k)$ using secret key SK.

3. Challenge Construction.
   1. Choose random $s, s_1, \ldots, s_\ell \in \mathbb{Z}_p$ and set $\Omega = \mathsf{e}(g_1, g_2)^{ys}$.
   2. For $1 \le i \ne j \le \ell$
      set $X_i = g_1^{t_{i,z_{0,i}}(s-s_i)}$ and $Z_i = g_1^{s_i v_{i,z_{0,i}}}$.
   3. set $X_j = g_1^{t_{j,z_{\eta,i}}(s-s_j)}$ and $Z_j = g_1^{s_j v_{j,z_{\eta,j}}}$.
   4. Set $\tilde{X} = (\Omega, (X_i, Z_i)_{i=1}^\ell)$.

4. Query Phase II. Identical to Query Phase I.

5. **return** $\mathcal{A}(\tilde{X})$.

We will use the writing $\mathsf{SemanticExp}(1^n, 1^\ell, z_0, z_1, \eta; s, s_1, \ldots, s_\ell)$ to denote the tuple $\tilde{X}$ computed by $\mathsf{SemanticExp}(1^n, 1^\ell, z_0, z_1, \eta)$ using $s, s_1, \ldots, s_\ell$ as random values. We will denote by $p_\eta^{\mathcal{A}}(z_0, z_1)$ the probability that experiment $\mathsf{SemanticExp}_{\mathcal{A}}(1^n, 1^\ell, z_0, z_1, \eta)$ returns η. Notice that, since z_0 and z_1 differ only in position j, then in $\mathsf{SemanticExp}_{\mathcal{A}}(1^n, 1^\ell, z_0, z_1, 0)$ adversary $\mathcal{A}$ receives a valid encrypted attribute vector $\tilde{X}$ for attribute vector z_0 whereas in $\mathsf{SemanticExp}_{\mathcal{A}}(1^n, 1^\ell, z_0, z_1, 1)$ adversary $\mathcal{A}$ receives $\tilde{X}$ for attribute vector z_1. Next we prove that, under the Double Decision Linear assumption, for all attribute vectors z_0, z_1 which differ only in position j, the difference $|p_0^{\mathcal{A}}(z_0, z_1) - p_1 \ell^{\mathcal{A}}(z_0, z_1)|$ is negligible. This implies the match concealing semantic security of the scheme.

Lemma 6. *Assume DDLinear holds. Then, for any j, for any attribute strings z_0 and z_1 which differ only in position j, and for any adversary $\mathcal{A}$,*

$$|p_0^{\mathcal{A}}(z_0, z_1) - p_1^{\mathcal{A}}(z_0, z_1)|$$

is negligible.

Proof. Suppose that there exist PPT adversary $\mathcal{A}$ and attribute vector z_0, z_1 for which $|p_0^{\mathcal{A}}(z_0, z_1) - p_1^{\mathcal{A}}(z_0, z_1)|$ is non-negligible. We assume without loss of generality that, for $i \neq j$, we have $z_{0,i} = z_{1,i} = 0$ and that $z_{0,j} = 0$ and $z_{1,j} = 1$. We next construct a PPT adversary $\mathcal{B}$ for the experiment DDLExp. $\mathcal{B}$ takes in input $[\mathcal{I}, U_1 = g_1^{u_1}, \hat{U}_1 = g_2^{u_1}, U_2 = g_1^{u_2}, U_4 = g_1^{u_4}, U_5 = g_1^{u_5}, U_{245} = g_2^{u_2 u_4 u_5}, U_{145} = g_2^{u_1 u_4 u_5}, U_{125} = g_2^{u_1 u_2 u_5}, U_{124} = g_2^{u_1 u_2 u_4}, U = g_1^u, Z, Z_0]$, and depending on whether $Z = g_1^{u_2(u-u_3)}$ and $Z_0 = g_1^{u_1 u_3}$ or $Z = g_1^{u_5(u-u_3)}, Z_0 = g_1^{u_4 u_3}$, simulates experiment $\mathsf{SemanticExp}(1^n, 1^\ell, z_0, z_1, 0)$ or $\mathsf{SemanticExp}(1^n, 1^\ell, z, 1)$ for $\mathcal{A}$. We next describe algorithm $\mathcal{B}$.

Initialization Phase. $\mathcal{B}$ simulates the key-generation phase by choosing random $y' \in \mathbb{Z}_p$ and sets $Y = \mathsf{e}(U_1^{y'}, g_2)$. This implicitly sets $y = u_1 y'$. $\mathcal{B}$ chooses random $t'_{i,0}, v'_{i,0}, t'_{i,1}, v'_{i,1} \in \mathbb{Z}_p$, for $i \neq j$, and then computes values $T_{i,0}, T_{i,1}, V_{i,0}$, and $V_{i,1}$ as follows.

$$T_{i,0} = g_1^{t'_{i,0}}, T_{i,1} = U_1^{t'_{i,1}}, V_{i,0} = g_1^{v'_{i,0}}, \text{ and } V_{i,1} = U_1^{v'_{i,1}}.$$

These settings implicitly define $t_{i,0} = t'_{i,0}$, $t_{i,1} = u_1 \cdot t'_{i,1}$, $v_{i,0} = v'_{i,0}$, and $v_{i,1} = u_1 \cdot v'_{i,1}$ which in turn define values $\bar{T}_{i,0}, \bar{T}_{i,1}, \bar{V}_{i,0}$, and $\bar{V}_{i,1}$. Then, $\mathcal{B}$ computes $T_{j,0}, T_{j,1}, V_{j,0}$, and $V_{j,1}$ by setting

$$T_{j,0} = U_2, T_{j,1} = U_5, V_{j,0} = U_1, \text{ and } V_{j,1} = U_4,$$

thus implicitly setting $t_{j,0} = u_2$, $t_{j,1} = u_5$, $v_{j,0} = u_1$, and $v_{j,1} = u_4$ which in turn define values $\bar{T}_{j,0}, \bar{T}_{j,1}, \bar{V}_{j,0}$ and $\bar{V}_{j,1}$.

After this step key $\mathsf{SK} = (\mathcal{I}, Y, y, \mathsf{SK}_1, \ldots, \mathsf{SK}_\ell)$ with $\mathsf{SK}_i = (T_{i,0}, T_{i,1}, V_{i,0}, V_{i,1}, \bar{T}_{i,0}, \bar{T}_{i,1}, \bar{V}_{i,0}, \bar{V}_{i,1})$ is implicitly defined even though $\mathcal{B}$ does not completely know SK. Notice that SK has the same distribution as a key given in output by $\mathsf{MasterKeyGen}$.

Answering Queries. $\mathcal{B}$ answers $\mathcal{A}$'s Enc queries for vector x by executing procedure Enc. Notice that Enc only needs values $T_{i,b}$'s and $V_{i,b}$'s which are known to $\mathcal{B}$ from the previous step. To describe how $\mathcal{B}$ answers $\mathcal{A}$'s KeyGen queries for vector k, we distinguish the following cases.

Case 1: $k_j \neq \star$. In this case there exists index $h \in S_k$ such that $k_h = 1$, for otherwise we would have $\mathsf{Match}(z_0, k) \neq \mathsf{Match}(z_1, k)$. Then, for $i \in S_k$, B chooses random values $a'_i \in \mathbb{Z}_p$, and sets $a' = \sum_{i \in S_k \setminus \{j, h\}} a'_i$. For $i \in S_k \setminus \{j, h\}$, $\mathcal{B}$ computes R_i and W_i as follows. If $k_i = 0$, then $\mathcal{B}$ sets

$$R_i = \hat{U}_1^{a'_i / t'_{i,k_i}} \quad \text{and} \quad W_i = \hat{U}_1^{a'_i / v'_{i,k_i}}$$

else $\mathcal{B}$ sets

$$R_i = g_2^{a_i'/t_{i,k_i}'} \quad \text{and} \quad W_i = g_2^{a_i'/v_{i,k_i}'}.$$

$\mathcal{B}$ then computes R_j and W_j as follows. If $k_j = 0$, then $\mathcal{B}$ sets

$$R_j = U_{145}^{a_j'} \quad \text{and} \quad W_j = U_{245}^{a_j'},$$

else $\mathcal{B}$ sets

$$R_j = U_{124}^{a_j'} \quad \text{and} \quad W_j = U_{125}^{a_j'}.$$

Finally, $\mathcal{B}$ sets

$$R_h = U_{245}^{-a_j'/t_{h,k_h}'} g_2^{(y'-a')/t_{h,k_h}'} \quad \text{and} \quad W_h = U_{245}^{-a_j'/v_{h,k_h}'} g_2^{(y'-a')/v_{h,k_h}'}.$$

$\mathcal{B}$ returns $\tilde{K} = (R_i, W_i)_{i \in S_k}$.

We next show that, even though $\mathcal{B}$ does not have complete access to SK, $\tilde{K}$ has the same distribution of the output of the KeyGen procedure on input SK and k.

Set $a_i = u_1 a_i'$, for $i \in S_k \setminus \{h, j\}$, $a_j = u_1 u_2 u_4 u_5 a_j'$, and $a_h = u_1 y' - u_1 u_2 u_4 u_5 a_j' - u_1 a'$. Then, for $i \in S_k \setminus \{j, h\}$ such that $k_i = 0$ we have

$$R_i = \hat{U}_1^{a_i'/t_{i,k_i}'} = g_2^{u_1 a_i'/t_{i,k_i}'} = g_2^{a_i/t_{i,k_i}'} = \bar{T}_{i,0}^{a_i}.$$

Similarly, for $i \in S_k \setminus \{j, h\}$ such that $k_i = 1$,

$$R_i = g_2^{a_i'/t_{i,k_i}'} = g_2^{u_1 a_i'/u_1 t_{i,k_i}'} = g_2^{a_i/u_1 t_{i,k_i}'} = \bar{T}_{i,1}^{a_i}.$$

Similarly, we have in both cases that $W_i = \bar{V}_{i,k_i}^{a_i}$. Furthermore, if $k_j = 0$ we have

$$R_j = U_{145}^{a_j'} = g_2^{u_1 u_4 u_5 a_j'} = g_2^{u_1 u_2 u_4 u_5 a_j'/u_2} = g_2^{a_j/u_2} = \bar{T}_{j,0}^{a_j}.$$

Similarly, for $k_j = 1$ and for W_j. Finally, we have

$$\begin{aligned}
R_h &= U_{245}^{-a_j'/t_{h,1}'} g_2^{(y'-a')/t_{h,1}'} \\
&= g_2^{(-u_2 u_4 u_5 a_j' + y' - a')/t_{h,1}'} \\
&= g_2^{u_1(-u_2 u_4 u_5 a_j' + y' - a')/t_{h,1}} \\
&= g_2^{a_h/t_{h,1}} \\
&= \bar{T}_{h,1}^{a_h}.
\end{aligned}$$

To conclude notice that the a_i's are random under the constraint that their sum is $u_1 y' = y$ and thus the simulation is perfect.

Case 2: $k_j = \star$. In this case, for $i \in S_k$, $\mathcal{B}$ chooses random values $a_i' \in \mathbb{Z}_p$ which sum up to y', and computes R_i and W_i as follows. If $k_i = 0$, then $\mathcal{B}$ sets

$$R_i = \hat{U}_1^{a_i'/t_{i,k_i}'} \quad \text{and} \quad W_i = \hat{U}_1^{a_i'/v_{i,k_i}'}$$

else $\mathcal{B}$ sets

$$R_i = g_2^{a_i'/t_{i,k_i}'} \quad \text{and} \quad W_i = g_2^{a_i'/v_{i,k_i}'}.$$

If we set, for $i \in S_k$, $a_i = u_1 a_i'$, we have that if $k_i = 0$ then

$$R_i = \hat{U}_1^{a_i'/t_{i,k_i}'} = g_2^{u_1 a_i'/t_{i,k_i}'} = g_2^{a_i/t_{i,k_i}'} = g_2^{a_i/t_{i,k_i}} = \bar{T}_{i,0}^{a_i},$$

and if $k_i = 1$ then

$$R_i = g_2^{a_i'/t_{i,k_i}'} = g_2^{u_1 a_i'/u_1 t_{i,k_i}'} = g_2^{a_i/t_{i,k_i}} = \bar{T}_{i,1}^{a_i}.$$

Similarly, we have $W_i = \bar{V}_{i,k_i}^{a_i}$. We thus conclude that $\tilde{K} = \mathsf{KeyGen}(\mathsf{SK}, k; (a_i)_{i \in S_k})$. Moreover, the a_i's are independently and randomly chosen in $\mathbb{Z}_p$ under the constraint that their sum is $u_1 y' = y$. Hence, also in this case, $\tilde{K}$ is distributed according to $\mathsf{KeyGen}(\mathsf{SK}, k)$.

Challenge construction. When $\mathcal{B}$ is asked to provide encrypted attribute vector for z_0 or z_1, $\mathcal{B}$ constructs the tuple $\tilde{X} = (\Omega, (X_i, Z_i)_{i=1}^\ell)$ in the following way. $\mathcal{B}$ sets $\Omega = \mathsf{e}(U, \hat{U}_1)^{-y'}$, thus implicitly setting $s = u$. For $i \neq j$, $\mathcal{B}$ chooses random $s_i \in \mathbb{Z}_p$ and computes X_i and Z_i as

$$X_i = U^{t_{i,0}'} g_1^{-t_{i,0}' s_i} \quad \text{and} \quad Z_i = g_1^{v_{i,0}' s_i}.$$

Notice that the above settings implies

$$X_i = U^{t_{i,0}'} g_1^{-t_{i,0}' s_i} = g_1^{u t_{i,0}'} T_{i,0}^{-s_i} = T_{i,0}^{s-s_i} \quad \text{and} \quad Z_i = g_1^{v_{i,0}' s_i} = V_{i,0}^{s_i}.$$

Finally, X_j and Y_j are computed as

$$X_j = Z \quad \text{and} \quad Z_j = Z_0.$$

Finally $\mathcal{B}$ returns $\mathcal{A}$'s output.

Suppose that $Z = g_1^{u_2(u-u_3)}$, $Z_0 = g_1^{u_1 u_3}$ and $s_j = u_3$. Then, we have

$$X_j = U_2^{u-u_3} = T_{j,0}^{u-u_3} = T_{j,0}^{s-s_3} \quad \text{and} \quad Z_j = U_1^{u_3} = V_{j,0}^{u_3} = V_{j,0}^{s_3}$$

and thus $\tilde{X} = \mathsf{SemanticExp}(1^n, 1^\ell, z_0, z_1, 0; s, s_1, \ldots, s_\ell)$. Moreover s and the s_i's are random in $\mathbb{Z}_p$ and thus we can conclude that $\tilde{X}$ is distributed as in $\mathsf{SemanticExp}(1^n, 1^\ell, z_0, z_1, 1)$.

Suppose instead that $Z = g_1^{u_5(u-u_3)}$ and $Z_0 = g_1^{u_4 u_3}$, and sets $s_j = u_3$ as before. Then we have

$$X_j = U_5^{u-u_3} = T_{j,1}^{u-u_3} = T_{j,1}^{s-s_3} \quad \text{and} \quad Z_j = U_4^{u_3} = V_{j,1}^{u_3} = V_{j,1}^{s_3}$$

and thus $\tilde{X} = \mathsf{SemanticExp}(1^n, 1^\ell, z_0, z_1, 1; s, s_1, \ldots, s_\ell)$. Since s and the s_i's are random in $\mathbb{Z}_p$, we can conclude that the challenge received by $\mathcal{A}$ is distributed as in $\mathsf{SemanticExp}(1^n, 1^\ell, z, 1)$. Furthermore notice that setting $s = u$ and $y = u_1 y'$ then Ω has the correct distribution.

By the observations above, we can say that if $Z = g_1^{u_2(u-u_3)}$ and $Z_0 = g_1^{u_1 u_3}$, then $\mathcal{A}$'s view is the same as in $\mathsf{SemanticExp}(1^n, 1^\ell, z_0, z_1, 0)$; whereas, if $Z = g_1^{u_5(u-u_3)}$ and $Z_0 = g_1^{u_4 u_3}$, then $\mathcal{A}$'s view is the same as in $\mathsf{SemanticExp}(1^n, 1^\ell, z_0, z_1, 1)$. This contradicts the DDLinear assumption. $\qquad \square$

Simple hybrid arguments can extend the lemma to arbitrary z_0 and z_1 (and not just for vectors differing in one position).

Lemma 7. *Assume DDLinear. Then predicate encryption* (MasterKeyGen, Enc, KeyGen, Test) *is match concealing semantically secure.*

8   Larger Alphabets

Our constructions have been presented for binary attribute vectors. The extension to larger alphabets is straightforward. Specifically, for an alphabet Σ of size s we would have a master secret key consisting of an instance $\mathcal{I}$ and of one element of $\mathbb{G}_T$, $2 \cdot \ell \cdot s$ elements of $\mathbb{G}_1$, and $2 \cdot \ell \cdot s$ elements of $\mathbb{G}_2$. The length of the encrypted attribute vectors and of the keys are independent of the size of Σ and only depend on ℓ. We can make the length of the secret key SK independent from the size of Σ by employing a pseudo-random function $\mathbb{F}$. Specifically, we randomly select a k-bit string R and set $t_{i,\sigma} = \mathbb{F}_R(i\|\sigma)$ and $v_{i,\sigma} = \mathbb{F}_R(i\|\sigma)$ for $i = 1, \ldots, \ell$ and $\sigma \in \Sigma$.

Acknowledgements

The work of the authors has been supported in part by the European Commission through the EU ICT program under Contract ICT-2007-216646 ECRYPT II and through the FP6 program under contract FP6-1596 AEOLUS.

References

[BBG05]   Boneh, D., Boyen, X., Goh, E.-J.: Hierarchical identity based encryption with constant size ciphertext. In: Cramer, R. (ed.) EUROCRYPT 2005. LNCS, vol. 3494, pp. 440–456. Springer, Heidelberg (2005)

[BDOP04]   Boneh, D., Di Crescenzo, G., Ostrovsky, R., Persiano, G.: Public key encryption with keyword search. In: Cachin, C., Camenisch, J.L. (eds.) EUROCRYPT 2004. LNCS, vol. 3027, pp. 506–522. Springer, Heidelberg (2004)

[Boy08]   Boyen, X.: The uber-assumption family – a unified complexity framework for bilinear groups. In: Galbraith, S.D., Paterson, K.G. (eds.) Pairing 2008. LNCS, vol. 5209, pp. 39–56. Springer, Heidelberg (2008)

[BW06]   Boyen, X., Waters, B.: Anonymous hierarchical identity-based encryption (Without random oracles). In: Dwork, C. (ed.) CRYPTO 2006. LNCS, vol. 4117, pp. 290–307. Springer, Heidelberg (2006)

[BW07]   Boneh, D., Waters, B.: Conjunctive, subset, and range queries on encrypted data. In: Vadhan, S.P. (ed.) TCC 2007. LNCS, vol. 4392, pp. 535–554. Springer, Heidelberg (2007)

[GPSW06]   Goyal, V., Pandey, O., Sahai, A., Waters, B.: Attribute-Based Encryption for Fine-Grained Access Control for Encrypted Data. In: ACM CCS 2006, Alexandria, VA, USA, pp. 89–98. ACM Press, New York (2006)

[IP08]   Iovino, V., Persiano, G.: Hidden-vector encryption with groups of prime order. In: Galbraith, S.D., Paterson, K.G. (eds.) Pairing 2008. LNCS, vol. 5209, pp. 75–88. Springer, Heidelberg (2008)

[KSW08]   Katz, J., Sahai, A., Waters, B.: Predicate encryption supporting disjunctions, polynomial equations, and inner products. In: Smart, N.P. (ed.) EUROCRYPT 2008. LNCS, vol. 4965, pp. 146–162. Springer, Heidelberg (2008)

[Nao03]   Naor, M.: On cryptographic assumptions and challenges (invited talk). In: Boneh, D. (ed.) CRYPTO 2003. LNCS, vol. 2729, pp. 96–109. Springer, Heidelberg (2003)

[SBC+07]   Shi, E., Bethencourt, J., Chan, H., Song, D., Perrig, A.: Multi-Dimensional Range Query over Encrypted Data. In: 2007 IEEE Symposium on Security and Privacy, Oakland, CA. IEEE Computer Society Press, Oakland (2007)

[SSW09]   Shen, E., Shi, E., Waters, B.: Predicate privacy in encryption systems. In: Reingold, O. (ed.) TCC 2009. LNCS, vol. 5444, pp. 457–473. Springer, Heidelberg (2009)

Building Secure Networked Systems
with Code Attestation
(Invited Talk)

Adrian Perrig

CyLab / Carnegie Mellon University
Pittsburgh, USA
perrig@cmu.edu

Attestation is a promising approach for building secure systems. The recent development of a Trusted Platform Module (TPM) by the Trusted Computing Group (TCG) that is starting to be deployed in common laptop and desktop platforms is fueling research in attestation mechanisms. In this talk, we will present approaches on how to build secure systems with advanced TPM architectures. In particular, we have designed an approach for fine-grained attestation that enables the design of efficient secure distributed systems, and other network protocols. We demonstrate this approach by designing a secure routing protocol.

J.A. Garay, A. Miyaji, and A. Otsuka (Eds.): CANS 2009, LNCS 5888, p. 278, 2009.
© Springer-Verlag Berlin Heidelberg 2009

HPAKE : Password Authentication Secure against Cross-Site User Impersonation

Xavier Boyen

Stanford University
xb@cs.stanford.edu

Abstract. We propose a new kind of asymmetric mutual authentication from passwords with stronger privacy against malicious servers, lest they be tempted to engage in "cross-site user impersonation" to each other.

It enables a person to authenticate (with) arbitrarily many independent servers, over adversarial channels, using a memorable and reusable single short password. Beside the usual PAKE security guarantees, our framework goes to lengths to secure the password against brute-force cracking from privileged server information.

1 Introduction

Password-based authentication and key exchange is the process whereby a client achieves mutual authentication with a remote server over an adversarial channel, turning it into a virtual secure communication channel, on the basis of a short password that should be easy to memorize but not guess.

Shared-Password Authentication. (Symmetric) password-authenticated key exchange (PAKE) assumes that the password is shared between the client and the server. The threat in this case is that a (passive or active) outside attacker might try to impersonate either party to the other, or to eavesdrop on the communication taking place within the secure channel. Though such attacks cannot be prevented in an adversarial network, they can be made to require one fresh online authentication attempt for each password being tested. This is a solved problem: many PAKE protocols achieve this notion very efficiently.

Private-Password Authentication. Asymmetric password-authenticated key exchange (APAKE), by contrast, allows the password to be known to the client only. The server holds a long-term authentication token, related in some way to the password, but from which it is (presumably) hard to recover the password itself. In addition to the unavoidable online attack, a secondary threat of concern here is that a compromise of the server database might give to the attacker the means of impersonating its users to another server. Thwarting this threat means that it is safe for a client to to reuse the same password with multiple servers. This constitutes a very significant usabily improvement around human limitations.

J.A. Garay, A. Miyaji, and A. Otsuka (Eds.): CANS 2009, LNCS 5888, pp. 279–298, 2009.

Many elegant APAKE protocols have been proposed over the years, that deliver more or less optimally on all those requirements — provided that the password is not too weak, that is. Indeed, a general attack strategy for the evil insider in the APAKE model is, once it has obtained the server's database, to mount an offline dictionary attack to recover the client password: a server can always do that, simply by simulating the authentication protocol with itself posing as the target user. Since such attack cannot be prevented, the user's only recourse is to make the attack slower, which requires: (1) that the protocol itself be made intentionally slower; (2) and that the server implements it correctly. Both are undesirable requirements.

On Password Strength. We remark in passing that the threshold for what constitutes a "good" password is *much* higher in the insider threat model than the outsider one, even though there are many more potential outsider attackers. Online attacks, the only option for outsiders, are indeed inherently slow and can be artificially and arbitrarily rate-limited; they are also easy to detect and counter by locking up the account.

Thus, as long as some basic security requirements are met against outsider attacks, one is much better served by devoting one's energy to the prevention of insider attacks.

Misaligned Incentives. Unfortunately for the end-user, servers generally have little incentive to assist in this task, since (1) it would presumably make the protocol more costly, and (2) it would be giving into the suggestion they, the servers, cannot be trusted with the users' passwords.

At a fundamental level, the entity most harmed in case of a password breach would be the owner of the account, not the server providing the service. The user thus has a greater incentive to do something about this, e.g., accept a slower protocols if it can make the password safer. Alas, users generally have no power to dictate such a change. The only available option is generally to preprocess the password *outside* of the protocol before it starts. (Though better than nothing, a problem of this approach is that the preprocessing function must be non-parametric if one is not willing to accept a statefull client on which the parameters can be stored).

Equally worrisome, the APAKE model does not explicitly take into account the threat of *cross-site user impersonation*, where the server itself at site A turns rogue and attempts to impersonate the client at some other site B (based on the oft-fulfilled premise that the client picked identical or very similar passwords on both sites). Since Server A *itself* is the threat in this attack scenario, one cannot reasonably expect it willingly to fight against itself (unless an external mechanism such as a reputation at stake comes into play). It could also be a matter of denial; after all, website operators who are genuinely honest will most likely consider themselves trustworthy — regardless of whether the user trusts them or not.

Reinternalizing the Externality. In economic parlance, one would say that, in the insider threat model, the incentives of the parties are mis-aligned; their wishes are at odds with each other. The root of the problem is that there an economic *externality*: the client is the one who suffers if the server fails to protect its database adequately.

(One of) our goal is thus to design a protocol whose "pricing structure" re-internalizes this externality. But, first, we look at other simpler alternatives. (We only consider alternatives that require no custom data storage, secret or public, on the client side; with it, our problems would be solved.)

Client-side Preprocessing. The application of a complex transformation on the password, *e.g.*, hashing it many times before use, is the implicit customary defense against offline server threats. Preprocessing can be very useful, if done by the client, because it realigns the costs with the incentives. However, it also creates new problems of its own, depending whether its complexity is fixed or variable.

Fixed-cost preprocessing, *e.g.*, with a hash function or password-based key derivation function of fixed constant complexity, is easy to implement, but it is a rather blunt instrument that can be too slow and inconvenient in some situations, and not provide enough of a deterrent against attacks in others.

Parametric-cost preprocessing, *i.e.*, based upon a user-selected complexity parameter, poses another problem, which is that the parameter must be stored somewhere, and available for the client to retrieve whenever needed. [1]

The need to retrieve parameters is what makes parametric preprocessing problematic in practice, because it is generally not desirable to keep them in the clear, and give them to anyone who asks. Indeed, the user's choice of complexity parameter can itself provide very valuable information, e.g, to guide an attack toward a promising target. And hiding the parameter behind an extra layer of authentication is a circular non-solution that just moves the problem around.

A Host of Requirements. As we said, our goal is to realize a secure and "economically sound" password protocol, i.e., with all the usual APAKE security guarantees, plus a provision for the user to defend her password against dishonest servers the way she sees fit. Hence there must be a (secret, user-programmed, user-computed) "computational bottleneck" somewhere that renders insider offline attacks arbitrarily slower, but that does not penalize a honest server.

Intuitively, our "computational bottleneck", or programmable costly function, will have to satisfy the following requirements:

Client-owned bottleneck: As discussed, the only way to thwart offline insider attacks is to make the protocol slower, somewhere. The client must own this feature, since his or her interests are at stake.

Server-side independence: Not only should the server be oblivious to the selection of the client-side bottleneck, it should also be removed from its calculation, for obvious scalability reasons (whereas a human will authenticate

---

[1] The password authentication system of [42] is based precisely on that idea. It relies on a third-party central server for storing and recalling the cost parameter. Anyone can request and obtain this data.

to one site at a time, a machine may have to answer thousands such requests per second).

Cacheable preprocessing: Because the hard function may, by user choice, take a very large amount of time to compute, it would be nice if the result could be cached in secure storage, for future uses with the same server, for as long as the user deems it safe to keep it there.[2] This requires: (1) the output of the bottleneck function to be independent of any random ephemeral used in the protocol execution; (2) the authentication process to be "hot-started" at the point where the client has just finished evaluating the bottleneck function.

Zero client storage: Conversely, no persistent storage whatsoever should be required of the client. Especially, all secrets should fit in the user's mind (those being the password and nothing else). We specifically demand that the user be allowed to forget the value of the cost parameter once it has been programmed into the registration data sent to the server.

Secrecy of the parameter: In general, in security it is a good idea not to leak any information that is not explicitly needed, unless one can prove that such leak is benign. Leaking the cost parameter is certainly not benign, since it might tell what the important targets are, and reveal other password usage pattern of the user.

Secrecy of the parameter's retrieval: Allowing an attacker to learn the bottleneck parameter can be very damaging, but even more so to let it learn whether the user has recovered it correctly. Depending on the leakage mechanism, e.g., if it comes from the protocol itself, then a dishonest server could use it to mount an offline dictionary attack that entirely bypasses the hard function. Neither party should learn whether a retrieval attempt succeeded, before the protocol actually reaches the accepting state.

Contribution. To address all the issues we raised, we propose the notion of Hardened Password-Authenticated Key Exchange (HPAKE), which integrates a user-programmable hard function with the above properties into an authentication protocol with PAKE and APAKE security.

With it, users will thus be able to reuse the same passwords at various sites, without having to trust that the server or the network is behaving nicely. The benefits over existing solutions, such as APAKE are especially pronounced in the case of weak passwords that would be easy to crack were they used in a regular APAKE protocol.

The architecture of HPAKE is easy to explain generically; it is based on three existing cryptographic primitives used as black boxes; all of them in fact have been known since the dawn of cryptography, except for the preferred instantiation of the user-programmable hard function which is recent.

---

[2] It is indeed a good strategy for long-term passwords of last resort to pick them very memorable, and thus very weak, and rely on a very high cost parameter to defeat offline attacks.

Setup Assumption. Before proceeding, we should mention that there are a lot of ways to attack password-based systems, many of which do not depend on the protocol used (key loggers and social engineering attacks being two common examples). Our objective is to provide the highest level of security for multi-site password authentication, under the common-sense assumption that neither the human user nor the electronic device acting as the client on his or her behalf (and on which the password will be seized), leaks any information to the outside world other than as specified by the protocol.

On the other hand, we stress once again that we make none of the following all-too-common assumptions: a public-key infrastructure (PKI), a preexisting one-sided authentication mechanism (such as SSL with a root CA), a client-side data storage device of any kind (whether private and/or authentic or neither), or any tamper-proof client hardware that has somehow become tied to the user (including physically unclonable functions or PUF).

2   Related Work

All password-based remote authentication and key exchange protocols can be divided into two broad categories, depending on the nature of the secrets held by the client and the server:

A. **Shared-password authentication,** where both parties share the same secret. Since there is no password privacy there is no possibility of password reuse. For completeness, we mention:
   - cleartext passwords, even if transmitted over an encrypted link *à la* SSL;
   - symmetric challenge-response authentication using nonces and hashes;
   - various *ad hoc* password-only protocols using public-key techniques;
   - most cryptographic password-authenticated key exchange PAKE protocols (see below).

B. **Password-private authentication,** where the secrets are asymmetric. The client proves possession of the password to a server that proves knowledge of a derived secret. There are:
   Stateful schemes, whose clients keep state or carry custom data beside the password, *e.g.*:
   - preregistered public keys, where the password unlocks a signing key;
   - multi-factor systems, *e.g.*, involving biometric or hardware credentials;
   - client-side "password managers" unlocked by a meta-password;
   - any authentication system that uses lists of one-time credentials.
   Stateless schemes, where the only client custom data is a small secret password. Such protocols truly enable "untethered" roaming for a human user. The only examples are:
   - augmented password-authenticated key exchange (APAKE, see below);
   - our HPAKE protocol, which is better hardened against malicious servers.

AKE. Key exchange (or key agreement) protocols from high-entropy secrets date back from the original Diffie-Hellman protocol [19]. Authenticated key exchange (AKE) further ensures that the two parties are mutually authenticated, *i.e.*, that they have the proper long-term secrets, and thus that no impersonation is taking place. Since achieving AKE from a shared high-entropy secret is all but trivial, mentions of AKE in the literature truly refer to "asymmetic" authentication (AAKE), where each party has its own private secret and has registered the corresponding public key with the other party. This notion of AAKE has been progressively refined and perfected over the years; see for example [20,3,7,15,32,30,37,17,31]. (We make the distinction between AKE and AAKE to emphasize the fact that later on we may elect to use one or the other.)

PAKE. For low-entropy human-memorable secrets, the grandfather of PAKE protocols is the Encrypted Key Exchange (EKE) scheme proposed by Bellovin and Merritt [4], and which can arguably be traced further back to the notion of Privacy Amplification [6]. In both cases, the goal was to take a short shared secret, and boost it into a cryptographically strong one by a public discussion process over an open channel [36]. The EKE protocol provided a particularly efficient way to do so, with (implicit) mutual authentication of the parties. It also jump-started a fruitful line of research, which led to many results including new definitions [3,1,25], increased efficiency [26,33], and/or provable security properties [11,2]. More recently, there has been a surge of interest in the construction of PAKE protocols with better proofs of security that avoid the random-oracle model, *e.g.*, in favor of the common reference string model; we mention the first reasonably efficient such protocol [29,13], and a simpler and faster variant [27]. Although by far most of the constructions are based on a Discrete Log assumption such as Diffie-Hellman or variations thereof, there are protocols based on the RSA assumption [34] or the Phi-hiding assumption [23].

APAKE. Although the EKE protocol of Bellovin and Merritt originally required both parties to know the password, it was soon followed by an asymmetric version called "Augmented" EKE, by the same authors [5], who had realized the impracticality of requiring users to remember independent passwords for different environments. However, it is not until much later that this concern has been addressed again, first in [25] under somewhat stringent operating conditions, then more practically in [2] and in a sequence of papers [11,33] which culminated in the so-called Omega-method [24] for "augmenting" any given symmetric PAKE protocol. Another way to deal with the threat of server corruption and password exposure is to use multiple servers in a threshold scheme, which is the solution adopted in [35], though this requires the user to believe that the servers are not colluding.

KDFs. Many approaches have been proposed to address the problem of offline dictionary attacks, whether for static storage, or in the context of an authentication protocol. Most of these proposals involve the use of password alternatives which are supposedly harder to brute-force without human assistance; we mention the interactive grid-like password system PassMaze [12], schemes based on visual recognition [39], sequences of challenges and responses [40], and solutions to "captcha"-like problems that are far easier for humans to solve than for computers [14]. In the context of traditional alphanumeric passwords, the method of choice to thwart guessing attacks remains the deliberately slow key derivation functions in the original Unix password log-on, made programmable in [41], and perfected into the secretly user-programmable halting key derivation functions of [9]. These (H)KDFs are somewhat related to the proofs of work used in other contexts [21,28,18].

3   Architecture

The generic HPAKE protocol is shown on Figure 1. We now informally explain what it does. In Section 4 we give more details on its components.

General Overview. In our system, the user and the server hold asymmetric credentials to authenticate each other: for the user, it is her password, for the server, it is a long-term authentication token obtained from the user when she initially registered. The user selects the cost parameter associated with that password/token pair during the initial registration with the server. The password is concealed from the server, and so is the cost parameter (see below). Once the registration is completed, the user can forget everything (*e.g.*, the token given to the server, and the cost parameter) except the password.

Later, when the user wishes to establish a secure session with the server, she sends a (blind) commitment to the server. The server responds with some ciphertext that depends on the commitment. The client uses some of that ciphertext as input to the hard function, and performs the computation (which may take a while). If she committed to the correct password, the hard function output will let her decrypt the rest of the ciphertext into a copy of the long-term authentication token held by the server. Based on this, the two parties can then mutually authenticate each other and set up a secure channel.

User Programmability and Parameter Secrecy. There are good reasons for letting the user select the complexity parameter associated with his password; but it is equally important to prevent anyone from learning this value prematurely (i.e., not until they have successfully completed the authentication).

At the same time, such value must be stored somewhere, since we cannot ask the user to remember it from memory (the only thing he should be asked to remember being the password).

This requirement of a user-programmable computational bottleneck whose cost parameter is hidden from everyone and yet implicitly stored, *requires* a specific kind of unpredictable function: one that halts (after the prescribed cost expenditure) *only* on the correct input — and that on all other inputs proceeds indefinitely without ever giving back any hint that its input might have been wrong. We refer to such functions as "(selectively) halting functions".

Selectively Halting Functions. Such notion of halting function is closely related to that of Halting Key Derivation Function (HKDF) used in [9] to boost the

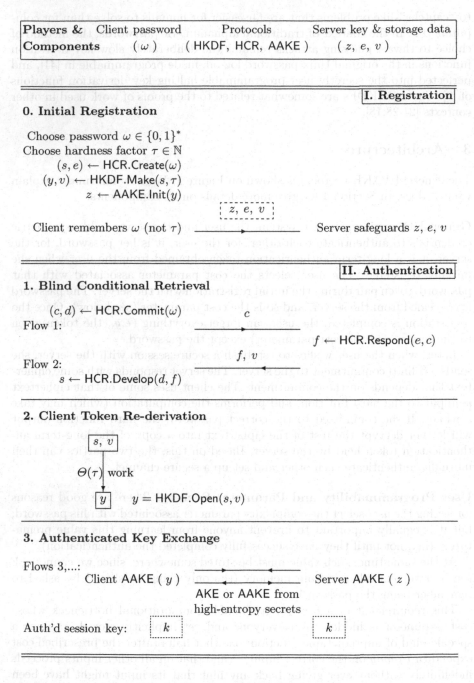

Players &	Client password	Protocols	Server key & storage data
Components	(ω)	(HKDF, HCR, AAKE)	(z, e, v)

I. Registration

0. Initial Registration

Choose password $\omega \in \{0,1\}^*$
Choose hardness factor $\tau \in \mathbb{N}$
$\quad (s, e) \leftarrow$ HCR.Create(ω)
$\quad (y, v) \leftarrow$ HKDF.Make(s, τ)
$\quad z \leftarrow$ AAKE.Init(y)

$\boxed{z, e, v}$

Client remembers ω (not τ) $\longrightarrow$ Server safeguards z, e, v

II. Authentication

1. Blind Conditional Retrieval

$\quad (c, d) \leftarrow$ HCR.Commit(ω)

Flow 1: $\qquad \xrightarrow{\quad c \quad}$

$\qquad\qquad\qquad\qquad\qquad\qquad f \leftarrow$ HCR.Respond(e, c)

Flow 2: $\qquad \xleftarrow{\quad f, v \quad}$

$\quad s \leftarrow$ HCR.Develop(d, f)

2. Client Token Re-derivation

$\boxed{s, v}$

$\Theta(\tau)$ work

$\boxed{y} \qquad y =$ HKDF.Open(s, v)

3. Authenticated Key Exchange

Flows 3,...:
$\qquad$ Client AAKE (y) $\xrightarrow{\quad}\xleftarrow{\quad}$ Server AAKE (z)

$\qquad\qquad\qquad$ AKE or AAKE from
$\qquad\qquad\qquad$ high-entropy secrets

Auth'd session key: $\boxed{k}$ $\qquad\qquad\qquad \boxed{k}$

Fig. 1. The generic HPAKE protocol

security of stand-alone password-based encryption. Indeed, conditionally halting functions such as HKDFs have another surprising benefit (which was the main point of [9]): they provide more security than any key derivation function KDF whose computational cost is known, for the same cost and the same password.

(Precisely, it is shown in [9] that a game-theoretically optimal attacker who has no idea about the programmed cost parameter must expend about $3.59\times$ more work than if it knew it, e.g., if it were facing a regular KDF with an explicit parameter.)

Incorporating stand-alone Halting Key Derivation Functions (HKDF) into our two-party key exchange protocol requires some precautions, because we want the *client* to compute it, but the *server* to store most of its input (since the client is memory-constrained, and the server mostly time-constrained). A fundamental and unavoidable problem with HKDFs is that they can serve as a password test predicate, since by definition they halt only on the correct input, which is a testable behavior. The consequence is that we will need a way to transport that data from server to client without exposing it to outside attackers, with the main complication being that the client will not have been authenticated yet by the time it needs the HKDF data.

Security by Obscurity? We emphatically stress that this notion of concealing a secretly programmed cost parameter from the adversary is not "security by obscurity", because all parties are deprived of the secret parameter, including the user who may safely forget the choice once it has been made and registered.

4   Components

We now give more details on the three cryptographic functions used in HPAKE.

Secure Registration. We note that the registration phase is special and not truly part of the protocol. It requires a secure channel which can stem from a face-to-face meeting or from a trusted PKI (which need not be used again in the actual protocol execution). Registration exists so that a user and a server who have never been in contact can start somewhere.

4.1   HKDF : Halting Key Derivation Functions

"Halting Key Derivation Functions" were originally defined in [9] to derive strong keys from weak passwords in a rate-limiting manner, to be used in a stand-alone password-based encryption system.

Here, we use HKDFs slightly differently: to map one secret random string (the retrieved secret s) into another (the client-side token y), in a manner that can be made as computationally expensive as the user wishes by selecting a suitable value of the parameter τ.

The primitive consists of two algorithms, HKDF.Make and HKDF.Open:

HKDF.Make takes as input a secret $s \in S$, a parameter $\tau \in \mathbb{N}$, and random coins, and returns a random token $y \in Y$ and its ciphertext $v \in V$.

HKDF.Open takes as input a secret $s \in S$ and a ciphertext $v \in V$, and, either returns a token $y \in Y$, or fails to halt in polynomial time.

We briefly recall the security requirements from [9]. For a random execution of Make, it must be infeasible to find, in polynomial time in the security parameter, a tuple (s', s, τ, y) such that $y = \mathrm{Open}(s', \mathrm{Make}(s, \tau))$ and $s \neq s'$. Furthermore, finding a tuple (s, τ, y) such that $y = \mathrm{Open}(s, \mathrm{Make}(s, \tau))$ must require $\Theta(\tau)$ units of time and memory, barring which no information about the correct y must be obtained from v, s, τ.

For concreteness, we give an HKDF construction adapted from [9].

HKDF.Make : $(s, \tau) \mapsto (v, k)$ $\qquad$ HKDF.Open : $(s, v) \mapsto k$

$\quad r \leftarrow \{0,1\}^{\ell}$ $\qquad\qquad\qquad\qquad$ parse v as (r, h)
$\quad z \leftarrow \mathrm{Hash}(s, r)$ $\qquad\qquad\qquad\quad$ $z \leftarrow \mathrm{Hash}(s, r)$
$\quad$ FOR $i := 1, ..., \tau$ or UNTIL user signal $\qquad$ FOR $i := 1, ..., \infty$
$\qquad z_i \leftarrow z$ $\qquad\qquad\qquad\qquad\qquad\quad$ $z_i \leftarrow z$
$\qquad$ REPEAT q times $\qquad\qquad\qquad\quad$ REPEAT q times
$\qquad\quad j \leftarrow 1 + (z \bmod i)$ $\qquad\qquad\quad$ $j \leftarrow 1 + (z \bmod i)$
$\qquad\quad z \leftarrow \mathrm{Hash}(z, z_j)$ $\qquad\qquad\qquad$ $z \leftarrow \mathrm{Hash}(z, z_j)$
$\quad v \leftarrow (r, \mathrm{Hash}(z_1, z))$ $\qquad\qquad$ IF $\mathrm{Hash}(z_1, z) = h$ BREAK
$\quad k \leftarrow \mathrm{Hash}(z, r)$ $\qquad\qquad\qquad\quad$ $k \leftarrow \mathrm{Hash}(z, r)$

The constant q is a design parameter that determines the ratio between the time and space requirements. It is not critical and wide range of values are acceptable for this parameter [9].

The primary purpose of using HKDFs is to let the user impose a computational cost without revealing it to the server or storing extra parameters locally.

The secondary benefit of HKDFs is that they are always at least as difficult to crack as a regular KDF of equal computational cost, and usually more depending on how wide or how far of the attacker's distribution of τ is compared to the user's choice.

We refer the reader to [9] for a full analysis and explanation of these phenomena. Suffice it to say that, in the best case, HKDFs provide a constant security multiplier of 3.59 (or 1.84 bits) over comparable KDFs, and in the worst case the multiplier is 1 (or 0 bit). In other words, HKDFs are never worse, and usually better than regular KDFs of same cost. To reap those benefits, the user-selected cost parameter must not be known exactly to the attacker, which is why it is important to let the user choose it, perhaps haphazardly, on a case-by-case basis.

4.2 HCR: Hidden Credential Retrieval

"Hidden Credential Retrieval" [10], the next ingredient, is a very simple cryptographic abstraction that allows a stateless client to retrieve some high-entropy secret s from a ciphertext e on remote storage server, based on a low-entropy password ω, in the safest possible way over an insecure channel. A feature of HCR is that it also protects the user data s and password ω against a curious server: the server only has in its custody a blinded string or ciphertext e, from which it is information-theoretically impossible to recognize either s or ω without also knowing the other. Furthermore, no party is to learn from the HCR protocol whether the user successfully retrieved the string s: in case of incorrect password, a junk string is silently recovered instead.

HCR was first formalized and utilized in [10] as a stand-alone protocol, though similar notions have been implicitly proposed much earlier, in different contexts. Notably, the notion of blind signature, coupled to some mild additional conditions (single-round signing and uniqueness of the unblinded signatures) already fulfilled in Chaum's original paper [16], subsumes that of HCR.

To define it, we consider three entities: a Preparer $\mathcal{P}$ that selects the retrieval password ω and the random string s to be stored; a Querier $\mathcal{Q}$ that knows the password ω and seeks to retrieve s; and a Responder $\mathcal{R}$ that acts as the storage server, prepared by the preparer and responding to queries from the querier. Both $\mathcal{P}$ and $\mathcal{Q}$ are meant to embody the same user, but we must separate the two to account for the possibility that the user does not need to remember s once it has finished to set up the server $\mathcal{R}$. The protocol consists of four algorithms:

HCR.Create, used by the Preparer $\mathcal{P}$, takes as input a reference password ω, and ouputs a plaintext s and a ciphertext e. The plaintext and ciphertext have uniform marginal distributions in some fixed sets S and Z respectively (that is, both s and e are marginally, but not jointly, independent of ω).

HCR.Commit, used by the Querier $\mathcal{Q}$, takes as input a query password ω, and outputs a commitment c and some private information d. The commitment is uniform in some set C and statistically independent of the query password.

HCR.Respond, used by the Responder $\mathcal{R}$, takes as input a ciphertext e and a commitment c, and outputs a response f in some set F.

HCR.Develop, used by the Querier $\mathcal{Q}$, takes as input the private data d and the response f, and outputs a plaintext s in the set S.

We refer to [10] for the formal security model of this primitive, and the various ways to construct it, but note that HCR can be constructed immediately from (very old) existing constructions such as unique blind signatures, including Chaum's [16] and Boldyreva's [8]. The Ford-Kalisky server-assisted password generation protocol from [22] is also an instantiation of HCR (though the use that Ford and Kalisky proposed for their scheme was different).

For illustration purposes, we describe the Ford-Kalisky version which is a bit simpler, but the Boldyreva signature would do just as well. Let $\mathbb{G}$ be a cyclic group of prime order p, and let Hash $: \{0,1\}^* \to \mathbb{G}$ be a hash function into $\mathbb{G}$.

HCR.Create $: \omega \mapsto (e, s)$. On input a registration password $\omega \in \{0,1\}^*$, output a storage-server string $e \in_\$ \mathbb{F}_p^\times$ and a user plaintext $s \leftarrow \mathsf{Hash}(\omega)^e$.

HCR.Commit $: \omega \mapsto (c, d)$. Given any candidate password $\omega \in \{0,1\}^*$, output a private decommitment $d \in_\$ \mathbb{F}_p^\times$ and a public commitment $c \leftarrow \mathsf{Hash}(\omega)^d$.

HCR.Respond : $(e, c) \mapsto f$. Given the ciphertext $e \in \mathbb{F}_p^\times$ and a commitment $c \in \mathbb{G}$, output the (deterministic) blind response $f \leftarrow c^e$.

HCR.Develop : $(d, f) \mapsto s$. Given an ephemeral $d \in \mathbb{F}_p^\times$ and a response $f \in \mathbb{G}$, output the retrieved (but unverified) user plaintext $s \leftarrow f^{1/d}$.

4.3   AKE: Authenticated Key Exchange

"(Asymmetric) Authenticated Key Exchange" is our third and final ingredient. Although it may seem strange to require an (A)AKE to build an HPAKE, there is no circularity given that AKE or AAKE from high-entropy keys is quite easy and very well known. In our description, the AKE shared secret, or the AAKE conjugate secrets, are the client-side token y and the server-side token z (such that $y = z$ for AKE or $y \neq z$ for AAKE, respectively).

Choosing an AAKE scheme for this stage instead of AKE (*i.e.*, with asymmetric secrets), will result in resistance to the compromise of the server database, even for authentication to the same server. That is, even with knowledge of all the server secrets including z, impersonating the client to the server itself will still require finding the password (and thus cracking the HKDF). The AAKE server token is initialized at registration time by the client; we generically wrote $z = $ AAKE.Init(y), but in practice y and z will be returned together by a key generation algorithm. Efficient AAKE schemes include [15] or the very compact MQV [32] on elliptic curves.

Alternatively, for increased server-side efficiency it is possible to use a symmetric AKE scheme instead. In this case, the client and server tokens are the same: $z = y$, though they will still vary from one server to the next even under the same password. We this choice, we forgo resistance to server database compromise against the same server, but we still get all the other security properties of HPAKE, including password secrecy and resistance to cross-site impersonation attacks. (Indeed, an attacker who learns $y = z$ for a specific client-server pair will be able to impersonate that client to that same server, but not to any other server, and without learning the password.)

For concreteness, we give an explicit "folklore" symmetric AKE protocol built purely from hash functions modeled as random oracles [38]. It is a very efficient three-flow AKE protocol where the client and server send each other fresh random nonces n_c and n_s, and verify their correct reception and create a session key by hashing them with the secret key $y = z$ they share.

AKE.1 : $C \rightarrow S$

C picks a fresh random nonce c_c and sends it to S:

AKE.2 : $C \leftarrow S$

Using S's stored copy of y and the received values $\hat{c}_c$ of c_c, S sends a fresh random nonce c_s and the value $a_s \leftarrow $ Hash(y, c_c, c_s) to C.

AKE.3 : $C \rightarrow S$

Using C's reconstructed copy $\hat{y}$ of y and the received values $\hat{c}_s$ and $\hat{a}_s$ of c_s and a_s, C verifies the equality, $\hat{a}_s \overset{?}{=} $ Hash$(\hat{y}, c_c, \hat{c}_s)$. If true, C accepts the session and sends $a_c \leftarrow $ Hash(y, c_s, a_s) to the server.

AKE : session key

Using S's stored copy of y and the received value $\hat{a}_c$ of a_c, S verifies that, $\hat{a}_c \overset{?}{=} \mathsf{Hash}(y, c_s, a_s)$. If true, C accepts the session. At this point, if both parties have accepted, they share a mutually authenticated random session key given by, $k \leftarrow \mathsf{Hash}(y, \hat{a}_c, a_s) = \mathsf{Hash}(\hat{y}, a_c, \hat{a}_s)$.

4.4 Consolidation of Flows

The above AKE protocol requires three flows (or half-rounds). If we add the two flows for HCR, that makes five flows for the complete HPAKE protocol.

However, it is possible and easy to interleave and consolidate the HCR and AKE messages so that HPAKE as a whole only requires three flows.

The idea is for the client eagerly to send Flow 1 of AKE along with the HCR commitment c in Phase 1 of HPAKE. The server then sends Flow 2 of AKE along with the HCR response f back to the client. The client performs the HKDF hard-function calculation in Phase 2 of HPAKE, and, once the token y is decrypted, sends the final Flow 3 of AKE, thereby completing Phase 3 of HPAKE in one additional flow instead of three.

It is easy to see that the first two flows of AKE are independent of the HCR phases on the protocol. In the random-oracle model, it is even acceptable to reuse the HCR commitment c directly as the AKE client nonce c_c, thereby saving a little extra bit of bandwidth.

The only drawback of this flow consolidation is that the server needs to pre-serve the AKE state across Phases 1–3 of the full protocol, while Phase 2 may by design take a long time for the client to complete (unless the client is caching a copy of y, which is explicitly allowed). By contrast, in the plain unconsolidated protocol, the server can remain stateless until Phase 3.

5 Security

Theorem 1. *Let χ be a security parameter, such that all hash functions have at least $\ell \geq 2\chi$ bits of output in the random-oracle model. Let $|\mathcal{D}| \ll 2^\chi$ be the size of the password dictionary. Assume that the HCR and AKE subprotocols are χ-bit secure in $\mathbb{G}$, that is, they yield to PPT computational adversaries with time-advantage product $TA \geq 2^\chi$ only. Suppose that $|\mathcal{D}| \ll 2^\chi$, i.e., the password dictionary size is the weak link. Then, in the random-oracle model, the advantage Adv of an polynomial-time adversary $\mathcal{A}$ at distinguishing from random a secure channel established by uncorrupted parties (either by causing one party to accept a new session with $\mathcal{A}$, or by stealing an already established session), is, $\forall k \in \mathbb{N}$:*

- *For an outsider $\mathcal{A}$ sending a total of q messages to the user and any number of honest servers:*

$$\mathsf{Adv}_{\mathcal{A}} \leq \frac{q}{|\mathcal{D}|} + o(1/\chi^k) \ .$$

- *For an insider $\mathcal{A}$ sending a total of q messages to the user and any number of honest servers, and making t queries to the random oracle used in the* *function* HKDF.Open *(expressed in the same unit as the hardness parameter* τ *used in* HKDF.Make *when registering with the insider):*

$$\mathsf{Adv}_{\mathcal{A}} \leq \frac{2t}{|\mathcal{D}|\,\tau} + \frac{q}{|\mathcal{D}|} + o(1/\chi^k) \qquad \textit{in the general case ;}$$

$$\mathsf{Adv}_{\mathcal{A}} \leq \frac{2t}{3.59\,|\mathcal{D}|\,\tau} + \frac{q}{|\mathcal{D}|} + o(1/\chi^k) \qquad \textit{in cases where :}$$

- *the amount of memory available to $\mathcal{A}$ is $\leq o(|\mathcal{D}|\,\tau)$ (which is always true in practice by a wide margin); and,*
- *either, the parameter τ is drawn by the user from a distribution of density $\sim \tau^{-1-\epsilon}$, or, the attacker $\mathcal{A}$ believes that it is not in its interest to try to guess τ (which is generally the case by a game-theoretic argument, see [9] for details).*

5.1   Interpretation

Theorem 1 expresses two very different bounds, depending on whether the user and server(s) are together facing a third-party attacker, or whether the user is facing a malicious server.

Against Outsiders. The advantage of outsiders, $\frac{q}{|\mathcal{D}|} + o(1/\chi^k)$, is the usual bound for PAKE and APAKE protocols. It corresponds (up to a negligible term) to the unavoidable online attack where the outsider tries to impersonate the user to the server (or *vice versa*) by trying out one password candidate at a time.

Since q is the number of *online* queries, and thus necessarily quite small, the security margin against outsiders will remain acceptable even for very small dictionaries $\mathcal{D}$, and thus very weak passwords. The banking industry, for example, is content to protect user accounts with four-digit PINs, thus with just 13 bits of entropy, by locking the account after three incorrect attempts.

Against Insiders. The advantage of insiders (*i.e.*, corrupt servers) is the same as outsiders plus an additional term, $\frac{2t}{B|\mathcal{D}|\tau}$, that accounts for the possibility that insiders have to mount an offline attack against the password. Here, τ if the user-selected complexity parameter, and B is a small constant ($B = 1$ or $B = 3.59$ depending on whether the "halting principle" is not, or is, applicable [9], i.e., whether τ is adequately uncertain to the attacker). Together, the dictionary size $|\mathcal{D}|$ and the user-programmable complexity parameter τ constitute the main actionable defenses at the user's disposal to thwart an insider offline attack. (Having $B > 1$ is merely a useful side-effect of enforcing the secrecy of τ, though the latter is already desirable in itself, as discussed previously.)

The offline attack, though it requires insider knowledge to be feasible, is far more dangerous than the online attack already available to outsiders. It is dangerous because, in an offline attack, the numerator t is out of the control of the user (or any honest server). It depends only on the adversary's resources and can therefore be quite large; in particular, $t \gg q$.

It is useful to take a very concrete example to illustrate this point. Suppose that the user, Alice, has a single password, which she uses everywhere, and changes every four months (10^7 seconds). Suppose also that one of the web sites where she has an account is a sham, and wishes to dedicate an enterprise-class computer farm (10^5 CPUs) to the single task of attempting to recover Alice's password. The attacker thus has a window of $10^{12} \approx 2^{40}$ CPU-seconds at his disposal before the password becomes useless. For comparison, Alice's password will succumb with probability $p = \frac{1}{2}$, in each of the following five scenarios ($|\mathcal{D}|$ = dictionary size; τ = user-selected hash complexity):

A. $|\mathcal{D}| = 2^{61}$ (61 bits) and fixed $\tau = 2^{-21}$ ($0.5\mu s$):
   i.e., a strong password (13 random letters) with a computer-instantaneous hash (*e.g.*, SHA1);
B. $|\mathcal{D}| = 2^{51}$ (51 bits) and fixed $\tau = 2^{-11}$ ($500\mu s$):
   i.e., a strong password (11 random letters) with a number-theoretic hash (*e.g.*, on curves);
C. $|\mathcal{D}| = 2^{38}$ (38 bits) and public parameter $\tau = 2^2$ ($4s$):
   i.e., an ok password (8 random letters) with a human-noticeable hash (such as a 4-sec KDF);
D. $|\mathcal{D}| = 2^{38}$ (38 bits) and *secret* user-selected $\tau = 2^0$ ($1s$):
   i.e., same password (8 random letters) with a human-instantaneous hash (here, 1-sec HKDF);
E. $|\mathcal{D}| = 2^{24}$ (24 bits) and *secret* user-selected $\tau = 2^{14}$ ($5h$):
   i.e., a very memorable but very weak "backup" password (5 random letters) protected by a very expensive hash (5-hours HKDF, taking, *e.g.*, 17 minutes to compute on a 16-core client).

(In all scenarios, the password lengths are for lowercase-only random letters, *i.e.*, a 26-symbol alphabet.)

Case A corresponds to the practice of simply hashing the password (possibly with some site-dependent non-secret information) before use, in a regular PAKE protocol.

Case B corresponds to most AEKE and APAKE protocol implementations, where the KDF is an inherent part of the protocol, and subject to number-theoretical constraints (such as compatibility with efficiently verifiable zero-knowledge proofs of knowledge of the password).

Cases C and D correspond to our HPAKE protocol with everyday settings, where the difference between the two is that in the former the hardness factor τ is a known parameter of the system, while in the latter it is chosen by the user in a somewhat unpredictable way (to the adversary).

Case E corresponds to the use of HPAKE with a last-resort backup password that ought never to be used, but must be very easy to remember in case it is ever needed, for instance because the user forgot her regular password. Because highly memorable passwords are also easy to guess, a very large value of τ is desirable to maintain a sufficient margin of security against insider attacks. (How large τ should be, depends on the actual strenght of the backup password, which is known to the user only. This case illustrates why τ must be kept secret.)

Those examples clearly show the superiority of HPAKE over previous PAKE and APAKE protocols in that it allows much weaker passwords to be safely reused, both in an everyday situation (*e.g.*, comparing Case D *vs.* Case B), as well as in a last-resort backup situation (for which none of the existing protocols offers a viable solution).

6   User Interface

Provided that text passwords are used, the client-side user interface (UI) does not require any special hardware: a keypad is all that is required, with perhaps a one-bit display to indicate that the hard function computation is in progress. There are however two important software requirements:

6.1   Trusted Local Password Entry

All the precautions we took to protect the user password and ensure its reusability are moot if an attacker ever manages to bypass the HPAKE protocol, *e.g.*, by tricking the user into entering the password directly into web form.

Software Solutions. A software solution, specific to internet transactions, would require native HPAKE support from the browser, and ideally from the operating system, so that password-entry prompts can be made distinctive enough to be easy to recognize as genuine by the user. *E.g.*, some browsers already attempt to make HTTP-Auth password dialog boxes look unlike regular browser windows; and the Windows operating system requires a Ctrl-Alt-Del attention sequence to escape any running application before a login password can be keyed.

Hardware Solutions : Commodity *vs.* Custom. The safest way to reduce the possibility of password exposure, is to seize it not on a general-purpose computer, but on a dedicated hardware device in the possession of the user.

"Pocket password calculators" have been used for decades by the banking industry for signing high-value electronic transactions, and more recently for generating one-time passwords to gain access to corporate VPNs. Such devices have a small keypad for entering a user PIN, but almost always also contain a custom user-specific private key, which makes them difficult to replace and also sensitive to theft and hardware key recovery attacks.

It is easy to imagine similar keypad-equipped hardware for securely entering one's HPAKE password and for performing all related HPAKE computations, possibly interfacing with a host computer connected to the internet. This would ensure that the password is never exposed, even in case of full compromise of the host computer. A key advantage over earlier "password calculators" is that an HPAKE device would be completely commoditized and contain no user-specific information. User would thus not need to worry about losing the, or having them stolen.

6.2   Real-Time User-Driven Cancellation

Because the HKDF component in HPAKE will not halt spontaneously on all inputs, the client-side UI must include an special button to allow the user to take corrective actions (and optionally, during the registration phase, to make it easier for the user to select the value of τ).

During registration, a "finish" button may serve as a simple and intuitive device for selecting the hardness parameter τ: the user would simply let HKDF.Make run for a while and then click on the finish button, which will cause the system to set τ to the current value of the HKDF loop iterator. The user need not be shown the value of τ, since she has no use for it (except perhaps a vague recollection of what kind of delay she chose, if she suspects she might forget her password).

During authentication, a "cancel" button must be available to let the user stop the process. Since the HKDF.Open function is designed to run forever when called on the wrong inputs, it is up to the user to stop it manually when she realizes that she entered a wrong password. Having a cancel button is always a good idea, since delays can occur for many reasons (*e.g.*, network congestion).

7   Conclusion

The sad reality is that people are not using passwords the way protocol designers and security experts wish they were. It is therefore natural to ask for an authentication protocol that remains as secure as possible under such stringent usage conditions.

Ideally, people should be able to conduct all their online business with a single easy-to-remember password, no matter how numerous or how untrustworthy the web sites they wish to authenticate with.

Just as importantly, the ideal protocol should need zero client-side long-term storage (other than the password), to lessen the security impact in case of loss or theft; this is especially important when traveling. This make a very compelling case for "reusable-password stateless roaming authentication", especially since by far the safest place to keep a password is in one's memory, where there is not much room for more.

Existing password authentication protocols are generally not safe when related passwords are used in multiple contexts. Protocols of the APAKE family come very close, but are still vulnerable to offline dictionary attacks by insiders, unless the password is strong, because they take no measure to limit the rate of such attacks.

Various client-side stop-gap measures have also been proposed, but they invariably have steep additional requirements: for example, browser-based "password managers" require long-term storage on the client side; whereas "anti-phishing" add-ons (intended to save you from mistakenly sending your password to an evildoer on a blacklist) make the tacit assumption that the DNS system and the PKI authorities used in SSL can be trusted. PKI-based solutions generally require the storage of at least one authentic certificate on the client, too.

Our HPAKE approach is certainly not perfect. However, it has a crucial combination of benefits over the existing alternatives: (1) client and servers have asymmetric secrets; (2) authentication is mutual; (3) no need for any client-side storage; (4) the password is a user secret and can be reused with other servers; (5) outsider attacks can do no better than online password guessing; (6) servers with access to the user registration data can always brute-force the user's password offline, but the presence of a hard function will greatly slow down such attacks; (7) the hard function is user-programmed, giving the user full control over it; (8) the hard function is user-computed, ensuring that it will be applied effectively; (9) the server-side protocol is independent of the hard function, it is lightweight and scales very well.

Our HPAKE protocol is but one example of a possible construction; there are certainly others. Ours has the advantage of being very simple and efficient, but relies heavily (and, in fact, almost exclusively) on the random-oracle model for its security. We leave it as an open question to find other realizations that avoid random oracles but are still reasonably efficient.

We conclude with an obvious but important word of caution: the reusability of weak passwords that HPAKE enables only applies within the confines of HPAKE (and HKDF [9], for local encryption applications). Reusing an HPAKE password on an unsecured web form will void all security guarantees that our cryptography sought to offer.

References

1. Bellare, M., Canetti, R., Krawczyk, H.: A modular approach to the design and analysis of authentication and key exchange protocols. In: Proceedings of the ACM Symposium on the Theory of Computing—STOC 1998. ACM Press, New York (1998)
2. Bellare, M., Pointcheval, D., Rogaway, P.: Authenticated key exchange secure against dictionary attacks. In: Preneel, B. (ed.) EUROCRYPT 2000. LNCS, vol. 1807, pp. 139–155. Springer, Heidelberg (2000)
3. Bellare, M., Rogaway, P.: Entity authentication and key distribution. In: Stinson, D.R. (ed.) CRYPTO 1993. LNCS, vol. 773, pp. 232–249. Springer, Heidelberg (1994)
4. Bellovin, S.M., Merritt, M.: Encrypted key exchange: Password-based protocols secure against dictionary attacks. In: IEEE Symposium on Security and Privacy— SP 1992, pp. 72–84. IEEE Press, Los Alamitos (1992)
5. Bellovin, S.M., Merritt, M.: Augmented encrypted key exchange. In: ACM Conference on Computer and Communications Security—CCS 1993, pp. 224–250. ACM Press, New York (1993)
6. Bennett, C.H., Brassard, G., Robert, J.-M.: Privacy amplification by public discussion. SIAM Journal of Computing 17(2) (1988)
7. Blake-Wilson, S., Johnson, D., Menezes, A.: Key agreement protocols and their security analysis. In: Darnell, M.J. (ed.) Cryptography and Coding 1997. LNCS, vol. 1355, pp. 30–45. Springer, Heidelberg (1997)

8. Boldyreva, A.: Threshold signatures, multisignatures and blind signatures based on the gap-Diffie-Hellman-group signature scheme. In: Desmedt, Y.G. (ed.) PKC 2003. LNCS, vol. 2567, pp. 31–46. Springer, Heidelberg (2003)
9. Boyen, X.: Halting password puzzles. In: USENIX Security Symposium— SECURITY 2007, pp. 119–134. The USENIX Association (2007)
10. Boyen, X.: Hidden credential retrieval from a reusable password. In: ACM Symposium on Information, Computer & Communication Security—ASIACCS 2009. ACM Press, New-York (2009)
11. Boyko, V., MacKenzie, P., Patel, S.: Provably secure password-authenticated key exchange using diffie-hellman. In: Preneel, B. (ed.) EUROCRYPT 2000. LNCS, vol. 1807, pp. 156–171. Springer, Heidelberg (2000)
12. Brown, D.R.L.: Prompted user retrieval of secret entropy: The passmaze protocol. Cryptology ePrint Archive, Report 2005/434 (2005), http://eprint.iacr.org/
13. Canetti, R., Halevi, S., Katz, J., Lindell, Y., MacKenzie, P.: Universally composable password-based key exchange. In: Cramer, R. (ed.) EUROCRYPT 2005. LNCS, vol. 3494, pp. 404–421. Springer, Heidelberg (2005)
14. Canetti, R., Halevi, S., Steiner, M.: Mitigating dictionary attacks on password-protected local storage. In: Dwork, C. (ed.) CRYPTO 2006. LNCS, vol. 4117, pp. 160–179. Springer, Heidelberg (2006)
15. Canetti, R., Krawczyk, H.: Analysis of key-exchange protocols and their use for building secure channels. In: Pfitzmann, B. (ed.) EUROCRYPT 2001. LNCS, vol. 2045, pp. 453–474. Springer, Heidelberg (2001)
16. Chaum, D.: Blind signatures for untraceable payments. In: Advances in Cryptology—CRYPTO 1982, pp. 199–203 (1982)
17. Choo, K.-K.R., Boyd, C., Hitchcock, Y.: Examining indistinguishability-based proof models for key establishment protocols. In: Roy, B. (ed.) ASIACRYPT 2005. LNCS, vol. 3788, pp. 585–604. Springer, Heidelberg (2005)
18. Dean, D., Stubblefield, A.: Using client puzzles to protect TLS. In: USENIX Security Symposium—SECURITY 2001 (2001)
19. Diffie, W., Hellman, M.: New directions in cryptography. IEEE Transactions on Information Theory 22(6), 644–654 (1976)
20. Diffie, W., van Oorschot, P., Wiener, M.: Authentication and authenticated key exchanges. Designs, Codes and Cryptography 2, 107–125 (1992)
21. Dwork, C., Naor, M.: Pricing via processing or combating junk mail. In: Brickell, E.F. (ed.) CRYPTO 1992. LNCS, vol. 740, pp. 139–147. Springer, Heidelberg (1993)
22. Ford, W., Kaliski Jr., B.S.: Server-assisted generation of a strong secret from a password. In: Proc. IEEE 9th Int. Workshop on Enabling Technologies: Infrastructure for Collaborative Enterprises, pp. 176–180. IEEE Computer Society Press, Los Alamitos (2000)
23. Gentry, C., MacKenzie, P., Ramzan, Z.: Password authenticated key exchange using hidden smooth subgroups. In: ACM Conference on Computer and Communications Security—CCS 2005, pp. 299–309. ACM Press, New York (2005)
24. Gentry, C., MacKenzie, P., Ramzan, Z.: A method for making password-based key exchange resilient to server compromise. In: Dwork, C. (ed.) CRYPTO 2006. LNCS, vol. 4117, pp. 142–159. Springer, Heidelberg (2006)
25. Halevi, S., Krawczyk, H.: Public-key cryptography and password protocols. In: ACM Conference on Computer and Communications Security—CCS 1998, pp. 122–131. ACM Press, New York (1998)
26. Jablon, D.: Strong password-only authenticated key exchange. Computer Communication Review (1996)

27. Jiang, S., Gong, G.: Password based key exchange with mutual authentication. In: Handschuh, H., Hasan, M.A. (eds.) SAC 2004. LNCS, vol. 3357, pp. 267–279. Springer, Heidelberg (2004)
28. Juels, A., Brainard, J.: Client puzzles: A cryptographic defense against connection depletion attacks. In: Proceedings of NDSS 1999, pp. 151–165 (1999)
29. Katz, J., Ostrovsky, R., Yung, M.: Efficient password-authenticated key exchange using human-memorable passwords. In: Pfitzmann, B. (ed.) EUROCRYPT 2001. LNCS, vol. 2045, pp. 475–494. Springer, Heidelberg (2001)
30. Krawczyk, H.: HMQV: A high-performance secure Diffie-Hellman protocol. In: Shoup, V. (ed.) CRYPTO 2005. LNCS, vol. 3621, pp. 546–566. Springer, Heidelberg (2005)
31. LaMacchia, B., Lauter, K., Mityagin, A.: Stronger security of authenticated key exchange. In: Susilo, W., Liu, J.K., Mu, Y. (eds.) ProvSec 2007. LNCS, vol. 4784, pp. 1–16. Springer, Heidelberg (2007)
32. Law, L., Menezes, A., Qu, M., Solinas, J., Vanstone, S.A.: An efficient protocol for authenticated key agreement. Designs, Codes and Cryptography 28(2), 119–134 (2003)
33. MacKenzie, P.: More efficient password-authenticated key exchange. In: Naccache, D. (ed.) CT-RSA 2001. LNCS, vol. 2020, pp. 361–377. Springer, Heidelberg (2001)
34. MacKenzie, P., Patel, S., Swaminathan, R.: Password-authenticated key exchange based on RSA. In: Okamoto, T. (ed.) ASIACRYPT 2000. LNCS, vol. 1976, pp. 599–613. Springer, Heidelberg (2000)
35. MacKenzie, P., Shrimpton, T., Jakobsson, M.: Threshold password-authenticated key exchange. Journal of Cryptology 19(1), 27–66 (2006)
36. Maurer, U.: Information-theoretically secure secret-key agreement by not authenticated public discussion. In: Fumy, W. (ed.) EUROCRYPT 1997. LNCS, vol. 1233, pp. 209–225. Springer, Heidelberg (1997)
37. Menezes, A.: Another look at HMQV. Cryptology ePrint Archive, Report 2005/205 (2005), http://eprint.iacr.org/
38. Menezes, A.J., van Oorschot, P.C., Vanstone, S.A.: Handbook of Applied Cryptography. CRC Press, Boca Raton (1997)
39. Naor, M., Pinkas, B.: Visual authentication and identification. In: Kaliski Jr., B.S. (ed.) CRYPTO 1997. LNCS, vol. 1294, pp. 322–336. Springer, Heidelberg (1997)
40. Pinkas, B., Sander, T.: Securing passwords against dictionary attacks. In: ACM Conference on Computer and Communications Security—CCS 2002, pp. 161–170. ACM Press, New York (2002)
41. Provos, N., Mazières, D.: A future-adaptable password scheme. In: USENIX Technical Conference—USENIX 1999 (1999)
42. Yee, K.-P., Sitaker, K.: Passpet: Convenient password management and phishing protection. In: Symposium On Usable Privacy and Security—SOUPS 2006. ACM Press, New York (2006)

An Efficient and Provably Secure Cross-Realm Client-to-Client Password-Authenticated Key Agreement Protocol with Smart Cards

Wenting Jin[1] and Jing Xu[2]

[1] State Key Laboratory of Information Security,
Graduate University of Chinese Academy of Sciences, Beijing, P.R. China
jinwenting@is.iscas.ac.cn
[2] State Key Laboratory of Information Security,
Institute of Software, Chinese Academy of Sciences, Beijing, P.R. China
xujing@is.iscas.ac.cn

Abstract. *Cross-realm* client-to-client password-authenticated key agreement (C2C-PAKA) protocols provide an authenticated key exchange between two clients of different realms, who only share their passwords with their own servers. Recently, several such *cross-realm* C2C-PAKA protocols have been suggested in the *private-key (symmetric)* setting, but all of these protocols are found to be vulnerable to *password-compromise impersonation* attacks. In this paper, we propose our innovative C2C-PAKA-SC protocol in which smart cards are first utilized in the *cross-realm* setting so that it can resist all types of common attacks including *password-compromise impersonation* attacks and provide improved efficiency. Moveover, we modify the original formal security model to adapt our proposed protocol and present a corresponding security proof.

Keywords: Password-authenticated key agreement, Cross-realm, Client-to-client, Smart cards, Provable security.

1 Introduction

Password-based authentication is a popular method for user authentication in the *client-server* model because of its easy-to-memorize property. Password-authenticated key agreement (PAKA) protocols are usually designed to provide secure authentication and key exchange between two entities who have a pre-shared password. Most password-authenticated key agreement protocols [12-14] in the literature are based on the *single-server* model which assumes each client has a secret password shared with a common server. The main advantage of this setting is that it provides two clients in the same realm with the capability of generating a common session key while only requiring them to remember their distinct passwords.

However, with dynamic diversity and development of communication environments such as mobile networks, home networking, ubiquitous networking and etc., it is considered as one of main concerns to establish a secure channel between

J.A. Garay, A. Miyaji, and A. Otsuka (Eds.): CANS 2009, LNCS 5888, pp. 299–314, 2009.
© Springer-Verlag Berlin Heidelberg 2009

clients registered in different servers. Such protocols are more popularly known as *cross-realm* C2C-PAKA protocols. In fact, each realm would have its own trusted server to manage clients' passwords and provide service for them. In this paper, we mainly focus on authentication settings involving *cross-realm*. For ease of notation, we will simply call these C2C-PAKAs for the rest of this paper.

1.1 Related Works and Our Contribution

Related works include [1-10]. In 2002, Byun *et al.* [1] first proposed a *cross-realm* C2C-PAKA protocol. Unfortunately, the protocol was found to be flawed. Chen [2] first pointed out that one malicious server in the *cross-realm* setting could mount a *dictionary* attack to obtain the password of a client who belongs to the other realm. Similarly, Wang *et al.* [3] showed *dictionary* attacks on the same protocol. Subsequently, Kim *et al.* [4] pointed out that the original protocol was susceptible to the *Denning-Sacco* attack and proposed an improved C2C-PAKA protocol. However, Phan *et al.* [5] suggested two *unknown key-share* attacks on the improved C2C-PAKA protocol [4]. Yoon *et al.* [6] also pointed out that Kim *et al.*'s protocol [4] was susceptible to a *one-way man-in-the-middle* attack and a *password-compromise impersonation* attack, and presented an enhancement. Nevertheless, all these protocols [1-6] were designed with heuristic security analysis. The first provably secure *cross-realm* C2C-PAKA protocols were independently proposed by Byun *et al.* [7] and Yin *et al.* [8], respectively. But quickly, Phan *et al.* [9] found that both protocols cannot withstand *undetectable on-line dictionary* attacks by any adversary. Recently, Feng *et al.* [10] showed that Byun *et al.*'s protocol [7] was insecure against the *password-compromise impersonation* attack and furthermore pointed out that it appeared infeasible to make any countermeasures against *password-compromise impersonation* attacks in the *private-key* setting. Thus they proposed a new provably secure protocol based on the *public-key (asymmetric)* setting which could avoid such attacks.

Indeed, up until now, there is no C2C-PAKA protocol in the *private-key* setting which is secure against *password-compromise impersonation* attacks. On the other hand, low-efficiency is the most serious disadvantage in the *public-key* setting. In this paper, aiming at resistance against all types of common attacks including *password-compromise impersonation* attacks and not reducing the efficiency, we initiate a challenge on contriving one satisfying protocol in the *private-key* setting. We first present a smart card based *cross-realm* C2C-PAKA protocol, named C2C-PAKA-SC. Compared with Byun *et al.*'s protocol [7] and Feng *et al.*'s protocol [10], our protocol reduces the computation and communication cost. We also define a new and clear security model for C2C-PAKA-SC protocols, which combines all the corresponding security properties. In addition, we prove our C2C-PAKA-SC protocol is secure under the well-known computational assumptions.

1.2 Organization

The remainder of this paper is organized as follows. In Section 2, we discuss the requirements of a secure C2C-PAKA-SC protocol. Our new *cross-realm*

C2C-PAKA-SC protocol is proposed in Section 3, along with its efficiency analysis. In Section 4, we introduce the formal model for C2C-PAKA-SC protocols and provide the detailed security proof for our proposed protocol. Finally, we conclude in Section 5.

2   Properties of C2C-PAKA-SC Protocol

Our C2C-PAKA-SC protocol is normally built as follows. In the registration phase, each server issues a unique smart card to its each client through an authenticated and secure environment where all parties are assumed to be honest and perform exactly according to the protocol specification. If a client registers with a server successfully, we say that the client is in the realm of the server. Once the registration phase is completed, two clients of different realms could authenticate each other and generate an agreed-upon session key with the help of smart cards and two servers through the login-and-authentication phase. In this phase, the communication channel is no longer considered to be secure. That is, an adversary $\mathscr{A}$ has total controlled over the communication channel and he may intercept, insert, delete, or modify any message in the channel. In addition, referring to the unique property of two-factor authentication mechanisms including passwords and smart cards which was discussed in [15], we suppose $\mathscr{A}$ may either steal a client's smart card and then extracts the information from it, or obtain the client's password, but not both of them. In other words, a client's password and smart card cannot be both compromised.

It is desirable for normal C2C-PAKA-SC protocols to possess the following security attributes:

(1) *Forward secrecy.* If passwords of the entities or long-term private keys of the smart cards are compromised, the secrecy of previous session keys is not affected.
(2) *Password-compromise impersonation resilience.* Compromising the password of client *Alice* in one realm should not enable an outside adversary to share the session key with *Alice* by masquerading as any other clients belonging to another realm.
(3) *Unknown key-share resilience.* Client *Alice* should not be coerced into sharing a key with client *Carol* when in fact she thinks that she is sharing the key with client *Bob*.
(4) *Dictionary attacks resilience.* The password of any client in one realm should be strongly protected against *dictionary* attacks of other entities (clients or/and servers) belonging to another realm, even if his smart card is stolen.

3   Our C2C-PAKA-SC Protocol

As stated above, all existing C2C-PAKA protocols in the *private-key* setting are vulnerable to *password-compromise impersonation* attacks. The weaknesses result from the fact that client *Alice* in one realm could not distinguish between interactions with the other honest client in another realm or with an adversary, if the password of *Alice* is compromised. Unfortunately, in the *private-key* setting, it appears infeasible to make any countermeasures against *password-compromise impersonation* attacks, or else *off-line dictionary* attacks would be exploited. However, such attacks may be avoided in the smart card setting in which additional security assumptions are required. In this section, we present a new smart card based C2C-PAKA protocol.

3.1   Protocol Description

There are four participants involved in our C2C-PAKA-SC protocol: *Clients* = {*Alice*; *Bob*}, and *Servers* = {S_A; S_B}, where *Alice* is a client in the realm of server S_A, and *Bob* is a client in the realm of server S_B. We assume that the key K is pre-distributed between S_A and S_B. To initialize, the system selects large prime number p and q such that $p = 2q + 1$. The detailed steps of the C2C-PAKA-SC protocol are described in the following subsections.

3.1.1   Registration Phase

In this phase (outlined in Fig. 1), client *Alice* registers with her server S_A and finally obtains a smart card SC_A through a secure channel. Then the protocol proceeds in the following steps:

(1) Client *Alice* chooses her identity ID_A and password PW_A. *Alice* then submits the registration request $r_A = \{ID_A, PW_A\}$ to her server S_A through a secure channel.

Client *Alice*	Server S_A
Select ID_A, PW_A	
$\xrightarrow{\quad r_A = \{ID_A, PW_A\} \quad}$	
	$R_A = h(ID_A)^x + h(PW_A) \bmod p$
	$SC_A : \{ID_A, R_A, h_1(\cdot), p\}$
$\xleftarrow{\quad SC_A \quad}$	

Client *Bob*	Server S_B
select ID_B, PW_B	
$\xrightarrow{\quad r_B = \{ID_B, PW_B\} \quad}$	
	$R_B = h(ID_B)^y + h(PW_B) \bmod p$
	$SC_B : \{ID_B, R_B, h_2(\cdot), p\}$
$\xleftarrow{\quad SC_B \quad}$	

Fig. 1. Registration Phase

(2) Server S_A chooses its secret key $x \in Z_q^*$ and an appropriate one-way hash function $h_1(\cdot) : \{0,1\}^* \to Z_p^*$. Upon receiving r_A, S_A computes $R_A = h_1(ID_A)^x + h(PW_A) \bmod p$.

(3) S_A stores $\{ID_A, R_A, h_1(\cdot), p\}$ into a smart card SC_A and issues it to *Alice*. In addition, S_A discards all of the information of *Alice*, only hold its own secret key x.

Similarly, client *Bob* interacts with his server S_B as above steps. Finally, *Bob* obtains his smart card SC_B which contains $\{ID_B, R_B, h_2(\cdot), p\}$ and S_B only keeps the value of its secret key y.

3.1.2 Login-And-Authentication Phase

In this phase (outlined in Fig. 2), two clients of different realms perform the following steps for mutual authentication with their respective servers' help. Moreover, a session key sk is agreed by two clients. Then the protocol proceeds in the following steps:

(1) Client *Alice* attaches her smart card SC_A to a device reader and inputs her identity ID_A and her password PW_A. Then, the device chooses a random $\alpha \in Z_q^*$ and computes $R_A' = (R_A - h_1(PW_A))^\alpha \bmod p$, $W_A = h_1(ID_A)^\alpha \bmod p$, $C_A = h_1(T_1 \| R_A' \| W_A \| ID_A)$, where T_1 is a time stamp. Finally, *Alice* sends the message $\{ID_A, ID_B, T_1, C_A, W_A\}$ to S_A.

(2) Upon receiving the message from *Alice*, S_A checks whether ID_A and T_1 are both valid. If so, S_A computes $R_A'' = W_A{}^x \bmod p$, and checks whether C_A equals $h_1(T_1 \| R_A'' \| W_A \| ID_A)$. If so, S_A believes *Alice* is authenticated. S_A then computes $K_A = h_1(R_A'' \oplus T_2)$, where T_2 is a time stamp. S_A also randomly chooses $k \in Z_q^*$ and computes $V_A = [k, ID_A, ID_B]_{K_A}$, $Ticket_B = [k, ID_A, ID_B, L]_K$, where K is the key between S_A and S_B, L is $Ticket_B$'s lifetime, and $[X]_K$ means the encryption of a message X using a symmetric key K. Finally, S_A sends $\{V_A, Ticket_B, T_2, L\}$ to *Alice*.

(3) Upon receiving the message from S_A, *Alice* checks whether the time stamp T_2 is valid. If so, she computes $K_A' = h_1(R_A' \oplus T_2)$, uses it to decrypt V_A, and obtains ID_A, ID_B and k. She also checks whether ID_A and ID_B are both correct. Then, *Alice* randomly chooses $a \in Z_q^*$, computes $E_a = g^a \| MAC_k(g^a)$, and forwards $\{ID_A, E_a, Ticket_B\}$ to client *Bob*.

(4) Upon receiving the message from *Alice*, *Bob* attaches his smart card SC_B to a device reader and inputs his ID_B and PW_B. Then, the device chooses a random $\beta \in Z_q^*$ and computes $R_B' = (R_B - h_2(PW_B))^\beta \bmod p$, $W_B = h_2(ID_B)^\beta \bmod p$, and $C_B = h_2(T_3 \| R_B' \| W_B \| ID_B)$, where T_3 is the current time. Finally, *Bob* sends the message $\{Ticket_B, T_3, C_B, W_B\}$ to S_B.

(5) Upon receiving the message from *Bob*, S_B first checks whether T_3 is valid. If so, it decrypts $Ticket_B$ by using the value K to obtain k, L, ID_A and ID_B. Then S_B checks whether L, ID_A and ID_B are valid. If so, S_B computes $R_B'' = W_B{}^y \bmod p$ and checks whether the received C_B equals $h_2(T_3 \| R_B'' \| W_B \| ID_B)$. If so, S_B believes *Bob* is authenticated. Then S_B computes $K_B = h_2(R_B'' \oplus T_4)$, $V_B = [k, ID_A, ID_B]_{K_B}$, where T_4 is the current time. Finally, S_B sends $\{V_B, T_4\}$ to *Bob*.

$S_A(x, K)$	$Alice(SC_A, PW_A)$	$Bob(SC_B, PW_B)$	$S_B(y, K)$

$$\text{Input } ID_A, PW_A$$

$$\text{Choose } \alpha \in Z_q^*$$

$$R'_A = (R_A - h_1(PW_A))^\alpha \bmod p$$

$$W_A = h_1(ID_A)^\alpha \bmod p$$

$$C_A = h_1(T_1 \| R'_A \| W_A \| ID_A)$$

$$\xrightarrow{\{ID_A, ID_B, T_1, C_A, W_A\}}$$

$$\text{Verify } T_1, ID_A$$

$$R''_A = W_A{}^x \bmod p$$

$$C_A \stackrel{?}{=} h_1(T_1 \| R''_A \| W_A \| ID_A)$$

$$\text{Choose } k \in Z_q^*$$

$$K_A = h_1(R''_A \oplus T_2)$$

$$V_A = [k, ID_A, ID_B]_{K_A}$$

$$Ticket_B = [k, ID_A, ID_B, L]_K$$

$$\xrightarrow{\{V_A, Ticket_B, T_2, L\}}$$

$$\text{Verify } T_2$$

$$K'_A = h_1(R'_A \oplus T_2)$$

$$\text{Decrypt } V_A$$

$$\text{Verify } ID_A, ID_B$$

$$\text{Choose } a \in Z_q^*$$

$$E_a = g^a \| MAC_k(g^a)$$

$$\xrightarrow{\{ID_A, E_a, Ticket_B\}}$$

$$\text{Input } ID_B, PW_B$$

$$\text{Choose } \beta \in Z_q^*$$

$$R'_B = (R_B - h_2(PW_B))^\beta \bmod p$$

$$W_B = h_2(ID_B)^\beta \bmod p$$

$$C_B = h_2(T_3 \| R'_B \| W_B \| ID_B)$$

$$\xrightarrow{\{Ticket_B, T_3, C_B, W_B\}}$$

$$\text{Verify } T_3$$

$$\text{Decrypt } Ticket_B$$

$$R''_B = W_B{}^y \bmod p$$

$$C_B \stackrel{?}{=} h_2(T_3 \| R''_B \| W_B \| ID_B)$$

$$K_B = h_2(R''_B \oplus T_4)$$

$$V_B = [k, ID_A, ID_B]_{K_B}$$

$$\xleftarrow{\{V_B, T_4\}}$$

$$\text{Check } T_4$$

$$K'_B = h_2(R'_B \oplus T_4)$$

$$\text{Decrypt } V_B$$

$$\text{Check } ID_A, ID_B$$

$$\text{Choose } b \in Z_q^*$$

$$E_b = g^b \| MAC_k(g^b)$$

$$\xleftarrow{\{E_b\}}$$

$$sk = h_3(ID_A \| ID_B \| g^a \| g^b \| g^{ab})$$

$$\xleftarrow{\hspace{1cm}} \text{- - - - - - - -} \xrightarrow{\hspace{1cm}}$$

Fig. 2. Login-And-Authentication Phase

(6) Upon receiving the message from S_B, *Bob* first checks whether T_4 is valid. If so, *Bob* computes $K_B' = h_2(R_B' \oplus T_4)$ and decrypts V_B to obtain ID_A, ID_B and k. He then checks whether ID_A and ID_B are both correct. If so, *Bob* randomly chooses $b \in Z_q^*$, computes $E_b = g^b \| MAC_k(g^b)$ and sends E_b to *Alice*.

(7) Upon receiving E_b, *Alice* checks the integrity of g^b. Finally, both *Alice* and *Bob* can compute the agreed session key $sk = h_3(ID_A \| ID_B \| g^a \| g^b \| g^{ab})$.

3.2 Performance Analysis of Our Protocol

In this subsection, we evaluate the performance of our C2C-PAKA-SC protocol in the login-and-authentication phase since the cost during the registration phase can be pre-computed *off-line*. With respect to the efficiency, we compare our C2C-PAKA-SC protocol with Byun *et al.*'s [7] and Feng *et al.*'s [10] in Table 1.

Table 1 shows that, for the clients, our C2C-PAKA-SC protocol introduces six modular exponentiations which could be pre-computed *off-line*, two extra modular exponentiations, and two symmetric decryption operations. For the servers, our C2C-PAKA-SC protocol only introduces two modular exponentiations, three symmetric encryption operations, and one symmetric decryption operation. Thus, the computation complexity of our C2C-PAKA-SC protocol is lower than those of [7] and [10].

Another advantage of our protocol is its low communication complexity. Our protocol takes only three rounds of message exchange (recall Fig. 2), while Byun *et al.*'s protocol [7] and Feng *et al.*'s protocol [10] both take five rounds of message exchange. Therefore, our protocol is more efficient than those proposed in [7] and [10].

Table 1. Comparisons of performance

		Our Protocol	[7]	[10]
Modular	clients	6 Pre + 2	4 Pre + 4	2 Pre + 5
exponentiation	servers	2	2 Pre + 2	3 Pre + 4
Symmetric	clients	N/A	2 Pre + 2	2 Pre + 1
encryption	servers	3	5	1 Pre + 1
Symmetric	clients	2	4	2
decryption	servers	1	5	3
Asymmetric	clients	N/A	N/A	N/A
operation	servers	N/A	N/A	6
Number of rounds		3	5	5
Note: "Pre" denotes pre-computed operation				

4   Security Analysis of Proposed Protocol

In this section, we introduce the formal security model for C2C-PAKA-SC protocols, and then present a security proof for our proposed protocol.

4.1   Security Model

In this subsection, we introduce a formal security model, which is mainly adopted from Bellare *et al.* [11]. In addition, we formally define the special security requirements for C2C-PAKA-SC protocols.

1. Communication model

We denote A, B as two clients belonging to two different realms. Client A shares her password PW_A and smart card SC_A with server S_A, and client B shares his password PW_B and smart card SC_B with another server S_B. S_A and S_B share the common key K which is pre-distributed between them by using a 2-party key exchange protocol. When client C enrolls in his server S, S stores R_C into a smart card and issues it to client C, where R_C is an (injective) transformation of the client's password PW_C and its server's secret key SK_S. In the end of the registration phase, S discards the information about the password of client C and the value R_C, only holds its own secret key SK_S. Additionally, all clients' passwords are chosen from the same small dictionary $\mathcal{D}$ whose distribution is D_{pw}. Each participant may has several instances involved in the execution of the protocol. We denote participant U's (maybe a client or a server) i-th instances as U^i. We denote the set of all clients as $\mathcal{C}$, and denote the set of all servers as $\mathcal{S}$. We denote the i-th instance of the protocol executed by entity $C(S)$ as C^i (S^i).

The C2C-PAKA-SC protocol is an interactive protocol among four participants' instances: A^i, B^j, S_A^s, S_B^t. During the execution of the protocol, an adversary $\mathscr{A}$ could interact with protocol participants via several oracle queries, which model adversary's possible attacks in the real execution. All possible oracle queries are listed in the following:

- *Execute*(A^i, B^j, S_A^s, S_B^t): This query returns transcripts of an honest execution between participants. This query models a passive adversary who simply eavesdrops on an execution of the C2C-PAKA-SC protocol.
- *Reveal*(C^i): This query models the possibility that an adversary gets session keys. After querying the oracle, the client instance C^i's session key is returned to the adversary.
- *Corrupt*(C^i, a): This query models the possibility that the adversary corrupts a client C^i. In our adversary model, we assume that the adversary $\mathscr{A}$ can get either the password or the smart card belonging to the same client, but not both of them. We define this property as follows:
    - ·If $a=1$, it outputs the password PW_C of C.
    - ·If $a=2$, it outputs information stored in the smart card.

- *SendClient*(C^i, m): This oracle query is used to simulate active attacks against the client. After querying the oracle, a message m is sent to the client instance C^i. At the end, client instance C^i's response is forwarded to the adversary.
- *SendServer*(S^i, m): This oracle query is used to simulate active attacks against the server. After querying the oracle, a message m is sent to the server instance S^i. At the end, server instance S^i's response is forwarded to the adversary.
- *Test*(C^i): This oracle query is not used to simulate adversary's attacks, but to define session key's semantic security. After querying the oracle, the session key of C^i or a random number will be returned according to a predefined random bit b. If $b=1$, the adversary would learn the session key of C^i; otherwise the adversary only learns a random number with the same length. This query can be called only once.

Partnering: As other formal models [7,8,11], The definition of partnering uses the session identification (*sid*). More specifically, two instances C^i and C^j are said to be partners if the following conditions are satisfied: (1) Both C^i and C^j accept; (2) Both C^i and C^j own the same sid; (3) C^i is C^j's partner and vice-verse; and (4) No instance other than C^i and C^j accepts with a partner identity equal to C^i and C^j.

Freshness: We say an instance C^i is *fresh* if the following conditions hold: (1) It has accepted and generated a valid session key; (2) No *Reveal* queries have been made to C^i or its partner; (3) Strictly less than 2 *Corrupt* queries have been made to C^i and its partner.

2. Security definition

A secure *cross-realm* C2C-PAKA-SC protocol should satisfy four security requirements: (1) the session key cannot be distinguished from a random number by an outside malicious adversary; (2) Any server does not know the session key between two clients; (3) Even if one client A's smart card is obtained by the other client B, B cannot learn A's password; and (4) Clients' passwords are not revealed to other servers except for their own servers.

Semantic Security Against A Malicious Outside Adversary. In the *Test* query, we require the adversary cannot tell whether the response received from the *Test* oracle is the session key or a random number. In other words, the adversary cannot guess the random bit b used in the *Test* query with non-negligible probability larger than $1/2$.

We say the adversary succeeds if he correctly guesses the value of b. Let $Succ^{ake}$ denote the event that the malicious outside adversary succeeds. Let $\mathcal{D}$ be client's password dictionary. For any adversary $\mathcal{A}$, we define his advantage $Adv_{\mathcal{D}}^{ake}(\mathcal{A})$ as

$$Adv_{\mathcal{D}}^{ake}(\mathcal{A}) = 2Pr[Succ^{ake}] - 1$$

$$Adv_{\mathcal{D}}^{ake}(t, R) = max\{Adv_{\mathcal{D}}^{ake}(\mathcal{A})\}$$

where the maximum is over all adversaries with time complexity at most t and using at most R times oracle queries.

We say our C2C-PAKA-SC protocol is semantically secure against a malicious outside adversary if the advantage $Adv_{\mathcal{D}}^{ake}(t, R)$ is only negligibly larger than $O(q_s)/|D|$, where q_s is the number of active sessions, and $|D|$ is the size of the password dictionary.

Key Privacy Against A Passive Server. We require that no information about the session key is revealed to servers. In the login-and-authentication phase, an active server can easily use its secret key SK_S to elicit an authenticated value which is equal to its client's value computed by using the password and the value R_C stored in smart card and could query the *Execute, SendClient, SendServer* and *Test* oracles, so it is always able to impersonate one of its clients and exchange a session key with another client by active attacks. As a result, we cannot require an active server cannot learn the session key.

The passive server S could query only two oracles: *Execute* and *Test*. We say S succeeds if it correctly guesses the value of the random bit b used in the *Test* query. Let $Succ^{kp}$ denote the event that the passive server succeeds. Let $\mathcal{D}$ be client's password dictionary. For any passive server $S \in \mathcal{S}$, we define its advantage $Adv_{\mathcal{D}}^{kp}(S)$ as

$$Adv_{\mathcal{D}}^{kp}(S) = 2Pr[Succ^{kp}] - 1$$

$$Adv_{\mathcal{D}}^{kp}(t, R) = max\{Adv_{\mathcal{D}}^{kp}(S)\}$$

where the maximum is over all adversaries with time complexity at most t and querying oracles at most R times. We say the C2C-PAKA-SC protocol is key private against a passive server if the advantage $Adv_{\mathcal{D}}^{kp}(t, R)$ is negligible.

Password Protection Against A Malicious Client. The *Test* oracle query is used to define the session key's security, which is not considered in current security notion. So the malicious client C does not have access to the *Test* query. Let $Succ^{pw-mc}(C)$ be the event that the malicious client C can successfully learn the honest client's password. The advantage of a malicious client $C \in \mathcal{C}$ are defined to be $Adv_{\mathcal{D}}^{pw-mc}(C)$ as

$$Adv_{\mathcal{D}}^{pw-mc}(C) = Pr[Succ^{pw-mc}(C)]$$

$$Adv_{\mathcal{D}}^{pw-mc}(t, R) = max\{Adv_{\mathcal{D}}^{pw-mc}(C)\}$$

We say our C2C-PAKA-SC protocol satisfies password protection against a malicious client if the advantage $Adv_{\mathcal{D}}^{pw-mc}(t, R)$ is only negligibly larger than $O(q_s)/|D|$, where q_s is the number of active sessions, and $|D|$ is the size of the password dictionary.

Password Protection Against A Malicious Server. Like the notion of password protection against a malicious client, for any malicious server $S \in \mathcal{S}$, we define it advantage $Adv_{\mathcal{D}}^{pw-ms}(S)$ as

$$Adv_{\mathcal{D}}^{pw-ms}(S) = Pr[Succ^{pw-ms}(S)]$$

$$Adv_{\mathcal{D}}^{pw-ms}(t, R) = max\{Adv_{\mathcal{D}}^{pw-ms}(S)\}$$

We say our C2C-PAKA-SC protocol satisfies password protection against a malicious server if the advantage $Adv_{\mathcal{D}}^{pw-ms}(t, R)$ is only negligibly larger than $O(q_s)/|D|$, where q_s is the number of active sessions, and $|D|$ is the size of the password dictionary.

3. Computational assumptions and cryptographic primitives

We adopt some computational assumptions and cryptographic primitives required in our *cross-realm* C2C-PAKA-SC protocol, some of them are similar to those proposed in [7]. So we introduce them in brief. Let $\mathbb{G}$ be a finite cyclic group of prime order p generated by an element g.

Discrete Logarithm (DL) Assumption. We denote by $Adv_{\mathbb{G}}^{dl}(\mathcal{A})$ the probability that $\mathcal{A}$ succeeds in computing x from (g, g^x) and have $Adv_{G}^{dl}(T_{dl}) = \max\{Adv_{\mathbb{G}}^{dl}(\mathcal{A})\}$, where the maximum is over all the adversaries $\mathcal{A}$ running in time at most T_{dl}. The *DL* assumption is that the value of $Adv_{G}^{dl}(T_{dl})$ is negligible.

Decisional Diffie-Hellman (DDH) Assumption. *DDH* assumption means that given (g, g^x, g^y), no probabilistic polynomial time algorithm can distinguish g^{xy} from a random element of $\mathbb{G}$ with non-negligible probability. We define $Adv_{\mathbb{G}}^{ddh}(\mathcal{A})$ as the probability that the adversary $\mathcal{A}$ could distinguish g^{xy} from a random element of $\mathbb{G}$, and define $Adv_{\mathbb{G}}^{ddh}(T_{ddh})$ as the maximum value of $Adv_{\mathbb{G}}^{ddh}(\mathcal{A})$ over all $\mathcal{A}$ with time complexity at most T_{ddh}.

Secure Message Authentication Code Under Chosen Message Attack. A message authentication code $MAC = (Key, Tag, Ver)$ is composed of a MAC key generation algorithm Key, a MAC generation algorithm Tag and a MAC verification algorithm Ver. A secure MAC should prevent existential forgeries under chosen-message attacks (CMA) if adversaries have access to the *generation* and *verification* oracles. The maximal value $Adv_{MAC}^{cma}(T_{mac})$ of the advantage $Adv_{MAC}^{cma}(\mathcal{A})$ with at most T_{mac} time complexity and at most q_t and q_v queries to its MAC *generation* and *verification* oracles, respectively, is a negligible function of the parameters above.

Secure Symmetric Encryption under Chosen Cipher Attack. A scheme symmetric scheme $SE = (E, D)$ is composed of an encryption algorithm E and a decryption algorithm D. We imagine an adversary $\mathcal{A}_{se}$ that runs in two stages. In the Find stage, the adversary endeavors to come up with a pair of equal-length plaintexts (x_0, x_1), whose encryption the adversary wants to tell apart. In the Guess stage, the adversary is given a random ciphertext y for one of x_0 and x_1. The adversary wins if it correctly identifies which plaintext goes with y. The maximal value $Adv_{SE}^{cca}(T_{se})$ of the advantage $Adv_{SE}^{cca}(\mathcal{A})$ with at most T_{se} time complexity and at most q_e and q_d queries to the *encryption* and *decryption* oracles, respectively, is a negligible function of the parameters above.

4.2 Security Proof

Theorem 4.1. *Let $\mathcal{A}$ be a probabilistic polynomial time adversary against semantic security of the proposed C2C-PAKA-SC protocol P within a time bound*

t. $\mathscr{A}$ can make q_{send} Send queries, q_{exe} Execute queries, q_E encryption queries for ideal cipher E, q_e encryption queries for symmetric encryption E, q_t tag queries, and q_v verification queries. Then

$$Adv_D^{ake}(t,R) \le \frac{q_E^2 + q_{h_1}^2 + q_{h_2}^2 + q_{h_3}^2}{(q-1)} + 4Adv_{\mathbb{G}}^{dl}(T_{dl}) + 6Adv_{SE}^{cca}(T_{se}, q_e, q_d)$$
$$+ 4Adv_{MAC}^{cma}(T_{mac}, q_t, q_v) + 2Adv_{\mathbb{G}}^{ddh}(T_{ddh}) + \frac{q_{send}}{|\mathcal{D}|} \tag{1}$$

where $|\mathcal{D}|$ is the size of the password space, q is a prime order of a cyclic group $\mathbb{G}$, $T_{se} = T_{dl} = T_{ddh} \le t + q_{send}(\tau_G + \tau_E)$, $T_{mac} \le t + \tau_{\mathscr{I}}(q_t + q_v)$, and τ_G, τ_E, $\tau_{\mathscr{I}}$ are computational time for an exponentiation, ideal encryption E, message authentication code, respectively.

Proof. Our proof defines a sequence of hybrid experiments, starting with the real attack and ending in an experiment in which the adversary has no advantage. Each experiment addresses a different security aspect. The detailed proof of Theorem 4.1 can be found in the Appendix A.

Theorem 4.2. *In our cross-realm C2C-PAKA-SC protocol, a passive server cannot learn the session key between two clients as long as the DDH assumption holds in the group $\mathbb{G}$. Formally,*

$$Adv_D^{kp}(t,R) \le 2Adv_{\mathbb{G}}^{ddh}(T_{ddh}) \tag{2}$$

where $T_{ddh} \le t + q_{exe}\tau_G$ and τ_G is computational time for an exponentiation.

Proof. See Appendix B.

Theorem 4.3. *In our cross-realm C2C-PAKA-SC scheme, the malicious client B cannot learn the client A's password as long as the DL assumption holds in the group $\mathbb{G}$. Formally,*

$$Adv_D^{pw-mc}(t,R) \le \frac{q_{send}}{|\mathcal{D}|} + 2Adv_{\mathbb{G}}^{dl}(T_{dl}) \tag{3}$$

where $|\mathcal{D}|$ is the size of the password space, $T_{dl} \le t + (q_{send} + q_{exe})\tau_G$ and τ_G is computational time for an exponentiation.

Proof. See Appendix C.

Theorem 4.4. *In our cross-realm C2C-PAKA-SC scheme, the malicious server S_B cannot learn the client A's password as long as the DL assumption holds in the group $\mathbb{G}$. Formally,*

$$Adv_D^{pw-ms}(t,R) \le \frac{q_{send}}{|\mathcal{D}|} + 2Adv_{\mathbb{G}}^{dl}(T_{dl}) \tag{4}$$

where $|\mathcal{D}|$ is the size of the password space, $T_{dl} \le t + (q_{send} + q_{exe})\tau_G$ and τ_G is computational time for an exponentiation.

Proof. The proof is similar to that of Theorem 4.3, omitted.

5   Conclusion

In this paper, we proposed an efficient and provably secure C2C-PAKA-SC protocol, which simultaneously utilizes passwords and smart cards in the *cross-realm* setting. We showed that in the *private-key* setting, our proposed C2C-PAKA-SC protocol is the foremost protocol which can resist all types of common attacks, especially *password-compromise impersonation* attacks. Moveover, our protocol is more efficient because many operations can be pre-computed *off-line*. To achieve a provable security for our proposed protocol, we defined a security model which is more appropriate to C2C-PAKA-SC protocols and provided a formal security proof in detail.

Acknowledgement

This work was supported by the National Grand Fundamental Research (973) Program of China under Grant No. 2007CB311202 and the National Natural Science Foundation of China (NSFC) under Grants No.60673083 and No.60873197.

References

1. Byun, J.W., Jeong, I.R., Lee, D.H., Park, C.: Password-authenticated key exchange between clients with different passwords. In: Deng, R.H., Qing, S., Bao, F., Zhou, J. (eds.) ICICS 2002. LNCS, vol. 2513, pp. 134–146. Springer, Heidelberg (2002)
2. Chen, L.: A weakness of the password-authenticated key agreement between clients with different passwords scheme, ISO/IEC JTC 1/SC27 N3716
3. Wang, S., Wang, J., Xu, M.: Weakness of a password-authenticated key exchange protocol between clients with different passwords. In: Jakobsson, M., Yung, M., Zhou, J. (eds.) ACNS 2004. LNCS, vol. 3089, pp. 414–425. Springer, Heidelberg (2004)
4. Kim, J., Kim, S., Kwak, J., Won, D.: Cryptoanalysis and improvements of password authenticated key exchange scheme between clients with different passwords. In: Laganá, A., Gavrilova, M.L., Kumar, V., Mun, Y., Tan, C.J.K., Gervasi, O. (eds.) ICCSA 2004. LNCS, vol. 3044, pp. 895–902. Springer, Heidelberg (2004)
5. Phan, R.C.-W., Goi, B.-M.: Cryptanalysis of an improved client-to-client password-authenticated key exchange (C2C-PAKE) scheme. In: Ioannidis, J., Keromytis, A.D., Yung, M. (eds.) ACNS 2005. LNCS, vol. 3531, pp. 33–39. Springer, Heidelberg (2005)
6. Yoon, E.J., Yoo, K.Y.: A secure password-authenticated key exchange between clients with different passwords. In: Shen, H.T., Li, J., Li, M., Ni, J., Wang, W. (eds.) APWeb Workshops 2006. LNCS, vol. 3842, pp. 659–663. Springer, Heidelberg (2006)
7. Byun, J.W., Lee, D.H., Lim, J.I.: EC2C-PAKA: An efficient client-to-client passwordauthenticated key agreement. Information Science 177, 3995–4013 (2007)
8. Yin, Y., Bao, L.: Secure cross-realm C2C-PAKE protocol. In: Batten, L.M., Safavi-Naini, R. (eds.) ACISP 2006. LNCS, vol. 4058, pp. 395–406. Springer, Heidelberg (2006)

9. Phan, R.C.-W., Goi, B.-M.: Cryptanalysis of two provably secure cross-realm C2C-PAKE protocols. In: Barua, R., Lange, T. (eds.) INDOCRYPT 2006. LNCS, vol. 4329, pp. 104–117. Springer, Heidelberg (2006)
10. Feng, D.G., Xu, J.: A New Client-to-Client Password-Authenticated Key Agreement Protocol. In: Chee, Y.M., Li, C., Ling, S., Wang, H., Xing, C. (eds.) Proc. IWCC 2009. LNCS, vol. 5557, pp. 63–76. Springer, Heidelberg (2009)
11. Bellare, M., Pointcheval, D., Rogaway, P.: Authenticated key exchange secure against dictionary attacks. In: Preneel, B. (ed.) EUROCRYPT 2000. LNCS, vol. 1807, pp. 139–155. Springer, Heidelberg (2000)
12. Abdalla, M., Pointcheval, D.: Interactive diffie-hellman assumptions with applications to password-based authentication. In: S. Patrick, A., Yung, M. (eds.) FC 2005. LNCS, vol. 3570, pp. 341–356. Springer, Heidelberg (2005)
13. Abdalla, M., Fouque, P., Pointcheval, D.: Password-based authenticated key exchange in the three-party setting. In: Vaudenay, S. (ed.) PKC 2005. LNCS, vol. 3386, pp. 65–84. Springer, Heidelberg (2005)
14. Lin, C., Sun, H., Steiner, M., Hwang, T.: Three-party encrypted key exchange without server public-keys. IEEE Communications Letters 5(12), 497–499 (2001)
15. Yang, G., Wong, D.S., Wang, H., Deng, X.: Two-factor mutual authentication based on smart cards and passwords. J. Comput. System Sci. 74, 1160–1172 (2008)

Appendix A: Security Proof of Theorem 4.1

Our proof defines a sequence of hybrid experiments, starting with the real attack and ending in an experiment in which the adversary has no advantage. For each experiment Exp_n, we define an event $Succ_n$ corresponding to the case in which the adversary correctly guesses the bit b involved in the $Test$ query. By using each difference of probability, we finally get the result of Theorem 4.1. Several experiments are similar to Byun $et\ al.$'s implement[7], so we explain them briefly.

Experiment Exp_0. This experiment corresponds to the real attack in the random oracle model. By definition, we have

$$Adv_{\mathcal{D}}^{ake}(t, R) = 2Pr[Succ_0] - 1 \qquad (5)$$

Experiment Exp_1. In this experiment, we simulate $\mathcal{H}_1$, $\mathcal{H}_2$, $\mathcal{H}_3$, $\mathcal{E}$ and $\mathcal{D}$ oracles. It is necessary to maintain $\mathcal{H}_1$, $\mathcal{H}_2$, $\mathcal{H}_3$, $\mathcal{E}$ and $\mathcal{D}$ tables (denoted by $T_{\mathcal{H}_1}$, $T_{\mathcal{H}_2}$, $T_{\mathcal{H}_3}$, $T_{\mathcal{E}}$ and $T_{\mathcal{D}}$, respectively.) during the simulation of the above oracles. We also simulate all instances of players for the $Send$, $Reveal$, and $Execute$ queries. The deductive process is similar to [7], the result is :

$$|Pr[Succ_0] - Pr[Succ_1]| \leq \frac{q_E^2 + q_{h_1}^2 + q_{h_2}^2 + q_{h_3}^2}{2(q - 1)} \qquad (6)$$

Experiment Exp_2. In this experiment, we replace a real value W_A with a random value W_A'. The other environments are identical to the previous experiment. By using $\mathscr{A}$'s advantage with respect to a session key in the $Exp1$ and $Exp2$, we construct a polynomial time algorithm $\mathscr{A}_{dl}$ to break the DL assumption. And we can perform it again when replacing the real W_B with a random value W_B'. Obviously, we can get

$$|Pr[Succ_1] - Pr[Succ_2]| = 2Adv_{\mathbb{G}}^{dl}(T_{dl}) \qquad (7)$$

Experiment Exp_3. In this experiment, we replace a real key k in the $Ticket_B$ with a random key r_k. The other environments are identical to the previous experiment. By exploiting $\mathscr{A}$ which tries to get the advantage of a session key in the $Exp2$ and $Exp3$, we can construct a polynomial time algorithm $\mathscr{A}_{se}$ to break the security of the symmetric encryption. This algorithm is designed by the similar method of [7]. Because there are three symmetric encryptions, so the formula is

$$|Pr[Succ_2] - Pr[Succ_3]| = 3Adv_{SE}^{cca}(T_{se}, q_e, q_d) \qquad (8)$$

Experiment Exp_4. The goal of this experiment is to construct an algorithm for the MAC adversary by using $\mathscr{A}$. This experiment is completely same to [7]. There is

$$|Pr[Succ_3] - Pr[Succ_4]| \leq 2Adv_{MAC}^{cma}(T_{mac}, q_t, q_v) \qquad (9)$$

Experiment Exp_5. In this experiment, we consider a random DDH triple (U, V, Z) where $U = g^u$, $V = g^v$, and $Z = g^r$. The random DDH triple is injected into the protocol, then the triple is used for generating a target session key. This experiment is perfectly identical to [7]. The aim is to construct a polynomial time algorithm $\mathscr{A}_{ddh}$ to break the DDH assumption. We get

$$|Pr[Succ_4] - Pr[Succ_5]| = Adv_{\mathbb{G}}^{ddh}(T_{ddh}) \qquad (10)$$

In [7], there is rigorous analysis about the successful probability of the Exp_5, denoted by $Adv_{Exp_5}^{sk}(\mathscr{A})$. In our C2C-PAKA-SC protocol, whether the $Corrupt(C^i, 2)$ query has been made or not, the *on-line password guessing* attack is unavoidable. The advantage $Adv_{Exp_5}^{sk}(\mathscr{A})$ is bounded by $q_{send}/|\mathcal{D}|$ where q_{send} is the maximum number of $Send$ queries. Therefore, We have

$$Pr[Succ_5] = \frac{1}{2}Adv_{Exp_5}^{sk}(\mathscr{A}) + \frac{1}{2} \leq \frac{q_{send}}{2|\mathcal{D}|} + \frac{1}{2} \qquad (11)$$

Consequently from (5) to (11), we come to a conclusion in Theorem 4.1.

Appendix B: Security Proof of Theorem 4.2

A passive server only has access to the $Execute$ and $Test$ oracles, so the proof is similar to the Exp_5 in the proof of semantic security of C2C-PAKA-SC. Let $\mathcal{S}_A$ be an adversary against the key privacy of C2C-PAKA-SC whose time complexity is at most t, we would construct another adversary $\mathscr{A}_{ddh}$ for DDH assumption using $\mathcal{S}_A$. $\mathscr{A}_{ddh}$ runs the adversary $\mathcal{S}_A$ in this environment and provides answers for all oracle queries with $\mathscr{A}_{ddh}$'s input and parameters. To deal with the security of the key privacy respect to a server, we only consider the last flow of C2C-PAKA-SC.

1. For the *Execute* query, $\mathscr{A}_{ddh}$ finally get (g, g^a, g^b) and computes the session key as described in the protocol.
2. For the *Test* query, $\mathscr{A}_{ddh}$ responses the session key to $\mathcal{S}_A$ when $b = 1$, or responses a random number to $\mathcal{S}_A$ when $b = 0$.

After all interaction, $\mathscr{A}_{ddh}$ sets his answer as the answer of the adversary $\mathcal{S}_A$. Now we analyze $\mathscr{A}_{ddh}$'s advantage. If $z = g^{ab}$, the simulation of $\mathscr{A}_{ddh}$ is perfect. Hence, The probability that $\mathscr{A}_{ddh}$ outputs 1 is exactly $1/2 + 1/2 \cdot Adv_{\mathbb{G}}^{kp}(\mathcal{S}_A)$. If z is a random number, the session key computed is a random number. Hence, the probability that $\mathscr{A}_{ddh}$ outputs 1 is exactly $1/2$. So

$$Adv_{\mathbb{G}}^{ddh}(\mathscr{A}) = \frac{1}{2} + \frac{1}{2} Adv_{\mathcal{D}}^{kp}(\mathcal{S}_A) - \frac{1}{2} \tag{12}$$

Therefore, we get the result of Theorem 4.2.

Appendix C: Security Proof of Theorem 4.3

We consider the following two cases: (1) the malicious client B hasn't made the $Corrupt(A^i, 2)$ query. In other word, he didn't obtain the honest client A's any information in her smart card; (2) the malicious client B has already made the $Corrupt(A^i, 2)$ query, that means he obtained A's information in her smart card. Obviously, in the case (2), the malicious client B has obtained A's smart card and some transcripts generated in the communicational process, if he still cannot learns A's password by using these useful information, he must not do it in the case (1). So it is advisable for us to only consider the case (2), we suppose the malicious client B has already obtained A's smart card and wants to learn A's password.

Firstly, the malicious client B makes the *Corrupt* $(A^i, 2)$ query to honest A to get the information in her smart card. Secondly, the client B makes the *Send* or *Execute* query and learns R_A, W_A, C_A with respect to PW_A. Obviously, if B learns the value of α, he can guess PW_A by using $C_A = h_1(T_1 \| (R_A - h_1(PW_A))^\alpha \| W_A \| ID_A)$. The maximal probability of obtaining α from W_A is $Adv_{G}^{dl}(T_{dl})$. Subsequently, we get

$$Adv_{\mathcal{D}}^{pw-mc}(B) = Adv_{\mathbb{G}}^{dl}(\mathscr{A}) \tag{13}$$

In addition, because the *on-line password guessing* attack is unavoidable, this advantage of $Adv_{\mathcal{D}}^{pw-mc}(B)$ is bounded by $q_{send}/|\mathcal{D}|$ where q_{send} is the maximum number of *Send* queries. Hence, Theorem 4.3 is concluded.

Ensuring Authentication of Digital Information Using Cryptographic Accumulators

Christophe Tartary

Institute for Theoretical Computer Science
Tsinghua University
Beijing, 100084
People's Republic of China
ctartary@mail.tsinghua.edu.cn

Abstract. In this paper, we study the broadcast authentication problem for both erasure and adversarial networks. Two important concerns for authentication protocols are the authentication delay and the packet overhead. In this paper, we address those points by proposing two schemes based on cryptographic accumulators. Our first scheme is developed for erasure channels and its packet overhead is less than the length of a digest most of the time. This makes our construction one of the least expensive protocols for this network model. Even if the sender processes the stream slightly in delay, the receivers can authenticate packets on-the-fly. Our second scheme is designed for adversarial networks. We show that our packet overhead is less than for the construction by Karlof et al. in 2004 and the protocol by Tartary and Wang in 2006 which are two recent efficient schemes dealing with adversarial networks.

Keywords: Stream Authentication, Polynomial Reconstruction, Erasure Channel, Adversarial Channel, Cryptographic Accumulator.

1 Introduction

In this early XXI century, communication networks have expended to such an extent that most human beings are daily connected to them. They are used for many applications such as video-conferences, pay-TV and air traffic control to name a few. A generalized way to distribute information through these networks is broadcasting. However, large-scale broadcasts have the drawback that lost content cannot be retransmitted as the size of the communication group would imply that a single deletion could lead to an overwhelming number of redistribution requests at the sender end. Furthermore, the communication network can be under the influence of malicious users altering the data stream[1]. As a consequence, the security of a broadcasting protocol depends on the properties of the communication network as well as the computational power of the adversaries. In this work, we present authentication protocols secure against computationally bounded opponents.

---

[1] In broadcasting, the sequence of information sent into the network is called *stream*.

J.A. Garay, A. Miyaji, and A. Otsuka (Eds.): CANS 2009, LNCS 5888, pp. 315–333, 2009.

The goal of streaming is to distribute continuous data such as stock market information. Therefore, the digital content obtained at the receiver end must be authenticated within a short period of delay upon reception. Moreover, many applications transmit private or sensitive information. Thus, non-repudiation of the stream source needs to be provided.

Network bandwidth availability and computational power of end-users are two primary concerns for a stream authentication protocol. Indeed, large packets may create a congestion of the network information flow while receivers with small computational resources will need more time to authenticate data delaying the stream play. Thus, when designing a protocol for stream authentication, one should aim at minimizing both the packet [2] overhead and the computational cost of authenticating information.

The multicast stream authentication problem has been widely studied [6]. Non-repudiation of the sender is provided using a digital signature. However, signing each data packet is not a practical solution as such a cryptographic primitive is generally expensive to generate and/or verify. Thus, a usual approach consists of generating a single signature and amortizing its communication and computation overheads over several packets using hash functions for instance.

In order to deal with erasures, Perrig et al. [26, 27], Challal et al. [7], Golle and Modadugu [10] as well as Miner and Staddon [18] appended the hash of each packet to several followers according to specific patterns. They all modeled the packet loss behavior of the network by k-state Markov chains [9] and they obtained bounds on the packet authentication probability. Nevertheless, the drawbacks of these schemes are twofold. First, they are degrading[3] as some received data packets may not be authenticated. Second, they rely on the reception of signed packets which cannot be guaranteed over networks such as the Internet where the User Datagram Protocol only provides a best effort delivery of information. These two issues restrict the range of applications for the previous protocols.

To overcome the issue of signature reliable delivery, a common approach is to split the signature into k smaller parts where only ℓ of them $(\ell < k)$ are sufficient to recover it. Signature dispersion can be achieved via various techniques: Park et al. [23, 24] as well as Park and Cho [25] used the Information Dispersal Algorithm [28], Al-Ibrahim and Pieprzyk [1] combined linear equations and polynomial interpolation, Pannetrat and Molva [22] utilized erasure codes whereas Desmedt and Jakimoski [8] employed cover-free families [30]. It should be noticed that each of those authentication schemes is non-degrading as well.

The major shortcoming of the previous constructions is that none of them tolerates a single packet injection. This is a central problem when data is distributed over large public networks since it is likely to have some unreliable nodes.

Using an algorithm developed by Guruswami and Sudan called Poly-Reconstruct to solve the polynomial reconstruction problem [11], Lysyanskaya et al. [14] constructed a non-degrading authentication protocol exhibiting $O(1)$ signature verification queries

---

[2] Since the stream size is large, it is divided into small fixed-size entities called *packets*.

[3] An authentication scheme is said to be *non-degrading* if every receiver can authenticate all the data packets he obtained. Otherwise, the scheme is said to be *degrading*.

per block[4] as a function of the block length n. Their construction was extended by Tartary and Wang [32] who used a Maximal Distance Separable (MDS) code to allow total recovery of all n data packets. In this paper, we denote this latter construction as TWMDS. The augmented packets[5] of TWMDS are $\Omega(\log_2 n)$-bit long as the underlying field used for polynomial operations must have at least n distinct points. Note that the same situation occurs in [14].

Another approach was followed by Karlof et al. in [12] when designing PRABS. This protocol combines an erasure code and an accumulator [4] based on a Merkle hash tree [17] to deal with injections. As TWMDS, PRABS only requires $O(1)$ signature verification queries per block. However, its packet overhead is $\Theta(\log_2(n))$-bit long as each augmented packet carries $\lceil \log_2(n) \rceil$ hashes. Nonetheless, the implementations done in [31] tend to infer that, for practical use, PRABS' overhead is larger than TWMDS'.

There exist several cryptographic accumulators. The advantage of using a construction based on hash functions is that aggregation and membership verification are fast contrary to [4, 20]. In [4], checking whether an element was accumulated costs as much as verifying a RSA signature whereas, in [20], it requires two pairing evaluations which is even slower [2, 5].

Nyberg's probabilistic accumulator is also based on hash functions [21]. Recently, Yum et al. proposed an improvement allowing to reduce the probability of false membership [35]. In this paper, we present two non-degrading authentication protocols based on this new accumulator, MDS codes and Poly-Reconstruct. Our first scheme is developed for erasure channels. Its overhead is smaller than [22, 23, 24, 25] and it allows each receiver to process information on-the-fly after a short part of the stream has been received. In particular, immediate data authentication can be achieved. Our second protocol is designed for adversarial networks as TWMDS and PRABS. It allows complete recovery of the data stream as TWMDS and we show on implementations that its overhead is smaller than TWMDS' and PRABS' in many situations. Another point worth noting is that our implementations also reinforce the intuition that TWMDS has smaller overhead than PRABS which has only been studied on a particular case so far ($n = 1000$) [31].

This paper is organized as follows. In the next section, we present the mathematical tools needed for the understanding of this paper. In particular, we recall the accumulator construction from [35] which plays a central role in our work. In Sect. 3, we present our authentication protocol for erasure channels. Our scheme for adversarial networks is studied in Sect. 4. The last section summarizes our contributions to the broadcast authentication problem.

2   Preliminaries

In this section, we present the network models and erasure correcting codes used in this paper. We also quote the polynomial reconstruction problem which plays an

---

[4] In order to be processed, packets are gathered into fixed-size sets called *blocks*.

[5] We call *augmented packets* the elements sent into the network. They generally consist of the original data packets with some redundancy used to prove the authenticity of the element.

important role for our authentication scheme over adversarial channels. Finally, we recall the cryptographic accumulator construction developed in [35].

2.1 Network Models

We consider that the communication network is under the control of an opponent $\mathcal{O}$.

Erasure Channels. In this model, $\mathcal{O}$ is simply an eavesdropper. Therefore, no injections of malicious packets occur. In other words, any packet collected by the receiver is authentic. We can assume that both sender and receivers have a buffering capacity of n consecutive packets and that at most t packets can be erased over a scope of n elements. This model generalizes the concept of bursts where, in the bursty model, the length of the longest burst occurring in the network is $t = n - 2$ (one packet must been receive on each side of the burst). An illustration is given as Fig. 1.

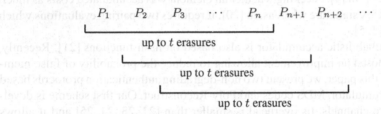

Fig. 1. Erasure Channel Model for Streaming

It should be noted that the bursty model has been used to analyze many authentication protocols [10, 18, 26, 27]. This is justified by the work of Yajnik et al. [34] who exhibited that the loss pattern of the Internet was bursty in nature. Notice that our model also encompasses [1, 8] as it does not require the t erasures to appear as a burst.

Adversarial Channels. In this case, $\mathcal{O}$ who can drop and rearrange packets of his choice as well as inject bogus data into the network [16]. Without loss of generality, we can assume that a reasonable number of original augmented packets reaches the receivers and not too many incorrect elements are injected by $\mathcal{O}$. We split the data stream into blocks of n packets: $P_1, \ldots, P_n$. In this settings, we introduced two parameters: $\alpha\,(0 < \alpha \leq 1)$ (the *survival* rate) and $\beta\,(\beta \geq 1)$ (the *flood* rate). It is assumed that at least a fraction α and no more than a multiple β of the number of augmented packets are received. This means that at least $\lceil \alpha n \rceil$ original augmented packets are received amongst a total which does not exceed $\lfloor \beta n \rfloor$ elements. The use of these two parameters to model $\mathcal{O}$ first appeared in [14] and was subsequently used in [32].

2.2 Correction of Deletions

Since the communication network is a priori unreliable, it is likely that some packets do not reach all the receivers. As in [32], we will use a linear correcting code to overcome this issue. A linear code of length N, dimension K and minimum distance D is denoted $[N, K, D]$.

Theorem 1 ([15]). *Any* $[N, K, D]$ *code satisfies:* $D - 1 \leq N - K$.

Since any $[N, K, D]$ code can correct up to $D - 1$ erasures [36], such a code can correct at most $N - K$ erasures. To maximize the efficiency of our protocols, we are interested in codes correcting exactly $N - K$ erasures. These codes are called *Maximum Distance Separable* (MDS) codes [15]. TWMDS is also based on this family of codes.

2.3 Reconstructing Polynomials

The Polynomial Reconstruction Problem (PRP) is the following mathematical problem.

Polynomial Reconstruction Problem
Input: Integers D, T and N points $\{(x_i, y_i)\}_{i \in \{1,...,N\}}$ where $x_i, y_i \in F$ for a field F.
Output: All univariate polynomials $P(X) \in F[X]$ of degree at most D such that $y_i = P(x_i)$ for at least T values of $i \in \{1, \ldots, N\}$.

Guruswami and Sudan developed an algorithm called Poly-Reconstruct to solve the PRP [11]. We modify it as in [32] where that new version was denoted MPR. Let $\mathbb{F}_{2^q}$ be the field of the polynomial coefficients. Every element of $\mathbb{F}_{2^q}$ can be represented as a polynomial of degree at most $q - 1$ over $\mathbb{F}_2$. Operations in $\mathbb{F}_{2^q}$ are performed modulo an irreducible polynomial $Q(X)$ over $\mathbb{F}_2$ having degree q [13].

Algorithm 1. MPR

Input: The maximal degree K of the polynomial $Q(X)$, the minimal number N of agreeable points, T points $\{(x_i, y_i), 1 \leq i \leq T\}$ and the polynomial $Q(X)$ of degree q.
1. If there are no more than $\sqrt{KN}$ distinct points then the algorithm stops.
2. Using $Q(X)$, run Poly-Reconstruct on the T points to get the list of all polynomials of degree at most K over $\mathbb{F}_{2^q}$ passing through at least N of the points.
3. Given the list $\{L_1(X), \ldots, L_\mu(X)\}$ obtained at Step 2. For each polynomial $L_i(X) := \mathcal{L}_{i,0} + \ldots + \mathcal{L}_{i,K} X^K$ where $\forall i \in \{0, \ldots, K\} \mathcal{L}_{i,j} \in \mathbb{F}_{2^q}$, form the elements: $\mathcal{L}_i := \mathcal{L}_{i,0} \| \cdots \| \mathcal{L}_{i,K}$.
Output: $\{\mathcal{L}_1, \ldots, \mathcal{L}_\mu\}$: list of candidates.

2.4 Cryptographic Accumulators

In [35], Yum et al. proposed a modified version of Nyberg's cryptographic accumulator [21]. A list $\{x_1, \ldots, x_m\}$ is aggregated into an accumulated value $\mathcal{A}$ using Algorithm 2.

One verifies the membership of an element $\tilde{x}$ to the list $\{x_1, \ldots, x_m\}$ using Algorithm 3.

Yum et al. have shown that Algorithm 3 was a YES-bias Monte-Carlo algorithm [29]. They demonstrated that the value of the bias was:

$$f(\epsilon, k) := \left[1 - \epsilon\left(1 - \frac{\epsilon}{r}\right)^{km}\right]^k$$

based on the *Random Oracle* (RO) model for h and h'.

Algorithm 2. ACCUMULATE

Input: Two cryptographic hash functions h and h' outputting $(r\,d)$-bit long and $(k\,\log_2(r))$-bit long digests respectively, a security parameter ϵ, a list of elements to be aggregated $\{x_1, \ldots, x_m\}$.

/* Digests Computation */

1. Compute the digests $h(x_i) := y_{i,1}\|\cdots\|y_{i,r}$ where each $y_{i,j}$ is d-bit long for $i \in \{1, \ldots, m\}$.
2. Compute the digests $h'(x_i) := y'_{i,1}\|\cdots\|y'_{i,r}$ where each $y'_{i,j}$ is $\log_2(r)$-bit long for $i \in \{1, \ldots, m\}$.

/* Binary Strings Generation */

3. For $i \in \{1, \ldots, m\}$, create the string $b_{i,1}\|\cdots\|b_{i,r}$ as follows:
   3.1. Set $b_{i,j} = 1$ for $j \in \{1, \ldots, r\}$.
   3.2. For $\tau \in \{1, \ldots, k\}$, do the following:
      3.2.1. Set: $j = y'_{i,\tau} + 1$.
      3.2.2. Set: $b_{i,j} = 0$ if $\frac{y_{i,j}}{2^d-1} \le \epsilon$.

/* Accumulated Value */

4. Compute the binary products: $\forall j \in \{1, \ldots, r\}\ a_j := \prod_{i=1}^{m} b_{i,j}$.

Output: $\mathcal{A} := (a_1, \ldots, a_r)$: accumulated value for the list $\{x_1, \ldots, x_m\}$.

Algorithm 3. MEMBERSHIP

Input: Two cryptographic hash functions h and h' outputting $(r\,d)$-bit long and $(k\,\log_2(r))$-bit long digests respectively, a security parameter ϵ, the accumulated value $\mathcal{A} = (a_1, \ldots, a_r)$ corresponding to the list $\{x_1, \ldots, x_m\}$ and a candidate element $\tilde{x}$.

/* Digests Computation */

1. Compute the digest $h(\tilde{x}) := \tilde{y}_1\|\cdots\|\tilde{y}_r$ where each $\tilde{y}_j$ is d-bit long.
2. Compute the digest $h'(\tilde{x}) := \tilde{y}'_1\|\cdots\|\tilde{y}'_r$ where each $\tilde{y}'_j$ is $\log_2(r)$-bit long.

/* Binary Strings Generation */

3. Create the string $\tilde{b}_1\|\cdots\|\tilde{b}_r$ as follows:
   3.1. Set $\tilde{b}_j = 1$ for $j \in \{1, \ldots, r\}$.
   3.2. For τ in $\{1, \ldots, n\}$, do the following:
      3.2.1. Set: $j = \tilde{y}'_\tau + 1$.
      3.2.2. Set: $\tilde{b}_j = 0$ if $\frac{\tilde{y}'_j}{2^d-1} \le \epsilon$.

/* Accumulated Value */

4. For $j \in \{1, \ldots, r\}$, do the following:
   If $(\tilde{b}_j = 1$ and $a_j = 1)$ then Return NO.
5. Return YES.

Output: Decide whether $\tilde{x}$ belongs to $\{x_1, \ldots, x_m\}$.

The issue in [35] is that Yum et al. only provide an asymptotic analysis of $f(\epsilon, k)$. Indeed, they substituted $\left(1 - \frac{\epsilon}{r}\right)^{k\,m}$ by $\exp(-\frac{k\,m\,\epsilon}{r})$. However, it is unlikely that a very large number of elements m be accumulated so that this approximation holds.

Fortunately, we can still get some information on how to choose ϵ. Indeed, the partial derivative $\frac{\partial f}{\partial \epsilon}$ is negative. This involves:

$$\forall \epsilon \in [0, 1] \quad f(\epsilon, k) \geq f(1, k)$$

Thus, it is suggested to choose $\epsilon = 1$. In this situation, the bias of the algorithm gets:

$$f(1, k) = \left[1 - \left(1 - \frac{1}{r} \right)^{km} \right]^k$$

As observed in [35], setting $\epsilon = 1$ allows us to completely remove h from the structure of the accumulator. That is, only the cryptographic hash function h' is needed. Given this observation, we assume in the remaining of this paper that $\epsilon = 1$.

Remark 1. The use of a cryptographic hash function to instantiate the RO model is frequent [33]. In 2007, the National Institute of Standards and Technology (NIST) set a competition for a new cryptographic hash algorithm SHA-3 [19]. One of the requirement that the candidates must satisfy is to support pseudo-random functions, in particular, the HMAC construction [3].

3 Stream Authentication over Erasure Channels

In the remaining of this paper, we work with a unforgeable S-bit long digital signature $(\text{Sign}_{\text{SK}}, \text{Verify}_{\text{PK}})$ [29] the key pair of which (SK,PK) is created by a generator KeyGen and a cryptographic hash function h' outputing $\mathcal{H}'$-bit long digests with $\mathcal{H}' = k \log_2(r)$.

3.1 Authentication Protocol

The stream is a continuous flow of information. First, the sender generates the signature σ on the digest $h'(P_1)$ of the first stream packet. He then encodes the concatenation $\sigma \| h'(P_1)$ using a MDS code of length n and dimension $n - t$. The corresponding codeword is denoted $(C_1 \cdots C_n)$ where each C_i is $\lceil \frac{S + \mathcal{H}'}{n - t} \rceil$- bit long.

Second, the sender buffers the first n packets $P_1, \ldots, P_n$ as list $\mathcal{L}_1$. He computes the accumulated value $\mathcal{A}_1$ of $\mathcal{L}_1$ and builds the augmented packet: $\text{AP}_1 := 1 \| P_1 \| \mathcal{A}_1 \| C_1$.

Third, when a new stream packet P_{n+j-1} ($j \geq 2$) is available, the sender builds the list $\mathcal{L}_j := \{P_j, \ldots, P_{n+j-1}\} \cup \{h'(P_1)\}$. He computes the corresponding accumulated value $\mathcal{A}_j$ and builds the augmented packet: $\text{AP}_j := j \| P_j \| \mathcal{A}_j \| C_{[j]}$ where $[j]$ denotes the unique integer in $\{1, \ldots, n\}$ congruent to j modulo n. In particular: $[n] = [2n] = [3n] = \cdots = n$. We notice that the delay at the sender is n packets as it sends into the network AP_j after P_{n+j-1} be available.

The receiver buffers the first $n - t$ packets $\text{AP}_{r_1}, \ldots, \text{AP}_{r_{n-t}}$ he collects. He can recover the whole codeword $(C_1 \cdots C_n)$ from them and then the signature σ on $h'(P_1)$. This allows to authenticate the $n - t$ accumulated values thanks to $h'(P_1)$ aggregated in them. Those values can in turn be used to authenticate all the received packets. The receiver buffers the accumulated value $\mathcal{A}_{r_{n-t}}$.

When the receiver gets the $(n - t + 1)^{\text{th}}$ packet $AP_{r_{n-t+1}}$, then it can be authenticated using $\mathcal{A}_{r_{n-t}}$. The receiver buffers $\mathcal{A}_{r_{n-t+1}}$ and the process repeats throughout the stream. One notices that any receiver can verify the authenticity of any packet on-the-fly from the $(n - t + 1)^{\text{th}}$ received packet.

The packet overhead of the scheme is $r + \lceil \frac{S+\mathcal{H}'}{n-t} \rceil$ bits.

3.2   Analysis of the Protocol

Security. We have the following theorem the proof of which is in Appendix A.

Theorem 2. *Our authentication scheme is a non-degrading authentication protocol. The sender processes data with a delay of n packets throughout streaming while the receiver can authenticate packets on-the-fly from the $(n - t + 1)^{\text{th}}$ received element.*

Remark 2. A single signature is needed to ensure non-repudiation of the **whole** stream.

Remark 3. One can notice that, when n is fixed, the lower t is, the larger the delay gets. This might seem to be surprising at first but having low t's implies having small redundancy for the codeword coordinates as $n - t$ is large. That is why one requires more codeword information to reconstruct $(C_1 \cdots C_n)$. The trade-off delay/overhead is an efficiency trade-off.

Packet Overhead. An important point to notice is that the value $f(1, k)$ does not have any impact on the security of our protocol for erasure channels. Therefore, the only restriction that we have to take into account is $0 < k \leq r$ as this is necessary to construct the accumulator. We minimize the overhead of our construction by tuning the pair (r, k) so that the bit size $\mathcal{H}'$ of the digest output by h' is $k \log_2(r)$. More precisely, we need to choose r as:

$$r_{\min} := \min\{R \in \mathbb{N} : (0 < K \leq R \text{ and } K \in \mathbb{N} \text{ and } \mathcal{H}' = K \log_2(R))\}$$

In the case of the SHA-3 competition, NIST has required that the new hash function provides message digests of $224, 256, 384$ and 512 bits at least [19]. In this situation, the optimal choice for r is given in Table 1.

Table 1. Optimal choice for the parameter r

$\mathcal{H}'$	224	256	384	512
$r_{\min}$	128	256	64	256

We plotted the behavior of the packet overhead when the ratio $\frac{t}{n}$ represented 10%, $30\%, 50\%, 70\%$ and 90% for n varying between 100 and 1000 as Fig. 2. We chose to use a 1024-bit long signature to illustrate this result (i.e. $S = 1024$).

We see that in most cases, our packet overhead is less than a digest long. In particular, it is less than in [22, 23, 24, 25].

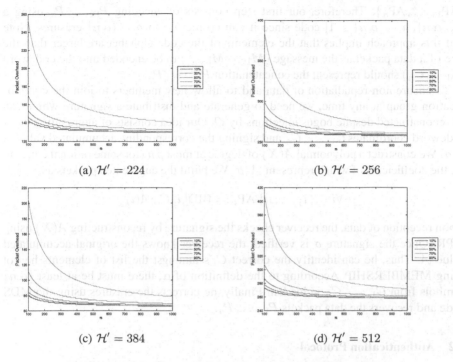

(a) $\mathcal{H}' = 224$

(b) $\mathcal{H}' = 256$

(c) $\mathcal{H}' = 384$

(d) $\mathcal{H}' = 512$

Fig. 2. Overhead of our authentication protocol for erasure channels

Remark 4. Desmedt and Jakimoski's scheme has small overhead as well [8]. Their result is based on optimal choices for cover-free families. The issue is that those optimal families have been shown to exist but they have yet to be constructed as the underlying result by Stinson et al. is a non-constructive proof of existence [30].

4 Stream Authentication over Adversarial Channels

In this channel model, $\mathcal{O}$ can inject bogus data packets into the network. In this situation, we will process the whole data stream per block of n packets $P_1, \ldots, P_n$. Each of these blocks is located within the whole stream using an identification value BID. This approach is used in the different schemes designed for adversarial networks quoted in Sect. 1 including TWMDS and PRABS. This is to be opposed to the on-the-fly authentication process from the $(n - t + 1)^{\text{th}}$ packet at the receiver for our protocol presented in Sect. 3.1.

4.1 Scheme Overview

Due to erasure of information, we want to generate n augmented packets $AP_1, \ldots, AP_n$ such that we can reconstruct all packets $P_1, \ldots, P_n$ from any $\lceil \alpha n \rceil$-subset of

$\{\mathrm{AP}_1, \ldots, \mathrm{AP}_n\}$. Therefore, our first step consists of encoding $P_1, \ldots, P_n$ using a $[n, \lceil \alpha n \rceil, n - \lceil \alpha n \rceil + 1]$ code since it can correct up to $n - \lceil \alpha n \rceil$ erasures. Note that this approach implies that the elements of the code alphabet are larger than the size of a data packet as the message $(M_1 \cdots M_{\lceil \alpha n \rceil})$ to be encoded into the codeword $(C_1 \cdots C_n)$ should represent the concatenation $P_1 \| \cdots \| P_n$.

To ensure non-repudiation of data and to allow new members to join the commu-nication group at any time, we need to generate and distribute a signature which can be reconstructed despite bogus injections by $\mathcal{O}$. Our idea consists of aggregating the n codeword coordinates $C_1, \ldots, C_n$ and signing the corresponding accumulated value $\mathcal{A}$ as σ. We construct a polynomial $A(X)$ of degree at most ρn (for some rational constant ρ), the coefficients of which represent $\mathcal{A} \| \sigma$. We build the augmented packets as:

$$\forall i \in \{1, \ldots, n\} \, \mathrm{AP}_i := \mathrm{BID} \| i \| C_i \| A(i)$$

Upon reception of data, the receiver checks the signature by reconstructing $A(X)$ using MPR. Once the signature σ is verified, the receiver knows the original accumulated value $\mathcal{A}$. Thus, he can identify the correct C_i's amongst the list of elements he got using MEMBERSHIP. According to the definition of α, there must be at least $\lceil \alpha n \rceil$ symbols from $C_1, \ldots, C_n$ in his list. Finally, he corrects the erasures using the MDS code and recovers the data packets $P_1, \ldots, P_n$.

4.2 Authentication Protocol

We assume that the values α and β are rational numbers so that we can represent them over a finite number of bits. In order to run Poly-Reconstruct as a subroutine of MPR, we have to choose a parameter $\rho \in (0, \frac{\alpha^2}{\beta})$. Notice that ρ has to be rational since ρn is an integer. Without loss of generality, one can consider that the value ρ is uniquely determined when n, α and β are known. Table 2 summarizes the scheme parameters which are assumed to be publicly known. The bit size $\mathcal{S}$ of the signature and its public key PK are also publicly known. They do not appear in Table 2 as they are considered as general parameters. Note that, once r is known, then k is uniquely determined since the digests of h' are $(k \log_2(r))$-bit long.

Table 2. Public parameters for our authentication scheme over adversarial channels

n: Block length	A list of irreducible polynomials over $\mathbb{F}_2$
α: Survival rate	β: Flood rate
$\mathcal{P}$: Bit size of data packets	r, k: Parameters of the accumulator hash function h'

The sender of data process the stream as in Algorithm 4. Note that the list of irre-ducible polynomials is used at Step 1 and Step 3. Furthermore, since any element of $\mathbb{F}_{2^{\tilde{q}}}$ can be represented as $\lambda_0 Y^0 + \lambda_1 Y_1 + \ldots + \lambda_{\tilde{q}-1} Y^{\tilde{q}-1}$ where each λ_i belongs to $\mathbb{F}_2$, we can define the first n elements as $(0, \ldots, 0)$, $(1, 0, \ldots, 0)$, $(0, 1, 0, \ldots, 0)$, $(1, 1, 0, \ldots, 0)$ and so on until the binary decomposition of $n - 1$ (Step 3).

Upon reception of data, the receivers use Algorithm 5 to authenticate information.

Algorithm 4. AUTHENTICATOR

Input: The secret key SK, the block number BID, Table 2 and n data packets $P_1, \ldots, P_n$.

/* Packet Encoding */

1. Parse $P_1 \| \cdots \| P_n$ as $M_1 \| \cdots \| M_{\lceil \alpha n \rceil}$ after padding. Encode the message $(M_1 \cdots M_{\lceil \alpha n \rceil})$ into the codeword $(C_1 \cdots C_n)$ using the MDS code over $\mathbb{F}_{2^q}$ with $q = \left\lceil \frac{n\,\mathcal{P}}{\lceil \alpha n \rceil} \right\rceil$.

/* Signature Generation and Representation */

2. Compute the accumulated value: $\mathcal{A} = \text{ACCUMULATE}(C_1, \ldots, C_n)$. Construct the block signature as: $\sigma = \text{Sign}_{\text{SK}}(h'(\text{BID}\|n\|\alpha\|\beta\|\mathcal{P}\|r\|\mathcal{A}))$.

3. Denote ξ the smallest element of $\mathbb{N}$ such that:

$$\left\lceil \frac{r + \mathcal{S} + \xi}{\rho n + 1} \right\rceil \geq \lceil \log_2 n \rceil \qquad (1)$$

Denote $\tilde{q}$ the left hand side of Inequality (1). Write $\mathcal{A}\|\sigma$ as the concatenation $a_0\| \cdots \|a_{\rho n}$ of $(\rho n + 1)$ elements of $\mathbb{F}_{2^{\tilde{q}}}$ after suitable padding. Form the polynomial $A(X) := a_0 + \cdots + a_{\rho n} X^{\rho n}$ and evaluate it at the first n points of $\mathbb{F}_{2^{\tilde{q}}}$: $\forall i \in \{1, \ldots, n\}$ $y_i := A(i)$.

/* Construction of Augmented Packets */

4. Build the augmented packets as:

$$\forall i \in \{1, \ldots, n\} \quad \text{AP}_i := \text{BID}\|i\|C_i\|y_i$$

Output: $\{\text{AP}_1, \ldots, \text{AP}_n\}$: set of augmented packets.

4.3 Analysis of the Protocol

Security. As the channel model allows an adversary to inject bogus elements into the network, we adopt the same security definition as in [32].

Definition 1. *The collection of algorithms* (KeyGen, AUTHENTICATOR, DECODER) *constitutes a* secure *and* (α, β)*-correct probabilistic multicast authentication scheme if no probabilistic polynomial-time opponent* $\mathcal{O}$ *can win with a non-negligible probability the following game:*

  i) A key pair (SK,PK) *is generated by* KeyGen.
  ii) $\mathcal{O}$ *is given: (a) The public key* PK *and (b) Oracle access to* AUTHENTICATOR *(but* $\mathcal{O}$ *can only issue at most one query with the same block identification tag* BID).
  iii) $\mathcal{O}$ *outputs* $(\text{BID}, n, \alpha, \beta, \mathcal{P}, r, \text{RP})$.

$\mathcal{O}$ *wins if one of the following happens:*

  a) (violation of the correctness property) $\mathcal{O}$ *succeeds to output* RP *such that even if it contains* $\lceil \alpha n \rceil$ *packets (amongst a total not exceeding* $\lfloor \beta n \rfloor$ *elements) of some authenticated packet set* AP_i *for block identification tag* BID *and parameters* $n, \alpha, \beta, \mathcal{P}$, *the decoder fails to authenticate all the correct packets.*

Algorithm 5. DECODER

Input: The public key PK, the block number BID, Table 2 and the set of received packets RP.

/* Signature Verification */

1. Write the packets as $BID_i\|j_i\|\widehat{C}_{j_i}\|\widehat{y}_{j_i}$ and discard those having $BID_i \neq BID$ or $j_i \notin \{1,\ldots,n\}$. Denote N the number of remaining elements. If $(N < \lceil \alpha n \rceil$ or $N > \lfloor \beta n \rfloor)$ then the algorithm stops.

2. Rename the remaining elements as $\{\widehat{AP}_1,\ldots,\widehat{AP}_N\}$ and write each element as: $\widehat{AP}_i = BID\|j_i\|\widehat{C}_{j_i}\|\widehat{y}_{j_i}$ where $j_i \in \{1,\ldots,n\}$. Compute $\tilde{q}$ as in Step 3 of AUTHENTICATOR. Get the irreducible polynomial of degree $\tilde{q}$ from the sender's public list and run MPR on the set $\{(j_i,\widehat{y}_{j_i}), 1 \leq i \leq N\}$ to get a list $\{c_1,\ldots,c_\mu\}$ of candidates for signature verification. If MPR rejects that set then the algorithm stops.

3. Initialize $\widehat{\mathcal{A}} = \emptyset$. While the list has not been exhausted (and the signature not verified yet), pick c_i and write it as: $\mathcal{A}_i\|\sigma_i$ after removing the pad where $\mathcal{A}_i$ is r-bit long. If $\text{Verify}_{\text{PK}}(h'(\text{BID}\|n\|\alpha\|\beta\|\mathcal{P}\|r\|\mathcal{A}_i),\sigma_i) = \text{TRUE}$ then set $\widehat{\mathcal{A}} = \mathcal{A}_i$ and break out the loop. Otherwise, increment i by 1 and start again the While loop.

/* Codeword Reconstruction */

4. If $\mathcal{A}' = \emptyset$ then the algorithm stops. Otherwise, set $C'_k := \emptyset$ for all $k \in \{1,\ldots,n\}$. For each $\widehat{AP}_i$ written as at Step 2, if $\text{MEMBERSHIP}(h',1,\widehat{\mathcal{A}},\widehat{C}_{j_i}) = \text{TRUE}$ then $C'_{j_i} = \widehat{C}_{j_i}$.

5. If $(C'_1 \cdots C'_n)$ has less than $\lceil \alpha n \rceil$ non-empty symbols then the algorithm stops. Otherwise, denote it into the message $(M'_1 \cdots M'_{\lceil \alpha n \rceil})$.

/* Packet Recovery */

6. If the decoding fails then the algorithm stops. Otherwise, remove the pad from $M'_1\| \cdots \|M'_{\lceil \alpha n \rceil}$ and write the remaining string as $P'_1\| \cdots \|P'_n$ where each P'_i is $\mathcal{P}$ bits long.

Output: $\{P'_1,\ldots,P'_n\}$: set of authenticated packets.

---

b) (violation of the security property) $\mathcal{O}$ succeeds to output RP such that the decoder outputs $\{P'_1,\ldots,P'_n\}$ that was never authenticated by AUTHENTICATOR for the value BID and parameters $n,\alpha,\beta,\mathcal{P}$.

Remark 5. A protocol with is secure and (α,β)-correct is non-degrading.

We have the following theorem the proof of which can be found as Appendix B.

Theorem 3. *If our authentication scheme is either insecure or not (α,β)−correct, then one can create a genuine element passing successfully* MEMBERSHIP.

Packet Overhead. Due to Theorem 3, we have to choose r in order to reduce the value of the YES-bias $f(1,k)$ as much as possible to ensure the security of the authentication protocol.

Since the packet overhead is:

$$\left(\frac{n\mathcal{P}}{\lceil \alpha n \rceil} - \mathcal{P}\right) + \left\lceil\frac{r+\mathcal{S}+\xi}{\rho n + 1}\right\rceil \text{ bits}$$

we look for the smallest value of r such that:

$$f(1,k) \leq t_{\text{bias}}$$

where t_{bias} is the threshold value for the bias. Note that this minimal value for r is independent from both rates α and β.

Table 3 represents the minimal values of r for n between 100 and 1000 when $t_{\text{bias}} = 10^{-10}$. As in Sect. 3.2, we used four digest length sizes for h': $224, 256, 384, 512$.

In order to provide a fair comparison with PRABS, we have to slightly modify Karlof et al.'s construction so that it also allows recovery of the whole data stream as TWMDS and our construction. This is at the cost of using a MDS code which leads to the additional overhead for PRABS of $\frac{n\,\mathcal{P}}{\lceil \alpha\,n \rceil} - \mathcal{P}$ bits. Therefore, in our implementations, the packet overhead for PRABS becomes:

$$\left(\frac{n\,\mathcal{P}}{\lceil \alpha\,n \rceil} - \mathcal{P} \right) + \mathcal{H}' \lceil \log_2(n) \rceil \text{ bits}$$

Table 3. Minimal values of r for $\mathcal{H}'$-bit long digests

		n		
		100	277	278	1000
	224	16384		65536	
$\mathcal{H}'$	256		65536		
	384		65536		
	512		65536		

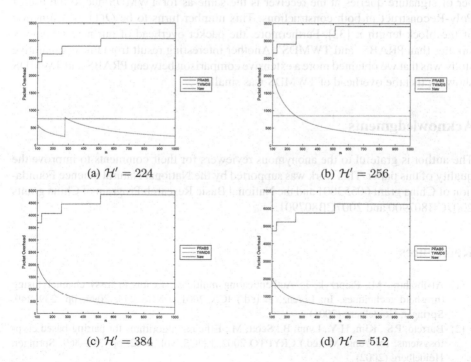

(a) $\mathcal{H}' = 224$

(b) $\mathcal{H}' = 256$

(c) $\mathcal{H}' = 384$

(d) $\mathcal{H}' = 512$

Fig. 3. Overhead comparison between PRABS, TWMDS and our scheme when $\alpha = 0.9$ and $\beta = 1.25$

We performed comparisons for $\alpha \in \{0.5, 0.75, 0.8, 0.9\}$ and $\beta = \{1.1, 1.25, 1.5, 2\}$ for the four digest sizes $\mathcal{H}' \in \{224, 256, 384, 512\}$. We chose $\rho = \frac{\alpha}{2\beta^2}$ as suggested in [32]. Due to space limitations, we can only depict the case $\alpha = 0.9$ and $\beta = 1.25$ as Fig. 3.

Our results show that, in most situations, the overhead of our new scheme is much smaller than PRABS' and TWMDS'. This is particularly acute when n gets large. Furthermore, those implementations demonstrate that the overhead of TWMDS is smaller than PRABS'. This extends the comparative survey between PRABS and TWMDS done so far which was only focused on the case $n = 1000$ [31, 32].

5   Conclusion

In this paper, we presented two protocols for the broadcast authentication problem using a modified version of Nyberg's accumulator due to Yum et al. Our first scheme was related to erasure channels. We showed that its packet overhead was less than the length of a digest and, in particular, far less than [22, 23, 24, 25]. Even if the sender processes the stream in delay of n packets, the receivers can authenticate packets on-the-fly from the $(n - t + 1)^{th}$ received element to the end of the stream (if any). In addition, a single signature is needed to provide the non-repudiation of the whole stream. Our second scheme was designed for adversarial networks. It is obvious that the number of signature queries at the receiver is the same as for TWMDS due to the use of Poly-Reconstruct in both constructions. This number turns to be $O(1)$ as a function of the block length n [32]. Furthermore, the packet overhead of our new scheme is smaller than PRABS' and TWMDS'. Another interesting result from this comparative study was that we obtained more a extensive comparison between PRABS and TWMDS showing that the overhead of TWMDS was smaller.

Acknowledgments

The author is grateful to the anonymous reviewers for their comments to improve the quality of this paper. This work was supported by the National Natural Science Foundation of China grant 60553001 and the National Basic Research Program of China grants 2007CB807900 and 2007CB807901.

References

[1] Al-Ibrahim, M., Pieprzyk, J.: Authenticating multicast streams in lossy channels using threshold techniques. In: Lorenz, P. (ed.) ICN 2001. LNCS, vol. 2094, pp. 239–249. Springer, Heidelberg (2001)

[2] Barreto, P.S., Kim, H.Y., Lynn, B., Scott, M.: Efficient algorithms for pairing-based cryptosystems. In: Yung, M. (ed.) CRYPTO 2002. LNCS, vol. 2442, pp. 354–369. Springer, Heidelberg (2002)

[3] Bellare, M., Canetti, R., Krawczyk, H.: Message authentication using hash functions - the HMAC construction. In: RSA Laboratories' CryptoBytes, vol. 2 (Spring 1996)

[4] Benaloh, J., de Mare, M.: One-way accumulators: A decentralized alternative to digital signatures. In: Helleseth, T. (ed.) EUROCRYPT 1993. LNCS, vol. 765, pp. 274–285. Springer, Heidelberg (1994)

[5] Boneh, D., Lynn, B., Shacham, H.: Short signatures from the Weil pairing. Journal of Cryptology 17(4), 297–319 (2004)

[6] Challal, Y., Bettahar, H., Bouabdallah, A.: A taxonomy of multicast data origin authentication: Issues and solutions. IEEE Communications Surveys and Tutorials 6(3), 34–57 (2004)

[7] Challal, Y., Bouabdallah, A., Bettahar, H.: H_2A: Hybrid hash-chaining scheme for adaptive multicast source authentication of media-streaming. Computer & Security 24(1), 57–68 (2005)

[8] Desmedt, Y., Jakimoski, G.: Non-degrading erasure-tolerant information authentication with an application to multicast stream authentication over lossy channels. In: Abe, M. (ed.) CT-RSA 2007. LNCS, vol. 4377, pp. 324–338. Springer, Heidelberg (2006)

[9] Fu, J.C., Lou, W.Y.W.: Distribution Theory of Runs and Patterns and its Applications. World Scientific Publishing, Singapore (2003)

[10] Golle, P., Modadugu, N.: Authenticating streamed data in the presence of random packet loss. In: Symposium on Network and Distributed Systems Security, San Diego, USA, pp. 13–22. Internet Society (February 2001)

[11] Guruswami, V., Sudan, M.: Improved decoding of Reed-Solomon and algebraic-geometric codes. IEEE Transactions on Information Theory 45(6), 1757–1767 (1999)

[12] Karlof, C., Sastry, N., Li, Y., Perrig, A., Tygar, J.D.: Distillation codes and applications to DoS resistant multicast authentication. In: 11th Network and Distributed Systems Security Symposium, San Diego, USA (February 2004)

[13] Lidl, R., Niederreiter, H.: Introduction to Finite Fields and their Applications (Revised Edition). Cambridge University Press, Cambridge (2000)

[14] Lysyanskaya, A., Tamassia, R., Triandopoulos, N.: Multicast authentication in fully adversarial networks. In: IEEE Symposium on Security and Privacy, Oakland, USA, pp. 241–253. IEEE Press, Los Alamitos (2003)

[15] MacWilliams, F.J., Sloane, N.J.A.: The Theory of Error-Correcting Codes. North-Holland, Amsterdam (1977)

[16] Menezes, A.J., van Oorschot, P.C., Vanstone, S.A.: Handbook of Applied Cryptography. CRC Press, Boca Raton (1996)

[17] Merkle, R.: A certified digital signature. In: Brassard, G. (ed.) CRYPTO 1989. LNCS, vol. 435, pp. 218–238. Springer, Heidelberg (1990)

[18] Miner, S., Staddon, J.: Graph-based authentication of digital streams. In: IEEE Symposium on Security and Privacy, Oakland, USA, pp. 232–246. IEEE Press, Los Alamitos (2001)

[19] National Institute of Standards and Technology. Announcing request for candidate algorithm nominations for a new cryptographic hash algorithm (SHA-3) family. Federal Register 72(212), 62212 – 62220 (November 2007)

[20] Nguyen, L.: Accumulators from bilinear pairings and applications. In: Menezes, A. (ed.) CT-RSA 2005. LNCS, vol. 3376, pp. 275–292. Springer, Heidelberg (2005)

[21] Nyberg, K.: Fast accumulated hashing. In: Gollmann, D. (ed.) FSE 1996. LNCS, vol. 1039, pp. 83–87. Springer, Heidelberg (1996)

[22] Pannetrat, A., Molva, R.: Authenticating real time packet streams and multicasts. In: 7th International Symposium on Computers and Communications, Taormina, Italy. IEEE Computer Society, Los Alamitos (2002)

[23] Park, J.M., Chong, E.K.P., Siegel, H.J.: Efficient multicast packet authentication using signature amortization. In: IEEE Symposium on Security and Privacy, Oakland, USA, pp. 227–240. IEEE Press, Los Alamitos (2002)

[24] Park, J.M., Chong, E.K.P., Siegel, H.J.: Efficient multicast stream authentication using erasure codes. ACM Transactions on Information and System Security 6(2), 258–285 (2003)

[25] Park, Y., Cho, Y.: The eSAIDA stream authentication scheme. In: Laganá, A., Gavrilova, M.L., Kumar, V., Mun, Y., Tan, C.J.K., Gervasi, O. (eds.) ICCSA 2004. LNCS, vol. 3046, pp. 799–807. Springer, Heidelberg (2004)

[26] Perrig, A., Canetti, R., Tygar, J., Song, D.: Efficient authentication and signing of multicast streams over lossy channels. In: IEEE Symposium on Security and Privacy, Oakland, USA, pp. 56–73. IEEE Press, Los Alamitos (2000)

[27] Perrig, A., Tygar, J.D.: Secure Broadcast Communication in Wired and Wireless Networks. Kluwer Academic Publishers, Dordrecht (2003)

[28] Rabin, M.O.: Efficient dispersal of information for security, load balancing, and fault tolerance. Journal of the Association for Computing Machinery 36(2), 335–348 (1989)

[29] Stinson, D.R.: Cryptography: Theory and Practice, 3rd edn. Discrete Mathematics and Its Applications. Chapman & Hall/CRC (2006)

[30] Stinson, D.R., Wei, R., Zhu, L.: Some new bounds for cover-free families. Journal of Combinatorial Theory, Series A 90(1), 224–234 (2000)

[31] Tartary, C.: Authentication for Multicast Communication. PhD thesis, Department of Computing - Macquarie University (October 2007)

[32] Tartary, C., Wang, H.: Achieving multicast stream authentication using MDS codes. In: Pointcheval, D., Mu, Y., Chen, K. (eds.) CANS 2006. LNCS, vol. 4301, pp. 108–125. Springer, Heidelberg (2006)

[33] van Tilborg, H.C.A.: Encyclopedia of Cryptography and Security. Springer, Heidelberg (2005)

[34] Yajnik, M., Moon, S., Kurose, J., Towsley, D.: Measurement and modeling of the temporal dependence in packet loss. In: IEEE INFOCOM 1999, vol. 1, pp. 345–352. IEEE Press, Los Alamitos (1999)

[35] Yum, D.H., Seo, J.W., Lee, P.J.: Generalized combinatoric accumulator. IEICE Transactions on Information and Systems E91-D(5), 1489–1491 (2008)

[36] Zanotti, J.-P.: Le code correcteur C.I.R.C,
http://zanotti.univ-tln.fr/enseignement/divers/chapter3.html

A   Proof of Theorem 2

Denote $AP_{r_1}, \ldots, AP_{r_{n-t}}$, the first $n - t$ augmented packets obtained by the receiver. Due to our channel model, we have $r_{n-t} - r_1 < n$.

As a consequence, $[r_1], \ldots, [r_{n-t}]$ are $n - t$ distinct values from $\{1, \ldots, n\}$. Thus, the receiver can uniquely identify $C_{[r_1]}, \ldots, C_{[r_{n-t}]}$ to their $n - t$ corresponding values from $C_1, \ldots, C_n$ by using the mapping $x \mapsto [x]$ over the values $r_1, \ldots, r_{n-t}$ contained within $AP_{r_1}, \ldots, AP_{r_{n-t}}$.

Using the correction capacity of the MDS code, the receiver can recover the codeword $(C_1 \cdots C_n)$ and then its corresponding message $(M_1 \cdots M_{n-t})$. This message easily leads to $\sigma \| h'(P_1)$ as this string represents the first $\mathcal{S} + \mathcal{H}'$ bits of $M_1 \| \cdots \| M_{n-t}$. Finally, the receiver verifies the authenticity of the signature σ on $h'(P_1)$ using $\mathrm{Verify}_{\mathrm{PK}}$.

So far, the receiver only authenticated $h'(P_1)$. It should be noticed that, since we are working over an erasure channel, it is sufficient to show that $h'(P_1)$ has been accumulated within the value $\mathcal{A}_{r_1}$ to authenticate this value and therefore every single element aggregated within (using MEMBERSHIP). We have two cases to consider.

1. $r_1 = 1$: The receiver obtained the first augmented AP_1 of the stream. In this case, he can verify the authenticity of AP_1 by computing $h'(P_1)$. He authenticates $\mathcal{A}_1$ using $h'(P_1)$ and MEMBERSHIP. The remaining $n - t - 1$ augmented packets $AP_{r_2}, \ldots,$ $AP_{r_{n-t}}$ belong to $\{AP_1, \ldots, AP_n\}$. Since $\mathcal{L}_1 = \{P_1, \ldots, P_n\}$, the validity of $P_{r_2}, \ldots,$ $P_{r_{n-t}}$ can be checked using MEMBERSHIP and $\mathcal{A}_1$.

2. $r_1 \geq 2$: The receiver did not receive AP_1. This case can also be seen as the receiver joining the communication group after the beginning of streaming.

Since the receiver authenticated $h'(P_1)$ thanks to the digital signature, he can check the authenticity of $\mathcal{A}_{r_1}$ using $h'(P_1)$ and MEMBERSHIP.

The remaining $n - t - 1$ augmented packets $AP_{r_2}, \ldots AP_{r_{n-t}}$ can be authenticated using $\mathcal{A}_{r_1}$ and MEMBERSHIP since $P_{r_2}, \ldots, P_{r_{n-t}}$ have been aggregated into $\mathcal{A}_{r_1}$ since: $r_1 < r_2 < \cdots < r_{n-t} \leq r_1 + (n - 1)$.

Up to this point, we showed that the receiver could authenticate the first $n - t$ augmented packets he got. In order to terminate the demonstration of this theorem, it remains to prove that the receiver can authenticate (on-the-fly) all the following augmented packets he obtains: $AP_{r_{n-t+1}}, AP_{r_{n-t+2}}, \ldots$

Consider $AP_{r_{n-t+1}}$. The accumulated value $\mathcal{A}_{r_{n-t}}$ is contained within the authenticated augmented packet $AP_{r_{n-t}}$. This value represents the aggregation of list $\mathcal{L}_{r_{n-t}}$ which includes the set $\{P_{r_{n-t}}, P_{r_{n-t+1}}, \ldots, P_{r_{n-t+(n-1)}}\} \cup \{h'(P_1)\}$. Therefore, $\mathcal{A}_{r_{n-t}}$ can be used to authenticate P_1 using MEMBERSHIP since: $r_{(n-t)+1} - r_{n-t} \leq t + 1 \leq n - 1$.

Once $\mathcal{A}_{r_{(n-t)+1}}$ is authenticated, the receiver can discard $\mathcal{A}_{r_{n-t}}$ and buffer $\mathcal{A}_{r_{(n-t)+1}}$ which will be used to authenticate $AP_{r_{(n-t)+2}}$ and so on.

This recursive process shows that the receiver can authenticate every packet he obtains, that is to say, the scheme is non-degrading.

B Proof of Theorem 3

Assume that the scheme is either insecure or not (α, β)-correct. By definition, a probabilistic polynomial time opponent $\mathcal{O}$ can break the scheme security or correctness with a non-negligible probability $\pi(k)$ where k is the security parameter setting up the digital signature and the hash function. Note that, since h' is a cryptographic hash function in the RO model, it is assumed to be collision-resistant. We must have either cases:

1. With probability at least $\pi(k)/2$, $\mathcal{O}$ breaks the scheme correctness.
2. With probability at least $\pi(k)/2$, $\mathcal{O}$ breaks the scheme security.

It should be noticed that since $\pi(k)$ is a non-negligible function of k, so is $\pi(k)/2$.

In both cases, we will demonstrate that $\mathcal{O}$ can turn an attack against either the correctness or the security of the scheme in polynomial time into forging an element $\widehat{C}$ passing successfully MEMBERSHIP in polynomial time as well.

<u>Point 1.</u> For this attack, $\mathcal{O}$ will have access to the signing algorithm Sign_{SK} (but $\mathcal{O}$ will not have access to SK itself). He can use the public key PK as well as the hash

function h'. $\mathcal{O}$ will be allowed to run AUTHENTICATOR whose queries are written as $(\text{BID}_i, n_i, \alpha_i, \beta_i, \mathcal{P}_i, r_i, \text{DP}_i)$ where DP_i is the set of n_i data packets to be authenticated. As said in Sect. 4.2, the knowledge of r_i determines the value of k_i as the digest output by h' are $(k_i \log_2(r_i))$-bit long with $\mathcal{H}' = k_i \log_2(r_i)$. In order to get the corresponding output, the signature is obtained by querying Sign_{SK} as a black-box at Step 2 of AUTHENTICATOR.

According to our hypothesis, $\mathcal{O}$ broke the correctness of the authentication protocol. This means that, following the previous process, $\mathcal{O}$ obtained values $\text{BID}, n, \alpha, \beta, \mathcal{P}, r$ and a set of received packets RP such that:

- $\exists i : (\text{BID}, n, \alpha, \beta, \mathcal{P}, r) = (\text{BID}_i, n_i, \alpha_i, \beta_i, \mathcal{P}_i, r_i)$.
  Denote $\text{DP} = \{P_1, \ldots, P_n\}(= \text{DP}_i)$ the n data packets associated with this query and AP the response given to $\mathcal{O}$. In particular, we denote σ the signature corresponding to DP and generated as in Step 2 of AUTHENTICATOR.
- $|\text{RP} \cap \text{AP}| \geq \lceil \alpha n \rceil$ and $|\text{RP}| \leq \lfloor \beta n \rfloor$.
- $\{P_1', \ldots, P_n'\} = \text{DECODER}(\text{PK}, \text{BID}, n, \alpha, \beta, \mathcal{P}, r, \text{RP})$ where $P_j' \neq P_j$ for some $j \in \{1, \ldots, n\}$.

It should be noticed that the values $n, \alpha, \beta, \mathcal{P}, r$ as well as PK are publicly known.

Since the digital signature is unforgeable and the hash function is collision resistant, it is impossible to obtain either a forgery (digital signature) or a collision (hash function) in polynomial time with non-negligible probability $\pi(k)/2$. This observation will be used to reduce the security of the authentication scheme to the security of the accumulator.

Since $|\text{RP} \cap \text{AP}| \geq \lceil \alpha n \rceil$ and $|\text{RP}| \leq \lfloor \beta n \rfloor$, Step 1 of DECODER ends successfully. The consistency of Poly-Reconstruct involves that the list returned by MPR at Step 2 contains the element $\mathcal{A} \| \sigma$ corresponding to DP after removing the pad. It should be noticed that the pad length can be uniquely determined from the public values as $\widetilde{q} = \left\lceil \frac{r + S + \xi}{\rho n + 1} \right\rceil$ (see Inequality (1)).

As the digital signature is unforgeable and the hash function is collision resistant, the pair message/signature going through the verification process at Step 3 corresponds to DP. Therefore, at the end of that step, we have:

$$\widehat{\mathcal{A}} = \mathcal{A}$$

At the beginning of Step 4, the receiver has recovered the accumulated value $\mathcal{A}$ corresponding to the original codeword $(C_1 \cdots C_n)$ related to DP.

Assume that $\mathcal{O}$ cannot forge any $\widehat{C} \notin \{C_1, \ldots, C_n\}$ passing successfully MEMBERSHIP.

In this case, only elements from $\text{RP} \cap \text{AP}$ will successfully pass MEMBERSHIP. As a consequence, at the end of Step 4, we get:

$$\forall i \in \{1, \ldots, n\} \ C_i' \in \{\emptyset, C_i\}$$

and at least $\lceil \alpha n \rceil$ values C_i'''s are non-empty.

Thus, at Step 5, $(C'_1 \cdots C'_n)$ is first corrected into $(C_1 \cdots C_n)$ and then decoded as $(M_1 \cdots M_{\lceil \alpha n \rceil})$. Finally, at the end of Step 6, we have: $\forall i \in \{1, \ldots, n\}$ $P'_i = P_i$. We obtain a contradiction with our original hypothesis which stipulated:

$$\exists j \in \{1, \ldots, n\} \quad P'_j \neq P_j$$

Therefore, $\mathcal{O}$ was able to construct a new value $\widehat{C}$ passing MEMBERSHIP successfully with non-negligible probability in polynomial time.

Point 2. We consider the same kind of reduction as in Point 1. The opponent $\mathcal{O}$ breaks the security of the scheme if one of the following holds:

I. AUTHENTICATOR was never queried on input BID, $n, \alpha, \beta, \mathcal{P}, r$ and the decoding algorithm DECODER does not reject RP, i.e. $\{P'_1, \ldots, P'_n\} \neq \emptyset$ where: $\{P'_1, \ldots, P'_n\} = \text{DECODER}(\text{PK}, \text{BID}, n, \alpha, \beta, \mathcal{P}, r, \text{RP})$.

II. AUTHENTICATOR was queried on input BID, $n, \alpha, \beta, \mathcal{P}, r$ for some data packets $\text{DP} = \{P_1, \ldots, P_n\}$. Nevertheless, the output of DECODER verifies $P'_j \neq P_j$ for some $j \in \{1, \ldots, n\}$.

Case I. Since DECODER output some non-empty packets, Step 3 had to terminate successfully. Thus, it has been found a pair $(h'(\text{BID}\|n\|\alpha\|\beta\|\mathcal{P}\|r\|\mathcal{A}), \sigma)$ such that:

$$\text{Verify}_{\text{PK}}(h'(\text{BID}\|n\|\alpha\|\beta\|\mathcal{P}\|r\|\mathcal{A}), \sigma) = \text{TRUE}$$

If $\mathcal{O}$ never queried AUTHENTICATOR for block tag BID, then either the previous pair is a forgery of the digital signature or $\text{BID}\|n\|\alpha\|\beta\|\mathcal{P}\|r\|\mathcal{A}$ collides with one of the queries $\text{BID}_i\|n_i\|\alpha_i\|\beta_i\|\mathcal{P}_i\|r_i\|\mathcal{A}_i$ made by $\mathcal{O}$ for the hash function h'. Since none of those cases can occur in polynomial time with non-negligible probability, we get a contradiction in this situation.

If $\mathcal{O}$ queried AUTHENTICATOR for block tag BID then denote $(\text{BID}, \hat{n}, \hat{\alpha}, \hat{\beta}, \hat{\mathcal{P}}, \hat{r})$ his query. By hypothesis, we have:

$$(\hat{n}, \hat{\alpha}, \hat{\beta}, \hat{\mathcal{P}}, \hat{r}) \neq (n, \alpha, \beta, \mathcal{P}, r)$$

We conclude as above. That is to say that we get a contradiction with the security of either the digital signature of the hash function.

Case II. We have the same situation as in Point 1.

MIBS: A New Lightweight Block Cipher[*]

Maryam Izadi, Babak Sadeghiyan, Seyed Saeed Sadeghian,
and Hossein Arabnezhad Khanooki

{izadi_maryam,basadegh,ssadeghian,qa_had}@aut.ac.ir

Abstract. In this paper, we propose a new lightweight 64-bit block cipher, which we call MIBS, suitable for resource-constrained devices, such as low-cost RFID tags. We also study its hardware implementation efficiency, as well as its security. The hardware implementation of MIBS requires 1400 gates on 0.18 μm technology, which is less than 2000 gates limit for low-cost RFID tags. We also show MIBS is secure against differential and linear cryptanalysis.

Keywords: Block Cipher, Lightweight, Low-cost RFID Tags, Resource-Constrained Devices.

1 Introduction

Radio frequency identification (RFID) is a technology for automated identification of objects and people. Although, this technology appeared quite a long time ago, it is recently used in wide range of applications due to technical improvements and dramatic cost decrease. There are security and privacy challenges concerning this technology. Cryptographic solutions require high computing resources and come with extra costs. Development of hardware efficient security primitive for resource-constrained devices such as low-cost RFID tags is a challenging task that recently is being more dealt with[1]. Gate constraints for security of low-cost tags are about 200-2000 gates, that is less than what is necessary for standard cryptographic primitives, so existing cryptographic algorithms can be hardly implemented under such resource constraint. In this paper, we propose a new light-weight block cipher to satisfy this requirement, and at the same time, it has the necessary security. In our design we have adopted several components which are already presented in other ciphers. The paper is organized as follows. Related works are described in section 2. In section 3, we present MIBS block cipher. In section 4, the design rationale of cipher is discussed. In section 5 we analyse the security of MIBS. The study of its hardware efficiency follows in section 6. We give a conclusion in section 7.

2 Related Works

In recent years, the lightweight cryptography for RFID tags has attracted much attention. Feldhofer et al. [2] have presented a hardware implementation of

---

[*] This project has been supported in part by Iran Telecommunication Research Center.

J.A. Garay, A. Miyaji, and A. Otsuka (Eds.): CANS 2009, LNCS 5888, pp. 334–348, 2009.

Advanced Encryption Standard (AES), with a gate count of 3595. Poschmann et al. [3] designed a lightweight variant of the Data Encryption Standard (DES) called DESL, which makes use of only one S-box mapping and can therefore be more compact than DES. Their implementation fits in 1848 gates. DESXL is an another version of DES that strengthened DESL with a key size of 184 bits and a hardware size of approximately 2168 gates [4]. PRESENT [5] is a 64-bit block cipher with a key length of 80 or 128 and consists of 31 rounds which has reasonable layout size for constrained environments such as low-cost RFID tags. Hardware implementation of PRESENT requires 1570 gates. But recently, Rolfes et al. [6] present a serialized architecture of the PRESENT that requires only 1000 gates. Other compact block ciphers like mCRYPTON [7], HIGHT [8], SEA [9], and PUFFIN [10] are also proposed, but they require more area to implement than PRESENT implementation. The very compact block cipher which we proposed has a reasonable area complexity. MIBS is a 32 rounds Feistel cipher with a block length of 64-bit, where two key lengths of 64-bit and 80-bit are supported.

3   MIBS Block Cipher

MIBS uses a Feistel structure with data block length of 64-bit and key lengths of 64-bit or 80-bit and consists of 32 rounds. The round structure is shown in Fig. 1. For applications that require moderate security levels, such as low-cost RFID tags, 64-bit security is adequate. In practice, there is a tradeoff between hardware efficiency and security. The F-function, depicted in Fig. 2, operates on half a block (32 bits), representing it into eight nibbles, and it consists of four stages: key addition, non-linear substitution layer, linear mixing layer, and nibble-wise permutation.

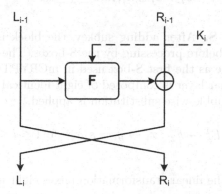

Fig. 1. Encrypt round of MIBS

Table 1. S-box mapping[7]

0	1	2	3	4	5	6	7	8	9	10	11	12	13	14	15
S	4	15	3	8	13	10	12	0	11	5	7	14	2	6	1	9

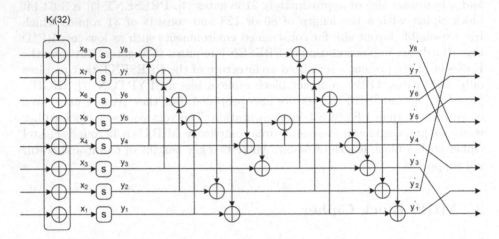

Fig. 2. The i-th round function of MIBS

Key addition. Current state $s_{31}, s_{30}, ..., s_0$, which is input to the F-function, is combined with a round subkey $k^i = k^i_{31}, k^i_{30}, ..., k^i_0 for 1 \leq i \leq 32$, using a bit-wise XOR operation. Since XOR is well-suited to hardware implementation, all subkeys are bitwise XORed with data before substitution layer.

$$s_j = s_j \oplus k^i_j, for 0 \leq j \leq 31$$

Substitution layer S. After adding subkey, the block is divided into eight nibbles $x_8, x_7, ..., x_1$, before processing by the S-boxes. The 4×4 S-box used in our cipher is the same as the first S-box used in mCRYPTON and is shown in Table 1. The non-linear layer is composed of eight identical 4×4 S-boxes, so in this transformation nibble-wise substitution is applied.

$$S : F_2^4 \rightarrow F_2^4 : x_i \rightarrow y_i = s(x_i), for 1 \leq i \leq 8$$

Mixing layer M. The linear transformation mixes eight nibbles as follows:

$$M : (GF(2)^4)^8 \rightarrow (GF(2)^4)^8, (y_8, y_7, ..., y_1) \rightarrow (y'_8, y'_7, ..., y'_1) \Leftrightarrow$$

Table 2. Permutation mapping

	1	2	3	4	5	6	7	8
P	2	8	1	3	6	7	4	5

$$y'_1 = y_2 \oplus y_3 \oplus y_4 \oplus y_5 \oplus y_6 \oplus y_7$$
$$y'_2 = y_1 \oplus y_3 \oplus y_4 \oplus y_6 \oplus y_7 \oplus y_8$$
$$y'_3 = y_1 \oplus y_2 \oplus y_4 \oplus y_5 \oplus y_7 \oplus y_8$$
$$y'_4 = y_1 \oplus y_2 \oplus y_3 \oplus y_5 \oplus y_6 \oplus y_8$$
$$y'_5 = y_1 \oplus y_2 \oplus y_4 \oplus y_5 \oplus y_6$$
$$y'_6 = y_1 \oplus y_2 \oplus y_3 \oplus y_6 \oplus y_7$$
$$y'_7 = y_2 \oplus y_3 \oplus y_4 \oplus y_7 \oplus y_8$$
$$y'_8 = y_1 \oplus y_3 \oplus y_4 \oplus y_5 \oplus y_8$$

Permutation layer P. Finally, the eight nibble outputs from the mixing layer are arranged according to Table 2. Each nibble is moved to a new position by P.

Key schedule for 64-bit key. The design principle of MIBS key schedule is adopted from the design principle of PRESENT key schedule. Our key schedule, generates 32-bit round key k^i, for $0 \leq i \leq 31$, from 64-bit user key K (represented as $k_{63}, k_{62}, ..., k_0$). We denote the key state of the i-th round as $state^i$. The key state for each round is updated as follows.

$$state^0 = \text{user-key}$$
$$state^i = state^i \ggg 15$$
$$state^i = \text{S-box}(state^i_{[63:60]}) || state^i_{[59:0]}$$
$$state^i = state^i_{[63:16]} || state^i_{[15:11]} \oplus \text{Round-Counter} || state^i_{[10:0]}$$
$$k^i = state^i_{[63:32]}$$

where $\ggg$ means rotation to right, $[i : j]$ indicates the i-th to the j-th bits are involved in the operation, and $||$ denotes concatenation. Also we use the same S-box as in the F-function. The round key k^i is the 32 left most bits of the current state.

Key schedule for 80-bit key. The key K is first initialized with the user key, and updates as follows.

$$state^0 = \text{user-key}$$
$$state^i = state^i \ggg 19$$
$$state^i = \text{S-box}(state^i_{[79:76]}) || \text{S-box}(state^i_{[75:72]}) || state^i_{[71:0]}$$
$$state^i = state^i_{[79:19]} || state^i_{[18:14]} \oplus \text{Round-Counter} || state^i_{[13:0]}$$
$$k^i = state^i_{[79:48]}$$

After that, the round key k^i is the 32 left most bits of the key state. Test vectors for MIBS with 64 bit and 80 bit key are provided in the Appendix I.

4   Design Rationale

4.1   The Cipher Structure

MIBS is based on Feistel structure with an SPN round function. A large proportion of block ciphers have used this scheme since the US Federal Government adopted the DES. Moreover, DES has endured various attacks for over 20 years, even though its round function is very simple. Since Feistel construction operates on half of the block length in each iteration, therefore the size of code or circuitry required to implement it is nearly halved. Thus we use Feistel network as an overall structure with the purpose of minimizing computational resources, which certainly is one of the most important considerations in hardware design for tiny ubiquitous devices.

4.2   Round Function

For round function we selected the Substitution-Permutation Network (SPN). The SPN structure is directly based on the concepts of confusion and diffusion. The confusion component is a nonlinear substitution and the diffusion component is a linear mixing which is used for diffusing the cryptographic characteristics of substitution layer.

The substitution layer. The most important objective in designing a block cipher targeted to embedded applications such as RFID tags, is to achieve low complexity in hardware while providing sufficient security. Consequently an appropriate substitution layer of such a block cipher should meets the above balance. Although, large S-boxes can achieve better security but even in software, large S-boxes require high storage cost and they are far worse in hardware. On the other hand, too small S-boxes can hardly achieve suitable security. We observed the gate count increases exponentially with the size of S-box. As a result, we decided to use 4×4 S-boxes with regard to hardware efficiency and at the same time adequate security. Also existing lightweight block ciphers like PRESENT, and mCRYPTON have used 4×4 S-boxes too. The S-box used in MIBS block cipher is the same as the S0 mapping applied in mCRYPTON [7].

The linear transformation. In order to construct a fast and strong block cipher, we design a round function that is secure against differential and linear cryptanalysis and yield small values for the maximum differential and linear probabilities p, q. Kanda et al. [11], proposed a search algorithm for constructing an optimal linear transformation layer by using the matrix representation in order to minimize probabilities p, q as much as possible. They determined an optimal linear transformation layer among many candidates which has a lower computational complexity, which we used in MIBS. Additionally they showed that any linear transformation following a non-linear layer consists of 8 parallel S-boxes, can not have branch number more than 5. The branch number is the minimum number of active S-boxes in two consecutive rounds of a non-trivial differential characteristic or non-trivial linear trail [12]. In this context, by Optimal we mean that the maximum differential and linear probabilities p, q are as small as possible. Similar linear transformations is used also in E2[13] and Camellia[14] block ciphers. The linear layer M, which we call mixing layer, is represented using only 16 nibble-wise XORs that is suitable for computational efficiency. For security against differential and linear cryptanalysis, the branch number of layer M is optimal. Consequently, the mixing layer piles up the number of active S-boxes every two rounds to minimize the maximum differential and linear probabilities.

5 Security Analysis

5.1 Differential and Linear Cryptanalysis

preliminaries. Two well-known attacks applicable against block ciphers are differential cryptanalysis, introduced by Biham and Shamir [15], and linear cryptanalysis proposed by Matsui[16]. Because of wide applicability of both attacks to numerous block ciphers, resistant against them should be considered in the design of block ciphers. The complexity of each attack is defined by the number of active S-boxes involved and their differential characteristic or linear approximation probabilities. Kanda et al. [17] show the minimum number of active S-boxes in differential and linear attacks for Feistel ciphers with SPN round function which is presented below.

Definition 1. *For any given* $\Delta_x, \Delta_y, \Gamma_x, \Gamma_y \in GF(2)^m$, *the differential and linear probabilities of each S-box are defined as:*

$$DP^{S_i}(\Delta_x \rightarrow \Delta_y) = \frac{\#\{x \in GF(2)^m | S_i(x) \oplus S_i(x \oplus \Delta_x) = \Delta_y\}}{2^m}$$

$$LP^{S_i}(\Gamma_y \rightarrow \Gamma_x) = (2 \times \frac{\#\{x \in GF(2)^m | x \Gamma_x = S_i(x) \Gamma_y\}}{2^m} - 1)^2$$

Where $x \cdot \Gamma_x$, *denotes the parity (0 or 1) of bitwise product of* x *and* Γ_x.

Definition 2. *The maximum differential and linear probabilities of S-boxes are defined as:*

$$p_s = \max_i \max_{\Delta x \neq 0, \Delta y} DP^{S_i}(\Delta x \rightarrow \Delta y)$$

$$q_s = \max_i \max_{\Gamma x, \Gamma y \neq 0} LP^{S_i}(\Gamma y \rightarrow \Gamma x)$$

Definition 3 ([11]). *A differential active S-box is defined as an S-box given a non-zero input difference, while a linear active S-box is defined as an S-box given a non-zero output mask value.*

As we mentioned earlier, the security against differential and linear cryptanalysis is evaluated using the branch number, and branch number is defined as follow[12].

Definition 4. *The differential branch number B_d is defined as:*

$$B_d = \min_{\Delta x \neq 0} (H_w(\Delta x) + H_w(\theta(\Delta x)))$$

where Δx is an input difference into the diffusion layer and $\theta(\Delta x)$ is an output difference from the layer. H_w denotes the number of non-zero nibbles as defined in [17].

In our case that the mixing transformation is bijective, the differential branch number B_d, and linear branch number B_l are identical ($B = B_d = B_l$).

Definition 5. *The minimum number of differential active S-boxes of the r-round Feistel cipher with SPN round function, is defined as:*

$$D^{(r)} = \min_{(\Delta x^{(0)}, \Delta x^{(1)}, \ldots, \Delta x^{(r+1)}) \neq (0, 0, \ldots, 0)} \sum_{i=1}^{r} H_w(\Delta x^{(i)})$$

where $H_w(\Delta x^{(i)})$ is the number of the ith-round differential active S-boxes.

Theorem 1 ([17]). *The minimum number of differential active s-boxes $D^{(4r)}$ for 4r round Feistel ciphers with SPN round function satisfies $D^{(4r)} \geq r \times B + \lfloor r/2 \rfloor$.*

Theoretical Analysis. Theorem 1 also holds for $L^{(r)}$, the number of non-zero nibbles in linear approximation of round r, because both non-linear and linear layers are bijective.

The maximum differential and linear probabilities of the S-boxes are $p_s = q_s = 2^{-2}$, and the branch number of the linear transformation is 5. According to theorem 1, the lower bound of the number of active S-boxes with respect to linear and differential cryptanalysis is as follows: $D^{(32)} \geq 8 \times 5 + 4 \Rightarrow D^{(32)} \geq 44$.

Therefore, the upper bound of maximum differential characteristic probability is $(2^{-2})^{44} = 2^{-88}$, and the bias of linear approximation according to pilling-up lemma, is $2^{43} \times (2^{-2})^{44} = 2^{-45}$. As a result our suggested number of rounds is a conservative choice and we have a fair amount of security margin.

Experimental Analysis

Differential cryptanalysis. In the previous section we presented the theoretical bound for Differential Cryptanalysis. Here we present the best non-trivial 4-round differential characteristic we have found. Table 3 illustrates this 4-round characteristic. The left and right columns represent the non-zero nibbles for each round. This characteristic has the least number of active S-boxes i.e 6, and results in the 4-round characteristic probability of 2^{-15}. If we assume it as an iterative characteristic, we can deduce that the 32-round characteristic should not have probability better than $(2^{-15})^8 = 2^{-120}$.

Table 3. 4 rounds differential characteristic

Round Number	Left	Right	Probability
Input	10000000	10001110	
1	00100001	10000000	1/8
2	01000010	00100001	1/16
3	00000100	01000010	1/64
4	00100101	00000100	1/4
		active S-boxes = 6	Total = 2^{-15}

Table 4. 4 rounds linear approximation

Round Number	Left	Right	Number of Active S-boxes	Bias
1	00000001	00000100	5	2^{-6}
2	00000100	11011101	1	2^{-2}
3	11011101	00000000	0	1
4	00000000	11011101	1	2^{-2}
			Total = 7	Total = 2^{-8}

Linear cryptanalysis. Here we show the best linear approximation we have found. The best linear approximation for 4-round MIBS is illustrated in Table 4. The left and right columns represent the Non-zero nibbles for input of each round.

It has 7 active S-boxes with the best bias in each S-box which is 2^{-2} ,that results in the 4-round approximation with bias $(2^{-2})^7 \times 2^6 = 2^{-8}$. So it yields that the bias for 32-round approximation is at least $(2^{-8})^8 \times 2^7 = 2^{-57}$, which requires $\lambda.2^{114}$ known plaintext, where λ is a small factor that is used for better success probability.

5.2   Multiple Linear Cryptanalysis

The idea of using multiple approximation in linear cryptanalysis is first introduced by Kaliski and Robshaw [18], their method has the restriction that only same bits of key can be used in approximations. In 2004 Biryukov et al. [19] introduced the general statistical framework which does not possess that restriction. As a result of their method the data complexity of attack becomes proportional to the capacity of approximations. Time complexity of attack is equal to the time for encrypting the known plaintexts and for each encryption, updating the counters for the approximations. Multidimensional approximations without the assumption of independence are introduced by Hermelin and Nyberg [20], which allows us to build 2^m approximations from m independent approximations. We have found several approximations with good bias and several can be obtained by changing the first round input mask and last round output mask. By taking that into account we can use 16 independent approximations and build 2^{16} approximations, so at the best case that all of the combined approximations have the maximum bias we may be able to reduce the data complexity by the order of 2^{16} at the cost of updating 2^{16} counters for each known plaintext. So the data complexity of linear cryptanalysis can be reduced to $\lambda.2^{98}$.

5.3   Other Variants of Linear Cryptanalysis

Differential-Linear is a method for connecting a differential characteristic to linear approximation, which is introduced by Langford and Hellman [21] and later enhanced for probabilistic differential characteristic by Dunkleman and Biham [22]. It is useful when we have differential and linear characteristics with high probability for small number of rounds. However as we dont have such characteristics, this attack is not applicable to the full rounds. Non-linear cryptanalysis [23] is not applicable to MIBS since it is usually useful when we have large s-boxes,and is used in outer rounds of the cipher. So the improvement by this attack does not pose a threat to MIBS because of large security margin. Bi-Linear cryptanalysis[24] which is proposed for feistel schemes is not a large improvement to the attack so this is also not a practical attack against MIBS.

5.4   Algebraic Attack

Algebraic attack is a method for the cryptanalysis of ciphers, which was first presented by Courtois and Pieprzyk [25] to analyze AES. The attack aims to recover the secret key through solving an overdefined system of multivariate algebraic equations. A block cipher, which consists small S-boxes, may be represented as many equations with small number of variables. By solving these multivariate equations the key of the block cipher may be found, but the problem of solving a system of multivariate quadratic equations is in general NP-hard. Several methods for solving such systems of equations has been proposed for the special cases of overdefined and sparse systems[25,26], although some flaws in all such techniques are claimed in [27,28]. Anyway, any 4×4 bit S-box can be represented as 21 quadratic equations of 8 input/output bit variables over GF(2) [25]. MIBS-64 consists of $n = (32 \times 8) + 32 = 288$ S-boxes, as there are 8 S-boxes in each round of the 32-round cipher, and one S-box in each round of key scheduling. Thus, the cipher can be described with 6048 ($= 288 \times 21$) quadratic equations of 2304 ($= 288 \times 8$) variables. MIBS-80 has 32 S-boxes more than MIBS-64, so the number of quadratic equations is 6720 with 2560 variables.

According to [25], an estimation of the complexity of XSL attack on a block cipher can be calculated with work factor. For MIBS-64, W.F. is accordingly estimated as follows:

$$WF \approx \Gamma^\omega \cdot \left((\text{Block size}) \cdot (\text{Number of rounds})^2\right)^{\omega\lceil \frac{t}{r} \rceil}$$

$$= (2^6)^{2.37} \cdot \left((64) \cdot (\frac{32}{2})^2\right)^{2.37\lceil \frac{37}{21} \rceil}$$

$$= 2^{80.58}$$

Which is greater than 2^{64} operations needed for exhaustive search, making the attack impractical.

5.5 Related Key Attack

Slide attack [29] and related-key [30] attack are a form of cryptanalysis which use some weakness of key schedule. The attacker can observe the operation of a cipher under several different keys whose values are initially unknown, but where some mathematical relationship connecting the keys is known to the attacker. The design rationale of the key schedule of MIBS is similar to the key schedule of PRESENT. Since key schedule uses the round-dependent counter and a non-linear operation to mix the contents of the key register K, it is secure against these attacks.

6 Hardware Implementation

MIBS block ciphers is designed for very efficient hardware implementations, and each component is carefully constructed with hardware implementations in mind. In order to check hardware complexity, MIBS was implemented in a standard cell library based on TSMC $0.18\mu m$ CMOS technology. The block cipher is described in Verilog and simulated using ModelSim SE PLUS 6.2b. The synthesis is only done for encryption using typical transistors with the aim of area optimization by LEONARDO SPECTRUM 2005a.82. The data path of MIBS is depicted in Fig. 3. Each round consists of key addition, substitution layer, mixing layer, permutation layer, and right data addition. The substitution layer is composed of eight 4×4 S-boxes which are used in parallel. 4×4 S-box is implemented with simple combinational logic of 4-bit Boolean function. The key addition, and the

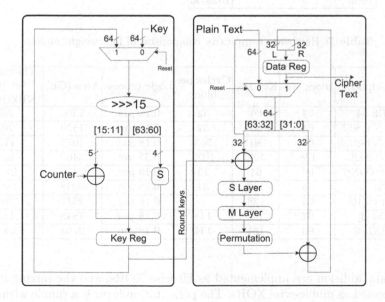

Fig. 3. Data path of MIBS-64

Table 5. Hardware complexity of MIBS-64

Module (Round Function)	GE	Module (Key Schedule)	GE
Data Register	384.68	Key Register	384.68
Substitution Layer	192	S-box	24
Key Xor	85.44	Right Rotation	0
Mixing Layer	170.84	Counter Xor	13.35
Permutation Layer	0	Total Key Schedule	422.03
Right Data Xor	85.44		
Total Round Function	918.4		
Control Unit	46		
Other Logics	8.74		
Total	1395.17		

Table 6. Hardware complexity of MIBS-80

Module (Round Function)	GE	Module (Key Schedule)	GE
Data Register	384.68	Key Register	484.46
Substitution Layer	192	S-box	48
Key Xor	85.44	Right Rotation	0
Mixing Layer	170.84	Counter Xor	13.35
Permutation Layer	0	Total Key Schedule	545.81
Right Data Xor	85.44		
Total Round Function	918.4		
Control Unit	46		
Other Logics	19.35		
Total	1529.56		

Table 7. Hardware complexity comparison of lightweight ciphers

Block ciphers	Block size	Key size	Cycles per block	Logic process	Area (GE)	Throughput at 100 KHZ(Kbps)
MIBS-64	64	64	32	0.18 μm	1396	200
PRESENT-80[5]	64	80	32	0.18 μm	1570	200
PRESENT-80[6]	64	80	563	0.18 μm	1075	11.4
AES-128[2]	128	128	1032	0.35 μm	3400	12.4
mCRYPTON[7]	64	64	13	0.13 μm	2420	492.3
HIGHT[8]	64	128	34	0.25 μm	3048	188.2
PUFFIN[10]	64	128	-	0.18 μm	2577	194
DESL[3]	64	64	144	0.18 μm	1848	44.4
DESXL[4]	64	184	144	0.18 μm	2168	44.4

right data addition are implemented as bit-wise XORs, and the mixing layer is implemented as nibble-wise XORs. The permutation layer is a simple wiring and does not have extra gates. The round keys used for each round function can be generated on-the-fly and, hence, there is no need to store all the round keys.

The implemented MIBS requires 32 clock cycles to encrypt a 64 bit plain text with 64 bit key (a single round per clock cycle), which result in throughput of 200 kilobit per second considering 100 KHz clock. MIBS implementation requires 1396 gates (2-input NAND gates). Table 5 shows the detailed gate counts of each component. The estimated area for MIBS with 80-bit key is about 1530 gates which is illustrated in table 6. A comparison for the hardware efficiency of MIBS and other lightweight block ciphers is shown in Table 7.

7 Conclusion

In this paper, we have presented a new lightweight block cipher MIBS with a 64-bit block length and 64/80-bit key lengths. Our goal in the design of MIBS was to provide security for resource-constrained applications, such as low-cost RFID tags, while having a lower hardware complexity in comparison with other compact block ciphers. MIBS is based on Feistel structure with SPN round function. We use Feistel network as an overall structure with the purpose of minimizing computational resources which is one of the important considerations in hardware design for tiny ubiquitous devices. For round function we selected the Substitution-Permutation Network. We use 4×4 S-boxes with regard to hardware efficiency and at the same time adequate security. The diffusion layer, which we named mixing layer M, is composed of 16 nibble-wise XORs, while for security against differential and linear cryptanalysis, its branch number is optimal. The hardware implementation of MIBS-64 requires 1400 gates on 0.18 μm technology, which is less than 2000 gates limit for low-cost RFID tags. We also studied the security of MIBS against several known attacks, where it showed adequate security margins. MIBS is a secure block cipher, while having a lower gate counts than PRESENT implemented in [5]. Although serialized implementation of PRESENT with lower gate counts is reported in [6], but we have not yet designed a serialized implementation for MIBS. Such an implementation remains as a future work.

References

1. Calmels, B., Canard, S., Girault, M., Sibert, H.: Low-cost cryptography for privacy in RFID systems. In: Domingo-Ferrer, J., Posegga, J., Schreckling, D. (eds.) CARDIS 2006. LNCS, vol. 3928, pp. 237–251. Springer, Heidelberg (2006)
2. Feldhofer, M., Dominikus, S., Wolkerstorfer, J.: Strong authentication for RFID systems using the AES algorithm. In: Joye, M., Quisquater, J.-J. (eds.) CHES 2004. LNCS, vol. 3156, pp. 85–140. Springer, Heidelberg (2004)
3. Leander, G., Paar, C., Poschmann, A., Schramm, K.: New lightweight DES variants. In: Biryukov, A. (ed.) FSE 2007. LNCS, vol. 4593, pp. 196–210. Springer, Heidelberg (2007)
4. Poschmann, A., Leander, G., Schramm, K., Paar, C.: A family of light-weight block ciphers based on DES suited for RFID applications. In: Proceedings of FSE (2007)

5. Bogdanov, A., Knudsen, L.R., Leander, G., Paar, C., Poschmann, A., Robshaw, M.J.B., Seurin, Y., Vikkelsoe, C.: PRESENT: An ultra-lightweight block cipher. In: Paillier, P., Verbauwhede, I. (eds.) CHES 2007. LNCS, vol. 4727, pp. 450–466. Springer, Heidelberg (2007)
6. Rolfes, C., Poschmann, A., Leander, G., Paar, C.: Ultra-Lightweight Implementations for Smart Devices-Security for 1000 Gate Equivalents. In: Grimaud, G., Standaert, F.-X. (eds.) CARDIS 2008. LNCS, vol. 5189, pp. 89–103. Springer, Heidelberg (2008)
7. Lim, C.H., Korkishko, T.: mCrypton – A lightweight block cipher for security of low-cost RFID tags and sensors. In: Song, J.-S., Kwon, T., Yung, M. (eds.) WISA 2005. LNCS, vol. 3786, pp. 243–258. Springer, Heidelberg (2006)
8. Hong, D., Sung, J., Hong, S., Lim, J., Lee, S., Koo, B.S., Lee, C., Chang, D., Lee, J., Jeong, K., Kim, H., Kim, J., Chee, S.: HIGHT: A new block cipher suitable for low-resource device. In: Goubin, L., Matsui, M. (eds.) CHES 2006. LNCS, vol. 4249, pp. 46–59. Springer, Heidelberg (2006)
9. Standaert, F.X., Piret, G., Gershenfeld, N., Quisquater, J.J.: SEA: A scalable encryption algorithm for small embedded applications. In: Domingo-Ferrer, J., Posegga, J., Schreckling, D. (eds.) CARDIS 2006. LNCS, vol. 3928, pp. 222–236. Springer, Heidelberg (2006)
10. Cheng, H., Heys, H., Wang, C.: PUFFIN: A Novel Compact Block Cipher Targeted to Embedded Digital Systems. In: Proceedings of the 2008 11th EUROMICRO Conference on Digital System Design Architectures, Methods and Tools, pp. 383–390. IEEE Computer Society, Washington (2008)
11. Kanda, M., Takashima, Y., Matsumoto, T., Aoki, K., Ohta, K.: A strategy for constructing fast round functions with practical security against differential and linear cryptanalysis. In: Tavares, S., Meijer, H. (eds.) SAC 1998. LNCS, vol. 1556, p. 264. Springer, Heidelberg (1999)
12. Rijmen, V., Daemen, J., Preneel, B., Bosselaers, A., Win, E.D.: The cipher SHARK. In: Gollmann, D. (ed.) FSE 1996. LNCS, vol. 1039, pp. 99–111. Springer, Heidelberg (1996)
13. Kanda, M., Moriai, S., Aoki, K., Ueda, H., Takashima, Y., Ohta, K., Matsumoto, T.: E2–A New 128-Bit Block Cipher. IEICE TRANSACTIONS on Fundamentals of Electronics, Communications and Computer Sciences 83(1), 48–59 (2000)
14. Aoki, K., Ichikawa, T., Kanda, M., Matsui, M., Moriai, S., Nakajima, J., Tokita, T.: Camellia: A 128-bit block cipher suitable for multiple platforms design andanalysis. In: Stinson, D.R., Tavares, S. (eds.) SAC 2000. LNCS, vol. 2012, pp. 39–56. Springer, Heidelberg (2001)
15. Biham, E., Shamir, A.: Differential cryptanalysis of the full 16-round DES. In: Brickell, E.F. (ed.) CRYPTO 1992. LNCS, vol. 740, pp. 487–496. Springer, Heidelberg (1993)
16. Matsui, M.: Linear cryptanalysis method for DES cipher. In: Helleseth, T. (ed.) EUROCRYPT 1993. LNCS, vol. 765, pp. 386–397. Springer, Heidelberg (1994)
17. Kanda, M.: Practical security evaluation against differential and linear cryptanalyses for feistel ciphers with SPN round function. In: Stinson, D.R., Tavares, S. (eds.) SAC 2000. LNCS, vol. 2012, pp. 324–338. Springer, Heidelberg (2001)
18. Kaliski Jr., B.S., Robshaw, M.J.B.: Linear cryptanalysis using multiple approximations. In: Desmedt, Y.G. (ed.) CRYPTO 1994. LNCS, vol. 839, pp. 26–39. Springer, Heidelberg (1994)
19. Biryukov, A., Canniere, C.D., Quisquater, M.: On multiple linear approximations. In: Franklin, M. (ed.) CRYPTO 2004. LNCS, vol. 3152, pp. 1–22. Springer, Heidelberg (2004)

20. Hermelin, M., Cho, J.Y., Nyberg, K.: Multidimensional linear cryptanalysis of reduced round serpent. In: Mu, Y., Susilo, W., Seberry, J. (eds.) ACISP 2008. LNCS, vol. 5107, pp. 203–215. Springer, Heidelberg (2008)
21. Langford, S.K., Hellman, M.E.: Differential-linear cryptanalysis. In: Desmedt, Y.G. (ed.) CRYPTO 1994. LNCS, vol. 839, pp. 17–25. Springer, Heidelberg (1994)
22. Biham, E., Dunkelman, O., Keller, N.: Enhancing differential-linear cryptanalysis. In: Zheng, Y. (ed.) ASIACRYPT 2002. LNCS, vol. 2501, pp. 587–592. Springer, Heidelberg (2002)
23. Knudsen, L.R., Robshaw, M.J.B.: Non-linear approximations in linear cryptanalysis. In: Maurer, U.M. (ed.) EUROCRYPT 1996. LNCS, vol. 1070, pp. 224–236. Springer, Heidelberg (1996)
24. Courtois, N.T.: Feistel schemes and bi-linear cryptanalysis. In: Franklin, M. (ed.) CRYPTO 2004. LNCS, vol. 3152, pp. 23–40. Springer, Heidelberg (2004)
25. Courtois, N.T., Pieprzyk, J.: Cryptanalysis of block ciphers with overdefined systems of equations. In: Zheng, Y. (ed.) ASIACRYPT 2002. LNCS, vol. 2501, pp. 267–287. Springer, Heidelberg (2002)
26. Courtois, N., Klimov, A., Patarin, J., Shamir, A.: Efficient algorithms for solving overdefined systems of multivariate polynomial equations. In: Preneel, B. (ed.) EUROCRYPT 2000. LNCS, vol. 1807, pp. 392–407. Springer, Heidelberg (2000)
27. Cid, C., Leurent, G.: An analysis of the xsl algorithm. In: Roy, B. (ed.) ASIACRYPT 2005. LNCS, vol. 3788, pp. 333–352. Springer, Heidelberg (2005)
28. Diem, C.: The xl-algorithm and a conjecture from commutative algebra. In: Lee, P.J. (ed.) ASIACRYPT 2004. LNCS, vol. 3329, pp. 323–337. Springer, Heidelberg (2004)
29. Biryukov, A., Wagner, D.: Slide attacks. In: Knudsen, L.R. (ed.) FSE 1999. LNCS, vol. 1636, pp. 245–259. Springer, Heidelberg (1999)
30. Biham, E.: New types of cryptanalytic attacks using related keys. Journal of Cryptology 7(4), 229–246 (1994)

Appendix I

Test vectors of MIBS for each key lenght are given here. The data are expressed in hexadecimal form.

Table 8. Test vectors for 64 bit key

Plaintext	Key	Ciphertext
00000000 00000000	00000000 00000000	6D1D3722 E19613D2
00000000 00000001	00000000 00000000	D79C5610 0851488A
00000000 00000000	FFFFFFFF FFFFFFFF	E538379F 99337F4A
00000000 00000001	FFFFFFFF FFFFFFFF	EF0840A9 4FCC2EAF
FFFFFFFF FFFFFFFF	00000000 00000000	66F21F5B 1F96D626
FFFFFFFF FFFFFFFE	00000000 00000000	5D86E9E2 96B4527F
FFFFFFFF FFFFFFFF	FFFFFFFF FFFFFFFF	595263B9 3FFE6E18
FFFFFFFF FFFFFFFE	FFFFFFFF FFFFFFFF	598CE962 22A34BDE

Table 9. Test vectors for 80 bit key

Plaintext	Key	Ciphertext
00000000 00000000	00000000 00000000 0000	F575004B 83ABA59F
00000000 00000001	00000000 00000000 0000	C80A965F 0969BB70
00000000 00000000	FFFFFFFF FFFFFFFF FFFF	F2144A89 F33C2AF0
00000000 00000001	FFFFFFFF FFFFFFFF FFFF	7A443766 74739625
FFFFFFFF FFFFFFFF	00000000 00000000 0000	DE2860FD B436725E
FFFFFFFF FFFFFFFE	00000000 00000000 0000	4617D4EB 1CE9E088
FFFFFFFF FFFFFFFF	FFFFFFFF FFFFFFFF FFFF	3185C8A3 5B51EB23
FFFFFFFF FFFFFFFE	FFFFFFFF FFFFFFFF FFFF	FC835FF 013970A5

Distinguishing and Second-Preimage Attacks on CBC-Like MACs[*]

Keting Jia[1], Xiaoyun Wang[1,2], Zheng Yuan[2,3], and Guangwu Xu[2,4]

[1] Key Laboratory of Cryptologic Technology and Information Security,
Ministry of Education, Shandong University, Jinan 250100, China
ktjia@mail.sdu.edu.cn
[2] Institute for Advanced Study, Tsinghua University, Beijing 100084, China
xiaoyunwang@tsinghua.edu.cn
[3] Beijing Electronic Science and Technology Institute, Beijing 100070, China
zyuan@mail.tsinghua.edu.cn
[4] Department of Electrical Engineering and Computer Science,
University of Wisconsin-Milwaukee, USA
gxu4uwm@uwm.edu

Abstract. This paper first presents a new distinguishing attack on the CBC-MAC structure based on block ciphers in cipher block chaining (CBC) mode. This attack detects a CBC-like MAC from random functions. The second result of this paper is a second-preimage attack on the CBC-MAC, which is an extension of the attack of Brincat and Mitchell. The attack also covers MT-MAC, PMAC and MACs with three-key enciphered CBC mode. Instead of exhaustive search, both types of attacks are of birthday attack complexity.

Keywords: CBC, MAC, Distinguishing attack, Second preimage attack.

1 Introduction

A message authentication code (MAC), also known as a keyed hash function, is a short piece of information used to authenticate both the source of a message and its integrity without the use of any additional mechanisms. A MAC algorithm takes as input a secret key and an arbitrary-length message to be authenticated, and outputs a short tag. As an important cryptographic primitive, MACs have been widely used in practice. The applications include internet communication protocols, e-commerce, e-banking etc. The cryptanalytic model of MACs usually involves three participants: a sender, a receiver and an adversary. The sender and the receiver have agreed on a secret key (or a set of keys). Prior to sending a message, the sender uses a MAC algorithm to produce an authentication tag (or MAC) from the message and the secret key. On receipt, the receiver verifies the message and the corresponding MAC by the same calculation using their shared secret key. The goal of the adversary is to trick the receiver into

[*] Supported by the National Natural Science Foundation of China (NSFC Grant No. 60525201) and 973 Project (No.2007CB807902).

J.A. Garay, A. Miyaji, and A. Otsuka (Eds.): CANS 2009, LNCS 5888, pp. 349–361, 2009.

accepting a message that was not sent by the sender. There are three main types of constructions for MAC algorithms: the construction based on block ciphers (OMAC, CBC-MAC and PMAC etc), the construction based on cryptographic hash functions (HMAC, NMAC, MDx-MAC etc) and the construction based on universal hashing. In this paper, we shall focus on the first type of construction.

The Security of MAC Algorithms. One of the most important requirements for a MAC is that, given a massage M and a l_k-bit secret key K, the computation of the MAC value $\text{MAC}_K(M)$ should be easy. However, it should be computationally infeasible to find $\text{MAC}_K(M)$ without knowing K. Security threats to a MAC algorithm include:

Existential Forgery. An adversary is able, without initial knowledge of K, to get a corresponding MAC C for any message M, which has not been MACed by the legitimate MAC generator. The message M may not have any particular meaning.

Selective Forgery. An adversary is able to determine the MAC for a message of his choice.

Second Preimage Attack. Second preimage attack is sometimes referred to as weak collision attack. If an adversary observes M and the corresponding MAC C, he can construct a message $M' \neq M$ with $\text{MAC}_K(M') = C$ without initial knowledge of K. Ideally, the relation $\Pr[\text{MAC}_K(M') = C] = 2^{-l_m}$ should hold in this case, where l_m is the length of the tag.

Universal Forgery. An adversary is able to find a MAC for every given message. This attack is much more powerful than previous cases.

Key Recovery Attack. A key recovery attack is more devastating than forgery. In this case an adversary is able to recover K itself, and thus can perform arbitrary forgeries. Ideally, any attack allowing key recovery requires about 2^{l_k} operations. Verification of such an attack requires $\lceil \frac{l_k}{l_m} \rceil$ text-MAC pairs.

Related Work. CBC-MAC is a technique for constructing a message authentication code from a block cipher through CBC mode. Most standards and applications use the CBC-MAC, such as [1,2,11]. Bellare et al.[3] formally examine the security of this construction for messages with fixed length. Variants of the CBC-MAC for variable length messages were then proposed, examples include EMAC [5], XCBC [6], TMAC [14], OMAC [13]and CMAC [16] etc. PMAC [7] was proposed by Black and Rogaway for parallel processing. Recently some provably secure MACs from differentially-uniform permutations were brought forward by Minematsu and Tsunoo [15]. Dodis et.al. [10] introduced a new mode of operation for block ciphers and length-preserving MACs.

In [18], Preneel and van Oorschot proposed a general forgery attack on all iterative MACs using the birthday attack. In [19], they presented a key recovery attack on the retail MAC [2] based on DES, which requires $2^{32.5}$ known text-MAC pairs and $3 \cdot 2^{56}$ off-line computations to find the 112-bit key. Knudesn [12] presented a forgery attack on CBC-MAC based on n-bit block cipher with

$2^{1+(n-l_m)/2}$ known text-MAC pairs and two known texts, where l_m is the truncation length of the tag. In [9], Coppersmith and Mitchell proposed a key recovery attack against the MacDES. This attack was further improved by Coppersmith, Knudsen and Mitchell [8]. The improved attack was applied to MacDES involves prefixing the data to be MACed with its length. In [4], Brincat and Mitchell gave new CBC-MAC forgery attacks. Recently, new techniques to identify the underlying hash functions of MACs were developed by Wang et al. in [22,21]. In [22], Wang et al. presented distinguishing attacks on HMAC/NMAC-MD5 and MD5-MAC, partial subkey recovery of MD5-MAC can be achieved. Their distinguisher makes use of internal near-collisions, which leaks more information than internal collisions. A distinguisher based on the internal near-collision was shown in [23]. Built upon that, a forgery attack on ALRED construction and equivalent subkey recovery attack on its AES-based instance Alpha-MAC were suggested.

Our Contribution. This paper explores a new distinguishing attack on MACs, which can be used to distinguish MACs based on block ciphers in CBC mode from random functions. Another main result of the paper is to construct a second preimage attack on the CBC-like MACs, which can be regarded as an extension of the attacks proposed by Brintcat and Mitchell [4]. Besides the cases discussed in [4], our attack also applies to MT-MAC [15], PMAC [7] and MACs with three-key enciphered CBC mode [10]. Our second preimage attack can be achieved with birthday attack complexity.

Our approach is to utilize the CBC structure, and turn the complexity of the second preimage attack into birthday attack complexity. More specifically, given a two-block message $x_1\|x_2$, we want to create another message $x_1'\|x_2'$ with the same MAC, i.e., $E_K(x_1') \oplus x_2' = E_K(x_1) \oplus x_2$. From the CBC operation, we get $E_K(x_1) \oplus x_2' = E_K(x_1') \oplus x_2$. We are able to choose x_1', x_2' at random to get a collision–$\mathrm{MAC}_K(x_1\|x_2') = \mathrm{MAC}_K(x_1'\|x_2)$. In particular, we explore the second preimage attack on the CBC-MAC, EMAC, XCBC, TMAC, OMAC, CMAC, PC-MAC, MT-MAC and PMAC etc.

Organization of the Paper. This paper is organized as follows. In Section 2, after basic notations are reviewed, we give brief descriptions of some MACs based on block ciphers in CBC mode. Section 3 shows a new method for distinguishing a CBC-like MAC from a random function. A second preimage attack on the CBC-like MACs is introduced in Section 4. Finally, we summary our results in Section 5.

2   Preliminaries and Notations

In this section, we first list some notations used in this paper, and then give brief descriptions of the relative MAC algorithms.

2.1 Notations

$M\|N$	:	the concatenation of two messages M and N
$\|M\|$	:	the length of M
K	:	the secret key
C	:	the output of a MAC taking a secret key K and a message M as input
x_i	:	the ith block of a message
y_i	:	the internal state after the ith iteration of a MAC
0^i	:	the strings of i 0s.
$\cdot$	:	the multiplication in the field $GF(2^n)$
l_m	:	the length of the MAC output
Σ^n	:	$\{0,1\}^n$, the set of all strings of length n

2.2 The CBC-Like MAC Algorithms

MAC algorithms based on block ciphers are of great practical significance. CBC-MAC is a well-known method to generate a MAC based on a block cipher in CBC mode. In general, we will call MAC algorithms based on block ciphers in CBC mode the *CBC-like* MAC algorithms.

CBC-MAC is used to compress the message M of a fixed length mn with a secret key K, where n is the length of a block, and m is the number of blocks. More precisely, the CBC-MAC is defined as:

$$y_0 = 0,$$
$$y_i = E_K(y_{i-1} \oplus x_i), \ i = 1, \ldots, m,$$
$$C = CBC_K(M) = f(y_m),$$

where f is a truncation function. For messages with fixed length, Bellare, Kilian, and Rogaway [3] established the security of the CBC-MAC. However, a well known fact says that the CBC-MAC without truncation is not secure for variable-length messages. In fact, suppose that an adversary has known $C = CBC_K(M)$, $C' = CBC_K(M')$, then for any single block Y, the messages $M\|Y$ and $M'\|(Y \oplus (C \oplus C'))$ have the same MAC. Several variants of CBC-MAC have been proposed for variable length messages in order to avoid this attack.

EMAC. EMAC is the encrypted CBC MAC, first proposed by Bosselaers et al. as RIPE-MAC in [5]. It is obtained by encrypting the CBC-MAC value by the block cipher encryption E again with a new key K_2. That is,

$$C = EMAC_{K_1,K_2}(M) = E_{K_2}(CBC_{K_1}(M)),$$

where K_1 is the key of the CBC-MAC and $CBC_{K_1}(M)$ is the CBC-MAC value of M without truncation. The result of EMAC is C. Petrank and Rackoff [20] proved that EMAC is secure if the message length is a positive multiple of n. It is remarked that, however, EMAC requires two key scheduling of the corresponding encryption E.

XCBC. Black and Rogaway [6] suggested some simple variants of the CBC-MAC i.e., ECBC, FCBC and XCBC. These modified CBC-MACs can be used to MAC messages of arbitrary lengths efficiently. The most typical construction in [6] is XCBC which requires only one key scheduling of the built-in block cipher E. In general, XCBC takes three keys: one block cipher key K_1, and two n-bit keys K_2 and K_3. If the message length is the multiple of n, set $K = K_2$, $P = M$. Otherwise, set $K = K_3$, $P = M \| 10^i$, where $i = n - 1 - (|M| \mod n)$. Write $P = x_1 \| \cdots \| x_m$, $y_0 = 0$. Then $y_i = E_{K_1}(x_i \oplus y_{i-1})$, for $i = 1, \ldots, m - 1$. The XCBC value is $C = E_{K_1}(x_m \oplus y_{m-1} \oplus K)$.

TMAC. Kurosawa and Iwata introduced two-key CBC-MAC (TMAC) in [14]. TMAC takes two keys with $(l_k + n)$ bits in total, one block cipher key K_1 with l_k bits and the other key K_2 with n bits. TMAC is obtained from XCBC by replacing (K_2, K_3) with $(K_2 \cdot u, K_2)$, where u is some non-zero constant.

OMAC. OMAC, also proposed by Iwata and Kurosawa [13], is a generic name for OMAC1 and OMAC2. OMAC1 is obtained from XCBC by replacing (K_2, K_3) with $(L \cdot u, L \cdot u^2)$ for some non-zero constant u in $GF(2^n)$, where L is given by $L = E_K(0^n)$. OMAC2 is constructed in a similar manner by using $(L \cdot u, L \cdot u^{-1})$. Note that $L \cdot u$, $L \cdot u^{-1}$ and $L \cdot u^2 = (L \cdot u) \cdot u$ can be computed efficiently by one shift operation and one conditional XOR from L, L and $L \cdot u$, respectively.

CMAC. CMAC [16] is equivalent to OMAC1, which is recommended by NIST.

PC-MAC and MT-MAC. PC-MAC and MT-MAC were proposed by Minematsu and Tsunoo [15] for higher performance for applications. Both are based on truncated block ciphers. PC-MAC is a very efficient periodic CBC-like construction. MT-MAC is an efficient MAC with provable security based on the modified tree hash (MTH). See [15] for more details.

Three-Key Enciphered CBC Mode. Three-key enciphered CBC mode [10] was constructed using the preserving fixed input length MAC by Dodis et.al. To describe it, some notations are needed. For $f_1, f_2 : \Sigma^n \to \Sigma^n$, let $g[f_1, f_2](x_2 \| x_1) = f_1(x_1) \oplus f_2(x_2)$. The compression function $G[f_1, f_2]$ is defined as

$$G[f_1, f_2](x_1 \| \cdots \| x_t) = g[f_1, f_2](x_t \| g[f_1, f_2](\cdots g[f_1, f_2](x_2 \| x_1) \cdots)).$$

The three key enciphered CBC construction uses three length-preserving functions $f_1, f_2, f_3 : \Sigma^n \to \Sigma^n$ and takes a variable-length input $M = x_1 \| \cdots \| x_t$ (w.l.o.g., we assume the length to be a multiple of n; if not, then append a bit "1" followed by the minimum "0"s to achieve this). The mode is defined as:

$$H[f_1, f_2, f_3](x_1 \| \cdots \| x_t) = f_3(G[f_1, f_2](x_1 \| \cdots \| x_t)).$$

Enhanced enciphered CBC, suggested in [10], can be used for the "unkeyed" settings (RO and CRHF) as well as the "keyed" settings (PRF and MAC).

The mode is denoted as $H^*[\pi_1, \pi_2, \pi_3]$. The MAC can be regarded as the basic enciphered CBC mode $H[f_1, f_2, f_3]$ with length-preserving functions f_1, f_2, f_3, where $f_i = \pi_i(x) \oplus x$ for $i=1, 2$, and $f_3(x) = \pi_3(x) \oplus \pi_3^{-1}(x)$.

3   Distinguishing Attack on CBC-Like MACs

Preneel and van Oorschot [18] proposed a general method to distinguish an iterative MAC algorithm from a random function, which is based on an internal collision searched by the birthday attack. However, we note the free-collision property of the CBC-like MACs with only one-block as a variable. This follows from the fact that the built-in block cipher is a permutation. So we utilize the free-collision property to distinguish a block-cipher-based MAC from a random function.

More specifically, we have the following observation.

Proposition 1. *Let $n, m \geq 2$ be integers. Given the block cipher $E: \Sigma^n \to \Sigma^n$ and a secret key K, the values of CBC-MAC without truncation for the messages $x_1 \| x_2 \| \cdots \| x_m$ and $x_1' \| x_2 \| \cdots \| x_m$ must be different provided that $x_1 \neq x_1'$, where x_1' and $x_i, i = 1, 2, \ldots, m$ are single-block messages.*

Proof. Let
$$y_1 = E_K(x_1) \text{ and } y_1' = E_K(x_1').$$
It is immediate that $y_1 \neq y_1'$ since $E_K(\cdot)$ is a permutation on Σ^n.

For $i = 2, \ldots, m$, y_i and y_i' can be recursively defined as
$$y_i = E_K(x_i \oplus y_{i-1}) \text{ and } y_i' = E_K(x_i \oplus y_{i-1}').$$

First, we claim that $y_2 \neq y_2'$. Otherwise, since E is a permutation, we would have
$$x_2 \oplus y_1 = x_2 \oplus y_1',$$
which forces that $y_1 = y_1'$. This is a contradiction.

Inductively, we have
$$y_i \neq y_i' \qquad \text{for } i = 3, \ldots, m.$$

In particular, the inequality $y_m \neq y_m'$ holds, and hence
$$C \neq C',$$
when there is no truncation on y_m and y_m'. $\square$

Without taking into account of the truncation, it is obvious that the CBC-MAC is a permutation on a single message block with other blocks fixed.

If there is a truncation function on the final output of the MACs, we can use the method in [18] to detect the internal collision (collision before the truncation) from all the collisions. For simplicity, we assume that there is no truncation function on the final output of the MACs (i.e., $l_m = n$) in the rest of our discussion.

To perform the distinguishing attack, the adversary is given an oracle $C = CBC_K(\cdot)$, and makes $2^{(n+1)/2}$ queries with m-block messages that have the same last $m - 1$ blocks. The following is the explicit structure S:

$$S = \{ M^i \mid M^i = x_1^i \| x_2 \| \cdots \| x_m, i = 1, 2, \ldots, 2^{(n+1)/2} \},$$

where x_1^i's are different random single-block messages. If there is a collision in the structure, the MAC algorithm is a random function. Otherwise it is the CBC-MAC.

This attack requires about $2^{(n+1)/2}$ chosen messages. By the birthday paradox, the success probability is 0.63.

Most variations of the CBC-MAC are obtained by modifying the padding method of the message and encrypting the value of the CBC-MAC with another key or the same key again. So these MACs are still permutations on single block (the other blocks are fixed). Therefore, the above distinguishing technique can also be used to attack other block-cipher-based MACs mentioned earlier, such as EMAC, XCBC, TMAC and OMAC. For these cases, we only need messages with one or two blocks to query the corresponding oracle.

4   Second Preimage Attack on CBC-Like MACs

Second preimage resistance is sometimes referred to as weak collision resistance. It means that if an adversary obtains M and the corresponding tag C, it must be computationally infeasible to construct another message M' s.t. $\mathrm{MAC}_K(M') = C$, without knowledge of K. The ideal complexity of finding the M' is 2^{l_m}. In ACISP 2001, Brincat and Mitchell introduced a second-preimage attack with birthday attack complexity. In this section, we extend this attack to the cases that include CBC-like MACs and PMAC, which proposed by Black and Rogaway as a parallelizable block-cipher mode of operation for message authentication [7].

4.1   The Attack of Brincat and Mitchell

The new forgery attack on CBC-MAC of Brincat and Mitchell [4] is a kind of second-preimage attack. To perform this attack, the attacker constructs two structures:

$$S_1 = \{ M^i \mid M^i = M_1 \| \cdots \| M_q \| X^i \| F_1 \| \cdots \| F_r, i = 1, \ldots, 2^{n/2} \},$$
$$S_2 = \{ M^i \mid M^i = Y^i \| F_1 \| \cdots \| F_r, i = 1, \ldots, 2^{n/2} \},$$

where $M_1, \ldots, M_q$ are arbitrary n-bit blocks, q is a positive integer, $F_1, F_2, \ldots, F_r$ are arbitrary but fixed n-bit blocks, and $X^i (1 \leq i \leq 2^{n/2})$ are n-bit blocks and pairwise distinct, so are $Y^i (1 \leq i \leq 2^{n/2})$. Then the attacker queries the corresponding MACs with elements in S_1 and S_2. It is argued that with high probability, a MAC of some element in S_1 matches that of an element in S_2. In other words, there are positive integers k, j such that $CBC(M_1\| \cdots \|M_q\|X^j\|F_1\| \cdots \|F_r)$ $=CBC(Y^k\|F_1\| \cdots \|F_r)$. Since the n−bit blocks $F_1, \ldots, F_r$ are the same for the two messages, it is immediate that $CBC^*(M_1\| \cdots \| M_q\|X^j) = CBC^*(Y^k)$. Here we use $CBC^*(X)$ to denote the computation of the MAC on message X without the output transformation. This yields $CBC^*(M_1\| \cdots \|M_q) = X^j \oplus Y^k$. As a result, if the attacker knows the MAC value C for the message $Z\|P_1\| \cdots \|P_t(t \geq 1)$, then the attacker knows that the MAC for the message $M_1\| \cdots \|M_q\|X^j \oplus Y^k \oplus Z\|P_1\| \cdots \|P_t$ is also C.

4.2   The Second-Preimage Attack on the CBC-MAC

In this subsection, we propose another second-preimage attack on the CBC-MAC. In this attack, we can construct a message that is different from the given message (with at least two blocks) and also with the birthday attack complexity.

Let us start with the following statement.

Proposition 2. *Let $n, m > 2$ be integers, $E: \Sigma^n \rightarrow \Sigma^n$ be the block cipher encryption, and K the secret key for the CBC-MAC. Then the CBC-MAC values of $x_1\|x_2\|x_3\| \cdots \|x_m$ and $x_1'\|x_2'\|x_3\| \cdots \|x_m$ are the same if and only if $E_K(x_1)\oplus x_2 = E_K(x_1')\oplus x_2'$, where x_1', x_2' and $x_i, i = 1, 2, \ldots, m$ are single-block messages.*

Proof. The proof is straightforward.

Note that the CBC-MAC value of $x_1\|x_2\|x_3\| \cdots \|x_m$ is given by

$$E_K(E_K(\cdots (E_K(E_K(E_K(x_1) \oplus x_2) \oplus x_3) \cdots) \oplus x_{m-1}) \oplus x_m).$$

Similarly, the CBC-MAC value of $x_1'\|x_2'\|x_3\| \cdots \|x_m$ is

$$E_K(E_K(\cdots (E_K(E_K(E_K(x_1') \oplus x_2') \oplus x_3) \cdots) \oplus x_{m-1}) \oplus x_m).$$

It is then easy to see that these two values are the same if and only if

$$E_K(x_1) \oplus x_2 = E_K(x_1')\oplus x_2'. \qquad \square$$

Suppose that the adversary is given an oracle $C = CBC_K(\cdot)$. Assume that the adversary has intercepted a message $M_0 = x_1\|x_2\| \cdots \|x_m$ and its MAC C_0. A collision is shown in Fig. 1.

The following procedure can be used to forge the MAC C_0 by finding another message M'.

1. Construct two structures

$$S_1 = \{ M_1^i \mid M_1^i = x_1\|x_2^i\|x_3\| \cdots \|x_m, i = 1, \ldots, 2^{(n+1)/2} \},$$
$$S_2 = \{ M_2^i \mid M_2^i = x_1^i\|x_2\|x_3\| \cdots \|x_m, i = 1, \ldots, 2^{(n+1)/2} \},$$

    where $x_2^1, x_2^2, \ldots$ are distinct random message blocks that are different from x_2, and so are $x_1^1, x_1^2, \ldots$ from x_1.
2. Query the oracle for the messages in the two structures and obtain $C_1^i = CBC_K(M_1^i)$, $C_2^i = CBC_K(M_2^i)$, where i=1, 2, ..., $2^{(n+1)/2}$.

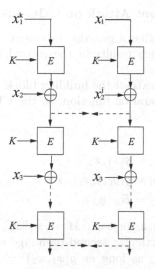

Fig. 1. A Collision of the CBC-MAC

3. Search a collision by the birthday attack. Assume w.l.o.g. that $C_1^k=C_2^j$. By Proposition 2,

$$E_{K_1}(x_1) \oplus x_2^j = E_{K_1}(x_1^k) \oplus x_2,$$

then

$$E_{K_1}(x_1) \oplus x_2 = E_{K_1}(x_1^k) \oplus x_2^j.$$

So the message $M' = x_1^k\|x_2^j\|x_3\|\cdots\|x_m$ has the same MAC value with the message M.

Given the message $M_0 = x_1\|x_2\|\cdots\|x_m$, it is not difficult to see that our technique still works for the following structures

$$S_1 = \{\, M_1^i \mid M_1^i = x_1\|\cdots\|x_j\|x_{j+1}^i\|x_{j+2}\|\cdots\|x_m,\ i = 1,\ldots, 2^{(n+1)/2} \,\},$$

where $x_{j+1}^1, x_{j+1}^2, \ldots$ are distinct random message blocks that are different from x_{j+1}; and

$$S_2 = \{\, M_2^i \mid M_2^i = X^i\|x_{j+1}\|x_{j+2}\|\cdots\|x_m, i = 1,\ldots, 2^{(n+1)/2} \,\},$$

where X^i's are distinct random j-block messages, and different from $x_1\|\cdots\|x_j$.

A collision can be obtained by the birthday attack, i.e., $M_1^k = M_2^l$. So we can deduce $E_K(y_j \oplus x_{j+1}^k) = E_K(y_j' \oplus x_{j+1})$, where $y_j' = CBC^*(X^l)$, thus $E_K(y_j \oplus x_{j+1}) = E_K(y_j' \oplus x_{j+1}^k)$. The messages $X^l\|x_{j+1}^k\|x_{j+2}\|\cdots\|x_m$ and M have the same tag.

This attack is apparently working for the following MACs with arbitrary message length such as EMAC, XCBC, TMAC, OMAC, CMAC, PC-MAC and MT-MAC.

4.3 A General Statement Attack on CBC-Like MACs

In this subsection, we describe a general statement about the building blocks of MAC constructions, which results in the second preimage attack on more CBC-like MACs.

By the above attack, we extract the building block: $g(x, y) = E_K(x) \oplus y$ from the CBC-MAC as a new iteration function, so the CBC-MAC for the message $M = x_1 \| x_2 \| \cdots \| x_m$ is

$$y_2 = g(x_1, x_2),$$
$$y_i = g(y_{i-1}, x_i), \ i = 3, \ldots, m,$$
$$y_{m+1} = E_K(y_m).$$

It can be seen that, given a message $M = x_1 \| x_2 \| \cdots \| x_i \| x_{i+1} \| \cdots \| x_m$ and its CBC-MAC value, a particular second-preimage can be of the form $M' = x_1' \| x_2' \| \cdots \| x_i \| x_{i+1} \| \cdots \| x_m$ as long as $g(x_1, x_2) = g(x_1', x_2')$. So, the second preimage attack for the CBC-MAC can be reduced to the second preimage attack on the equation $g(x, y) = E_K(x) \oplus y$, where x, y are independent block variables.

Definition 1. *Let $b_1, b_2 \geq n$, $f_1 : \Sigma^{b_1} \to \Sigma^n$ and $f_2 : \Sigma^{b_2} \to \Sigma^n$ be two maps. Define $g_{f_1, f_2} : \Sigma^{b_1} \times \Sigma^{b_2} \to \Sigma^n$ as*

$$g_{f_1, f_2}(x, y) = f_1(x) \oplus f_2(y),$$

where $x \in \Sigma^{b_1}$ and $y \in \Sigma^{b_2}$.

For notational simplicity, we shall write g for g_{f_1, f_2} hereafter.

Proposition 3. *Given $x_1, y_1 \in \Sigma^b$, there exists an algorithm to search (x_2, y_2) such that $g(x_2, y_2) = g(x_1, y_1)$ with birthday attack complexity $2^{(n+3)/2}$. The probability of success is 0.63.*

Proof. We present Algorithm 1 similar to Yuval's birthday attack algorithm [24].

Algorithm 1. Find another pair (x_2, y_2) to make $g(x_2, y_2) = g(x_1, y_1)$.

INPUT   : x_1, y_1, g.
OUTPUT   : x_2, y_2.
1. $S_1 \leftarrow \emptyset$
2. For $i \leftarrow 0$ to $2^{(n+1)/2}$ do
    Choose $x_1^i \notin \{x_1, x_1^0, \ldots, x_1^{i-1}\}$ at random and compute $z_1^i \leftarrow g(x_1^i, y_1)$
    $S_1 \leftarrow S_1 \bigcup \{(x_1^i, z_1^i)\}$
    End For
3   For $i \leftarrow 0$ to $2^{(n+1)/2}$ do
    Choose $y_1^i \notin \{y_1, y_1^0, \ldots, y_1^{i-1}\}$ at random and compute $z_2^i \leftarrow g(x_1, y_1^i)$
    If $z_2^i = z_1^k$ where z_1^k is the second component of an element of S_1
        Return (x_1^k, y_1^i)
    End For

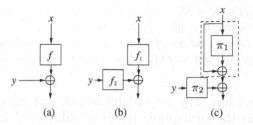

Fig. 2. Three Types of Building Blocks to MAC Iterative Construction

To show that this is a successful attack, we need to check two things. First, the forgery is valid, i.e., $g(x_2, y_2) = g(x_1, y_1)$. Second, the messages are different, i.e., $(x_2, y_2) \neq (x_1, y_1)$.

If a pair (x_2, y_2) is obtained from the Algorithm 1, $(x_2, y_2) \neq (x_1, y_1)$ is obvious.

Next, we prove $g(x_2, y_2) = g(x_1, y_1)$. By the algorithm, we have $g(x_2, y_1) = g(x_1, y_2)$, i.e.,

$$f_1(x_2) \oplus f_2(y_1) = f_1(x_1) \oplus f_2(y_2).$$

This implies that $f_1(x_2) \oplus f_2(y_2) = f_1(x_1) \oplus f_2(y_1)$, and hence

$$g(x_2, y_2) = g(x_1, y_1).$$

Step 2 needs $2^{(n+1)/2}$ computations of the function g, and Step 3 has the same complexity. Therefore, the complexity of the algorithm is $2^{(n+3)/2}$. The probability of success is

$$1 - \prod_{i=1}^{2^{(n+1)/2}-1} (1 - \frac{i}{2^n}) \approx 1 - e^{-1} \approx 0.63. \qquad \square$$

By our observation, there are three types of building blocks for MAC iterative constructions that are consistent with the general statement. See Fig. 2.

1. The first type of building block is defined as $g(x, y) = E_K(x) \oplus y$, where x, y are independent variables. The building block is available to CBC-MAC, EMAC, XCBC, OMAC, PC-MAC etc. It is obvious that $f_1(x) = E_K(x)$, and $f_2(y) = y$. See Fig 2 (a). We would like to point out that the CFB-MAC[1] is the type, whose iteration function is defined as follows:

$$y_i = g(y_{i-1}, x_i) = E_K(y_{i-1}) \oplus x_i.$$

2. The second type is given by $g(x, y) = f_1(x) \oplus f_2(y)$, where x, y are independent variables. The building block $g(x, y)$ comes from the MAC based on the three-key enciphered CBC mode and PMAC. See Fig. 2 (b).
3. The third type is $g(x, y) = (\pi_1(x) \oplus x) \oplus \pi_2(y)$, which is corresponding to the MAC with the enhanced three-key enciphered CBC mode. Here, $f_1(x) = \pi_1(x) \oplus x$, and $f_2(y) = \pi_2(y)$. See Fig. 2 (c).

Proposition 4. *Let C be a MAC algorithm such that $C(x\|y\|m_f) = f(g(x,y))$, where f is a permutation, $g(x,y) = f_1(x) \oplus f_2(y)$, x and y are independent (block) variables, m_f is a concatenation of blocks. Then there exists an algorithm to get the second-preimage for any message of at least two blocks with birthday attack complexity.*

Proof. Let $M = x\|y\|m_f$, x, y are variable blocks, and m_f is a fixed concatenation of blocks. By the assumption, the MAC value of M is a permutation of $g(x,y)$. By Proposition 3, we can get (x', y'), s.t. $g(x', y') = g(x, y)$ with birthday attack complexity. Since f is a permutation, then $f(g(x', y')) = f(g(x, y))$, i.e., $C(M') = C(M)$, where $M' = x'\|y'\|m_f$. $\qquad\square$

It is clear that, the general statement is the core of the second preimage attack, which is applicable to most CBC-like MACs, including the recently proposed MAC based on three-key enciphered CBC mode and its enhanced version etc.

5 Conclusion

This paper first explores a new distinguishing attack on CBC-like MACs. The distinguisher can detect a CBC-like MAC from a random function. The other result of this paper is to show a second preimage attack on a variety of MACs based on block ciphers in CBC mode. In particular, we have proved that the MACs such as CBC-MAC, EMAC, XCBC, TMAC, OMAC, CMAC, PC-MAC, MT-MAC, three-key enciphered CBC mode and PMAC etc. are all vulnerable to the second preimage attack.

Acknowledgements. We would like to thank the reviewers for their very helpful comments on the paper.

References

1. ANSI X9.9 (revised): Financial Institution Message Authentication (wholesale), American Bankers Association (1986)
2. ANSI X9.19: Financial Institution Retail Message Authentication, American Bankers Association (1986)
3. Bellare, M., Kilian, J., Rogaway, P.: The Security of the Cipher Block Chaining Message Authentication Code. In: Desmedt, Y.G. (ed.) CRYPTO 1994. LNCS, vol. 839, pp. 341–358. Springer, Heidelberg (1994)
4. Brincat, K., Mitchell, C.J.: New CBC-MAC Forgery Attacks. In: Varadharajan, V., Mu, Y. (eds.) ACISP 2001. LNCS, vol. 2119, pp. 3–14. Springer, Heidelberg (2001)
5. Bosselaers, A., Preneel, B. (eds.): RIPE 1992. LNCS, vol. 1007. Springer, Heidelberg (1995)
6. Black, J., Rogaway, P.: CBC MACs for Arbitrary-Length Messages: The Three-Key Constructions. In: Bellare, M. (ed.) CRYPTO 2000. LNCS, vol. 1880, pp. 197–215. Springer, Heidelberg (2000)

7. Black, J., Rogaway, P.: A Block-Cipher Mode of Operation for Parallelizable Message Authentication. In: Knudsen, L.R. (ed.) EUROCRYPT 2002. LNCS, vol. 2332, pp. 384–397. Springer, Heidelberg (2002)
8. Coppersmith, D., Knudsen, L.R., Mitchell, C.J.: Key Recovery and Forgery Attacks on the MacDES MAC Algorithm. In: Bellare, M. (ed.) CRYPTO 2000. LNCS, vol. 1880, pp. 184–196. Springer, Heidelberg (2000)
9. Coppersmith, D., Mitchell, C.J.: Attacks on MacDES MAC algorithm. Electronics Letters 35, 1626–1627 (1999)
10. Dodis, Y., Pietrzak, K., Puniya, P.: A New Mode of Operation for Block Ciphers and Length-Preserving MACs. In: Smart, N.P. (ed.) EUROCRYPT 2008. LNCS, vol. 4965, pp. 198–219. Springer, Heidelberg (2008)
11. ISO/IEC 9797-1, Information technology Security techniques Message Authentication Codes (MACs) Part 1: Mechanisms using a block cipher. International Organization for Standardization, Genève, Switzerland (1999)
12. Knudsen, L.R.: Chosen-text Attack on CBC-MAC. Electronic Letters 33(1) (1997)
13. Iwata, T., Kurosawa, K.: OMAC: One-Key CBC MAC. In: Johansson, T. (ed.) FSE 2003. LNCS, vol. 2887, pp. 129–153. Springer, Heidelberg (2003)
14. Kurosawa, K., Iwata, T.: TMAC: Two-Key CBC MAC. In: Joye, M. (ed.) CT-RSA 2003. LNCS, vol. 2612, pp. 265–273. Springer, Heidelberg (2003)
15. Minematsu, K., Tsunoo, Y.: Provably Secure MACs from Differentially-Uniform Permutations and AES-Based Implementations. In: Robshaw, M.J.B. (ed.) FSE 2006. LNCS, vol. 4047, pp. 226–241. Springer, Heidelberg (2006)
16. NIST, Recommendation for Block Cipher Modes of Operation: The CMAC Mode for Authentication. NIST Special Publication 800-38B (2005)
17. Preneel, B., Knudsen, L.R.: MacDES: MAC algorithm based on DES. Electronic Letters 33(1) (1997)
18. Preneel, B., van Oorschot, P.C.: MDx-MAC and Building Fast MACs from Hash Functions. In: Coppersmith, D. (ed.) CRYPTO 1995. LNCS, vol. 963, pp. 1–14. Springer, Heidelberg (1995)
19. Preneel, B., van Oorschot, P.C.: Key Recovery Attack on ANSI X9.19 Retail MAC. Electronic Letters 32(17) (1996)
20. Petrank, E., Rackoff, C.: CBC MAC for Real-Time Data Sources. J. Cryptology 13(3), 315–338 (2000)
21. Wang, X., Wang, W., Jia, K., Wang, M.: New Distinguishing Attack on MAC using Secret-Prefix Method. In: Dunkelman, O. (ed.) FSE 2009. LNCS, vol. 5665, pp. 363–374. Springer, Heidelberg (2009)
22. Wang, X., Yu, H., Wang, W., Zhang, H., Zhan, T.: Cryptanalysis on HMAC/NMAC-MD5 and MD5-MAC. In: Joux, A. (ed.) EUROCRYPT 2009. LNCS, vol. 5479, pp. 121–133. Springer, Heidelberg (2009)
23. Yuan, Z., Jia, K., Wang, W., Wang, X.: Distinguishing and Forgery Attacks on Alred and Its AES-based Instance Alpha-MAC (2008), http://eprint.iacr.org/2008/516
24. Yuval, G.: How to Swindle Rabin. Cryptologia 3, 187–189 (1979)

Improving the Rainbow Attack by Reusing Colours

Martin Ågren, Thomas Johansson, and Martin Hell

Dept. of Electrical and Information Technology, Lund University,
P.O. Box 118, 221 00 Lund, Sweden

Abstract. Hashing or encrypting a key or a password is a vital part in most network security protocols. The most practical generic attack on such schemes is a time memory trade-off attack. Such an attack inverts any one-way function using a trade-off between memory and execution time. Existing techniques include the Hellman attack and the rainbow attack, where the latter uses different reduction functions ("colours") within a table.

This work investigates the possibility of reusing colours, i.e., repeating the reduction functions, in the rainbow attack. We show how this outperforms the Hellman and the rainbow attack in a model of fixed resources. We try to characterize exactly when this improvement appears and in such a case the choice of an optimal number of colours.

1 Introduction

Almost all network security protocols have as an essential building block the use of a hash function or a symmetric cipher to hash or encrypt some known value, call it a key, a password or a plaintext. The function simply acts like a one-way function, with the property that knowing an output value, it is computationally difficult to find the input that maps to the given value.

In general, the key space could have been chosen very large (say 128 bits), and it is essentially impossible to apply any generic attack to invert the one-way function. However, there are several scenarios where the key space would be much smaller. One such example is when the key is selected by a user as a password. Another example could be in protocols where resources are very constrained, say protocols for RFID and similar devices. Yet another class of examples would be old security protocols that are still running; one such example is GSM.

In essence, there is a very strong motivation for studying generic attacks. Apart from the trivial approach of an exhaustive key search, the most well known and well studied class of generic attacks is the Time Memory Trade-Off (TMTO) attacks. TMTO attacks provide a middle ground between brute force and table lookup attacks on various cryptographic primitives. The size of e.g., the key, a stream cipher's internal state or a set of possible passwords gives the number N of possible values and determines the complexity which can be balanced in terms of time and memory. In 1980, Hellman [1] described one such attack and several improvements and tweaks have been suggested since. The

J.A. Garay, A. Miyaji, and A. Otsuka (Eds.): CANS 2009, LNCS 5888, pp. 362–378, 2009.

concept has also been extended by Biryukov and Shamir [2] into Time Memory
Data Trade-Off (TMDTO) attacks, where another trade-off parameter is added
to the mix. However, this type of attack will not be studied in this paper.

The basic problem that TMTO attacks attempt to solve is that of inverting
a one-way function $f : \Omega \rightarrow \Omega_c$. While it is always a theoretical possibility to
try out all the possible values to see which one matches the value we attempt to
invert, in practice this usually takes too long. One could instead pre-calculate
and sort a table, which would allow a fast lookup, but the memory requirements
would be very large. Thus, it would be interesting to somehow balance the time
and memory requirements.

The Hellman attack creates *chains* by repeatedly applying the one-way function to different starting points. Each chain could consist of t different points, and
is represented on disk using only the starting points and endpoints. By creating a
large number of chains we could expect to cover a large part of the search space.
During the online phase of the attack, we traverse the chains, searching among
the endpoints. One modification to the original idea is that of using *distinguished
points* [3] which saves on disk accesses.

Another significant idea is to apply different *reduction functions* after each
successive application of the one-way function, in order to avoid the impact of
collisions. This latter approach is called *the rainbow attack* [4] since the different functions used in each column of the table could be compared to different
"colours" that together make up a rainbow. The improvement offered by the
rainbow attack is only a relatively small reduction. Furthermore, it has been
shown [5] that there can be no substantial improvements compared to the currently known attacks, but only small reductions in required complexity.

Nevertheless, since the TMTO attack applies to all sorts of cryptographic
primitives and it is the most practical attack for many schemes, even improvements giving small complexity reductions are very important. The rainbow attack quickly became famous, even if it only provided at most a halving of complexity compared to the Hellman attack using distinguished points.

This paper will investigate a generalized version of the rainbow attack, namely
what happens when the colours are systematically repeated within the chains. It
will become clear that for a certain chain length, the number of colours (number
of reduction functions we use before starting to reuse them) will affect the online
attack time and the probability of success when trying to invert a value. The
repeated pattern of colours will provide a speed-up of the attack that allows it
to perform better than the rainbow attack under some circumstances. We will
try to characterize exactly when such an improvement appears and in such a
case the choice of an optimal number of colours.

The analysis will be done in a model where we assume the attacker to have
fixed resources. This means that the attacker has a fixed memory of M_0 words
(each word of size suitable for representing one chain) and we allow a certain
computational complexity T_0 (measured in evaluations of f) in the online phase.
Then we are interested in maximizing the probability of success in the attack,
call it P_{succ}, assuming that the attacker has access to one hashed value y, where

$y = f(x)$. The precomputation is large, but similar for all different attacks, so we do not consider this part in detail. To the best of our knowledge, the performance of TMTO attacks under fixed resources has not been considered before.

As a motivating example, we might consider an attempt to attack a password from an average equipped attacker. Using up to twelve alphanumerical characters, we have $N \approx 2^{72}$ passwords. An attacker equipped with a few large disks could have access to $M_0 = 2^{36}$ words of memory and we might allow him to do $T_0 = 2^{54}$ computational operations (evaluations of f) in the online phase, either on an FPGA or in a small network of computers. The time memory trade-off attacks we know can achieve only a small success probability in this scenario. How do we get the highest success probability P_{succ} for an attack using these given resources?

This is certainly an interesting practical question. One can buy rainbow tables for various search spaces online [6]. However, these tables are very limited in the set of characters (numeric or lowercase) or in the password lengths (seven characters) that they attack. If one wants to attack stronger passwords, one simply has to accept a low success probability unless one wants to use completely unrealistic hardware.

The paper is outlined as follows. In Section 2 we give an introduction to time memory trade-off attacks and introduce some basic notation. Section 3 presents the thin rainbow attack, which is the focus of this paper. Then we present some initial properties in Section 4 before continuing to a comparison between the Hellman attack, the original rainbow attack and the thin rainbow attack in Section 5. The paper is concluded in Section 6.

2   TMTO Attacks

The family of TMTO attacks dealt with in this paper has evolved from an attack introduced by Hellman almost thirty years ago. We start by supplementing the one-way function $f : \Omega \rightarrow \Omega_c$ with a reduction function $h : \Omega_c \rightarrow \Omega$ which maps an output from f to a new input. The name is not always entirely accurate since we might have $\Omega = \Omega_c$, but is kept for historical reasons.

By defining $g = h \circ f$, the *round function* g can be applied repeatedly to a *starting point* $SP \in \Omega$ to create an *endpoint* $EP = g(\ldots g(SP) \ldots)$ where g is applied t times. This chain can be represented in memory by just preserving SP and EP. By creating a long *table* consisting of m chains for random starting points, and sorting the chains on the endpoints, we have allowed for a fast table lookup.

During the online phase, we are given $y \in \Omega_c$ and want to find a[1] corresponding x such that $y = f(x)$. We calculate the first *candidate endpoint* $\hat{x}_{t-1} = h(y)$, thus assuming that y occurs as y_{t-1}^i in some chain i and letting $\hat{x}_{t-1}$ be the endpoint resulting from it. By searching for $\hat{x}_{t-1}$ among the endpoints, we might

---

[1] We might be looking for a password with a certain hash, in which case it is not interesting that we find the "correct" password as long as we find *some* password that gives the desired hash value.

get an alarm. If so, we use the corresponding starting point SP_i to reconstruct the value $\hat{x}$ immediately prior to y_{t-1}^i in the chain. This could be x. It could also be a *false alarm* due to colliding rows (in turn caused by the reduction function mapping several elements in Ω_c to a certain element in Ω), meaning that $\hat{x}$ does not yield $f(\hat{x}) = y$. Unless we find x, we calculate another candidate endpoint

$$\hat{x}_{t-2} = g(h(y)) = g(\hat{x}_{t-1}).$$

We continue calculating candidate endpoints as

$$\hat{x}_{t-i} = g(\dots (g(h(y)))\dots) = g(\hat{x}_{t-i+1}), \quad i = 3, 4, \dots, t$$

until we find x or until we have calculated $\hat{x}_0$ without result.

The two parameters that govern the number of points in the table, m and t, can be chosen according to the *TMTO curve* [1]. This equation gives the number of points that can be added to the table before expecting (too many) duplicates. Let T and M be the time complexity of the online attack and the memory complexity needed for storing the tables, respectively. Then, using the birthday paradox, the curve of the original Hellman table can be found as $N^2 = TM^2$ [1] and similarly, each TMTO attack has an associated TMTO curve. Hellman suggested using t tables, with different reduction functions h_i, in order to avoid the problem of colliding chains where a repeated point causes all subsequent points to be repeated, wasting memory in some sense. An implementation improvement to the original idea is that of using *distinguished points* [3]. The chains are allowed to be of variable length and the endpoints are those where some of the bits, say the last b bits, are all zero. Such values occur with probability $\lambda = 2^{-b}$, so the chains will have an expected length of 2^b.

In 2003 Oechslin proposed the rainbow attack [4] as another way to avoid this problem. In this attack, each column of the table is created using a different reduction function, i.e., $EP = g_{t-1}(\dots (g_0(SP)))$. For a repeated point to cause further points to be repeated, it will have to occur using the same round function. Since the problem of collisions of chains has disappeared, this allows the rainbow method to use a single table sized t times larger than one of the Hellman tables. See Figure 2 for an overview of a matrix, its chains and their points.

3   The Thin Rainbow Attack

One apparent drawback of the rainbow scheme is that calculating $\hat{x}_{t-i}$ cannot make use of the recently calculated $\hat{x}_{t-i+1}$. While each new candidate endpoint in the Hellman attack required only one application of g, each new candidate endpoint in the rainbow attack will require more calculations than the previous one. This makes the online time complexity

$$T = 1 + 2 + \dots + t \approx \frac{t^2}{2}.$$

However, this does not compare negatively with the Hellman attack once we realize that the Hellman attack uses t tables where each will require t applications

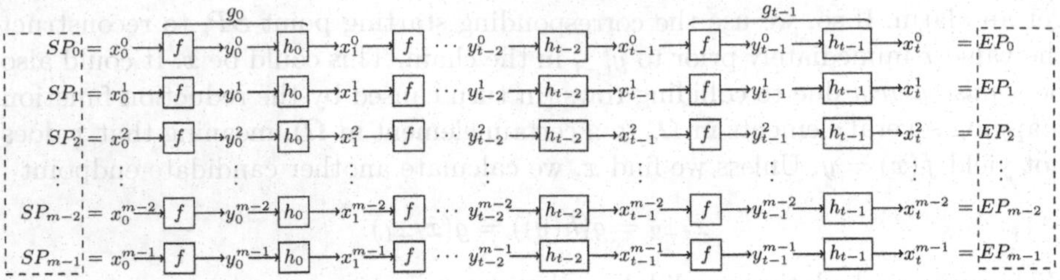

Fig. 1. A general rainbow matrix consisting of m rows and t columns. Only the starting and endpoints (shown in dashed rectangles) are stored on disk. A Hellman table uses $h_i = h, \forall i$, while a rainbow table features different reduction functions, $h_i \neq h_j, \forall i, j \neq i$.

of g, giving $T = t^2$. Indeed, the TMTO curve derived for the rainbow attack is $N^2 = 2TM^2$, which has been taken as evidence of the superiority of the rainbow attack. There has however been some discussion [5] as to whether the rainbow attack really is as superior as was originally claimed.

We now examine a generalized version of the rainbow attack, which is called the *thin rainbow attack*. This attack reuses colours in a repeated sequence: we make a "mini-chain" of S colours which we repeat L times,

$$\underbrace{g_0 g_1 \cdots g_{S-1}}\ \underbrace{g_0 g_1 \cdots g_{S-1}} \cdots \underbrace{g_0 g_1 \cdots g_{S-1}},$$

i.e., $h_i = h_j, i \equiv j \pmod{S}$. Of course, $S = t$ gives an ordinary rainbow chain while $S = 1$ is the Hellman case with one table. We have found that the idea was mentioned in [5] as a TMDTO attack. The name "thin" stems from the fact that the colours are repeated in many thin stripes.

As noted in [5], the candidate endpoints in the thin rainbow attack can be calculated faster than in the rainbow attack when the chains are equally long, $SL = t$. The observation is that if the *sequence function* is defined as

$$g_{\text{Seq}} = g_{S-1} \circ g_{S-2} \circ \dots \circ g_0,$$

most of the candidate endpoints can be calculated using previously calculated ones. Since every point in a chain can be identified by the sequence it is in and the location within that sequence, using $0 \leq j < S$ gives

$$\hat{x}_{iS+j} = \underbrace{g_{\text{Seq}} \circ g_{\text{Seq}} \circ \dots \circ g_{\text{Seq}}}_{L-i-1} \circ g_{S-1} \circ \dots \circ g_{j+1} \circ h_j(y)$$

$$= g_{\text{Seq}}(\underbrace{g_{\text{Seq}} \circ \dots \circ g_{\text{Seq}}}_{L-i-2} \circ g_{S-1} \circ \dots \circ g_{j+1} \circ h_j(y))$$

$$= g_{\text{Seq}}(\hat{x}_{(i+1)S+j}). \tag{1}$$

The drawback of the thin rainbow attack is that the coverage is likely to be smaller; while a reoccurring value in a rainbow table only leads to further reoccurring values if the duplicate value appears in the exact same column, there is now an entire congruence class of columns where any of the L columns after which g_i is applied will cause succeeding values to not add anything to the coverage of the table.

Investigating the online behaviour reveals that the number of chain walk steps that need to be made in the worst case scenario (not counting those that are due to false alarms) is

$$T = 1 + 2 + \ldots + S + (L - 1)S \cdot S$$
$$= \frac{S(S+1)}{2} + (L - 1)S^2 \approx LS^2. \tag{2}$$

Given a chain length $t = SL$ we have $T = tS$ so it might be tempting to deduce that $S = 1$ (a single Hellman table) gives the fastest online operation. However, to find the actual online time of the thin rainbow attack we would also need to investigate the number of false alarms, e.g., as in [7], and take into account the different number of tables employed by the various methods. A more detailed investigation of the proposed TMTO method is needed to see if it can provide some improved performance.

4   The Deteriorating Coverage of the Thin Rainbow Attack

Such an investigation has been done in the model of fixed resources and our findings are presented in this section. The thin rainbow attack has been compared to other TMTO attacks applied on fixed memory M_0 and complexity T_0. Throughout the paper, simulations are based on password hashing using MD5.[2]

We start by studying the development of the tables' coverage, i.e., we count how many new points are added with each column of a table. Given a rainbow table column count t and using the same column count for thin rainbow tables, we get Figure 2, where it is clear how the number of unique points in each column naturally develops in a similar manner for the first S columns, while the subsequent slope of the curve is steeper for a thin rainbow table than it is for an original rainbow table.

Since the area under each curve can be considered to represent the number of points represented by the table, it is evident how the thin rainbow table does not allow for as many successful inversions as the original rainbow table.

As a consequence of this, the thin rainbow attack is not as successful as the rainbow attack using equally long chains — however, it is much faster due to property (1). While keeping m constant, the coverage of the thin rainbow table could be improved by increasing $t = SL$, creating longer chains. Increasing SL makes the online time longer, but when the length has been increased so much

---

[2] While this hash function is broken in its own right, none of its deficiencies have been exploited here — it is simply used as a one-way function. There is no perceived need to rerun the simulations using other hash functions as the conclusions would be the same.

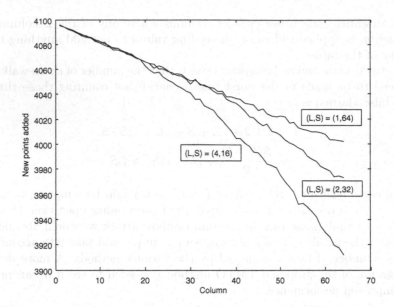

Fig. 2. With $m = 2^{12}$ rows, the number of new points added with each column has been plotted for an original rainbow table, corresponding to $(L, S) = (1, 64)$, and thin rainbow tables $((L, S) = (2, 32)$ and $(L, S) = (4, 16))$. The curves naturally start out the same way for 16 columns but when the thin tables start reusing colours, the number of added points gets lower than in the rainbow table. Results are averaged over 30 tables.

as to make the attack take equally long time as the original rainbow attack, it is our hope that the coverage will actually outperform that of the original rainbow attack. The memory consumption will not be affected.

In the next section, we will allow the thin rainbow attack to use longer chains, and see where that leads us.

5   Detailed Performance Analysis under Fixed Resources

We consider the following problem: What probability of success P_{succ} can we reach using M_0 chains and T_0 function evaluations? We apply a very strict policy to T_0: once the threshold is exceeded, we abandon all calculations, no matter if we are currently investigating an alarm or if we are calculating a candidate end point. We allow the M_0 chains to be split among l tables of m chains each.

In our model we consider an attacker who is given a hashed password after an initial pre-computation phase and wants to find the password within a certain time. After building a set of tables, we have repeated the online attack on a large number of hashes in order to establish the success probability P_{succ}. We have performed many such experiments and averaged the results.

Note that our setting does not translate directly into that of TMDTO attacks. In the latter attacks, we might e.g., be trying to find the internal state in a stream cipher [2]. If we can find the initial state during a certain time instance, there is no point in continuing the search in order to find the state at a later time. Thus, we do not care *how many* points we can invert, all that we are interested in is inverting *one* out of many. The attacker in our setting investigated here, however, is certainly interested in having a success probability P_{succ} that is as high as possible. Given x passwords, our attacker can expect to reconstruct xP_{succ} of them.

5.1   Algorithms and Parameters Considered

In order to investigate the chances for such an attack in real-world circumstances, we have limited the resources available to our attacker. After deciding upon resources M_0 and T_0, we have compared the following algorithms, while trying out various parameter choices:

- **The Hellman algorithm:** The tables have been allowed to use any column count t and any chain count m (and thus the table count l such that $ml = M_0$).
- **The rainbow algorithm:** Similarly, the tables have been allowed to use any colour count t and any chain count m (and the table count l producing $ml = M_0$).
- **The thin rainbow algorithm:** These tables have been allowed to use any colour count S and any sequence count L. As above, we have allowed any value of m and the corresponding l. Keep in mind that the Hellman and the rainbow attacks can both be obtained as special cases of the thin rainbow attack, so the thin rainbow attack is guaranteed to be able to perform at least as good as the other algorithms. In order to make a meaningful comparison we have thus excluded the cases $L = 1$ and $S = 1$ from the values used with the thin rainbow attack.

For each algorithm, the best success probability has been included in our results. All other calculations have been discarded. This has been done for various search space sizes N. Thus, for each choice of N, we have three data points which tell us the success probability P_{succ} for the three algorithms when they perform "as good as possible".

The search has been done by first searching table one completely, then table two, etc. until the last table is searched. Note that our fixed time resource may be exhausted once we have searched some tables, causing the latter tables to never be looked at. We have investigated whether this is the case, but it has turned out that parameter choices that cause all tables to be searched have performed better. Still, we allow the attacker to do whatever she pleases with the memory and time alloted and if, for some reason, the best approach she can

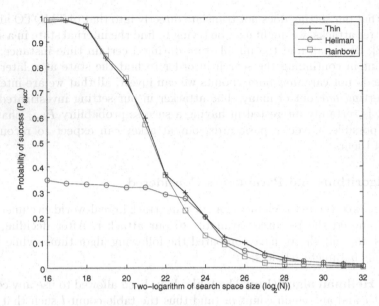

Fig. 3. The probability of success for the different algorithms as a function of the size of the search space. $M = 2^{12}$ and $T = 2^{18}$.

come up with is to only use half of the memory, then we must allow her to do just that.

Also note that we disregard the difference in bits needed to store the tables. As pointed out in [5], the starting points of fewer, larger tables cannot be compressed as much as equally many starting points in a larger number of smaller tables. On the other hand [8,9], the compression of the endpoints behaves the other way round. Even with a very small number of tables, these effects taken together put the difference in number of chains that fit into the tables at just a few per cent.

5.2   Comparing TMTO Algorithms under Fixed Resources

Figure 3 illustrates the behaviour when $M_0 = 2^{12}$ and $T_0 = 2^{18}$. The Hellman, rainbow and thin rainbow algorithms have all been applied to a password from various domains, where the size of the domain is indicated on the x axis. The y axis shows the success probability P_{succ} obtained. Note that the smaller search spaces could be searched exhaustively in the allowed time.

Studying Figure 3 we find a pattern: at smaller domain sizes, the rainbow attack and the thin rainbow attack are much more similar (and better) than the Hellman attack. In this region, the original rainbow attack is slightly better, while for larger domains, the thin rainbow attack has the upper hand. This is perhaps more evident in the zoomed-in version, Figure 4. The behaviour for even larger values of N is depicted in Figure 5 and we can see that the trend continues.

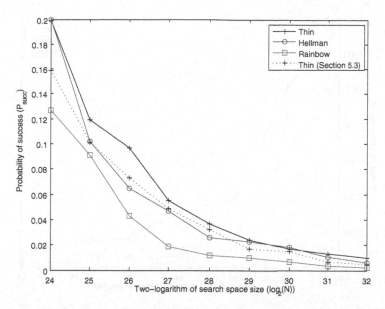

Fig. 4. The probability of success for the different algorithms as a function of the size of the search space. $M = 2^{12}$ and $T = 2^{18}$. The dotted line illustrates the behaviour when we choose parameters for the thin rainbow algorithm as in Section 5.3.

We naturally ask ourselves to which extent this is merely an effect of the parameters used in this specific case. To investigate this, we have studied other choices of (M_0, T_0) for which $T_0 = M_0^{3/2}$, i.e., $(M_0, T_0) = (2^8, 2^{12})$ and $(M_0, T_0) = (2^{10}, 2^{15})$. Our findings are depicted in Figures 7 and 8 in Appendix A. Since the behaviour is largely the same, we have reason to believe that similar effects will be appearing for e.g., $(M_0, T_0) = (2^{36}, 2^{54})$. We believe that $T_0 = M_0^{3/2}$ is a suitable case to investigate considering current possibilities regarding resources.

Regarding the failures of the Hellman attack on smaller search spaces, we believe that these occur due to the relatively large cost of collisions and investigations of false alarms. Using a large column count t would give many collisions and false alarms, so m needs to be small. This in turn makes the table count l large, increasing the online time. We attribute the slightly irregular behaviour in e.g., Figure 9 (Appendix A) to the smaller values of M_0 and T_0 which make random effects much more prominent.

Now that we have found there is reason to believe that increasing the size of the search space changes the choice of algorithm from the rainbow algorithm to the thin rainbow algorithm or even the Hellman algorithm, we ask ourselves just when this change occurs. Inserting $T_0 = M_0^{3/2}$ into the rainbow TMTO curve yields $N = \sqrt{2}M_0^{7/4} = \sqrt{2}T_0^{7/6}$. Clearly, the point where the thin rainbow algorithm becomes better than the rainbow one lies beyond this. Suggesting $N = T_0^{5/4}$ as the point of break-even seems to be supported by the data.

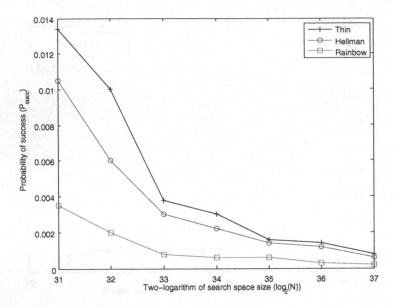

Fig. 5. The probability of success for the different algorithms as a function of the size of the search space. $M = 2^{12}$ and $T = 2^{18}$.

We thus predict that the case $(M_0, T_0, N) = (2^{36}, 2^{54}, 2^{72})$ discussed in Section 1 can be handled with much better performance using the thin rainbow algorithm than with the rainbow algorithm.

Appendix B contains similar figures for configurations with other relations between M_0 and T_0, namely $T_0 = M_0$ and $T_0 = M_0^2$. These figures show the same basic behaviour of the different algorithms.

5.3   Determining the Optimal Parameters

The TMTO curve for a thin rainbow table with m rows is $mL^2S = N$ (see [5]) while the time complexity for searching it is LS^2 (disregarding false alarms) as given by (2). Keeping in mind that we have l tables with $M_0 = lm$, we find that $mL^2S = N$ and $LS^2/m = T_0/M_0$. Investigating these equations for the choices that turned out to be optimal above, we find that

$$mL^2S = c_1N, \quad 1 \leq c_1 \leq 4, \tag{3}$$

$$LS^2/m = c_2T_0/M_0, \quad 4^{-1} \leq c_2 \leq 1. \tag{4}$$

In other words, we fill the tables slightly more than suggested by the TMTO curve, and we keep the theoretical time consumption slightly below that suggested by a step count analysis (thus allowing a certain fraction of steps due to false alarms).

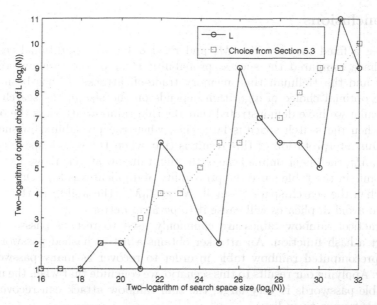

Fig. 6. With $M_0 = 2^{12}$ and $T_0 = 2^{18}$, the most successful choice of L is displayed for each search space size. The dotted line indicates the value of L chosen in in Section 5.3.

This gives us a rough idea about how to choose the parameters m, L and S for other choices of T_0 and M_0. There is still one thing troubling us — we have three unknown variables, L, S and m, but only two equations, (3) and (4). Studying Figure 6 we get an idea regarding the optimal choice of L. The figure illustrates the parameters that turned out to be the best for the case $(M_0, T_0) = (2^{12}, 2^{18})$.

If we assume that the parameter L increases from 1 to 10 as N goes from 18 to 32 (see Figure 6) and choose S and m to satisfy the above criteria, we obtain the dotted line in Figure 4. Our guess at the behaviour of L is motivated by the trend in Figure 6, but also the observation that larger L (and smaller S) make the thin chain look more like Hellman chains and less like rainbow chains. Since the Hellman algorithm apparently performs better than the rainbow algorithm with increasing values of N, we choose to make the thin chains resemble Hellman chains to a greater extent.

Certainly, the parameters produced in this manner do not give the best success probability possible. However, it does perform better than the rainbow algorithm and sometimes better than the best instance of the Hellman algorithm found.

A really determined attacker might be able to decide on a few promising configurations of (m, L, S) and build a single table of each before investigating their online behaviour, both with respect to success probability and (just as importantly) time consumption. If the number of tables l is somewhat large, the increase in precomputation due to tables being produced and later discarded in favour of another configuration should be within a factor 2 or so.

6   Conclusions

Assuming a fixed memory size M_0 and fixed online computational resources T_0, we have compared the success probability P_{succ} of the rainbow, the thin rainbow and the Hellman time memory trade-off attacks. It has been shown that the optimal choice of algorithm depends on the size of the search space. In particular we have demonstrated that the thin rainbow attack is the optimal choice when the search space is large, i.e., when the probability of success is small. Our interpretation of this result is that when the search space is large, $N^2 \gg T_0 M_0^2$, the faster online time of the thin rainbow attack allows us to cover more points in the table since the probability of duplicates is low. On the other hand, when the search space is small, $N^2 \approx T_0 M_0^2$, the ability of the rainbow attack to avoid duplicates will cause it to perform better.

In practice, rainbow tables are commonly used to recover passwords, i.e., to invert a hash function. An attacker obtains a list of hashed passwords and uses a precomputed rainbow table in order to recover as many passwords as possible. Applying our results to this scenario, we conclude that when the number of possible passwords is very large, the thin rainbow attack can recover more passwords from a given list.

We have given approximate values for the parameters L, S and m in order to make the attack as successful as possible, but studying the details surrounding these parameters remains an open problem.

References

1. Hellman, M.: A cryptanalytic time-memory trade-off. IEEE Transactions on Information Theory IT-26, 401–406 (1980)
2. Biryukov, A., Shamir, A.: Cryptanalytic time/memory/data tradeoffs for stream ciphers. In: Okamoto, T. (ed.) ASIACRYPT 2000. LNCS, vol. 1976, pp. 1–13. Springer, Heidelberg (2000)
3. Denning, D.: Cryptography and data security, p.100. Out of Print (1982)
4. Oechslin, P.: Making a faster cryptanalytic time-memory trade-off. In: Boneh, D. (ed.) CRYPTO 2003. LNCS, vol. 2729, pp. 617–630. Springer, Heidelberg (2003)
5. Barkan, E., Biham, E., Shamir, A.: Rigorous bounds on cryptanalytic time/memory tradeoffs. In: Dwork, C. (ed.) CRYPTO 2006. LNCS, vol. 4117, pp. 1–21. Springer, Heidelberg (2006)
6. project-rainbowcrack.com, RainbowCrack — crack hashes with rainbow tables, Website (2009), Retrieved 2009-06-11,
   http://project-rainbowcrack.com/index.htm
7. Hong, J.: The cost of false alarms in Hellman and rainbow tradeoffs. Cryptology ePrint Archive, Report 2008/362 (2008), http://eprint.iacr.org/
8. Biryukov, A., Shamir, A., Wagner, D.: Real time cryptanalysis of A5/1 on a PC. In: Schneier, B. (ed.) FSE 2000. LNCS, vol. 1978, pp. 1–13. Springer, Heidelberg (2001)
9. Avoine, G., Junod, P., Oechslin, P.: Time-memory trade-offs: False alarm detection using checkpoints. In: Maitra, S., Veni Madhavan, C.E., Venkatesan, R. (eds.) INDOCRYPT 2005. LNCS, vol. 3797, pp. 183–196. Springer, Heidelberg (2005)

A   Other Results with $T = M^{3/2}$

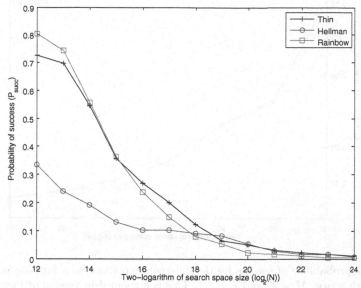

Fig. 7. The probability of success for the different algorithms as a function of the size of the search space. $M_0 = 2^8$ and $T_0 = 2^{12}$. This is a full version of Figure 9.

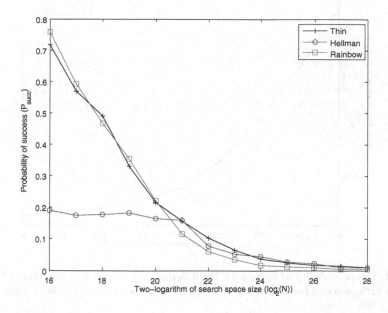

Fig. 8. The probability of success for the different algorithms as a function of the size of the search space. $M_0 = 2^{10}$ and $T_0 = 2^{15}$. This is a full version of Figure 10.

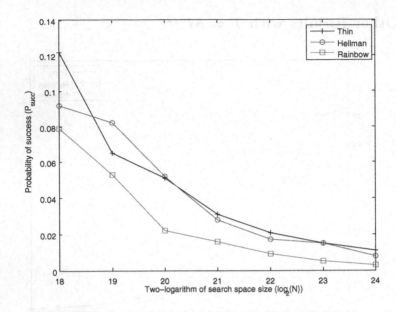

Fig. 9. The probability of success for the different algorithms as a function of the size of the search space. $M = 2^8$ and $T = 2^{12}$. This is a zoomed-in version of Figure 7.

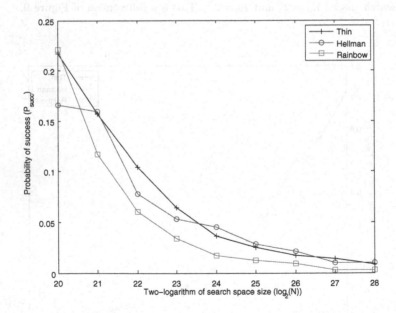

Fig. 10. The probability of success for the different algorithms as a function of the size of the search space. $M = 2^{10}$ and $T = 2^{15}$. This is a zoomed-in version of Figure 8.

B   Other Relations between Memory and Time

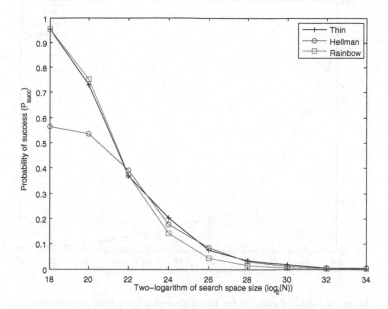

Fig. 11. The probability of success for the different algorithms as a function of the size of the search space. $M_0 = 2^{14}$ and $T_0 = 2^{14}$. This is a full version of Figure 13.

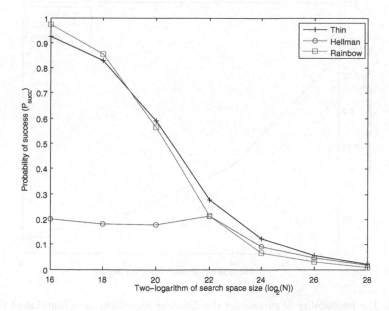

Fig. 12. The probability of success for the different algorithms as a function of the size of the search space. $M_0 = 2^{10}$ and $T_0 = 2^{20}$. This is a full version of Figure 14.

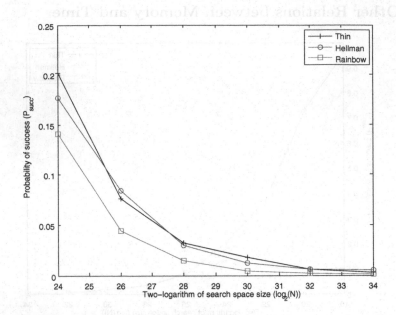

Fig. 13. The probability of success for the different algorithms as a function of the size of the search space. $M_0 = 2^{14}$ and $T_0 = 2^{14}$. This is a zoomed-in version of Figure 11.

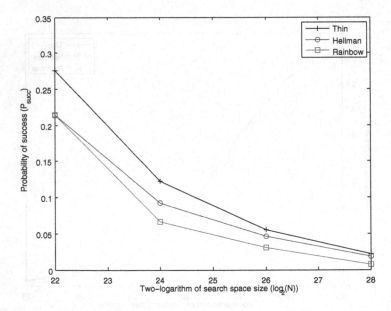

Fig. 14. The probability of success for the different algorithms as a function of the size of the search space. $M_0 = 2^{10}$ and $T_0 = 2^{20}$. This is a zoomed-in version of Figure 12.

Side Channel Cube Attack on PRESENT*

Lin Yang[1,2], Meiqin Wang[1], and Siyuan Qiao[1]

[1] Key Laboratory of Cryptologic Technology and Information Security,
Ministry of Education, Shandong University, Jinan 250100, China
linyang_sdu@mail.sdu.edu.cn,
mqwang@sdu.edu.cn,
sy_qiao@mail.sdu.edu.cn
[2] Department of Physics, Tsinghua University, Beijing 100084, China

Abstract. As an ultra-lightweight block cipher, PRESENT is presented by A. Bogdanov et al. in CHES 2007. In this paper, we detect the non-random properties in the first four rounds of PRESENT based on cube attack proposed by Shamir et al. By analyzing the features of the S-box and pLayer, we give the side channel cube attack on PRESENT. If any output bit of the third round is leaked, we can recover the total 80-bit key. Especially, for the leaked bit 1, bit 2 or bit 3 of the output bits in the third round, we can also recover 80-bit key with lower computing complexity compared to other leaked bits, and our attack requires 2^{15} chosen plaintexts and 2^{32} 31-round PRESENT encryptions.

Keywords: Cryptanalysis, Cube attack, Side channel attack, PRESENT.

1 Introduction

Tiny computing devices such as RFID tags and sensor networks will be very popular in future. In order to guarantee security in extremely constrained environments, many ultra-lightweight block ciphers have been developed such as PRESENT [1], TEA [5,6], MCRYPTON [7], HIGHT [8], SEA [9] and CGEN [10]. PRESENT, a hardware-optimized block cipher, is proposed by A. Bogdanov et al. in CHES 2007 which consists of 31 rounds with a SP-network. The block size is 64-bit, and the key size can be 80-bit and 128-bit for PRESENT-80 and PRESENT-128 respectively. Compared with other ultra-lightweight block ciphers, it has the lowest implementation costs.

In EUROCRYPT 2009, Itai and Shamir proposed a new type of algebraic attacks named cube attack [2]. Cube attack is a generic key derivation attack [2,3,4]. It can be used to attack any cryptosystem in which even a single bit can be represented by a low degree multivariate polynomial in the key and plaintext variables. Standard side channel attacks concentrate on how to obtain the side

* Supported by National Natural Science Foundation of China (NSFC Grant No. 60525201), 973 Project (No. 2007CB807902) and China Hi-Tech program (No. 2006AA01Z420).

J.A. Garay, A. Miyaji, and A. Otsuka (Eds.): CANS 2009, LNCS 5888, pp. 379–391, 2009.

information. However, in this paper, we only concentrate on exploiting the leaked information. We apply the data mode of leakage attacks in Figure 1 of [4].

In the standard attacks on block ciphers, only the plaintexts and ciphertexts are avaiable. But in the side channel attacks, we can obtain either the plaintexts and the ciphertexts or the internal state information for the intermediate rounds. Under the attack model, Shamir et al. presented the side channel cube attack on some block ciphers such as Serpent [2]. For the special property of PRESENT, we will exploit the cube attack to cryptanalyze PRESENT.

As we know, there are several cryptanalysis for reduced-round of PRESENT. In [13], the algebraic cryptanalysis for 5-round of PRESENT is shown, and the integral attacks on 5, 6 and 7 rounds of PRESENT are given in [14]. In [12], the differential cryptanalysis of 16-round PRESENT is presented. Based on the differential characteristics in [12], a further analysis for 17-round PRESENT is given integrating differential cryptanalysis and algebraic cryptanalysis [15].

By analyzing the relationship between the inter-media state in the first few rounds with the plaintext bits and key bits for PRESENT, we found some important properties which can be utilized by the side channel cube attack. In this paper, we will present how to recover 80-bit key of PRESENT even if only one bit in the special position of the third round is leaked. As a result, with 2^{15} chosen plaintexts, 2^{16} 64-bit key candidates can be obtained by solving a system of linear equations. We can recover all 80-bit key with 2^{32} 31 round PRESENT encryptions.

The paper is organized as follows. Section 2 introduces the PRESENT block cipher and Section 3 gives the brief description of cube attack. We present the non-random properties for reduced-round of PRESENT and give the side channel cube attack on PRESENT-80 in Section 4. Finally, Section 5 concludes this paper.

2   Description of PRESENT

PRESENT is a 31-round ultra-lightweight block cipher. The block length is 64-bit. The cipher is described in Figure 1. As in Serpent, there are three stages involved in each round of PRESENT. The first stage is addRoundKey described as follows,

$$b_j \rightarrow b_j \oplus K_i^j,$$

where b_j, $(0 \leq j \leq 63)$ is the current state bit and K_i^j, $(1 \leq i \leq 32, 0 \leq j \leq 63)$ is the jth subkey bit of round key K_i. The second stage is sBoxLayer which consists of 16 parallel versions of the 4-bit to 4-bit S-box, which is given in Table 1. The

Table 1. Table of S-box

x	0	1	2	3	4	5	6	7	8	9	A	B	C	D	E	F
S[x]	C	5	6	B	9	0	A	D	3	E	F	8	4	7	1	2

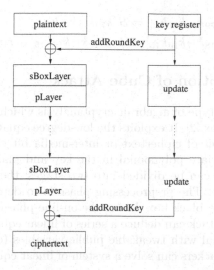

Fig. 1. 31-round PRESENT Encryption Algorithm

third stage is the bit permutation pLayer, which is given by Table 2. The bit i is moved to the bit $P(i)$ by pLayer.

Table 2. Table of pLayer

i	0	1	2	3	4	5	6	7	8	9	10	11	12	13	14	15
$P(i)$	0	16	32	48	1	17	33	49	2	18	34	50	3	19	35	51
i	16	17	18	19	20	21	22	23	24	25	26	27	28	29	30	31
$P(i)$	4	20	36	52	5	21	37	53	6	22	38	54	7	23	39	55
i	32	33	34	35	36	37	38	39	40	41	42	43	44	45	46	47
$P(i)$	8	24	40	56	9	25	41	57	10	26	42	58	11	27	43	59
i	48	49	50	51	52	53	54	55	56	57	58	59	60	61	62	63
$P(i)$	12	28	44	60	13	29	45	61	14	30	46	62	15	31	47	63

The key size of PRESENT can be taken as 80-bit or 128-bit. We will crypt-analyze 80 bits version, so only the key schedule algorithm for 80 bits version is given in the following.

- Step 1: The 80-bit user key will be stored in a key register K denoted as $K = k_{79}k_{78} \ldots k_0$.
- Step 2: for $i = 1$ to 32 do
  - Extract 64-bit subkeys K_i as the round key for the ith round,

$$K_i = K_i^{63}K_i^{62} \ldots K_i^0 = k_{79}k_{78} \ldots k_{16}.$$

  - Update key register $K = k_{79}k_{78} \ldots k_0$ as follows,

$$[k_{79}k_{78} \ldots k_1k_0] = [k_{18}k_{17} \ldots k_{20}k_{19}],$$

$$[k_{79}k_{78}k_{77}k_{76}] = S[k_{79}k_{78}k_{77}k_{76}],$$
$$[k_{19}k_{18}k_{17}k_{16}k_{15}] = [k_{19}k_{18}k_{17}k_{16}k_{15}] \oplus roundcounter.$$

3   Brief Description of Cube Attack

Cube attack is a new type of algebraic cryptanalysis which considered the cryptosystem as a black box [2]. It exploits the low-degree equations in cryptosystem which even a single bit of ciphertext or inter-media bit can be represented as a low-degree multivariate polynomial in the key and plaintext variables. The cryptanalytic process can be divided into two stages: the preprocessing phase and the on-line phase. The preprocessing phase is to determine which queries should be made to the black box during the on-line phase of the attack. In the on-line phase, cube attack can deduce a series of linear equations by querying to a black box polynomial with tweakable public variables (e.g. chosen plaintexts attack). Then the attackers can solve a system of linear equations to recover the secret key bits.

Any output bit can be represented as a multivariate master polynomial $p(k_1, \ldots, k_n, v_1, \ldots, v_m)$ over $GF(2)$. The variables include secret variables k_i (key bits) and public variables v_i (plaintext bits in block ciphers and MACs, IV bits in stream ciphers). Let the degree of the polynomial be d.

In the preprocessing phase, the attacker chooses t_I randomly which can be indexed by the subset $I \subseteq \{1, \ldots, m\}$ of the multiplied public variables. The index of the subset I is defined as cube index. The polynomial can be represented as follows,

$$p(k_1, \ldots, k_n, v_1, \ldots, v_m) = t_I \cdot p_{S(I)} + q(k_1, \ldots, k_n, v_1, \ldots, v_m),$$

$p_{S(I)}$ is called the superpoly of I in p. The polynomial is divided into two parts $p_{S(I)}$ and q by t_I.

To demonstrate these notions, let

$$p(k_1, k_2, k_3, v_1, v_2, v_3) = v_1 v_2 k_1 + v_1 v_2 k_3 + v_1 k_2 k_3 + v_1 v_2 v_3 + k_1 k_2 + v_3 + 1$$

be a polynomial of degree 3 in 3 secret variables and 3 public variables. Let $I = \{1, 2\}$ be an index subset of the public variables which size is 2. We can represent p as:

$$p(k_1, k_2, k_3, v_1, v_2, v_3) = v_1 v_2 (k_1 + k_3 + v_3) + (v_1 k_2 k_3 + k_1 k_2 + v_3 + 1),$$

where

$$t_I = v_1 v_2,$$
$$p_{S(I)} = k_1 + k_3 + v_3,$$
$$q(k_1, k_2, k_3, v_1, v_2, v_3) = v_1 k_2 k_3 + k_1 k_2 + v_3 + 1.$$

We assign all the public variables 0/1 values. Then the $p_{S(I)}$ becomes a polynomial including secret variables only. A maxterm of p is a term t_I such that

$deg(p_{S(I)}) \equiv 1$, (We only consider the degree of the secret variables.) i.e. the supperpoly of I in p is a linear polynomial which is not a constant. The $p_{S(I)}$ corresponding to the maxterm calls maxterm equation. According to Theorem. 1 of [2], Sum p that $t_I \in \{0,1\}^{d-1}$, then

$$\sum_{t_I \in \{0,1\}^{d-1}} p = \sum_{t_I \in \{0,1\}^{d-1}} (t_I p_{S(I)} + q) = p_{S(I)}.$$

The public variables do not involve in the subset I, we can set them to be constant. For convenience we set them to be zero. Since the key can be chosen in this phase, it is easy to check whether a superpoly is linear by linear tests [16]. We choose secret variable vectors $\mathbf{x}, \mathbf{y} \in \{0,1\}^n$ randomly, and verify the equation $p_{S(I)}[\mathbf{0}] + p_{S(I)}[\mathbf{x}] + p_{S(I)}[\mathbf{y}] = p_{S(I)}[\mathbf{x}+\mathbf{y}]$. The test always succeeds if $p_{S(I)}$ is linear. The attackers repeat the test N times, and a non-linear superpoly can be accepted with probability 2^{-N}. The preprocessing phase is not key-dependant and perform once per cryptosystem. In this phase the attackers find many maxterms and their equations.

In on-line phase, the attackers choose plaintexts to get a system of linear equations and solve the equations to recover the key. The superpoly can be evaluated by summing over every possible assignment to its maxterm. Assuming the degree of the maxterm is $d-1$, each sum requires 2^{d-1} evaluations of the derived polynomials.

4 Cube Attack on PRESENT

4.1 The Non-randomness of PRESENT

In cryptographic algorithms, each output bit can be regard as a polynomial of plaintext and key bits. We identify the non-random properties on the polynomial for PRESENT. In general, as the input variables, the plaintext bits and the subkey bits are confused so completely that the multivariate polynomial of the ciphertext bit should be random polynomial. Firstly we give the definition of a random polynomial [2].

Definition 1. *A random polynomial of degree d in $n + m$ variables is a polynomial $p \in \mathbb{P}_n^{n+m}$ such that each possible term of degree at most d is independently chosen to occur with probability 0.5.*

If the polynomial in a cryptosystem doesn't satisfy this definition, it means the cryptosystem is not random [3] which will incur a potential flaw in the cryptosystem. We have identified the polynomial after a few rounds of PRESENT is not random.

We denote the 4 input bits of S-box as x_0, x_1, x_2 and x_3 respectively, and the 4 output bits of S-box as y_0, y_1, y_2 and y_3 respectively. We present the boolean functions of S-box in Table 3. According to Table 3, the highest degree of the four boolean functions is 3, and specially the degree of the boolean function for y_0 is 2.

Table 3. Boolean Functions of S-box

Output Bit	Boolean Function
y_0	$x_0 + x_2 + x_1x_2 + x_3$
y_1	$x_1 + x_0x_1x_2 + x_3 + x_1x_3 + x_0x_1x_3 + x_2x_3 + x_0x_2x_3$
y_2	$1 + x_0x_1 + x_2 + x_3 + x_0x_3 + x_1x_3 + x_0x_1x_3 + x_0x_2x_3$
y_3	$1 + x_0 + x_1 + x_1x_2 + x_0x_1x_2 + x_3 + x_0x_1x_3 + x_0x_2x_3$

Table 4. Distribution of Degree for State bits

Position	0	1	2	3	4	5	6	7	8	9	10	11	12	13	14	15
Degree	8	12	12	12	12	18	18	18	12	18	18	18	12	18	18	18
Position	16	17	18	19	20	21	22	23	24	25	26	27	28	29	30	31
Degree	12	18	18	18	18	27	27	27	18	27	27	27	18	27	27	27
Position	32	33	34	35	36	37	38	39	40	41	42	43	44	45	46	47
Degree	12	18	18	18	18	27	27	27	18	27	27	27	18	27	27	27
Position	48	49	50	51	52	53	54	55	56	57	58	59	60	61	62	63
Degree	12	18	18	18	18	27	27	27	18	27	27	27	18	27	27	27

The basic approach is to exploit the relationship between plaintext bits and the state information in the first few rounds. PRESENT produces 64-bit state variable after each round. Particularly, any state bit after the first round can be represented as a simple polynomial involving 4-bit plaintext and 4-bit subkey. Further more, the output bits from the same S-box are related with the same plaintext bits and subkey bits. In this way, each state bit can be represented as a polynomial with the state bits and the subkey bits of the top one round. So we can derive the polynomials of the state bits round by round. Based on the boolean functions of the S-box and the pLayer permutation, we can calculate a distribution table of the degree for different bits before the pLayer in the third round in Table 4.

According to definition 1, if we use cube index with the size less than $d-1$, the superpoly is most likely non-linear. We test the randomness of the multivariate polynomial after the third round by cube test (a randomness test with cube attack) [3]. We take the cube index of consecutive c plaintext bits$(c < d-1)$, e.g. bit 0, 1, ... , $c-1$. For each cube index, we produce 64 corresponding superpolys for 64 state bits. Then we count the number of the non-linear superpolys for the cube index. If the polynomials for the state bits are random polynomials, the number of non-linear superpolys is close to 64. The degree of the polynomials in Table 4 is bigger than 8. If we use the cube indices with size less than 6, the corresponding superpolys are mostly non-linear.

We choose $c = 2$ and $c = 4$ respectively as the degree of cube index for cube test in the third round, and the obtained distributions of the number of non-linear superpolys are listed in Fig. 2. At the same time, we choose $c = 4$ and $c = 6$ respectively as the degree of cube index for cube test in the fourth round, and the obtained distribution of the number of non-linear superpolys are listed in Fig. 3. In Fig. 2 and Fig. 3, the x-axis denotes the first bit for cube index, and the y-axis denotes the number of non-linear superpolys for 64 state bits we

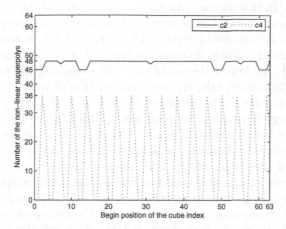

Fig. 2. Distribution of Biases ($c = 2$, $c = 4$) in Round 3

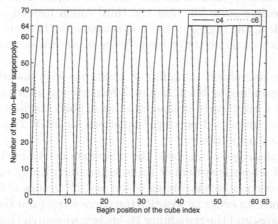

Fig. 3. Distribution of Biases ($c = 4$, $c = 6$) in Round 4

get. From the figures, the non-randomness after the third round and the fourth round will incur that the size of some cube index is very low.

4.2 Side Channel Cube Attack on PRESENT-80

Given an explicit representation of the multivariate polynomial, all the maxterms can be deduced easily. Due to the number of terms of the polynomial increases rapidly after several rounds, it is difficult to find the representation and store it. If one variables of the term is a key variable, and others of it are public variables, the maxterm may be identified from it. For each round, we reserve these terms involving a key variable and the terms only involving public variables. Then in the next round we calculate the partial terms in the polynomial of the state bit which related to the reserved terms only. And we discard the terms involving more than one key variables. In this way, we can control the number of the terms in the first three rounds. The maxterms can be deduced from the reduced polynomial.

From the non-randomness of Section 4.1 for PRESENT, it is easy to recover partial key bits from a single state bit in the first or second round. But after the fourth round, the lowest degree of the polynomials of the 64 state bits is 24. And most polynomials take the boolean function of the S-box with degree 3, so the degree can get to the theoretical degree $d = 3^4 = 81$. The cubes with comparatively large dimension are required in order to get maxterms for any state bit after the fourth round. It is not efficient to use cube attack.

If we proceed the side channel attack for the first round or the second round, the complexity of chosen plaintexts is minimised. However the number of key bits we can recover for the first round or the second round would be very few. For the remained key bits, we search them exhaustively and it will procude much more complexity. At the same time, if we proceed the attack on the 4-round, the maxterms will involve much more plaintext bits, which will also lead much more complexity. So we proceed the side channel attack for the third round.

According to the side channel attacks mode in Figure 1 of [4], we need to obtain an intermediate state bit after the third round instead of a bit in each round. From Table 4, the degree of the polynomials for the state bit 0, 1, 2 and 3 are lower than others, so we try to identify the maxterms corresponding to the three bits.

In the preprocessing phase, we found 48 maxterms with linearly independent superpolys corresponding to bit 1, 2 and 3 respectively after the third round, and 32 maxterms with linearly independent superpolys corresponding to bit 0. We show the cube indicse and maxterm equations for bit 1, 2 and 3 in Table 6, Table 7 and Table 8 respectively.

In the on-line phase, we can choose the plaintexts determined in the preprocessing phase and solve the system of linear equations with 48 equations.

For the system of equations involving 64-bit key, 2^{16} groups of 64 bits key candidates are obtained. In order to identify the correct key from the 2^{16} groups of 64-bit key candidates and the remaining 16-bit key, we search them exhaustively with 2^{32} full 31-round PRESENT encryptions.

Since the cube index size is different, we choose the number of plaintexts as follows,

$$N_{CP} = 12 \cdot 2^2 + 12 \cdot 2^5 + 12 \cdot 2^8 + 12 \cdot 2^{11} \approx 2^{15}.$$

Although $2^{16} \times 2^{16}$ key candidates should be searched to assure the right key. Noticing the feature of the system of linear equations, it can be split into 16

Table 5. Sorts of the bits in 3-round

Class 1	the bits of S-box 0, 4, 8, 12
Class 2	the first bits of S-box 1, 2, 3
Class 3	the other bits of S-box 1, 2, 3
Class 4	the first bits of rest S-boxes
Class 5	the other bits of rest S-boxes

Table 6. 48 Maxterms and Maxterm Equations in Bit 1

Cube Indexes	Maxterm Equations
0,1	$X_{18} + X_{19}$
0,2	$X_{17} + X_{19}$
1,2	X_{16}
12,13	$X_{30} + X_{31}$
12,14	$X_{29} + X_{31}$
13,14	X_{28}
48,49	$X_{66} + X_{67}$
48,50	$X_{65} + X_{67}$
49,51	$1 + X_{64}$
60,61	$X_{78} + X_{79}$
60,62	$X_{77} + X_{79}$
61,62	X_{76}
4,5,8,9,10	$X_{22} + X_{23}$
4,6,8,9,10	$X_{21} + X_{23}$
5,6,8,9,10	X_{20}
4,5,6,8,9	$X_{26} + X_{27}$
4,5,6,8,10	$X_{25} + X_{27}$
4,5,6,9,10	X_{24}
52,53,56,57,58	$X_{70} + X_{71}$
52,54,56,57,58	$X_{69} + X_{71}$
53,54,56,57,58	X_{68}
52,53,54,56,57	$X_{74} + X_{75}$
52,53,54,56,58	$X_{73} + X_{75}$
52,53,54,58,59	$1 + X_{72}$
16,17,36,37,38,40,41,42	$X_{34} + X_{35}$
16,18,36,37,38,40,41,42	$X_{33} + X_{35}$
17,18,36,37,38,40,41,42	X_{32}
28,29,36,37,38,40,41,42	$X_{46} + X_{47}$
28,30,36,37,38,40,41,42	$X_{45} + X_{47}$
29,30,36,37,38,40,41,42	X_{44}
20,21,22,24,25,26,32,33	$X_{50} + X_{51}$
20,21,22,24,25,26,32,34	$X_{49} + X_{51}$
20,21,22,24,25,26,33,34	X_{48}
20,21,22,24,25,26,44,45	$X_{62} + X_{63}$
20,21,22,24,25,26,44,46	$X_{61} + X_{63}$
20,21,22,24,25,26,45,46	X_{60}
20,21,24,25,26,36,37,38,40,41,42	$X_{38} + X_{39}$
20,22,24,25,26,36,37,38,40,41,42	$X_{37} + X_{39}$
21,22,24,25,26,36,37,38,40,41,42	X_{36}
20,21,22,24,25,36,37,38,40,41,42	$X_{42} + X_{43}$
20,21,22,24,26,36,37,38,40,41,42	$X_{41} + X_{43}$
20,21,22,25,26,36,37,38,40,41,42	X_{40}
20,21,22,24,25,26,36,37,40,41,42	$X_{54} + X_{55}$
20,21,22,24,25,26,36,38,40,41,42	$X_{53} + X_{55}$
20,21,22,24,25,26,37,38,40,41,42	X_{52}
20,21,22,24,25,26,36,37,38,40,41	$X_{58} + X_{59}$
20,21,22,24,25,26,36,37,38,40,42	$X_{57} + X_{59}$
20,21,22,24,25,26,36,37,38,41,42	X_{56}

Table 7. 48 Maxterms and Maxterm Equations in Bit 2

Cube Indexes	Maxterm Equations
0,1	$1 + X_{19}$
0,3	$1 + X_{17} + X_{18}$
1,3	$1 + X_{16}$
12,13	$1 + X_{31}$
12,15	$1 + X_{29} + X_{30}$
13,15	$1 + X_{28}$
48,49	$1 + X_{67}$
48,51	$1 + X_{65} + X_{66}$
49,51	$1 + X_{64}$
60,61	$1 + X_{79}$
60,63	$1 + X_{77} + X_{78}$
61,63	$1 + X_{76}$
4,5,8,9,11	$1 + X_{23}$
4,7,8,9,11	$1 + X_{21} + X_{22}$
5,7,8,9,11	$1 + X_{20}$
4,5,7,8,9	$1 + X_{27}$
4,5,7,8,11	$1 + X_{25} + X_{26}$
4,5,7,9,11	$1 + X_{24}$
52,53,56,57,59	$1 + X_{71}$
52,55,56,57,59	$1 + X_{69} + X_{70}$
53,55,56,57,59	$1 + X_{68}$
52,53,55,56,57	$1 + X_{75}$
52,53,55,56,59	$1 + X_{73} + X_{74}$
52,53,55,57,59	$1 + X_{72}$
16,17,36,37,39,40,41,43	$1 + X_{35}$
16,19,36,37,39,40,41,43	$1 + X_{33} + X_{34}$
17,19,36,37,39,40,41,43	$1 + X_{32}$
28,29,36,37,39,40,41,43	$1 + X_{47}$
28,31,36,37,39,40,41,43	$1 + X_{45} + X_{46}$
29,31,36,37,39,40,41,43	$1 + X_{44}$
20,21,23,24,25,27,32,33	$1 + X_{51}$
20,21,23,24,25,27,32,35	$1 + X_{49} + X_{50}$
20,21,23,24,25,27,33,35	$1 + X_{48}$
20,21,23,24,25,27,44,45	$1 + X_{63}$
20,21,23,24,25,27,44,47	$1 + X_{61} + X_{62}$
20,21,23,24,25,27,45,47	$1 + X_{60}$
20,21,24,25,27,36,37,39,40,41,43	$1 + X_{39}$
20,23,24,25,27,36,37,39,40,41,43	$1 + X_{37} + X_{38}$
21,23,24,25,27,36,37,39,40,41,43	$1 + X_{36}$
20,21,23,24,25,36,37,39,40,41,43	$1 + X_{43}$
20,21,23,24,27,36,37,39,40,41,43	$1 + X_{41} + X_{42}$
20,21,23,25,27,36,37,39,40,41,43	$1 + X_{40}$
20,21,23,24,25,27,36,37,40,41,43	$1 + X_{55}$
20,21,23,24,25,27,36,39,40,41,43	$1 + X_{53} + X_{54}$
20,21,23,24,25,27,37,39,40,41,43	$1 + X_{52}$
20,21,23,24,25,27,36,37,39,40,41	$1 + X_{59}$
20,21,23,24,25,27,36,37,39,40,43	$1 + X_{57} + X_{58}$
20,21,23,24,25,27,36,37,39,41,43	$1 + X_{56}$

Table 8. 48 Maxterms and Maxterm Equations in Bit 3

Cube Indexes	Maxterm Equations
0,2	$X_{17} + X_{19}$
0,3	$X_{17} + X_{18}$
1,2	$1 + X_{16}$
12,14	$X_{29} + X_{31}$
12,15	$X_{29} + X_{30}$
13,14	$1 + X_{28}$
48,50	$X_{65} + X_{67}$
48,51	$X_{65} + X_{66}$
49,51	X_{64}
60,62	$X_{77} + X_{79}$
60,63	$X_{77} + X_{78}$
61,62	$1 + X_{76}$
4,5,8,9,10	$X_{22} + X_{23}$
4,6,8,9,10	$X_{21} + X_{23}$
5,6,8,9,10	$1 + X_{20}$
4,5,6,8,9	$X_{26} + X_{27}$
4,5,6,8,10	$X_{25} + X_{27}$
4,5,6,9,10	$1 + X_{24}$
52,53,56,57,58	$X_{70} + X_{71}$
52,54,56,57,58	$X_{69} + X_{71}$
53,54,56,57,58	$1 + X_{68}$
52,53,54,56,57	$X_{74} + X_{75}$
52,53,54,56,58	$X_{73} + X_{75}$
52,53,54,58,59	X_{72}
16,17,36,37,38,40,41,42	$X_{34} + X_{35}$
16,18,36,37,38,40,41,42	$X_{33} + X_{35}$
17,18,36,37,38,40,41,42	$1 + X_{32}$
28,29,36,37,38,40,41,42	$X_{46} + X_{47}$
28,30,36,37,38,40,41,42	$X_{45} + X_{47}$
29,30,36,37,38,40,41,42	$1 + X_{44}$
20,21,22,24,25,26,32,33	$X_{50} + X_{51}$
20,21,22,24,25,26,32,34	$X_{49} + X_{51}$
20,21,22,24,25,26,33,34	$1 + X_{48}$
20,21,22,24,25,26,44,45	$X_{62} + X_{63}$
20,21,22,24,25,26,44,46	$X_{61} + X_{63}$
20,21,22,24,25,26,45,46	$1 + X_{60}$
20,21,24,25,26,36,37,38,40,41,42	$X_{38} + X_{39}$
20,22,24,25,26,36,37,38,40,41,42	$X_{37} + X_{39}$
21,22,24,25,26,36,37,38,40,41,42	$1 + X_{36}$
20,21,22,24,25,36,37,38,40,41,42	$X_{42} + X_{43}$
20,21,22,24,26,36,37,38,40,41,42	$X_{41} + X_{43}$
20,21,22,25,26,36,37,38,40,41,42	$1 + X_{40}$
20,21,22,24,25,26,36,37,40,41,42	$X_{54} + X_{55}$
20,21,22,24,25,26,36,38,40,41,42	$X_{53} + X_{55}$
20,21,22,24,25,26,37,38,40,41,42	$1 + X_{52}$
20,21,22,24,25,26,36,37,38,40,41	$X_{58} + X_{59}$
20,21,22,24,25,26,36,37,38,40,42	$X_{57} + X_{59}$
20,21,22,24,25,26,36,37,38,41,42	$1 + X_{56}$

small systems of linear equations. Each one can solve 2 solutions for 4 key bits. So we can use negligible memory to store the 2^{16} 64-bit key candidates.

So the data complexity for our attack is 2^{15} chosen plaintexts, the time complexity is about 2^{32} full 31-round encryptions and the memory requirements can be negligible.

We can exploit any bit in 3-round to recover the key also, but the complexity of bit 1, 2, 3 is the lowest. Since the intention of pLayer is to change the position after sBoxLayer, we sort the bits into 5 classes by different S-boxes and output positions of the S-box.

For each bit in class 1, 32 equations can be deduced to recover 32 key bits and the remained key bits can be searched exhaustively with 2^{48} encryptions. For the bits in other classes can be utilized to derive 48 equations. For the output bit 1, 2 and 3 of the third round in class 2, the number of chosen plaintexts is 2^{15}. For the bits in class 3, class 4 and class 5, the number of chosen plaintexts are about $24 \cdot 2^8 + 24 \cdot 2^{11} \approx 2^{16}$, $24 \cdot 2^8 + 24 \cdot 2^{17} \approx 2^{22}$ and $48 \cdot 2^{26} \approx 2^{32}$ respectively and the time complexity is about 2^{32} full 31-round encryptions which is same as that of the class 2.

5   Conclusion

The key points for a successful cube attack are how to gain a single output bit represented as a low-degree polynomial and how to determine cube indices leading to maxterms and maxterm equations efficiently. In order to identify the low-degree state bits, we utilize the side channel attack. The original cube attack searches the cube indices with random walk as the multivariate polynomial is randomness. However we identify the low-degree state bits by analyzing the property of round function of PRESENT. The term discarded technique is utilized to reduce the scale of the polynomial. Then we can search maxterms in the reduced polynomial efficiently in the third round.

According to our analysis, side channel cube attack is efficient to attack PRESENT-80. We have identified the polynomials after a few rounds is not random. Then we recover the 80 key bits with 2^{15} chosen plaintexts and 2^{32} 31-round PRESENT encryptions with a special leaked output bit (bit 1, 2 or 3) of the third round. Moreover, any other output bit of the third round can be utilized to proceed the cube attack with more computing complexity.

There are some problems left open to this attack. Now many types of side channel attacks are not accurate. In another word, when the side channel attack access the cryptosystem to obtain data, the error data with noise may be received. The cube attack is very sensitive to the error. And [4] gave the analysis for error correct cube attack to random polynomials. But it is need further research for non-random polynomials.

Acknowledgements. We would like to thank the reviewers for their very helpful comments on the paper.

References

1. Bogdanov, A., Knudsen, L.R., Leander, G., Paar, C., Poschmann, A., Robshaw, M.J.B., Seurin, Y., Vikkelsoe, C.: PRESENT: An Ultra-Lightweight Block Cipher. In: Paillier, P., Verbauwhede, I. (eds.) CHES 2007. LNCS, vol. 4727, pp. 450–466. Springer, Heidelberg (2007)
2. Dinur, I., Shamir, A.: Cube Attacks on Tweakable Black Box Polynomials. In: Joux, A. (ed.) EUROCRYPT 2009. LNCS, vol. 5479, pp. 278–299. Springer, Heidelberg (2009)
3. Aumasson, J.-P., Dinur, I., Meier, W., Shamir, A.: Cube Testers and Key Recovery attacks on Reduced-Round MD6 and Trivium. In: Dunkelman, O. (ed.) FSE 2009. LNCS, vol. 5665, pp. 1–22. Springer, Heidelberg (2009)
4. Dinur, I., Shamir, A.: Side Channel Cube Attacks on Block Ciphers. Cryptology ePrint Archive. Report 2009/127 (2009)
5. Wheeler, D., Needham, R.: TEA, a Tiny Encryption Algorithm. In: Preneel, B. (ed.) FSE 1994. LNCS, vol. 1008, pp. 363–366. Springer, Heidelberg (1995)
6. Wheeler, D., Needham, R.: TEA extensions (October 1997), Also Correction to XTEA (October 1998), www.ftp.cl.cam.ac.uk/ftp/users/djw3/
7. Lim, C., Korkishko, T.: mCrypton - A Lightweight Block Cipher for Security of Low-cost RFID Tags and Sensors. In: Song, J.-S., Kwon, T., Yung, M. (eds.) WISA 2005. LNCS, vol. 3786, pp. 243–258. Springer, Heidelberg (2006)
8. Hong, D., Sung, J., Hong, S., Lim, J., Lee, S., Koo, B.-S., Lee, C., Chang, D., Lee, J., Jeong, K., Kim, H., Kim, J., Chee, S.: HIGHT: A New Block Cipher Suitable for Low-Resource Device. In: Goubin, L., Matsui, M. (eds.) CHES 2006. LNCS, vol. 4249, pp. 46–59. Springer, Heidelberg (2006)
9. Standaert, F.-X., Piret, G., Gershenfeld, N., Quisquater, J.-J.: SEA: A Scalable Encryption Algorithm for Small Embedded Applications. In: Domingo-Ferrer, J., Posegga, J., Schreckling, D. (eds.) CARDIS 2006. LNCS, vol. 3928, pp. 222–236. Springer, Heidelberg (2006)
10. Robshaw, M.J.B.: Searching for Compact Algorithms: cgen. In: Nguyên, P.Q. (ed.) VIETCRYPT 2006. LNCS, vol. 4341, pp. 37–49. Springer, Heidelberg (2006)
11. Xun, Y.: On Boolean Function Expressions of Sbox (1995) (in chinese)
12. Wang, M.Q.: Differential Cryptanalysis of Reduced-Round PRESENT. In: Vaudenay, S. (ed.) AFRICACRYPT 2008. LNCS, vol. 5023, pp. 40–49. Springer, Heidelberg (2008)
13. Courtois, N.T., Debraize, B.: Specific S-Box Criteria in Algebraic Attacks on Block Ciphers with Several Known Plaintexts. In: Lucks, S., Sadeghi, A.-R., Wolf, C. (eds.) WEWoRC 2007. LNCS, vol. 4945, pp. 100–113. Springer, Heidelberg (2008)
14. Z'aba, M.R., Raddum, H., Henricksen, M., Dawson, E.: Bit-Pattern Based Integral Attack. In: Nyberg, K. (ed.) FSE 2008. LNCS, vol. 5086, pp. 363–381. Springer, Heidelberg (2008)
15. Albrecht, M., Cid, C.: Algebraic Techniques in Diferential Cryptanalysis. In: Dunkelman, O. (ed.) FSE 2009. LNCS, vol. 5665, pp. 193–208. Springer, Heidelberg (2009)
16. Blum, M., Luby, M., Rubinfeld, R.: Self-testing/correcting with applications to numerical problems. Journal of Computer and System Sciences 47, 549–595 (1993)

Algebraic Attack on the MQQ Public Key Cryptosystem

Mohamed Saied Emam Mohamed[1], Jintai Ding[2], Johannes Buchmann[1],
and Fabian Werner[1]

[1] TU Darmstadt, FB Informatik
Hochschulstrasse 10, 64289 Darmstadt, Germany
{mohamed,buchmann}@cdc.informatik.tu-darmstadt.de, fw@cccmz.de
[2] Department of Mathematical Sciences, University of Cincinnati
Cincinnati OH 45220, USA
jintai.ding@uc.edu

Abstract. In this paper, we present an efficient attack on the multivariate Quadratic Quasigroups (MQQ) public key cryptosystem. Our cryptanalysis breaks the MQQ cryptosystem by solving a system of multivariate quadratic polynomial equations using both the MutantXL algorithm and the F_4 algorithm. We present the experimental results that show that MQQ systems is broken up to size n equal to 300. Based on these results we show also that MutantXL solves MQQ systems with much less memory than the F_4 algorithm implemented in Magma.

Keywords: Algebraic Cryptanalysis, MQQ public key cryptosystem, MutantXL algorithm, F_4 algorithm.

1 Introduction

The intractability of solving mathematical problems is the security basis for many public key cryptosystems. One example are multivariate cryptosystems which are based on the problem of solving large systems of multivariate polynomial equations over finite fields.

The first multivariate public key cryptosystem was the Matsumoto-Imai scheme in [12] which was broken by Patarin in [14]. Other systems like the Hidden Field Equation (HFE) by Patarin [15] and the unbalanced Oil and Vineger (UOV) by Kipnis, Patarin and Goubin in [11] were attacked by Faugère et al. [7] and Wolf et al. [2,16] respectively.

At the American Conference on Applied Mathematics 2008 (MATH08), Gligoroski et al. presented a new multivariate public key encryption scheme referred to as Multivariate Quadratic Quasigroups (MQQ), which, according to the authors, is as fast as highly efficient block ciphers. MQQ is parameterized by the number of variables n in the multivariate polynomials that are used in the public key. The inventors of MQQ claim that for $n \geq 140$ the security level of MQQ is at least $2^{\frac{n}{2}}$, means an attacker must perform at least $2^{\frac{n}{2}}$ elementary operations to discover the plaintext.

J.A. Garay, A. Miyaji, and A. Otsuka (Eds.): CANS 2009, LNCS 5888, pp. 392–401, 2009.

In this paper, we present an algebraic attack that breaks the MQQ scheme. We present experiments that show that MQQ is easily broken for n up to 300. For our algebraic attack we use the MutantXL algorithm, which was published by Ding et al. [4], and improved by Mohamed et al. [13]. We slightly adjusted the MutantXL implementation to make it more efficient when attacking MQQ.

We also use Magma's implementation of the F_4 algorithm [6] for this attack. The result is, that F_4 can also successfully attack MQQ, but MutantXL turns out to use significantly less space than Magma's F_4. For example, if $n = 200$ MutantXL requires 6.3 Gigabytes of memory while F_4 needs 17.7 Gigabytes. After the initial publication of our attack on the IACR eprint archive, we were informed that the F_4 attack on MQQ was independently discovered by Ludovic Perret.

The paper is organized as follows. In Section 2 we review the MQQ cryptosystems. In Section 3 we describe the MutantXL algorithm and its adaptation to MQQ. Section 4 contains our experimental results and shows how to cryptanalyze MQQ. Finally we conclude our paper in Section 5.

2 MQQ Cryptosystem

The MQQ public-key cryptosystem is a standard multivariate public key cryptosystem that is constructed using quasigroup string transformation performed on a class of quasigroups. The security parameter is a positive integer n which is the number of variables and polynomials used in the public key. The authors of MQQ proposed the length of $n \geq 140$ for a conjectured security level of $2^{\frac{n}{2}}$. In this Section we present an overview of the MQQ cryptosystem. A more detailed explanation is found in [9,10].

Definition 1. *Let $Q = \{a_1, \ldots, a_n\}$ be a finite set of n elements. A quasigroup $(Q, *)$ is a groupoid satisfying the law*

$$(\forall a, b \in Q)(!\exists x, y \in Q)(a * x = b \quad \wedge \quad y * a = b) \tag{1}$$

The unique solutions to these equations are written $x = a \backslash_* b$ and $y = b /_* a$ where $\backslash_*$ and $/_*$ are called a left parastrophe and a right parastrophe of $*$ respectively. The basic quasigroup string transformation, called e-transformation is defined as follows [8]:

Definition 2. *A quasigroup e-transformation of a string $S = (s_0, \ldots, s_{k-1}) \in Q^k$ with a leader $l \in Q$ is the function $e_l : Q \times Q^k \to Q^k$ defined as $T = e_l(S)$, $T = (t_0, \ldots, t_{k-1})$ such that*

$$t_i = \begin{cases} l * s_0 & i = 0 \\ t_{i-1} * s_i & 1 \leq i \leq k - 1 \end{cases} \tag{2}$$

Consider the case where each element $a \in Q$ has a unique d-bit representation $x_1, \ldots, x_d \in \{0, 1\}$ such that $a = x_1 x_2 \ldots x_d$. The binary operation $*$ of the finite quasigroups $(Q, *)$ is equivalent to a vector valued operation $*_{vv} : \{0, 1\}^{2d} \to \{0, 1\}^d$ defined as:

$$a * b = c \Leftrightarrow *_{vv}(x_1, \ldots, x_d, y_1, \ldots, y_d) = (z_1, \ldots, z_d)$$

where $x_1 \ldots x_d$, $y_1 \ldots y_d$, and $z_1 \ldots z_d$ are binary representations of a, b, and c respectively.

Lemma 1. *For every quasigroup $(Q, *)$ of order 2^d and for each d-bit representation of Q there is a unique vector valued operation $*_{vv}$ and d uniquely determined arrays of length 2d of boolean functions $f_1, \ldots, f_d$ such that $\forall a, b, c \in Q$*

$$a * b = c \Leftrightarrow *_{vv}(X^d, Y^d) = (f_1(X^d, Y^d), \ldots, f_d(X^d, Y^d))$$

where $X^d = x_1, \ldots, x_d$, $Y^d = y_1, \ldots, y_d$.

Each $k-bit$ boolean function $f(x_1, \ldots, x_k)$ has the following algebraic normal form (ANF):

$$ANF(f) = c_0 + \sum_{1 \leq i \leq k} c_i x_i + \sum_{1 \leq i \leq j \leq k} c_{i,j} x_i x_j + \ldots, \tag{3}$$

where $c_0, c_i, c_{i,j}, \ldots \in \{0, 1\}$. The degrees of the boolean functions f_i are one of the complexity factors of the quasigroup $(Q, *)$.

Definition 3. *A quasigroup $(Q, *)$ of order 2^d is called multivariate quadratic quasigroup (MQQ) of type $Quad_{d-k}Lin_k$ if exactly $d - k$ of the polynomials f_i are quadratic and k of them are linear, where $0 \leq k \leq d$.*

The authors of [9,10] provide a heuristic algorithm to generate MQQs of order 2^d and of type $Quad_{d-k}Lin_k$. The public and the private keys are constructed as follows.

A system $P' = (p_1, \ldots, p_n)$ of quadratic polynomials over $\mathbb{F}_2$ in n variables is generated using uniformly and randomly selected quasigroups $*_1, \ldots, *_8$ as described in Table 1. That system represents a map $P' : \mathbb{F}_2^n \to \mathbb{F}_2^n$. The matrices $T, S \in \mathbb{F}_2^{(n,n)}$ are selected uniformly at random. The public key is the map

$$P = T \circ P' \circ S$$

which can also be represented by n quadratic polynomials in n variables over $\mathbb{F}_2$. The secret key consists of the 10-tuple $(T, S, *_1, \ldots, *_8)$.

The plain text space is $\mathbb{F}_2^n$. A plaintext $x = (x_1, \ldots, x_n) \in \mathbb{F}_2^n$ is encrypted by computing

$$c = P(x)$$

Decrypting means solving the multivariate system $P(x) = c$. If the secret key is known then we can decrypt using the form

$$x = S^{-1} P'^{-1} T^{-1}(c)$$

where P'^{-1} is computed by using the left parastrophes $\backslash_*$ of the quasigroups $*_1, \ldots, *_8$.

The parameter that was suggested for the practical applications of the MQQ scheme is $n(= 140, 160, 180, 200, \ldots)$, where n is the bit length of the encrypted block.

Table 1. Definition of the the nonlinear mapping P'

Input: Integer n, where $n = 5k, k \geq 28$
Output: Eight quasigroups $*_1, \ldots, *_8$ and n multivariate quadratic polynomials P'
1. Randomly generate n Boolean functions $L = (f_1, \ldots, f_n)$ of n variables $x = (x_1, \ldots, x_n)$;
2. Represent a vector L as a string $L = X_1 \ldots X_k$, where X_i are vectors of dimension 5;
3. Generate several MQQs of type $Quad_4 Lin1$ and $Quad_5 Lin_0$;
The algorithm of generating MQQs is described in [9,10] Table 2.
4. Randomly choose $*_1, *_2 \in Quad_4 Lin_1$ and $*_3, *_4, *_5, *_6, *_7, *_8 \in Quad_5 Lin_0$;
5. Define a $(k-1) - tuple$ $I = (i_1, \ldots, i_{k-1})$ where $i_j \in \{1, \ldots, 8\}$ such that, $1, 2$ are
repeated 8 times in I, without loss of generality let $i_1, \ldots, i_8 \in \{1, 2\}$.
6. Compute $y = Y_1 \ldots Y_k$ where $Y_1 = X_1, Y_{j+1} = X_j *_{i_j} X_{j+1}$, for $j = 1, 2, \ldots, k-1$;
7. Set a vector $Z = Y_1 \|Y_{2,1}\|Y_{3,1}\| \ldots \|Y_{8,1}$ that has 13 components as linear Boolean
functions, where $Y_{j,1}$ means the first coordinate of the vector Y_j;
8. Transform Z by the bijection Dobbertin: $W = Dob(Z)$;
9. Set $Y_1 = (W_1, W_2, W_3, W_4, W_5), Y_{2,1} = W_6, \ldots, Y_{8,1} = W_{13}$;
10. Return y as n multivariate quadratic polynomials $P' = \{p'(x_1, \ldots, x_n)), i = 1, \ldots, n\}$
and the eight Quasigroups $*_1, \ldots, *_8$;

3   MutantXL

MutantXL is an efficient algorithm for solving systems of multivariate polynomial equations that have only one solution. It is a variant of the XL algorithm [3] uses mutant strategy [5].

Let F be a finite field and q be its cardinality. We consider the ring

$$R = F[x_1, \ldots, x_n]/(x_1^q - x_1, \ldots, x_n^q - x_n)$$

of functions over F in the n variables $x_1, \ldots, x_n$. Here $x_i^q - x_i = 0, 1 \leq i \leq n$ are the so-called field equations. In R, each element is uniquely expressed as a polynomial where each x_i has degree less than q. Let the monomials of R are ordered by the graded lexicographical order $<_{glex}$.

Let P be a finite set of polynomials in R. Given a degree bound D, the XL algorithm is simply based on extending the set of polynomials P by multiplying each polynomial in P by all the possible monomials such that the resulting polynomials have degree less than or equal to D. Then, by using linear algebra, XL computes a row echelon form of the extended set P. XL uses univariate polynomials in this row echelon form of P to solve $P(\underline{x}) = 0$ at least partially. If the system can not be solved, D is increased.

In [5,4], it was pointed out that during the linear algebra step, certain polynomials of degrees lower than expected appear. These polynomials are called mutants. The mutant strategy is to give mutants a predominant role in the process of solving the system. The precise definition of mutants is as follows:

Let I be the ideal generated by the finite set of polynomials P. An element f in I can be written as

$$f = \sum_{p \in P} f_p p \qquad (4)$$

where $f_p \in R$. The maximum degree of $f_p p$, $p \in P$, is the *level* of this representation. The level of f is the minimum level of all of its representations. The polynomial f is called *mutant* with respect to P if $\deg(p)$ is less than its level.

We describe the MutantXL algorithm and its adaptation to MQQ. The input of MutantXL is a set P. The output of MutantXL is a vector $x = (x_1, \ldots, x_n) \in \mathbb{F}^n$ such that $p_i(x) = 0$, $1 \leq i \leq m$. The MutantXL algorithm executes the following steps:

- *Initialize*: Set the degree bound D to the maximum degree of the polynomials in P, set the elimination degree d to the minimum degree of the polynomials in P, and set the set of mutants M to the empty set.

- *Eliminate*: Compute the row echelon form of the set $P_d = \{p \in P : deg(p) \leq d\}$. Here polynomials are identified with their coefficient vectors as explained in [6].

- *Solve*: If there are univariate polynomials in P, then determine the values of the corresponding variables. If this solves the system return the solution and terminate, otherwise substitute the values of the variables in P, set D to $max\{deg(p) : p \in P\}$, set d to D, and go back to *Eliminate*.

- *ExtractMutants*: Add all the new elements of P_d, that have degree $< d$, to M.

- *MultiplyMutants*: If M is not empty, then multiply a necessary number of mutants that have degree $k = min\{deg(p): p \in M\}$ by all monomials of degree one, remove the multiplied polynomials from M, add the new polynomials obtained to P, set d to $k + 1$, and go back to *Eliminate*. The necessary number of mutants are numerically computed as in [13].

- *Extend*: Extend P by adding all polynomials that are obtained by multiplying the degree D elements in P by all monomials of degree one. Increment D by one, set d to D and go back to *Eliminate*.

We explain why and how we adapted MutantXL to MQQ. When MutantXL is not successful in solving the system $P = 0$ with a certain degree bound, the algorithm increments the degree bound by one and extends P in the *Extend* step. For MQQ it turned out that *Extend* yields more polynomials than required to solve the system. In the *Extend* step we therefore only multiply the degree D polynomials from the original system. We do not use the degree D polynomials that were generated from mutants.

4   Experimental Results and Analysis

We performed several experiments to attack MQQ systems built using the algorithm described in [9,10]. These systems come from decrypting ciphertext using

the public key but not the secret key. The MQQ inventors supplied us with
a few MQQ systems that are not sufficient for the analysis. However, we used
them to confirm our implementation of the MQQ cryptosystem. We generated
some systems for $n = 60, 80, \ldots, 300$ that were created using MQQs of type
$Quad_4Lin_1$ and $Quad_5Lin_0$ as in Table 1. According to [9,10], these correspond
to $30, 40, \ldots, 150$ bits of security. Our experiment setup has a Sun X4440 server,
with four "Quad-Core AMD Opteron^{TM} Processor 8356" CPUs and 128 GB of
main memory. Each CPU is running at 2,3 GHz. MutantXL code at the moment
uses only one out of the 16 cores. We used both our MutantXL variant and the
Magma's implementation of the F_4 algorithm (version V2.13-10).

Table 2 shows the results of our attacks. There we list the number n of variables and equations, the maximum required memory in Megabytes, the maximum matrix size, and the executed time in seconds. It is clear from Table 2 that
all systems up to $n = 300$ were successfully attacked by MutantXL as well as
Magma's implementation of F_4.

Figure 1(a) compares the maximum number of polynomials used in case of
MutantXL and Magma's F_4. We noticed from it that the MutantXL algorithm
solves the MQQ systems with smaller number of polynomial equations than
Magma's F_4. Conversely, Figure 1(b) shows that the number of monomials of
Magma's F_4 is smaller than MutantXL. This is due to the special selection
strategy used by the F_4 algorithm, while MutantXL multiplies polynomials of the
initial system by all monomials up to certain degree D. For the MQQ systems,
all the quadratic monomials appear in the initial system. In this case, all the
monomials up to degree D will appear in the enlarged system.

Table 3 and Table 4 show the steps of solving an MQQ system for $n = 200$
using MutantXL and Magma's F_4, respectively. In Table 3, for each step we show
the elimination degree (d), the matrix size, the rank of the matrix (Rank), the

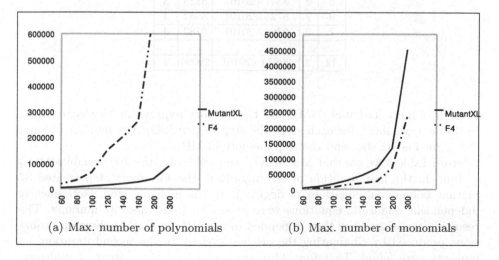

(a) Max. number of polynomials   (b) Max. number of monomials

Fig. 1. Comparison between MutantXL and F_4 for MQQ

Table 2. Performance of MutantXL versus F_4

n	MutantXL			F_4		
	Memory MB	Max Matrix	Time in in sec.	Memory MB	Max Matrix	Time in in sec.
60	1.6	3714×3605	8	88.8	18835×35918	4
80	70.1	6830×85401	23	217.5	33267×32863	10
100	212.7	10649×166751	76	538	63258×62697	55
120	498.2	14387×288101	283	1819	149077×148234	298
140	1109	20192×457451	556	2909	200397×199391	873
160	2281	26937×682801	1283	4364	262244×261130	1366
200	6437	39497×1333501	16694	18198	699138×697280	17186
300	47952	88647×4500251	237362	111160	2339710×2336171	387754

Table 3. MutantXL: Results for MQQ-200

Step	D	Matrix Size	Rank	NM
1	2	200×20101	200	37
2	2	7600×20101	6897	0
3	3	39497×1333501	39497	130
4	2	7427×20101	7347	5
5	2	8347×20101	8125	3
6	2	8725×20101	8584	4
7	2	9384×20101	9183	4
...	...	...	...	...
42	2	20874×20101	20094	4

number of mutants found (NM), and the memory required in Megabyte (MB). In Table 4 we show, for each step, the step degree (SD), the number of pairs (NP), the matrix size, and the step memory in MB.

From Table 3 we see that MutantXL can easily solve the 200 variables MQQ system. In the first iteration of the algorithm, the *Eliminate* step created 37 mutant polynomial equations of degree 1. In the multiply step, 6697 linearly independent quadratic equations were generated from these 37 mutants. The resulting equations were then appended to the 163 quadratic polynomial equations produced by eliminating the original system. In the second iteration, no mutants were found. Therefore, MutantXL extended the system by multiplying only the 163 quadratic equations producing 32600 cubic equations. In the

Table 4. Magma-F_4: Results for MQQ-200

Step	SD	NP	Matrix Size
1	2	192	200×20101
2	2	163	6897×20101
3	3	4640	26732×721928
4	2	84	914×13366
5	2	182	1948×13366
6	2	130	2596×13233
7	2	148	3409×13189
...	...	...	...
42	3	58891	699138×697280

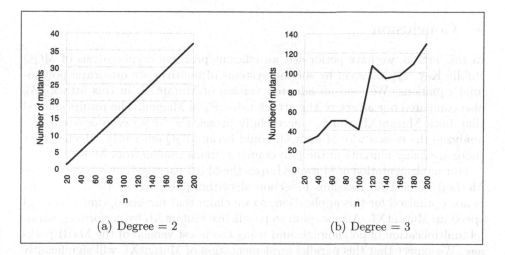

(a) Degree = 2   (b) Degree = 3

Fig. 2. Relation between n and the number of mutants obtained

third iteration, MutantXL eliminated the extended system thus generating 128 quadratic mutants and two linear mutants. Further iteration steps continuously generate linear mutants as shown in Table 3 until some of these mutants are univariate which finally leads to solving the system.

It was indeed observed that all MQQ systems offer enough algebraic information (in the sense that it finds enough mutants) to the MutantXL algorithm such that it was always able to solve the system having a critical degree of 3. Figure 2(a) shows the direct linear relation between the size of the initial system n and the number of linear mutants obtained from it. On the other hand Figure 2(b) shows the relation between n and the number of mutants obtained from the extended degree 3 system. Both figures point out two main drawbacks of the MQQ system: First, the initial systems contains linear equations, which is easily discovered in the first step of MutantXL. Second, at degree 3, the system keeps producing mutants until the system is solved. This explains why MQQ systems are solved at degree 3 and therefore can be easily defeated.

We are going to estimate the security level of MQQ against attacks using MutantXL. The MQQ systems that we have broken are solved by MutantXL and F_4 at degree $D = 3$ as explained above. It is reasonable to assume that all the MQQ systems can be defeated at degree $D = 3$. For MutantXL the memory resources are basically measured by the matrix size. From the linear relation explained in Figure 2(a), we can easily estimate the number of linear mutants obtained in the first step. This will enable us to calculate the maximum number of polynomials. For MutantXL, all the monomials appear. We claim that MutantXL can attack MQQ cryptosystems up to $n = 365$ using the same memory resources of the architecture specified above (128 GB). In this case, the expected matrix size is 133590×8104461 which needs $\simeq 126$ Gigabyte of space in less than 10 days.

5   Conclusion

In this article, we have performed an efficient practical cryptanalysis of MQQ Public Key cryptosystem by solving systems of multivariate quadratic polynomial equations. We used an adapted version of MutantXL in this attack. We also compared our attack to the attack using F_4 in Magma. The results showed that both MutantXL and F_4 successfully attack the MQQ cryptosystem. We analyzed the reason why MutantXL could break MQQ efficiently, which is that there are many mutants in the polynomial systems coming from MQQ.

Our implementation of MutantXL uses the M4RI package that is based on the Method of the Four Russians Inversion algorithm published by Bard [1]. M4RI is not optimized for this application, so we claim that further optimization will speed up MutantXL. Also we plan to parallelize MutantXL by performing parallel multiplication of polynomials and using the latest version of the M4RI package. We expect that this parallel implementation of MutantXL will significantly improve its speed performance. Also Magma's implementation of F_4 does not use mutants, so we plan to combine the mutant strategy with the F_4 algorithm.

Acknowledgment

We would like to thank Danilo Gligoroski for supplying us with some MQQ systems generated by his implementation.

References

1. Bard, G.V.: Accelerating cryptanalysis with the Method of Four Russians. Report 251, Cryptology ePrint Archive (2006)
2. Braeken, A., Wolf, C., Preneel, B.: A study of the security of Unbalanced Oil and Vinegar signature schemes. In: Menezes, A. (ed.) CT-RSA 2005. LNCS, vol. 3376, pp. 29–43. Springer, Heidelberg (2005)
3. Courtois, N., Klimov, A., Patarin, J., Shamir, A.: Efficient Algorithms for Solving Overdefined Systems of Multivariate Polynomial Equations. In: Preneel, B. (ed.) EUROCRYPT 2000. LNCS, vol. 1807, pp. 392–407. Springer, Heidelberg (2000)

4. Ding, J., Buchmann, J., Mohamed, M.S.E., Moahmed, W.S.A., Weinmann, R.-P.: MutantXL. In: Proceedings of the 1st international conference on Symbolic Computation and Cryptography (SCC 2008), Beijing, China, pp. 16–22. LMIB (April 2008)
5. Ding, J., Gower, J.E., Schmidt, D.S.: Zhuang-Zi: A New Algorithm for Solving Multivariate Polynomial Equations over a Finite Field. Technical Report 038, Cryptology ePrint Archive (2006)
6. Faugère, J.-C.: A new efficient algorithm for computing Gröbner bases (F4). Journal of Pure and Applied Algebra 139(1-3), 61–88 (1999)
7. Faugère, J.-C., Joux, A.: Algebraic Cryptoanalysis of Hidden Field Equation (HFE) Cryptosystems Using Gröbner Bases. In: Proceedings of the International Association for Cryptologic Research 2003, pp. 44–60. Springer, Heidelberg (2003)
8. Gligoroski, D.: Candidate One-Way Functions and One-Way Permutations Based on Quasigroup String Transformations. Report 352, Cryptology ePrint Archive (2005)
9. Gligoroski, D., Markovski, S., Knapskog, S.J.: Multivariate Quadratic Trapdoor Functions Based on Multivariate Quadratic Quasigroups. In: Proceedings of The American Conference on Applied Mathematics (MATH 2008), Cambridge, Massachusetts, USA (March 2008)
10. Gligoroski, D., Markovski, S., Knapskog, S.J.: Public Key Block Cipher Based on Multivariate Quadratic Quasigroups. Report 320, Cryptology ePrint Archive (2008)
11. Kipnis, A., Hotzvim, H.S.H., Patarin, J., Goubin, L.: Unbalanced oil and vinegar signature schemes. In: Stern, J. (ed.) EUROCRYPT 1999. LNCS, vol. 1592, pp. 206–222. Springer, Heidelberg (1999)
12. Matsumoto, T., Imai, H.: Public Quadratic Polynomial-Tuples for Efficient Signature-Verification and Message-Encryption. In: Günther, C.G. (ed.) EUROCRYPT 1988. LNCS, vol. 330, pp. 419–453. Springer, Heidelberg (1988)
13. Mohamed, M.S.E., Mohamed, W.S.A.E., Ding, J., Buchmann, J.: MXL2: Solving Polynomial Equations over GF(2) using an Improved Mutant Strategy. In: Buchmann, J., Ding, J. (eds.) PQCrypto 2008. LNCS, vol. 5299, pp. 203–215. Springer, Heidelberg (2008)
14. Patarin, J.: Cryptanalysis of the Matsumoto and Imai Public Key Scheme. In: Coppersmith, D. (ed.) CRYPTO 1995. LNCS, vol. 963, pp. 248–261. Springer, Heidelberg (1995)
15. Patarin, J.: Hidden Fields Equations (HFE) and Isomorphisms of Polynomials (IP): two new families of Asymmetric Algorithms. In: Maurer, U.M. (ed.) EUROCRYPT 1996. LNCS, vol. 1070, pp. 33–48. Springer, Heidelberg (1996)
16. Wolf, C., Preneel, B.: Superfluous keys in multivariate quadratic asymmetric systems. In: Vaudenay, S. (ed.) PKC 2005. LNCS, vol. 3386, pp. 275–287. Springer, Heidelberg (2005)

Construction of Rotation Symmetric Boolean Functions with Maximum Algebraic Immunity

Shaojing Fu[1,2], Chao Li[1,3], Kanta Matsuura[2], and Longjiang Qu[1]

[1] Department of Mathematics and System Science, National University of
Defence Technology, Changsha, China
[2] Institute of Industrial Science, University of Tokyo, 4-6-1 Komaba,
Meguro-ku, Tokyo 153-8505, Japan
[3] State Key Laboratory of Information Security, Beijing, China

Abstract. Rotation symmetric Boolean functions (RSBFs) which are
invariant under circular translation of indices have been used as components of different cryptosystems. In this paper, we study the construction
of RSBFs with maximum algebraic immunity. First, a new construction
of RSBFs on odd number of variables with maximum possible Algebraic
Immunity is given. Then by using the relationship between some flats
and support of a n-variables Boolean function f, we prove that a construction of RSBFs on even number of variables has maximum possible
Algebraic Immunity. Furthermore, we study the nonlinearity of functions
by our construction.

Keywords: Rotation Symmetry; Boolean Function; Algebraic Immunity; Nonlinearity.

1 Introduction

The subject of Boolean functions is well established and constitutes a corner-
stone of cryptography and coding theory, and Boolean functions are the basic
building blocks of the most cryptosystems. The study of the different cryptographic properties of Boolean functions is important because of the strong connections between known cryptanalytic attacks and these properties. Recently,
rotation symmetric Boolean functions (RSBFs) have attracted attention due to
their simplicity-invariant under rotation transform-for efficient computation.

An n-variable boolean function which is invariant under the action of the
cyclic group C_n is called rotation symmetric Boolean functions. These functions
have been analyzed in [18] where the authors studied the nonlinearity of these
Boolean functions and found encouraging results. This study has been extended
in [19,20,13,9]. And important properties of RSBFs have been demonstrated. On
the other hand, Pieprzyk and Qu studied RSBFs as components in the rounds
of a hashing algorithm [14] and research in this direction was later continued
in [7].

In recent years algebraic attacks [1,5,6] have become an important tool in
cryptanalysis of stream and block cipher systems. A new cryptographic property

J.A. Garay, A. Miyaji, and A. Otsuka (Eds.): CANS 2009, LNCS 5888, pp. 402–412, 2009.

for designing Boolean functions to resist this kind of attacks, called algebraic immunity(AI), has been introduced [8,12]. Since then several classes of Boolean functions with large AI have been investigated and constructed in order to against the algebraic attack [3,8,10,12,15,16,17].

It is Sarkar and Maitra who first present a theoretical construction of RSBFs on odd number of variables with maximum possible AI [21], in the construction, n-variable majority function is considered and its outputs are toggled at the inputs of the orbits of size $\lceil n/2 \rceil$ and $\lfloor n/2 \rfloor$ respectively. We here work in the direction at construction of RSBFs with maximum AI. In our construction of odd-variable functions with maximum AI, n-variable majority function is considered and its outputs are toggled at the inputs of the orbits of size $\geq \lceil n/2 \rceil$ and $\lfloor n/2 \rfloor$ respectively. We also give a construction of RSBFs on even number of variables has maximum AI, and we prove the constructed even-variable RSBFs are of better nonlinearity than the existing theoretical constructions with maximum AI.

The paper is organized as follows. Section 2 provides basic definitions and notations. In Section 3, a construction of RSBFs on odd number of variables with maximum possible AI is given. In Section 4, we present a construction of RSBFs on even number of variables with maximum possible AI by using the relationship between some flats and support of a n-variables Boolean function f. The nonlinearity of constructed even-variable RSBFs are studied in Section 5. Section 6 concludes this paper.

2 Preliminaries

Let $\mathbb{F}_2$ be the binary finite field, the vector space of dimension n over $\mathbb{F}_2$ is denoted by $\mathbb{F}_2^n$. A Boolean function on n variables may be viewed as a mapping from $\mathbb{F}_2^n$ into $\mathbb{F}_2$. A Boolean function $f(x_1, x_2, \cdots, x_n)$ is also interpreted as the output column of its truth table, that is, a binary string of length 2^n having the form:

$$\{f(0,0,\cdots,0),\ \ f(0,0,\cdots,1),\ \ \cdots,\ f(1,1,\cdots,1)\}.$$

The weight of f is the number of ones in its output column, and is denoted by $wt(f)$. The support of f is the set $\{x|f(x) = 1\}$ and is denoted by $supp(f)$. The support of a vector $x = (x_1, \cdots, x_n)$ is $supp(x) = \{i | x_i \neq 0\}$.

Definition 1. *An n-variable function f is balanced iff $wt(f) = 2^{n-1}$.*

Let us denote the addition operator over $\mathbb{F}_2$ by $+$. An n-variable function $f(x_1, \cdots, x_n)$ can be seen as a multivariate polynomial over $\mathbb{F}_2$, that is,

$$f(x_1, \cdots, x_n) = a_0 + \sum_{i=1}^{n} a_i x_i + \sum_{1 \leq i < j \leq n} a_{i,j} x_i x_j + \cdots + a_{1,2,\cdots,n} x_1 x_2 \cdots x_n$$

where the coefficients $a_0, a_i, a_{i,j}, \cdots, a_{1,2,\cdots,n}$ are in $\mathbb{F}_2$. This representation of f is called the algebraic normal form (ANF) of f. And the algebraic normal form of a

Boolean function is a unique representation of this function in $\mathbb{F}_2[x_1, .., x_n]/(x_1^2 - x_1, \cdots, x_n^2 - x_n)$.

Let $x = (x_1, \cdots, x_n)$ and $w = (w_1, \cdots, w_n)$ belong to $\mathbb{F}_2^n$ and $x \cdot w = x_1 w_1 + x_2 w_2 + \cdots + x_n w_n$. The Walsh transform of an n-variable function f is a real valued function defined as

$$W_f(w) = \sum_{x \in \mathbb{F}_2^n} (-1)^{f(x)+x \cdot w}.$$

The nonlinearity of f is defined as

$$NL(f) = 2^{n-1} - \frac{1}{2} \max_{u \in \mathbb{F}_2^n} |W_f(u)| \tag{1}$$

A nonzero n-variable Boolean function g is called an annihilator of an n-variable Boolean function f if $f * g = 0$. We denote the set of all annihilators of f by $AN(f)$.

Definition 2. *For $f \in B_n$, the algebraic immunity(AI) of f is the minimum degree of non-zero functions $g \in B_n$ such that $g * f = 0$ or $g * (f+1) = 0$. Namely,*

$$AI(f) = min\{deg(g)|0 \neq g \in AN(f) \cup AN(1+f)\}$$

If $x_i \in \mathbb{F}_2$ for any $1 \leq i \leq n$, and $0 \leq k \leq n-1$. We define

$$\rho_n^k(x_i) = \begin{cases} x_{i+k}, & \text{if } i+k \leq n, \\ x_{i+k-n}, & \text{if } i+k > n. \end{cases}$$

Let $x = (x_1, \cdots, x_n) \in \mathbb{F}_2^n$, then we can extend the definition of ρ_n^k on tuples and monomials as follows:

$$\rho_n^k(x_1, \cdots, x_n) = (\rho_n^k(x_1), \cdots, \rho_n^k(x_n)),$$

and

$$\rho_n^k(x_{i_1} x_{i_2} \cdots) = \rho_n^k(x_{i_1}) \rho_n^k(x_{i_2}) \cdots .$$

Definition 3. *$f(x_1, x_2, \cdots, x_n)$ is called Rotation Symmetric if for each input $x = (x_1, x_2, \cdots, x_n) \in \mathbb{F}_2^n$, and for any k, $0 \leq k \leq n-1$,*

$$f(\rho_n^k(x_1, x_2, \cdots, x_n)) = f(x_1, x_2, \cdots, x_n).$$

Note that there are 2^n different input values corresponding to a function. Let us define $G_n(x_1, x_2, \cdots, x_n) = \{\rho_n^k(x_1, x_2, \cdots, x_n)|0 \leq k \leq n-1\}$, that is, the orbit of $(x_1, x_2, \cdots, x_n)$ under the action of ρ_n^k, $0 \leq k \leq n-1$. It is clear that G_n generates a partition of the set $\mathbb{F}_2^n$. Let g_n be the cardinality of the partition. Using Burnside lemma, it can be shown (see [19]) the number of n-variable RSBFs is

$$2^{g_n}, \text{ where } g_n = \frac{1}{n} \sum_{t|n} \phi(t) 2^{\frac{n}{t}},$$

where $\phi(.)$ is Euler's phi-function.

The orbit of an input can be represented by its representative element which is the lexicographically first element belonging to the orbit. The representative elements are again arranged lexicographically as $\Lambda_1 > \cdots > \Lambda_{g_n}$. Note that for any n, $\Lambda_1 = (1, 1, \cdots, 1)$ and $\Lambda_{g_n} = (0, 0, \cdots, 0)$. Thus an n-variable RSBF f can be represented by the length string $f(\Lambda_1), \cdots, f(\Lambda_{g_n})$ which we call *RSTT* of f and denote it by $RSTT_f$.

In [20] it was shown that the Walsh spectrum of an RSBF f takes the same value for all elements belonging to the same orbit, i.e., $W_f(u) = W_f(v)$ if $v \in G_n(u)$. Therefore the Walsh spectrum of f can be represented by the g_n length vector $(W_f(\Lambda_1), \cdots, W_f(\Lambda_{g_n}))$.

Definition 4. *Let us defined n-variables Boolean function F as follows,*

$$F(x) = \begin{cases} 1, wt(x) < \lceil \frac{n}{2} \rceil, \\ 0, wt(x) \geq \lceil \frac{n}{2} \rceil. \end{cases}$$

F is called the majority function.

3   Construction of Odd-Variables RSBFs with Maximum AI

Let us start with a few available results on n-variable Boolean functions with maximum AI when n is odd.

Theorem 1. *[10]Let n be a odd number, then the algebraic immunity of the majority function F is $AI(F) = \lceil \frac{n}{2} \rceil$.*

In [2], A. Canteaut has observed the following:

Proposition 1. *Let n be a odd number, and f be an n-variable Boolean function, if f has no non-zero annihilator of degree at most $\lfloor \frac{n}{2} \rfloor$, then $AI(f) = \lceil \frac{n}{2} \rceil$.*

For $y = (y_1, \cdots, y_n)$, we denote $x^y = x_1^{y_1} \cdots x_n^{y_n}$, We present in the sequel our new construction.

Construction 1

1. Take $n \geq 5$, n odd.
2. Take Λ_p such that $wt(\Lambda_p) = \frac{n-1}{2}$.
3. Choose Λ_q such that $|G_n(\Lambda_p)| = |G_n(\Lambda_q)| \triangleq N$, and for each $(\lambda_1, \cdots, \lambda_N) \neq 0$, the equation $\sum_{i=0}^{N-1} \lambda_i x^{\rho^i(\Lambda_q)} = 1$ has at least one solution in the set $G_n(\Lambda_q)$.
4. Construct

$$T(x) = \begin{cases} F(x) + 1, x \in G_n(\Lambda_p) \cup G_n(\Lambda_q), \\ F(x), otherwise. \end{cases}$$

Theorem 2. *The function T in Construction 1 is an n-variable RSBF with maximum AI.*

Proof. By Proposition 1, We only need to prove that T has no annihilators with degree less than $\lceil n/2 \rceil$.

Let g be a non-zero annihilators of T with degree less than $\lceil n/2 \rceil$, then $supp(g) \subset supp(T+1)$, and g can be represented as the following polynomial,

$$g(x) = \sum_{(\tau_1, \cdots, \tau_n) \in supp(g)} (x_1 + \tau_1 + 1) \cdots (x_n + \tau_n + 1)$$

If $wt(\tau_1, \cdots, \tau_n) \geq \lceil n/2 \rceil$, then the degree of each term of $(x_1 + \tau_1 + 1) \cdots (x_n + \tau_n + 1)$ is not less than $\lceil n/2 \rceil$. Since $\deg(g) < \lceil n/2 \rceil$, there must be some vectors with weight less than $\lceil n/2 \rceil$ in the support of g.

From the construction, we have $supp(T+1) = \{x | wt(x) \geq \lceil n/2 \rceil\} \cup G_n(\Lambda_p) \setminus G_n(\Lambda_q)$, which indicates that

$$G_n(\Lambda_p) \cap supp(g) \neq \emptyset.$$

and

$$g(x) = \sum_{(\tau_1, \cdots, \tau_n) \in G_n(\Lambda_p) \cap supp(g)} (x_1 + \tau_1 + 1) \cdots (x_n + \tau_n + 1)$$

Then g can be formed as

$$g(x) = \sum_{i=0}^{N-1} \lambda_i x^{\rho^i(\Lambda_q)}, \quad where \quad \prod_{i=0}^{N-1} \lambda_i \neq 0.$$

From the assumptions, there are at least one vector in $G_n(\Lambda_q)$ such that $g(x) = 1$, then

$$G_n(\Lambda_q) \cap supp(g) \neq \emptyset \Rightarrow G_n(\Lambda_q) \cap supp(T+1) \neq \emptyset \Rightarrow G_n(\Lambda_q) \subset supp(T+1).$$

which contradict with the assumptions, so T has no annihilators with degree less than $\lceil n/2 \rceil$.

Hence, we finish our proof.   □

4 Construction of Even-Variables RSBFs with Maximum AI

In this section, we study the construction of even-variable RSBFs with maximum algebraic immunity.

It is known that, if a function has degree strictly less than k and if it is null on a flat of dimension at least k, except maybe at one vector of this flat, then it must be null on the whole flat. Carlet exploit this idea for the annihilators of a boolean function f and $f + 1$, and obtain the following result.

Lemma 1. *[4] Let k be any positive integer such that $k \leq \lceil n/2 \rceil$. A sufficient condition for a function f to have no non-zero annihilator of degree strictly less* *than k is that there exists a sequence of flats (i.e. of affine subspaces of $\mathbb{F}_2^n$) $(A_i)_{1 \leq i \leq r}$ of dimensions at least k, such that:*

$$\begin{cases} \forall i \leq r, \ |A_i \setminus [\cup_{i^* \leq i} A_{i^*} \cup supp(f)]| \leq 1, \\ \mathbb{F}_2^n \setminus supp(f) \subseteq \cup_{i \leq r} A_i. \end{cases}$$

Now we present our construction

Construction 2

1. Take $n \geq 6$.
2. Take Λ_p such that $wt(\Lambda_p) \leq \frac{n}{2} - 1$.
3. Choose Λ_q such that $|G_n(\Lambda_p)| = |G_n(\Lambda_q)|$, $wt(\Lambda_q) = \frac{n}{2}$, and for each $x \in G_n(\Lambda_p)$, there is a unique $y \in G_n(\Lambda_q)$ such that $supp(x) \subseteq supp(y)$.
4. Construct

$$R(x) = \begin{cases} F(x) + 1, x \in G_n(\Lambda_p) \cup G_n(\Lambda_q), \\ F(x), otherwise. \end{cases}$$

Henceforth, we will consider R as the function on n-variables obtained from Construction 1. We have the following theorem.

Theorem 3. *The function R in construction 2 is an n-variable RSBF with maximum AI.*

Proof. it comes from a given property of binomial coefficients that $\sum_{i=0}^{n/2} \binom{n}{i} = \sum_{i=n/2}^{n} \binom{n}{i}$, we denote this value by M.

1. Let $a_1, a_2, \cdots, a_M$ be an ordering of the set of all vectors of weight at least $n/2$ in $\mathbb{F}_2^n$, the order being by increasing weights (with any order for vectors of the same weight). We take for A_i the vector spaces $\{x \in \mathbb{F}_2^n | supp(x) \subseteq supp(a_i)\}$. Then when $1 \leq i \leq \binom{n}{n/2}$

$$\begin{cases} A_i \setminus supp(R) = \{a_i\}, a_i \notin \Lambda_q \\ A_i \setminus supp(R) = \{a_i^*\}, a_i^* \in \Lambda_p, a_i \in \Lambda_q, supp(a_i^*) \subseteq supp(a_i). \end{cases}$$

when $i > \binom{n}{n/2}$

$$A_i \setminus [\cup_{i^* \leq i} A_{i^*} \cup supp(R)] = \{a_i\}$$

From Lemma 1, we prove that $R(x)$ have no non-zero annihilator of degree strictly less than $n/2$.

2. Let $b_1, b_2, \cdots, b_M$ be an ordering of the set of all vectors of weight at most $n/2$ in $\mathbb{F}_2^n$, the order being by decreasing weights (with any order for vectors of the same weight). We take for A_i the vector spaces $\{x \in \mathbb{F}_2^n | supp(a_i) \subseteq supp(x)\}$. Then when $1 \leq i \leq \binom{n}{n/2}$

$$\begin{cases} A_i \setminus supp(R+1) = \varnothing, b_i \notin \Lambda_q \\ A_i \setminus supp(R+1) = \{b_i\}, b_i \in \Lambda_q. \end{cases}$$

when $i > \binom{n}{n/2}$

$$\begin{cases} A_i \setminus [\cup_{i^* \leq i} A_{i^*} \cup supp(R+1)] = \{b_i\}, b_i \notin \Lambda_p \\ A_i \setminus [\cup_{i^* \leq i} A_{i^*} \cup supp(R+1)] = \varnothing, b_i \in \Lambda_p. \end{cases}$$

From Lemma 1, we prove that $R(x) + 1$ have no non-zero annihilator of degree strictly less than $n/2$. □

Example 1. Take $n = 6$. Consider $\Lambda_p = (1, 1, 0, 0, 0, 0)$ and $\Lambda_q = (1, 1, 0, 1, 0, 0)$ and generate the orbits

$G_n(\Lambda_p) = (1, 1, 0, 0, 0, 0), (0, 1, 1, 0, 0, 0), (0, 0, 1, 1, 0, 0), (0, 0, 0, 1, 1, 0),$
$(0, 0, 0, 0, 1, 1), (1, 0, 0, 0, 0, 1).$
$G_n(\Lambda_q) = (1, 1, 0, 1, 0, 0), (0, 1, 1, 0, 1, 0), (0, 0, 1, 1, 0, 1), (1, 0, 0, 1, 1, 0),$
$(0, 1, 0, 0, 1, 1), (1, 0, 1, 0, 0, 1).$

Here, for each $x \in G_n(\Lambda_p)$, there is a unique $y \in G_n(\Lambda_q)$ such that $supp(x) \subseteq supp(y)$. Therefore by Theorem 3, the function $R(x)$ is a 6-variable RSBF with Maximum AI.

5   The Nonlinearity of the Constructed RSBFs

In this dection, we will study the nonlinearity of functions by our construction 2. First, we give the results relating algebraic immunity and the nonlinearities of a Boolean function which was presented in [3].

Lemma 2. *[3] if f has maximum algebraic immunity, then*

$$NL(f) \geq \begin{cases} 2^{n-1} - \binom{n-1}{\frac{n-1}{2}}, n \text{ odd} \\ 2^{n-1} - \binom{n-1}{\frac{n}{2}-1} - \binom{n-1}{\frac{n}{2}}, n \text{ even.} \end{cases}$$

Now we study the Walsh spectra of the majority function F which have very nice combinatorial properties related to Krawtchouk polynomial.

Krawtchouk polynomial [11] of degree i is defined by

$$K_i(x, n) = \sum_{j=0}^{i} (-1)^j \binom{x}{j} \binom{n-x}{i-j}.$$

It is known that for a fixed w, such that $wt(w) = k$,

$$\sum_{wt(w)=i} (-1)^{w \cdot x} = K_i(k, n).$$

Let us now list some known results about Krawtchouk polynomial $K_i(x, n)$.

Lemma 3. *We have,*

1.

$$K_i(k,n) = (-1)^k K_{n-i}(k,n),$$

2.

$$K_i(k,n) = (-1)^i K_i(n-k,n),$$

3.

$$K_i(n/2,n) = \begin{cases} 0, & i \text{ odd}, \\ (-1)^{i/2} \binom{n/2}{i/2}, & i \text{ even}. \end{cases}$$

Lemma 4. *The Walsh transform of the function F is as follows,*

$$W_F(w) = \begin{cases} K_{n/2}(wt(w),n), & wt(w) \text{ even}, \\ -2 \sum_{i=0}^{n/2-1} K_i(wt(w),n), & wt(w) \text{ odd}, \end{cases}$$

Proof. If $wt(w)$ is even, then

$$W_F(w) = \sum_{0 \le wt(x) < n/2} (-1)^{w \cdot x + 1} + \sum_{n/2 \le wt(x) \le n} (-1)^{w \cdot x}$$

$$\Rightarrow W_F(w) = - \sum_{i=0}^{n/2-1} K_i(wt(w),n) + \sum_{i=n/2}^{n} K_i(wt(w),n)$$

$$\Rightarrow W_F(w) = K_{n/2}(wt(w),n).$$

If $wt(w)$ is odd, then

$$W_F(w) = \sum_{0 \le wt(x) < n/2} (-1)^{w \cdot x + 1} + \sum_{n/2 \le wt(x) \le n} (-1)^{w \cdot x}$$

$$\Rightarrow W_F(w) = - \sum_{i=0}^{n/2-1} K_i(wt(w),n) + \sum_{i=n/2}^{n} K_i(wt(w),n)$$

$$\Rightarrow W_F(w) = -2 \sum_{i=0}^{n/2-1} K_i(wt(w),n). \qquad \square$$

The following corollary is clear.

Corollary 1. *We have,*

$$W_F(w) = \begin{cases} -\binom{n}{n/2}, & wt(w) = 1, \\ (-1)^{n/2}\binom{n}{n/2}, & wt(w) = n. \end{cases}$$

Theorem 4. *The nonlinearity of the function R is not less than $2^{n-1} - \frac{1}{2}\binom{n}{n/2}$.*

Proof. Let us first find the relation between the values of $W_R(\Lambda_k)$ and $W_F(\Lambda_k)$.

$$W_R(\Lambda_k) = \sum_{x \in G_n(\Lambda_p)} (-1)^{R(x)+x \cdot \Lambda_k} + \sum_{x \in G_n(\Lambda_q)} (-1)^{R(x)+x \cdot \Lambda_k}$$

$$+ \sum_{x \notin G_n(\Lambda_p) \cup G_n(\Lambda_q)} (-1)^{R(x)+x \cdot \Lambda_k}$$

$$= \sum_{x \in G_n(\Lambda_p)} (-1)^{R(x)+x \cdot \Lambda_k+1} + \sum_{x \in G_n(\Lambda_q)} (-1)^{R(x)+x \cdot \Lambda_k+1}$$

$$+ \sum_{x \notin G_n(\Lambda_p) \cup G_n(\Lambda_q)} (-1)^{R(x)+x \cdot \Lambda_k}$$

$$= W_F(\Lambda_k) + 2 \sum_{x \in G_n(\Lambda_p)} (-1)^{x \cdot \Lambda_k} - 2 \sum_{x \in G_n(\Lambda_q)} (-1)^{x \cdot \Lambda_k}.$$

Now we have the following four cases.

1. If $wt(\Lambda_k) = 0$. It is clear that $|W_F(\Lambda_k)| = \binom{n}{n/2}$.
2. If $wt(\Lambda_k) = 1$. $\sum_{x \in G_n(\Lambda_p)}(-1)^{x \cdot \Lambda_k} = 2$, $\sum_{x \in G_n(\Lambda_q)}(-1)^{x \cdot \Lambda_k} = n - 2wt(\Lambda_q)$, from Corollary 1,

$$|W_R(\Lambda_k)| = |W_F(\Lambda_k) + 4 - 2(n - 2wt(\Lambda_q))|$$

$$= |-\binom{n}{n/2} + 4 - 2(n - 2wt(\Lambda_q))| \geq -\binom{n}{n/2} + 4.$$

3. If $wt(\Lambda_k) = n$, $\sum_{x \in G_n(\Lambda_p)}(-1)^{x \cdot \Lambda_k} = (-1)^{wt(\Lambda_p)} \cdot n$, $\sum_{x \in G_n(\Lambda_q)}(-1)^{x \cdot \Lambda_k} = (-1)^{wt(\Lambda_q)} \cdot n$. Then from Corollary 1,

$$W_R(\Lambda_k) = W_F(\Lambda_k) + 2(-1)^{wt(\Lambda_p)} \cdot n - 2(-1)^{wt(\Lambda_q)} \cdot n$$

$$\Rightarrow \begin{cases} W_R(\Lambda_k) \leq \binom{n}{n/2} - 2n + 2n, & n/2 \text{ odd}, \\ W_R(\Lambda_k) \geq -\binom{n}{n/2} + 2n - 2n, & n/2 \text{ even}. \end{cases}$$

4. If $2 \leq wt(\Lambda_k) \leq n-1$, $|\sum_{x \in G_n(\Lambda_p)}(-1)^{x \cdot \Lambda_k}| \leq n$, $|\sum_{x \in G_n(\Lambda_q)}(-1)^{x \cdot \Lambda_k}| \leq n \cdot wt(\Lambda_q)$. Then, from Lemma 4,
   if $wt(\Lambda_k)$ is odd,

$$|W_R(\Lambda_k)| \leq |W_F(\Lambda_k)| + 4n = |2 \sum_{i=0}^{n/2-1} K_i(wt(\Lambda_k), n)| + 4n$$

$$\leq 2 \sum_{i=0}^{n/2-1} |K_i(wt(\Lambda_k), n)| + 4n$$

$$\leq 2 \sum_{i=0}^{n/2-1} |K_i(2, n)| + 4n \leq \binom{n}{n/2}$$

if $wt(\Lambda_k)$ is even,

$$|W_R(\Lambda_k)| \le |W_F(\Lambda_k)| + 4n = |K_{n/2}(wt(\Lambda_k), n)| + 4n$$

$$= |\sum_{j=0}^{n/2}(-1)^j \binom{wt(\Lambda_k)}{j}\binom{n - wt(\Lambda_k)}{n/2 - j}| + 4n \le \binom{n}{n/2}.$$

then by Relation (1), we end our proof.   □

It is obvious that $2^{n-1} - \frac{1}{2}\binom{n-1}{n/2} > 2^{n-1} - \binom{n-1}{\frac{n}{2}-1} - \binom{n-1}{\frac{n}{2}}$, which indicates that the nonlinearity of functions is than better nonlinearity than the existing constructions with maximum AI.

6   Conclusion

In this paper, we firstly present a new construction of RSBFs on odd number of variables with maximum AI. We also give a construction of RSBFs on even number of variables has maximum AI, and it is shown that our constructed even-variable RSBFs are of better nonlinearity than the existing theoretical constructions with maximum AI. However, it is still an open problem to construct balanced RSBFs on even variables. Furthermore, there are still some problems need to be studied such as whether the constructed functions can achieve high degree and be robust against fast algebraic attacks.

Acknowledgments

This work is partially supported by the National Natural Science Foundation of China (No:60803156) and the open research fund of State Key Laboratory of Information Security(No: 01–07).

References

1. Armknecht, F.: Improving Fast Algebraic Attacks. In: Roy, B., Meier, W. (eds.) FSE 2004. LNCS, vol. 3017, pp. 65–82. Springer, Heidelberg (2004)
2. Canteaut, A.: Open problems related to algebraic attacks on stream ciphers. In: Ytrehus, Ø. (ed.) WCC 2005. LNCS, vol. 3969, pp. 120–134. Springer, Heidelberg (2006)
3. Carlet, C., Dalai, D.K., Gupta, K.C., Maitra, S.: Algebraic Immunity for Cryptographically Significant Boolean Functions: Analysis and Construction. IEEE Trans. Inform. Theory 52, 3105–3121 (2006)
4. Carlet, C.: A method of construction of balanced functions with optimum algebraic immunity, http://eprint.iacr.org/2006/149
5. Courtois, N., Pieprzyk, J.: Cryptanalysis of block ciphers with overdefined systems of equations. In: Zheng, Y. (ed.) ASIACRYPT 2002. LNCS, vol. 2501, pp. 267–287. Springer, Heidelberg (2002)
6. Courtois, N., Meier, W.: Algebraic attacks on stream ciphers with linear feedback. In: Biham, E. (ed.) EUROCRYPT 2003. LNCS, vol. 2656, pp. 345–359. Springer, Heidelberg (2003)

7. Cusick, T.W., Stanica, P.: Fast Evaluation, Weights and Nonlinearity of Rotation-Symmetric Functions. Discrete Mathematics 258, 289–301 (2002)
8. Dalai, D.K., Gupta, K.C., Maitra, S.: Results on Algebraic Immunity for Cryptographically Significant Boolean Functions. In: Canteaut, A., Viswanathan, K. (eds.) INDOCRYPT 2004. LNCS, vol. 3348, pp. 92–106. Springer, Heidelberg (2004)
9. Dalai, D.K., Maitra, S., Sarkar, S.: Results on rotation symmetric bent functions. In: Second International Workshop on Boolean Functions: Cryptography and Applications, BFCA 2006, pp. 137–156 (2006)
10. Dalai, D.K., Maitra, S., Sarkar, S.: Basic theory in construction of Boolean functions with maximum possible annihilator immunity. Des. Codes, Cryptogr. 40, 41–58 (2006)
11. MacWilliams, F.J., Sloane, N.J.A.: The Theory of Error Correcting Codes. North-Holland, Amsterdam (1977)
12. Meier, W., Pasalic, E., Carlet, C.: Algebraic attacks and decomposition of Boolean functions. In: Cachin, C., Camenisch, J.L. (eds.) EUROCRYPT 2004. LNCS, vol. 3027, pp. 474–491. Springer, Heidelberg (2004)
13. Maximov, A., Hell, M., Maitra, S.: Plateaued Rotation Symmetric Boolean Functions on Odd Number of Variables. In: First Workshop on Boolean Functions: Cryptography and Applications, BFCA 2005, pp. 83–104 (2005)
14. Pieprzyk, J., Qu, C.X.: Fast Hashing and Rotation-Symmetric Functions. Journal of Universal Computer Science 5, 20–31 (1999)
15. Qu, L.J., Li, C., Feng, K.Q.: A note on symmetric Boolean functions with maximum algebraic immunity in odd number of variables. IEEE Transactions on Information Theory 53, 2908–2910 (2007)
16. Qu, L.J., Li, C.: On the 2^m-variable Symmetric Boolean Functions with Maximum Algebraic Immunity. Science in China Series F-Information Sciences 51, 120–127 (2008)
17. Qu, L.J., Feng, G.Z., Li, C.: On the Boolean Functions with Maximum Possible Algebraic Immunity: Construction and A Lower Bound of the Count, http://eprint.iacr.org/2005/449
18. Stanica, P., Maitra, S.: Rotation symmetric Boolean functions-count and cryptographic properties. Discrete Mathematics and Applications 156, 1567–1580 (2008)
19. Stanica, P., Maitra, S.: A constructive count of rotation symmetric functions. Information Processing Letters 88, 299–304 (2003)
20. Stanica, P., Maitra, S., Clark, J.: Results on rotation symmetric bent and correlation immune Boolean functions. In: Roy, B., Meier, W. (eds.) FSE 2004. LNCS, vol. 3017, pp. 161–177. Springer, Heidelberg (2004)
21. Sarkar, S., Maitra, S.: Construction of rotation symmetric Boolean functions with maximun algebraic immunity on odd number of variables. In: Boztaş, S., Lu, H.-F. (eds.) AAECC 2007. LNCS, vol. 4851, pp. 271–280. Springer, Heidelberg (2007)

Multi-core Implementation of the Tate Pairing over Supersingular Elliptic Curves

Jean-Luc Beuchat[1], Emmanuel López-Trejo[2], Luis Martínez-Ramos[3],
Shigeo Mitsunari[4], and Francisco Rodríguez-Henríquez[3]

[1] Graduate School of Systems and Information Engineering, University of Tsukuba,
1-1-1 Tennodai, Tsukuba, Ibaraki, 305-8573, Japan
[2] Nehalem Platform Validation, Intel Guadalajara Design Center, Periférico Sur 7980
Edificio 4E, 45600 Tlaquepaque, Jalisco, México
[3] Computer Science Department, Centro de Investigación y de Estudios Avanzados
del IPN, Av. Instituto Politécnico Nacional No. 2508, 07300 México City, México
[4] Akasaka Twin Tower East 15F, 2-17-22 Akasaka, Minato-ku, Tokyo 107-0052,
Cybozu Labs, Inc.

Abstract. This paper describes the design of a fast multi-core library
for the cryptographic Tate pairing over supersingular elliptic curves. For
the computation of the reduced modified Tate pairing over $\mathbb{F}_{3^{509}}$, we
report calculation times of just 2.94 ms and 1.87 ms on the Intel Core2
and Intel Core i7 architectures, respectively. We also try to answer one
important design question that arises: how many cores should be utilized
for a given application?

Keywords: Tate pairing, η_T pairing, supersingular curve, finite field
arithmetic, multi-core.

1 Introduction

During the early years of this century it was generally assumed that computing
cryptographic bilinear pairings was a computationally expensive task. Taking as
a starting point the breakthrough introduced by Miller [23, 24], who proposed
the first iterative approach to compute a cryptographic pairing, several authors
focused their efforts on finding algorithmic improvements and shortcuts to further reduce the complexity of the so-called *Miller's Algorithm* [4, 9, 3, 19, 18, 29].
Those theoretical findings were experimentally validated by different means. At
first, it was thought that the rich parallelization potential shown by hardware
platforms could be exploited in order to produce faster and more compact pairing implementations. Through the years, this assumption has been confirmed in
many research works (see for instance [28, 20, 8] for a comprehensive bibliography). Nevertheless, with the only exception of the ASIP in [20], all hardware
accelerators reported in the open literature until today, have targeted low and
medium security levels.

On the other hand, in the last few years a second wave of authors have investigated the challenges associated to the efficient implementation of pairings

J.A. Garay, A. Miyaji, and A. Otsuka (Eds.): CANS 2009, LNCS 5888, pp. 413–432, 2009.

in software platforms [17, 12, 21, 16]. ¿From the results reported by those research works, it appears that software pairing libraries can sometimes compete with their hardware counterparts. Furthermore, yet another way to exploit parallelism can be instrumented when the multi-core architectures introduced just recently by Intel are targeted. Multi-core architectures can be seen as a massive way to obtain parallelism via the concurrent usage of powerful individual processors that are tightly interconnected.

To our knowledge, the only bilinear pairing library targeting a multi-core architecture was reported in [12]. After considering several scenarios, the authors came up with the not too optimistic conclusion that on a Core2 64-bit platform, the best option to parallelize the computation was to perform one pairing on each core. They state that, "if the requirement is for two pairing evaluations, the slightly moronic conclusion is that one can perform one pairing on each core [...], doubling the performance versus two sequential invocations of any other method that does not already use multi-core parallelism internally" [12].

This paper is devoted to the design of a software library for the cryptographic Tate pairing on supersingular elliptic curves defined over finite fields of characteristics two and three. After a careful selection of the field arithmetic (Section 2) and pairing algorithms (Section 3) we show that multi-core architectures can be effectively used to provide significant computational speedups. A single-core version of our software [1] computes the reduced modified Tate pairing in 11.19 and 7.59 ms for the extension fields $\mathbb{F}_{2^{1223}}$ and $\mathbb{F}_{3^{509}}$, respectively, on an Intel Quad Core running at 2.4 GHz. Speedups of approximately 2.6× are obtained when using the four cores available in the target architecture (Section 4).

2   Finite Field Arithmetic Using SSE

2.1   Characteristic Two Field

Frobenius and Inverse-Frobenius Operators. We define the binary extension field $\mathbb{F}_{2^m}$ as $\mathbb{F}_2[x]/(f(x))$, where $f(x)$ is an irreducible degree-m polynomial over $\mathbb{F}_2$. An arbitrary element $a \in \mathbb{F}_{2^m}$ is written as $a(x) = \sum_0^{m-1} a_i x^i$, where $a_i \in \mathbb{F}_2$ for $i = 0, 1, \ldots, m - 1$. Let us also assume that the extension degree m can be expressed as $m = 2u + 1$, with $u \geq 1$. Then, the Frobenius operator applied to a consists of computing $c = a^2 \bmod f(x)$, which can be obtained as

$$c = \sum_{i=0}^{u} a_i x^{2i} + \left(x^m \cdot \sum_{i=1}^{u} a_{u+i} x^{2i-1} \right) \bmod f(x) = a_L + x^m a_H \bmod f(x).$$

The field element c can be efficiently calculated in software by extracting the two half-length vectors a_L and a_H along with the computation of an $\frac{m}{2}$-bit multiplication by the per-field constant x^m. The subsequent reduction process modulo $f(x)$ is typically implemented in software by using XOR and shift operations.

---

[1] An open source code for benchmarking the library is available at
http://homepage1.nifty.com/herumi/crypt/pairing.html

The inverse-Frobenius operator of a is computed by determining the unique field element $b \in \mathbb{F}_{2^m}$ such that $b^2 = a$ holds. The element b can be computed in terms of the square root of the field constant x as

$$b = \sum_{i=0}^{\lfloor \frac{m-1}{2} \rfloor} a_{2i}x^i + \left(x^{\frac{1}{2}} \cdot \sum_{i=0}^{\lfloor \frac{m-3}{2} \rfloor} a_{2i+1}x^i \right) \bmod f(x) = a_{\text{even}} + x^{\frac{1}{2}} a_{\text{odd}} \bmod f(x).$$

The efficient computation in software of b defined as above is performed by extracting the even and odd bits of a into the half length vectors a_{even} and a_{odd}, respectively. This should be followed by multiplying the half length vector a_{odd} by the pre-computed constant $x^{\frac{1}{2}}$.

In the case that the irreducible polynomial happens to be a trinomial of the form $f(x) = x^m + x^n + 1$, where m and n are odd numbers, we have that $x^{\frac{1}{2}} = x^{\frac{m+1}{2}} + x^{\frac{n+1}{2}}$. Since $x^{\frac{1}{2}} a_{\text{odd}}$ has degree $m - 1$, it follows that we do not need to perform a polynomial modular reduction and hence the inverse Frobenius operator of an arbitrary element $a \in \mathbb{F}_{2^m}$ can be obtained by computing [10]

$$b = \sum_{i=0}^{\lfloor \frac{m-1}{2} \rfloor} a_{2i}x^i + \left(x^{\frac{m+1}{2}} + x^{\frac{n+1}{2}} \right) \cdot \sum_{i=0}^{\lfloor \frac{m-3}{2} \rfloor} a_{2i+1}x^i.$$

Multiplier. We implemented this arithmetic block by using a variation of the left-to-right *comb* multiplication scheme presented in [22], one of the fastest multiplier schemes for binary fields reported in the open literature.

Let w be the processor word size in bits. Then, the number of processor words required for storing an arbitrary element in the field $\mathbb{F}_{2^m}$ is $s = \lceil \frac{m}{w} \rceil$. From these definitions, authors in [22] found that the computational complexity of their algorithm (excluding the one associated to the reduction process) was of $s(\frac{m}{4})$ w-bit XOR operations and a total of $(w/4 - 1)$ 4-bit left shift operations of a $2s$-word vector. Additionally, their method makes use of a look-up table containing sixteen s-word entries, which is queried a total of $s(\frac{m}{4})$ times. The look-up table is pre-computed at a cost of three 1-bit left shift operations over an s-word vector and eleven s-word XOR operations.

In the case of the SSE instruction set, we have $w = 128$. Hence, we can invoke specialized instructions to perform any logic or arithmetic operation over a bank of 128-bit SSE register operators. It is also possible to manipulate the contents of the SSE registers by applying left/right shift/rotate operators over them. Those shift and rotate operations are executed with very high efficiency if the operand is shifted or rotated by a constant value multiple of 8 bits. This feature motivated us to propose a right-to-left comb multiplication scheme that trades all but one of the 4-bit left shift operations required by the multiplier in [22], with 8-bit right shift operations. In the rest of this subsection, we describe our formulation.

Let us define $n = 32 \cdot s$. It appears convenient to group the bit representation of a field element $a \in \mathbb{F}_{2^m}$ into 4-bit digits as follows:

$$a = (a_{m-1} \ldots a_1 a_0) \Leftrightarrow a = (A_{n-1} \ldots A_1 A_0),$$

where A_i, for $i = 0, 1, \ldots, n-1$, is defined as $A_i = \sum_{j=0}^{3} a_{4i+j} x^{4i+j}$. Notice that each 128-bit SSE register can store thirty-two such digits.[2]

As it will be discussed below, it appears convenient to rearrange the n digits of the field element a into a $2 \times 16s$ matrix Idx as

$$\text{Idx}[2][16s] = \begin{pmatrix} A_{n-1} \cdots A_{33} \ A_{31} \cdots A_3 \ A_1 \\ A_{n-2} \cdots A_{32} \ A_{30} \cdots A_2 \ A_0 \end{pmatrix}. \tag{1}$$

In order to calculate the product $c = a \cdot b$ we prepare first a 16-entry look-up table by pre-computing $\text{TblMul}[i] \leftarrow (i_3 x^3 + i_2 x^2 + i_1 x + i_0) \cdot b$, for $i = 0, 1, \ldots, 15$ and where $i = (i_3 i_2 i_1 i_0)_2$ is the binary expansion of i. Then, the polynomial product $c = ab$ can be computed as follows:

$$\begin{aligned} ab &= \sum_{i=0}^{n-1} A_i x^{4i} b = \sum_{i=0}^{16s-1} x^{8i} \left(A_{2i} + x^4 A_{2i+1} \right) b \\ &= \sum_{i=0}^{15} \sum_{j=0}^{s-1} x^{8i} \left(A_{32j+2i} + x^4 A_{32j+2i+1} \right) x^{128j} b \\ &= \sum_{i=0}^{15} \sum_{j=0}^{s-1} x^{8i} \left(\text{TblMul}[A_{32j+2i}] + x^4 \text{TblMul}[A_{32j+2i+1}] \right) x^{128j} \qquad (2) \\ &= \left(\sum_{i=0}^{15} \sum_{j=0}^{s-1} x^{8(i-16)} \text{TblMul}[\text{Idx}[1][16j+i]] x^{128j} \right) x^{128} + \\ &\quad \left(\sum_{i=0}^{15} \sum_{j=0}^{s-1} x^{8(i-16)} x^4 \text{TblMul}[\text{Idx}[0][16j+i]] x^{128j} \right) x^{128}. \end{aligned}$$

Note that in the last equality of Eq. (2) we used the matrix Idx as defined in Eq. (1), which allows us to recover the digits $A_{2(16j+i)}$ and $A_{2(16j+i)+1}$ as

$$\text{Idx}[1][16j+i] = A_{2(16j+i)} \quad \text{and} \quad \text{Idx}[0][16j+i] = A_{2(16j+i)+1}.$$

One can compute Eq. (2) as shown in Algorithm 1. It is worth stressing that:

- In step 5, we extract the bits of a in such a way that its digits A_i for $i = 0, \ldots, n-1$ are rearranged into a two dimensional array $\text{Idx}[2][16s]$ as described in Eq. (1).
- The shift operations of step 10, namely, $x^{128}, x^{256}, \ldots, x^{1152}$, can be performed at no cost because they correspond to shifts by entire 128-bit words.
- In step 12, a 1-byte right rotation is applied over the contents of the $2s$-word accumulator R. This rotation operation is invoked thirty-two times.
- In step 15, a 4-bit left rotation over the $2s$-word accumulator R is performed. This is the only left rotation by four bits included in the algorithm.

---

[2] Note that the $v = 128 - (m \bmod 128)$ most significant bits of the last SSE register should be filled with zeroes.

Algorithm 1. SSE implementation of a right-to-left comb multiplier over $\mathbb{F}_{2^m}$.

Input: $a, b \in \mathbb{F}_{2^m}$.
Output: $c = a \cdot b \bmod f(x) \in \mathbb{F}_{2^m}$.
1: **for** $i \leftarrow 0$ **to** 15 **do**
2:      Compute the binary expansion of $i = (i_3 i_2 i_1 i_0)_2$;
3:      TblMul$[i] \leftarrow (i_3 x^3 + i_2 x^2 + i_1 x + i_0) \cdot b(x)$;
4: **end for**

5: Idx$[2][16s] \leftarrow$ extractIdx(a);
6: $R \leftarrow 0$;

7: **for** $k \leftarrow 0$ **to** 1 **do**
8:    **for** $i \leftarrow 0$ **to** 15 **do**
9:       **for** $j \leftarrow 0$ **to** $s - 1$ **do**
10:          $R \leftarrow R + \text{TblMul}[\text{Idx}[k][i + 16 \cdot j]]x^{128 \cdot j}$;
11:       **end for**
12:       $R \leftarrow \text{rotRight\_Byte}(R, 1)$
13:    **end for**
14:    **if** $k = 0$ **then**
15:       $R \leftarrow \text{rotLeft\_bit}(R, 4)$;
16:    **end if**
17:    $R \leftarrow \text{rotLeft\_Byte}(R, 16)$;
18: **end for**

19: $c \leftarrow R \bmod f(x)$;

20: **return** c;

- In step 17, a left rotation by 16 bytes must be executed. This rotation is almost for free as it only implies the reassignment of the SSE registers.
- Finally in Step 19, a modular reduction with the polynomial $f(x)$ must be performed.

It is easy to verify that the computational cost of Algorithm 1 is of three 1-bit left shift operations over an s-word vector, $(32 + 11)s$ XOR operations, $32s$ queries to the look-up table TblMul, thirty-two 1-byte right rotations of a $2s$-word vector, one 4-bit left rotation of a $2s$-word vector, plus the computational cost of the reduction operation of Step 19. As a final remark we state that it is straightforward to generalize the multiplier of Algorithm 1 so that it can compute field multiplications over finite fields with characteristic $p > 2$.

Multiplicative Inverse. We compute the multiplicative inverse of an arbitrary field element $a \in \mathbb{F}_{2^m}$ by implementing the Almost Inverse Algorithm [26, 15], which is a variant of the binary extended Euclidean algorithm.

2.2   Characteristic Three Field

Addition and Subtraction. In 2002, Galbraith *et al.* [11] showed how to compute additions of two elements $a, b \in \mathbb{F}_3$ using 12 AND, OR, XOR and

NOT Boolean functions. That same year, Harrison *et al.* [17] noted that this operation could be computed using only 7 OR and XOR logical instructions. This was considered the minimal number of logical operations for this arithmetic operation until Kawahara *et al.* [21] presented in 2008 an expression that only requires 6 logical instructions. However, our experiments, performed on a multi-core processor environment, showed that the expression in [17] consistently yields a shorter computation time than the one associated to the expression in [21]. Hence, we decided to adopt the expression reported in [17], which is briefly described next. Each coefficient (trit) $a \in \mathbb{F}_3$ can be encoded as two bits a_h and a_l with $a = 2a_h + a_l$. The addition of two elements a, $b \in \mathbb{F}_3$ can then be computed as [17]

$$t = (a_l|b_h) \oplus (a_h|b_l), \quad c_l = t \oplus (a_h|b_h), \text{ and } \quad c_h = t \oplus (a_l|b_l).$$

As mentioned by the authors of [17], the order in which the above expression is evaluated has a major impact on the performance of its implementation in software. In fact, we use the following equivalent expression for computing $c = a + b$:

$$t = (a_l|a_h) \,\&\, (b_l|b_h), \quad c_l = t \oplus (a_l|b_l), \text{ and } \quad c_h = t \oplus (a_h|b_h).$$

Similarly, subtraction in $\mathbb{F}_3$ can be computed using only 7 instructions as

$$t = (a_l|a_h) \,\&\, (b_l|b_h), \quad c_l = t \oplus (a_l|b_h), \text{ and } \quad c_h = t \oplus (a_h|b_l).$$

Frobenius and Inverse-Frobenius Operators. Let $f(x)$ be an irreducible polynomial of degree m over $\mathbb{F}_3$. Then, the ternary extension field $\mathbb{F}_{3^m}$ can be defined as $\mathbb{F}_{3^m} \cong \mathbb{F}_3[x]/ (f(x))$. Let a be an arbitrary element of that field, which can be written in canonical basis as $a = \sum_{i=0}^{m-1} a_i x^i$, $a_i \in \mathbb{F}_3$. Assume that the extension degree m is an integer of the form $m = 3u + r$, with $u \geq 1$ and $r \in \{0, 1, 2\}$. Then, the Frobenius operator applied to a consists of computing $c = a^3$, which can be obtained as [1]

$$c = a^3 \bmod f(x) = (C_0 + x^m C_1 + x^{2m} C_2) \bmod f(x), \qquad (3)$$

where $C_0 = \sum_{i=0}^{u} a_i x^{3i}$, $C_1 = \sum_{i=1}^{u+r-1} a_{i+u} x^{3i-r}$, and $C_2 = \sum_{i=r}^{u+r-1} a_{i+2u} x^{3i-2r}$.

One can evaluate Eq. (3) by determining the constants x^m and x^{2m}, which are per-field constants. The inverse-Frobenius operator of a is computed by determining the unique field element $b \in \mathbb{F}_{3^m}$ such that $b^3 = a$ holds. The element b can be computed as [2]

$$b = \sum_{i=0}^{u-1+\lceil \frac{r}{2} \rceil} a_{3i} x^i + x^{\frac{1}{3}} \cdot \sum_{i=0}^{u-1+\lfloor \frac{r}{2} \rfloor} a_{3i+1} x^i + x^{\frac{2}{3}} \cdot \sum_{i=0}^{u-1} a_{3i+2} x^i \bmod f(x), \qquad (4)$$

Eq. (4) allows us to compute the inverse-Frobenius operator by performing two third-length polynomial multiplications with the per-field constants $x^{\frac{1}{3}}$ and $x^{\frac{2}{3}}$.

Table 1. Pre-computation look-up table when using the comb method for $a \in \mathbb{F}_{3^m}$ with a window size $v = 2$

Entry	Value	Entry	Value	Entry	Value
[00]	0	[10]	[01] $\ll$ 1	[20]	$\sim$ [10]
[01]	a	[11]	[10] + [01]	[21]	$\sim$ [12]
[02]	$\sim$ [01]	[12]	[10] + [02]	[22]	$\sim$ [11]

Multiplier. We use here the same comb method discussed previously. As we did in characteristic two, we selected a window size $v = 4$, which in characteristic three means that we have to pre-compute a look-up table containing $3^4 = 81$ entries. Due to the fact that almost half of the entries can be obtained by performing one single logical NOT, pre-computing such look-up table requires a moderate computational effort. As an example, consider the case where we want to build a look-up table for a given element $a \in \mathbb{F}_{3^m}$, with a size of $v = 2$. Then, we have to pre-compute $3^2 = 9$ entries. Table 1 shows how to obtain those 9 elements, where $\sim$ and $\ll$ stand for the logical negation and left shift operations, respectively. As it can be seen in Table 1, 4 out of 9 entries can be computed using logical negation only. We also need to compute two $\mathbb{F}_{3^m}$ additions, one initialization to zero and one assignment of the element a. In the case of $v = 4$, generating the 81-entry look up table requires a computational effort of 40 logical negations, 36 field additions, and 3 left-shift operations.

Multiplicative Inverse. In order to compute the multiplicative inverse d of a field element $b \in \mathbb{F}_{3^m}$, namely, $d = b^{-1} \bmod f(x)$, we used the ternary variant of the binary extended Euclidean algorithm reported in [17].

2.3 Field-Arithmetic Implementation Timings

We present in Table 2 a timing performance comparison of our field arithmetic library against the timings reported by Hankerson et al. in [16].[3] In both works, the libraries were executed on a Intel Core2 processor running at 2.4 GHz. It is noticed that our multipliers in characteristic two and three are faster than their counterparts in [16]. However, the field multiplier for 256-bit prime fields reported in [16] easily outperforms all the other four multipliers listed in Table 2.

3 Pairing Computation on Supersingular Curves in Characteristics Two and Three

In the following, we consider a supersingular elliptic curve $E/\mathbb{F}_{p^m}$ (where $p = 2$ or 3) with a distortion map ψ. The point at infinity is denoted by $\mathcal{O}$. Let ℓ be a large prime factor of $N = \#E(\mathbb{F}_{p^m})$, and suppose that the embedding degree of

---

[3] Our library was compiled using the MS Visual Studio 2008SP1 in 64-bit mode, and it was executed on the Windows XP 64-bit SP2 environment.

Table 2. A comparison of field arithmetic software implementations on an Intel Core2 processor (clock frequency: 2.4 GHz). All timings are reported in μs.

	Field	Prime/polynomial	x^p	$\sqrt[p]{x}$	Mult
Hankerson *et al.* [16]	$\mathbb{F}_{p256}$	256-bit prime Hamming weight 87	–	–	0.129
	$\mathbb{F}_{2^{1223}}$	$x^{1223} + x^{255} + 1$	0.250	0.208	3.417
	$\mathbb{F}_{3^{509}}$	$x^{509} - x^{318} - x^{191} + x^{127} + 1$	0.375	0.500	3.208
This work	$\mathbb{F}_{2^{1223}}$	$x^{1223} + x^{255} + 1$	0.200	0.312	2.266
	$\mathbb{F}_{3^{509}}$	$x^{509} - x^{318} - x^{191} + x^{127} + 1$	0.375	0.406	1.720

the curve k is larger than 1 and that there are no points of order ℓ^2 in $E(\mathbb{F}_{p^{km}})$. Let $f_{n,P}$, for $n \in \mathbb{N}$ and $P \in E(\mathbb{F}_{p^m})[\ell]$, be a family of normalized $\mathbb{F}_{p^{km}}$-rational functions with divisor $(f_{n,P}) = n(P) - ([n]P) - (n-1)(\mathcal{O})$. The modified Tate pairing of order ℓ is a non-degenerate and bilinear pairing given by the map

$$\hat{e} : E(\mathbb{F}_{p^m})[\ell] \times E(\mathbb{F}_{p^m})[\ell] \longrightarrow \mathbb{F}^*_{p^{km}}/(\mathbb{F}^*_{p^{km}})^\ell$$
$$(P, Q) \longmapsto f_{\ell,P}(\psi(Q)).$$

Note that $\hat{e}(P, Q)$ is defined as a coset of $(\mathbb{F}^*_{p^{km}})^\ell$. However, $\mathbb{F}\mathbb{F}^*_{p^{km}}/(\mathbb{F}^*_{p^{km}})^\ell$ is cyclic of order ℓ and isomorphic to the group of ℓ-th roots of unity $\mu_\ell = \{u \in \overline{\mathbb{F}}^*_{p^m} : u^\ell = 1\} \subseteq \mathbb{F}^*_{p^{km}}$. Hence, in order to obtain a unique representative, which is desirable for pairing-based protocols, it suffices to raise $f_{\ell,P}(\psi(Q))$ to the $(p^{km}-1)/\ell$-th power. This operation is often referred to as *final exponentiation*. We define the reduced modified Tate pairing as $\hat{e}_r(P, Q) = \hat{e}(P, Q)^{(p^{km}-1)/\ell}$.

3.1   Miller's Algorithm

Miller [23,24] proposed the first iterative approach to compute the function $f_{\ell,P}$. By proving the equality of the divisors, he showed that

$$f_{a+b,P} = f_{a,P} \cdot f_{b,P} \cdot \frac{l_{[a]P,[b]P}}{v_{[a+b]P}},$$

where $l_{[a]P,[b]P}$ is the equation of the line through $[a]P$ and $[b]P$ (or the tangent line if $[a]P = [b]P$), $v_{[a+b]P}$ is the equation of the vertical line through $[a+b]P$, and $f_{1,P}$ is a constant function (usually, $f_{1,P} = 1$). We obtain a double-and-add algorithm for computing the rational function $f_{n,P}$ in $\lfloor \log_2 n \rfloor$ iterations. A nice property of supersingular elliptic curves is that multiplication by p is a relatively easy operation: it involves only a few Frobenius maps and additions over $\mathbb{F}_{p^m}$ (see for instance [3] for details) and a p-ary expansion of ℓ seems perfectly suited to the computation of $f_{\ell,P}(\psi(Q))$.

Several researchers focused on shortening the loop of Miller's algorithm (see for instance [19,18,29] for a comprehensive bibliography). Barreto *et al.* [3] introduced the η_T pairing as *"an alternative means of computing the Tate pairing on certain supersingular curves"* [25, page 108]. They suggest to compute $\hat{e}_r(P, Q)$

using an order $T \in \mathbb{Z}$ that is smaller than ℓ. Their main result is a lemma giving a method to select T such that $\eta_T(P, Q)$ is a non-degenerate bilinear pairing [3]. In the case of characteristics two and three, they show that one can half the number of basic Miller's iterations by choosing $T = p^m - N$:

$$\eta_T(P, Q) = \begin{cases} f_{T,P}(\psi(Q)) \text{ if } T > 0, \text{ or} \\ f_{-T,-P}(\psi(Q)) \text{ if } T < 0. \end{cases}$$

It is worth noticing that T has a low p-adic Hamming weight and the computation of $[T]P$ (or $[-T]P$) requires $(m+1)/2$ multiplications by p and a single addition. It is therefore possible to pre-compute multiples of P by means of Frobenius maps and to parallelize Miller's algorithm on several cores. Multiplications over $\mathbb{F}_{p^{km}}$ are of course necessary to obtain $\hat{e}(P, Q)$ from the partial results computed on each core.

3.2 Reduced Modified Tate Pairing in Characteristic Two

We follow [6, Algorithm 1] to compute $\hat{e}(P, Q)$ on a supersingular curve $E/\mathbb{F}_{2^m}$, m an odd number and with embedding degree $k = 4$ given by $E : y^2 + y = x^3 + x + b$, where m is odd and $b \in \{0, 1\}$. The reduced modified Tate pairing is defined by [3, 6]:

$$\hat{e}_r(P, Q) = \eta_T([2^m]P, Q)^{\frac{2^{4m}-1}{N}}.$$

Loop unrolling does not allow one to reduce the number of multiplications over $\mathbb{F}_{2^m}$ and Miller's algorithm requires $(m-1)/2$ iterations which can be parallelized on several cores.[4] Final exponentiation consists of raising $\hat{e}(P, Q)$ to the exponent $M = \frac{2^{4m}-1}{N} = (2^{2m}-1)\cdot(2^m+1-\nu 2^{(m+1)/2})$ [3], where $\nu = (-1)^b$ when $m \equiv 1, 7$ (mod 8) and $\nu = (-1)^{1-b}$ in all other cases. We perform this operation according to a slightly optimized version of [6, Algorithm 3] (see Appendices A.1 and A.2 for technical details):

- **Raising to the $(2^m + 1)$-st power.** Raising the outcome of Miller's algorithm to the $(2^{2m} - 1)$-st power produces an element $U \in \mathbb{F}_{2^{4m}}$ of order $2^{2m} + 1$. This property allows one to save a multiplication over $\mathbb{F}_{2^{4m}}$ when raising U to the $(2^m + 1)$-st power compared to [6, page 304].
- **Raising to the $2^{\frac{m+1}{2}}$-th power.** Beuchat et al. [8] exploited the linearity of the Frobenius map in order to reduce the cost of successive cubings over $\mathbb{F}_{3^{6m}}$. The same approach can be straightforwardly applied to characteristic two: raising an element of $\mathbb{F}_{2^{4m}}$ to the 2^i-th power involves $4i$ squarings and at most four additions over $\mathbb{F}_{2^m}$.

---

[4] Shirase et al. propose a loop unrolling technique in reference [27] and claim that they reduce the computation time by 14.3%. However, they assume that additions and multiplications by small constants are almost free, and inverse Frobenius maps over $\mathbb{F}_{2^m}$ are m times more expensive than Frobenius maps. Such estimates do not hold in our context (see for instance Table 2) and we did not investigate further the approach by Shirase et al.

3.3   Reduced Modified Tate Pairing in Characteristic Three

We consider a supersingular curve $E/\mathbb{F}_{3^m}$ with embedding degree $k = 6$ defined by $E : y^2 = x^3 - x + b$, where m is coprime to 6 and $b \in \{-1, 1\}$. According to [3,6], we have

$$\hat{e}_r(P, Q) = \eta_T \left(\left[-\mu b 3^{\frac{3m-1}{2}} \right] P, Q \right)^{\frac{3^{6m}-1}{N}},$$

where $\mu = 1$ when $m \equiv 1, 11 \pmod{12}$, or $\mu = -1$ otherwise. There are several ways to compute the η_T pairing (see for instance [3, 7, 8]) and the choice of an algorithm depends on the target architecture. Here, we decided to minimize the number of arithmetic operations over $\mathbb{F}_{3^m}$ and applied the well-known loop unrolling technique [13] to [7, Algorithm 3] (technical details are provided in Appendix B). This approach allows us to save several multiplications over $\mathbb{F}_{3^m}$ compared to the original algorithm. Final exponentiation is carried out according to [7, 8].

4   Results and Comparisons

We list in Table 3 the timings achieved on an Intel Core2 processor by our library for low, medium and high security levels (66, 89, and 128 bits, respectively), including the performance obtained when using one, two, and four cores. Our library was compiled using the MS Visual Studio 2008SP1 in 64-bit mode, and it was executed on the Windows XP 64-bit SP2 environment. For comparison purposes, we also include in Table 3 the performance reported by Hankerson et al. [16], which is the fastest pairing library that we know of. The work by Grabher et al. [12] is also of interest as it is the only pairing library preceding this work that reports a multi-core platform implementation.

Table 4 shows the timings achieved by our library when implemented on an Intel core i7 multi-processor platform running at 2.9 GHz. Finally, in Table 5 we list some of the fastest hardware accelerators for the Tate pairing reported at low, medium, and high security levels.

Grabher et al. reported in [12] a multi-core implementation of the Ate pairing defined over a Barreto–Naehrig (BN) curve [5], when using a 256-bit prime. Since the BN curves have an embedding degree of $k = 12$, this implies a 128-bit security level. As shown in Table 3, our pairing implementation in characteristic three is faster than the prime field pairing library reported in [12].

In Table 4 we report a calculation time for the reduced modified Tate pairing of just 3.08 ms and 1.87 ms for characteristics two and three, respectively. This performance, that was obtained on a two Intel quad-core i7 multi-processor platform, appears to be the fastest pairing timings yet reported.

Although our software implementation outperforms several hardware architectures previously reported for low and medium levels of security, when we compare our results against the ones in [28, 8] (Table 5), we see that there still exists a large gap between software and hardware pairing implementations for

Table 3. Performance comparison of software implementations for pairings on an Intel Core2 processor

	Curve	Security [bits]	# of cores	Freq. [GHz]	Calc. time [ms]
This work	$E(\mathbb{F}_{3^{97}})$	66	1	2.6	0.15
	$E(\mathbb{F}_{3^{97}})$	66	2	2.6	0.09
This work	$E(\mathbb{F}_{3^{193}})$	89	1	2.6	0.98
	$E(\mathbb{F}_{3^{193}})$	89	2	2.6	0.55
Hankerson et al. [16]	$E(\mathbb{F}_{p^{256}})$	128	1	2.4	4.16
	$E(\mathbb{F}_{2^{1223}})$	128	1	2.4	16.25
	$E(\mathbb{F}_{3^{509}})$	128	1	2.4	13.75
Grabher et al. [12]	$E(\mathbb{F}_{p^{256}})$	128	1	2.4	9.71
	$E(\mathbb{F}_{p^{256}})$	128	2	2.4	6.01
This work	$E(\mathbb{F}_{2^{1223}})$	128	1	2.4	11.19
	$E(\mathbb{F}_{2^{1223}})$	128	2	2.4	6.72
	$E(\mathbb{F}_{2^{1223}})$	128	4	2.4	4.22
	$E(\mathbb{F}_{3^{509}})$	128	1	2.4	7.59
	$E(\mathbb{F}_{3^{509}})$	128	2	2.4	4.31
	$E(\mathbb{F}_{3^{509}})$	128	4	2.4	2.94

Table 4. Implementations timings for the reduced modified Tate pairing at the 128-bit security level on an Intel core i7 processor (clock frequency: 2.9 GHz)

Curve	# of cores	Calc. time [ms]	Curve	# of cores	Calc. time [ms]
$E(\mathbb{F}_{3^{509}})$	1	5.22	$E(\mathbb{F}_{2^{1223}})$	1	7.94
$E(\mathbb{F}_{3^{509}})$	2	3.16	$E(\mathbb{F}_{2^{1223}})$	2	4.53
$E(\mathbb{F}_{3^{509}})$	4	2.31	$E(\mathbb{F}_{2^{1233}})$	4	3.13
$E(\mathbb{F}_{3^{509}})$	8	1.87	$E(\mathbb{F}_{2^{1223}})$	8	3.08

Table 5. Some hardware accelerators for the Tate pairing

	Curve	Security [bits]	Platform	Area	Freq. [MHz]	Calc. time [ms]
Shu et al. [28]	$E(\mathbb{F}_{2^{239}})$	66	xc4vlx200	29920 slices	100	0.0365
Beuchat et al. [8]	$E(\mathbb{F}_{3^{97}})$	66	xc4vlx60-11	18683 slices	179	0.0048
Shu et al. [28]	$E(\mathbb{F}_{2^{457}})$	88	xc4vlx200	58956 slices	100	0.1
Beuchat et al. [8]	$E(\mathbb{F}_{3^{193}})$	89	xc4vlx100-11	47433 slices	167	0.01
Shu et al. [28]	$E(\mathbb{F}_{2^{557}})$	96	xc4vlx200	37931 slices	66	0.6758
Kammler et al. [20]	$E(\mathbb{F}_{p^{256}})$	128	130 nm CMOS	97 kGates	338	15.8

moderate security levels. The computation of the reduced modified Tate pairing over $\mathbb{F}_{3^{193}}$ on a Virtex-4 LX FPGA reported in [8] with a medium speed grade is for instance roughly fifty times faster than our software timings. Depending on the application, this speedup may justify the usage of large FPGAs which are now available in servers and supercomputers such as the SGI Altix 4700 platform.

5   Conclusion

In this work we presented the multi-core implementation of a software library that is able to compute the reduced modified Tate pairing on supersingular elliptic curves at a high speed. The sequential timings reported in this work are significantly faster than the ones achieved in [16] for pairings computed over characteristics two and three fields.

In the light of the results obtained here, one important design question that arises is: how many cores should be utilized by a given application? As discussed in the Appendices, our pairing library successfully parallelize the computation of Miller's algorithm. However, if we use n cores for the implementation of the Miller's algorithm, in order to combine the n partial products generated by the n parallel sub-loops executed in each core, we are forced to add $n - 1$ extra field multiplications over $\mathbb{F}_{p^{km}}$. Furthermore, due to the dependencies among the different operations involved in the final exponentiation step, this portion of the pairing has to be computed sequentially. As shown in Table 4, these two factors cause the acceleration achieved by an n-core implementation to be always less than the ideal $n\times$ speedup factor. From Table 4, we can see for instance that the acceleration provided by the eight-core implementation is modest compared with the timings achieved by the four-core one. On the other hand, when comparing the timings of the single-core and dual-core implementations, the acceleration factor is roughly $1.70\times$ for both, characteristics two and three.

Our future work includes the implementation of pairings for large characteristics on ordinary curves and the SSE implementation of our pairing library using the built-in carry-less 64-bit multiplier recently announced by Intel [14].

Acknowledgments

The authors would like to thank Jérémie Detrey and the anonymous referees for their valuable comments.

References

1. Ahmadi, O., Rodríguez-Henríquez, F.: Low complexity cubing and cube root computation over $\mathbb{F}_{3^m}$ in standard basis. Cryptology ePrint Archive, Report 2009/070 (2009)
2. Barreto, P.S.L.M.: A note on efficient computation of cube roots in characteristic 3. Cryptology ePrint Archive, Report 2004/305 (2004)
3. Barreto, P.S.L.M., Galbraith, S.D., Ó hÉigeartaigh, C., Scott, M.: Efficient pairing computation on supersingular Abelian varieties. Designs, Codes and Cryptography 42, 239–271 (2007)
4. Barreto, P.S.L.M., Kim, H.Y., Lynn, B., Scott, M.: Efficient algorithms for pairing-based cryptosystems. In: Yung, M. (ed.) CRYPTO 2002. LNCS, vol. 2442, pp. 354–368. Springer, Heidelberg (2002)

5. Barreto, P.S.L.M., Naehrig, M.: Pairing-friendly elliptic curves of prime order. In: Preneel, B., Tavares, S. (eds.) SAC 2005. LNCS, vol. 3897, pp. 319–331. Springer, Heidelberg (2006)
6. Beuchat, J.-L., Brisebarre, N., Detrey, J., Okamoto, E., Rodríguez-Henríquez, F.: A comparison between hardware accelerators for the modified tate pairing over $\mathbb{F}_{2^m}$ and $\mathbb{F}_{3^m}$. In: Galbraith, S.D., Paterson, K.G. (eds.) Pairing 2008. LNCS, vol. 5209, pp. 297–315. Springer, Heidelberg (2008)
7. Beuchat, J.-L., Brisebarre, N., Detrey, J., Okamoto, E., Shirase, M., Takagi, T.: Algorithms and arithmetic operators for computing the η_T pairing in characteristic three. IEEE Transactions on Computers 57(11), 1454–1468 (2008)
8. Beuchat, J.-L., Detrey, J., Estibals, N., Okamoto, E., Rodríguez-Henríquez, F.: Hardware accelerator for the Tate pairing in characteristic three based on Karatsuba–Ofman multipliers. In: Clavier, C., Gaj, K. (eds.) CHES 2009. LNCS, vol. 5747, pp. 225–239. Springer, Heidelberg (2009)
9. Duursma, I., Lee, H.S.: Tate pairing implementation for hyperelliptic curves $y^2 = x^p - x + d$. In: Laih, C.-S. (ed.) ASIACRYPT 2003. LNCS, vol. 2894, pp. 111–123. Springer, Heidelberg (2003)
10. Fong, K., Hankerson, D., López, J., Menezes, A.: Field inversion and point halving revisited. IEEE Transactions on Computers 53(8), 1047–1059 (2004)
11. Galbraith, S.D., Harrison, K., Soldera, D.: Implementing the Tate pairing. In: Fieker, C., Kohel, D.R. (eds.) ANTS 2002. LNCS, vol. 2369, pp. 324–337. Springer, Heidelberg (2002)
12. Grabher, P., Großschädl, J., Page, D.: On software parallel implementation of cryptographic pairings. In: SAC 2008. LNCS, vol. 5381, pp. 34–49. Springer, Heidelberg (2008)
13. Granger, R., Page, D., Stam, M.: On small characteristic algebraic tori in pairing-based cryptography. LMS Journal of Computation and Mathematics 9, 64–85 (2006)
14. Gueron, S., Kounavis, M.E.: Carry-less multiplication and its usage for computing the GCM mode. Intel Corporation White Paper (May 2009)
15. Hankerson, D., López Hernandez, J., Menezes, A.J.: Software implementation of elliptic curve cryptography over binary fields. In: Paar, C., Koç, Ç.K. (eds.) CHES 2000. LNCS, vol. 1965, pp. 1–24. Springer, Heidelberg (2000)
16. Hankerson, D., Menezes, A., Scott, M.: Software Implementation of Pairings. Cryptology and Information Security Series, ch. 12, pp. 188–206. IOS Press, Amsterdam (2009)
17. Harrison, K., Page, D., Smart, N.P.: Software implementation of finite fields of characteristic three, for use in pairing-based cryptosystems. LMS Journal of Computation and Mathematics 5, 181–193 (2002)
18. Hess, F.: Pairing lattices. In: Galbraith, S.D., Paterson, K.G. (eds.) Pairing 2008. LNCS, vol. 5209, pp. 18–38. Springer, Heidelberg (2008)
19. Hess, F., Smart, N., Vercauteren, F.: The Eta pairing revisited. IEEE Transactions on Information Theory 52(10), 4595–4602 (2006)
20. Kammler, D., Zhang, D., Schwabe, P., Scharwaechter, H., Langenberg, M., Auras, D., Ascheid, G., Leupers, R., Mathar, R., Meyr, H.: Designing an ASIP for cryptographic pairings over Barreto-Naehrig curves. Cryptology ePrint Archive, Report 2009/056 (2009)
21. Kawahara, Y., Aoki, K., Takagi, T.: Faster implementation of η_T pairing over $GF(3^m)$ using minimum number of logical instructions for GF(3)-addition. In: Galbraith, S.D., Paterson, K.G. (eds.) Pairing 2008. LNCS, vol. 5209, pp. 282–296. Springer, Heidelberg (2008)

22. López, J., Dahab, R.: High-speed software multiplication in $\mathbb{F}_{2^m}$. In: Roy, B.K., Okamoto, E. (eds.) INDOCRYPT 2000. LNCS, vol. 1977, pp. 203–212. Springer, Heidelberg (2000)
23. Miller, V.S.: Short programs for functions on curves (1986), http://crypto.stanford.edu/miller
24. Miller, V.S.: The Weil pairing, and its efficient calculation. Journal of Cryptology 17(4), 235–261 (2004)
25. Ó hÉigeartaigh, C.: Pairing Computation on Hyperelliptic Curves of Genus 2. PhD thesis, Dublin City University (2006)
26. Schroeppel, R., Orman, H., O'Malley, S.W., Spatscheck, O.: Fast key exchange with elliptic curve systems. In: Coppersmith, D. (ed.) CRYPTO 1995. LNCS, vol. 963, pp. 43–56. Springer, Heidelberg (1995)
27. Shirase, M., Takagi, T., Choi, D., Han, D., Kim, H.: Efficient computation of Eta pairing over binary field with Vandermonde matrix. ETRI Journal 31(2), 129–139 (2009)
28. Shu, C., Kwon, S., Gaj, K.: Reconfigurable computing approach for Tate pairing cryptosystems over binary fields. IEEE Transactions on Computers 58(9), 1221–1237 (2009)
29. Vercauteren, F.: Optimal pairings. Cryptology ePrint Archive, Report 2008/096 (2008)

A   Reduced Modified Tate Pairing in Characteristic Two

Table 6 summarizes the parameters of the supersingular curve considered to compute $\hat{e}_r(P, Q)$ in characteristic two. Noting $T' = -\nu T$ and $P' = [-\nu 2^m]P$, we have:

$$\hat{e}_r(P, Q) = f_{T', P'}(\psi(Q))^M$$

$$= \left(l_{P'}(\psi(Q)) \cdot \left(\prod_{j=0}^{\frac{m-1}{2}} g_{\left[2^{\frac{m-1}{2}-j}\right]P'}(\psi(Q))^{2^j} \right) \right)^M, \tag{5}$$

where, for all $V \in E(\mathbb{F}_{2^m})[\ell]$, l_V is the equation of the line corresponding to the addition of $[\nu]V$ with $\left[2^{\frac{m+1}{2}}\right]V$ and g_V is the rational function defined over $E(\mathbb{F}_{2^{4m}})[\ell]$ corresponding to the straight line in doubling V. More precisely, we have:

$$\left(g_{\left[2^{\frac{m-1}{2}-j}\right]P'}(\psi(Q)) \right)^{2^j} = (x_{P'}^{2^{-j}} + \alpha) \cdot (x_Q^{2^j} + \alpha) + y_{P'}^{2^{-j}} + y_Q^{2^j} + \beta +$$

$$(x_{P'}^{2^{-j}} + x_Q^{2^j} + \alpha)s + t, \text{ and}$$

$$l_{P'}(\psi(Q)) = g_{\left[2^{\frac{m-1}{2}}\right]P'}(\psi(Q)) + x_{P'}^2 + x_Q + \alpha + s.$$

Algorithm 2 describes the computation of $\hat{e}_r(P, Q)$ according to Eq. (5) (this algorithm is based on [6, Algorithm 1]). The equations of all straight lines can be pre-computed by storing all square roots of x_P and y_P, as well as all squares of x_Q and y_Q (lines 12 and 13 that can be computed in parallel on two cores). Then, one can split the execution of Miller's algorithm (lines 16 to 20) into several parts that are run concurrently. We perform the final exponentiation according to an improved version of [6, Algorithm 3] detailed in the following.

Table 6. Supersingular curves over $\mathbb{F}_{2^m}$

Underlying field	$\mathbb{F}_{2^m}$, where m is an odd integer.
Curve	$E : y^2 + y = x^3 + x + b$, with $b \in \{0, 1\}$.
Number of rational points	$N = 2^m + 1 + \nu 2^{(m+1)/2}$, with $$\delta = \begin{cases} b & \text{if } m \equiv 1, 7 \pmod 8, \\ 1 - b & \text{if } m \equiv 3, 5 \pmod 8, \end{cases}$$ and $\nu = (-1)^\delta$.
Embedding degree	$k = 4$
Distortion map	$\psi : E(\mathbb{F}_{2^m})[\ell] \longrightarrow E(\mathbb{F}_{2^{4m}})[\ell] \setminus E(\mathbb{F}_{2^m})[\ell]$ $(x, y) \longmapsto (x + s^2, y + sx + t)$ with $s \in \mathbb{F}_{2^2}$ satisfying $s^2 = s + 1$, and $t \in \mathbb{F}_{2^4}$ satisfying $t^2 = t + s$.
Tower field	$\mathbb{F}_{2^{4m}} = \mathbb{F}_{2^m}[s, t] \cong \mathbb{F}_{2^m}[X, Y]/(X^2 + X + 1, Y^2 + Y + X)$
Final exponentiation	$M = \left(2^{2m} - 1\right) \cdot \left(2^m + 1 - \nu 2^{(m+1)/2}\right)$
Parameters of Algorithm 2	$$\alpha = \begin{cases} 0 & \text{if } m \equiv 3 \pmod 4, \\ 1 & \text{if } m \equiv 1 \pmod 4, \end{cases}$$ $$\beta = \begin{cases} b & \text{if } m \equiv 1, 3 \pmod 8, \text{ and} \\ 1 - b & \text{if } m \equiv 5, 7 \pmod 8. \end{cases}$$

A.1 Raising an Element of Order $2^{2m} + 1$ to the $(2^m + 1)$-st Power over $\mathbb{F}_{2^{4m}}$

Let $F, U \in \mathbb{F}_{2^{4m}}$ and assume that $U = F^{2^{2m}-1}$. According to Fermat's little theorem, the result of raising to the $(2^{2m} - 1)$-st power produces an element of order $2^{2m} + 1$, i.e. $U^{2^{2m}+1} = 1$. Let us write

$$U = \underbrace{u_0 + u_1 s}_{U_0} + \underbrace{(u_2 + u_3 s)}_{U_1} t,$$

where $u_0, u_1, u_2, u_3 \in \mathbb{F}_{2^m}$ and $U_0, U_1 \in \mathbb{F}_{2^{2m}}$. Since $t^{2^{2m}} = 1 + t$, we have:

$$U^{2^{2m}+1} = (U_0 + U_1 t)(U_0 + U_1 t)^{2^{2m}} = U_0^2 + U_0 U_1 + U_1^2 s = 1.$$

Therefore,

$$u_0 u_2 + u_1 u_3 = u_0^2 + u_1^2 + u_3^2 + 1, \text{ and}$$
$$u_0 u_3 + u_1 u_2 + u_1 u_3 = u_1^2 + u_2^2.$$

Algorithm 2. Computation of the reduced modified Tate pairing in characteristic two.

Input: $P, Q \in \mathbb{F}_{2^m}[\ell]$.	
Output: $\hat{e}_r(P, Q) \in \mathbb{F}_{2^{4m}}^*$.	
1: $x_P \leftarrow x_P + 1$;	(1 XOR)
2: $y_P \leftarrow x_P + y_P + \alpha + \bar{\delta}$;	($\alpha + \bar{\delta}$ XOR, 1 A)
3: $u \leftarrow x_P + \alpha$; $v \leftarrow x_Q + \alpha$	(2α XOR)
4: $g_0 \leftarrow u \cdot v + y_P + y_Q + \beta$;	(1 M, 2 A, β XOR)
5: $g_1 \leftarrow u + x_Q$; $g_2 \leftarrow v + x_P^2$;	(1 S, 2 A)
6: $G \leftarrow g_0 + g_1 s + t$;	
7: $L \leftarrow (g_0 + g_2) + (g_1 + 1)s + t$;	(1 A, 1 XOR)
8: $F \leftarrow L \cdot G$;	(2 M, 1 S, 5 A, 2 XOR)
9: $x_P[0] \leftarrow x_P$; $y_P[0] \leftarrow y_P$;	
10: $x_Q[0] \leftarrow x_Q$; $y_Q[0] \leftarrow y_Q$;	
11: **for** $j = 1$ to $\frac{m-1}{2}$ **do**	
12: $x_P[j] \leftarrow \sqrt{x_P[j-1]}$; $x_Q[j] \leftarrow x_Q[j-1]^2$;	(1 R, 1 S)
13: $y_P[j] \leftarrow \sqrt{y_P[j-1]}$; $y_Q[j] \leftarrow y_Q[j-1]^2$;	(1 R, 1 S)
14: **end for**	
15: **for** $j = 1$ to $\frac{m-1}{2}$ **do**	
16: $u \leftarrow x_P[j] + \alpha$; $v \leftarrow x_Q[j] + \alpha$	(2α XOR)
17: $g_0 \leftarrow u \cdot v + y_P[j] + y_Q[j] + \beta$;	(1 M, 2 A, β XOR)
18: $g_1 \leftarrow u + x_Q[j]$;	(1 A)
19: $G \leftarrow g_0 + g_1 s + t$;	
20: $F \leftarrow F \cdot G$;	(6 M, 14 A)
21: **end for**	
22: **return** F^M;	

Let $\alpha = 0$ when $m \equiv 3 \pmod 4$ and $\alpha = 1$ when $m \equiv 1 \pmod 4$. Seeing that $s^{2^m} = s + 1$ and $t^{2^m} = t + s + \alpha + 1$, we obtain:

$$U^{2^m} = \begin{cases} (u_0 + u_1 + u_3) + (u_1 + u_2)s + (u_2 + u_3)t + u_3 st & \text{if } \alpha = 1, \\ (u_0 + u_1 + u_2) + (u_1 + u_2 + u_3)s + (u_2 + u_3)t + u_3 st & \text{if } \alpha = 0. \end{cases}$$

A first solution to compute U^{2^m+1} would be to multiply U^{2^m} by U. There is however a faster way to raise U to the power of $2^m + 1$. Defining $m_0 = u_0 u_1$, $m_1 = u_0 u_3$, $m_2 = u_1 u_2$, and $m_3 = u_2 u_3$, we have:

$$\begin{aligned} U^{2^m+1} &= (u_0 u_1 + u_0 u_3 + u_1 u_2 + u_0^2 + u_1^2) + \\ &\quad (\underbrace{u_0 u_2 + u_1 u_3}_{=u_0^2 + u_1^2 + u_3^2 + 1} + u_1 u_2 + u_2 u_3 + u_2^2 + u_3^2)s + \\ &\quad (u_0 u_3 + u_1 u_2 + u_2 u_3 + u_2^2 + u_3^2)t + (u_2 u_3 + u_2^2 + u_3^2)st \\ &= (m_0 + m_1 + m_2 + (u_0 + u_1)^2) + (m_2 + m_3 + (u_0 + u_1 + u_2)^2 + 1)s + \\ &\quad (m_1 + m_2 + m_3 + (u_2 + u_3)^2)t + (m_3 + (u_2 + u_3)^2)st, \end{aligned}$$

when $\alpha = 1$, and

$$
\begin{aligned}
U^{2^m+1} &= (\underbrace{u_0 u_2 + u_1 u_3}_{=u_0^2+u_1^2+u_3^2+1} + u_0 u_1 + u_1 u_2 + u_0^2 + u_1^2) + \\
&\quad (\underbrace{u_0 u_2 + u_1 u_3}_{=u_0^2+u_1^2+u_3^2+1} + u_0 u_3 + u_2 u_3 + u_2^2 + u_3^2)s + \\
&\quad (u_0 u_3 + u_1 u_2)t + (u_2 u_3 + u_2^2 + u_3^2)st \\
&= (m_0 + m_2 + u_3^2 + 1) + (m_1 + m_3 + (u_0 + u_1 + u_2)^2 + 1)s + \\
&\quad (m_1 + m_2)t + (m_3 + u_2^2 + u_3^2)st,
\end{aligned}
$$

when $\alpha = 0$. Thus, computing U^{2^m+1} involves only four multiplications, three squarings, and eleven additions over $\mathbb{F}_{2^m}$ (Algorithm 3). This approach allows us to save one multiplication over $\mathbb{F}_{2^m}$ compared to [6].

A.2   Computing $U^{2^{\frac{m+1}{2}}}$ over $\mathbb{F}_{2^{4m}}$

Let $U = u_0 + u_1 s + u_2 t + u_3 st \in \mathbb{F}_{2^{4m}}$. Noting that $s^{2^i} = s + \gamma_1$ and $t^{2^i} = t + \gamma_1 s + \gamma_2$, where $\gamma_1 = i \bmod 2$ and $\gamma_2 = \lfloor \frac{i}{2} \rfloor \bmod 2$, we obtain the following formula for U^{2^i}, depending on the value of i modulo 4:

Algorithm 3. Computation of U^{2^m+1} over $\mathbb{F}_{2^{4m}}$, where U is an element of order $2^{2m} + 1$.

Input: $U = u_0 + u_1 s + u_2 t + u_3 st \in \mathbb{F}_{2^{4m}}$ with $U^{2^{2m}+1} = 1$.
Output: $V = U^{2^m+1}$.

1: $m_0 \leftarrow u_0 \cdot u_1$; $m_1 \leftarrow u_0 \cdot u_3$; $m_2 \leftarrow u_1 \cdot u_2$; $m_3 \leftarrow u_2 \cdot u_3$;   (4 M)
2: $a_0 \leftarrow u_0 + u_1$; $a_1 \leftarrow a_0 + u_2$;   (2 A)
3: $s_1 \leftarrow a_1^2$;   (1 S)
4: **if** $\alpha = 1$ **then**
5:   $a_2 \leftarrow u_2 + u_3$; $a_3 \leftarrow m_1 + m_2$;   (2 A)
6:   $s_0 \leftarrow a_0^2$; $s_2 \leftarrow a_2^2$;   (2 S)
7:   $v_3 \leftarrow m_3 + s_2$;   (1 A)
8:   $v_2 \leftarrow v_3 + a_3$;   (1 A)
9:   $v_1 \leftarrow m_2 + m_3 + s_1 + 1$;   (3 A)
10:   $v_0 \leftarrow m_0 + a_3 + s_0$;   (2 A)
11: **else**
12:   $s_0 \leftarrow u_2^2$; $s_2 \leftarrow u_3^2$;   (2 S)
13:   $v_0 \leftarrow m_0 + m_2 + s_2 + 1$;   (3 A)
14:   $v_1 \leftarrow m_1 + m_3 + s_1 + 1$;   (3 A)
15:   $v_2 \leftarrow m_1 + m_2$;   (1 A)
16:   $v_3 \leftarrow m_3 + s_0 + s_2$;   (2 A)
17: **end if**
18: **return** $v_0 + v_1 s + v_2 t + v_3 st$;

$$U^{2^i} = (u_0 + \gamma_1 u_1 + \gamma_2 u_2 + \gamma_3 u_3)^{2^i} + (u_1 + \gamma_1 u_2 + \gamma_2 u_3)^{2^i} s$$
$$+ (u_2 + \gamma_1 u_3)^{2^i} t + u_3^{2^i} st,$$

where $\gamma_3 = 1$ when $i \bmod 4 = 1$, and $\gamma_3 = 0$ otherwise. According to the value of $(m+1)/2 \bmod 4$, the computation of $U^{2^{\frac{m+1}{2}}}$ requires $2m + 2$ squarings and at most four additions over $\mathbb{F}_{2^m}$.

B   Reduced Modified Tate Pairing in Characteristic Three

In the following, we consider the computation of the reduced modified Tate pairing in characteristic three on several cores. Table 7 summarizes the parameters of the supersingular curve. Noting $T' = -\mu b T$ and $P' = (x_{P'}, y_{P'}) = \left[3^{\frac{m-1}{2}}\right] P$, we have to compute:

$$\hat{e}_r(P, Q) = f_{T', P'}(\psi(Q))^M$$
$$= \left(l_{P'}(\psi(Q)) \cdot \left(\prod_{j=0}^{\frac{m-1}{2}} g_{\left[3^{\frac{m-1}{2}-j}\right]P'}(\psi(Q))^{3^j} \right) \right)^M, \qquad (6)$$

where, for all $V \in E(\mathbb{F}_{3^m})[\ell]$, l_V is the equation of the line corresponding to the addition of $[\mu b]V$ with $\left[3^{\frac{m+1}{2}}\right]V$, and g_V is the rational function introduced by Duursma and Lee [9] and having divisor $(g_V) = 3(V) + ([-3]V) - 4(\mathcal{O})$. Expanding everything, we obtain the following expressions:

$$l_{P'}(\psi(Q)) = y_Q \sigma + \lambda y_{P'}(x_{P'} + x_Q - \nu b - \rho), \text{ and}$$
$$g_{\left[3^{\frac{m-1}{2}-j}\right]P'}(\psi(Q))^{3^j} = -\lambda y_{P'}^{3^{-j}} y_Q^{3^j} \sigma - \left(x_{P'}^{3^{-j}} + x_Q^{3^j} - \nu b - \rho \right)^2.$$

It is worth noticing that the Duursma-Lee functions can be pre-computed by building a table of all cube roots of x_P and y_P as well as all cubes of x_Q and y_Q.

Beuchat *et al.* described an algorithm to compute $\hat{e}_r(P, Q)$ according to Eq. (6) (see [7, Algorithm 3]). They took advantage of the sparsity of $l_{P'}$ and g_V to reduce the cost of the first multiplication over $\mathbb{F}_{3^m}$ and needed therefore $\frac{m-1}{2}$ iterations of Miller's algorithm to accumulate the remaining products. We propose to optimize further the algorithm by computing two iterations at a time and obtain Algorithm 4 for the case where $\frac{m-1}{2}$ is even. When $\frac{m-1}{2}$ is odd, one has to restrict the loop on j from 1 to $\frac{m-3}{4}$, and perform the last product by means of an iteration of the original loop. Thanks to the sparsity of g_V, the cost of a double iteration is of 25 multiplications over $\mathbb{F}_{3^m}$, whereas two iterations of the original loop involve 28 multiplications [7]. The key observation is that the sparse multiplication over $\mathbb{F}_{3^{6m}}$ on line 15 requires only 8 multiplications over $\mathbb{F}_{3^m}$. Keeping in mind that our SSE implementation of multiplication over $\mathbb{F}_{3^m}$

Table 7. Supersingular curves over $\mathbb{F}_{3^m}$ (reprinted from [8])

Underlying field	$\mathbb{F}_{3^m}$, where m is coprime to 6.
Curve	$E : y^2 = x^3 - x + b$, with $b \in \{-1, 1\}$.
Number of rational points	$N = \#E(\mathbb{F}_{3^m}) = 3^m + 1 + \mu b 3^{(m+1)/2}$, with $$\mu = \begin{cases} +1 & \text{if } m \equiv 1, 11 \pmod{12}, \text{ and} \\ -1 & \text{if } m \equiv 5, 7 \pmod{12}. \end{cases}$$
Embedding degree	$k = 6$
Distortion map	$\psi : E(\mathbb{F}_{3^m})[\ell] \longrightarrow E(\mathbb{F}_{3^{6m}})[\ell] \setminus E(\mathbb{F}_{3^m})[\ell]$ $(x, y) \longmapsto (\rho - x, y\sigma)$ with $\sigma \in \mathbb{F}_{3^2}$ satisfying $\sigma^2 = -1$, and $\rho \in \mathbb{F}_{3^3}$ satisfying $\rho^3 = \rho + b$.
Tower field	$\mathbb{F}_{3^{6m}} = \mathbb{F}_{3^m}[\rho, \sigma] \cong \mathbb{F}_{3^m}[X, Y]/(X^3 - X - b, Y^2 + 1)$
Final exponentiation	$M = \left(3^{3m} - 1\right) \cdot \left(3^m + 1\right) \cdot \left(3^m + 1 - \mu b 3^{(m+1)/2}\right)$
Parameters of Algorithm 4	$$\lambda = \begin{cases} +1 & \text{if } m \equiv 7, 11 \pmod{12}, \\ -1 & \text{if } m \equiv 1, 5 \pmod{12}, \end{cases}$$ $$\nu = \begin{cases} +1 & \text{if } m \equiv 5, 11 \pmod{12}, \text{ and} \\ -1 & \text{if } m \equiv 1, 7 \pmod{12}. \end{cases}$$

involves a pre-computation step depending on the second operand, we designed Algorithm 5 where several multiplications over $\mathbb{F}_{3^m}$ share a common operand (lines 5 and 6).

Since lines 9 and 10 of Algorithm 4 do not present dependencies, the pre-computation of the Duursma-Lee functions can be performed in parallel on two cores. Then, one can split the execution of Miller's algorithm (lines 13 to 16) into several parts that are run concurrently.

Algorithm 4. Unrolled loop for computing the reduced modified Tate pairing in characteristic three when $\frac{m-1}{2}$ is even.

Input: $P, Q \in E(\mathbb{F}_{3^m})[\ell]$.
Output: $\hat{e}_r(P, Q) \in \mathbb{F}_{3^{6m}}^*$.

1: $x_P \leftarrow \sqrt[3]{x_P} - (\nu + 1)b;$ (1 R, 1 A when $m \equiv 5, 11 \bmod 12$)
2: $y_P \leftarrow \lambda \sqrt[3]{y_P};$ (1 R)
3: $y_Q \leftarrow -\lambda y_Q;$

4: $t \leftarrow x_P + x_Q;$ (1 A)
5: $R \leftarrow \lambda(y_P t - y_Q \sigma - y_P \rho) \cdot (-t^2 + y_P y_Q \sigma - t\rho - \rho^2);$ (6 M, 1 C, 6 A)

6: $x_P[0] \leftarrow x_P; y_P[0] \leftarrow y_P;$
7: $x_Q[0] \leftarrow x_Q; y_Q[0] \leftarrow y_Q;$
8: **for** $j = 1$ to $\frac{m-1}{2}$ **do**
9:     $x_P[j] \leftarrow \sqrt[3]{x_P[j-1]}; x_Q[j] \leftarrow x_Q[j-1]^3;$ (1 R, 1 C)
10:     $y_P[j] \leftarrow \sqrt[3]{y_P[j-1]}; y_Q[j] \leftarrow y_Q[j-1]^3;$ (1 R, 1 C)
11: **end for**

12: **for** $j = 1$ **to** $\frac{m-1}{4}$ **do**
13:     $t \leftarrow x_P[2j-1] + x_Q[2j-1];$ $u \leftarrow y_P[2j-1]y_Q[2j-1];$ (1 M, 1 A)
14:     $t' \leftarrow x_P[2j] + x_Q[2j];$ $u' \leftarrow y_P[2j]y_Q[2j];$ (1 M, 1 A)
15:     $S \leftarrow (-t^2 + u\sigma - t\rho - \rho^2) \cdot (-t'^2 + u'\sigma - t'\rho - \rho^2);$ (8 M, 13 A)
16:     $R \leftarrow R \cdot S;$ (15 M, 67 A)
17: **end for**
18: **return** $R^M;$

Algorithm 5. Computation of $(-t^2 + u\sigma - t\rho - \rho^2) \cdot (-t'^2 + u'\sigma - t'\rho - \rho^2)$.

Input: $t, u, t',$ and $u' \in \mathbb{F}_{3^m}$.
Output: $W = (-t^2 + u\sigma - t\rho - \rho^2) \cdot (-t'^2 + u'\sigma - t'\rho - \rho^2).$

1: $a_1 \leftarrow t + u; a_2 \leftarrow t' + u';$ (2 A)
2: $a_3 \leftarrow t + t'; a_4 \leftarrow u + u';$ (2 A)
3: $m_1 \leftarrow t \cdot t'; m_2 \leftarrow u \cdot u'; m_3 \leftarrow a_1 \cdot a_2;$ (3 M)
4: $w_3 \leftarrow m_1 + m_2 - m_3;$ (2 A)
5: $m_4 \leftarrow m_1 \cdot m_1; m_5 \leftarrow m_1 \cdot a_3; m_6 \leftarrow m_1 \cdot a_4;$ (3 M)
6: $m_7 \leftarrow a_3 \cdot a_3; m_8 \leftarrow a_3 \cdot w_3;$ (2 M)
7: $w_0 \leftarrow m_4 - m_2 + ba_3;$ (2 A)
8: $w_1 \leftarrow m_6 + m_8;$ (1 A)
9: $w_2 \leftarrow m_5 + a_3 + b;$ (2 A)
10: $w_4 \leftarrow m_7 - m_1 + 1;$ (2 A)
11: $w_5 \leftarrow -a_4;$

12: **return** $w_0 + w_1\sigma + w_2\rho + w_3\sigma\rho + w_4\rho^2 + w_5\sigma\rho^2;$

On the Complexity of Computing Discrete Logarithms over Algebraic Tori

Shuji Isobe[1], Eisuke Koizumi[1], Yuji Nishigaki[1,2], and Hiroki Shizuya[1]

[1] Graduate School of Information Sciences,
Tohoku University, Sendai, 980-8576 Japan
{iso,koizumi,shizuya}@cite.tohoku.ac.jp
[2] Presently with Development Planning Dept.,
Brother Industries, Ltd., Nagoya, 467-8562 Japan

Abstract. This paper studies the complexity of computing discrete logarithms over algebraic tori. We show that the order certified version of the discrete logarithm over general finite fields (OCDL, in symbols) reduces to the discrete logarithm over algebraic tori (TDL, in symbols) with respect to the polynomial-time Turing reducibility. This reduction means that if the integer factorization can be computed in polynomial time, then TDL is equivalent to the discrete logarithm DL over general finite fields with respect to the Turing reducibility.

Keywords: algebraic tori, order certified discrete logarithms, Turing reduction.

1 Introduction

It is a significant subject to construct compact and efficient cryptographic schemes for reducing the sizes of ciphertexts and public keys without influencing the security. One of ideas for achieving this is to find an efficiently computable compression map from a certain subgroup of $\mathbb{F}_{q^n}^{\times}$, the unit group of the finite field of q^n elements, to the affine space $\mathbb{F}_q^m$ of some dimension $m < n$. The length of messages, ciphertexts or signatures decreases from $n \log q$ bits to $m \log q$ bits if one applies the map to known cryptographic schemes based on the discrete logarithms, such as the Diffie-Hellman key agreement protocol [3] and the ElGamal encryption and signature schemes [5]. A typical example of compression maps is the trace map of fields. The trace maps of $\mathbb{F}_{q^2}$, $\mathbb{F}_{q^3}$ and $\mathbb{F}_{q^6}$ are applied in LUC [11], the Gong-Harn scheme [6] and XTR [7], respectively.

In the paper proposing CEILIDH, Rubin and Silverberg [8] pointed out that these compression maps can be described in the context of algebraic tori. They stated, in [8], that subgroups of $\mathbb{F}_{q^n}^{\times}$ applied in the above schemes can be regarded as (the quotient of) algebraic tori and that these maps are birational maps from

J.A. Garay, A. Miyaji, and A. Otsuka (Eds.): CANS 2009, LNCS 5888, pp. 433–442, 2009.

the algebraic tori to the affine spaces. Their observations imply that studying algebraic tori may give a lead in understanding properties and security of torus-based cryptographic schemes. A lot of research concerning the torus-based cryptography has worked on proposal for admissible parameters or security arguments against known attacks. On the other hand, there are few investigations on the torus-based cryptography from a complexity theoretic viewpoint so far.

In this paper, we study the computational complexity of the discrete logarithm problem over algebraic tori. We prove that the order certified version of the discrete logarithm over general finite fields reduces to the discrete logarithm problem over algebraic tori. If we assume that the prime factorizations can easily be computed, then this result means that computing discrete logarithms over algebraic tori is as hard as computing those over finite fields. Hence, we can consider the discrete logarithm problem over algebraic tori as a good candidate for the hard cryptographic primitives as long as we believe that the discrete logarithms over finite fields is hard to compute.

This paper is organized as follows: In Section 2, we refer to the definition of algebraic tori and some notations needed later. In Section 3, we introduce the discrete logarithm function of which we explore the computational complexity. We then state the main theorem, and prove it. Concluding remarks are given in Section 4.

2   Preliminaries

In this section, we introduce necessary notions and notations.

Let p be a prime, and set $q = p^n$ for $n \geq 1$. Then there exists a field of q elements. In particular, it is unique in a sense that if K and K' are fields of q elements, then $K \simeq K'$ holds. We denote by $\mathbb{F}_q$ the field of q elements.

When $q = p^n$, $\mathbb{F}_q$ is a Galois extension of degree n over its prime field $\mathbb{F}_p$. Then $\mathbb{F}_q/\mathbb{F}_p$ is a simple extension (e.g. [12, Section 6.10]), that is, there exists an element $\theta \in \mathbb{F}_q$ such that $\mathbb{F}_q = \mathbb{F}_p(\theta)$. Let $f = f(X)$ be the minimal polynomial of θ over $\mathbb{F}_p$. Then f is an irreducible polynomial over $\mathbb{F}_p$ of degree n, and we have a canonical isomorphism $\mathbb{F}_q = \mathbb{F}_p(\theta) \simeq \mathbb{F}_p[X]/(f)$, where $\mathbb{F}_p[X]$ denotes the ring of polynomials over $\mathbb{F}_p$ and (f) denotes the ideal of $\mathbb{F}_p[X]$ generated by f. On the other hand, let $f^* \in \mathbb{F}_p[X]$ be an irreducible polynomial of degree n. Then the residue ring $\mathbb{F}_p[X]/(f^*)$ is a field of p^n elements, and $\mathbb{F}_p(\theta^*) \simeq \mathbb{F}_p[X]/(f^*) \simeq \mathbb{F}_{p^n}$ follows for any root θ^* of f^*.

This observation suggests that one can identify $\mathbb{F}_{p^n}$ with the residue ring $\mathbb{F}_p[X]/(f)$ for any irreducible polynomial $f \in \mathbb{F}_p[X]$ of degree n. Then each element $\alpha \in \mathbb{F}_{p^n}$ can be expressed by a tuple $\alpha = (p, f, g)$, where g is a polynomial over $\mathbb{F}_p$ of degree less than n. Namely, $\alpha = (p, f, g)$ means that α is identified with the residue class $g + (f)$ in the residue ring $\mathbb{F}_p[X]/(f)$. Under this expression, the length of the expression for $\alpha \in \mathbb{F}_{p^n}$ is $O(n \log p)$. If p and f are well understood, then we denote the element $\alpha = (p, f, g)$ by the polynomial g itself. Let n_0 be a divisor of n. Then the subset

$$K_{n_0} = \{(p, f, g) \in \mathbb{F}_{p^n} \mid g^{p^{n_0}} = g \bmod f\} \tag{1}$$

is a subfield of $\mathbb{F}_{p^n}$ with p^{n_0} elements. Hence, for any divisor n_0 of n, $\mathbb{F}_{p^{n_0}}$ can be identified with the subfield K_{n_0}.

For each $n \in \mathbb{N}$, the n-th cyclotomic polynomial $\Phi_n(X)$ is defined by

$$\Phi_n(X) = \prod_{\zeta}(X - \zeta),$$

where the product is taken over all the n-th primitive roots ζ of 1 in the field $\mathbb{C}$ of complex numbers. $\Phi_n(X)$ is an irreducible polynomial over $\mathbb{Z}$ of degree $\varphi(n)$, where φ denotes the Euler function. For the cyclotomic polynomials, we refer to the following two equations [4]:

$$X^n - 1 = \prod_{d \mid n} \Phi_d(X) \tag{2}$$

and

$$\Phi_n(X) = \prod_{d \mid n}(X^d - 1)^{\mu(n/d)}, \tag{3}$$

where $\prod_{d \mid n}$ denotes the product that is taken over all the positive divisors d of n, and μ denotes the Möbius function.

We next introduce the notion of algebraic tori [8]. Let L/K be a Galois extension of degree n, and let $\mathrm{Gal}(L/K)$ be the Galois group of L/K, that is, the group of all the K-automorphisms of L. Note that $\mathrm{Gal}(L/K)$ is a group of order n. For each element $\alpha \in L$, its *norm* $N_{L/K}(\alpha)$ is defined by

$$N_{L/K}(\alpha) = \prod_{\sigma \in \mathrm{Gal}(L/K)} \sigma(\alpha).$$

The *algebraic torus* $T_{L/K}$ for L/K is defined to be the set of all the elements $\alpha \in L$ such that $N_{L/M}(\alpha) = 1$ for all intermediate fields M with $K \subseteq M \subsetneq L$. If $q = p^m$ for some prime p, $L = \mathbb{F}_{q^n}$ and $K = \mathbb{F}_q$, then we write $T_{L/K} = T_n(\mathbb{F}_q) = T_n(\mathbb{F}_{p^m})$. We refer to the following proposition [8].

Proposition 1 ([8]). $T_n(\mathbb{F}_q)$ *is a subgroup of* $\mathbb{F}_{q^n}^\times$ *of order* $\Phi_n(q)$.

Since $\mathbb{F}_{q^n}^\times$ is a cyclic group, $T_n(\mathbb{F}_q)$ is characterized as the unique subgroup of $\mathbb{F}_{q^n}^\times$ of order $\Phi_n(q)$. Equivalently, we have

$$T_n(\mathbb{F}_q) = \left\{\alpha \in \mathbb{F}_{q^n}^\times \mid \alpha^{\Phi_n(q)} = 1\right\}. \tag{4}$$

Note that $\mathbb{F}_{q^n}$ is a subfield of $\mathbb{F}_{q^{n_1}}$ for any multiple n_1 of n. Hence, using an irreducible polynomial $f \in \mathbb{F}_q[X]$ of degree n_1, we can regard the algebraic torus $T_n(\mathbb{F}_q)$ as the subgroup of $\mathbb{F}_{q^{n_1}}^{\times}$ in a way that

$$T_n(\mathbb{F}_q) = \left\{ (p, f, g) \in \mathbb{F}_{q^{n_1}}^{\times} \mid g^{\Phi_n(q)} = 1 \bmod f \right\} \tag{5}$$

under the canonical isomorphism $\mathbb{F}_{q^{n_1}} \simeq \mathbb{F}_p[X]/(f)$.

In this paper, we use the notion of the polynomial-time Turing reducibility. We follow the definition presented in [10]. For a multivalued function ξ, a *refinement* of ξ is a function ξ_0 such that $\mathrm{dom}(\xi) = \mathrm{dom}(\xi_0)$, and $\xi(x) = y$ whenever $\xi_0(x) = y$, where dom denotes the domain of function. For any multivalued functions ξ and η, ξ *reduces* to η, denoted by $\xi \leq_{\mathrm{T}}^{\mathrm{P}} \eta$, if there exists a deterministic polynomial-time Turing machine M with oracle tapes such that for any single-valued refinement η_0 of η, M computes a single-valued refinement of ξ with the help of the oracle η_0. $\xi \equiv_{\mathrm{T}}^{\mathrm{P}} \eta$ means that both $\xi \leq_{\mathrm{T}}^{\mathrm{P}} \eta$ and $\eta \leq_{\mathrm{T}}^{\mathrm{P}} \xi$ hold.

3   Main Theorem

In this section, we present the main theorem on the complexity of computing discrete logarithms over algebraic tori.

3.1   Discrete Logarithm Functions

We start with introducing the discrete logarithm functions that we investigate.

The discrete logarithm function DL over finite fields maps each pair (y, g) of elements y and g in some finite field $\mathbb{F}_q$ to the exponents $x \in \mathbb{Z}_{q-1}$ such that $y = g^x$ holds in the field $\mathbb{F}_q$. Namely, DL is formally defined as follows: The domain $\mathrm{dom}(\mathrm{DL})$ is the set of all tuples (p, f, g, y) that express two elements $(p, f, g), (p, f, y) \in \mathbb{F}_{p^n}^{\times}$, where $n = \deg f$. For each tuple $(p, f, g, y) \in \mathrm{dom}(\mathrm{DL})$, we define

$$\mathrm{DL}(p, f, g, y) = x, \quad \text{where } x \in \mathbb{Z}_{p^n - 1} \text{ and } y = g^x,$$

where we write $y = g^x$ instead of $y = g^x \bmod f$ for simplicity. Note that $\mathrm{DL}(p, f, g, y)$ is defined to be $\perp$ if there exists no exponent x such that $y = g^x$. We denote by PDL the discrete logarithm function over prime fields. Namely, the domain $\mathrm{dom}(\mathrm{PDL})$ is the set of all tuples (p, g, y) of a prime p and two elements $g, y \in \mathbb{F}_p^{\times}$, and for each tuple $(p, g, y) \in \mathrm{dom}(\mathrm{PDL})$, $\mathrm{PDL}(p, g, y)$ is defined by

$$\mathrm{PDL}(p, g, y) = x, \quad \text{where } x \in \mathbb{Z}_{p-1} \text{ and } y = g^x.$$

$\mathrm{PDL}(p, g, y)$ is defined to be $\perp$ if there exists no exponent x such that $y = g^x$. PDL is a restriction of DL in a sense that $\mathrm{PDL}(p, g, y) = \mathrm{DL}(p, X, g, y)$.

The function TDL (DL on algebraic Tori) is defined to be a restriction of DL to the case where g and y are contained in algebraic tori. Formally, the domain dom(TDL) of TDL is the set of all tuples (p, m, n, f, g, y) satisfying $mn \mid \deg f$ and $(p, f, g), (p, f, y) \in T_m(\mathbb{F}_{p^n})$, where we identify $T_m(\mathbb{F}_{p^n})$ with the subgroup of $\mathbb{F}_{p^{\deg f}}^{\times}$ by Eq. (5). For each tuple $(p, m, n, f, g, y) \in$ dom(TDL), we define

$$\text{TDL}(p, m, n, f, g, y) = x, \quad \text{where } x \in \mathbb{Z}_{\Phi_m(p^n)} \text{ and } y = g^x.$$

$\text{TDL}(p, m, n, f, g, y)$ is defined to be $\perp$ if there exists no exponent x such that $y = g^x$. Since $T_m(\mathbb{F}_{p^n}) = \mathbb{F}_p^{\times}$ when $(m, n) = (1, 1)$, TDL involves PDL in a sense that $\text{PDL}(p, g, y) = \text{TDL}(p, 1, 1, X, g, y)$. It should be noted that since $T_1(\mathbb{F}_{p^n}) \simeq \mathbb{F}_{p^n}^{\times}$, TDL involves DL in a sense that $\text{DL}(p, f, g, y) = \text{TDL}(p, 1, \deg f, f, g, y)$. Therefore, allowing the case where $m = 1$ seems to be improper for capturing the complexity of discrete logarithms over algebraic tori in comparing with the discrete logarithms over general finite fields. In fact, if $m = 1$ is allowed, then one can obtain only the trivial relationship $\text{DL} \equiv_T^P \text{TDL}$. However, this tells nothing about the complexity-theoretic characterization of TDL. Hence, it looks proper that we exclude the case where $m = 1$. On the other hand, it looks natural to allow the case where $(m, n) = (1, 1)$, that is, $\text{PDL} \leq_T^P \text{TDL}$, since what our main result states is that the discrete logarithm over finite fields is equivalent to the combination of TDL and PDL. Therefore, we always require that either $m \geq 2$ or $(m, n) = (1, 1)$ holds for any input $(p, m, n, f, g, y) \in$ dom(TDL).

We should also note that DL is a multivalued function. In fact, for an input tuple (p, f, g, y), if the base g is not a generator of the group $\mathbb{F}_{p^n}^{\times}$, where $n = \deg f$, then the exponent x satisfying $y = g^x$ is not unique under modulo $p^n - 1$ but it is unique under modulo $\text{ord}(g)$, the order of g, as long as the exponent x exists. Similarly, PDL and TDL are multivalued functions. If the order $\text{ord}(g)$ of g in $\mathbb{F}_{p^n}^{\times}$ is known, then one can find the unique exponent $x_0 \in \mathbb{Z}_{\text{ord}(g)}$ satisfying $y = g^{x_0}$ from an exponent $x = \text{DL}(p, f, g, y)$ by $x_0 = x \mod \text{ord}(g)$. A simple way for efficiently computing the order $\text{ord}(g)$ is to give the prime factorization $\text{IF}(p^n - 1)$ of the order $\#\mathbb{F}_{p^n}^{\times} = p^n - 1$ as a part of the input tuple. Using the factorization $\text{IF}(p^n - 1)$, one can compute the order $\text{ord}(\alpha)$ of any given element $\alpha = (p, f, g) \in \mathbb{F}_{p^n}^{\times}$ in polynomial time in the expression length $\ell = O(n \log p)$ of α. Thus, we introduce the "order certified" version OCDL of DL as follows [9]: The domain dom(OCDL) is the set of all tuples $(p, f, g, y, \text{IF}(p^n - 1))$ such that $(p, f, g, y) \in$ dom(DL), where $n = \deg f$. For each tuple $(p, f, g, y, \text{IF}(p^n - 1)) \in$ dom(OCDL), $\text{OCDL}(p, f, g, y, \text{IF}(p^n - 1))$ is defined to be the unique exponent $x \in \mathbb{Z}_{\text{ord}(g)}$ such that $y = g^x$. It is obvious from the definition that $\text{OCDL} \leq_T^P \text{DL}$, and moreover $\text{OCDL} \equiv_T^P \text{DL}$ if $\text{IF}(p^n - 1)$ can be computed in polynomial time in ℓ.

3.2   Reductions

In this section, we prove the following main theorem.

Theorem. OCDL $\leq_T^P$ TDL.

Our strategy for the proof is to reduce the extension degree by successively applying the oracle TDL and finally one reduces OCDL to the discrete logarithm PDL over prime fields.

The goal is to construct a deterministic polynomial-time Turing machine M with oracle tapes such that, for any single-valued refinement TDL_0 of TDL, M computes OCDL with the help of the oracle TDL_0. Let w be any input string for M. Since the primarity test can be done in polynomial time [1] and the irreducibility test of polynomials can be done in polynomial time [2], one can determine in polynomial time whether or not w is contained in dom(OCDL). If $w \notin$ dom(OCDL), then M halts with output $\perp$. We assume that $w \in$ dom(OCDL), that is, w is of the form $w = (p, f, g, y, \mathrm{IF}(p^{k_0} - 1))$, where $k_0 = \deg f$. M has to output the exponent $x \in \mathbb{Z}_{\mathrm{ord}(g)}$ that satisfies $y = g^x$ with the help of the oracle TDL_0.

Using the prime factorization $\mathrm{IF}(p^{k_0} - 1)$, one can compute the order $\mathrm{ord}(\alpha)$ of any given element $\alpha \in \mathbb{F}_{p^{k_0}}^\times$ in polynomial time in the length of α. Therefore, one can determine in polynomial time in $\ell = O(k_0 \log p)$, the expression length of w, whether or not there exists the desired exponent x satisfying $y = g^x$ by verifying the divisibility relationship $\mathrm{ord}(y) \mid \mathrm{ord}(g)$. If there exists no such exponent x, then M outputs $\perp$ and halts.

We now assume that there exists the exponent x, that is, $\mathrm{ord}(y) \mid \mathrm{ord}(g)$ holds. We show the following lemma.

Lemma 2. *For any divisor k of k_0, the exponent $\mathrm{OCDL}(p, f, g, y, \mathrm{IF}(p^{k_0} - 1))$ can be computed in polynomial time in $\ell = O(k_0 \log p)$ with the help of the oracle TDL_0 whenever $(p, f, g) \in K_k^\times$, where K_k is defined in Eq. (1).*

Since k_0 is a divisor of k_0 itself, the theorem immediately follows from this lemma.

We now prove Lemma 2 by induction on divisors k of k_0. When $k = 1$, one can obtain the exponent x by calling the oracle TDL_0 with the query $(p, 1, 1, f, g, y)$. Note that we allow $(m, n) = (1, 1)$ for the instance of TDL.

Let $k > 1$ be any divisor of k_0, and inductively assume that one can obtain the exponent $x' = \mathrm{OCDL}(p, f, g', y', \mathrm{IF}(p^{k_0} - 1))$ in time polynomial in ℓ with the help of the oracle TDL_0 whenever $(p, f, g') \in K_{k'}^\times$ for any divisor $k' < k$ of k_0. One can find a prime factor p' of k in time polynomial in ℓ by a straightforward successive division algorithm. Set $k' = k/p'$. Using $\mathrm{IF}(p^{k_0} - 1)$, one can compute the prime factorization of $p^k - 1$ in polynomial time in ℓ. Since $p^k - 1 = (p^{k'})^{p'} - 1 = (p^{k'} - 1)\Phi_{p'}(p^{k'})$ from Eq. (2), one can also compute the prime factorizations of $\Phi_{p'}(p^{k'})$ and $p^{k'} - 1$ in time polynomial in ℓ. Let

$$p^k - 1 = p_1^{a_1} \cdots p_r^{a_r}, \quad \Phi_{p'}(p^{k'}) = p_1^{b_1} \cdots p_r^{b_r} \quad \text{and} \quad p^{k'} - 1 = p_1^{c_1} \cdots p_r^{c_r}$$

be the prime factorizations of $p^k - 1$, $\Phi_{p'}(p^{k'})$ and $p^{k'} - 1$, respectively, where $a_i = b_i + c_i$ for each $i = 1, \ldots, r$. Furthermore, one can compute the prime factorization of $\mathrm{ord}(g)$ in time polynomial in ℓ as $\mathrm{IF}(p^{k_0} - 1)$ is given. Since $g \in K_k^\times$, we have $\mathrm{ord}(g) \mid p^k - 1$, and hence, $\mathrm{ord}(g) = p_1^{d_1} \cdots p_r^{d_r}$ for some exponents $0 \le d_i \le a_i$. Renumbering the indices, we may assume without loss of generality that $b_i > d_i$ for $i = 1, \ldots, t$ and $b_i \le d_i$ for $i = t+1, \ldots, r$. We write

$$\Phi_{p'}(p^{k'}) = \beta_1 \beta_2, \qquad \beta_1 = \prod_{i=1}^{t} p_i^{b_i}, \quad \beta_2 = \prod_{i=t+1}^{r} p_i^{b_i}, \qquad (6)$$

$$\mathrm{ord}(g) = \delta_1 \delta_2, \qquad \delta_1 = \prod_{i=1}^{t} p_i^{d_i}, \quad \delta_2 = \prod_{i=t+1}^{r} p_i^{d_i}. \qquad (7)$$

When $b_i > d_i$ for all $i = 1, \ldots, r$, that is, $t = r$, we set $\beta_2 = \delta_2 = 1$, $\beta_1 = \Phi_{p'}(p^{k'})$ and $\delta_1 = \mathrm{ord}(g)$. Similarly, when $b_i \le d_i$ for all $i = 1, \ldots, r$, that is, $t = 0$, we set $\beta_1 = \delta_1 = 1$, $\beta_2 = \Phi_{p'}(p^{k'})$ and $\delta_2 = \mathrm{ord}(g)$. Let $\mu = \beta_1/\delta_1$, $\nu = \delta_2/\beta_2$,

$$g_0 = g^\nu \quad \text{and} \quad y_0 = y^\nu = g^{x\nu}. \qquad (8)$$

Then we have $\mu, \nu \in \mathbb{Z}$,

$$g_0^{\Phi_{p'}(p^{k'})} = g^{\nu \Phi_{p'}(p^{k'})} = g^{\beta_1 \delta_2} = (g^{\delta_1 \delta_2})^\mu = (g^{\mathrm{ord}(g)})^\mu = 1$$

and $y_0^{\Phi_{p'}(p^{k'})} = (g_0^{\Phi_{p'}(p^{k'})})^x = 1$. Therefore, it follows from Eq. (5) that $g_0, y_0 \in T_{p'}(\mathbb{F}_{p^{k'}})$. Calling the oracle TDL_0 with the query (p, p', k', f, g_0, y_0), one obtains an exponent $x_0 \in \mathbb{Z}_{\Phi_{p'}(p^{k'})}$ such that $y_0 = g_0^{x_0}$. Since $g^{x_0 \nu} = g_0^{x_0} = y_0 = g^{x\nu}$, we have $(x - x_0)\nu = u\,\mathrm{ord}(g)$, namely $x - x_0 = u\delta_1\beta_2$ for some integer $u \in \mathbb{Z}$.

Let

$$g_1 = g^{\delta_1 \beta_2} \quad \text{and} \quad y_1 = yg^{-x_0} = g^{x-x_0} = g^{u\delta_1\beta_2} = g_1^u. \qquad (9)$$

Then we have

$$\mathrm{ord}(g_1) = \frac{\mathrm{ord}(g)}{\gcd(\mathrm{ord}(g), \delta_1\beta_2)} = \frac{\delta_1\delta_2}{\gcd(\delta_1\delta_2, \delta_1\beta_2)} = \frac{\delta_1\delta_2}{\delta_1\beta_2} = \nu.$$

Since

$$\delta_1\beta_2(p^{k'} - 1) = \prod_{i=1}^{t} p_i^{c_i + d_i} \prod_{i=t+1}^{r} p_i^{b_i + c_i} = \prod_{i=1}^{t} p_i^{c_i + d_i} \prod_{i=t+1}^{r} p_i^{a_i}$$

$$= \mathrm{ord}(g) \prod_{i=1}^{t} p_i^{c_i} \prod_{i=t+1}^{r} p_i^{a_i - d_i},$$

that is, $\delta_1\beta_2(p^{k'} - 1)$ is a multiple of $\mathrm{ord}(g)$, we have $g_1^{p^{k'}-1} = g^{\delta_1\beta_2(p^{k'}-1)} = 1$. This implies that $g_1 \in K_{k'}^{\times}$, and hence it follows that $y_1 = g_1^u \in K_{k'}^{\times}$. By the induction hypothesis, one can find the exponent

$$x_1 = \mathrm{OCDL}(p, f, g_1, y_1, \mathrm{IF}(p^{k_0} - 1)) \in \mathbb{Z}_{\mathrm{ord}(g_1)} = \mathbb{Z}_{\nu}$$

in time polynomial in ℓ with making queries to the oracle TDL_0.

Using the exponent x_1, one can obtain the desired exponent $x = \mathrm{OCDL}(w)$ as follows. Since $g_1^{x_1} = y_1 = g_1^u$, we have $x_1 - u = v\,\mathrm{ord}(g_1) = v\nu$ for some integer $v \in \mathbb{Z}$. Setting

$$x' = x_0 + x_1\delta_1\beta_2, \tag{10}$$

we have

$$g^{x'} = g^{x_0}\left(g^{\delta_1\beta_2}\right)^{x_1} = g^{x_0}g_1^{x_1} = g^{x_0}y_1 = g^{x_0}yg^{-x_0} = y = g^x.$$

This implies that $x \equiv x' \pmod{\mathrm{ord}(g)}$, and one can obtain the desired exponent $x \in \mathbb{Z}_{\mathrm{ord}(g)}$ by $x = x' \bmod \mathrm{ord}(g)$. This completes the proof of Lemma 2.

We are now ready to construct a polynomial-time bounded deterministic oracle Turing machine M which reduces OCDL to TDL by the proof of Lemma 2. For an input string w, M works with the following procedure, where we omit the verification of the validity of the input w.

> **procedure** $\mathrm{OCDL}(p, f, g, y, \mathrm{IF}(p^{k_0} - 1))$
> find the minimum integer $k \geq 1$ such that $g^{p^k-1} = 1$;
> **if** $k = 1$ **then**
> **return** $\mathrm{TDL}_0(p, 1, 1, f, g, y)$;
> **else**
> find a prime factor p' of k and let $k' = k/p'$;
> compute β_1, β_2, δ_1 and δ_2 by Eqs. (6) and (7);
> compute g_0 and y_0 by Eq. (8);
> let $x_0 = \mathrm{TDL}(p, p', k', f, g_0, y_0)$ by calling the oracle TDL;
> compute g_1 and y_1 by Eq. (9);
> let $x_1 = \mathrm{OCDL}(p, f, g_1, y_1, \mathrm{IF}(p^{k_0} - 1))$ by recursively calling OCDL;
> **return** $x_0 + x_1\delta_1\beta_2 \bmod \mathrm{ord}(g)$;
> **endprocedure**

Each step can be done in time polynomial in ℓ, and the number of recursive calls is at most $O(\log k_0)$. Hence, M works in time polynomial in ℓ.

4   Concluding Remarks

In this paper, we have investigated the complexity of computing the discrete logarithms over algebraic tori in the context of Turing reductions, and we have shown that OCDL, the order certified discrete logarithm over finite fields, reduces to TDL, the discrete logarithm over algebraic tori. For simplicity, assume that the prime factorizations of $p^n - 1$ can be easily computed, that is, OCDL is equivalent to DL with respect to the Turing reducibility. Then our result implies that DL $\equiv_T^P$ TDL. Since $T_m(\mathbb{F}_{p^n})$ is a subgroup of $\mathbb{F}_{p^{mn}}^\times$ of order $\Phi_m(p^n)$, we have

$$\frac{\#T_m(\mathbb{F}_{p^n})}{\#\mathbb{F}_{p^{mn}}^\times} \leq \frac{cp^{n\varphi(m)}}{p^{mn} - 1} = O\left(q^{\varphi(m)-m}\right), \quad q = p^n$$

for some constant $c > 0$ and any sufficiently large $q = p^n$. Therefore, the algebraic tori are negligibly small compared to the unit group of the finite fields. In view of this observation, DL $\equiv_T^P$ TDL looks a strong reduction because it means that DL reduces to the discrete logarithms over negligibly small subgroups. Furthermore, we assume that PDL $\equiv_T^P$ TDL (Note that since PDL $\leq_T^P$ TDL, this is equivalent to TDL $\leq_T^P$ PDL). Then our result implies that DL $\leq_T^P$ PDL. Conversely, since TDL $\leq_T^P$ DL, DL $\leq_T^P$ PDL implies that TDL $\leq_T^P$ PDL. Therefore, our result implies that TDL $\leq_T^P$ PDL if and only if DL $\leq_T^P$ PDL. Since the prime field $\mathbb{F}_p$ is negligibly smaller than the extension $\mathbb{F}_{p^n}$ with n being larger, this is an unlikely event. Thus, as long as we believe that DL $\not\leq_T^P$ PDL, TDL is properly harder than PDL. Namely, TDL can be thought as a good candidate for the hard cryptographic primitives.

Acknowledgement. The authors thank Hirofumi Muratani and Atsushi Shimbo for their helpful suggestion.

References

1. Agrawal, M., Kayal, N., Saxena, N.: PRIMES is in P. Ann. of Math. 160(2), 781–793 (2004)
2. Cohen, H.: A Course in Computational Algebraic Number Theory. In: Graduate Text in Mathematics, vol. 138. Springer, Heidelberg (1993)
3. Diffie, W., Hellman, M.E.: New directions in cryptography. IEEE Trans. on Inform. Theory 22, 644–654 (1976)
4. van Dijk, M., Woodruff, D.: Asymptotically optimal communication for torus-based cryptography. In: Franklin, M. (ed.) CRYPTO 2004. LNCS, vol. 3152, pp. 157–178. Springer, Heidelberg (2004)
5. ElGamal, T.: A public key cryptosystem and a signature scheme based on discrete logarithms. IEEE Trans. on Inform. Theory 31, 469–472 (1985)
6. Gong, G., Harn, L.: Public-key cryptosystems based on cubic finite field extensions. IEEE Trans. on Inform. Theory 45, 2601–2605 (1999)
7. Lenstra, A.K., Verheul, E.R.: The XTR public key system. In: Bellare, M. (ed.) CRYPTO 2000. LNCS, vol. 1880, pp. 1–19. Springer, Heidelberg (2000)

 8. Rubin, K., Silverberg, A.: Torus-based cryptography. In: Boneh, D. (ed.) CRYPTO 2003. LNCS, vol. 2729, pp. 349–365. Springer, Heidelberg (2003)
 9. Sakurai, K., Shizuya, H.: A Structural Comparison of the Computational Difficulty of Breaking Discrete Log Cryptosystems. J. of Cryptology 11, 29–43 (1998)
10. Selman, A.: A taxonomy of complexity classes of functions. J. of Computer and System Sciences 48, 357–381 (1994)
11. Smith, M., Lennon, M.J.J.: LUC: A New Public Key System. In: Proceedings of IFIP TC11 Ninth International Conference on Information Security IFIP/Sec 1993, pp. 103–117. North-Holland, Amsterdam (1993)
12. van der Waerden, B.L.: Algebra, vol. I. Springer, New York (1991)

On the Usability of Secure Association of Wireless Devices Based on Distance Bounding

Mario Cagalj[1], Nitesh Saxena[2], and Ersin Uzun[3]

[1] FESB
University of Split, Croatia
mario.cagalj@fesb.hr
[2] Computer Science and Engineering
Polytechnic Institute of New York University
nsaxena@poly.edu
[3] Information and Computer Sciences
University of California, Irvine
euzun@ics.uci.edu

Abstract. When users wish to establish wireless communication between their devices, the channel needs to be bootstrapped first. Usually, the channel is desired to be authenticated and confidential, in order to mitigate any malicious control of or eavesdropping over the communication. When there is no prior security context, such as, shared secrets, common key servers or public key certificates, device association necessitates some level of user involvement into the process. A wide variety of user-aided security association techniques have been proposed in the past. A promising set of techniques require out-of-band communication between the devices (e.g., auditory, visual, or tactile). The usability evaluation of such techniques has been an active area of research.

In this paper, our focus is on the usability of an alternative method of secure association – *Integrity regions* (I-regions) [40] – based on distance bounding. I-regions achieves secure association by verification of entity proximity through time-to-travel measurements over ultrasonic or radio channels. Security of I-regions crucially relies on the assumption that human users can correctly gauge the distance between two communicating devices. We demonstrate, via a thorough usability study of the I-regions technique and related statistical analysis, that such an assumption does not hold in practice. Our results indicate that I-regions can yield high error rates, undermining its security and usability under common communication scenarios.

Keywords: Authentication, Distance Bounding, Usable Security, Wireless Networks.

1 Introduction

Short- and medium-range wireless communication, based on technologies such as Bluetooth and WiFi, is becoming increasingly popular and promises to remain so in the future. With this surge in popularity, come various security risks. Wireless communication channel is easy to eavesdrop upon and to manipulate, and therefore a fundamental security objective is to secure this communication channel. In this paper, we will use the

J.A. Garay, A. Miyaji, and A. Otsuka (Eds.): CANS 2009, LNCS 5888, pp. 443–462, 2009.

term "pairing" to refer to the operation of bootstrapping secure communication between two devices connected with a short-range wireless channel. The examples of pairing, from day-to-day life, include pairing of a WiFi laptop and an access point, a Bluetooth keyboard and a desktop.

One of the main challenges in secure device pairing is that, due to sheer diversity of devices and lack of standards, no global security infrastructure exists today and none is likely for the foreseeable future. Consequently, traditional cryptographic means (such as authenticated key exchange protocols) are unsuitable, since unfamiliar devices have no prior security context and no common point of trust.

A number of research directions have been undertaken by the research community to address the problem of pairing of *ad hoc* wireless devices. One valuable and well-established research direction is the use of auxiliary – also referred to as "out-of-band" (OOB) – channels, which are both perceivable and manageable by the human user(s) who own and operate the devices[1]. An OOB channel takes advantage of human sensory capabilities to authenticate human-imperceptible (and hence subject to Man-in-the-Middle or MitM attacks) information exchanged over the wireless channel. OOB channels can be realized using senses such as auditory, visual and tactile. Unlike the in-band (wireless) channel, the attacker can not remain undetected if it actively interferes with the OOB channel. A number of device pairing methods based on a variety of OOB channels have been proposed (we overview these methods later in Section 5; see [18] for a relevant survey). Usability evaluation of these methods is an active research area these days [18,16,14].

The focus of this paper is on an alternative approach to device pairing, called *Integrity regions (I-regions)*. I-regions is based on distance bounding [4] and can be implemented using ultrasonic or radio time-of-arrival ranging techniques. It relies on range measurements to prevent MitM attackers from inserting forged messages into the communication between the devices. Basically, the distance bounding technique allows a communicating device A to compute an upper bound of its (physical) distance d from another device it is being paired with. Note that the latter can be device B, with which the user of A intends to pair her device or it could be an MitM attacker. An MitM attack can be effectively foiled if the user controlling A can verify whether the actual distance between A and B is less than or equal to d and make sure that there is no other device (except B) at a distance less than or equal to d. Figure 1 illustrates an MitM attack scenario for I-regions. In the figure, *inter-device distance* denotes the actual physical distance between the two devices (i.e., between the user's phone and kiosk) and *attacker distance bound* is the actual physical distance between user's phone and attacker's device. In this example, attacker distance bound (6 ft) is larger than inter-device distance (3.5 ft), which indicates to the user an ongoing MitM attack. As defined in [40], an *integrity region* is a space centered at user's location, within which the user can confidently establish the presence (or absence) of other wireless devices.

Motivation and Contributions: In this paper, our focus is on the "User Layer" of the I-regions method. In I-regions, once A computes the upper bound of its distance d from B, and shows it on its screen, the user (controlling A) is required to perform two tasks: (1) determine if the perceived distance between A and B is not more than d, and (2)

---

[1] This has been the subject of recent standardization activities [37].

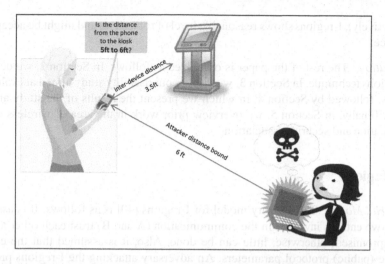

Fig. 1. An MitM Scenario for I-regions (user intends to pair her phone with the kiosk)

make sure if there is no other device (except B) at distance less than or equal to d, i.e., if B belongs to A's integrity region.[2] Clearly, a pre-requisite to the security of I-regions is users' ability of distance judgement. In other words, if users can not correctly perceive the distance shown on devices' screens as well as the distance between the two devices, the security of I-regions can not be guaranteed.

We hypothesize that users perception and interpretation of physical distances (needed to execute the first task mentioned above) is far from accurate. Consequently, I-regions is quite likely to result in both safe errors [38] (i.e, rejection of a valid pairing attempt) and more critically, fatal errors (i.e., acceptance of an MitM attack). In order to test our hypothesis and to evaluate I-regions in terms of efficiency (i.e., speed), robustness (i.e., error tolerance) and usability (i.e., System Usability Score of the method and user's self-confidence about distance judgement), we pursue a thorough and systematic usability study. We remark that such an experimental study was necessary to evaluate I-regions, which is akin to human behavior.

Based on the results of our study, I-regions can be termed quite efficient in terms of completion time. As hypothesized, however, in general (i.e., for arbitrary values of inter-device distance and attacker distance bound), I-regions exhibits poor robustness, with high likelihood of users committing both safe as well as fatal errors. This undermines the security of I-regions, either directly (i.e, in case of fatal errors) or indirectly (i.e, in case of safe errors). Thus, we can conclude that I-regions is not a suitable method for all possible pairing scenarios. However, for some specific values of inter-device distance (1 ft or 3.5 ft) in conjunction with attacker distance bound (at least 4.5 ft or 7

---

[2] The first manual task can be eliminated if device A is only allowed to accept pairing with devices located within a small, pre-determined distance (e.g., less than 1 meter). This would, however, severely limit the utility of I-regions only to scenarios where devices are in close proximity, and at the same time, damage usability by forcing user to move devices within a certain distance bound, which may not always be possible (such as in case of a wall-mounted access point or when two users are sitting across a long table in a meeting room).

ft, respectively), I-regions shows reasonable level of robustness and might be acceptable in practice.

Organization: The rest of the paper is organized as follows. In Section 2, we describe the I-regions technique. In Section 3, we discuss our usability study aimed at evaluating I-regions, followed by Section 4, in which we present the results of the study and our analysis. Finally, in Section 5, we overview prior work in the area of wireless device authentication and security association.

2   I-Regions

Adversarial Model: The security model for I-regions [40] is as follows. It is assumed that the two entities involved in the communication (A and B) trust each other and are not compromised; otherwise, little can be done. Also, it is assumed that the entities know the (public) protocol parameters. An adversary attacking the I-regions protocol is assumed to have full control on the wireless channel, namely, it can eavesdrop, delay, drop, replay and modify messages. The security notion for I-regions protocol in this setting is adopted from the model of authenticated key agreement due to Canneti and Krawczyk [6]. In this model, a multi-party setting is considered wherein a number of parties simultaneously run multiple/parallel instances of pairing protocols. In practice, however, it is reasonable to assume only two-parties running only a few serial/parallel instances of the pairing protocol. The security model does not consider denial-of-service (DoS) attacks. Note that on wireless channels, explicit attempts to prevent DoS attacks might not be useful because an adversary can simply launch an attack by jamming the wireless signal.

Protocol: The I-regions key exchange protocol, based on Diffie-Hellman (denoted DH-IR), unfolds as shown in Fig. 2. Both Alice and Bob calculate the commitment/opening pairs $((c_A, o_A)$ and $(c_B, o_B))$ for messages $m_A \leftarrow 0 \| g^{X_A} \| N_A$ and $m_B \leftarrow 1 \| g^{X_B} \| N_B$, respectively. Here, N_A and N_B are k bit long random strings and "0" and "1" are two public (and fixed) values that are used to break the symmetry and thus prevent a *reflection attack* [21]. In the first two messages, Alice and Bob exchange the commitments c_A and c_B. Then, in the following two messages they open the commitments by sending out o_A and o_B, respectively. It is important to stress that a given party opens his/her commitment only after having received the commitment value from the other party. The first four messages are exchanged over a radio link. Having received the commitment/opening pairs (c_A, o_A) and (c_B, o_B), Alice and Bob open the corresponding commitments and verify that "1" and "0" appear at the beginning of $\widehat{m}_B$ and $\widehat{m}_A$, respectively. If this verification is successful, Alice and Bob generate the authentication strings s_A and s_B. Note that the length of each of these strings is k. The main purpose of the last two messages in the DH-IR protocol is to allow Alice to compare s_A against the authentication string s_B generated by Bob, in a secure way. Thus, Alice sends a k-bit long random string N'_A to Bob and measures the time until she received the response from Bob. Bob responds with $R_B \leftarrow \widehat{N}'_A \oplus s_B$, where the sign hat denotes that the N'_A as transmitted by Alice may have been altered by the adversary. Alice receives $\widehat{R}_B$, where again the sign hat denotes that R_B as transmitted by Bob may have been altered by the adversary. At the same time, Alice calculates the distance d_A and verifies the corresponding integrity region for the presence of devices other than Bob's device (see Section 2). If this verification is successful, Alice knows that (with a high probability) Bob must have transmitted $\widehat{R}_B$, that is, $\widehat{R}_B = R_B$. Finally, if s_A equals $\widehat{R}_B \oplus N'_A$, Alice notifies Bob and they both accept the messages $\widehat{m}_A$ and $\widehat{m}_B$ (i.e., the corresponding DH public keys) as being authentic. Note that $\widehat{R}_B \oplus N'_A = s_B$ in case no attack takes place.

An adversary against the DH-IR protocol can only succeed with a probability at most 2^{-k}, as long as the commitment scheme used in the protocol is secure. To achieve a high level of security, k can be chosen to be arbitrarily long. For details regarding the security arguments of DH-IR, refer to [40].

Implementation: The DH-IR protocol can be implemented using two techniques: (1) using ultrasonic ranging (US) and (2) using radio (RF) ranging. Both exhibit equal security guarantees, but require different equipment attached to the devices.

Ultrasonic ranging requires time measurement precision only in hundreds of μ-seconds, but requires each device to be able to communicate via ultrasonic channel. Current ultrasonic ranging systems (e.g., Cricket motes [29,27]) can have centimeter precision ranging when the transceivers are perfectly aligned. However, the dependable

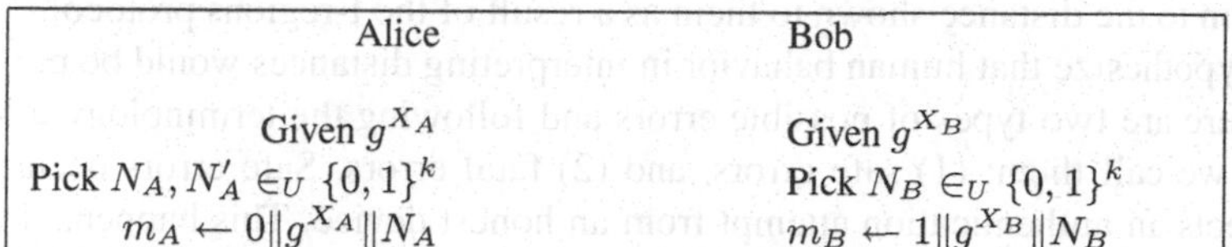

Alice		Bob
Given g^{X_A}		Given g^{X_B}
Pick $N_A, N'_A \in_U \{0,1\}^k$		Pick $N_B \in_U \{0,1\}^k$
$m_A \leftarrow 0\|g^{X_A}\|N_A$		$m_B \leftarrow 1\|g^{X_B}\|N_B$
$(c_A, o_A) \leftarrow \mathrm{commit}(m_A)$		$(c_B, o_B) \leftarrow \mathrm{commit}(m_B)$
	$\xrightarrow{\quad c_A \quad}$	
	$\xleftarrow{\quad c_B \quad}$	
	$\xrightarrow{\quad o_A \quad}$	$m_A \leftarrow \mathrm{open}(\widehat{c}_A, \widehat{o}_A)$
$\widehat{m}_B \leftarrow \mathrm{open}(\widehat{c}_B, \widehat{o}_B)$	$\xleftarrow{\quad o_B \quad}$	Verify 0 in $\widehat{m}_A$.
Verify 1 in $\widehat{m}_B$.		$s_B \leftarrow N_B \oplus \widehat{N}_A$
$s_A \leftarrow N_A \oplus \widehat{N}_B$		
(t_s^A) $\xrightarrow{\quad N'_A \quad}$		
(t_r^A) $\xleftarrow{\;-\;-R_B-\;-\;}$		$R_B \leftarrow \widehat{N}'_A \oplus s_B$
$d_A = s\left(t_r^A - t_s^A\right)$		
Verify $s_A \overset{?}{=} N'_A \oplus \widehat{R}_B$		

Alice verifies if distance between A and B is no more than d_A.
If verification OK, Alice and Bob accept the mutual authentication.

$\xrightarrow{\quad\quad}$: the wireless channel
$\xleftarrow{\;-\;-\;-\;}$: unidirectional ultra-sonic or ultra wide band channel
X_A, X_B: Diffie-Hellman exponents of devices A and B
$\mathrm{commit}()$ and $\mathrm{open}()$: functions of a commitment scheme

Fig. 2. DH-IR Key Agreement Protocol for I-regions

accuracy is about 0.5ft in practical applications when there are imperfections in the alignment of transceivers.

Radio ranging is more demanding and it requires devices with a high (nanosecond) precision-of-time measurement. To the best of our knowledge, the only commercial technique that achieves such precision, and achieves therefore a high precision-of-distance measurement, is Ultra Wide Band (UWB). In [9], Fontana has demonstrated that with UWB, distances can be measured with an error margin of up to 0.5 ft. Some protocols, e.g., the distance-bounding protocol of Brands and Chaum [4] propose some optimizations through which the cost of nanosecond processing of nodes can be reduced.

In both radio-frequency and ultra-sound solutions, the response time (the XOR operation and the reversion of the transceiver) of the challenged principal must be tightly bound and predictable. With current off-the-self components, ultrasonic ranging seems a more viable implementation of DH-IR and for both techniques the reasonable practical accuracy would be about 0.5ft in a typical use case for I-regions.

3 Usability Evaluation of I-Regions

A pre-requisite to the security as well as usability of I-regions is the ability of human users to correctly gauge and interpret the distance between two communicating devices in relation to the distance shown to them as a result of the I-regions protocol.

We hypothesize that human behavior in interpreting distances would be prone to errors. There are two types of possible errors and following the terminology introduced in [38], we call them: (1) safe errors, and (2) fatal errors. Safe errors occur when a user rejects an authentication attempt from an honest device. This happens if the user believes that the distance shown on her device is more than the (perceived) distance between the two devices. As the name suggests, safe errors might not directly undermine the security of I-regions, however, they have an adverse effect on its efficiency and thus usability. Once rejected, the user needs to re-execute another instance of the I-regions protocol by varying the distance between the two devices. This process needs to repeated iteratively until the user has sufficient confidence that she is indeed communicating with the intended device (and not with an attacker). This will clearly slow down the authentication process. In addition, this will lead to poor usability due to user annoyance and increased user burden. Moreover, in certain communication scenarios, it might not be possible to vary the distance between two devices (e.g., two users wanting to communicate in a meeting room). An adversary could also possibly take advantage of such a situation because a user who gets frustrated due to repeated authentication attempts is likely to accept even an attacked session, thus committing a fatal error (which we explain next).

Fatal errors occur when the user accepts an authentication attempt from an attacking device. This can happen if the user believes that the distance shown on her device is less than or equal to the (perceived) distance between the two intended devices. Fatal errors are clearly dangerous as the user's device will now be communicating with the attacker, even though the user believes her device is communicating with the intended device.

In order to test our hypothesis and to evaluate I-regions, we performed usability experiments. These experiments were simultaneously conducted at two different university campuses: Polytechnic Institute of NYU, USA and University of Split, Croatia.

1. *Efficiency*: time it takes to complete the I-regions method (at the usability layer).
2. *Robustness*: how often the I-regions method leads to safe and fatal errors, with varying inter-device distances.
3. *Usability*: how the method fares in system usability scale [5] and in terms of user confidence in judging distance.

3.1 Testing Apparatus

In our experiments, we used Nokia cell-phones as the testing devices. The models used in the U.S. were N73 and E61 and the model used in Europe was Nokia 6310.[3] We chose to use Nokia cell-phones as they are quite ubiquitous and familiar to many people.

Since our purpose was to test the I-regions method at the usability layer, we chose a simulated test set-up. Our implementation of the I-regions method mock-up was developed to run over the open-source comparative usability testing framework developed by Kostiainen et al. [17] (this framework has previously been used in comparative usability studies of device authentication methods [18]). We used the basic communication primitives as well as automated logging and timing functionalities as provided by this framework.

In terms of user experience, our mock-up closely approximates a real implementation. The two main differences are: (1) our version omits the rounds of the underlying DH-IR protocol, (2) the device only displays the syntactic distance measurement provided by the framework instead of measuring the distance using the packet trip time as in the DH-IR protocol. Notice that the first difference is completely transparent to users as the wireless (and if used, the ultra-sonic) channel is "human-imperceptible." The second difference was necessary to evaluate subjects' ability of comprehending distances and in order to measure resulting error rates.

3.2 Test Cases

We tested the usability of I-regions method with respect to 5 physical distance values, where the actual distances between the devices were set to 1, 2, 3.5, 5 and 6.5 ft. These distances were chosen to capture typical wireless device authentication scenarios. In most situations, the two devices can be within a distance of few feet (e.g., less than 3-4 ft). In some situations, however, it may not be possible to bring the two devices very close to each other (such as in case of a wall-mounted access point or when two users are sitting across a table in a meeting room). For each inter-device distance value, we created a total of 5 test-cases simulating normal scenarios (i.e., when no attacks occur and the maximum distance shown on device's screen is less than or equal to the physical inter-device distance) as well as attack scenarios (i.e., when a MitM attack is simulated

---

[3] See http://europe.nokia.com/phones/n73, http://europe.nokia.com/ A4142101 and http://europe.nokia.com/A4143044, respectively, for the specifications of Nokia phones N73, E61 and 6310.

and the maximum distance shown on device's screen is more that the physical inter-device distance). Two of these test cases simulated normal pairing scenarios, while the remaining three simulated attack scenarios, wherein the "attacker distance bound" (i.e., the simulated distance between attacker's device and user's own device) was kept as 1.5, 2.5 and 3.5 ft more than the inter-device distance (Figure 1 depicts an attack scenario). This was done to estimate safe error rates as well as fatal error rates with the attacker residing/hiding within a reasonable proximity of the two devices.

In our study, we only consider the MitM attack cases where the attacker's physical distance is farther than the intended device's. As explained in section 1, users also have to make sure that there is no other device at any distance less than or equal to the actual distance between the intended devices. We did not test users' ability to perform this task. We believe that such attacks, where the attacker is closer than the intended device, would be rare due to higher risk of detection by the user and attacker exposure.

3.3   Test Procedures

In our experiments, all participants were subject to the following procedures (in the given order):

Background Questionnaire: Subjects were asked to fill out a questionnaire through which they were polled for their age, gender and prior experience with device pairing.

Scenario Presentation: Subjects were asked to imagine that they had to send a confidential file from their smart phone to a co-worker's phone. In order to proceed with the file transfer, they needed to first securely pair the two devices.

Experimentation with the Method: Each subject was provided with a test device. The other device was held by the test administrator. The subject was then asked to perform the following procedure a number of times with varying distances between the two devices being paired.

1. Subject was instructed to move to a fixed test point/location previously marked for him/her by the test administrator. He/she was instructed not to move away from this point throughout the experiment.
2. Subject was then given brief and simple instructions on the I-regions pairing method, both textually on the device and orally by the test administrator.
3. After the test administrator set the physical distance between the subject and the administrator's device to one of the pre-defined distances for a given test case, the subject's device showed a (simulated) value for the lower and upper bound distances. Subject then indicated whether the actual physical distance between his device and test administrator's device was within the shown boundaries by pressing the button labeled with his answer.
4. Test administrator relocated the administrator device according to the next test case.

To avoid order effects (particularly due to learning and fatigue), the sequence of test cases was randomized. Also the distance marks/indicators used by the test administrator to correctly set the physical distance according to different test cases were obscured from test participants.

At the beginning of the experiment, we also provided the participants with the choice of distance measurement unit to be used during the experiment. Participants were given the choice of using either the metric (shown in meter and centimeters) or the British units (shown in feet). This was done in order to personalize the pairing method according to individual participants and facilitate better distance comprehension.

In every run of the experiment, the following measures of observable efficiency and robustness indicators were automatically recorded by the testing software: task performance time, fatal errors (if any) and safe errors (if any).

Post-Test Questionnaire: After completing the experiments, subjects completed the System Usability Scale (SUS) questionnaire [5], a widely used and highly reliable 10-item Likert scale that polls subjects' satisfaction with computer systems [2]. We used the original questions from [5], but replaced "system" with "method". Subjects also rated their confidence level on judging physical distances of 1-10 ft with 0.5 ft accuracy, i.e., the typical practical accuracy level provided by distance bounding techniques (as discussed in 2). This allowed us to measure the usability of I-regions as perceived by our participants.

3.4   Subjects

We recruited a total of 43 subjects for our study.[4] 20 of these users participated in our study at the US venue and the remaining 23 of them took part in our study in Croatia. Most participants were students and staff members from the respective universities that we conducted our test at. The subjects were recruited on a first-come-first-serve basis with no controlling or balancing on subject dependent variables such as age and gender. As a result, our sample consisted of a large fraction (85%) of young subjects in the 18-25 age group and a relatively smaller fraction (15%) belonging to the age group of 26-40. We also had a high proportion (70%) of male participants. An overwhelmingly high fraction (91.5%) of our subjects reported prior experience connecting two wireless devices (in response to one of the questions in the Background Questionnaire) and none of them reported any visual disability.

4   Test Results and Interpretations

As we described previously in Section 3.2, each subject participated in a total of 25 test cases (5 test cases each for 5 different values of inter-device distances). Through our tests, we collected data regarding following measures.

- **Within-subjects measures:** Task performance time, fatal error (categorical) and safe error (categorical).
- **Within-subjects factors:** Test-case varying with respect to (1) physical inter-device distance, (2) the attacker distance bound, and (3) normal or attack scenario.

---

[4] It is well-known that a usability study performed by at least 20 participants captures over 98% of usability related problems [8].

- **Between-subjects measures:** SUS-score, average task performance time, self-confidence ratings for gauging distances.
- **Between-subjects factors:** Age group, gender and prior experience with wireless device authentication methods.

In the remainder of this section, we discuss our results and interpretations. Unless stated otherwise, statistical significance is reported at the 5% level.

4.1   Overview of Results

Before delving into a detailed analysis of logged data, we provide a brief overview. Following observations were made from the initial analysis of the data collected from 43 subjects, each of whom completed 25 test cases.

- The mean task completion time, over all test cases, was 4.93 seconds with standard error of 0.84 seconds. Over the test cases simulating normal scenarios, the mean of task completion time was 5.02 seconds with standard error of 0.82 seconds. Figure 3(a) and 3(b) depict the average task completion time for different inter-device distances, calculated over normal test scenarios and over all test scenarios, respectively. When compared to the completion times for other pairing methods studied in [18,16], we find that I-regions is quite fast for all inter-device distances.
- Depending on the inter-device distance and simulated attacker distance bound, observed fatal error rate ranged from 9.5% to 78.5%. Over all executed test-cases, the observed fatal error rate was 42%. Figure 4 shows the average rate of fatal errors for different test cases. These numbers are alarmingly high, especially when the simulated bound for the attacker distance is close to the inter-device distance, i.e., 1.5 ft and 2.5 ft more than the inter-device distance. Recall that a fatal error leads to a successful MitM attack.
- Over all normal test scenarios, the observed rate of safe errors was 29%. Figure 5 shows the observed average safe error rates for different inter-device distances. Although safe error rates are smaller than fatal errors rates (as observed above), they are still quite high (more than 10% in all cases). We believe that safe error

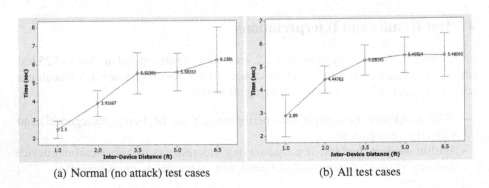

(a) Normal (no attack) test cases   (b) All test cases

Fig. 3. Task Completion Time vs. Inter-Device Distance

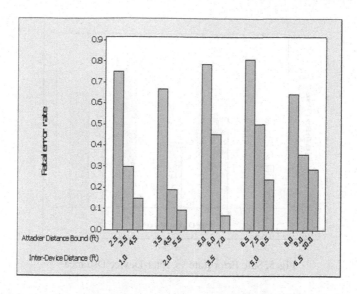

Fig. 4. Fatal Error Rates for Different Test Cases

rates higher than 10% are problematic since such errors undermine the usability, can cause user frustration and eventually lead to fatal errors.

- The mean SUS-score assigned by the subjects was 75 (out of 100) with standard deviation of 12.4. In general, this means that our subjects were reasonably happy with the method and felt that it is easy to use. This can be seen as a positive indication.
- The mean of all participant responses to the last question of the post-test questionnaire, i.e., the self-confidence level in guessing short distances with 0.5 ft accuracy, was 3.24 with standard deviation 1.14. This was rated on a 5-point Likert scale (1=strongly disagree, through 5=strongly agree). This implies that most of our users believed that their comprehension of physical distances was quite up to the mark.

4.2 Within-Subjects Analysis

We analyze the effect of test case on the efficiency and robustness of I-regions. Repeated measures analysis of variance and Chi-square tests revealed that the type of test case has a highly significant effect on task completion time as well as fatal and safe error rates. To better understand the effect of each independent variable of every test case, we look at them individually.

Physical inter-device distance: Analysis of variance revealed that actual physical distance between devices has a significant effect on task completion time. As shown in Figure 3(a), the mean task completion time, for normal scenarios, gradually increases from 2.50 seconds to 6.24 seconds as the inter-device distance increases from 1 ft to 6.5 ft. A similar pattern can be observed in Figure 3(b) for completion time over all test cases. This finding is intuitive as it is easier (and thus faster) to gauge the distance between devices that are closer compared to those that are farther.

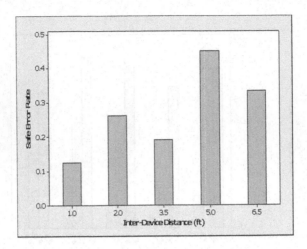

Fig. 5. Safe Error Rate vs. Inter-Device Distance

Chi-square tests revealed that inter-device distance also has an effect on the like-lihood of committing fatal and safe errors. However, the error rates are not directly correlated with inter-device distance as it was in the case of task completion time.

Normal vs. attack scenario: We did not find any significant difference between the mean completion timings of test cases corresponding to normal and attack scenarios.

Simulated attacker distance bound: The difference between the attacker's distance bound and the inter-device distance has a highly significant effect on fatal errors. As the difference increases from 1.5 ft to 3.5ft, the mean proportion of fatal errors drops from 0.73 to 0.17. This pattern can also be observed in Figure 4, irrespective of the inter-device distance. As expected, this means that fatal errors are more likely to occur when the attacker's distance is close to the actual inter-device distance. In other words, human subjects are expected to be less erratic in detecting a distant attacker.

Simulated attacker distance, on the other hand, did not have any significant effect on task completion time.

4.3   Between-Subject Analysis

Effect of gender: In conducted unpaired t-tests, we have not found any significant effect of gender on task completion time, and users' SUS-scores and self-confidence ratings.

According to the results of Chi-square tests, the effect of gender on fatal error rate and safe error rate were also not significant.

Effect of age: Our test sample consisted of subjects belonging to two age groups, namely 18-25 years and 26-40 years. Unpaired t-tests revealed that subjects belong-ing to the age group 26-40 take significantly longer time to complete the task compared to the subjects belonging to the 18-25 group (p=0.037). The means of task completion times were 7.128 and 4.744 seconds, respectively, corresponding to the two age groups. There was no significant effect of age, however, on users' SUS-scores and self-confidence ratings.

Chi-Square tests revealed that age has significant effect on fatal (p=0.041) and safe errors (p=0.038). The rate of making a fatal error was 0.44 for the 18-25 age group and 0.32 for the 26-40 age group. Similarly, the rate of safe errors in 18-25 and 26-40 age groups were 0.31 and 0.18 respectively.

A plausible explanation of the above findings is that our slightly older subjects were more conscious while completing the assigned tasks compared to their younger counterparts. This also helps to explain the higher task completion durations for the older group. Another possible reason could be that older subjects were perhaps more familiar with distance measurements.

Self-confidence in distance judgement: ANOVA tests revealed that participants' self-confidence ratings for accurately judging distance have a significant effect on task completion time. Although there was no obvious linear correlation between self-confidence and the completion, we observed that people with the highest confidence ratings tend to have shorter completion times. People with mid-range confidence ratings took the longest and subjects having the lowest confidence had the highest variance in completion times.

Self-confidence ratings also had a significant effect on fatal and safe errors. In both cases, the proportion of errors (Y-axis) are almost *bell shaped* with respect to the self-confidence ratings (X-axis). However, the variance is observed to be higher for subjects with low confidence ratings. For fatal errors, the mean proportions corresponding to self-confidence levels 1, 3 and 5 were 0.29, 0.49 and 0.34 respectively. For the same self-confidence levels, the respective corresponding mean safe-error proportions were 0.18, 0.30 and 0.11.

When we look at the error rates, the most surprising finding was that the mean error rates for subjects with lowest confidence ratings was smaller than the mean error rates for subjects with mid-range confidence ratings. Although hard to explain, this finding could be partly because some subjects rated themselves higher due to optimism and overconfidence biases. On the other hand, some subjects might also have become over-cautious while answering this question and under-rated their confidence level or performed better than they would normally do due to the observer effect (also known as the Hawthorne Effect [19]. This also helps to explain the higher variance observed within the task completion timings corresponding to the group of subjects with lowest self-confidence ratings.

4.4   Discussion of Combined Measures

A usable wireless device authentication method should perform well in terms of all three (not just one of the) measures, i.e., efficiency (task completion time), robustness (likelihood of committing safe and fatal errors) and usability (user ratings and self-confidence). As our analysis in prior subsections indicate, I-regions is certainly quite efficient and can be considered usable in terms of its SUS-score. However, *in general,*

I-regions has poor robustness, with high likelihood of safe as well as fatal errors. This means that I-regions might not be a practical method for arbitrary values of inter-device distance and attacker distance bound.

On the other hand, since I-regions exhibited, in spite of its manual nature, quite low task completion time and good usability ratings from the participants, we set out to further explore it. We were interested in investigating whether I-regions is robust (for practical purposes) for any specific values for inter-device distance and attacker distance bound. Looking at Figure 4, we find that fatal error rates are on a lower side (less than 10%), when the distance between attacker's device and user's device is 3.5 ft more than the inter-device distance, especially for inter-device distances of 2.0 ft and 3.5 ft. As mentioned previously, an error rate around 10% might be acceptable in practice for certain scenarios. Similarly, looking at Figure 5, mean safe error rate for inter-device distance of 1.0 ft is 12.5%, which might also be an acceptable fraction in practice, especially for scenarios where user can vary (reduce) the inter-device distance prior to re-executing the authentication process in case a safe error occurs.

We wanted to determine values (if any) of inter-device distance and attacker distance bound optimal with respect to safe errors, fatal errors and task completion time (all taken together). To this end, we first set out to check whether our efficiency, robustness and usability measures were independent of one another. Table 1 shows the correlation coefficients and their respective statistical significance (P-values). As shown in the table, none of the measures is sufficiently correlated with others that it could be justifiably omitted. However, it should be noted that the fatal and the safe errors are positively correlated. Although this correlation is modest, it still suggests that the subjects who accepted incorrect distances showed a tendency to reject correct distances (or vice versa).

In terms of efficiency, shorter inter-device distances results in better completion times. However, I-regions was quite efficient in general and completion times were almost always under 8 seconds. Compared to other pairing methods, the time required for I-regions is quite low and completion time differences among various inter-device distances were small. Thus, it is more appropriate to concentrate on the combined effect of fatal and safe errors on I-regions. Figure 6 shows this effect for varying inter-device distances and attacker distance bounds. Clearly, the distance values lying on the lower left are considered better. We can observe that although [inter-device distance, attacker

Table 1. Cross-Correlation of Different Measures

	Average Task Performance Time	SUS-Score	Fatal Error Rate
SUS-Score	-0.242 (0.129)	-	-
Fatal Error Rate	-0.182 (0.249)	0.111 (0.484)	-
Safe Error Rate	0.144 (0.365)	-0.157 (0.332)	0.393 (0.010)

Pearson correlation coefficient.
(P-Value)

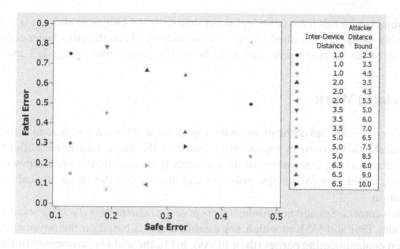

Fig. 6. Mean Fatal Error Rate vs. Mean Safe Error Rate for Different Test Cases

distance bound] values [1 ft, 2.5 ft] and [1 ft, 3.5 ft] have low safe error rates, they yield quite high fatal error rates and are thus not suitable. [1 ft, 4.5 ft] and [3.5 ft, 7ft] are the only tuples with reasonable safe and fatal error rates (although such rates may be still high for many practical applications). We can conclude, therefore, that inter-device distance of 1 ft and 3.5 ft, with the attacker distance bound no less than 4.5 ft and 7 ft, respectively, works the best for I-regions. These values might be suitable for certain pairing scenarios. However, for all other values, I-regions can be deemed impractical.

4.5 Summary of Results

Our (significant) findings can be summarized as follows:

- I-regions exhibits low task completion timings and rated as usable by the test participants.
- Most users were quite confident about their distance judgement ability.
- In general (i.e., for arbitrary values of inter-device distance and attacker distance bound), I-regions shows poor robustness, with high likelihood of users committing both safe as well as fatal errors. However, for some specific values of inter-device distance (1 ft and 3.5 ft) in conjunction with attacker distance bound (at least 4.5 ft and 7 ft), I-regions shows reasonable level of robustness and might be acceptable in practice.
- Fatal errors become less likely as the difference between attacker's distance bound and the inter-device distance increases.
- The task completion time has a tendency to increase as the inter-device distance increases.
- Older subjects (26-40 age group) commit less fatal and safe errors compared to their younger counterparts (18-25 age group). Older subject, on the other hand, took longer to complete the tasks.

– Subjects who felt most confident about their distance judgment abilities (i.e., those with a rating 5) committed less safe errors and completed the tasks faster compared to those having mid-range confidence levels (i.e., those with ratings 2, 3 and 4).

5   Related Work

Providing integrity and authentication over insecure wireless channels is an active area of research. This provision has mainly focused on the key establishment after which the integrity and the authenticity of the messages is ensured by the use of known cryptographic techniques. We review prior work in this area, in the chronological order of publication.

In this context, Stajano and Anderson propose the *resurrecting duckling* security policy model, [36] and [35], in which key establishment is based on the physical contact between communicating parties (their PDAs). In [1], the authors go one step further and relax the requirement that the location limited channel has to be secure against passive eavesdropping; they introduce the notion of a *location-limited channel* (e.g., an infrared link), which is used to exchange pre-authentication data and should be resistant to active attacks.

Another early approach involves image comparison. It encodes a small checksum data calculated over the exchanged data into images and asks the user to compare them on two devices. Prominent examples include "Snowflake" [10], "Random Arts Visual Hash" [26] and "Colorful Flag" [7]. Such methods, however, require both devices to have displays with sufficiently high resolution. A more practical approach, based on SAS protocols [25,20], suitable for simpler displays and LEDs has been investigated in [30] and [28].

More recent work [24] proposed the "Seeing-is-Believing" (SiB) pairing method. In SiB one device encodes a checksum data into a two-dimensional barcode which it displays on its screen and the other device "reads it" using a photo camera, operated by the user. For bidirectional authentication, the same procedure is executed once more with devices changing roles. A related approach has been explored in [31]. Like SiB, it uses the visual out-of-Band (OOB) channel but requires one device to have a continuous visual receiver, e.g., a light detector or a video camera. The other device must have at least one LED. The LED-equipped device transmits OOB data via blinking while the other receives it by recording the transmission and extracting information based on inter-blink gaps. The receiver device indicates success/failure to the user who, in turn, informs the other to accept or abort.

Another recent method is "Loud-and-Clear" (L&C) [11]. It uses the audio (acoustic) OOB channel along with vocalized MadLib sentences which represent the digest of information exchanged over the main wireless channel. There are two L&C variants: "Display-Speaker" and "Speaker-Speaker". In the latter the user compares two vocalized sentences and in the former – displayed sentence with its vocalized counterpart. Some follow-on work (HAPADEP [34,12]) considered pairing devices using only the audio channel. HAPADEP transmits cryptographic protocol messages over audio and requires the user to merely monitor device interaction for any extraneous interference.

Yet another approach: "Button-Enabled Device Authentication (BEDA)" [33,32] suggests pairing devices with the help of user button presses, thus utilizing the tactile OOB channel. This method has several variants: "LED-Button", "Beep-Button", "Vibration-Button" and "Button-Button". In the first two variants, based on the SAS protocol variant [31], the sending device blinks its LED (or vibrates or beeps) and the user presses a button on the receiving device. Each 3-bit block of the SAS string is encoded as the delay between consecutive blinks (or vibrations). As the sending device blinks (or vibrates), the user presses the button on the other device thereby transmitting the SAS from one device to another. In the Button-Button variant, the user simultaneously presses buttons on both devices and random user-controlled inter-button-press delays are used as a means of establishing a common secret using a password based key agreement protocol (e.g., [3]).

There are also other methods which require hardware that is less common. To briefly summarize a few. [15] suggested using ultrasound and [23] suggested using laser as the OOB channel. A very different OOB channel was considered in "Smart-Its-Friends" [13]: a common movement pattern is used to communicate a shared secret to both devices as they are shaken together by the user. A similar approach is taken in "Shake Well Before Use" [22].

A closely related approach to the method tested in this paper is introduced in [39]. Although practical for establishing secure connection between devices that are in very close proximity, [39] lacks the flexibility to accommodate various distances between devices. This limitation is due to the fact that it uses the environmental radio signal noise as the initiating shared secret between devices and the sensed noise is sufficiently similar only within close proximity. Moreover, the security of using radio noise as a location dependent secret is not well studied and currently unknown at best.

An experimental investigation [38] presented the results of a comparative usability study of simple pairing methods for devices with displays capable of showing a few digits. In the "Compare-and-Confirm" approach, the user simply compares two 4-, 6- or 8-digit numbers displayed by devices. In the "Select-and-Confirm" approach, one device displays to the user a set of (4-, 6- or 8-digit) numbers, the user selects the one that matches the number displayed by the other device. In the "Copy-and-Confirm" approach, the user copies a number from one device to the other. The last variant is "Choose-and-Enter" which asks the user to pick a "random" 4-to-8-digit number and enter it into both devices. All methods except "Choose-and-Enter" are based on SAS protocols and the latter is based on password based key agreement protocols e.g., [3].

Quite recently, more comprehensive studies of different pairing methods have been introduced in [18,16] and [14]. In [18], authors selected 13 pairing methods that they deem practical and comparatively investigated the security and usability of them. [16,14] also conducted similar studies but their main focus was usability rather than security. Unfortunately, distance bounding based pairing methods were not included into any of these studies and the usability of such methods left unknown. In this paper, we try to fill this gap left by the previous work and shed light on the usability of distance bounding based pairing methods.

6   Conclusion

In this paper, we presented the results of the first usability study of the I-regions technique. Based on our results, I-regions can be termed quite efficient and it is found to be usable by our subjects. However, in general (i.e., for arbitrary values of inter-device distance and attacker distance bound), I-regions exhibits poor robustness, with high likelihood of users committing both safe as well as fatal errors. This undermines the security of I-regions, either directly (i.e, in case of fatal errors) or indirectly (i.e, in case of safe errors). Thus, we can conclude that I-regions is not a suitable method for all communication scenarios. However, for some specific values of inter-device distance (1 ft or 3.5 ft) in conjunction with attacker distance bound (at least 4.5 ft or 7 ft), I-regions shows reasonable level of robustness and might be acceptable in practice.

Acknowledgements

We are thankful to Arun Kumar and Toni Perkovic for administering our usability studies at our US and Croatian locations, respectively. We also thank Yang Wang for his comments on an earlier version of this paper and CANS'09 anonymous reviewers for their feedback.

References

1. Balfanz, D., Smetters, D., Stewart, P., Wong, H.: Talking to Strangers: Authentication in Ad-Hoc Wireless Networks. In: Proceedings of the 9th Annual Network and Distributed System Security Symposium (NDSS) (2002)
2. Bangor, A., Kortum, P.T., Miller, J.T.: An empirical evaluation of the system usability scale. International Journal of Human-Computer Interaction 24(6), 574–594 (2008)
3. Boyko, V., MacKenzie, P., Patel, S.: Provably Secure Password-Authenticated Key Exchange Using Diffie-Hellman. In: Preneel, B. (ed.) EUROCRYPT 2000. LNCS, vol. 1807, pp. 156–171. Springer, Heidelberg (2000)
4. Brands, S., Chaum, D.: Distance-bounding protocols. In: Workshop on the theory and application of cryptographic techniques on Advances in cryptology, pp. 344–359. Springer-Verlag New York, Inc., Heidelberg (1994)
5. Brooke, J.: SUS: a quick and dirty usability scale. In: Jordan, P.W., Thomas, B., Weerd-meester, B.A., McClelland, A.L. (eds.) Usability Evaluation in Industry, Taylor and Francis, London (1996)
6. Canetti, R., Krawczyk, H.: Analysis of key-exchange protocols and their use for building secure channels. In: Pfitzmann, B. (ed.) EUROCRYPT 2001. LNCS, vol. 2045, pp. 453–474. Springer, Heidelberg (2001)
7. Ellison, C.M., Dohrmann, S.: Public-key support for group collaboration. ACM Transactions on Information and System Security 6(4), 547–565 (2003)
8. Faulkner, L.: Beyond the five-user assumption: Benefits of increased sample sizes in usability testing. Behavior Research Methods, Instruments, & Computers 35(3), 379–383 (2003)
9. Fontana, R.J.: Experimental Results from an Ultra Wideband Precision Geolocation System. Ultra-Wideband, Short-Pulse Electromagnetics (May 2000)
10. Goldberg, I.: Visual Key Fingerprint Code (1996), http://www.cs.berkeley.edu/iang/visprint.c
11. Goodrich, M., et al.: Loud and Clear: Human-Verifiable Authentication Based on Audio. In: International Conference on Distributed Computing Systems (2006)
12. Goodrich, M.T., et al.: Using audio in secure device pairing. International Journal of Security and Networks 4(1), 57–68 (2009)

13. Holmquist, L.E., et al.: Smart-its friends: A technique for users to easily establish connections between smart artefacts. In: Abowd, G.D., Brumitt, B., Shafer, S. (eds.) UbiComp 2001. LNCS, vol. 2201, pp. 116–122. Springer, Heidelberg (2001)
14. Kainda, R., et al.: Usability and security of out-of-band channels in secure device pairing protocols. In: Symposium On Usable Privacy and Security (SOUPS) (2009)
15. Kindberg, T., Zhang, K.: Validating and securing spontaneous associations between wireless devices. In: Information Security Conference, pp. 44–53 (2003)
16. Kobsa, A., et al.: Serial hook-ups: A comparative usability study of secure device pairing methods. In: Symposium On Usable Privacy and Security (SOUPS) (2009)
17. Kostiainen, K., Uzun, E.: Framework for comparative usability testing of distributed applications. In: Security User Studies: Methodologies and Best Practices Workshop (2007)
18. Kumar, A., et al.: Caveat Emptor: A Comparative Study of Secure Device Pairing Methods. In: IEEE International Conference on Pervasive Computing and Communications (PerCom) (2009)
19. Landsberger, H.A.: Hawthorne revisited. Cornell University Press (1968)
20. Laur, S., Nyberg, K.: Efficient mutual data authentication using manually authenticated strings. In: Pointcheval, D., Mu, Y., Chen, K. (eds.) CANS 2006. LNCS, vol. 4301, pp. 90–107. Springer, Heidelberg (2006)
21. Mao, W.: Modern Cryptography, Theory & Practice. Prentice Hall PTR, Englewood Cliffs (2004)
22. Mayrhofer, R., Gellersen, H.-W.: Shake Well Before Use: Authentication Based on Accelerometer Data. In: LaMarca, A., Langheinrich, M., Truong, K.N. (eds.) Pervasive 2007. LNCS, vol. 4480, pp. 144–161. Springer, Heidelberg (2007)
23. Mayrhofer, R., Welch, M.: A Human-Verifiable Authentication Protocol Using Visible Laser Light. In: International Conference on Availability, Reliability and Security (ARES), pp. 1143–1148 (2007)
24. McCune, J.M., Perrig, A., Reiter, M.K.: Seeing-is-believing: Using camera phones for human-verifiable authentication. In: IEEE Symposium on Security and Privacy (2005)
25. Pasini, S., Vaudenay, S.: SAS-Based Authenticated Key Agreement. In: Yung, M., Dodis, Y., Kiayias, A., Malkin, T.G. (eds.) PKC 2006. LNCS, vol. 3958, pp. 395–409. Springer, Heidelberg (2006)
26. Perrig, A., Song, D.: Hash visualization: a new technique to improve real-world security. In: International Workshop on Cryptographic Techniques and E-Commerce (1999)
27. Piontek, H., Seyffer, M., Kaiser, J.: Improving the accuracy of ultrasound-based localisation systems. Personal and Ubiquitous Computing 11(6), 439–449 (2007)
28. Prasad, R., Saxena, N.: Efficient device pairing using Human-comparable synchronized audiovisual patterns. In: Bellovin, S.M., Gennaro, R., Keromytis, A.D., Yung, M. (eds.) ACNS 2008. LNCS, vol. 5037, pp. 328–345. Springer, Heidelberg (2008)
29. Priyantha, N.B., Chakraborty, A., Balakrishnan, H.: The Cricket location-support system. In: Proceedings of the ACM/IEEE International Conference on Mobile Computing and Networking (MobiCom), pp. 32–43. ACM Press, New York (2000)
30. Roth, V., et al.: Simple and effective defense against evil twin access points. In: ACM conference on Wireless network security (WISEC), pp. 220–235 (2008)
31. Saxena, N., et al.: Extended abstract: Secure device pairing based on a visual channel. In: IEEE Symposium on Security and Privacy (2006)
32. Soriente, C., Tsudik, G., Uzun, E.: Secure pairing of interface constrained devices. International Journal of Security and Networks 4(1), 17–26 (2009)
33. Soriente, C., Tsudik, G., Uzun, E.: BEDA: Button-Enabled Device Association. In: International Workshop on Security for Spontaneous Interaction (IWSSI), UbiComp Workshop Proceedings (2007)

34. Soriente, C., Tsudik, G., Uzun, E.: HAPADEP: human-assisted pure audio device pairing. In: Information Security, pp. 385–400 (2008)

35. Stajano, F., Anderson, R.: The Resurrecting Duckling: Security Issues for Ad-hoc Wireless Networks. In: International Workshop on Security Protocols (1999)

36. Stajano, F.: Security for Ubiquitous Computing. John Wiley & Sons, Ltd., Chichester (2002)

37. Suomalainen, J., Valkonen, J., Asokan, N.: Security Associations in Personal Networks: A Comparative Analysis. In: Stajano, F., Meadows, C., Capkun, S., Moore, T. (eds.) ESAS 2007. LNCS, vol. 4572, pp. 43–57. Springer, Heidelberg (2007)

38. Uzun, E., Karvonen, K., Asokan, N.: Usability analysis of secure pairing methods. In: Dietrich, S., Dhamija, R. (eds.) FC 2007 and USEC 2007. LNCS, vol. 4886, pp. 307–324. Springer, Heidelberg (2007)

39. Varshavsky, A., et al.: Amigo: Proximity-Based Authentication of Mobile Devices. In: Krumm, J., Abowd, G.D., Seneviratne, A., Strang, T. (eds.) UbiComp 2007. LNCS, vol. 4717, pp. 253–270. Springer, Heidelberg (2007)

40. Čapkun, S., Čagalj, M.: Integrity regions: authentication through presence in wireless networks. In: WiSe 2006: Proceedings of the 5th ACM workshop on Wireless security (2006)

Short Hash-Based Signatures for Wireless Sensor Networks

Erik Dahmen[1] and Christoph Krauß[2]

[1] Technische Universität Darmstadt
Department of Computer Science
Hochschulstraße 10, 64289 Darmstadt, Germany
dahmen@cdc.informatik.tu-darmstadt.de
[2] Fraunhofer Institute for Secure Information Technology SIT
Department Network Security and Early Warning Systems
Parkring 4, 85748 Garching, Germany
christoph.krauss@sit.fraunhofer.de

Abstract. We present a hash-based signature scheme customized for wireless sensor networks. For message lengths required by instructions or queries from the base station or single measurements like the temperature, signature generation is 7 times faster and verification is 158 times faster than state-of-the-art implementations of ECDSA at the same security level. For message lengths sufficient for most sensor network applications, the signature generation time is comparable to ECDSA, while signature verification remains 20 times faster. Our scheme can be used to authenticate messages exchanged between sensor nodes, as well as for securing broadcast authentication. Our scheme minimizes the overhead introduced in the network by the signature verification done by each sensor before relaying the message.

Keywords: hash-based signature scheme, hash chain, wireless sensor network, Winternitz one-time signature scheme.

1 Introduction

Wireless sensor networks (WSNs) [1] can be deployed in many security- and safety-critical applications such as military surveillance, or medical applications such as patient health monitoring. Thus, securing WSNs is of paramount importance. A WSN can be viewed as a closed user group and therefore the application of symmetric cryptography seems sufficient. However, since sensor nodes are often deployed in an unattended or even hostile environment, an adversary may compromise a sensor node to access stored keys and compromise the security of the communication in the whole group [19].

This problem can be solved by using public key cryptography. For example, one can use the elliptic curve digital signature algorithm (ECDSA) to ensure the authenticity and integrity of the communication in a WSN. ECDSA is especially suited for WSNs because of its small signatures: 320 bit for a sufficiently high

J.A. Garay, A. Miyaji, and A. Otsuka (Eds.): CANS 2009, LNCS 5888, pp. 463–476, 2009.

security level of 80 bits. There exist several implementations of ECDSA on WSNs [16,27,30] providing this security level. Signature generation takes between 0.81 and 2.16 seconds and signature verification takes between 1.62 and 4.32 seconds on typical sensor hardware. This shows that especially the ECDSA signature verification introduces a significant overhead in terms of time and energy required to verify a signature.

This paper presents a hash-based digital signature scheme customized for WSNs that addresses this issue. For the same security level, our scheme provides nearly the same signature sizes as ECDSA. To achieve this, we exploit the property that messages to be signed mainly consist of only a few bits. Our scheme provides a trade-off between the signature generation / verification time and the maximum length of the message to be signed. If only 8-bit messages must be signed, signature generation is 7 times faster and signature verification is 158 times faster than ECDSA. In case of 14-bit messages, the signing time is comparable to ECDSA while signature verification remains 20 times faster. The signature size is 330 bit in both cases. These timings assume that our scheme is implemented on an Atmel ATMega128 [7] microcontroller and that the sensor node generates at most 2^{10} signatures with one key pair.

Being able to efficiently verify signatures is crucial for WSNs, because each node that relays a message should check its authenticity and integrity before doing so. The signature verification time is of main interest especially in broadcast authentication, where the signatures are generated by a powerful base station but signature verification is performed by resource constrained sensor nodes. While 8-bit messages can be used to transmit basic commands and single measurements like the temperature, 14-bit messages are sufficient for most other WSN applications.

Our scheme makes use of a variant of the Winternitz one-time signature scheme [11] and a hash chain for the authentication of the one-time verification keys. Compared to the usual approach of using a Merkle tree [25] for authenticating verification keys, applying a hash chain significantly reduces the signature size because no authentication path must be transmitted. For example, at security level 80, this saves 800 bits if at most 2^{10} signatures are generated. Similar to the Merkle construction, our scheme requires predetermining the number of signatures that can be generated with one key pair. Using a hash chain requires a verifier to receive all signatures generated by the signer, because the current signature is used to compute the authentication data for the next verification key. In our opinion, this is a reasonable assumption for most WSNs. Nevertheless, we discuss measures that can be taken in case a sensor node misses some signatures. Our scheme is similar to the construction proposed in [22]. The difference is that our construction explicitly makes use of a hash chain traversal algorithm. The benefit is, that the signer is no longer required to store all potentially used one-time key pairs to generate the hash chain. This reduces the memory requirements from "linear in the maximum number of signatures" to "logarithmic in the maximum number of signatures". This is one of the reasons why our scheme is also applicable on sensor nodes and not only for broadcast authentication of base station messages. The second novelty is the modified Winternitz construction that provides signature sizes currently only achievable by ECDSA.

The remainder of this paper is organized as follows: Section 2 summarizes related work. In Section 3, the new hash-based signature scheme is introduced. Section 4 analyses the security and Section 5 provides a performance evaluation of the signature scheme. Finally, Section 6 states the authors conclusion.

2   Related Work

Research in the early stages of securing WSNs was predominantly focused on symmetric encryption schemes. For example, the work presented in [26,17,35] uses a Message Authentication Code (MAC) for two party authentication. To establish the required symmetric keys, a variety of key establishment schemes have been proposed [13,6]. However, for the reasons mentioned above, symmetric cryptography is not applicable if the sensors are deployed in a hostile environment.

For this reason, research effort has been shifted on investigating the use of public key cryptography in resource constrained systems and WSNs. The goal is to overcome the key management issues of symmetric schemes and to provide a higher resilience against node compromise [33,15,16,23,32,12,3,4,27,31,2,30]. Public key cryptography is especially useful to authenticate broadcast messages from the base station since all verifying sensor nodes require only the public key of the base station. Gura et al. [16] presented a software implementation of the elliptic curve secp160r1 on an Atmel ATmega128 which is a commonly used CPU on sensor nodes such as the MICA2 [9]. An ECDSA sign operation takes 0.81 seconds and an ECDSA verify operation about 1.62 seconds. Similar values (0.89/1.77 seconds) for the MICA2 nodes are stated by Piotrowski et al. [27]. In [30], an ECC point multiplication using NIST k163 Koblitz curve over $GF(2^{163})$ is stated with 2.16 seconds (using binary field $GF(2^m)$) and 1.27 (using prime field $GF(p)$) on a MICA2 node. This implies that an ECDSA verify operation would take 4.32 (2.54) seconds. In [4], an implementation for $GF(2^{113})$ is presented which takes 6.74 seconds for one fixed point multiplication on a MICA2 node. However, the large execution times and energy requirements raise the question if public key cryptography is really applicable on sensor nodes. Especially the slow ECDSA signature verification poses a problem.

As alternative to public key cryptography, approaches based on symmetric primitives have been proposed. The seminal μTESLA protocol [26] provides an asymmetric mechanism by employing a delayed disclosure of symmetric keys. The drawback of this approach is the introduction of an authentication delay and the requirement for loose time synchronization.

Hash-based one-time signature schemes have the advantage that signature generation and verification are very efficient. At first glance, this makes them very interesting for WSNs. However, they suffer two drawbacks. First, each key pair can be used to sign at most one message. Thus, for each message, a new authentic verification key is required at the receivers side. This problem can be solved using Merkle's tree authentication scheme [25]. This approach uses a binary hash tree, also called Merkle tree, to reduce the validity of an arbitrary but fixed number of one-time verification keys to the validity of a single public key. Using Merkle's approach, the transmission of a potentially large authentication path becomes necessary. The second drawback is that hash-based signature schemes produce quite large signatures. Although the signature size can be reduced using the time-memory trade-off provided by the Winternitz one-time signature scheme [11], signature sizes comparable to ECDSA seem out of reach.

3 Short Hash-Based Signatures

We now describe the new digital signature scheme. It uses a variant of the Winternitz one-time signature scheme to sign the messages and a hash chain to authenticate the one-time verification keys.

Our scheme is parameterized by the three integers n, l, w. The parameter n denotes the security level. The parameter $l \geq 1$ determines the number of signatures that can be generated with one key pair. The parameter $w \geq 2$ denotes the maximum bit length of the messages that can be signed using this key pair, where w must be divisible by 2.

In the following, let $f : \{0,1\}^n \rightarrow \{0,1\}^n$ and $g : \{0,1\}^{4n} \rightarrow \{0,1\}^n$ be hash functions that map bit strings of length n and $4n$ to bit strings of length n, respectively. Further, let $prng : \{0,1\}^n \rightarrow \{0,1\}^n \times \{0,1\}^n$ be a pseudo random number generator that maps an n bit seed to an n bit pseudo random number and an updated seed, i.e. $prng(\text{seed}_{\text{in}}) = (\text{rand}, \text{seed}_{\text{out}})$.

Key pair generation. We first choose an initial seed $\psi_1 \in \{0,1\}^n$ and the end link $z_l \in \{0,1\}^n$ of the hash chain uniformly at random. Next, we generate the one-time signature keys $X_i = (x_i[0], x_i[1], x_i[2]) \in \{0,1\}^{(n,3)}$ for $i = 1, \ldots, l$ using the PRNG and the seed ψ_1.

$$(x_i[0], \psi_i') \leftarrow prng(\psi_i), \ (x_i[1], \psi_i'') \leftarrow prng(\psi_i'), \ (x_i[2], \psi_{i+1}) \leftarrow prng(\psi_i'') \quad (1)$$

Note that the seed is updated each time the PRNG is used. The next step is to compute the one-time verification keys $Y_i = (y_i[0], y_i[1], y_i[2]) \in \{0,1\}^{(n,3)}$ for $i = 1, \ldots, l$ and the hash chain links $z_i \in \{0,1\}^n$ for $i = 0, \ldots, l-1$. We repeat the following steps for $i = l, \ldots, 1$

$$\begin{aligned} y_i[0] \leftarrow f^{2^{w/2}-1}(x_i[0]), \quad & y_i[2] \leftarrow f^{2^{w/2+1}-2}(x_i[2]), \\ y_i[1] \leftarrow f^{2^{w/2}-1}(x_i[1]), \quad & z_{i-1} \leftarrow g(y_i[0] \parallel y_i[1] \parallel y_i[2] \parallel z_i) \end{aligned} \quad (2)$$

Here, $\parallel$ denotes the concatenation of bit strings and $f^k(x)$ means that the function f is applied k times to x.

Initially, the public key is the beginning link z_0 of the hash chain and the private key is the seed ψ_1 and the link z_1. Figure 1 shows how the hash chain is generated.

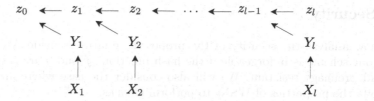

Fig. 1. Visualization of the hash chain

Signature generation. We describe the generation of the ith signature, $i \in \{1,\ldots,l\}$, of the w-bit message $m = m_1 \parallel m_2 \in \{0,\ldots,2^{w/2} - 1\}^2$. The signer knows the seed ψ_i required to generate the ith one-time signature key and the ith link of the hash chain z_i. For the first signature ($i = 1$), the signer knows ψ_1 and z_1 from key pair generation.

At first the signer generates $x_i[0], x_i[1], x_i[2]$ as described in (1). Doing so he obtains the updated seed ψ_{i+1} and stores it in the private key for the next signature $i + 1$. Next, he uses the message $m_1 \parallel m_2$ to compute the checksum

$$c \leftarrow 2^{w/2+1} - 2 - m_1 - m_2 \tag{3}$$

and finally generates $(\alpha_1, \alpha_2, \alpha_3)$, the one-time signature of m.

$$\alpha_1 \leftarrow f^{m_1}(x_i[0]), \quad \alpha_2 \leftarrow f^{m_2}(x_i[1]), \quad \alpha_3 \leftarrow f^c(x_i[2]) \tag{4}$$

In total, the signature of message m is given as $\sigma = (i, \alpha_1, \alpha_2, \alpha_3, z_i)$. After signing, the signer must compute the previous link of the hash chain z_{i+1} which is required for the next signature. This can be accomplished using an arbitrary hash chain traversal algorithm. For our analysis we use the algorithm presented in [34], which requires the computation of $\lceil \frac{1}{2} \log_2 l \rceil$ links in each round and needs to store $\lceil \log_2 l \rceil$ links and seeds.

Signature verification. The verification of a signature $\sigma = (i, \alpha_1, \alpha_2, \alpha_3, z_i)$ of message m works as follows. The verifier knows the $i - 1$th link of the hash chain z_{i-1}. For the first signature ($i = 1$), the link z_0 is the signers public key.

Signature verification consists of two steps. First, the verifier uses $\alpha_1, \alpha_2, \alpha_3$ to recompute the one-time verification key $Y_i = (\beta_1, \beta_2, \beta_3)$. Second, he checks whether the chain link z_{i-1} can be computed using this one-time verification key and the link z_i included in the signature.

$$\begin{aligned} \beta_1 &= f^{2^{w/2}-1-m_1}(\alpha_1), \quad \beta_3 = f^{2^{w/2+1}-2-c}(\alpha_3), \\ \beta_2 &= f^{2^{w/2}-1-m_2}(\alpha_2), \quad g(\beta_1 \parallel \beta_2 \parallel \beta_3 \parallel z_i) \overset{?}{=} z_{i-1} \end{aligned} \tag{5}$$

If the comparison is successful, the signature is accepted as valid. Finally, the verifier discards the link z_{i-1} and stores z_i for the verification of the next signature $i + 1$.

4 Security

We now analyze the security of the proposed signature scheme. We will show that our scheme is unforgeable if the hash functions f and g are preimage and second preimage resistant. We will also consider the case where an adversary exploits the properties of WSNs to perform attacks.

4.1 Security against Preimage and Second Preimage Attacks

The goal of this section is to show that breaking the proposed signature scheme requires breaking cryptographic properties of the used hash functions f and g. These cryptographic properties are preimage and second preimage resistant. For a detailed description of the various cryptographic properties of hash functions we refer to [28]. We begin by considering two attacks and show that performing them successfully requires the attacker to be able to compute preimages or second preimages. Then we discuss the provable security of our scheme.

An attacker has two possibilities attacking our scheme via breaking cryptographic properties of the used hash functions. The first possibility is trying to forge a one-time signature as follows. Assume the attacker is given a one-time signature $(\alpha_1, \alpha_2, \alpha_3)$ of message $m = m_1 \parallel m_2$ and wants to generate a valid one-time signature $\sigma' = (\alpha'_1, \alpha'_2, \alpha'_3)$ of a different message $m' = m'_1 \parallel m'_2$. There are two cases to consider.

Case 1: $(m'_1 < m_1$ or $m'_2 < m_2)$ To obtain a valid one-time signature, the attacker must compute $\alpha'_1 = f^{m'_1 - m_1}(\alpha_1)$ and $\alpha'_2 = f^{m'_2 - m_2}(\alpha_2)$. This requires the attacker to be able to compute preimages of the hash function f, since either $m'_1 - m_1 < 0$ or $m'_2 - m_2 < 0$ holds.

Case 2: $(m'_1 \geq m_1$ and $m'_2 \geq m_2)$ In this case, the signer can easily compute $\alpha'_1 = f^{m'_1 - m_1}(\alpha_1)$ and $\alpha'_2 = f^{m'_2 - m_2}(\alpha_2)$. However, in order to compute $\alpha'_3 = f^{c' - c}(\alpha_3)$, the attacker must be able to compute preimages of the hash function f because, according to Equation (3), the checksum c' of message m' is smaller than the checksum c of message m; $c' - c = m_1 - m'_1 + m_2 - m'_2 < 0$.

If the function f is preimage resistant, this attack cannot be performed. Note, that, as in the Winternitz scheme, the purpose of the checksum is to prevent this attack.

The second possibility an attacker has is to choose his own one-time signature key $X' = (x'[0], x'[1], x'[2]) \in \{0,1\}^{(n,3)}$ which allows him to sign arbitrary messages. The task of the attacker then is to include the corresponding one-time verification key $Y' = (y'[0], y'[1], y'[2]) \in \{0,1\}^{(n,3)}$, computed as in Equation (2), in the hash chain of the signer. He must therefore be able to find $z' \in \{0,1\}^n$, such that

$$g(y'[0] \parallel y'[1] \parallel y'[2] \parallel z') = z_{i-1} = g(y_i[0] \parallel y_i[1] \parallel y_i[2] \parallel z_i)$$

holds for some $i \in \{1, \ldots, l\}$. Here we assume that the values $y_i[0], y_i[1], y_i[2], z_i, z_{i-1}$ are known by the attacker, since he is able to compute them from the

ith signature generated by the signer as shown in Equation (5). An attacker who is able to successfully perform this attack is also able to compute second preimages of the hash function g. In fact, successfully performing this attack is more complicated than computing second preimages, since the first part of the second preimage is determined by the one-time verification key $Y' = (y'[0], y'[1], y'[2])$ and cannot be chosen freely. Hence, this attack cannot be performed if the function g is second preimage resistant.

In summary, based on these attacks we require f to be a preimage resistant hash function and g to be a second preimage resistant hash function. The best generic attack to break these cryptographic properties of hash functions with output length n bit is exhaustive search. In the following, we therefore assume that our scheme provides a security level of n bits, which means that an attacker on average requires 2^{n-1} evaluations of the hash functions to break the scheme.

Estimating the security level based on the best known attacks is sometimes considered dubious. Although we are confident that no attack better than the ones described above exists, there is no guarantee. To guarantee that no better attack exists, a security reduction is required. In hash-based cryptography there exist security reductions for Merkle's tree authentication scheme to the collision resistance of the used hash function and the Lamport–Diffie one-time signature scheme to the preimage resistance of the used hash function [10]. By slightly modifying Merkle's construction, a security reduction to the second preimage resistance of the used hash function is also possible [10]. Since our hash chain based construction can be interpreted as a linearized version of Merkle's tree authentication scheme, these security reductions can be adopted to our scheme. Estimating the security level of our signature scheme based on the security reduction increases the signature size by n bit, since the hash function g is now required to be collision resistant. However, no attack that requires the adversary to be able to compute collisions is known neither for Merkle's tree authentication scheme combined with the Winternitz scheme nor for our signature scheme. We therefore assume that our scheme has security level n bit.

4.2   WSN Specific Attacks

Next, we consider an adversary that tries to exploit some of the special properties of WSNs to perform attacks. Relevant are the wireless multihop communication and the possibility of node compromise.

One threat to WSNs are replay attacks, where an attacker replays previously sent messages, for example to order the sensor nodes to reboot. One great advantage of our scheme is that such attacks are not possible, since a sensor node can detect a replay attack using the index i enclosed in the signature. However, because of the wireless multihop communication, an adversary could perform a successful replay attack by first preventing the reception of messages (e.g., by jamming the wireless channel or by performing a selective forwarding attack [18]) and then replaying valid messages at a later point in time. This is a problem that all signature schemes have in common and is not unique to the scheme presented here. To mitigate this type of attack, timestamps could be added to the signature to create a time window where the message is valid. Another strategy an adversary can pursue is to prevent valid authentications. An adversary can perform Denial-of-Service attacks such as jamming or selective forwarding attacks to prevent the sensor node from receiving messages. This is hard to prevent and a general problem of the wireless multihop communication. However, in the case of our scheme, message loss prevents later messages to be authenticated. This might not only be a result of an attack but could also be caused by the unreliable wireless channel. To enable sensor nodes to resynchronize to the current verification link z_{i-1}, they could send a request to the base station which unicasts the link z_{i-1} to the respective sensor nodes. To secure this communication, a pairwise symmetric key could be used to generate MACs over the exchanged messages. This approach is reasonable if the sensor nodes are located nearby the sink. To decrease the communication overhead for farther located sensor nodes, each sensor node could buffer a certain number of received signatures σ and send them to neighboring nodes if required. Only if neighboring nodes cannot provide the required signatures σ to resynchronize, the base station is contacted.

If the scheme is used to authenticate messages exchanged between sensor nodes, an adversary could try to compromise a sensor node to access all keys and data stored on the node. This would enable the adversary to generate valid signatures originating from this node. However, in contrast to the symmetric key approach, the adversary can only generate signatures from the compromised node and not from other non-compromised sensor nodes. This is a general advantage of public key based signature schemes. One possibility to cope with node compromise is the use of tamper-resistant hardware which may be applicable in certain scenarios [20].

5   Performance Evaluation

This section deals with the performance of our signature scheme. We begin by estimating the cost and memory requirements. Then we explain how the functions f, g and $prng$ are implemented. Finally, we estimate timings and sizes for different parameters and compare our scheme with ECDSA.

5.1   Cost and Memory Requirements

The following formulae show the number of evaluations of f, g, and $prng$ required for computing a hash chain link, key pair generation, signature generation, and verification. Here, c_f, c_g, c_{prng} denotes the cost for one evaluation of $f, g, prng$, respectively. The cost for one link of the hash chain is given by Equations (1) and (2). Key pair generation requires l links to be computed. The initialization of the hash chain traversal algorithm [34] can be done during key pair generation without extra cost. Signing requires the computation of $\lceil \frac{1}{2} \log_2 l \rceil$ links for the traversal algorithm as well as $(2^{w/2+1} - 2)c_f$ and $3c_{prng}$ according to Equations (4) and (1). The verification cost is given by Equation (5).

$$c_{\text{link}} = \left(2^{w/2+2} - 4\right) c_f + c_g + 3\, c_{prng} \tag{6}$$

$$c_{\text{keygen}} = l \cdot c_{\text{link}} \tag{7}$$

$$c_{\text{sign}} = \lceil \tfrac{1}{2} \log_2 l \rceil\, c_{\text{link}} + \left(2^{w/2+1} - 2\right) c_f + 3\, c_{prng} \tag{8}$$

$$c_{\text{verify}} = \left(2^{w/2+1} - 2\right) c_f + c_g \tag{9}$$

The following formulae show the sizes of the signature, private key, and public key. The private key requires $2n$ bits to store the current seed ψ_i and hash chain link z_i as well as $2n\lceil \log_2 l \rceil$ bits to store the $\lceil \log_2 l \rceil$ links and seeds required by the traversal algorithm. The public key consists only of the last link of the hash chain received by the verifier. The signature requires $3n$ bits for the one-time signature, n bits for the hash chain link, and $\log_2 l$ bits for the index i.

$$s_{\text{privkey}} = 2n\left(\lceil \log_2 l \rceil + 1\right)\ \text{bit} \tag{10}$$

$$s_{\text{pubkey}} = n\ \text{bit} \tag{11}$$

$$s_{\text{signature}} = 4n + \log_2 l\ \text{bit} \tag{12}$$

5.2 Construction of f, g, and $prng$

We now describe concrete implementations of the three functions

$$f : \{0,1\}^n \rightarrow \{0,1\}^n \qquad prng : \{0,1\}^n \rightarrow \{0,1\}^n \times \{0,1\}^n$$
$$g : \{0,1\}^{4n} \rightarrow \{0,1\}^n \qquad \text{seed}_{\text{in}} \mapsto (\text{rand}, \text{seed}_{\text{out}}) \tag{13}$$

used in our scheme. For the construction of f and g we use AES in the Matyas-Meyer-Oseas (MMO) mode [24] and truncate the output to n bits. Let $E_k(m)$ denote an AES encryption of message m with key k, let $\oplus$ denote bitwise XOR, and let $\lfloor x \rfloor_n$ denote truncation of the bit string x to n bits. Also, let IV be an initialization vector. Then the functions f and g are constructed as follows.

$$f(x) = \lfloor E_{IV}(x) \oplus x \rfloor_n \qquad \begin{aligned} g(x_1, x_2, x_3, x_4) &= \lfloor E_{k_3}(x_4) \oplus x_4 \rfloor_n \\ \text{with } k_3 &= E_{k_2}(x_3) \oplus x_3, \\ k_2 &= E_{k_1}(x_2) \oplus x_2, \\ k_1 &= E_{IV}(x_1) \oplus x_1 \end{aligned} \tag{14}$$

Before applying the AES encryption, x is padded from the left with $128 - n$ zeroes. For the implementation of the PRNG we use the following construction. due to [14].

$$prng(\psi) = (f(\psi), f(\psi) + \psi + 1 \bmod 2^n) \tag{15}$$

This construction is already used for the signature key generation in implementations of the Merkle signature scheme [5,29].

5.3   Practical Performance

In the following, we estimate timings and sizes of our signature scheme for different maximum message lengths w and number of signatures l. Then we compare our scheme with ECDSA. We compare our scheme exclusively with ECDSA, because ECDSA is the only digital signature scheme that offers the same signature size as our scheme. Other hash-based signature schemes, like the Winternitz scheme, have much larger signature sizes. In the following, we choose $n = 80$ as output length of the hash functions. According to Section 4, this yields a security level comparable to elliptic curves over a 160-bit prime field.

The timings are obtained by first estimating the number of evaluations of f, g, and $prng$ required for the specific choice of parameters using Equations (6)-(9). This number is converted to actual timings using that one AES encryption takes 0.3 msec as measured by the authors of [29] on an Atmel ATMega128 microcontroller. We therefore assume that one evaluation of f, g, and $prng$ takes 0.3, 1.2, and 0.3 msec, respectively

Table 1 summarizes the signature generation and verification time for different choices of parameters. The security parameter n is fixed to 80 to provide a security level comparable to 160-bit elliptic curves. We state timings for up to $l = 2^{10}$ and $l = 2^{16}$ signatures and different maximum message lengths. We don't state timings for more than 2^{16} signatures, because on the one hand 2^{16} signatures are sufficient for typical sensor node applications and on the other hand, updating the private and public key more than 2^{16} times is not possible because this would exceed the life-span of the EEPROM where private and public

Table 1. Signature generation and verification timings of our signature scheme at security level 80 ($n = 80$)

(l, w)	t_{sign}	t_{verify}
$(2^{10}, 8)$	110.4 msec	10.2 msec
$(2^{16}, 8)$	170.7 msec	10.2 msec
$(2^{10}, 10)$	216.0 msec	19.8 msec
$(2^{16}, 10)$	333.9 msec	19.8 msec
$(2^{10}, 12)$	427.2 msec	39.0 msec
$(2^{16}, 12)$	660.3 msec	39.0 msec
$(2^{10}, 14)$	849.6 msec	77.4 msec
$(2^{16}, 14)$	1313.1 msec	77.4 msec
$(2^{10}, 16)$	1694.4 msec	154.2 msec
$(2^{16}, 16)$	2618.7 msec	154.2 msec
$(2^{10}, 18)$	3384.0 msec	307.8 msec
$(2^{16}, 18)$	5229.9 msec	307.8 msec
$(2^{10}, 20)$	6763.2 msec	615.0 msec
$(2^{16}, 20)$	10452.3 msec	615.0 msec
$(2^{10}, 22)$	13521.6 msec	1229.4 msec
$(2^{16}, 22)$	20897.1 msec	1229.4 msec

Table 2. Key and signature sizes of the new signature scheme at security level 80 ($n = 80$)

(l, w)	$s_{\text{signature}}$	s_{privkey}	s_{pubkey}
$(2^{10}, \cdot)$	330 bits	1760 bits	80 bits
$(2^{16}, \cdot)$	336 bits	2720 bits	80 bits

key are stored. Table 1 omits the time required for key pair generation on the sensor node, since it would take far to long. However, from previous timings for the Merkle scheme [5], we can deduce that key pair generation on a PC does take more than a few minutes. Table 2 shows the sizes of the keys and signatures, again for up to $l = 2^{10}$ and $l = 2^{16}$ signatures. Note that these sizes do not depend on the parameter w.

Table 2 shows that signatures of our scheme are almost as small as ECDSA signatures. This is because we need additional 10 to 16 bits to store the index of the signature. Although the size of the private key is large compared to ECDSA, it can be easily stored on the sensor node. Table 1 shows that the signature verification time does not depend on the parameter l. This is especially meaningful in the scenario where our scheme is used for broadcast authentication. Table 1 also clarifies the flexibility of our construction. If a sensor node needs to sign only 8-bit messages, e.g. if it measures only the temperature, our scheme drastically outperforms ECDSA both in signature generation (7 times faster) and verification (158 times faster). For the parameters $(l, w) = (2^{16}, 12)$ and $(l, w) = (2^{10}, 14)$ the signature generation time of our scheme is comparable to ECDSA but signature verification still is more than 20 times faster. If messages of bit length more than 14 have to be signed, signature generation of our scheme is less efficient than ECDSA. However, for message lengths up to 22-bit, signature verification of our scheme remains faster than ECDSA which is again meaningful for broadcast authentication. For the comparison with ECDSA, we use the timings due to Gura et al. [16], i.e. 0.81 seconds for signature generation and 1.62 seconds for signature verification. Finally, we remark that 16384 different messages can be encoded in 14 bits, which should me more than enough for most sensor network applications.

6   Conclusion

We present a hash-based digital signature scheme customized for wireless sensor networks that offers the same signature size as ECDSA at the same security level. We show explicit formulae to estimate the timings and memory requirements of our scheme for a given choice of parameters. Using AES timings measured on an Atmel ATMega128, we show that our scheme provides 7 times faster signature generation and 158 times faster verification than ECDSA for message lengths required by instructions or queries from the base station or single measurements like the temperature. In case of message lengths sufficient for most other WSN applications, the signature generation time is comparable to ECDSA, while signature verification remains 20 times faster.

Future research includes an implementation of our scheme to estimate the exact power consumption of signature generation and verification. In addition, we also plan an implementation on a platform providing hardware acceleration for AES such as the Atmel ATxmega 128A1 processor [8]. Results presented in [29] indicate that this would result in a speed up of our scheme by a factor of more than 3. We will also compare our implementations with hardware accelerated ECC implementations [21].

References

1. Akyildiz, I., Su, W., Sankarasubramaniam, Y., Cayirci, E.: A survey on sensor networks. IEEE Communications Magazine 40(8), 102–114 (2002)
2. Batina, L., Mentens, N., Sakiyama, K., Preneel, B., Verbauwhede, I.: Low-cost elliptic curve cryptography for wireless sensor networks. In: Buttyán, L., Gligor, V.D., Westhoff, D. (eds.) ESAS 2006. LNCS, vol. 4357, pp. 6–17. Springer, Heidelberg (2006)
3. Blaß, E.-O., Junker, H., Zitterbart, M.: Effiziente implementierung von public-key algorithmen für sensornetze. GI Jahrestagung (2), 140–144 (2005)
4. Blaß, E.-O., Zitterbart, M.: Towards acceptable public-key encryption in sensor networks. In: IWUC (2005)
5. Buchmann, J., Coronado, C., Dahmen, E., Döring, M., Klintsevich, E.: CMSS - an improved merkle signature scheme. In: Barua, R., Lange, T. (eds.) INDOCRYPT 2006. LNCS, vol. 4329, pp. 349–363. Springer, Heidelberg (2006)
6. Chan, H., Perrig, A., Song, D.: Random key predistribution schemes for sensor networks. In: Proceedings of the 2003 IEEE Symposium on Security and Privacy (2003)
7. ATMEL Corporation. Atmel atmega128 datasheet (2006),
   http://www.atmel.com/dyn/resources/prod_documents/doc2467.pdf
8. ATMEL Corporation. Atmel atxmega128a1 datasheet (2009),
   http://www.atmel.com/dyn/resources/prod_documents/doc8067.pdf
9. Crossbow Technology Inc. MICA2 datasheet,
   http://www.xbow.com/Products/Product_pdf_files/Wireless_pdf/
   MICA2_Datasheet.pdf
10. Dahmen, E., Okeya, K., Takagi, T., Vuillaume, C.: Digital signatures out of second-preimage resistant hash functions. In: Buchmann, J., Ding, J. (eds.) PQCrypto 2008. LNCS, vol. 5299, pp. 109–123. Springer, Heidelberg (2008)
11. Dods, C., Smart, N., Stam, M.: Hash based digital signature schemes. In: Smart, N.P. (ed.) Cryptography and Coding 2005. LNCS, vol. 3796, pp. 96–115. Springer, Heidelberg (2005)
12. Du, W., Wang, R., Ning, P.: An efficient scheme for authenticating public keys in sensor networks. In: Proceedings of the 6th ACM international symposium on Mobile ad hoc networking and computing (MobiHoc) (2005)
13. Eschenauer, L., Gligor, V.D.: A key-management scheme for distributed sensor networks. In: Proceedings of the 9th ACM conference on Computer and communications security (CCS) (2002)
14. Digital signature standard (DSS). FIPS PUB 186-2 (2007),
   http://csrc.nist.gov/publications/fips/

15. Gaubatz, G., Kaps, J.-P., Sunar, B.: Public key cryptography in sensor networks-revisited. In: Castelluccia, C., Hartenstein, H., Paar, C., Westhoff, D. (eds.) ESAS 2004. LNCS, vol. 3313, pp. 2–18. Springer, Heidelberg (2005)
16. Gura, N., Patel, A., Wander, A., Eberle, H., Shantz, S.C.: Comparing elliptic curve cryptography and RSA on 8-bit cPUs. In: Joye, M., Quisquater, J.-J. (eds.) CHES 2004. LNCS, vol. 3156, pp. 119–132. Springer, Heidelberg (2004)
17. Karlof, C., Sastry, N., Wagner, D.: TinySec: A link layer security architecture for wireless sensor networks. In: Proceedings of the 2nd international conference on Embedded networked sensor systems (SenSys) (2004)
18. Karlof, C., Wagner, D.: Secure routing in wireless sensor networks: attacks and countermeasures. In: Proceedings of the First IEEE International Workshop on Sensor Network Protocols and Applications (2003)
19. Krauß, C., Schneider, M., Eckert, C.: On handling insider attacks in wireless sensor networks. Information Security Technical Report 13, 165–172 (2008)
20. Krauß, C., Stumpf, F., Eckert, C.: Detecting node compromise in hybrid wireless sensor networks using attestation techniques. In: Stajano, F., Meadows, C., Capkun, S., Moore, T. (eds.) ESAS 2007. LNCS, vol. 4572, pp. 203–217. Springer, Heidelberg (2007)
21. Kumar, S.S., Paar, C.: Reconfigurable instruction set extension for enabling ECC on an 8-bit processor. In: Becker, J., Platzner, M., Vernalde, S. (eds.) FPL 2004. LNCS, vol. 3203, pp. 586–595. Springer, Heidelberg (2004)
22. Luk, M., Perrig, A., Whillock, B.: Seven cardinal properties of sensor network broadcast authentication. In: ACM Workshop on Security of Ad Hoc and Sensor Networks (SASN) (2006)
23. Malan, D.J., Welsh, M., Smith, M.D.: A public-key infrastructure for key distribution in tinyos based on elliptic curve cryptography. In: First IEEE International Conference on Sensor and Ad Hoc Communications and Networks (2004)
24. Menezes, A.J., Vanstone, S.A., Van Oorschot, P.C.: Handbook of Applied Cryptography. CRC Press, Boca Raton (1996)
25. Merkle, R.C.: A certified digital signature. In: Brassard, G. (ed.) CRYPTO 1989. LNCS, vol. 435, pp. 218–238. Springer, Heidelberg (1990)
26. Perrig, A., Szewczyk, R., Tygar, J.D., Wen, V., Culler, D.E.: SPINS: security protocols for sensor networks. Wirel. Netw. 8(5), 521–534 (2002)
27. Piotrowski, K., Langendoerfer, P., Peter, S.: How public key cryptography influences wireless sensor node lifetime. In: Proceedings of the fourth ACM workshop on Security of ad hoc and sensor networks (SASN) (2006)
28. Rogaway, P., Shrimpton, T.: Cryptographic hash-function basics: Definitions, implications, and separations for preimage resistance, second-preimage resistance, and collision resistance. In: Roy, B., Meier, W. (eds.) FSE 2004. LNCS, vol. 3017, pp. 371–388. Springer, Heidelberg (2004)
29. Rohde, S., Eisenbarth, T., Dahmen, E., Buchmann, J., Paar, C.: Fast hash-based signatures on constrained devices. In: Grimaud, G., Standaert, F.-X. (eds.) CARDIS 2008. LNCS, vol. 5189, pp. 104–117. Springer, Heidelberg (2008)
30. Szczechowiak, P., Oliveira, L.B., Scott, M., Collier, M., Dahab, R.: NanoECC: Testing the limits of elliptic curve cryptography in sensor networks. In: Verdone, R. (ed.) EWSN 2008. LNCS, vol. 4913, pp. 305–320. Springer, Heidelberg (2008)
31. Uhsadel, L., Poschmann, A., Paar, C.: Enabling full-size public-key algorithms on 8-bit sensor nodes. In: Stajano, F., Meadows, C., Capkun, S., Moore, T. (eds.) ESAS 2007. LNCS, vol. 4572, pp. 73–86. Springer, Heidelberg (2007)

32. Wander, A.S., Gura, N., Eberle, H., Gupta, V., Shantz, S.C.: Energy analysis of public-key cryptography for wireless sensor networks. In: Proceedings of the Third IEEE International Conference on Pervasive Computing and Communications (PERCOM) (2005)
33. Watro, R., Kong, D., Cuti, S.-f., Gardiner, C., Lynn, C., Kruus, P.: TinyPK: securing sensor networks with public key technology. In: Proceedings of the 2nd ACM workshop on Security of ad hoc and sensor networks (SASN) (2004)
34. Yum, D.H., Seo, J.W., Eom, S., Lee, P.J.: Single-layer fractal hash chain traversal with almost optimal complexity. In: CT-RSA 2009. LNCS, vol. 5473, pp. 325–339. Springer, Heidelberg (2009)
35. Zhu, S., Setia, S., Jajodia, S.: LEAP: efficient security mechanisms for large-scale distributed sensor networks. In: Proceedings of the 10th ACM conference on Computer and communications security (CCS) (2003)

Computing on Encrypted Data

Craig Gentry

IBM Research

Abstract. What if you want to query a search engine, but don't want to tell the search engine what you are looking for? Is there a way that you can encrypt your query, such that the search engine can process your query without your decryption key, and send back an (encrypted) response that is well-formed and concise (up to some upper bound on length that you specify)? The answer is yes, if you use a "fully homomorphic" encryption scheme. As another application, you can store your encrypted data in the "cloud", and later ask the server to retrieve only those files that contain a particular (boolean) combination of keywords, without the server being able to "see" either these keywords or your files.

We will present a recent fully homomorphic encryption scheme. In particular, we will highlight the main ideas of the construction, discuss issues concerning the scheme's performance, and mention other applications.

J.A. Garay, A. Miyaji, and A. Otsuka (Eds.): CANS 2009, LNCS 5888, p. 477, 2009.
© Springer-Verlag Berlin Heidelberg 2009

Fully Robust Tree-Diffie-Hellman Group Key Exchange

Timo Brecher[3], Emmanuel Bresson[2], and Mark Manulis[1]

[1] Cryptographic Protocols Group, TU Darmstadt & CASED, Germany
mark@manulis.eu
[2] DCSSI Crypto Lab, Paris, France
emmanuel@bresson.org
[3] INFODAS GmbH, Cologne, Germany
timo.brecher@googlemail.com

Abstract. We extend the well-known Tree-Diffie-Hellman technique used for the design of group key exchange (GKE) protocols with robustness, i.e. with resistance to faults resulting from possible system crashes, network failures, and misbehavior of the protocol participants. We propose a fully robust GKE protocol using the novel tree replication technique: our basic protocol version ensures security against outsider adversaries whereas its extension addresses optional insider security. Both protocols are proven secure assuming stronger adversaries gaining access to the internal states of participants. Our security model for robust GKE protocols can be seen as a step towards unification of some earlier security models in this area.

1 Introduction and Contributions

In *group key exchange (GKE)* protocols, users interact over a network to exchange contributions and finally compute a common group key which is suitable for subsequent cryptographic use. Outsider security encompasses scenarios in which all users are honest and the adversary is an external entity, trying to violate the privacy of the established key (indistinguishability). Outsider security is sufficient for many applications as it protects the communication between trusted users. However, with the increasing group size it is quite natural to assume that some users will not follow the protocol execution in a correct way. Insider security aims to define what security means if some users misbehave. Since the secrecy property of the key in this case becomes vacuous (nothing can prevent insiders from learning and disclosing group keys), the insider security goals usually focus on preventing the dishonest users from disrupting the protocol execution amongst the remaining honest users.

The Tree-Diffie-Hellman GKE protocol from [12,26], called TDH1, achieves security against both outsider and insider attacks. Yet, if any protocol participant fails then this protocol has to abort. In this work we provide two extensions of TDH1 towards **full robustness**. Our first protocol, R-TDH1, preserves the strong **outsider security** of TDH1 (in the standard model), whereas our second protocol, denoted IR-TDH1, combines robustness and **insider security** (in the random oracle model). Our constructions are based on a technical novelty called the *tree replication* technique. R-TDH1 is essentially as efficient as the underlying protocol TDH1, while IR-TDH1 needs some (rather inefficient) NIZK proofs.

J.A. Garay, A. Miyaji, and A. Otsuka (Eds.): CANS 2009, LNCS 5888, pp. 478–497, 2009.
© Springer-Verlag Berlin Heidelberg 2009

1.1 Related Work

Outsider Security In most cases (e.g. [9, 15, 8, 24, 11]), security is defined against an adversary that does not control any player (at least during the target session) resulting in the main requirement called *Authenticated Key Exchange (AKE)* security, which ensures the indistinguishability of the established group key from a random bit string. Usually, AKE-security comes along with forward-secrecy that deals with indistinguishability of keys computed in *earlier* sessions assuming that users can be corrupted at some later stage, and recently it has been updated to include outsider key compromise impersonation attacks [19]. Another requirement called *Mutual Authentication (MA)* has been formalized and studied for the honest users setting in [9, 8].

Insider Security. In 2005, Katz and Shin formalized[1] the notion of security against insider impersonation attacks and key agreement in [23], and Choo et al. defined resistance against unknown-key share attacks [18]. Another goal — *contributiveness* — has been identified in [3, 10, 12, 19] and is also related to non-malleability [21] or key control [31, 34]: briefly speaking, it prevents the adversary from "fixing" the value of the group key computed by honest users; this property states the main difference between key exchange and key transport. In particular, it also prevents key-replication attacks [28].

Robustness. Executions of GKE protocols that do not provide robustness are aborted if some deviation from the given protocol specification is detected by the users, e.g. when users are not able to send or receive messages or if some necessary verification steps fail. Amir et al. [1] were the first to consider robustness in GKE protocols; they merged the non-robust GKE protocol by Steiner et al. [38] with an underlying *group communication system (GCS)* [17]. However their protocol must be restarted in case of failures. Assuming authenticated channels, Cachin and Strobl [14] proposed and formally proved (in the framework of Reactive Simulatability [33]) an asynchronous GKE protocol by combining the GKE protocol by Burmester and Desmedt [5] with an additional *k-resilient consensus protocol* [16, 13]. Their protocol can tolerate only up to $n - 2k$ corruptions to remain forward secure (which is an upper-bound for the asynchronous setting). Desmedt et al. [21] considered an unauthenticated reliable broadcast setting and designed a provably secure scheme immune to outsider and insider attacks based on *verifiable secret sharing (VSS)*. They also explained how to use the authentication compiler by Katz and Yung [24] to sort out invalid messages and tolerate failures. Jarecki et al. [22] followed by Kim and Tsudik [27] used reliable broadcast/multicast setting to design robust GKE protocols proven secure against outsider adversaries. They also defined *full-robustness*: since a GKE protocol requires at least two users, the optimal criterion for robustness is the ability to tolerate up to $n - 2$ (out of n) failed users. We remark that solutions proposed in [14, 21] are not fully robust.

---

[1] First security definitions for insider attacks in non-robust GKE protocols were formalized in [23]. Later, [10, 12] merged these definitions as part of MA-security and formalized contributiveness. Then, [19] updated MA-security with insider key compromise impersonation.

Corruptions and Opening Attacks. Beside formal definitions of outsider and insider security, existing models for GKE protocols differ in the adversarial abilities to corrupt users: *weak corruptions* [9, 24, 21] allow the adversary to obtain users' long-lived keys, but not their internal states, whereas *strong corruptions* [36, 8, 38, 23, 29, 10, 12] reveal both secret types at the same time. The latter gain more and more on attention due to the significant advances in the field of malware and side-channel attacks used to recover information stored locally within hardware and software. Corruptions allow to model the requirement of *(strong) forward secrecy* [9, 8, 11, 12] whose goal is to ensure AKE-security "in the future". Strong corruptions have been recently refined with so-called *opening attacks* [12] that provide higher flexibility: they allow the adversary to get users' ephemeral secrets without necessarily obtaining their long-lived keys. The formal advantage of this refinement is that it can exclude subsequent impersonation attacks on the "opened" users who are then treated as *honest* (rather than *corrupted*). As an illustrative example, AKE-security can be extended to capture the leakage of users' internal states prior to the protocol execution; thus, GKE protocols which pre-compute their ephemeral secrets off-line (for better efficiency) may become insecure.

1.2   Organization of the Paper

In Section 2 we present our *tree replication* technique. Then in Sections 3 and 4, we present the protocols R-TDH1 and IR-TDH1, respectively. The security model is described in Section 5; interestingly, it can be viewed as an extension of [12, 19] towards consideration of (full) robustness, and at the same time as an extension of [21, 22] towards consideration of strong corruptions and opening attacks. In Section 6 we compare security and efficiency of R-TDH1 and IR-TDH1 with some earlier robust GKE protocols.

2   The Tree-Diffie-Hellman Protocol and Our Tree Replication Technique

First, we recall the basic steps of the non-robust protocol TDH1 (see [12, 26] for more details). Then, we introduce at a high-level our tree replication technique that achieves full robustness.

2.1   Overview of Basic TDH1

Preliminaries. The protocol makes use of a *linear* binary tree T_n: a full binary tree with one leaf at each level, except for the deepest one with two leaves, i.e. each node in T_n has a *label* $\langle l, v \rangle$, where $l \in [0, n-1]$ denotes its level and $v \in [0, 1]$ its *position* within that level. For each node $\langle l, v \rangle$, there are two associated values: a secret value, denoted $x_{l,v}$, and a public one, denoted $y_{l,v}$ and computed as $y_{l,v} = g^{x_{l,v}}$. Moreover, each secret value associated to an internal node is the Diffie-Hellman function of the public values associated to its children. In other words, for any l we have:

$$x_{l,0} = \mathsf{DH}(y_{l+1,0}, y_{l+1,1}) = g^{x_{l+1,0}x_{l+1,1}}$$

In order to be able to "chain" such operations, all operations are performed in a cyclic group $\mathbb{G}$ with generator g in which the classical Decisional Diffie-Hellman (DDH) assumption is assumed to be hard and for which there exists an *efficient*, bijective mapping from $\mathbb{G}$ to $\mathbb{Z}_{|\mathbb{G}|}$ (which is not the discrete logarithm!); this bijection is used to consider multiple-decker exponentiations: the result of an exponentiation (an element of $\mathbb{G}$) can be in turn re-interpreted as an exponent in $\mathbb{Z}_{|\mathbb{G}|}$. A suitable group $\mathbb{G}$ of prime order q generated by a quadratic residue g modulo a large safe prime number $p = 2q + 1$ has been described in [25, 26, 12]. In this group if some exponent x is uniform and random in $\mathbb{Z}_q$ then so is g^x in $\mathbb{G}$ and $\mathbb{G} = \mathbb{Z}_q$ (as sets).

Protocol Steps. The protocol is based on the following (intuitive) trick: users are associated to the leaf nodes. Each user U_i knows the secret value associated to its own node, and can reveal the corresponding public value y_i. Using these values, it is easy to check that the users associated to the deepest leaves have enough information to compute all values (both secret and public) associated to the internal nodes. And once all public values are available (that is, including public values associated to internal nodes), it is clear that every user can inductively compute the secret value associated to the root.

- Round 1. User U_i associated to leaf $\langle l, v \rangle$ chooses a secret $x_{l,v}$ and broadcasts $y_{l,v} := g^{x_{l,v}}$;
- Round 2. The goal of the protocol is to let each U_i compute the secret value $x_{0,0}$ associated to the root. Therefore, U_1 assigned to $\langle n - 1, 0 \rangle$ computes a set X_1 of secret values $x_{l,0}$ in its path up to the root $\langle 0, 0 \rangle$. Each $x_{l,0}$ can be seen as the output of the Diffie-Hellman function of $y_{l+1,0}$ and $y_{l+1,1}$. Note that for each internal node $\langle l, 0 \rangle$ user U_1 knows the public value $y_{l+1,1}$ broadcasted by some other user and the secret value $x_{l+1,0}$ by induction at level $l + 1$. Having computed the set X_1 user U_1 broadcasts each $y_{l,0} = g^{x_{l,0}}$. We emphasize, however, that $y_{0,0}$ is never made public —it is not used in the protocol;
- Group Key Derivation. Once user U_1 finishes, all other users $U_{i \neq 1}$ are also able to compute the secret values $x_{l,0}$ in their paths up to the root. Hence, every user finally learns $x_{0,0}$ and uses it to derive the group key.

2.2   The Tree Replication Technique

The original TDH1 protocol is not robust, in particular if the second round message is not delivered then all parties have to abort. In order to achieve robustness, we enhance TDH1 such that, even if some users fail (they halt and/or are not able to continue), the remaining users are still able to compute a common tree structure and to compute a common root secret. This feature is achieved through what we call the *tree replication* technique.

Intuitively, it means that every user is going to compute its own key tree structure, and act as if it would be in the position of U_1. Then, after some users failed, the "deepest" common structure will be used by all users to compute the root secret. At a high level the modification is as follows.

- Round 1. Each user U_i chooses its secret exponent $x_i \in_R \mathbb{G}$ and broadcasts its public value $y_i := g^{x_i}$. After this round all active users (i.e. those who do not fail)

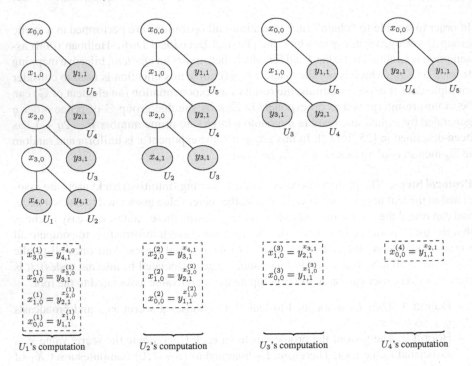

Fig. 1. Tree Replication Technique. Representation of each user's computation in the second protocol round. One of the trees will be used commonly by all active users to derive the group key.

will receive sent public values of other active users. Based on this information, they are assigned to tree leaves;

- Round 2. Each user U_i computes its own set X_i as visualized in Figure 1. Each set X_i is composed of secret values $x_{l,0}^{(i)}$, where the superscript "(i)" indicates a quantity which is computed by U_i only; when we write $x_{l,0}^{(i)}$ we mean that user U_i computes his own value of the variable named "$x_{l,0}$". Users U_i also broadcasts the public values $y_{l,0}^{(i)}$ corresponding to the exponents $x_{l,0}^{(i)}$ in X_i (and with the obvious exclusion of $y_{0,0}^{(i)}$);

- Group Key Derivation. For the computation of the secret root, all users choose the message broadcasted by the lowest-index alive user. Of importance is that all users who are still active choose the same broadcast message and compute the same secret value for the root $\langle 0, 0 \rangle$.

Remarks. Unlike in TDH1, the lowest-indexed user is not necessarily U_1: it can be U_2 if U_1 has failed and so on. Note also that X_n and X_{n-1} are empty sets: if U_{n-1} and U_n are the only remaining users, the protocol reduces to two-party Diffie-Hellman. Yet, U_n and U_{n-1} must still broadcast their "liveness" messages in the second round.

3   A Fully Robust Protocol with Strong *Outsider* Security: R-TDH1

Here we specify R-TDH1 that achieves full robustness while preserving the constant number of rounds and strong outsider security of TDH1 [12]. In what follows we assume that each user U_i has a long-lived key $LL_i = (sk_i, pk_i)$ generated by $\Sigma.\texttt{Gen}(1^\kappa)$, where $\Sigma = (\texttt{Gen}, \texttt{Sign}, \texttt{Verify})$ is an existentially unforgeable signature scheme. By $F := \{\{f_k\}_{k \in \{0,1\}^\kappa}\}_{\kappa \in \mathbb{N}}$ we denote a pseudo-random function ensemble. Our descriptions are provided from the perspective of one session with the initial set of participants $(U_1, \ldots, U_n)$.

Formal Description of R-TDH1. We assume that each U_i is initialized with a *partner id* pid_i encompassing the identities of all users participating in that session. In the beginning of each round each U_i will update own pid_i by removing users that are no longer active based on the messages it received. We assume that if at some stage $\text{pid}_i = \{U_i\}$ then U_i erases every secret information from it internal state $\texttt{state}_i$ and terminates without accepting.

Round 1. Each user U_i does the following:
- $r_i \in_R \{0,1\}^\kappa$; broadcast $U_i|1|r_i$.

Round 2. Each user U_i does the following:
- Remove from pid_i every U_j with missing messages;
- $\texttt{nonces}_i \leftarrow r_1|\ldots|r_{n'}$;
- $x_i \in_R \mathbb{G}; y_i \leftarrow g^{x_i}$;
- $\sigma_i \leftarrow \Sigma.\texttt{Sign}(sk_i, 2|y_i|\texttt{nonces}_i|\text{pid}_i)$;
- Broadcast $U_i|2|y_i|\sigma_i$.

Round 3. Each user U_i does the following:
- Remove from pid_i every U_j with missing or invalid messages;
- Remove nonces of failed oracles from $\texttt{nonces}_i$;
- Assigned remaining oracles to the leaves of T_n, where $n = |\text{pid}_i|$;
- $Y \leftarrow \{y_j\}_{1 \le j \le n}$;
- $x_{n-i,0}^{(i)} \leftarrow x_i$ (renaming);
- For $l = n - i - 1$ downto 0, iteratively compute a set X_i made of values: $x_{l,0}^{(i)} = y_{l+1,1}^{x_{l+1,0}^{(i)}}$;
- For $l = n - i - 1$ downto 1, compute a set $\hat{Y}_i$ made of values: $y_{l,0}^{(i)} = g^{x_{l,0}^{(i)}}$;
- $\sigma_i \leftarrow \Sigma.\texttt{Sign}(sk_i, 3|M|Y|\texttt{nonces}_i|\text{pid}_i)$ where $M = \hat{Y}_i$ if $i < n-1$ and $M =$'alive' if $i \ge n-1$;
- Broadcast $U_i|3|M|\sigma_i$.

Group Key Derivation. Each user U_i does the following:
- Remove from pid_i every U_j with missing messages or invalid signatures;
- Update $\texttt{nonces}_i$ by removing the nonces of failed oracles;
- Determine the *lowest-indexed* oracle U_γ;
- $x_{n-i,0} \leftarrow \begin{cases} y_\gamma^{x_i} & \text{if } i = \gamma + 1 \\ y_{n-i+1,0}^{(\gamma)}{}^{x_i} & \text{if } i > \gamma + 1 \end{cases}$
- For $l = n - i - 1$ downto 0: $x_{l,0} = y_{l+1,1}^{x_{l+1,0}}$
- $k_i \leftarrow f_{x_{0,0}^{(\gamma)}}(v)$;
- Erase every ephemeral secret information from $\texttt{state}_i$ and accept with k_i.

4   Securing R-TDH1 against Strong Insider Attacks

In this section we describe IR-TDH1, an extension of R-TDH1 which provides security against outsider *and* insider attacks. To do so, we use a special NIZK proof for the *equality of the double discrete logarithm and a (single) discrete logarithm* from [37,2].

4.1   NIZK Proof Lg²EqLg

Let $H : \{0,1\}^* \to \{0,1\}^\ell$ be a cryptographic hash function for some ℓ. Let $g, y, \tilde{y}_1, \tilde{y}_2$ be public elements of $\mathbb{G}$. A Non-Interactive Zero-Knowledge Proof for the statement $\log_g(\tilde{y}_1) = \log_y(\log_g(\tilde{y}_2))$ is denoted $\mathrm{Lg^2EqLg}[(x) : \tilde{y}_1 = g^x \wedge \tilde{y}_2 = g^{y^x}]$ and can be constructed as follows using the witness x: for $i \in [1, \ell]$, pick at random α_i in $\mathbb{G}$ and compute $t_{1,i} := g^{\alpha_i}$ and $t_{2,i} := g^{y^{\alpha_i}}$; then output $z := (c, s_1, \ldots, s_\ell)$ with

$$c := H(g|y|\tilde{y}_1|\tilde{y}_2|t_{1,1}|\ldots|t_{1,\ell}|t_{2,1}|\ldots|t_{2,\ell})$$

$$s_i := \begin{cases} \alpha_i & \text{if } c[i] = 0 \quad (c[i] \text{ is the } i\text{-th bit of } c) \\ \alpha_i - x & \text{otherwise.} \end{cases}$$

To verify z one simply checks whether $c \stackrel{?}{=} H(g|y|\tilde{y}_1|\tilde{y}_2|\bar{t}_{1,1}|\ldots|\bar{t}_{1,\ell}|\bar{t}_{2,1}|\ldots|\bar{t}_{2,\ell})$ where

$$\bar{t}_{1,i} := \begin{cases} g^{s_i} & \text{if } c[i] = 0 \\ \tilde{y}_1 g^{s_i} & \text{otherwise} \end{cases} \quad \text{and} \quad \bar{t}_{2,i} := \begin{cases} g^{y^{s_i}} & \text{if } c[i] = 0 \\ \tilde{y}_2^{y^{s_i}} & \text{otherwise.} \end{cases}$$

Using the Random Oracle Model (ROM) [4] one can show that $\mathrm{Lg^2EqLg}$ is secure; we denote by $\mathrm{Adv}^{\mathrm{zk}}_{\mathrm{Lg^2EqLg}}(\kappa)$ the maximum advantage of distinguishing a real proof from a simulated one, and by $\mathrm{Succ}^{\mathrm{snd}}_{\mathrm{Lg^2EqLg}}(\kappa)$ the probability of computing a valid proof for a false statement (*soundness*).

4.2   Description of IR-TDH1

Our protocol is an extension of R-TDH1. We add NIZK proofs in order to prevent corrupted users from sending bad values. This increases the costs of the protocol. Briefly speaking, the protocol is modified in the third round, as follows. In addition to computing X_i (a set of values), each user computes a set Z_i of NIZK proofs $\{z_l^{(i)}\}_{n-i-1 \geq l \geq 1}$:

– for $1 \leq l \leq n - i - 2$, proof $z_l^{(i)}$ proves that

$$x_{l+1,0}^{(i)} = \log_g\left(y_{l+1,0}^{(i)}\right) = \log_{y_{l+1,1}}\left(\log_g\left(y_{l,0}^{(i)}\right)\right)$$

– for $l = n - i - 1$, proof $z_l^{(i)}$ proves that

$$\begin{cases} x_{n-1,0} = \log_g\left(y_{n-1,0}\right) = \log_{y_{n-1,1}}\left(\log_g\left(y_{n-2,0}^{(1)}\right)\right) & \text{if } i = 1 \\ x_{n-i+1,1} = \log_g\left(y_{n-i+1,1}\right) = \log_{y_{n-i,1}}\left(\log_g\left(y_{n-i-1,0}^{(i)}\right)\right) & \text{if } i > 1 \end{cases}$$

We note that Z_{n-1} and Z_n computed by U_{n-1} and U_n are empty. The remaining of the protocol is identical to R-TDH1, however, in the Random Oracle Model, the key derivation is simplified as: $k_i := H'(x_{0,0}^{(\gamma)}|\mathrm{nonces}_i|\mathrm{pid}_i)$, where $x_{0,0}^{(\gamma)}$ is the common secret computed by active users.

5 Security of R-TDH1 and IR-TDH1

5.1 Security Model

Protocol Participants and Execution Model. In order to capture multiple sessions, we model each user U through different *instance oracles* Π_U^s. Then, the *session identifier* is of the form $\mathtt{sid} := U_{i_1}|s_{i_1}|\ldots|U_{i_n}|s_{i_n}$. We say that Π_U^s and $\Pi_{U'}^t$ are *partners* if there exists $\mathtt{sid}$ containing $U|s$ and $U'|t$ as substrings. For each Π_U^s we also define its *partner id* $\mathtt{pid}_U^s$ and the *internal state* $\mathtt{state}_U^s$ as mentioned in the description of the protocols. Once invoked Π_U^s turns into the *processing* stage where it communicates and updates $\mathtt{pid}_U^s$ by removing user identities of oracles that it treats as failed. As long as $|\mathtt{pid}_U^s| > 1$, the oracle is *active* (it continues the execution); at some point, it *accepts* with a *session key* k_U^s and terminates successfully. Otherwise ($|\mathtt{pid}_U^s| \leq 1$ or no acceptance), it terminates with a failure.

Communication. For the security of R-TDH1 and IR-TDH1 we consider a reliable broadcast channel without authentication and any ordering guarantees; this is similar to [21] and less restrictive than [22]. The protocol execution is organized in rounds, which are delimited by a local timer δ (within a round, events are asynchronous): that is participants expect to receive round messages before their timer expires. At the beginning of each round, the *adversary* $\mathcal{A}$ learns each round message to be broadcast. It can then block (refuse to deliver) some of these messages. Additionally, it can inject its own messages. Thus at some point, $\mathcal{A}$ will have a set of messages which it "puts" on the broadcast channel before the timer expires. We model network failures by considering user U as *disconnected* if no expected message containing U as sender's identity is put on the channel. Reliability of the broadcast channel means that all messages put on it are delivered to *all participants that are still connected* in that round. The actual delivery order is determined by the adversary. At the end of the round, each oracle Π_U^s updates its partner id based on the previously received messages: users from whom no message has been delivered are removed. Reliability of the channel implies consistency of updated partner ids.

Adversarial Queries. The adversary $\mathcal{A}$ is modeled as a PPT (*probabilistic polynomial-time Turing machine*), it is assumed to mount its attacks through the following queries:

   Initialize($\mathcal{S}$): for each user in the set $\mathcal{S}$ a new oracle Π_U^s is initialized and the resulting session id $\mathtt{sid}$ is given to $\mathcal{A}$.

   Invoke($\mathtt{sid}, \mathcal{S}'$), assuming that $\mathtt{sid}$ is a valid session id and $\mathcal{S}'$ is a set of initialized oracles ($\mathcal{S}' \subseteq \mathcal{S}$ where $\mathcal{S}$ led to the construction of $\mathtt{sid}$). In response, for each $U \in \mathcal{S}'$ the oracle Π_U^s turns into the processing stage and learns $\mathtt{pid}_U^s = \mathcal{S}$. Then, $\mathcal{A}$ is given the first protocol message m computed by each Π_U^s and the round timer δ is started. The separation between *Initialize* and *Invoke* allows opening attacks against honest oracles prior to the first protocol round. It also allows $\mathcal{A}$ to decide which of these oracles should proceed with the execution. We require that the *Invoke* query can be invoked only once with a given argument.

Broadcast(sid, *m*): In this query the message *m* is supposed to contain the identity of its sender *U*. The *Broadcast* queries for the current round are collected in the order of their occurrence; at the end of the round, message *m* is delivered to all connected oracles Π_U^s in sid (i.e., oracles such that *U* is part of *m* for some collected *Broadcast*(sid, *m*) query). $\mathcal{A}$ can also provide several messages *m* that include the same sender *U*. It is the task of the protocol to determine which of these messages should be processed or dropped. At the end of the round, $\mathcal{A}$ receives messages to be sent by the connected oracles in the next round, and each oracle Π_U^s updates its set pid_U^s according to the protocol specification.

Corrupt(*U*): $\mathcal{A}$ obtains LL_U. This allows impersonation attacks, in which $\mathcal{A}$ can "talk" on the network pretending to be *U*.

AddUser(*U*, Λ), where Λ contains the registration information and a long-lived key LL_U: in response, a new user with that long-lived key is added to $\mathcal{U}$. This query (which is missing in [12]) allows $\mathcal{A}$ to register new users whose behavior it will fully control.

RevealState(Π_U^s): $\mathcal{A}$ obtains ephemeral secrets stored in state_U^s (which may also be empty, if erased). This query models opening attacks [12].

RevealKey(Π_U^s): $\mathcal{A}$ obtains k_U^s (only if Π_U^s has already accepted).

Terminology. We say *U* is *corrupted* or *malicious* if LL_U is known to $\mathcal{A}$, either via *Corrupt*(*U*) or *AddUser*(*U*, Λ); if no such queries have been asked then *U* is *honest*. This terminology also refers to the oracles of *U*. However, an opening attack is not sufficient to make Π_U^s malicious.

Definition of Robust Group Key Exchange. We can now formally specify what a (fully) robust GKE protocol is.

Definition 1 (Robust GKE Protocol). *A robust group key exchange (RGKE) protocol* P *consists of a key generation algorithm* KeyGen, *and a protocol* Setup:

- P.KeyGen(1^κ): *On input a security parameter* 1^κ, *each user is given* LL_U.
- P.Setup($\mathcal{S}$): *On input a set* $\mathcal{S} \subseteq \mathcal{U}$ *a new oracle* Π_U^s *is created for each* $U \in \mathcal{S}$. *A probabilistic interactive protocol is executed between these oracles such that at the end all active oracles (those that have not failed and are still connected) accept with the session group key and terminate.*

P *is* correct *if all active oracles that are honest accept the same session group key.* P *is* fully robust *if it can tolerate all oracles dishonest except two.*

Strong Outsider Security. The *outsider* security of GKE protocols can be expressed through AKE-security. In this section we define its strong version revising the one from [12] to address robustness.

We use the classical query *Test*(Π_U^s) to model AKE-security: in response, a bit *b* is privately flipped and $\mathcal{A}$ is given k_U^s if *b* = 1 or a random string if *b* = 0. The difficulty is in defining how to use this query; the notion of freshness aims at excluding trivial and meaningless attacks. We provide the following four conditions: condition (a) excludes prevents $\mathcal{A}$ from introducing new users; condition (b) allows $\mathcal{A}$ to corrupt some user of the attacked session but $\mathcal{A}$ must remain passive on behalf of that user's oracle until the session is complete, this models outsider key compromise impersonation attacks [19] for robust protocols; condition (c) follows from [12] and allows $\mathcal{A}$ to inspect internal states of participants before and after the attacked session but not during it; condition (d) prevents $\mathcal{A}$ from obtaining the key directly via the *RevealKey* query.

Definition 2. (Oracle Freshness for RGKE) *In a session* sid *of* P *an oracle* Π_U^s *that has accepted is* fresh *if* all *of the following holds:*

(a) *no U' included in* sid *has been added by $\mathcal{A}$ via a corresponding AddUser query,*
(b) *if some U' (incl. $U' = U$) from* sid *has been asked Corrupt(U') prior to the acceptance of Π_U^s then any message m with sender's identity U' asked via a Broadcast(sid, m) query must have been produced by the corresponding oracle $\Pi_{U'}^t$, partnered with Π_U^s,*
(c) *neither Π_U^s nor any of its partners has been asked for a query RevealState before they terminated,*
(d) *neither Π_U^s nor any of its partners is asked for a query RevealKey after having accepted and terminated.*

Definition 3. (Strong AKE-Security for RGKE) *Let* P *be a correct RGKE protocol and b a uniformly chosen bit. Consider an adversary $\mathcal{A}$ against the AKE-security of* P. *We define the adversarial game* $\mathsf{Game}_{\mathcal{A},\mathrm{P}}^{\mathsf{ake}-b}(\kappa)$ *as follows:*

- *$\mathcal{A}$ interacts via queries;*
- *at some point $\mathcal{A}$ asks a Test query to an oracle Π_U^s which is (and remains) fresh;*
- *$\mathcal{A}$ continues interacting via queries;*
- *when $\mathcal{A}$ terminates, it outputs a bit, which is set as the output of the game.*

We define: $\mathsf{Adv}_{\mathcal{A},\mathrm{P}}^{\mathsf{ake}}(\kappa) := \left| 2\Pr[\mathsf{Game}_{\mathcal{A},\mathrm{P}}^{\mathsf{ake}-b}(\kappa) = b] - 1 \right|$

and denote with $\mathsf{Adv}_{\mathrm{P}}^{\mathsf{ake}}(\kappa)$ *the maximum advantage over all PPT adversaries $\mathcal{A}$. We say that a RGKE protocol* P *provides* strong AKE-security *if this advantage is negligible.*

Strong Insider Security. We now revisit the strong *insider* security definitions, that is MA-security and contributiveness, from [12] to address robustness.

In the next definition, condition (a) models robustness since it requires that every honest, non-failed participant accepts, provided there exists other participants that have not failed as well. In condition (b) we model mutual authentication in the sense that no user accepts the group key until it is assured of the active participation of the other users; this takes into account insider key compromise impersonation attacks [19] as $\mathcal{A}$ can obtain the long-lived key of a user, as long as it remains passive with respect to that user's oracle. Finally, condition (c) models key confirmation and requires that session group keys accepted by any two participants are identical.

Definition 4. (Strong MA-Security for RGKE) *Let* P *a correct RGKE protocol and* $\mathcal{A}$ *an adversary who is allowed to query Initialize, Invoke, Broadcast, AddUser, Corrupt, RevealKey and RevealState. We denote this interaction as* $\mathrm{Game}^{\mathrm{ma}}_{\mathcal{A},\mathrm{P}}(\kappa)$. *We say that* $\mathcal{A}$ *wins in some session* sid *if at the end of that session* **one** *of the following conditions is satisfied:*

(a) *there is an honest oracle* Π^s_U *(with* $U|s$ *part of* sid*) which terminated without having accepted some key but for which other partners exist (i.e.,* $|\mathrm{pid}^s_U| > 1$*),*

(b) *there are two partnered oracles* Π^s_U *and* $\Pi^t_{U'}$, *such that* Π^s_U *has accepted and* $U' \in \mathrm{pid}^s_U$ *is uncorrupted but* $\Pi^t_{U'}$ *has not been invoked via Invoke*(sid, $\cdot$)*,*

(c) *there are two honest partnered oracles* Π^s_U *and* $\Pi^t_{U'}$, *which have accepted and* $k^s_U \neq k^t_{U'}$.

The maximum probability of this event is denoted $\mathrm{Succ}^{\mathrm{ma}}_{\mathrm{P}}(\kappa)$; *we say that a RGKE protocol* P *provides* strong MA-security *if this probability is negligible.*

The following requirement of strong contributiveness resists key control attacks by which a malicious subset of (at most $n-1$) users aims to predetermine the resulting value of the group key [34]. This is in contrast to the *non-malleability* property [21], which ensures uniform distribution of group keys in the presence of malicious participants but in a weaker model in which no opening attacks exist (as discussed in [12]).

Definition 5. (Strong Contributiveness for RGKE) *Let* P *be a correct RGKE protocol and* $\mathcal{A}$ *an adversary operating in two stages (*prepare *and* attack*) and having access to the queries Initialize, Invoke, Broadcast, AddUser, Corrupt, RevealKey and RevealState. We define the following game* $\mathrm{Game}^{\mathrm{con}}_{\mathcal{A},\mathrm{P}}(\kappa)$*:*

– $\mathcal{A}$(prepare) *interacts via queries and outputs some* $\tilde{k} \in \{0,1\}^\kappa$, *and some state information* ζ;

– *A set* Ψ *is built, consisting of all session ids* sid *for which a query Invoke*(sid, $\mathcal{S}'$) *has been asked during the* prepare *stage;*

– $\mathcal{A}$(attack, ζ) *continues interacting via queries and outputs some oracle identifier* $U|s$.

The adversary $\mathcal{A}$ *wins in* $\mathrm{Game}^{\mathrm{con}}_{\mathcal{A},\mathrm{P}}(\kappa)$ *if* **all** *of the following holds:*

(a) Π^s_U *is honest, has terminated accepting* $\tilde{k}$, *and there is no* sid $\in \Psi$ *which contains its identifier* $U|s$.

(b) *There are at most* $n-1$ *corrupted oracles that are partnered with* Π^s_U.

$$\text{We define:} \quad \mathrm{Succ}^{\mathrm{con}}_{\mathcal{A},\mathrm{P}}(\kappa) := \Pr[\mathcal{A} \text{ wins in } \mathrm{Game}^{\mathrm{con}}_{\mathcal{A},\mathrm{P}}(\kappa)]$$

and denote with $\mathrm{Succ}^{\mathrm{con}}_{\mathrm{P}}(\kappa)$ *the maximum probability of this event over all PPT adversaries* $\mathcal{A}$; *we say* P *provides* strong contributiveness *if this probability is negligible in* κ.

Since Ψ contains identifiers of sessions that have been invoked during the prepare stage, the requirement that no sid $\in \Psi$ should contain $U|s$ excludes a trivial attack by which $\mathcal{A}$ chooses $\tilde{k}$ as a key computed in some session invoked during the prepare stage.

5.2 Security Results

The following theorems proven in appendix show that R-TDH1 is secure against strong outsiders and that IR-TDH1 is additionally secure against strong insider attacks.

As the underlying number-theoretic assumption we use the well-known Square-Exponent Decisional Diffie-Hellman (SEDDH) assumption [35, 22], i.e. the following probability is assumed to be negligible:

$$\text{Adv}_{\mathbb{G}}^{\text{SEDDH}}(\kappa) =_{\mathcal{A}'}^{\max} \left| \Pr_a \left[\mathcal{A}'(g, g^a, g^{a^2}) = 1 \right] - \Pr_{a,b} \left[\mathcal{A}'(g, g^a, g^b) = 1 \right] \right|.$$

Theorem 1. *If Σ is existentially unforgeable under chosen message attacks, if F is pseudo-random, and $\mathbb{G}$ is SEDDH-hard then R-TDH1 provides strong AKE-security, and*

$$\text{Adv}_{\text{R-TDH1}}^{\text{ake}}(\kappa) \leq \frac{Nq_s^2}{2^{\kappa-1}} + 2N^2\text{Succ}_{\Sigma}^{\text{euf-cma}}(\kappa) + q_sN^2\text{Adv}_{\mathbb{G}}^{\text{SEDDH}}(\kappa) + 2q_s\text{Adv}_F^{\text{prf}}(\kappa).$$

Theorem 2. *If Σ is existentially unforgeable under chosen message attacks, Lg^2EqLg is zero-knowledge, and $\mathbb{G}$ is SEDDH-hard then IR-TDH1 provides strong AKE-security in ROM, and*

$$\text{Adv}_{\text{IR-TDH1}}^{\text{ake}}(\kappa) \leq \frac{Nq_s^2}{2^{\kappa-1}} + 2N\text{Succ}_{\Sigma}^{\text{euf-cma}}(\kappa) + q_sN^2\left(\text{Adv}_{\text{Lg}^2\text{EqLg}}^{\text{zk}}(\kappa) + \text{Adv}_{\mathbb{G}}^{\text{SEDDH}}(\kappa)\right).$$

Theorem 3. *If Σ is existentially unforgeable under chosen message attacks and if Lg^2EqLg is sound then IR-TDH1 provides strong MA-security in ROM, and*

$$\text{Succ}_{\text{IR-TDH1}}^{\text{ma}}(\kappa) \leq \frac{Nq_s^2}{2^{\kappa}} + N^2\text{Succ}_{\Sigma}^{\text{euf-cma}}(\kappa) + \frac{q_sN^2}{2}\text{Succ}_{\text{Lg}^2\text{EqLg}}^{\text{snd}}(\kappa).$$

Theorem 4. *IR-TDH1 provides strong contributiveness in ROM, and*

$$\text{Succ}_{\text{IR-TDH1}}^{\text{con}}(\kappa) \leq \frac{Nq_s^2 + 2Nq_s + q_{H'}^2}{2^{\kappa}}.$$

6 Comparison with Prior Work

At a glance, Table 1 describes how R-TDH1 and IR-TDH1 fit into the current state of the art of provably secure RGKE protocols in terms of security, robustness and complexity: (i) we indicate whether AKE-security, MA-security and contributiveness (CON) is achieved, and for which strength of corruptions, (ii) we indicate the maximum of users that may fail without disrupting the protocol execution (fully robust protocols have robustness of $n - 2$), and (iii) we compare the broadcast complexity and the total number of operations *per* user.

To ensure fair comparison we adopt [14] to the reliable broadcast setting as described in [22] and add authentication costs to non-authenticated protocols from [21, 22] based on the technique from [24].

We highlight that R-TDH1 and the reliable broadcast version of [14] are the only RGKE protocols that have been formally proven to achieve strong outsider security.

Table 1. Security, Robustness, and Complexity of R/IR-TDH1 and other *Robust* GKE Protocols

RGKE Prot.	Out-/Insider Security				Robustness	Complexity		
	AKE	MA	CON	Model	max Faults $(k \leq)$	Rounds	Broadcast	Ops
adopted [14]	strong	-	-	STD	$n - 2$	2	$O(n^2)$	$O(n)$
[21]	weak	weak	weak	STD	$n/2 - 1$	7	$O(nk)$	$O(n)$
BD-RGKA [22]	weak	-	-	STD	$n - 2$	2	$O(n^3)$	$O(n^2)$
RGKA [22]	weak	-	-	STD	$n - 2$	2	$O(n^2)$	$O(n)$
t-RGKA [22]	weak	-	-	STD	$2t - 1$	2	$O(nt)$	$O(t)$
RGKA' [22]	weak	-	-	STD	$n - 2$	$O(\delta)$	$O(n \log n)$	$O(n)$
R-TDH1	strong	-	-	STD	$n - 2$	3	$O(n^2)$	$O(n)$
IR-TDH1	strong	strong	strong	ROM	$n - 2$	3	$O(n^2 l)$	$O(nl)$
TDH1 [12]	strong	strong	strong	STD	0	3	$O(n)$	$O(n)$

The out-/insider security entries reflect the formally proven properties of the protocols, though it might be possible to amend the protocols from [14, 21, 22] to achieve strong outsider and insider security using techniques that are close to those proven secure for R-TDH1 and IR-TDH1.

The protocols from [22] are proven under consideration of weak corruptions only. In terms of complexity and robustness R-TDH1 is similar to both RGKA and the modified version of [14].

We proved that IR-TDH1 provides strong outsider and insider security, while [22, 14] did not address insider security, and [21] did not consider strong corruptions. Compared to R-TDH1, IR-TDH1 has loss in the broadcast and computation complexity by factor $O(l)$ where l ranges from the number of users that do not fail $(n - k)$ to 1. Finally, we notice that compared to the original TDH1 from [12] the use of our tree replication technique achieves full robustness but increases the communication complexity by factors $O(n)$ and $O(nl)$, respectively.

7   Conclusion

This paper introduced two fully robust versions of the Tree-Diffie-Hellman protocol TDH1 from [12] based on our novel tree replication technique. R-TDH1 preserves the strong outsider security of the original protocol, whereas IR-TDH1 presents the first construction of a fully robust GKE protocol that remains resilient to strong insider attacks. We proved both protocols in a security model which is of independent interest as it combines strengths of several previous modeling approaches.

As mentioned, some existing robust GKE protocols can also be modified to achieve insider security using NIZK proofs, however, this would require random oracles as well. Hence, designing a fully robust GKE protocol with strong outsider and insider security in the standard model remains an interesting open problem.

References

1. Amir, Y., Nita-Rotaru, C., Schultz, J.L., Stanton, J.R., Kim, Y., Tsudik, G.: Exploring Robustness in Group Key Agreement. In: Proc. of ICDCS 2001, pp. 399–408. IEEE CS, Los Alamitos (2001)

2. Ateniese, G., Song, D.X., Tsudik, G.: Quasi-Efficient Revocation in Group Signatures. In: Blaze, M. (ed.) FC 2002. LNCS, vol. 2357, pp. 183–197. Springer, Heidelberg (2003)
3. Ateniese, G., Steiner, M., Tsudik, G.: Authenticated Group Key Agreement and Friends. In: Proc. of ACM CCS 1998, pp. 17–26. ACM Press, New York (1998)
4. Bellare, M., Rogaway, P.: Random Oracles are Practical: A Paradigm for Designing Efficient Protocols. In: Proc. of ACM CCS 1993, pp. 62–73. ACM Press, New York (1993)
5. Burmester, M., Desmedt, Y.: A secure and efficient conference key distribution system. In: De Santis, A. (ed.) EUROCRYPT 1994. LNCS, vol. 950, pp. 275–286. Springer, Heidelberg (1995)
6. Boyd, C., Mathuria, A.: Protocols for Authentication and Key Establishment. Springer, Heidelberg (2003)
7. Bresson, E., Chevassut, O., Pointcheval, D.: Provably Authenticated Group Diffie-Hellman Key Exchange — The Dynamic Case. In: Boyd, C. (ed.) ASIACRYPT 2001. LNCS, vol. 2248, pp. 290–390. Springer, Heidelberg (2001)
8. Bresson, E., Chevassut, O., Pointcheval, D.: Dynamic Group Diffie-Hellman Key Exchange under Standard Assumptions. In: Knudsen, L.R. (ed.) EUROCRYPT 2002. LNCS, vol. 2332, pp. 321–336. Springer, Heidelberg (2002)
9. Bresson, E., Chevassut, O., Pointcheval, D., Quisquater, J.-J.: Provably Authenticated Group Diffie-Hellman Key Exchange. In: ACM CCS 2001, pp. 255–264. ACM Press, New York (2001)
10. Bresson, E., Manulis, M.: Malicious Participants in Group Key Exchange: Key Control and Contributiveness in the Shadow of Trust. In: Xiao, B., Yang, L.T., Ma, J., Muller-Schloer, C., Hua, Y. (eds.) ATC 2007. LNCS, vol. 4610, pp. 395–409. Springer, Heidelberg (2007)
11. Bresson, E., Manulis, M., Schwenk, J.: On Security Models and Compilers for Group Key Exchange Protocols. In: Miyaji, A., Kikuchi, H., Rannenberg, K. (eds.) IWSEC 2007. LNCS, vol. 4752, pp. 292–307. Springer, Heidelberg (2007)
12. Bresson, E., Manulis, M.: Securing Group Key Exchange against Strong Corruptions. In: Proc. of ASIACCS 2008, pp. 249–261. ACM Press, New York (2008)
13. Cachin, C., Kursawe, K., Shoup, V.: Random Oracles in Constantinople: Practical Asynchronous Byzantine Agreement using Cryptography. In: Proc. of PODC 2000, pp. 123–132. ACM Press, New York (2000)
14. Cachin, C., Strobl, R.: Asynchronous Group Key Exchange with Failures. In: Proc. of PODC 2004, pp. 357–366. ACM Press, New York (2004)
15. Canetti, R., Krawczyk, H.: Analysis of Key-Exchange Protocols and Their Use for Building Secure Channels. In: Pfitzmann, B. (ed.) EUROCRYPT 2001. LNCS, vol. 2045, pp. 453–474. Springer, Heidelberg (2001)
16. Canetti, R., Rabin, T.: Fast Asynchronous Byzantine Agreement with Optimal Resilience. In: STOC 1993, pp. 42–51. ACM Press, New York (1993)
17. Chockler, G.V., Keidar, I., Vitenberg, R.: Group Communication Specifications: A Comprehensive Study. ACM Computing Surveys 33(4), 427–469 (2001)
18. Choo, K.-K.R., Boyd, C., Hitchcock, Y.: Examining Indistinguishability-Based Proof Models for Key Establishment Protocols. In: Roy, B. (ed.) ASIACRYPT 2005. LNCS, vol. 3788, pp. 585–604. Springer, Heidelberg (2005)
19. Choudary Gorantla, M., Boyd, C., González Nieto, J.M.: Modeling Key Compromise Impersonation Attacks on Group Key Exchange Protocols. In: PKC 2009. LNCS, vol. 5443, pp. 105–123. Springer, Heidelberg (2009)
20. Crescenzo, G.D., Ferguson, N., Impagliazzo, R., Jakobsson, M.: How to Forget a Secret. In: Meinel, C., Tison, S. (eds.) STACS 1999. LNCS, vol. 1563, pp. 500–509. Springer, Heidelberg (1999)
21. Desmedt, Y.G., Pieprzyk, J., Steinfeld, R., Wang, H.: A Non-Malleable Group Key Exchange Protocol Robust Against Active Insiders. In: Katsikas, S.K., López, J., Backes, M., Gritzalis, S., Preneel, B. (eds.) ISC 2006. LNCS, vol. 4176, pp. 459–475. Springer, Heidelberg (2006)

22. Jarecki, S., Kim, J., Tsudik, G.: Robust Group Key Agreement using Short Broadcasts. In: Proc. of ACM CCS 2007, pp. 411–420. ACM, New York (2007)
23. Katz, J., Shin, J.S.: Modeling Insider Attacks on Group Key Exchange Protocols. In: Proc. of ACM CCS 2005, pp. 180–189. ACM Press, New York (2005)
24. Katz, J., Yung, M.: Scalable Protocols for Authenticated Group Key Exchange. In: Boneh, D. (ed.) CRYPTO 2003. LNCS, vol. 2729, pp. 110–125. Springer, Heidelberg (2003)
25. Kim, Y., Perrig, A., Tsudik, G.: Group Key Agreement Efficient in Communication. IEEE Transactions on Computers 53(7), 905–921 (2004)
26. Kim, Y., Perrig, A., Tsudik, G.: Tree-Based Group Key Agreement. ACM Transactions on Information and System Security 7(1), 60–96 (2004)
27. Kim, J., Tsudik, G.: Survival in the Wild: Robust Group Key Agreement in Wide-Area Networks. In: ICISC 2008. LNCS, vol. 5461, pp. 66–83. Springer, Heidelberg (2008)
28. Krawczyk, H.: HMQV: A High-Performance Secure Diffie-Hellman Protocol. In: Shoup, V. (ed.) CRYPTO 2005. LNCS, vol. 3621, pp. 546–566. Springer, Heidelberg (2005)
29. LaMacchia, B., Lauter, K., Mityagin, A.: Stronger Security of Authenticated Key Exchange. In: Susilo, W., Liu, J.K., Mu, Y. (eds.) ProvSec 2007. LNCS, vol. 4784, pp. 1–16. Springer, Heidelberg (2007)
30. Manulis, M.: Provably Secure Group Key Exchange. PhD thesis, Ruhr Univ. Bochum (2007)
31. Mitchell, C.J., Ward, M., Wilson, P.: Key Control in Key Agreement Protocols. Electronic Letters 34(10), 980–981 (1998)
32. Pedersen, T.P.: Non-interactive and Information-Theoretic Secure Verifiable Secret Sharing. In: Feigenbaum, J. (ed.) CRYPTO 1991. LNCS, vol. 576, pp. 129–140. Springer, Heidelberg (1992)
33. Pfitzmann, B., Waidner, M.: A Model for Asynchronous Reactive Systems and its Application to Secure Message Transmission. In: IEEE S&P 2001, pp. 184–200. IEEE CS, Los Alamitos (2001)
34. Pieprzyk, J., Wang, H.: Key Control in Multi-party Key Agreement Protocols. In: CCC/PCS 2003, vol. 23 (2003)
35. Sadeghi, A.-R., Steiner, M.: Assumptions Related to Discrete Logarithms: Why Subtleties Make a Real Difference. In: Pfitzmann, B. (ed.) EUROCRYPT 2001. LNCS, vol. 2045, pp. 244–261. Springer, Heidelberg (2001)
36. Shoup, V.: On Formal Models for Secure Key Exchange (Version 4). TR-RZ 3120, IBM Research (November 1999)
37. Stadler, M.: Publicly Verifiable Secret Sharing. In: Maurer, U.M. (ed.) EUROCRYPT 1996. LNCS, vol. 1070, pp. 90–99. Springer, Heidelberg (1996)
38. Steiner, M.: Secure Group Key Agreement. PhD thesis, Saarland University (2002)

A   Security Proofs

In our proofs, similar to [8, 12], we assume that $\Sigma.\mathsf{Sign}$ is executed under the same protection mechanism as sk_i so that any randomness used to compute the signature will not be revealed in response to a *RevealState* query.

A.1   Proof of Strong AKE-Security of R-TDH1 (Theorem 1)

Proof (Sketch). We define a sequence of games: $\mathbf{G}_i$, $i = 0, \ldots, 5$ (whereby $\mathbf{G}_4$ is a sequence of $n-1$ hybrid games where n is the number of invoked oracles in the attacked session) with the adversary $\mathcal{A}$ against the strong AKE-security of R-TDH1. In each game $\mathsf{Win}_i^{\mathsf{ake}}$ denotes the event that the bit b' output by $\mathcal{A}$ is identical to the randomly chosen bit b in Game $\mathbf{G}_i$.

Game G_0. This is the real game $\text{Game}_{A,\text{R-TDH1}}^{\text{ake}-b}(\kappa)$ where a *simulator* Δ simulates the execution of the protocol and answers all queries of A.

Recall that the $\textit{Test}(U_i|s)$ query is asked to a fresh oracle $\Pi_{U_i}^s$ which has accepted, and that A then receives either a random string or a session group key $k_{U_i}^s$. Our definition of the oracle freshness restricts A from legal participation in the attacked session and from opening oracles that are partnered with $\Pi_{U_i}^s$ until these oracles terminate (having erased ephemeral secrets as required in R-TDH1).

Game G_1. This game is identical to Game G_0 except that Δ aborts and b' is set at random if any two honest oracles identified by $U|s$ and $U|s'$ that have been invoked for two different sessions choose the same nonce in Round 1. Since there are at most N users and at most q_s sessions we get $|\Pr[\text{Win}_1^{\text{ake}}] - \Pr[\text{Win}_0^{\text{ake}}]| \leq Nq_s^2/2^\kappa$. This game ensures the uniqueness of $\texttt{nonces}$ computed by each honest oracle Π_U^s in Round 2 over all invoked sessions.

Game G_2. In this game the only exception is that Δ aborts and b' is set at random if A asks a query of the form $\textit{Broadcast}(\texttt{sid},U|t|m|\sigma)$ such that $t \in \{2,3\}$, the session id $\texttt{sid}$ contains $U_i|s$ and there is an oracle of sender U which is partnered with $\Pi_{U_i}^s$ and is still treated as active during the t-th round, and σ is a valid signature on m, that has not been previously output by that oracle of U prior to a query $\textit{Corrupt}(U)$.

In other words the simulation fails if A outputs a successful forgery of the signature. A classical reductionist argument (e.g. [19]) can be used to construct a forger algorithm against Σ such that $|\Pr[\text{Win}_2^{\text{ake}}] - \Pr[\text{Win}_1^{\text{ake}}]| \leq N^2\text{Succ}_\Sigma^{\text{euf}-\text{cma}}(\kappa)$.

Since the concatenation $\texttt{nonces}|\texttt{pid}$ is part of every signed protocol message sent by the oracles this game prevents successful replay attacks.

Game G_3. In this game we add the following rule: Δ chooses $q^* \in [1,q_s]$ and aborts if the $\textit{Test}$ query does not occur in the q^*-th session. Let Q be the event that this guess for q^* is correct and $\Pr[Q] = 1/q_s$. Then, similar to the AKE-security proof of TDH1 in [12] we get $\Pr[\text{Win}_2^{\text{ake}}] = q_s\left(\Pr[\text{Win}_3^{\text{ake}}] - \frac{1}{2}\right) + \frac{1}{2}$.

Game $G_{4,j}$ for $j = 1,\ldots,n-1$. Each Game $G_{4,j}$ is composed of two Sub-Games $G_{4,j,1}$ and $G_{4,j,2}$.

Sub-Game $G_{4,j,1}$. In this (sub-)game Δ is given a tuple from the real SEDDH distribution, i.e., $(g, A = g^a, B = g^{a^2})$ for some unknown $a \in G$, and embeds it into the simulation of the q^*-th session as follows. In Round 2 for each of n' active oracles Π_i the simulator defines the public value $y_i := A^{\alpha_i}$ for some random $\alpha_i \in_R G$. In Round 3 for every remaining active oracle Π_i assigned to the leaf node $\langle n-1, 0\rangle$ (if $i = 1$) or $\langle n-i+1, 1\rangle$ (if $i > 1$), the iterative computation of the values $x_{c,0}^{(i)}$ and $y_{c,0}^{(i)}$, for $n-i-1 \geq c \geq 0$ is modified according to the following three rules (see also Fig. 2 for an example):

<u>Rule 1:</u> For $c = n-i-1$ downto $n-i-j+1$: the computation of $x_{c,0}^{(i)}$ is ignored and $y_{c,0}^{(i)}$ is defined to $A^{\alpha_{c,0}^{(i)}}$ for a randomly chosen $\alpha_{c,0}^{(i)}$; note that the rule is vacuous for $j = 1$.

<u>Rule 2:</u> For $c = n - i - j$, define $x_{c,0}^{(i)} := B^{\alpha_{c+1,0}^{(i)}\alpha_{c+1,1}}$ and $y_{c,0}^{(i)} = g^{x_{c,0}^{(i)}}$, where

$$\alpha_{c+1,0}^{(i)} = \begin{cases} \alpha_i & \text{if } j = 1 \text{ (value chosen in Round 2)} \\ \alpha_{c+1,0}^{(i)} & \text{as chosen in Rule 1, if } j > 1 \end{cases}$$

$$\alpha_{c+1,0}^{(i)} = \alpha_{i+1} \text{ (value chosen in Round 2)}$$

<u>Rule 3:</u> For $c = n - i - j - 1$ downto 0, the computation is done normally: $x_{c,0}^{(i)} = y_{c+1,1}^{(i)} \quad x_{c+1,0}^{(i)}$ and $y_{c,0}^{(i)} = g^{x_{c,0}^{(i)}}$.

Sub-Game $G_{4,j,2}$. In this game Δ is given a tuple from the random SEDDH distribution, i.e., $(g, A = g^a, B = g^b)$ for some unknown $a, b \in \mathbb{G}$. Since the simulator performs the same steps as defined for the Sub-Game $G_{4,j,1}$ the only difference between them is that in $G_{4,j,1}$ $B = g^{a^2}$ and in $G_{4,j,2}$ $B = g^b$. In each Game $G_{4,j}$ the value B is embedded exactly $n - j$ times (that is once per set X_i for $n - j \geq i \geq 1$ using the re-randomization exponent $\alpha_{c+1,0}^{(i)}\alpha_{c+1,1}$ whose factor $\alpha_{c+1,0}^{(i)}$ is different for each i). Since $n \leq N$ the probability difference between $G_{4,j,2}$ and $G_{4,j,1}$ can be upper-bounded by $(N - j)\mathsf{Adv}_{\mathbb{G}}^{\mathsf{SEDDH}}(\kappa)$.

Further we stress that by construction in Sub-Game $G_{4,1,1}$ (the first sub-game in the sequence) the distribution of secret values in each X_i is identical to G_3. And since j is a running variable from 1 to $n - 1$ and $n \leq N$ we can upper-bound the probability difference between Game $G_{4,n-1}$ (which ends with Sub-Game $G_{4,n-1,2}$) and G_3 as follows:

$$|\Pr[\mathsf{Win}_{4,n-1}^{\mathsf{ake}}] - \Pr[\mathsf{Win}_{3}^{\mathsf{ake}}]| \leq \sum_{j=1}^{N-1} (N - j)\mathsf{Adv}_{\mathbb{G}}^{\mathsf{SEDDH}}(\kappa).$$

The consequence of $G_{4,n-1}$ is that among different sets $X_i = \{x_{c,0}^{(i)}\}_{l \geq c \geq 0}$ computed by Δ in the q^*-th session all values $x_{c,0}^{(i)}$ are random and independent in $\mathbb{G} = \mathbb{Z}_q$ (the equality is due to the construction of $\tilde{\mathbb{G}}$ [12]). In particular, this implies that the value $x_{0,0}^{(\gamma)}$ used by every active oracle Π_i to derive the group key k_i in the q^*-th session is uniformly distributed in $\{0,1\}^\kappa$ (since κ is the length of q).

Game G_5. In this game Δ replaces in the q^*-th session f by a truly random function. Hence, k_i computed by every active oracle Π_i including the one for which the *Test* query is asked is uniformly distributed, and $|\Pr[\mathsf{Win}_{5}^{\mathsf{ake}}] - \Pr[\mathsf{Win}_{4,n-1}^{\mathsf{ake}}]| \leq \mathsf{Adv}_{F}^{\mathsf{prf}}(\kappa)$. And since k_i is uniform: $\Pr[\mathsf{Win}_{5}^{\mathsf{ake}}] = 1/2$. Combining the previous equations, we obtain the desired inequality for $\mathsf{Adv}_{\mathsf{R-TDH1}}^{\mathsf{ake}}(\kappa)$. negligible advantage.

A.2 Proof of Strong AKE-Security of IR-TDH1 (Theorem 2)

Proof (Sketch). This proof is identical to the proof of Theorem 1 except that we need to show that $\mathsf{Lg}^2\mathsf{EqLg}$ proofs computed by every honest oracle Π_i within Z_i do not reveal any additional information to the outsider adversary that asks the *Test* query. For this we need to plug in an additional game prior to Sub-Game $G_{4,1,1}$ (the first sub-game of $G_{4,1}$) in which Δ simulates the $\mathsf{Lg}^2\mathsf{EqLg}$ proofs $\{z_l^{(i)}\}_l$ in Z_i computed by each active Π_i in Round 3. It is clear that the simulation of $\mathsf{Lg}^2\mathsf{EqLg}$ proofs can be done via the classical technique of programmable random oracles. Hence, we omit the details.

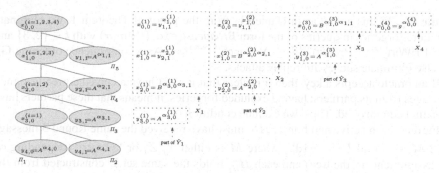

Fig. 2. Snapshot of $G_{4,2,1}$ and $G_{4,2,2}$ with oracles Π_i, $1 \le i \le 5$. In $G_{4,2,1}$: $(A = g^a, B = g^{a^2})$. In $G_{4,2,2}$: $(A = g^a, B = g^b)$. <u>Left side:</u> Δ embeds A into y_{l_i,v_i}. <u>Right side:</u> Δ follows the defined rules, i.e., Rule 1: it defines randomized $y_{3,0}^{(1)}$, $y_{2,0}^{(2)}$, and $y_{1,0}^{(3)}$ leaving corresponding $x_{3,0}^{(1)}$, $x_{2,0}^{(2)}$, and $x_{1,0}^{(3)}$ undefined, uses $\alpha_{0,0}^{(4)}$ to define $x_{0,0}^{(4)}$ (note that $x_{0,0}^{(4)}$ is already randomized at the end of Game $G_{4,1,2}$); Rule 2: embeds B in each second value of each X_i ($i = 1, 2, 3$); Rule 3: computes all subsequent values within X_i as specified in R-TDH1.

Assuming that n oracles remain active in Round 3 the number of simulated $\mathrm{Lg^2EqLg}$ proofs within each Z_j is $n - j - 1$ (remember, oracles Π_{n-1} and Π_n do not compute any proofs). Since $j \le n - 2$ and $n \le N$ we can upper-bound the probability difference between $G_{4,1,1}$ and this game by $\sum_{j=1}^{N-2}(N-j-1)\mathrm{Adv}_{\mathrm{Lg^2EqLg}}^{\mathrm{zk}}(\kappa)$.

Since the group key derivation in IR-TDH1 is performed through the random oracle H' we can omit Game G_5. The session group key k_i computed by Π_i to which the *Test* query is asked is already uniform at the end of Game $G_{4,n-1}$. It is easy to see that the combination of probability upper-bounds for the difference of sequence games gives the desired inequality for $\mathrm{Adv}_{\mathrm{IR\text{-}TDH1}}^{\mathrm{ake}}(\kappa)$.

A.3   Proof of Strong MA-Security of IR-TDH1 (Theorem 3)

Proof (Sketch). In the following games, event $\mathrm{Win}_i^{\mathrm{ma}}$ means that $\mathcal{A}$ wins in G_i.

Game G_0. This is the real game $\mathrm{Game}_{\mathrm{IR\text{-}TDH1}}^{\mathrm{ma}}(\kappa)$ played between Δ and $\mathcal{A}$. Recall that $\mathcal{A}$ wins if at some point there is a session sid for which the last protocol round is finished and one of its honest oracles $\Pi_{U_i}^s$ that does not fail after this round: (a) does not accept a group key although other active partners exist, or (b) accepts a group key without being assured that all other honest oracles that remain active after this last round have been invoked, or (c) accepts a different group key.

We observe that, by construction of IR-TDH1, every honest oracle $\Pi_{U_i}^s$ which has been invoked to participate in some session sid and during the group key derivation phase holds $\mathrm{pid}_{U_i}^s$ consisting of at least two identities (from which one is U_i) will compute the session group key $k_{U_i}^s$, and, thus accept. Hence, the probability that condition (a) ever occurs is 0. Therefore, in the following we focus on conditions (b) and (c).

Game G_1. In this game, as in Game G_1 from the proof of Theorem 1, Δ aborts if any two honest oracles identified by $U|s$ and $U|s'$ that have been invoked for two different sessions choose the same nonce in Round 1. Thus, $|\Pr[\mathrm{Win}_1^{\mathrm{ma}}] - \Pr[\mathrm{Win}_0^{\mathrm{ma}}]| \le Nq_s^2/2^\kappa$.

Game G_2. In this game, as in Game G_2 from the proof of Theorem 1, we eliminate signature forgeries in queries of the form $Broadcast(\text{sid}, U|t|m|\sigma)$ with $t \in \{2,3\}$ and get $|\Pr[\text{Win}_2^{ma}] - \Pr[\text{Win}_1^{ma}]| \le N^2 \text{Succ}_\Sigma^{euf-cma}(\kappa)$. As a consequence of Game G_1 we also eliminate successful replay attacks.

If an oracle accepts a key, then it is clear that it must have received correctly signed messages from its partners; having excluded forgeries, it means that these partners have actually been invoked. Thus, we exclude condition (b).

Further, each active and honest $\Pi_{U_i}^s$ must have received the same Round 3 message $U_j|3|M|\sigma_j'$ for all $U_j \in \text{pid}_{U_i}^s$ where M is either $\hat{Y}_j|Z_j$ or 'alive' (depending on the assignments in the tree) and each $\Pi_{U_i}^s$ holds the same set Y constructed from the received Round 2 messages (which is also signed by σ_j').

Game G_3. In this game Δ aborts if on behalf of any two partnered honest oracles $\Pi_{U_i}^s$ and $\Pi_{U_j}^t$ that are active during the group key derivation phase Δ computes two different values for $x_{0,0}^{(\gamma)}$ which should be used by $\Pi_{U_i}^s$ and $\Pi_{U_j}^t$ to derive the session group key.

Assume that this failure event occurs and let $U_\gamma|3|\hat{Y}_\gamma|Z_\gamma|\sigma_\gamma'$ be the Round 3 message received by $\Pi_{U_i}^s$ and $\Pi_{U_j}^t$. In line with the notations used in IR-TDH1 upon the construction of T_n we denote $\Pi_{U_i}^s$ as Π_i (assigned to the leaf node $\langle n-i+1, 1\rangle$) and $\Pi_{U_j}^s$ as Π_j (assigned to the leaf node $\langle n-j+1, 1\rangle$) and assume w.l.o.g. that $i < j$. We know that $\gamma < i$ since Π_γ has the lowest index.

Since oracles Π_i and Π_j receive the same Round 3 message (due to the broadcast channel) and since in previous games we have excluded any impersonation attacks on honest oracles, we conclude that if both oracles compute different values for $x_{0,0}^{(\gamma)}$ then Π_γ must be malicious.

Now we focus on the computation of $x_{0,0}^{(\gamma)}$ by Π_j (still assuming that $j > i$). The first value computed by Π_j in the key derivation phase using its secret exponent $x_{n-j+1,1}^{(\gamma)}$ and $y_{n-j+1,0}^{(\gamma)} \in \hat{Y}_\gamma$ is $x_{n-j,0}^{(\gamma)}$. Also Π_i computes this value, however, using the secret exponent $x_{n-j+1,0}^{(\gamma)}$ and $y_{n-j+1,1} \in Y_i$. Since the computation of $x_{0,0}^{(\gamma)}$ is deterministic and both oracles Π_i and Π_j use identical sets Y (and $Y_i \subset Y$) we follow that Π_i and Π_j compute different values for $x_{0,0}^{(\gamma)}$ only if they compute different values for $x_{n-j,0}^{(\gamma)}$.

Since Π_j honestly uses its secret exponent $x_{n-j+1,1}^{(\gamma)}$ its computed value for $x_{n-j,0}^{(\gamma)}$ is different from that computed by Π_i only if $y_{n-j+1,0}^{(\gamma)} \in \hat{Y}_\gamma$ used by Π_j does not have the form $g^{x_{n-j+1,0}^{(\gamma)}}$ where $x_{n-j+1,0}^{(\gamma)}$ is the exponent used by Π_i. In turn, Π_i assigned to $\langle n-i+1, 1\rangle$ computes $x_{n-j+1,0}^{(\gamma)}$ through iteration starting with the computation of $x_{n-i,0}^{(\gamma)}$ for which it honestly uses its secret exponent $x_{n-i+1,1}$.

By construction, if $y_{n-j+1,0}^{(\gamma)} \in \hat{Y}_\gamma$ used by Π_j does not have the required form $g^{x_{n-j+1,0}^{(\gamma)}}$, then the set Z_γ contains at least one forged proof. Moreover this proof can be discovered by Δ as follows. Δ uses $x_{n-i+1,1}$, $\hat{Y}_\gamma$, and Y to iteratively compute each $x_{l,0}^{(\gamma)}$ for $n-i \ge l \ge n-j+1$ and the corresponding $y_{l,0}^{(\gamma)} := g^{x_{l,0}^{(\gamma)}}$. Then, Δ sequentially checks whether each computed $y_{l,0}^{(\gamma)}$ is the same as the one included in $\hat{Y}_\gamma$ until it finds the first one which is different. At least one such value must exist;

otherwise, oracles would have computed identical values for $x_{n-j,0}^{(\gamma)}$. When Δ finds the first such value $y_{l,0}^{(\gamma)}$ the corresponding $\mathtt{Lg^2EgLg}$ proof $z_l^{(\gamma)}$ must be a forgery since it claims that $y_{l,0}^{(\gamma)} = g^{\frac{x_{l+1,0}^{(\gamma)}}{y_{l+1,1}}}$, a false statement.

Thus, if the described failure event occurs then Δ is able to output a forged $\mathtt{Lg^2EgLg}$ proof. Since for each oracle Π_j with $j \le n-2$ there are exactly $(n-j-1)$ $\mathtt{Lg^2EgLg}$ proofs we can upper-bound

$$| \Pr[\mathsf{Win}_3^{\mathsf{ma}}] - \Pr[\mathsf{Win}_2^{\mathsf{ma}}]| \le q_\mathsf{s} \sum_{j=1}^{N-2} (N-j-1)\mathsf{Succ}_{\mathtt{Lg^2EqLg}}^{\mathsf{snd}}(\kappa).$$

As a consequence of this game any two honest oracles $\Pi_{U_i}^s$ and $\Pi_{U_j}^t$ that are partnered with respect to some $\mathtt{sid}$ and remain active during the key derivation phase compute identical values for $x_{0,0}^{(\gamma)}$ and thus, accept with identical session group keys $k_{U_i}^s = k_{U_j}^s$. This excludes condition (c). Hence, $\Pr[\mathsf{Win}_3^{\mathsf{ma}}] = 0$. Combining the previous equations we obtain the desired inequality for $\mathsf{Succ}_{\mathtt{IR\text{-}TDH1}}^{\mathsf{ma}}(\kappa)$.

A.4 Proof of Strong Contributiveness of IR-TDH1 (Theorem 4)

Proof. In the following games, event $\mathsf{Win}_i^{\mathsf{con}}$ means that $\mathcal{A}$ wins in $\mathbf{G}_i$.

Game $\mathbf{G}_0$. This is the real game $\mathsf{Game}_{\mathcal{A},\mathtt{IR\text{-}TDH1}}^{\mathsf{con}}(\kappa)$, in which the honest players are simulated by Δ. Recall that $\mathcal{A}$ wins if after the stage **prepare** it returned $\tilde{k}$ and if in the stage **attack** the honest oracle $\Pi_{U_i}^s$ accepts $\tilde{k}$.

Game $\mathbf{G}_1$. In this game, similar to the previous proofs, Δ aborts if any two honest oracles identified by $U|s$ and $U|s'$ that have been invoked for two different sessions choose the same nonce in Round 1. Thus, $| \Pr[\mathsf{Win}_1^{\mathsf{con}}] - \Pr[\mathsf{Win}_0^{\mathsf{con}}]| \le Nq_\mathsf{s}^2/2^\kappa$. This implies that in every session in which U_i participates through some oracle $\Pi_{U_i}^s$ which remains active during the group key derivation phase the concatenation of random nonces $\mathtt{nonces}_i$ held by $\Pi_{U_i}^s$ contains a fresh nonce r_i. Since the concatenation preserves the lexicographic order of user identities in $\mathtt{pid}_{U_i}^s$ the concatenation $\mathtt{nonces}_i|\mathtt{pid}_{U_i}^s$ used in addition to $x_{0,0}^{(\gamma)}$ as input for H' to derive $k_{U_i}^s$ is unique for each session.

Game $\mathbf{G}_2$. In this game Δ aborts if $\mathcal{A}(\mathbf{prepare})$ returned some $\tilde{k}$ which it did not receive from Δ in response to some query to the random oracle H' but which is computed by Δ on behalf of some honest $\Pi_{U_i}^s$ as $k_{U_i}^s$ (as output of H') invoked during the **attack** stage. Since H' is modeled as a random oracle the probability that this event occurs for any honest oracle and any invoked session is given by the probability for the random guess of the output of H', so that $| \Pr[\mathsf{Win}_2^{\mathsf{con}}] - \Pr[\mathsf{Win}_1^{\mathsf{con}}]| \le Nq_\mathsf{s}/2^\kappa$.

Therefore, $\mathcal{A}$ wins in this game only if it queried H' on some input m during the **prepare** stage and received $\tilde{k}$ in response, which is then accepted by $\Pi_{U_i}^s$ as $k_{U_i}^s$.

Case $m = x_{0,0}^{(\gamma)}|\mathtt{nonces}_i|\mathtt{pid}_{U_i}^s$: The probability of $\mathcal{A}$ to win is given by the guess of r_i, i.e. $Nq_\mathsf{s}/2^\kappa$ (due to Game $\mathbf{G}_1$).

Case $m \ne x_{0,0}^{(\gamma)}|\mathtt{nonces}_i|\mathtt{pid}_{U_i}^s$: The probability of $\mathcal{A}$ to win is upper-bounded by $q_{H'}^2/2^\kappa$ by the birthday paradox.

Thus, $\Pr[\mathsf{Win}_2^{\mathsf{con}}] = (Nq_\mathsf{s} + q_{H'}^2)/2^\kappa$. Combining previous equations gives us the desired inequality for $\mathsf{Succ}_{\mathtt{IR\text{-}TDH1}}^{\mathsf{con}}(\kappa)$.

Group Signatures with Verifier-Local Revocation and Backward Unlinkability in the Standard Model

Benoît Libert[1] and Damien Vergnaud[2],*

[1] Université Catholique de Louvain, Microelectronics Laboratory, Crypto Group
Place du Levant, 3 – 1348 Louvain-la-Neuve – Belgium
[2] École normale supérieure – C.N.R.S. – I.N.R.I.A.
45, Rue d'Ulm – 75230 Paris CEDEX 05 – France

Abstract. Group signatures allow users to anonymously sign messages in the name of a group. Membership revocation has always been a critical issue in such systems. In 2004, Boneh and Shacham formalized the concept of group signatures with *verifier-local revocation* where revocation messages are only sent to signature verifiers (as opposed to both signers and verifiers). This paper presents an efficient verifier-local revocation group signature (VLR-GS) providing *backward unlinkability* (*i.e.* previously issued signatures remain anonymous even after the signer's revocation) with a security proof in the standard model (*i.e.* without resorting to the random oracle heuristic).

Keywords: Group signatures, verifier-local revocation, bilinear maps, backward unlinkability, standard model.

1 Introduction

The *group signature* primitive, as introduced by Chaum and van Heyst in 1991 [17], allows members of a group to sign messages, while hiding their identity within a population group members administered by a group manager. At the same time, it must be possible for a tracing authority holding some trapdoor information to "open" signatures and find out which group members are their originator. A major issue in group signatures is the revocation of users whose membership should be cancelled: disabling the signing capability of misbehaving members (or honest users who intentionally leave the group) without affecting remaining members happens to be a highly non-trivial problem. In 2004, Boneh and Shacham [11] formalized the concept of group signatures with *verifier-local*

* The first author acknowledges the Belgian National Fund for Scientific Research (F.R.S.-F.N.R.S.) for their financial support and the BCRYPT Interuniversity Attraction Pole. The second author is supported by the European Commission through the IST Program under Contract ICT-2007-216646 ECRYPT II and by the French *Agence Nationale de la Recherche* through the ANR 07-TCOM-013-04 PACE Project.

J.A. Garay, A. Miyaji, and A. Otsuka (Eds.): CANS 2009, LNCS 5888, pp. 498–517, 2009.
© Springer-Verlag Berlin Heidelberg 2009

revocation where revocation messages are only sent to signature verifiers (as opposed to both signers and verifiers). This paper describes the first efficient verifier-local revocation group signature scheme providing *backward unlinkability* (*i.e.*, previously issued signatures remain anonymous even after the signer's revocation) whose proof of security does not hinge upon the random oracle heuristic.

1.1 Related Work

GROUP SIGNATURES. Many group signatures were proposed in the nineties, the first provably coalition-resistant proposal being the famous ACJT scheme [2] proposed by Ateniese, Camenisch, Joye and Tsudik in 2000. The last few years saw the appearance of new constructions using bilinear maps [9,32,21,19]. Among these, the Boneh-Boyen-Shacham scheme [9] was the first one to offer signatures shorter than 200 bytes using the *Strong Diffie-Hellman assumption* [7]. Its security was analyzed using random oracles [5] in the model of Bellare, Micciancio and Warinschi [4] (BMW) which captures all the requirements of group signatures in three well-defined properties.

The BMW model, which assumes static groups where no new member can be introduced after the setup phase, was independently extended by Kiayias and Yung [28] and Bellare-Shi-Zhang [6] to a dynamic setting. In these models (that are very close to each other), efficient pairing-based schemes were put forth by Nguyen and Safavi-Naini [32], Furukawa and Imai [21] and, later on, by Delerablée and Pointcheval [19]. In dynamically growing groups, Ateniese *et al.* [1] also proposed a construction without random oracles offering a competitive efficiency at the expense of a security resting on interactive assumptions that are not efficiently falsifiable [31]. Another standard model proposal was put forth (and subsequently improved [12]) by Boyen-Waters [13] in the static model from [4] under more classical assumptions. Groth [23] described a scheme with constant-size signatures without random oracles in the dynamic model [6] but signatures were still too long for practical use. Later on, Groth showed [24] a fairly practical random-oracle-free group signature with signature length smaller than 2 kB and full anonymity (*i.e.*, anonymity in a model where the adversary is allowed to open anonymous signatures at will) in the model of [6].

VERIFIER-LOCAL REVOCATION. Membership revocation has always been a critical issue in group signatures. The simplest solution is to generate a new group public key and provide unrevoked signers with a new signing key, which implies the group master to send a secret message to each individual signer as well as to broadcast a public message to verifiers. In some settings, it may not be convenient to send a new secret to signers after their inclusion in the group. In *verifier-local revocation group signatures* (VLR-GS), originally suggested in [15] and formalized in [11], revocation messages are only sent to verifiers (making the group public key and the signing procedure independent of which and how many members were excluded). The group manager maintains a (periodically updated) revocation list (RL) which is used by all verifiers to perform the revocation test and make sure that signatures were not produced by a revoked member.

The RL contains a token for each revoked user. The verification algorithm accepts all signatures issued by unrevoked users and reveals no information about which unrevoked user issued the signature. However, if a user is revoked, his signatures are no longer accepted. It follows that signatures from a revoked member become linkable: to test that two signatures emanate from the same revoked user, one can simply verify signatures once using the RL before the alleged signer's revocation and once using the post-revocation RL. As a result, users who deliberately leave the group inevitably lose their privacy.

The property of *backward unlinkability*, first introduced in [36] in the context of key-evolving group signatures, ensures that signatures that were generated by a revoked member *before* his revocation remain anonymous and unlinkable. This property is useful when members who voluntarily leave the group wish to retain a certain level of privacy. When users' private keys get stolen, preserving the anonymity of their prior signatures is also definitely desirable.

Boneh and Shacham [11] proposed a VLR group signature using bilinear maps in a model inspired from [4]. In [33], Nakanishi and Funabiki extended Boneh-Shacham group signatures and devised a scheme providing backward unlinkability. They proved the anonymity of their construction under the Decision Bilinear Diffie-Hellman assumption [10]. In [34], the same authors suggested another backward-unlinkable scheme with shorter signatures. Other pairing-based VLR-GS constructions were put forth in [38,39]

Traceable signatures [27], that also have pairing-based realizations [32,18], can be seen as extensions of VLR-GS schemes as they also admit an implicit tracing mechanism. They provide additional useful properties such as the ability for signers to claim (and prove) the authorship of anonymously generated signatures or the ability for the group manager to reveal a trapdoor allowing to publicly trace all signatures created by a given user. This primitive was recently implemented in the standard model [30]. However, it currently does not provide a way to trace users' signatures per period: once the tracing trapdoor of some group member is revealed, all signatures created by that member become linkable. In some situations, it may be desirable to obtain a fine-grained traceability and only trace signatures that were issued in specific periods. The problem of VLR-GS schemes with backward unlinkability can be seen as the one of tracing some user's signatures from a given period onwards while preserving the anonymity and the unlinkability of that user's signatures for earlier periods. The solution described in this paper readily extends to retain the anonymity of signatures produced during past *and* future periods.

1.2   Contribution of the Paper

All known constructions of group signatures with verifier local revocation (with or without backward unlinkability) make use of the Fiat-Shamir paradigm [20] and thus rely on the random oracle methodology [5], which is known not to provide more than heuristic arguments in terms of security. Failures of the random oracle model were indeed reported in several papers such as [16,22]. When first analyzed in the random oracle model, cryptographic primitives thus deserve further efforts towards securely instantiating them without appealing to the random oracle idealization.

The contribution of this paper is to describe a new VLR-GS scheme with backward unlinkability in the standard model. Recently, Groth and Sahai [25] described powerful non-interactive proof systems allowing to prove that a number of committed variables satisfy certain algebraic relations. Their techniques notably proved useful to design standard model group signatures featuring constant signature size [12,23,24].

Extending the aforementioned constructions to obtain VLR-GS schemes with backward unlinkability is not straightforward. The approach used in [34], which can be traced back to Boneh-Shacham [11], inherently requires to use programmable random oracles, the behavior of which currently seems impossible to emulate in the standard model (even with the techniques developed in [26]). Another approach used in [33] looks more promising as it permits traceability with backward unlinkablity without introducing additional random oracles. This technique, however, does not interact with the Groth-Sahai toolbox in a straightforward manner as it typically requires non-interactive zero-knowledge (NIZK) proofs for what Groth and Sahai called *pairing product equations*. The problem that we face is that proving the required anonymity property of VLR-GS schemes entails to simulate a NIZK proof for such a pairing-product equation at some step of the reduction. As pointed out in [25], such non-interactive proofs are only known to be simulatable in NIZK under specific circumstances that are not met if we try to directly apply the technique of [33].

To address the above technical difficulty, we use the same revocation mechanism as [33] but use a slightly stronger (but still falsifiable [31]) assumption in the proof of anonymity: while Nakanishi and Funabiki rely the Decision Bilinear Diffie-Hellman assumption, we rest on the hardness of the so-called Decision Tripartite Diffie-Hellman problem, which is to distinguish g^{abc} from random given (g, g^a, g^b, g^c). Our contribution can be summarized as showing that the implicit tracing mechanism of [33] can be safely applied to the Boyen-Waters group signature [12] to make it backward-unlinkably revocable. This property comes at the expense of a quite moderate increase of signature sizes w.r.t. [12]. The main price to pay is actually to use a slightly stronger assumption than in [33] in the security proof.

2   Preliminaries

2.1   Verifier-Local Revocation Group Signatures

This section presents the model of VLR group signatures with backward unlinkability proposed in [33] which extends the Boneh-Shacham model [11] of VLR group signatures.

Definition 1. *A VLR group signature scheme with backward unlinkability consists of the following algorithms:*

Keygen(λ, N, T): *is a randomized algorithm taking as input a security parameter* $\lambda \in \mathbb{N}$ *and integers* $N, T \in \mathbb{N}$ *indicating the number of group members and the number of time periods, respectively.*

Its output consists of a group public key gpk, *a* N-*vector of group members' secret keys* gsk $= (\text{gsk}[1], \ldots, \text{gsk}[N])$ *and a* $(N \times T)$-*vector of revocation tokens* grt $= (\text{grt}[1][1], \ldots, \text{grt}[N][T])$, *where* grt$[i][j]$ *indicates the token of member* i *at time interval* j.

Sign$(\text{gpk}, \text{gsk}[i], j, M)$: *is a possibly randomized algorithm taking as input, the group public key* gpk, *the current time interval* j, *a group member's secret key* gsk$[i]$ *and a message* $M \in \{0, 1\}^*$. *It outputs a group signature* σ.

Verify$(\text{gpk}, j, RL_j, \sigma, M)$: *is a deterministic algorithm taking as input* gpk, *the period number* j, *a set of revocation tokens* RL_j *for period* j, *a signature* σ, *and the message* M. *It outputs either "*valid*" or "*invalid*". The former output indicates that* σ *is a correct signature on* M *at interval* j *w.r.t.* gpk, *and that the signer is not revoked at interval* j.

For all $(\text{gpk}, \text{gsk}, \text{grt}) = $ **Keygen**(λ, N, T), *all* $j \in \{1, \ldots, T\}$, *all* RL_j, *all* $i \in \{1, \ldots, N\}$ *and any message* $M \in \{0, 1\}^*$, *it is required that if* grt$[i][j] \notin RL_j$ *then:*

$$\textbf{Verify}(\text{gpk}, j, RL_j, \textbf{Sign}(\text{gpk}, \text{gsk}[i], j, M), M) = \text{``}valid\text{''}.$$

Remark 1. As mentioned in [11], any such group signature scheme has an associated *implicit tracing* algorithm that allows tracing a signature to the group member who generated it using the vector grt as the tracing key: on input a valid message-signature pair (M, σ) for period j, the opener can determine which user was the author of σ by successively executing the verification algorithm on (M, σ) using the vector of revocation tokens (*i.e.*, with $RL_j = \{\text{grt}[i][j]\}_{i \in \{1, \ldots, N\}}$) and outputting the first index $i \in \{1, \ldots, N\}$ for which the verification algorithm returns "invalid" whereas verifying the same pair (M, σ) with $RL_j = \emptyset$ yields the answer "valid".

From a security standpoint, VLR group signatures with backward unlinkability should satisfy the following properties:

Definition 2. *A VLR-GS with backward unlinkability has the* **traceability** *property if no probabilistic polynomial time (PPT) adversary* $\mathcal{A}$ *has non-negligible advantage in the following game.*

1. *The challenger* $\mathcal{C}$ *runs the setup algorithm to produce a group public key* gpk, *a group master secret* gsk *and a vector* grt *of revocation tokens. It also defines a set of corrupt users* U *which is initially empty. The adversary* $\mathcal{A}$ *is provided with* gpk *and* grt *while* $\mathcal{C}$ *keeps* gsk *to itself.*
2. $\mathcal{A}$ *can make a number of invocations to the following oracles:*

   Signing oracle: *on input of a message* M, *an index* $i \in \{1, \ldots, N\}$ *and a period number* j, *this oracle responds with a signature* σ *generated on behalf of member* i *for period* j.

Corruption oracle: *given an index $i \in \{1, \ldots, N\}$, this oracle reveals the private key $\mathsf{gsk}[i]$ of member i which is included in the set U.*

3. *$\mathcal{A}$ eventually comes up with a signature $\sigma^\star$ on a message $M^\star$, a period number $j^\star$ and a set of revocation tokens $RL^\star_{j^\star}$.*

The adversary $\mathcal{A}$ is declared successful if

- **Verify**$(\mathsf{gpk}, j^\star, RL^\star_{j^\star}, \sigma^\star, M^\star) = $ *"valid".*
- *The execution of the implicit tracing algorithm on input of revocation tokens $(\mathsf{grt}[1][j^\star], \ldots, \mathsf{grt}[N][j^\star])$, ends up in one of the following ways:*
    - *$\sigma^\star$ traces to a member outside the coalition $U \backslash RL^\star_{j^\star}$ that did not sign $M^\star$ during period $j^\star$*
    - *the tracing fails.*

$\mathcal{A}$'s advantage in breaking traceability is measured as

$$\mathbf{Adv}^{\text{trace}}_{\mathcal{A}}(k) := \Pr[\mathcal{A} \text{ is successful}],$$

where the probability is taken over the coin tosses of $\mathcal{A}$ and the challenger.

This definition slightly weakens the original one [33] that captures the strong unforgeability requirement (*i.e.*, the message-signature pair $(M^\star, \sigma^\star)$ must be different from that of any signing query during period $j^\star$). Due to the use of publicly randomizable non-interactive witness indistinguishable proofs, we need to settle for the usual flavor of unforgeability according to which the message $M^\star$ must not have been queried for signature during the target period $j^\star$.

Definition 3. *A VLR-GS with backward unlinkability provides* **BU-anonymity** *if no PPT adversary $\mathcal{A}$ has non-negligible advantage in the following game.*

1. *The challenger $\mathcal{C}$ runs* **Keygen**(λ, n, T) *to produce a group public key gpk, a master secret gsk and a vector grt of revocation tokens. The adversary $\mathcal{A}$ is given gpk but is denied access to grt and gsk.*
2. *At the beginning of each period, $\mathcal{C}$ increments a counter j and notifies $\mathcal{A}$ about it. During the current time interval j, $\mathcal{A}$ can adaptively invoke the following oracles:*

    Signing oracle: *on input of a message M and an index $i \in \{1, \ldots, n\}$, this oracle outputs a signature σ generated for member i and period j.*
    Corruption oracle: *for an adversarially-chosen $i \in \{1, \ldots, n\}$, this oracle reveals member i's private key $\mathsf{gsk}[i]$.*
    Revocation oracle: *given $i \in \{1, \ldots, n\}$, this oracle outputs member i's revocation token for the current period j.*

3. *At some period $j^\star \in \{1, \ldots, T\}$, $\mathcal{A}$ comes up with a message M and two distinct user indices $i_0, i_1 \in \{1, \ldots, n\}$ such that neither i_0 or i_1 has been corrupt. Moreover, they cannot have been revoked before or during period $j^\star$. At this stage, $\mathcal{C}$ flips a fair coin $d^\star \xleftarrow{R} \{0,1\}$ and generates a signature $\sigma^\star$ on M on behalf of user $i_{d^\star}$ which is sent as a challenge to $\mathcal{A}$.*

4. $\mathcal{A}$ is granted further oracle accesses as in phase 2. Of course, she may not query the private key of members i_0, i_1 at any time. On the other hand, she may obtain their revocation tokens for time intervals after j^*.
5. Eventually, $\mathcal{A}$ outputs $d' \in \{0, 1\}$ and wins if $d' = d^*$.

The advantage of $\mathcal{A}$ in breaking BU-anonymity is defined as $\mathbf{Adv}_{\mathcal{A}}^{\text{bu-anon}}(k) := |\Pr[d' = d^*] - 1/2|$, where the probability is taken over all coin tosses.

2.2   Bilinear Maps and Complexity Assumptions

BILINEAR GROUPS. Groups $(\mathbb{G}, \mathbb{G}_T)$ of prime order p are called *bilinear groups* if there is an efficiently computable mapping $e : \mathbb{G} \times \mathbb{G} \to \mathbb{G}_T$ such that:

1. $e(g^a, h^b) = e(g, h)^{ab}$ for any $(g, h) \in \mathbb{G} \times \mathbb{G}$ and $a, b \in \mathbb{Z}$;
2. $e(g, h) \neq 1_{\mathbb{G}_T}$ whenever $g, h \neq 1_{\mathbb{G}}$.

In such groups, we will need three non-interactive (and thus falsifiable [31]) complexity assumptions.

Definition 4. *In a group* $\mathbb{G} = \langle g \rangle$ *of prime order* $p > 2^\lambda$, *the* **Decision Linear Problem** *(DLIN) is to distinguish the distributions* $(g, g^a, g^b, g^{ac}, g^{bd}, g^{c+d})$ *and* $(g, g^a, g^b, g^{ac}, g^{bd}, g^z)$, *with* $a, b, c, d \xleftarrow{R} \mathbb{Z}_p^*$, $z \xleftarrow{R} \mathbb{Z}_p^*$. *The* **Decision Linear Assumption** *posits that, for any PPT distinguisher* $\mathcal{D}$,

$$\mathbf{Adv}_{\mathbb{G}, \mathcal{D}}^{\text{DLIN}}(\lambda) = |\Pr[\mathcal{D}(g, g^a, g^b, g^{ac}, g^{bd}, g^{c+d}) = 1|a, b, c, d \xleftarrow{R} \mathbb{Z}_p^*]$$
$$- \Pr[\mathcal{D}(g, g^a, g^b, g^{ac}, g^{bd}, g^z) = 1|a, b, c, d \xleftarrow{R} \mathbb{Z}_p^*, \ z \xleftarrow{R} \mathbb{Z}_p^*]| \in \mathsf{negl}(\lambda).$$

This problem amounts to deciding whether vectors $\vec{g_1} = (g^a, 1, g)$, $\vec{g_2} = (1, g^b, g)$ and $\vec{g_3}$ are linearly dependent or not. It has been used [25] to construct efficient non-interactive proof systems.

We also rely on a variant, introduced by Boyen and Waters [12], of the Strong Diffie-Hellman assumption [7].

Definition 5 ([12]). *In a group* $\mathbb{G}$ *of prime order* p, *the* ℓ-**Hidden Strong Diffie-Hellman problem** *(ℓ-HSDH) is, given elements* $(g, \Omega = g^\omega, u) \xleftarrow{R} \mathbb{G}^3$ *and* ℓ *distinct triples* $(g^{1/(\omega+s_i)}, g^{s_i}, u^{s_i})$ *with* $s_1, \ldots, s_\ell \xleftarrow{R} \mathbb{Z}_p^*$, *to find another triple* $(g^{1/(\omega+s)}, g^s, u^s)$ *such that* $s \neq s_i$ *for* $i \in \{1, \ldots, \ell\}$.

We also rely on the following intractability assumption suggested for the first time in [10, Section 8].

Definition 6. *In a prime order group* $\mathbb{G}$, *the* **Decision Tripartite Diffie-Hellman Assumption** *(DTDH) is the infeasibility of deciding if* $\eta = g^{abc}$ *on input of* (g, g^a, g^b, g^c, η), *where* $a, b, c \xleftarrow{R} \mathbb{Z}_p^*$. *The advantage function* $\mathbf{Adv}_{\mathbb{G}, \mathcal{D}}^{\text{DTDH}}(\lambda)$ *of any PPT distinguisher* $\mathcal{D}$ *is defined analogously to the DLIN case.*

The above assumption is a bit stronger than the widely accepted Decision Bilinear Diffie-Hellman assumption according to which the distributions

$$\{(g, g^a, g^b, g^c, e(g, g)^{abc})|a, b, c, \xleftarrow{R} \mathbb{Z}_p\} \text{ and } \{(g, g^a, g^b, g^c, e(g, g)^z)|a, b, c, z \xleftarrow{R} \mathbb{Z}_p\}$$

are computationally indistinguishable. Yet, the DTDH problem is still believed to be hard in groups with a bilinear map where the DDH problem is easy.

2.3 Groth-Sahai Proof Systems

In the following notations, for equal-dimension vectors or matrices A and B containing group elements, $A \odot B$ stands for their entry-wise product (*i.e.* it denotes their Hadamard product).

When based on the DLIN assumption, the Groth-Sahai (GS) proof systems [25] use a common reference string comprising vectors $\vec{g_1}, \vec{g_2}, \vec{g_3} \in \mathbb{G}^3$, where $\vec{g_1} = (g_1, 1, g), \vec{g_2} = (1, g_2, g)$ for some $g_1, g_2 \in \mathbb{G}$. To commit to group elements $X \in \mathbb{G}$, one sets $\vec{C} = (1, 1, X) \odot \vec{g_1}^r \odot \vec{g_2}^s \odot \vec{g_3}^t$ with $r, s, t \xleftarrow{R} \mathbb{Z}_p^*$. When the proof system is configured to give perfectly sound proofs, $\vec{g_3}$ is chosen as $\vec{g_3} = \vec{g_1}^{\xi_1} \odot \vec{g_2}^{\xi_2}$ with $\xi_1, \xi_2 \xleftarrow{R} \mathbb{Z}_p^*$. Commitments $\vec{C} = (g_1^{r+\xi_1 t}, g_2^{s+\xi_2 t}, X \cdot g^{r+s+t(\xi_1+\xi_2)})$ are then Boneh-Boyen-Shacham (BBS) ciphertexts that can be decrypted using $\alpha_1 = \log_g(g_1)$, $\alpha_2 = \log_g(g_2)$. In the witness indistinguishability (WI) setting, vectors $\vec{g_1}, \vec{g_2}, \vec{g_3}$ are linearly independent and $\vec{C}$ is a perfectly hiding commitment. Under the DLIN assumption, the two kinds of CRS are computationally indistinguishable.

To commit to a scalar $x \in \mathbb{Z}_p$, one computes $\vec{C} = \vec{\varphi}^x \odot \vec{g_1}^r \odot \vec{g_2}^s$, with $r, s \xleftarrow{R} \mathbb{Z}_p^*$, using a CRS comprising vectors $\vec{\varphi}, \vec{g_1}, \vec{g_2}$. In the soundness setting $\vec{\varphi}, \vec{g_1}, \vec{g_2}$ are linearly independent (typically $\vec{\varphi} = \vec{g_3} \odot (1, 1, g)$ where $\vec{\varphi} = \vec{g_1}^{\xi_1} \odot \vec{g_2}^{\xi_2}$) whereas, in the WI setting, choosing $\vec{\varphi} = \vec{g_1}^{\xi_1} \odot \vec{g_2}^{\xi_2}$ gives a perfectly hiding commitment since $\vec{C}$ is always a BBS encryption of $1_{\mathbb{G}}$.

To prove that committed variables satisfy a set of relations, the GS techniques replace variables by commitments in each relation. The whole proof consists of one commitment per variable and one proof element (made of a constant number of group elements) per relation.

Such proofs are easily obtained for pairing-product relations, which are of the type

$$\prod_{i=1}^{n} e(\mathcal{A}_i, \mathcal{X}_i) \cdot \prod_{i=1}^{n} \prod_{j=1}^{n} e(\mathcal{X}_i, \mathcal{X}_j)^{a_{ij}} = t_T, \tag{1}$$

for variables $\mathcal{X}_1, \ldots, \mathcal{X}_n \in \mathbb{G}$ and constants $t_T \in \mathbb{G}_T$, $\mathcal{A}_1, \ldots, \mathcal{A}_n \in \mathbb{G}$, $a_{ij} \in \mathbb{G}$, for $i, j \in \{1, \ldots, n\}$. Efficient proofs also exist for multi-exponentiation equations

$$\prod_{i=1}^{m} \mathcal{A}_i^{y_i} \cdot \prod_{j=1}^{n} \mathcal{X}_j^{b_j} \cdot \prod_{i=1}^{m} \prod_{j=1}^{n} \mathcal{X}_j^{y_i \gamma_{ij}} = T, \tag{2}$$

for variables $\mathcal{X}_1, \ldots, \mathcal{X}_n \in \mathbb{G}$, $y_1, \ldots, y_m \in \mathbb{Z}_p$ and constants $T, \mathcal{A}_1, \ldots, \mathcal{A}_m \in \mathbb{G}$, $b_1, \ldots, b_n \in \mathbb{Z}_p$ and $\gamma_{ij} \in \mathbb{G}$, for $i \in \{1, \ldots, m\}, j \in \{1, \ldots, n\}$.

In both cases, proofs for quadratic equations cost 9 group elements. Linear pairing-product equations (when $a_{ij} = 0$ for all i, j) take 3 group elements each. Linear multi-exponentiation equations of the type $\prod_{j=1}^{n} \mathcal{X}_j^{b_j} = T$ (resp. $\prod_{i=1}^{m} \mathcal{A}_i^{y_i} = T$) demand 3 (resp. 2) group elements.

Multi-exponentiation equations admit zero-knowledge proofs at no additional cost. On a simulated CRS (prepared for the WI setting), a trapdoor makes it is possible to simulate proofs without knowing witnesses and simulated proofs are identically distributed to real proofs.

On the other hand, pairing-product equations are not known to always have zero-knowledge proofs. Proving relations of the type (1) in NIZK usually comes at some expense since auxiliary variables have to be introduced and proof sizes are not necessarily independent of the number of variables. If $t_T = 1_{\mathbb{G}_T}$ in relation (1), the NIZK simulator can always use $\mathcal{X}_1 = \cdots = \mathcal{X}_n = 1_{\mathbb{G}}$ as witnesses. If t_T equals $\prod_{j=1}^{n'} e(g_j, h_j)$ for known group elements $g_1, \ldots, g_{n'}, h_1, \ldots, h_{n'} \in \mathbb{G}$, the simulator can prove that

$$\prod_{i=1}^{n} e(\mathcal{A}_i, \mathcal{X}_i) \cdot \prod_{i=1}^{n} \cdot \prod_{j=1}^{n} e(\mathcal{X}_i, \mathcal{X}_j)^{a_{ij}} = \prod_{j=1}^{n'} e(g_j, \mathcal{Y}_j) \qquad (3)$$

and that introduced variables $\mathcal{Y}_1, \ldots, \mathcal{Y}_{n'}$ satisfy the linear equations $\mathcal{Y}_j = h_j$ for $j \in \{1, \ldots, n'\}$. Since linear equations are known to have NIZK proofs and the proof of relation (3) can be simulated using witnesses $\mathcal{X}_1 = \cdots = \mathcal{X}_n = \mathcal{Y}_1 = \cdots = \mathcal{Y}_{n'} = 1_{\mathbb{G}}$. When t_T is an arbitrary element of $\mathbb{G}_T$, pairing-product equations are currently not known to have NIZK proofs at all.

3   A Scheme in the Standard Model

3.1   Description of the Scheme

In notations hereafter, it will be useful to define the coordinate-wise pairing $E : \mathbb{G} \times \mathbb{G}^3 \to \mathbb{G}_T^3$ such that, for any $h \in \mathbb{G}$ and any vector $\vec{g} = (g_1, g_2, g_3) \in \mathbb{G}^3$, $E(h, \vec{g}) = (e(h, g_1), e(h, g_2), e(h, g_3))$. As in [25], we will also make use of a symmetric bilinear map $F : \mathbb{G}^3 \times \mathbb{G}^3 \to \mathbb{G}_T$ defined in such a way that, for any vectors $\vec{X} = (X_1, X_2, X_3) \in \mathbb{G}^3$ and $\vec{Y} = (Y_1, Y_2, Y_3) \in \mathbb{G}^3$, we have $F(\vec{X}, \vec{Y}) = \tilde{F}(\vec{X}, \vec{Y})^{1/2} \cdot \tilde{F}(\vec{Y}, \vec{X})^{1/2}$, where $\tilde{F} : \mathbb{G}^3 \times \mathbb{G}^3 \to \mathbb{G}_T^9$ is a non-commutative bilinear mapping that sends $(\vec{X}, \vec{Y})$ onto the matrix $\tilde{F}(\vec{X}, \vec{Y})$ of entry-wise pairings (i.e., containing $e(X_i, Y_j)$ in its entry (i, j)).

Also, for any $z \in \mathbb{G}_T$, $\iota_T(z)$ denotes the 3×3 matrix containing z in position $(3, 3)$ and 1 everywhere else. For group elements $X \in \mathbb{G}$, the notation $\iota(X)$ will denote the vector $(1, 1, X) \in \mathbb{G}^3$.

The group manager holds a public key $(g, \Omega = g^\omega, A = e(g, g)^\alpha, u)$, where (α, γ) is the private key. As in the Boyen-Waters construction [12], group members' private keys consist of triples $(K_1, K_2, K_3) = ((g^\alpha)^{1/(\omega + s_i)}, g^{s_i}, u^{s_i})$, where s_i uniquely identifies the group member. Messages can be signed by creating tuples $(S_1, S_2, S_3, S_4) = (K_1, K_2, K_3 \cdot F(m)^r, g^r)$, where r is a random exponent and $F : \{0, 1\}^* \to \mathbb{G}$ is a Waters-like hash function [37].

The revocation mechanism of [33] consists in introducing a vector $(h_1, \ldots, h_T)$ of group elements, where T is the number of time periods, that allow to form revocation tokens for each user: the revocation token of user i for period j is obtained as $\mathrm{grt}[i][j] = h_j^{s_i}$. When user i must be revoked at stage j, the group manager can simply add $\mathrm{grt}[i][j]$ to the revocation list RL_j of period j. When user i signs a message during stage j, he is required to include a pair $(T_1, T_2) = (g^\delta, e(h_j, g^{s_i})^\delta)$ in the signature and append a proof that $(g, T_1 = g^\delta, K_2 = g^{s_i}, h_j, T_2)$ satisfy the forementioned relation and that T_2 is indeed the "Bilinear Diffie-Hellman value" $e(h_j, g^{s_i})^\delta$ associated with (g, T_1, K_2, h_j).

Keygen(λ, N, T): for security parameters λ and $n \in \text{poly}(\lambda)$, choose bilinear groups $(\mathbb{G}, \mathbb{G}_T)$ of order $p > 2^\lambda$, with $g, h_1, \ldots, h_T, u \xleftarrow{R} \mathbb{G}$. Select $\alpha, \omega \xleftarrow{R} \mathbb{Z}_p^*$ and set $A = e(g, g)^\alpha$, $\Omega = g^\omega$. Select $\bar{v} = (v_0, v_1, \ldots, v_n) \xleftarrow{R} \mathbb{G}^{n+1}$. Choose vectors $\mathbf{g} = (\vec{g_1}, \vec{g_2}, \vec{g_3})$ such that $\vec{g_1} = (g_1, 1, g) \in \mathbb{G}^3$, $\vec{g_2} = (1, g_2, g) \in \mathbb{G}^3$, and $\vec{g_3} = \vec{g_1}^{\xi_1} \cdot \vec{g_2}^{\xi_2}$, with $g_1 = g^{\alpha_1}, g_2 = g^{\alpha_2}$ and $\alpha_1, \alpha_2 \xleftarrow{R} \mathbb{Z}_p^*, \xi_1, \xi_2 \xleftarrow{R} \mathbb{Z}_p$. Finally, select a collision-resistant hash function $H : \{0, 1\}^* \to \{0, 1\}^n$. The group public key is defined to be

$$\text{gpk} := \left(g, \ h_1, \ldots, \ h_T, \ A = e(g, g)^\alpha, \ \Omega = g^\omega, \ u, \ \bar{v}, \ \mathbf{g}, \ H \right)$$

while the group manager's private key is $(\alpha, \omega, \alpha_1, \alpha_2)$. User i is assigned the group signing key $\text{gsk}[i] = (K_1, K_2, K_3) = \left((g^\alpha)^{\frac{1}{\omega+s_i}}, g^{s_i}, u^{s_i}\right)$ and his revocation token for period $j \in \{1, \ldots, T\}$ is defined as $\text{grt}[i][j] := h_j^{s_i}$.

Sign$(\text{gpk}, \text{gsk}[i], j, M)$: given $\text{gsk}[i] = (K_1, K_2, K_3) = \left((g^\alpha)^{\frac{1}{\omega+s_i}}, g^{s_i}, u^{s_i}\right)$, to sign a message M during period j, the signer $\mathcal{U}_i$ first computes a hash value $m = m_1 \ldots m_n = H(j\|M) \in \{0, 1\}^n$ and conducts the following steps.

1. Choose $\delta, r \xleftarrow{R} \mathbb{Z}_p^*$ and first compute

$$T_1 = g^\delta \qquad\qquad T_2 = e(h_j, K_2)^\delta \qquad\qquad (4)$$

  as well as

$$\theta_1 = K_1 = (g^\alpha)^{1/(\omega+s_i)} \qquad\qquad (5)$$
$$\theta_2 = K_2 = g^{s_i} \qquad\qquad (6)$$
$$\theta_3 = K_3 \cdot F(m)^r = u^{s_i} \cdot F(m)^r \qquad\qquad (7)$$
$$\theta_4 = g^r \qquad\qquad (8)$$
$$\theta_5 = h_j^\delta, \qquad\qquad (9)$$

  where $F(m) = v_0 \cdot \prod_{k=1}^n v_k^{m_k}$.
2. Commit to group elements θ_ℓ, for $\ell \in \{1, \ldots, 5\}$. For $\ell \in \{1, \ldots, 5\}$, choose $r_\ell, s_\ell, t_\ell \xleftarrow{R} \mathbb{Z}_p^*$ and set $\vec{\sigma_\ell} = (1, 1, \theta_\ell) \cdot \vec{g_1}^{r_\ell} \cdot \vec{g_2}^{s_\ell} \cdot \vec{g_3}^{t_\ell}$.
3. Give NIWI proofs that committed variables $\theta_1, \ldots, \theta_4$ satisfy

$$e(\theta_1, \Omega \cdot \theta_2) = A \qquad\qquad (10)$$
$$e(\theta_3, g) = e(u, \theta_2) \cdot e(F(m), \theta_4) \qquad\qquad (11)$$

Relation (10) is a quadratic pairing product equation (in the Groth-Sahai terminology) over variables θ_1, θ_2. Such a relation requires a proof consisting of 9 group elements that we denote by $\pi_1 = (\vec{\pi}_{1,1}, \vec{\pi}_{1,2}, \vec{\pi}_{1,3})$. Relation (11) is a linear pairing product equation over the variables $\theta_2, \theta_3, \theta_4$. The corresponding proof, that we denote by $\pi_2 = (\pi_{2,1}, \pi_{2,2}, \pi_{2,3}) \in \mathbb{G}^3$, consists of 3 group elements.

5. Give NIZK proofs that committed variables θ_2 and θ_5 satisfy

$$T_2 = e(\theta_2, \theta_5) \tag{12}$$
$$e(h_j, T_1) = e(g, \theta_5) \tag{13}$$

These are two linear pairing product equations over the variables θ_2 and θ_5 and proving them in NIZK requires to introduce an auxiliary variable θ_6. Proving (13) is achieved by proving in NIZK that $e(\theta_6, T_1) = e(g, \theta_5)$ and $\theta_6 = h_j$. The proof for (13) thus comprises an auxiliary commitment $\vec{\sigma_6} = \iota(h_j) \odot \vec{g_1}^{r_6} \odot \vec{g_2}^{s_6} \odot \vec{g_3}^{t_6}$ to $\theta_6 = h_j$ and proofs that relations

$$e(\theta_6, T_1) = e(g, \theta_5) \tag{14}$$
$$e(\theta_6, g) = e(h_j, g) \tag{15}$$

are simultaneously satisfied. These relations are all pairing-product equations. Relation (12) is quadratic and costs 9 group elements to prove. We will call this proofs $\pi_3 = (\vec{\pi}_{3,1}, \vec{\pi}_{3,2}, \vec{\pi}_{3,3})$. Relations (14)-(15) are linear and only require 3 group elements each. The corresponding proofs are denoted by $\pi_4 = (\pi_{4,1}, \pi_{4,2}, \pi_{4,3})$ and $\pi_5 = (\pi_{5,1}, \pi_{5,2}, \pi_{5,3})$.

The signature consists of $\sigma = (T_1, T_2, \vec{\sigma_1}, \ldots, \vec{\sigma_6}, \pi_1, \pi_2, \pi_3, \pi_4, \pi_5)$.

Verify$(j, M, \sigma, \mathsf{gpk}, RL_j)$: parse σ as $(T_1, T_2, \vec{\sigma_1}, \ldots, \vec{\sigma_6}, \pi_1, \pi_2, \pi_3, \pi_4, \pi_5)$ and return "valid" if and only if all proof are valid and σ passes the revocation test:

1. We abstracted away the construction of proof elements $\pi_1, \pi_2, \pi_3, \pi_4, \pi_5$ for clarity. To explain to proof of anonymity, it will be useful to outline what verification equations look like: namely, $\pi_1, \pi_2, \pi_3, \pi_4, \pi_5$ must satisfy

   1) $F(\vec{\sigma_1}, \iota(\Omega) \cdot \vec{\sigma_2}) = \iota_T(A) \odot F(\vec{g_1}, \vec{\pi}_{1,1}) \odot F(\vec{g_2}, \vec{\pi}_{1,2}) \odot F(\vec{g_3}, \vec{\pi}_{1,3})$
   2) $E(g, \vec{\sigma_3}) = E(u, \vec{\sigma_2}) \odot E(F(m), \vec{\sigma_4})$
      $$\odot E(\pi_{2,1}, \vec{g_1}) \odot E(\pi_{2,2}, \vec{g_2}) \odot E(\pi_{2,3}, \vec{g_3})$$
   3) $F(\vec{\sigma_2}, \vec{\sigma_5}) = F(\iota(T_2)) \odot F(\vec{\pi}_{3,1}, \vec{g_1}) \odot F(\vec{\pi}_{3,2}, \vec{g_2}) \odot F(\vec{\pi}_{3,3}, \vec{g_3})$
   4) $E(T_1, \vec{\sigma_6}) = E(\iota(g), \vec{\sigma_5}) \odot E(\pi_{4,1}, \vec{g_1}) \odot E(\pi_{4,2}, \vec{g_2}) \odot E(\pi_{4,3}, \vec{g_3})$
   5) $E(g, \vec{\sigma_6}) = E(h_j, \iota(g)) \odot E(\pi_{5,1}, \vec{g_1}) \odot E(\pi_{5,2}, \vec{g_2}) \odot E(\pi_{5,3}, \vec{g_3})$

2. The signer must not be revoked at period j: for all $B_{ij} = h_j^{s_i} \in RL_j$,

   $$T_2 \neq e(B_{ij}, T_1) \tag{16}$$

As in all VLR-GS schemes, there is an implicit tracing algorithm that can determine which group member created a valid signature using the vector of revocation tokens (and the revocation test (16)) which acts as a tracing key. We observe that, if necessary, the group manager is able to explicitly open the signature in $O(1)$ time by performing a BBS-decryption of $\vec{\sigma_2}$ using the trapdoor information α_1, α_2.

As far as efficiency goes, signatures consist of 46 elements of $\mathbb{G}$ and 1 element of $\mathbb{G}_T$. If we consider an implementation using symmetric pairings with a 256-bit group order and also assume that elements of $\mathbb{G}_T$ have a 1024-bit representation (with symmetric pairings and supersingular curves, such pairing-values can even be compressed to the third of their length as suggested in [35]), we obtain signatures of about 1.56 kB.

3.2   Security

When proving the BU-anonymity property, it seems natural to use a sequence of games starting with the real attack game and ending with a game where T_2 is replaced by a random element of $\mathbb{G}_T$ so as to leave no advantage to the adversary while avoiding to affect the adversary's view provided the Decision Bilinear Diffie-Hellman (DBDH) assumption holds. The problem becomes to simulate (using a fake common reference string) the NIZK proof that (g, T_1, h_j, K_2, T_2) forms a bilinear Diffie-Hellman tuple. Since T_2 is a given element of $\mathbb{G}_T$ in the proof, there is apparently no way to simulate the proof for relation (12).

As a natural workaround to this problem, we use the Decision Tripartite Diffie-Hellman assumption instead of the DBDH assumption in the last transition of the sequence of games.

Theorem 1 (BU-anonymity). *The scheme satisfies the backward unlinkable anonymity assuming that the Decision Linear problem and the Decision Tripartite Diffie-Hellman problem are both hard in* $\mathbb{G}$. *More precisely, we have*

$$\mathbf{Adv}_{\mathcal{A}}^{\text{bu-anon}}(\lambda) \leq T \cdot N \cdot \left(2 \cdot \mathbf{Adv}_{\mathbb{G}}^{\text{DLIN}}(\lambda) + \mathbf{Adv}_{\mathbb{G}}^{\text{DTDH}}(\lambda)\right) \tag{17}$$

where N is the maximal number of users and T is the number of time periods.

Proof. The proof is a sequence of games organized in such a way that even an unbounded adversary has no advantage in the final game while the first one is the real attack game as captured by definition 3. Throughout the sequence, we call S_i the event that the adversary wins and her advantage is $Adv_i = |\Pr[S_i] - 1/2|$.

Game 1: the challenger $\mathcal{B}$ sets up the scheme by choosing random exponents

$$\omega, \alpha, \alpha_1, \alpha_2, \xi_1, \xi_2 \xleftarrow{R} \mathbb{Z}_p^*$$

and setting g^ω and $A = e(g, g)^\alpha$. It also sets $u = g^\gamma$ for a randomly chosen $\gamma \xleftarrow{R} \mathbb{Z}_p^*$ and picks $h_1, \ldots, h_T \in \mathbb{G}$ as well as vectors $\vec{v} \in \mathbb{G}^{n+1}$, and defines $\vec{g_1} = (g_1 = g^{\alpha_1}, 1, g)$, $\vec{g_2} = (1, g_2 = g^{\alpha_2}, g)$, $\vec{g_3} = \vec{g_1}^{\xi_1} \odot \vec{g_2}^{\xi_2}$. Using ω, α, it generates users' private keys and answers all queries as in the real game. At the challenge phase, the adversary chooses two unrevoked and uncorrupted users $i_0^\star, i_1^\star$ and is given a challenge signature $\sigma^\star$ on behalf of signer $i_{d^\star}^\star$. Eventually, she outputs a guess $d' \in \{0, 1\}$ and her advantage is $Adv_1 = |\Pr[S_1] - 1/2|$, where S_1 denotes the event that $d' = d^\star$.

Game 2: we modify the simulation and let the simulator $\mathcal{B}$ pick two indices $i^\star \in \{1, \ldots, N\}, j^\star \xleftarrow{R} \{1, \ldots, T\}$ at the outset of the simulation. In the challenge phase, $\mathcal{B}$ aborts if $\mathcal{A}$'s chosen pair $(i_0^\star, i_1^\star)$ does not contain $i^\star$ or if $\mathcal{A}$ does not choose to be challenged for period $j^\star$. It also fails if $i^\star$ is ever queried for corruption or if it is queried for revocation before or during period $j^\star$. Assuming that $\mathcal{B}$ is lucky when drawing $i^\star, j^\star$ (which is the case with probability $(2/N) \cdot (1/T)$ since $i^\star$ and $j^\star$ are independent of $\mathcal{A}$'s view), the introduced failure event does not occur. We can write $Adv_2 = 2 \cdot Adv_1/(NT)$.

Game 3: we introduce a new rule that causes $\mathcal{B}$ to abort. At the challenge step, we have $i^\star \in \{i_0^\star, i_1^\star\}$ unless the failure event of Game 2 occurs. The new rule is the following: when $\mathcal{B}$ flips $d^\star \xleftarrow{R} \{0,1\}$, it aborts if $i_{d^\star}^\star \neq i^\star$. With probability $1/2$, this rule does not apply and we have $Adv_3 = 1/2 \cdot Adv_2$.

Game 4: we modify the setup phase and consider group elements $Z_1 = g^{z_1}$, $Z_2 = g^{z_2}$ that are used to generate the public key gpk and users' private keys. Namely, for $j \in \{1, \ldots, T\} \backslash \{j^\star\}$, $\mathcal{B}$ chooses $\mu_j \xleftarrow{R} \mathbb{Z}_p^*$ and defines $h_j = g^{\mu_j}$ whereas it sets $h_{j^\star} = Z_2$. Also, $\mathcal{B}$ chooses $\nu \xleftarrow{R} \mathbb{Z}_p^*$ and sets $A = e(g, Z_1 \cdot g^\omega)^\nu$ (so that α is implicitly fixed as $\alpha = \nu(z_1 + \omega)$). Private keys of users $i \neq i^\star$ are calculated as $(K_1, K_2, K_3) = ((Z_1 \cdot g^\omega)^{\nu/(\omega+s_i)}, g^{s_i}, u^{s_i})$, for a random $s_i \xleftarrow{R} \mathbb{Z}_p^*$ and using ω. Since $\mathcal{B}$ knows s_i for each $i \neq i^\star$, it can compute revocation tokens $B_{ij} = h_j^{s_i}$ for users $i \neq i^\star$ in any period.

The group signing key of the expected target user $i^\star$ is set as the triple $(K_1, K_2, K_3) = (g^\nu, Z_1, Z_1^\gamma)$, which implicitly defines $s_{i^\star} = z_1 = \log_g(Z_1)$. We note that, for periods $j \neq j^\star$, the revocation tokens $h_j^{s_{i^\star}}$ are also computable as $Z_2^{\mu_j}$. On the other hand, the token $h_{j^\star}^{s_{i^\star}} = g^{z_1 z_2}$ is not computable from Z_1, Z_2. However, unless the abortion rule of Game 2 occurs, $\mathcal{A}$ does not query it. Although $\mathcal{B}$ does not explicitly use $z_1 = \log_g(Z_1)$ and $z_2 = \log_g(Z_2)$, it still knows all users' private keys and it can use them to answer signing queries according to the specification of the signing algorithm. It comes that $\mathcal{A}$'s view is not altered by these changes and we have $\Pr[S_4] = \Pr[S_3]$.

Game 5: we bring a new change to the setup phase and generate the CRS $(\vec{g_1}, \vec{g_2}, \vec{g_3})$ by setting $\vec{g_3} = \vec{g_1}^{\xi_1} \odot \vec{g_2}^{\xi_2} \odot \iota(g)^{-1}$ instead of $\vec{g_3} = \vec{g_1}^{\xi_1} \odot \vec{g_2}^{\xi_2}$. We note that vectors $\vec{g_1}, \vec{g_3}, \vec{g_3}$ are now linearly independent. Any noticeable change in the adversary's behavior is easily seen[1] to imply a statistical test for the Decision Linear problem so that we can write $|\Pr[S_5] - \Pr[S_4]| = 2 \cdot \mathbf{Adv}^{\mathrm{DLIN}}(\mathcal{B})$.

Game 6: we modify the generation of the challenge signature and use the trapdoor (ξ_1, ξ_2) of the CRS to simulate NIZK proofs. We suppose that $\mathcal{B}$ knows values $(Z_1, Z_2, Z_3) = (g^{z_1}, g^{z_2}, g^{z_3})$ and $\eta = g^{z_1 z_2 z_3}$. Elements Z_1 and Z_2 are used to define the group public key as in Game 4 whereas Z_3 will be used to

---

[1] Indeed, $\Pr[\mathcal{B}(g_1, g_2, g_1^{\xi_1}, g_2^{\xi_2}, g^{\xi_1+\xi_2}) = 1]$ and $\Pr[\mathcal{B}(g_1, g_2, g_1^{\xi_1}, g_2^{\xi_2}, g^{\xi_1+\xi_2-1}) = 1]$ are both within distance $\mathbf{Adv}^{\mathrm{DLIN}}(\mathcal{B})$ from $\Pr[\mathcal{B}(g_1, g_2, g_1^{\xi_1}, g_2^{\xi_2}, g^z) = 1]$, where z is random.

create the challenge signature on behalf of user $i^\star$ for period $j^\star$. To this end, $\mathcal{B}$ first implicitly defines $\delta = z_3$ by setting

$$T_1 = Z_3 \qquad\qquad T_2 = e(g, \eta).$$

Elements $\theta_1, \ldots, \theta_4$ are committed to as specified by the scheme and π_1, π_2 are calculated accordingly. This time however, $\vec{\sigma_5}$ is calculated as a commitment to $1_\mathbb{G}$: namely, $\vec{\sigma_5} = \vec{g_1}^{r_5} \odot \vec{g_2}^{s_5} \odot \vec{g_3}^{t_5}$, where $r_5, s_5, t_5 \overset{R}{\leftarrow} \mathbb{Z}_p^*$. Then, $\mathcal{B}$ generates a proof $\pi_3 = (\vec{\pi}_{3,1}, \vec{\pi}_{3,2}, \vec{\pi}_{3,3})$ satisfying

$$F(\vec{\sigma_2}, \vec{\sigma_5}) = F(\iota(g), \iota(\eta)) \odot F(\vec{\pi}_{3,1}, \vec{g_1}) \odot F(\vec{\pi}_{3,2}, \vec{g_2}) \odot F(\vec{\pi}_{3,3}, \vec{g_3}). \qquad (18)$$

Such an assignment can be obtained as

$$\vec{\pi}_{3,1} = \vec{\sigma_2}^{r_5} \odot \iota(\eta)^{-\xi_1} \qquad \vec{\pi}_{3,2} = \vec{\sigma_2}^{s_5} \odot \iota(\eta)^{-\xi_2} \qquad \vec{\pi}_{3,3} = \iota(\eta) \odot \vec{\sigma_2}^{t_5}.$$

We note that the value $\theta_5 = h_{j^\star}^\delta = g^{z_2 z_3}$ is not used by $\mathcal{B}$. To simulate the proof π_3 that $T_2 = e(\theta_2, \theta_5)$ without knowing θ_5, the simulator takes advantage of the fact that $T_2 = e(g, \eta)$ for known $g, \eta \in \mathbb{G}$ (and simulating such a proof would not have been possible if T_2 had been a given element of $\mathbb{G}_T$). To simulate proofs $\pi_4 = (\pi_{4,1}, \pi_{4,2}, \pi_{4,3})$, $\pi_5 = (\pi_{5,1}, \pi_{5,2}, \pi_{5,3})$ that relations (14)-(15) are both satisfied, $\mathcal{B}$ generates π_4 as if it were a real proof using the variable assignment $\theta_5 = \theta_6 = 1_\mathbb{G}$ that obviously satisfies $e(\theta_6, T_1) = e(g, \theta_5)$ (and $\vec{\sigma_6} = \vec{g_1}^{r_6} \odot \vec{g_2}^{s_6} \odot \vec{g_3}^{t_6}$ is thus computed as a commitment to $1_\mathbb{G}$). As for π_5, the assignment

$$\pi_{5,1} = g^{r_6} \cdot h_j^{-\xi_1} \qquad \pi_{5,2} = g^{s_6} \cdot h_j^{-\xi_2} \qquad \pi_{5,3} = g^{t_6} \cdot h_j.$$

is easily seen to satisfy the last verification equation

$$E(g, \vec{\sigma_6}) = E(h_j, \iota(g)) \odot E(\pi_{5,1}, \vec{g_1}) \odot E(\pi_{5,2}, \vec{g_2}) \odot E(\pi_{5,3}, \vec{g_3})$$

since $\vec{g_3} = \vec{g_1}^{\xi_1} \odot \vec{g_2}^{\xi_2} \odot \iota(g)^{-1}$. Simulated proofs π_4, π_5 are then randomized as explained in [25] to be uniform in the space of valid proofs and achieve perfect witness indistinguishability. Simulated proofs are perfectly indistinguishable from real proofs and $\Pr[S_6] = \Pr[S_5]$.

Game 7: is identical to Game 6 but we replace η (that was equal to $g^{z_1 z_2 z_3}$ in Game 6) by a random group element. It is clear that, under the DTDH assumption, this change does not significantly alter $\mathcal{A}$'s view. We thus have $|\Pr[S_7] - \Pr[S_6]| \leq \mathbf{Adv}_{\mathbb{G}, \mathcal{B}}^{\mathrm{DTDH}}(\lambda)$.

In Game 7, it is easy to see that $\Pr[S_7] = 1/2$. Elements T_1 and T_2 are indeed completely independent of $s_{i^\star} = z_1$ (and thus of $i^\star$). Moreover, in the WI setting, all commitments $\vec{\sigma_1}, \ldots, \vec{\sigma_5}$ are perfectly hiding and proofs $\pi_1, \ldots, \pi_5$ reveal no information on underlying witnesses.

When gathering probabilities, we obtain the upper bound (17) on $\mathcal{A}$'s advantage in Game 1. $\qquad\qquad\square$

Theorem 2 (Traceability). *The scheme satisfies the full non-traceability assuming that the N-Hidden Strong Diffie-Hellman problem is hard in* $\mathbb{G}$. *More precisely, we have*

$$\mathbf{Adv}_{\mathcal{A}}^{\text{trace}}(\lambda) \leq 4 \cdot n \cdot N \cdot q_s \cdot \left(1 - \frac{(N-1)}{p}\right)^{-1}$$
$$\cdot \left(\mathbf{Adv}^{N\text{-HSDH}}(\lambda) + \mathbf{Adv}^{\text{CR}}(n)\right) \quad (19)$$

where N is maximum of the number of the adversary signature queries and the maximal number of users and T is the number of time periods.

Proof. The proof is very similar to the proof of full traceability in the Boyen-Waters [12] group signature. One difference is that [12] reduces the full traceability property of their scheme to the unforgeability of a 2-level hierarchical signature [29]. To prove this result, Boyen and Waters restricted the message space (where the element s_i, that uniquely identifies the group member is the group signature, must be chosen) to a relatively small interval at the first level.

In our proof of anonymity, we need elements s_i to be uniformly chosen in $\mathbb{Z}_p^*$. Therefore, we cannot directly link the security of our scheme to that of the 2-level hierarchical signature of [12] and a direct proof is needed (but it is simply obtained using the techniques from [12]). Namely, two kinds of forgeries must be considered as in [12]:

- **Type I forgeries** are those for which the implicit tracing algorithm fails to identify the signer using the vector of revocation tokens for the relevant period j^*.
- **Type II forgeries** are those for which the implicit tracing algorithm incriminates a user outside the coalition and that was not requested to sign the message M^* during period j^*.

The two kinds of adversaries are handled separately in lemmas 1 and 2.

To conclude the proof, we consider an algorithm $\mathcal{B}$ that guesses the kind of forgery that $\mathcal{A}$ will come up with. Then, $\mathcal{B}$ runs the appropriate HSDH solver among those described in previous lemmas. If the guess is correct, $\mathcal{B}$ solves the HSDH problem with the success probability given in the lemmas. Since this guess is correct with probability $1/2$, we obtain the claimed security bound.   □

Lemma 1. *If N is the maximal number of users, any Type I forger $\mathcal{A}$ has no advantage than* $\mathbf{Adv}_{\mathcal{A}}^{\text{Type-I}}(\lambda) \leq \mathbf{Adv}^{N\text{-HSDH}}(\lambda)$.

Proof. The proof is close to the one of lemma A.1 in [12]. The simulator $\mathcal{B}$ is given a N-HSDH instance consisting of elements $(g, \Omega = g^\omega, u)$ and triples $\{(A_i, B_i, C_i) = (g^{1/(\omega+s_i)}, g^{s_i}, u^{s_i})\}_{i=1,\ldots,N}$.

The simulator picks $\alpha, \beta_0, \ldots, \beta_n \xleftarrow{R} \mathbb{Z}_p^*$ and sets $v_i = g^{\beta_i}$, for $i = 0, \ldots, n$. Vectors $\vec{g_1}, \vec{g_2}, \vec{g_3}$ are chosen as $\vec{g_1} = (g_1 = g^{\alpha_1}, 1, g)$, $\vec{g_2} = (1, g_2 = g^{\alpha_2}, g)$ and $\vec{g_3} = \vec{g_1}^{\xi_1} \odot \vec{g_2}^{\xi_2}$, for randomly chosen $\alpha_1, \alpha_2, \xi_1, \xi_2 \xleftarrow{R} \mathbb{Z}_p^*$, in such a way that the CRS $\mathbf{g} = (\vec{g_1}, \vec{g_2}, \vec{g_3})$ provides perfectly sound proofs for which $\mathcal{B}$ retains the extraction trapdoor $(\alpha_1 = \log_g(g_1), \alpha_2 = \log_g(g_2))$. Finally, $\mathcal{B}$ generates $(h_1, \ldots, h_T) \in \mathbb{G}^T$ as $h_j = g^{\zeta_j}$, for $j = 1, \ldots, T$, with $\zeta_1, \ldots, \zeta_T \xleftarrow{R} \mathbb{Z}_p^*$. Then, $\mathcal{B}$ starts interacting with the Type I adversary $\mathcal{A}$ who is given the group public key gpk $:=$ $(g, A = e(g,g)^\alpha, h_1, \ldots, h_T, \Omega, u, \overline{v}, \mathbf{g})$ and the vector of revocation tokens grt, which $\mathcal{B}$ generates as $\text{grt}[i][j] = h_j^{s_i} = B_i^{\zeta_j}$. The simulation proceeds as follows:

- when $\mathcal{A}$ decides to corrupt user $i \in \{1, \ldots, N\}$, $\mathcal{B}$ returns the HDSH triple (A_i, B_i, C_i).
- when $\mathcal{A}$ queries a signature from user $i \in \{1, \ldots, N\}$ for a message M, $\mathcal{B}$ uses the private key $(K_1, K_2, K_3) = (A_i, B_i, C_i)$, to generate the signature by following the specification of the signing algorithm.

When $\mathcal{A}$ outputs her forgery $(M^\star, j^\star, \sigma^\star)$, $\mathcal{B}$ uses elements α_1, α_2 to decrypt $\vec{\sigma}_i^\star$, for indices $i \in \{1, \ldots, 5\}$, and obtain $\theta_1^\star = (g^\alpha)^{1/(\omega + s^\star)}$, $\theta_2^\star = g^{s^\star}$ as well as $\theta_3^\star = u^{s^\star} \cdot (v_0 \cdot \prod_{k=1}^n v_k^{m_k})^r$ and $\theta_4^\star = g^r$. From these values, $\mathcal{B}$ can extract $u^{s^\star}$ since it knows the discrete logarithm $\log_g(v_0 \cdot \prod_{k=1}^n v_k^{m_k}) = \beta_0 + \sum_{k=1}^n m_k \beta_k$, where $m_1 \ldots m_n = H(j^\star \| M^\star) \in \{0, 1\}^n$. Since $\sigma^\star$ is a Type I forgery, the implicit tracing algorithm must fail to identify one of the group members $\{1, \ldots, N\}$. The perfect soundness of the proof system implies that $s^\star \notin \{s_1, \ldots, s_N\}$ and $(\theta_1^{\star 1/\alpha}, \theta_2^\star, u^{s^\star})$ is necessarily an acceptable solution. $\qquad\square$

Lemma 2. *The scheme is secure against Type II forgeries under the $(N-1)$-HSDH assumption. The advantage of any Type II adversary $\mathcal{A}$ is at most*

$$\mathbf{Adv}_{\mathcal{A}}^{\text{Type-II}}(\lambda, n) \leq 2 \cdot n \cdot N \cdot q_s \cdot \left(1 - \frac{(N-1)}{p}\right)^{-1} \cdot \left(\mathbf{Adv}^{(N-1)\text{-HSDH}}(\lambda) + \mathbf{Adv}^{\text{CR}}(n)\right)$$

where N and q_s stand for the number of users and the number of signing queries, respectively, and the last term accounts for the probability of breaking the collision-resistance of H.

Proof. The proof is based on lemma A.2 in [12]. Namely, the simulator $\mathcal{B}$ receives a $(N-1)$-HSDH input comprising $(g, \Omega = g^\omega, u)$ and a set of triples $\{(A_i, B_i, C_i) = (g^{1/(\omega + s_i)}, g^{s_i}, u^{s_i})\}_{i=1, \ldots, N-1}$.

To prepare the public key gpk, the simulator $\mathcal{B}$ picks a random index $\nu \xleftarrow{R} \{0, \ldots, n\}$, as well as $\rho_0, \ldots, \rho_n \xleftarrow{R} \mathbb{Z}_p^*$ and integers $\beta_0, \ldots, \beta_n \xleftarrow{R} \{0, \ldots, 2q_s - 1\}$. It sets $v_0 = u^{\beta_0 - 2\nu q_s} \cdot g^{\rho_0}$, $v_i = u^{\beta_i} \cdot g^{\rho_i}$ for $i = 1, \ldots, n$. It also defines $h_1, \ldots, h_T$ by setting $h_j = g^{\zeta_j}$, with $\zeta_j \xleftarrow{R} \mathbb{Z}_p^*$, for $j = 1, \ldots, T$. It finally chooses vectors $\mathbf{g}$ as specified by the setup algorithm to obtain perfectly sound proofs.

Before starting its interaction with the Type II forger $\mathcal{A}$, $\mathcal{B}$ initializes a counters $ctr \leftarrow 0$ and chooses an index $i^\star \xleftarrow{R} \{1, \ldots, N\}$ as a guess for the honest user on behalf of which $\mathcal{A}$ will attempt to generate a forgery. The simulation proceeds by handling $\mathcal{A}$'s queries in the following way.

Queries: at the first time that user $i \in \{1, \ldots, N\}$ is involved in a signing query or a corruption query, $\mathcal{B}$ does the following:

- if the query is a corruption query, $\mathcal{B}$ halts and declares failure if $i = i^\star$ as it necessarily guessed the wrong user $i^\star$. Otherwise, it increments ctr and returns the triple $(A_{ctr}, B_{ctr}, C_{ctr})$ as a private key for user (K_1, K_2, K_3).

- if the query is a signing query for period $j \in \{1, \ldots, T\}$,
    - if $i \neq i^\star$ $\mathcal{B}$ increments ctr and answers the query by running the signing algorithm using the private key $(K_1, K_2, K_3) = (A_{ctr}, B_{ctr}, C_{ctr})$.
    - if $i = i^\star$, $\mathcal{B}$ chooses $t^\star \overset{R}{\leftarrow} \mathbb{Z}_p^*$ at random and implicitly defines a triple $(K_1^\star, K_2^\star, K_3^\star) = (g^{1/t^\star}, g^{t^\star} \cdot \Omega^{-1}, *)$, where $*$ is a placeholder for an unknown group element (note that this implicitly defines $s^* = t^\star - \omega$). Then, $\mathcal{B}$ computes $m_1 \ldots m_n = H(j\|M) \in \{0,1\}^n$. At this stage, it is convenient to write $F(m_1 \ldots m_n) = v_0 \cdot \prod_{k=1}^{n} v_k^{m_k}$ as $F(m_1 \ldots m_n) = u^J \cdot g^K$ where $J = \beta_0 - 2\nu q_s + \sum_{j=1}^{n} \beta_j m_j$, $K = \rho_0 + \sum_{j=1}^{n} \rho_j m_j$. If $J = 0$, $\mathcal{B}$ aborts. Otherwise, it can pick $r \overset{R}{\leftarrow} \mathbb{Z}_p^*$ and compute a pair

$$\left(\theta_3 = u^{t^\star} \cdot F(m_1 \ldots m_n)^r \cdot \Omega^{\frac{K}{J}}, \ \theta_4 = g^r \cdot \Omega^{\frac{1}{J}} \right),$$

which can be re-written as $(\theta_4 = u^{t^\star - \omega} \cdot F(m_1 \ldots m_n)^{\tilde{r}}, \theta_5 = g^{\tilde{r}})$ if we define $\tilde{r} = r + \omega/J(\mathsf{m})$. This pair then allows generating a suitably anonymized signature. In particular, since $\mathcal{B}$ knows $\theta_2 = K_2^\star = g^{t^\star} \cdot \Omega^{-1}$, it is able to compute $T_2 = e(h_j, K_2^\star)^\delta$ and $T_1 = g^\delta$ for a random $\delta \overset{R}{\leftarrow} \mathbb{Z}_p^*$.

When subsequent queries involve the same user i, $\mathcal{B}$ responds as follows (we assume that corruption queries are distinct):

- For corruption queries on users $i \in \{1, \ldots, N\}$ that were previously involved in signing queries, $\mathcal{B}$ aborts if $i = i^\star$. Otherwise, it knows the private key (K_1, K_2, K_3) (that was used to answer signing queries) and hands it to $\mathcal{A}$.
- For signing queries, $\mathcal{B}$ uses the same values as in the first query involving the user $i \in \{1, \ldots, N\}$. If $i \neq i^\star$, $\mathcal{B}$ uses the same triple $(A_{ctr}, B_{ctr}, C_{ctr})$. In the case $i = i^\star$, $\mathcal{B}$ re-uses the pair $(K_1^\star, K_2^\star) = (g^{1/t^\star}, g^{t^\star} \cdot \Omega^{-1})$ and proceeds as in the first query involving $i^\star$ (but uses a fresh random exponent r).

Forgery: the game ends with the adversary outputting message $M^\star$ together with a type II forgery $\sigma^\star = (T_1^\star, T_2^\star, \vec{\sigma_1}^\star, \ldots, \vec{\sigma_6}^\star, \pi_1^\star, \ldots, \pi_5^\star)$ for some period $j^\star \in \{1, \ldots, T\}$. By assumption, the implicit tracing algorithm must point to some user who did not sign $M^\star$ at period $j^\star$. Then, $\mathcal{B}$ halts and declares failure if $\sigma^\star$ does not trace to user $i^\star$. Since the chosen index $i^\star$ was independent of $\mathcal{A}$'s view, with probability $1/N$, $\mathcal{B}$'s guess turns out to be correct. Then, the perfect soundness of the proof system implies that $\vec{\sigma_2}^\star$ is a BBS encryption of $K_2^\star$. Then, $\mathcal{B}$ computes $\mathsf{m}^\star = m_1 \ldots m_n = H(j^\star \| M^\star)$. If user $i^\star$ signed a message M at period j such that $(j, M) \neq (j^\star, M^\star)$ but $H(j\|M) = H(j^\star\|M^\star)$, $\mathcal{A}$ was necessarily able to generate a collision on H. Otherwise, the perfect soundness of the proof system implies that $\vec{\sigma_3}^\star$ and $\vec{\sigma_4}^\star$ decrypt into

$$\theta_3^\star = u^{t^\star - \omega} F(\mathsf{m}^\star)^r \qquad \theta_4^\star = g^r$$

for some $r \in \mathbb{Z}_p^*$ and where $F(\mathsf{m}^\star) = v_0 \cdot \prod_{k=1}^{n} v_k^{m_k} = u^{J^\star} \cdot g^{K^\star}$ and $s_\star = t_{i^\star} - \omega$. Then, $\mathcal{B}$ aborts if $J(\mathsf{m}^\star) = \beta_0 + \sum_{j=1}^{n} \beta_j m_j - 2\nu q_s \neq 0$. Otherwise, $\mathcal{B}$ can compute $u^{s^\star}$ and thereby obtains a full tuple $\left(g^{1/(\omega + s^\star)}, g^{s^\star}, u^{s^\star} \right)$ where $s^\star = t^\star - \omega$ differs from $s_1, \ldots, s_{N-1}$ with probability at least $1 - (N-1)/p$ (since the value t^* was chosen at random).

$\mathcal{B}$'s probability not to abort throughout the simulation can assessed as in [37,12]. More precisely, one can show that $J \neq 0$ in all signing queries with probability greater than $1/2$. Conditionally on the event that $\mathcal{B}$ does not abort before the forgery stage, the probability to have $J^* = 0$ is then shown to be at least $1/(2nq_s)$ (see [37,12] for details). $\square$

3.3   A Variant with Shorter Group Public Keys

As described in this section, the scheme suffers from a group public key of size $O(T)$, which makes it impractical when the number of time periods is very large. In the random oracle model $h_1, \ldots, h_T$ could be derived from a random oracle. However, avoiding the dependency on T in the group public key size is also possible without resorting to random oracles. This can be achieved using the techniques introduced in [7] in the context of identity-based encryption.

The vector $(h_1, \ldots, h_T)$ is replaced by a triple $(h, h_0, h_1) \in \mathbb{G}^3$ and the revocation token of user i at period $j \in \{1, \ldots, T\}$ is defined to be the pair $(B_{ij1}, B_{ij2}) = (h^{s_i} \cdot F(j)^\rho, g^\rho)$, where $\rho \xleftarrow{R} \mathbb{Z}_p^*$ and $F(j) = h_0 \cdot h_1^j$ is the selectively-secure identity-hashing function of Boneh and Boyen [7]. Since the revocation token (B_{ij1}, B_{ij2}) satisfies the relation $e(B_{ij1}, g) = e(h, g^{s_i}) \cdot e(F(j), B_{ij2})$, we have $e(B_{ij1}, g^\delta) = e(h, g^{s_i})^\delta \cdot e(F(j)^\delta, B_{ij2})$ for any $\delta \in \mathbb{Z}_p^*$.

Therefore, in each signature σ, the pair (T_1, T_2) is superseded by a triple $(T_1, T_2, T_3) = (g^\delta, F(j)^\delta, e(h, K_2)^\delta)$ (so that the verifier needs the check that $e(T_1, F(j)) = e(g, T_2)$) whereas $\vec{\sigma}_5$ becomes a commitment to $\theta_5 = h^\delta$ and the NIZK proof for relation (13) is replaced by a proof that $e(h, T_1) = e(g, \theta_5)$. At step 2 of the verification algorithm, the revocation test then consists in testing whether $e(T_1, B_{ij1}) = T_3 \cdot e(T_2, B_{ij2})$ for revocation tokens $\{(B_{ij1}, B_{ij2})\}_{i \in RL_j}$. Using the technique of [7] to generate tokens for periods $j \in \{1, \ldots, T\} \setminus \{j^*\}$, it can be checked that everything goes through in the proof of anonymity.

4   Conclusion

We described a simple way to provide Boyen-Waters group signatures with an efficient verifier local revocation mechanism with backward unlinkability.

The scheme can be easily extended so as to provide exculpability (and prevent the group manager from signing on behalf of users) using a dynamic joining protocol such as the one of [30]. It would be interesting to turn the scheme into a traceable signature [27] supporting fine-grained (*i.e.* per period) user tracing while leaving users the ability to claim their signatures.

References

1. Ateniese, G., Camenisch, J., Hohenberger, S., de Medeiros, B.: Practical Group Signatures without Random Oracles. Cryptology ePrint Archive, Report 2005/385 (2005), http://eprint.iacr.org/2005/385

2. Ateniese, G., Camenisch, J., Joye, M., Tsudik, G.: A Practical and Provably Secure Coalition-Resistant Group Signature Scheme. In: Bellare, M. (ed.) CRYPTO 2000. LNCS, vol. 1880, pp. 255–270. Springer, Heidelberg (2000)
3. Ateniese, G., Song, D., Tsudik, G.: Quasi-efficient revocation of group signatures. In: Blaze, M. (ed.) FC 2002. LNCS, vol. 2357, pp. 183–197. Springer, Heidelberg (2003)
4. Bellare, M., Micciancio, D., Warinschi, B.: Foundations of Group Signatures: Formal Definitions, Simplified Requirements, and a Construction Based on General Assumptions. In: Biham, E. (ed.) EUROCRYPT 2003. LNCS, vol. 2656, pp. 614–629. Springer, Heidelberg (2003)
5. Bellare, M., Rogaway, P.: Random Oracles are Practical: A Paradigm for Designing Efficient Protocols. In: 1st ACM Conference on Computer and Communications Security, pp. 62–73. ACM Press, New York (1993)
6. Bellare, M., Shi, H., Zhang, C.: Foundations of Group Signatures: The Case of Dynamic Groups. In: Menezes, A. (ed.) CT-RSA 2005. LNCS, vol. 3376, pp. 136–153. Springer, Heidelberg (2005)
7. Boneh, D., Boyen, X.: Short Signatures Without Random Oracles. In: Cachin, C., Camenisch, J.L. (eds.) EUROCRYPT 2004. LNCS, vol. 3027, pp. 56–73. Springer, Heidelberg (2004)
8. Boneh, D., Boyen, X.: Efficient Selective-ID Secure Identity-Based Encryption Without Random Oracles. In: Cachin, C., Camenisch, J.L. (eds.) EUROCRYPT 2004. LNCS, vol. 3027, pp. 223–238. Springer, Heidelberg (2004)
9. Boneh, D., Boyen, X., Shacham, H.: Short Group Signatures. In: Franklin, M. (ed.) CRYPTO 2004. LNCS, vol. 3152, pp. 41–55. Springer, Heidelberg (2004)
10. Boneh, D., Franklin, M.: Identity based encryption from the Weil pairing. SIAM J. of Computing 32(3), 586–615 (2003); Extended abstract in Crypto 2001. LNCS, vol. 2139, pp. 213–229 (2001)
11. Boneh, D., Shacham, H.: Group signatures with verifier-local revocation. In: ACM-CCS 2004, pp. 168–177. ACM Press, New York (2004)
12. Boyen, X., Waters, B.: Full-Domain Subgroup Hiding and Constant-Size Group Signatures. In: Okamoto, T., Wang, X. (eds.) PKC 2007. LNCS, vol. 4450, pp. 1–15. Springer, Heidelberg (2007)
13. Boyen, X., Waters, B.: Compact Group Signatures Without Random Oracles. In: Vaudenay, S. (ed.) EUROCRYPT 2006. LNCS, vol. 4004, pp. 427–444. Springer, Heidelberg (2006)
14. Bresson, E., Stern, J.: Efficient Revocation in Group Signatures. In: Kim, K.-c. (ed.) PKC 2001. LNCS, vol. 1992, pp. 190–206. Springer, Heidelberg (2001)
15. Brickell, E.: An efficient protocol for anonymously providing assurance of the container of the private key. Submission to the Trusted Computing Group (April 2003)
16. Canetti, R., Goldreich, O., Halevi, S.: The random oracle methodology, revisited. Journal of the ACM 51(4), 557–594 (2004)
17. Chaum, D., van Heyst, E.: Group Signatures. In: Davies, D.W. (ed.) EUROCRYPT 1991. LNCS, vol. 547, pp. 257–265. Springer, Heidelberg (1991)
18. Choi, S.G., Park, K., Yung, M.: Short Traceable Signatures Based on Bilinear Pairings. In: Yoshiura, H., Sakurai, K., Rannenberg, K., Murayama, Y., Kawamura, S.-i. (eds.) IWSEC 2006. LNCS, vol. 4266, pp. 88–103. Springer, Heidelberg (2006)
19. Delerablée, C., Pointcheval, D.: Dynamic fully anonymous short group signatures. In: Nguyên, P.Q. (ed.) VIETCRYPT 2006. LNCS, vol. 4341, pp. 193–210. Springer, Heidelberg (2006)
20. Fiat, A., Shamir, A.: How to prove yourself: Practical Solutions to Identification and Signature Problems. In: Odlyzko, A.M. (ed.) CRYPTO 1986. LNCS, vol. 263, pp. 186–194. Springer, Heidelberg (1987)

21. Furukawa, J., Imai, H.: An Efficient Group Signature Scheme from Bilinear Maps. In: Boyd, C., González Nieto, J.M. (eds.) ACISP 2005. LNCS, vol. 3574, pp. 455–467. Springer, Heidelberg (2005)
22. Goldwasser, S., Tauman-Kalai, Y.: On the (In)security of the Fiat-Shamir Paradigm. In: FOCS 2003, pp. 102–115 (2003)
23. Groth, J.: Simulation-Sound NIZK Proofs for a Practical Language and Constant Size Group Signatures. In: Lai, X., Chen, K. (eds.) ASIACRYPT 2006. LNCS, vol. 4284, pp. 444–459. Springer, Heidelberg (2006)
24. Groth, J.: Fully anonymous group signatures without random oracles. In: Kurosawa, K. (ed.) ASIACRYPT 2007. LNCS, vol. 4833, pp. 164–180. Springer, Heidelberg (2007)
25. Groth, J., Sahai, A.: Efficient non-interactive proof systems for bilinear groups. In: Smart, N.P. (ed.) EUROCRYPT 2008. LNCS, vol. 4965, pp. 415–432. Springer, Heidelberg (2008)
26. Hofheinz, D., Kiltz, E.: Programmable Hash Functions and Their Applications. In: Wagner, D. (ed.) CRYPTO 2008. LNCS, vol. 5157, pp. 21–38. Springer, Heidelberg (2008)
27. Kiayias, A., Tsiounis, Y., Yung, M.: Traceable Signatures. In: Cachin, C., Camenisch, J.L. (eds.) EUROCRYPT 2004. LNCS, vol. 3027, pp. 571–589. Springer, Heidelberg (2004)
28. Kiayias, A., Yung, M.: Group Signatures: Provable Security, Efficient Constructions and Anonymity from Trapdoor-Holders. Cryptology ePrint Archive: Report 2004/076 (2004), http://eprint.iacr.org/2004/076
29. Kiltz, E., Mityagin, A., Panjwani, S., Raghavan, B.: Append-Only Signatures. In: Caires, L., Italiano, G.F., Monteiro, L., Palamidessi, C., Yung, M. (eds.) ICALP 2005. LNCS, vol. 3580, pp. 434–445. Springer, Heidelberg (2005)
30. libert, B., Yung, M.: Efficient Traceable Signatures in the Standard Model. In: Shacham, H. (ed.) Pairing 2009. LNCS, vol. 5671, pp. 187–205. Springer, Heidelberg (2009)
31. Naor, M.: On Cryptographic Assumptions and Challenges. In: Boneh, D. (ed.) CRYPTO 2003. LNCS, vol. 2729, pp. 96–109. Springer, Heidelberg (2003)
32. Nguyen, L., Safavi-Naini, R.: Efficient and Provably Secure Trapdoor-Free Group Signature Schemes from Bilinear Pairings. In: Lee, P.J. (ed.) ASIACRYPT 2004. LNCS, vol. 3329, pp. 372–386. Springer, Heidelberg (2004)
33. Nakanishi, T., Funabiki, N.: Verifier-local revocation group signature schemes with backward unlinkability from bilinear maps. In: Roy, B. (ed.) ASIACRYPT 2005. LNCS, vol. 3788, pp. 533–548. Springer, Heidelberg (2005)
34. Nakanishi, T., Funabiki, N.: A Short Verifier-Local Revocation Group Signature Scheme with Backward Unlinkability. In: Yoshiura, H., Sakurai, K., Rannenberg, K., Murayama, Y., Kawamura, S.-i. (eds.) IWSEC 2006. LNCS, vol. 4266, pp. 17–32. Springer, Heidelberg (2006)
35. Scott, M., Barreto, P.S.L.M.: Compressed Pairings. In: Franklin, M. (ed.) CRYPTO 2004. LNCS, vol. 3152, pp. 140–156. Springer, Heidelberg (2004)
36. Song, D.X.: Practical forward secure group signature schemes. In: ACM-CCS 2001, pp. 225–234. ACM Press, New York (2001)
37. Waters, B.: Efficient identity-based encryption without random oracles. In: Desmedt, Y.G. (ed.) PKC 2003. LNCS, vol. 2567. Springer, Heidelberg (2002)
38. Zhou, S., Lin, D.: A Shorter Group Signature with Verifier-Location Revocation and Backward Unlinkability. Cryptology ePrint Archive: Report 2006/100 (2006), http://eprint.iacr.org/2006/100
39. Zhou, S., Lin, D.: Shorter Verifier-Local Revocation Group Signatures from Bilinear Maps. In: Pointcheval, D., Mu, Y., Chen, K. (eds.) CANS 2006. LNCS, vol. 4301, pp. 126–143. Springer, Heidelberg (2006)

Relinkable Ring Signature

Koutarou Suzuki[1], Fumitaka Hoshino[2], and Tetsutaro Kobayashi[1]

[1] NTT Information Sharing Platform Laboratories, NTT Corporation,
3-9-11 Midori-cho, Musashino-shi, Tokyo, 180-8585 Japan
suzuki.koutarou@lab.ntt.co.jp, kobayashi.tetsutaro@lab.ntt.co.jp
[2] IPA : Information-Technology Promotion Agency, Japan
2-28-8, Hon-Komagome, Bunkyo-ku, Tokyo, 113-6591 Japan
f-hoshi@ipa.go.jp

Abstract. In this paper, we propose the concept of a relinkable ring signature, which is a ring signature with ring reformation function, i.e., a signer can delegate ring reformation ability separately from signing ability to his/her proxy. The relinkable ring signature can be applicable to proxy ring reformation, anonymization of past-generated signature, or ring signature for dynamic group. We also propose a concrete relinkable ring signature scheme that uses pairing in the random oracle model.

Keywords: ring signature, anonymity, pairing.

1 Introduction

Ring signature, where a signer can sign anonymously on behalf of a group, the ring members, without a setup procedure or group manager, was introduced in [RST01]. The signer generates a ring signature for a message using his/her secret key and the public keys of all the ring members. Thus, by the moment of ring signature generation, the ring members need to be determined and their public keys need to be provided.

In this paper, we propose the concept of a *relinkable ring signature*: an extension of a ring signature where ring members do not need to be determined by the moment of ring signature generation and the ring members of the generated ring signature can be changed at a later point. Compared with the usual ring signature scheme, a relinkable ring signature scheme has a *relink algorithm* that can change the ring members of a given signature by using the relink key rk after the signature is generated using the signing key sk. The relink key rk is weaker than the signing key sk and can change the ring members but cannot change the message and the real signer, i.e., using relink key rk, one can create a new ring signature for the same message with different ring members that include the same real signer from the existing ring signature, but one cannot create a new ring signature for a new message and a new real signer. Thus, using the relinkable ring signature, one can separately select a message and ring members.

The relinkable ring signature provides restricted anonymity in comparison with the usual ring signature, i.e., it guarantees only computational anonymity,

J.A. Garay, A. Miyaji, and A. Otsuka (Eds.): CANS 2009, LNCS 5888, pp. 518–536, 2009.

while the usual ring signature guarantees unconditional anonymity. Moreover, using the relink key rk, one can check whether or not the real signer of a ring signature is the signer corresponding to the the relink key rk by changing the ring members to a set that consists of only the corresponding signer. Thus, the relinkable ring signature is not suitable for applications that highly require anonymity, e.g., voting or whistle-blowing. However, the restricted anonymity of the relinkable ring signature can be utilized for the following applications.

Proxy Ring Formation: By providing relink key rk to the signer's proxy, the signer can delegate the ring reformation ability to the proxy separately from the signing ability. This is useful for a signer with small computational resources. For instance, one can securely store the signing key sk in a tamper-resistant IC card that has only small computational resources and store the relink key rk in a PC that has large computational resources and access to PKI. The IC card computes a ring signature with ring members including only the signer by using the signing key sk, then the PC reforms the ring members of the ring signature by using the relink key rk and public keys of the other ring members from PKI. By this, one can isolate a ring formation process whose computational cost is heavy, i.e., is proportional to the number of ring members, and delegate it to a PC with large computational resources.

Anonymization of Past-generated Signature: In the case that one publicizes a document with a signature, he/she needs to ensure the privacy of the signer, e.g., in the case of publication of a governmental document by a "freedom of information act". To hide the content of the document, one can use a sanitizing signature [SBZ01]. To hide the signer, one can use the relinkable ring signature, i.e., one can anonymize a past-generated signature by using relink key rk. The signer submits the document, the relinkable ring signature on it with ring members including only the signer, and relink key rk. When the document is publicized, one can anonymize the signature by reforming the ring members using relink key rk.

Ring Signature for Dynamic Group: In a ring signature scheme [RST01], a signer can sign anonymously on behalf of the ring members and there is no group manager. However, this would not be suitable for a dynamic group whose members change, since after the public key of a member is removed, one can no longer verify stored past-generated ring signatures whose ring members include the removed member. The proposed relinkable ring signature can resolve this problem as follows. When a new member joins the group, the new member registers his/her public key to the PKI of the group and passes his/her relink key to the relink manager. When a member leaves the group, the PKI removes the public key of that member and the relink manager removes that member from the ring members of stored past-generated ring signatures by using that member's relink key.

Private Key Exposure Attack: The usual ring signature scheme that has unconditional anonymity is susceptible to a private key exposure attack, i.e., once a signer makes his/her private key public, all ring signatures whose ring members include that signer become meaningless because anyone can use the publicized

private key to generate the signature. The relinkable ring signature is not susceptible to a private key exposure attack because it has computational anonymity
and one can exclude the signer who exposed his/her private key from the ring
members of a signature by using relink key rk.

Convertible Ring Signature: One can gradually decrease the anonymity of past-
generated ring signatures by decreasing the number of ring members of the signature by using relink key rk. In an extreme case, one can convert a past-generated
ring signature to non-anonymous signature by making the ring members include
only the signer. This is similar to a convertible ring signature [LWH05], though
one has a conversion key for each signature in the convertible ring signature.

Ring Signature and Group Signature: In a group signature scheme [CvH91], there
is a group manager who can revoke the anonymity of signatures. In contrast, in
a ring signature scheme [RST01], there is no group manager. As an intermediary
between the ring and group signatures, the revocable ring signature [LLM+07]
was invented. The relinkable ring signature can also be considered an intermediary between the ring and group signatures, where a signer can generate a
signature without a setup procedure and the signer's proxy who has relink key
rk can revoke anonymity.

The proposed relinkable ring signature scheme is secure in the random oracle model, and uses groups with efficiently computable pairing on a non-
supersingular elliptic curve known as an MNT curve [MNT01], where there is
no efficiently computable distortion map. More precisely, we use three assumptions in groups G_1, G_2, and G_3, where there is efficiently computable pairing
$e : G_1 \times G_2 \rightarrow G_3$ but no efficiently computable distortion map $\psi : G_1 \rightarrow G_2$.
These groups are studied by [SHUK03, GPS08] and used in some existing
schemes [Sco02, ACdM05, BGdMM05, ACHdM05] where the XDH (eXternal
Diffie-Hellman) assumption, i.e., the DDH problem in the group G_1 is intractable, is used.

In Section 2, we describe bilinear groups on a non-supersingular elliptic curve
and three assumptions used in the proposed scheme. In Section 3, we define
the relinkable ring signature. In Section 4, we show the proposed relinkable ring
signature scheme using pairing, prove its security, and estimate its efficiency. In
Section 5, we conclude the paper.

2   Bilinear Group and Assumptions

In this section, we describe groups with efficiently computable pairing but without an efficiently computable distortion map, we also describe three assumptions
on these groups used in the proposed relinkable ring signature scheme.

We use the pairing on a non-supersingular elliptic curve known as an MNT
curve [MNT01], where no distortion map is known. By using an MNT curve, we
can construct cyclic groups G_1, G_2, and G_3 of prime order p, which are called a
bilinear groups, and a polynomial-time computable bilinear non-degenerate map
called pairing

$$e : G_1 \times G_2 \to G_3.$$

Let $g \in G_1$ be a generator of G_1 and $\hat{g} \in G_2$ be a generator of G_2. See Appendix B for a detailed construction.

On MNT curves [MNT01], no polynomial-time computable homomorphism $\psi : G_1 \to G_2$, called a distortion map, is known. See Appendix B for details.

These bilinear groups are studied by [SHUK03, GPS08], and used in some existing schemes [Sco02, ACdM05, BGdMM05, ACHdM05] where the XDH (eXternal Diffie-Hellman) assumption, i.e., the DDH problem in the group G_1 is intractable, is used.

We now state the three assumptions on these groups that are used in the proposed relinkable ring signature scheme as follows.

For adversary A, we define advantage

$$\text{Adv}^{\text{skewCDH}}(A) = \Pr[g \in_U G_1, \hat{g} \in_U G_2, \alpha \in \mathbb{Z}_p, A(g, g^\alpha, \hat{g}) = \hat{g}^\alpha],$$

where the probability is taken over the choices of $g, \hat{g}, \alpha$ and the coin tosses of A.

Definition 1 (Skew CDH Assumption from G_1 to G_2). *We assume that for all polynomial-time adversary A, advantage $\text{Adv}^{\text{skewCDH}}(A)$ is negligible in security parameter k.*

For adversary A, we define advantage

$$\text{Adv}^{\text{hintedCDH}}(A) = \Pr[g \in_U G_1, \hat{g} \in_U G_2, \alpha, \beta \in \mathbb{Z}_p, A(g, g^\alpha, g^\beta, \hat{g}, \hat{g}^\alpha, \hat{g}^\beta) = g^{\alpha\beta}],$$

where the probability is taken over the choices of $g, \hat{g}, \alpha, \beta$ and the coin tosses of A.

Definition 2 (Hinted CDH Assumption in G_1). *We assume that for all polynomial-time adversary A, advantage $\text{Adv}^{\text{hintedCDH}}(A)$ is negligible in security parameter k.*

We denote by $D_1 = \{(g, h, g', h') \in G_1^4 | \log_g h = \log_{g'} h'\}$ the set of DDH tuple, and by $D_0 = \{(g, h, g', h') \in G_1^4\}$ the set of random tuple. For adversary A, we define advantage

$$\text{Adv}^{\text{DDH}}(A) = |\Pr[b \in_U \{0,1\}, X \in_U D_b : A(X) = b] - 1/2|$$

where the probability is taken over the choices of b, X and the coin tosses of A.

Definition 3 (DDH Assumption in G_1). *We assume that for all polynomial-time adversary A, advantage $\text{Adv}^{\text{DDH}}(A)$ is negligible in security parameter k.*

Notice that if there exists polynomial-time computable distortion map $\psi : G_1 \to G_2$, "Skew CDH Assumption from G_1 to G_2" and "DDH Assumption in G_1" are not true and "Hinted CDH Assumption in G_1" is equivalent to "CDH Assumption in G_1".

3 Relinkable Ring Signature

In this section, we provide the definition of relinkable ring signature. *Anonymity* means, informally, that adversary cannot distinguish test signature is generated by signer 0 or by signer 1, where the adversary knows all secret and relink keys except of signer 0 and 1. *Traceability* means, informally, that the real signer of signature generated by adversary can be determined uniquely, where the adversary knows all secret and relink keys. *Unforgeability* means, informally, that adversary cannot create new forged signature and cannot modify a signature from signing oracle, where the adversary does not know secret and relink keys. *Relinker unforgeability* means, informally, that adversary cannot create new forged signature and cannot modify message and real signer of a signature from signing oracle, where the adversary knows relink keys. Our definition of anonymity does not adopt *adversarially-chosen keys* and *full key exposure* [BKM06], since exposure of revoke keys trivially breaks anonymity. Our definitions of unforgeability and relinker unforgeability adopt *insider corruption* [BKM06].

3.1 Definition of Relinkable Ring Signature

We provide the definition of the relinkable ring signature scheme. In this scheme there are two secret keys: signing key sk by which signer can generate a ring signature, and relink key rk by which relinker can reform the ring member of generated ring signature.

We denote the set of signers $N = \{0, 1, ...\}$. We also denote subset of signers $L \subset N$ that is called ring.

Syntax. A relinkable ring signature scheme is a tuple of four algorithms $\Sigma = (\text{Gen}, \text{Sig}, \text{Ver}, \text{Rel})$, s.t.

- Gen, the key generation algorithm, is a probabilistic polynomial-time algorithm that takes security parameter $k \in \mathbb{N}$, and outputs secret, relink, and public key (sk, rk, pk):

$$\text{Gen}(k) \to (sk, rk, pk).$$

We denote by (sk_i, rk_i, pk_i) the public, secret, and relink key of the i-th signer.
- Sig, the signing algorithm, is a probabilistic polynomial-time algorithm that takes secret key sk_i, ring $L \subset N$ s.t. $i \in L$, set of public keys of L, and message $m \in \{0, 1\}^*$, and outputs signature σ:

$$\text{Sig}(sk_i, L, (pk_j)_{j \in L}, m) \to \sigma.$$

- Ver, the signature verification algorithm, is a probabilistic polynomial-time algorithm that takes ring $L \subset N$, set of public keys of L, message $m \in \{0, 1\}^*$, and signature σ, and outputs a bit $0/1$ that means reject/accept, respectively:

$$\text{Ver}(L, (pk_j)_{j \in L}, m, \sigma) \to 0/1.$$

– Rel, the relinking algorithm, is a probabilistic polynomial-time algorithm that takes relink key rk_i, rings $L, L' \subset N$ s.t. $i \in L, L'$, sets of public keys of $L \cup L'$, message $m \in \{0,1\}^*$, and signature σ, and outputs new signature σ':

$$\mathrm{Rel}(rk_i, L, L', (pk_j)_{j \in L \cup L'}, m, \sigma) \to \sigma'/"reject".$$

A relinkable ring signature scheme satisfies the following correctness.

Correctness. For every $i \in N$, every $L_1, ..., L_J \subset N$ s.t. $i \in L_1, ..., L_J$, and every $m \in \{0,1\}^*$, if $\mathrm{Gen}(k) \to (sk_i, rk_i, pk_i)$, $\mathrm{Sig}(sk_i, L_1, (pk_l)_{l \in L_1}, m) \to \sigma_1$, $\mathrm{Rel}(rk_i, L_j, L_{j+1}, (pk_l)_{l \in L_j \cup L_{j+1}}, m, \sigma_j) \to \sigma_{j+1}$ for $j = 1, ..., J - 1$, it holds with overwhelming probability that $\mathrm{Ver}(L_j, (pk_l)_{l \in L_j}, m, \sigma_j) = 1$ for $j = 1, ..., J$.

We first define the following three oracles called by adversary in the games of security definitions. We then define the following four security notions of relinkable ring signature: anonymity, unforgeability, relinker unforgeability, and traceability.

Key Registration Oracle $KO(i, rk_i, pk_i)$. Let $N = \{0, 1, ..\}$ be set of registered signers. Let $RK = \{rk_1, rk_2, ...\}$ and $PK = \{pk_1, pk_2, ...\}$ be set of registered relink keys and public keys. Adversary generates secret, relink, and public keys $(sk_i, rk_i, pk_i) \leftarrow Gen(k)^1$ and send index of key i, relink keyrk_i, and public key pk_i to key registration oracle KO. Key registration oracle KO registers the keys sent from the adversary, i.e., appends i to N, rk_i to RK, and pk_i to PK.

Signing Oracle $SO(i, L, m)$. Adversary sends index of signing key i, ring $L \subset N$, and message m to signing oracle SO. Signing oracle SO generates signature $\mathrm{Sig}(sk_i, L, (pk_j)_{j \in L}, m) \to \sigma$ and return signature σ.

Relink Oracle $RO(L, L', m, \sigma)$. Adversary sends ring $L, L' \subset N$, message m, and signature σ to relink oracle RO. If $\mathrm{Ver}(L, (pk_j)_{j \in L}, m, \sigma) \neq 1$, return "reject". Relink oracle RO finds i^* s.t. $\mathrm{Ver}(\{i^*\}, pk_i, m, \mathrm{Rel}(rk_i, L, \{i^*\}, (pk_j)_{j \in L}, m, \sigma)) = 1$. Here, such i^* is unique because of Traceability. If $i^* \in L, L'$ does not hold, return "reject". Relink oracle RO generates relinked signature $\mathrm{Rel}(rk_{i^*}, L, L', (pk_j)_{j \in L \cup L'}, m, \sigma) \to \sigma'$ and return relinked signature σ'.

Anonymity. We define the anonymity of a relinkable ring signature scheme Σ. We consider the following game of adversary D against Σ.

At the beginning of the game, simulator chooses a random bit $b \in \{0, 1\}$, and generates secret, relink, and public keys $(sk_i, rk_i, pk_i) \leftarrow Gen(k)$ $(i = 0, 1)$ and registers (i, rk_i, pk_i) $(i = 0, 1)$. D takes pk_0, pk_1 as input, and performs the following steps.[2]

---

[1] To guarantee correct key generation, PKI may require zero-knowledge proof of correctness of keys, when user registers his/her keys.

[2] We assume that D tries to distinguish signature generated by 0 or 1 w.l.o.g.

D may make queries to key registration oracle KO, signing oracle SO, and relink oracle RO. D is allowed to execute these oracle calls polynomially many times at any moment.

D sends $L^* \subset N$ s.t. $\{0,1\} \subset L^*$ and m^* to the challenge oracle CO, and can obtain signature $\sigma^* \leftarrow \mathrm{Sig}(sk_b, (pk_j)_{j \in L^*}, m^*)$. D is allowed to execute this once at any moment.

Finally, D outputs a bit b'.

D cannot ask to relink oracle RO, if

- σ is σ^* or its (polynomially many times) relinked signatures,
- and $L' \cap \{0,1\} = \{0\}$ or $\{1\}$.

When the game is defined in the random oracle model, D may access the random oracle polynomially many times at any moment.

We define the advantage of D against Σ as

$$\mathrm{Adv}_{\Sigma}^{\mathrm{anon}}(D) = \left| \Pr \left[\begin{matrix} b \in \{0,1\}, (sk_i, rk_i, pk_i) \leftarrow \mathrm{Gen}(k) \ (i = 0, 1), \\ b' \leftarrow D^{KO,SO,RO,CO}(pk_0, pk_1) \end{matrix} : b = b' \right] - \frac{1}{2} \right|$$

where the probability is taken over the choice of bit b, keys (sk_i, rk_i, pk_i) $(i = 0, 1)$, and the coin tosses of KO, SO, RO, CO and D.

Definition 4. *We say that relinkable ring signature scheme Σ is anonymous, if for every probabilistic polynomial-time adversary D the advantage $\mathrm{Adv}_{\Sigma}^{\mathrm{anon}}(D)$ is negligible in k.*

Traceability. We define the traceability of a relinkable ring signature scheme Σ. We consider the following game of adversary F against Σ.

F performs the following steps.

F may make queries to key registration oracle KO. F is allowed to execute these oracle calls polynomially many times at any moment.

Finally, F outputs (L^*, m^*, σ^*).

We say F wins the game if

- $\mathrm{Ver}(L^*, (pk_j)_{j \in L^*}, m^*, \sigma^*) = 1$,
- $\#\{i : i \in L^*, \mathrm{Ver}(\{i\}, pk_i, m^*, \mathrm{Rel}(rk_i, L^*, \{i\}, (pk_j)_{j \in L^*}, m, \sigma^*)) = 1\} \neq 1$.

When the game is defined in the random oracle model, F may access the random oracle polynomially many times at any moment.

We define the advantage of F against Σ as

$$\mathrm{Adv}_{\Sigma}^{\mathrm{trace}}(F) = \Pr \left[(L^*, m^*, \sigma^*) \leftarrow F^{KO}() : F \text{ wins.} \right]$$

where the probability is taken over the choice of the coin tosses of KO and F.

Definition 5. *We say that relinkable ring signature scheme Σ is traceable, if for every probabilistic polynomial-time adversary F the advantage $\mathrm{Adv}_{\Sigma}^{\mathrm{trace}}(F)$ is negligible in k.*

If a scheme is traceable, for given (L^*, m^*, σ^*) generated by probabilistic polynomial-time adversary F, there exists unique signer $i \in L^*$ s.t. $\mathrm{Ver}(\{i\}, pk_i, m^*, \mathrm{Rel}(rk_i, L^*, \{i\}, (pk_j)_{j \in L^*}, m, \sigma^*)) = 1$ except negligible probability. We denote the unique signer by $i^* \in L^*$.

Unforgeability. We define the unforgeability of a relinkable ring signature scheme Σ. We consider the following game of adversary F against Σ.

At the beginning of the game, simulator generates secret, relink, and public keys $(sk_i, rk_i, pk_i) \leftarrow Gen(k)$ $(i = 0, ..., n-1)$ and registers (i, rk_i, pk_i) $(i = 0, ..., n-1)$. F takes $(pk_j)_{j=0,...,n-1}$ as input, and performs the following steps.

F may make queries to key registration oracle KO, signing oracle SO, and relink oracle RO. F is allowed to execute these oracle calls polynomially many times at any moment.

Finally, F outputs (L^*, m^*, σ^*).

We say F wins the game if

- $\mathrm{Ver}(L^*, (pk_j)_{j \in L^*}, m^*, \sigma^*) = 1$,
- $L^* \subset \{0, ..., n-1\}$,
- $((i^*, L^*, m^*), \sigma^*)$ never appears in oracle query and reply list of SO,
- and $((i^*, L, L^*, m^*, \sigma), \sigma^*)$ never appears in oracle query and reply list of RO for any L and σ.

When the game is defined in the random oracle model, F may access the random oracle polynomially many times at any moment.

We define the advantage of F against Σ as

$$\mathrm{Adv}_{\Sigma}^{\mathrm{unforge}}(F) = \Pr \left[\begin{array}{l} (sk_i, rk_i, pk_i) \leftarrow Gen(k) \ (i = 0, ..., n-1), \\ (L^*, m^*, \sigma^*) \leftarrow F^{KO,SO,RO}((pk_j)_{j=0,...,n-1}) \end{array} : F \text{ wins.} \right]$$

where the probability is taken over the choice of keys (sk_i, rk_i, pk_i) $(i = 0, ..., n-1)$ and the coin tosses of KO, SO, RO and F.

Definition 6. *We say that relinkable ring signature scheme Σ is unforgeable, if for every probabilistic polynomial-time adversary F the advantage $\mathrm{Adv}_{\Sigma}^{\mathrm{unforge}}(F)$ is negligible in k.*

Relinker Unforgeability. We define the relinker unforgeability of a relinkable ring signature scheme Σ. We consider the following game of adversary F against Σ.

At the beginning of the game, simulator generates secret, relink, and public keys $(sk_i, rk_i, pk_i) \leftarrow Gen(k)$ $(i = 0, ..., n-1)$ and registers (i, rk_i, pk_i) $(i = 0, ..., n-1)$. F takes $(rk_i, pk_i)_{i=0,...,n-1}$ as input, and performs the following steps.

F may make queries to key registration oracle KO, signing oracle SO. F is allowed to execute these oracle calls polynomially many times at any moment.

Finally, F outputs (L^*, m^*, σ^*).

We say F wins the game if

- $\mathrm{Ver}(L^*, (pk_j)_{j \in L^*}, m^*, \sigma^*) = 1$,
- $L^* \subset \{0, ..., n-1\}$,
- and $((i^*, L, m^*), \sigma)$ never appears in oracle query and reply list of SO for any L and σ.

When the game is defined in the random oracle model, F may access the random oracle polynomially many times at any moment.

We define the advantage of F against Σ as

$$\mathrm{Adv}_{\Sigma}^{\mathrm{relink}}(F) = \Pr \left[\begin{array}{l} (sk_i, rk_i, pk_i) \leftarrow \mathrm{Gen}(k) \ (i = 0, ..., n-1), \\ (L^*, m^*, \sigma^*) \leftarrow F^{KO,SO}((rk_i, pk_i)_{i=0,...,n-1}) \end{array} : F \text{ wins.} \right]$$

where the probability is taken over the choice of keys (sk_i, rk_i, pk_i) $(i = 0, ..., n-1)$ and the coin tosses of KO, SO and F.

Definition 7. *We say that relinkable ring signature scheme Σ is relinker unforgeable, if for every probabilistic polynomial-time adversary F the advantage $\mathrm{Adv}_{\Sigma}^{\mathrm{relink}}(F)$ is negligible in k.*

4   Proposed Relinkable Ring Signature Scheme

In this section, we propose a relinkable ring signature scheme using pairing, prove its security, and estimate its efficiency.

4.1   Intuition of the Proposed Scheme

The following scheme is the interactive protocol which our ring signature is based on.

1. Prover P and verifier V have common input $g, y = g^x, h, w \in G_1$ and $\hat{g} \in G_2$. Prover P has witness $\hat{y} = \hat{g}^x \in G_2$.
2. Prover P chooses random $r \in_U \mathbb{Z}_p$ and sends $a = e(g^r, \hat{g}) \in G_3$ and $b = e(h^r, \hat{g}) \in G_3$ to verifier V.
3. Verifier V chooses random $c \in_U \mathbb{Z}_p$ and sends it to prover P.
4. Prover P sends $\hat{z} = \hat{g}^r \hat{y}^{-c} \in G_2$ to verifier V.
5. Verifier V checks $a = e(g, \hat{z})e(y, \hat{g})^c$ and $b = e(h, \hat{z})e(w, \hat{g})^c$.

This protocol is a variant of well known Chaum-Pedersen's protocol [CP92]. By using the Chaum-Pedersen's protocol, the prover who knows x which satisfies $y = g^x \wedge w = h^x$ can give an interactive proof of knowledge about x, and he is able to prove that $(g, y, h, w) \in$ DDH-tuple. Instead of the discrete logarithm x, we employ the corresponding group element of G_2 as the knowledge to prove, that is the prover who knows $\hat{y} \in G_2$ such that $e(y, \hat{g}) = e(g, \hat{y}) \wedge e(w, \hat{g}) = e(h, \hat{y})$ can give an interactive proof of knowledge about $\hat{y}$. Therefore we can separate the ability to prove that $(g, y, h, w) \in$ DDH-tuple from the discrete logarithm x, namely we can use $\hat{y}$ as a relink key, and can use x as a signing key. Furthermore this protocol inherits honest verifier zero-knowledgeness, language soundness, and knowledge soundness which are closely related to the security of our ring signature scheme. Finally we combine this protocol with the Cramer-Damgård-Schoenmakers' standard technique (proof of partial knowledge) [CDS94] to construct our ring signature scheme.

4.2   Proposed Relinkable Ring Signature Scheme

The proposed relinkable ring signature scheme is as follows.

Let G_1, G_2, G_3 be a multiplicative cyclic group with prime order p. Let $g \in G_1$ and $\hat{g} \in G_2$ be generators of G_1 and G_2. Let $e : G_1 \times G_2 \to G_3$ be pairing. Let $k \in \mathbb{N}$ be a security parameter that is the bit length of group element. Let $H : \{0,1\}^* \to G_1$ and $H' : \{0,1\}^* \to \mathbb{Z}_p$ be distinct hash functions that are modeled as random oracles in the security statements below. We denote by $N = \{0, ..., n-1\}$ the set of n signers.

Key Generation. Gen takes security parameter k, randomly chooses $x_i \in_U \mathbb{Z}_p$, and outputs secret, relink, public keys $(sk_i = x_i, rk_i = \hat{y}_i = \hat{g}^{x_i}, pk_i = y_i = g^{x_i})$ for i-th signer.

Signing. Sig takes i-th secret key $sk_i = x_i$, ring $L \subset N$ s.t. $i \in L$, public keys $(pk_i = y_i)_{i \in L}$, and message m, and outputs signature σ as follows.

1. Choose random $r \in_U \{0,1\}^l$, compute $h = H(r,m) \in G_1$, $w = h^{x_i} \in G_1$.
2. Generate a (non-interactive) zero-knowledge proof for language
   $\{(g, (y_i)_{i \in L}, h, w) \mid \exists i \in L, \log_g y_i = \log_h w\}$ as follows.
   (a) For i, choose random $r_i \in_U \mathbb{Z}_p$, compute $a_i = e(g^{r_i}, \hat{g}), b_i = e(h^{r_i}, \hat{g}) \in G_3$.
   (b) For all $j \in L \setminus \{i\}$, choose random $c_j \in_U \mathbb{Z}_p$, $\hat{z}_j \in_U G_2$, compute $a_j = e(g, \hat{z}_j)e(y_j, \hat{g})^{c_j}, b_j = e(h, \hat{z}_j)e(w, \hat{g})^{c_j} \in G_3$.
   (c) Compute $c_i = H'(L, h, w, (a_i)_{i \in L}, (b_i)_{i \in L}) - \sum_{j \neq i} c_j \mod p$.
   (d) Compute $\hat{z}_i = \hat{g}^{r_i} \hat{y}_i^{-c_i} \in G_2$.
3. Output signature $\sigma = (r, w, (c_i)_{i \in L}, (\hat{z}_i)_{i \in L})$.

Verification. Ver takes ring $L \subset N$, public keys $(pk_i = y_i)_{i \in L}$, message m, and signature $\sigma = (r, w, (c_i)_{i \in L}, (\hat{z}_i)_{i \in L})$, and outputs bit 0/1 as follows.

1. Check $y_i, w \in G_1$, $c_i \in \mathbb{Z}_p$, $\hat{z}_i \in G_2$ for all $i \in L$, otherwise reject.
2. Compute $h = H(r,m)$.
3. Compute $a_i = e(g, \hat{z}_i)e(y_i, \hat{g})^{c_i}, b_i = e(h, \hat{z}_i)e(w, \hat{g})^{c_i} \in G_3$ for all $i \in L$.
4. Check that $H'(L, h, w, (a_i)_{i \in L}, (b_i)_{i \in L}) = \sum_{i \in L} c_i \mod p$, otherwise reject.
5. Output accept if all checks above are passed, otherwise output reject.

Relinking. Rel takes i-th relink key $rk_i = \hat{y}_i$, rings $L, L' \subset N$ s.t. $i \in L, L'$, public keys $(pk_i = y_i)_{i \in L \cup L'}$, message m, and signature $\sigma = (r, w, (c_i)_{i \in L}, (\hat{z}_i)_{i \in L})$, and outputs signature σ' as follows.

1. Verify signature σ, otherwise reject.
2. Compute $h = H(r,m)$.
3. Check $e(h, \hat{y}_i) = e(w, \hat{g})$, otherwise reject.

Table 1. The comparison of costs of the proposed relinkable ring signature scheme and ring signature scheme [AOS02]. Here, n is the number of group member, T_{exp} is the time to compute exponential in G, T_{pair} is the time to compute pairing, $L_G, L_{\mathbb{Z}_p}$ are the lengths of elements of $G, \mathbb{Z}_p$, respectively.

	proposed scheme	ring signature [AOS02]
Signing costs	$2nT_{exp} + 4nT_{pair}$	nT_{exp}
Verification costs	$2nT_{exp} + 4nT_{pair}$	$2nT_{exp}$
Relinking costs	$2nT_{exp} + 4nT_{pair}$	–
Signature size	$(n+1)(L_G + L_{\mathbb{Z}_p})$	$2nL_{\mathbb{Z}_p}$

4. Generate a (non-interactive) zero-knowledge proof for language
   $\{(g, (y_i)_{i \in L'}, h, w) \mid \exists i \in L', \log_g y_i = \log_h w\}$ as follows.
   (a) For i, choose random $r_i \in_U \mathbb{Z}_p$, compute $a_i = e(g^{r_i}, \hat{g}), b_i = e(h^{r_i}, \hat{g}) \in G_3$.
   (b) For all $j \in L' \setminus \{i\}$, choose random $c'_j \in_U \mathbb{Z}_p$, $\hat{z}'_j \in_U G_2$, compute
       $a_j = e(g, \hat{z}'_j)e(y_j, \hat{g})^{c'_j}, b_j = e(h, \hat{z}'_j)e(w, \hat{g})^{c'_j} \in G_3$.
   (c) Compute $c'_i = H'(L', h, w, (a_i)_{i \in L'}, (b_i)_{i \in L'}) - \sum_{j \neq i} c'_j \mod p$.
   (d) Compute $\hat{z}'_i = \hat{g}^{r_i} \hat{y}_i^{-c'_i} \in G_2$.
5. Output signature $\sigma' = (r, w, (c'_i)_{i \in L'}, (\hat{z}'_i)_{i \in L'})$.

4.3   Security

The proposed relinkable ring signature scheme satisfies anonymity, unforgeability, relinker unforgeability, and traceability.

Theorem 1. *The proposed scheme satisfies*

1. *anonymity if we assume DDH problem in G_1 is intractable and H and H' are random oracle,*
2. *unforgeability if we assume skew CDH problem from G_1 to G_2 is intractable and H and H' are random oracle,*
3. *relinker unforgeability if we assume hinted CDH problem in G_1 is intractable and H and H' are random oracle,*
4. *and traceability if we assume H and H' are random oracle.*

The proofs of theorem is provided in Appendix C.

4.4   Efficiency

The comparison of costs of the proposed relinkable ring signature scheme and existing discrete logarithm based ring signature scheme [AOS02] is provided in Table1. Although the proposed relinkable ring signature scheme has complexity of same order in n as existing ring signature scheme [AOS02], the proposed scheme needs more pairing operations and modular exponentiations.

5   Conclusion

In this paper, we proposed the concept of the relinkable ring signature, which is a ring signature with ring reformation, i.e., a signer can delegate ring reformation ability separately from signing ability. The security of relinkable ring signature is defined by anonymity, unforgeability, relinker unforgeability, and traceability. We also proposed a concrete relinkable ring signature scheme that uses pairing.

Acknowledgments

The authors would like to thank the anonymous referees for their valuable comments.

References

[ACdM05]   Ateniese, G., Camenisch, J., de Medeiros, B.: Untraceable RFID tags via insubvertible encryption. In: Atluri, V., Meadows, C., Juels, A. (eds.) ACM Conference on Computer and Communications Security, pp. 92–101. ACM, New York (2005)

[ACHdM05]   Ateniese, G., Camenisch, J., Hohenberger, S., de Medeiros, B.: Practical group signatures without random oracles. Cryptology ePrint Archive: 2005/385 (2005)

[AOS02]   Abe, M., Ohkubo, M., Suzuki, K.: 1-out-of-n Signatures from a Variety of Keys. In: Zheng, Y. (ed.) ASIACRYPT 2002. LNCS, vol. 2501, pp. 415–432. Springer, Heidelberg (2002)

[BB04]   Boneh, D., Boyen, X.: Short signatures without random oracles. In: Cachin, C., Camenisch, J. (eds.) EUROCRYPT 2004. LNCS, vol. 3027, pp. 56–73. Springer, Heidelberg (2004)

[BGdMM05]   Ballard, L., Green, M., de Medeiros, B., Monrose, F.: Correlation-resistant storage. Technical Report TR-SP-BGMM-050705, Johns Hopkins University, CS Dept, 2005 (2005)

[BKM06]   Bender, A., Katz, J., Morselli, R.: Ring signatures: Stronger definitions, and constructions without random oracles. In: Halevi, S., Rabin, T. (eds.) TCC 2006. LNCS, vol. 3876, pp. 60–79. Springer, Heidelberg (2006)

[BLS02]   Barreto, P.S.L.M., Lynn, B., Scott, M.: Constructing elliptic curves with prescribed embedding degrees. In: Cimato, S., Galdi, C., Persiano, G. (eds.) SCN 2002. LNCS, vol. 2576, pp. 257–267. Springer, Heidelberg (2003)

[CDS94]   Cramer, R., Damgård, I., Schoenmakers, B.: Proofs of partial knowledge and simplified design of witness hiding protocols. In: Desmedt, Y.G. (ed.) CRYPTO 1994. LNCS, vol. 839, pp. 174–187. Springer, Heidelberg (1994)

[CP92]   Chaum, D., Pedersen, T.P.: Wallet databases with observers. In: Brickell, E.F. (ed.) CRYPTO 1992. LNCS, vol. 740, pp. 89–105. Springer, Heidelberg (1993)

[CvH91]   Chaum, D., van Heyst, E.: Group signatures. In: Davies, D.W. (ed.)
             EUROCRYPT 1991. LNCS, vol. 547, pp. 257–265. Springer, Heidelberg
             (1991)
[DEM05]   Dupont, R., Enge, A., Morain, F.: Building curves with arbitrary small
             mov degree over finite prime fields. J. Cryptology 18(2), 79–89 (2005)
[GPS08]   Galbraith, S.D., Paterson, K.G., Smart, N.P.: Pairings for cryptogra-
             phers. Discrete Applied Mathematics 156(16), 3113–3121 (2008)
[JN01]   Joux, A., Nguyen, K.: Separating decision diffie-hellman from diffie-
             hellman in cryptographic groups. Cryptology ePrint Archive: 2001/003
             (2001)
[LLM+07]   Liu, D.Y.W., Liu, J.K., Mu, Y., Susilo, W., Wong, D.S.: Revocable ring
             signature. J. Comput. Sci. Technol. 22(6), 785–794 (2007)
[LWH05]   Lee, K.C., Wei, H., Hwang, T.: Convertible ring signature. IEE Proceed-
             ings of Communications 152(4), 411–414 (2005)
[MNT01]   Miyaji, A., Nakabayashi, M., Takano, S.: New explicit conditions of el-
             liptic curve traces for fr-reduction. IEICE Transactions on Fundamen-
             tals E84-A(5), 1234–1243 (2001)
[RST01]   Rivest, R.L., Shamir, A., Tauman, Y.: How to leak a secret. In: Boyd,
             C. (ed.) ASIACRYPT 2001. LNCS, vol. 2248, pp. 552–565. Springer,
             Heidelberg (2001)
[SB06]   Scott, M., Barreto, P.S.L.M.: Generating more mnt elliptic curves. Des.
             Codes Cryptography 38(2), 209–217 (2006)
[SBZ01]   Steinfeld, R., Bull, L., Zheng, Y.: Content extraction signatures. In: Kim,
             K. (ed.) ICISC 2001. LNCS, vol. 2288, pp. 285–304. Springer, Heidelberg
             (2002)
[Sco02]   Scott, M.: Authenticated id-based key exchange and remote log-in with
             simple token and pin number. Cryptology ePrint Archive: 2002/164
             (2002)
[Sho97]   Shoup, V.: Lower bounds for discrete logarithms and related problems.
             In: Fumy, W. (ed.) EUROCRYPT 1997. LNCS, vol. 1233, pp. 256–266.
             Springer, Heidelberg (1997)
[SHUK03]   Saito, T., Hoshino, F., Uchiyama, S., Kobayashi, T.: Candidate one-
             way functions on non-supersingular elliptic curves. Technical Report of
             IEICE, ISEC 2003-65 (2003)
[Ver01]   Verheul, E.R.: Evidence that XTR is more secure than supersingular el-
             liptic curve cryptosystems. In: Pfitzmann, B. (ed.) EUROCRYPT 2001.
             LNCS, vol. 2045, pp. 195–210. Springer, Heidelberg (2001)

A   Security of Skew and Hinted CDH Assumptions

Security in Generic Group Model: First, we prove security of the skew
CDH problem and the hinted CDH problem in generic groups in the sense of
[Sho97]. Our proof is essentially the same as the proof of CDH problem in [Sho97]
except the complicated group settings of the pairing. We employ the settings of
[BB04] which gives a generic proof of the q-SDH assumption on the type 2 curve
in [GPS08].

In the generic group model, elements of $G_1 = \langle g \rangle$, $G_2 = \langle \hat{g} \rangle$, and G_3 are
encoded as unique random strings. Let $\xi_i : G_i \to \{0,1\}^*$ be a random encoding

function of G_i for $i = 1, 2, 3$. The adversary $\mathcal{A}$ can make the following oracle calls

- the group operation in each of G_1, G_2 and G_3,
- the bilinear pairing $e : G_1 \times G_2 \to G_3$, and
- the projection $\phi : G_2 \to G_1$.

These oracles takes encoding(s) of the input element(s) and answers encoding of the output element. Notice that we consider projection just for generality.

We have the following propositions about security of skew and hinted CDH assumptions in the generic group model. Due to lack of space, we omit here the proofs of these propositions that only follow [Sho97, BB04].

Proposition 1. *Let $\mathcal{A}$ be an algorithm that solves the skew CDH problem in the generic group model, making a total of at most q queries to the oracles computing the group action in G_1, G_2, G_3, the oracle computing the projection ϕ, and the oracle computing the bilinear pairing e. If $a \in \mathbb{Z}_p$ and ξ_1, ξ_2, ξ_3 are chosen at random, then the probability ϵ that $\mathcal{A}(\xi_1(g), \xi_1(g^a), \xi_2(\hat{g}))$ outputs $\xi_2(\hat{g}^a)$, is bounded by $\epsilon \leq O(q^2/p)$*

Proposition 2. *Let $\mathcal{A}$ be an algorithm that solves the hinted CDH problem in the generic group model, making a total of at most q queries to the oracles computing the group action in G_1, G_2, G_3, the oracle computing the projection ϕ, and the oracle computing the bilinear pairing e. If $a, b \in \mathbb{Z}_p$ and ξ_1, ξ_2, ξ_3 are chosen at random, then the probability ϵ that $\mathcal{A}(\xi_1(g), \xi_1(g^a), \xi_1(g^b), \xi_2(\hat{g}), \xi_2(\hat{g}^a), \xi_2(\hat{g}^b))$ outputs $\xi_1(g^{ab})$, is bounded by $\epsilon \leq O(q^2/p)$*

Reduction Security: Second, we show reductions of the skew CDH problem and the hinted CDH problem to other known problems, assuming that there exists projection $\phi : G_2 \to G_1$, i.e., the type 2 curve in [GPS08].

Proposition 3. *If there exists probabilistic polynomial-time algorithm $A_s : G_1^2 \times G_2 \to G_2$ that solves the skew CDH problem on (G_1, G_2), there exists probabilistic polynomial-time algorithm $A_c : G_1^3 \to G_1$ that solves a variant of the CDH problem on G_1 (named chosen generator CDH).*

Proof. Let $g = \phi(\hat{g})$. $A_c(g, g^a, g^b)$ outputs $\phi(A_s(g, g^b, A_s(g, g^a, \hat{g})))$. □

Proposition 4. *If there exists probabilistic polynomial-time algorithm $A_h : G_1^3 \times G_2^3 \to G_1$ that solves the hinted CDH problem on (G_1, G_2), there exists probabilistic polynomial-time algorithm $A_b : G_2^4 \to G_3$ that solves the BDH problem on G_2.*

Proof. $A_b(\hat{g}, \hat{g}^a, \hat{g}^b, \hat{g}^c)$ outputs $e(A_h(\phi(\hat{g}), \phi(\hat{g}^a), \phi(\hat{g}^b), \hat{g}, \hat{g}^a, \hat{g}^b), \hat{g}^c)$. □

B   Selection of Bilinear Group

Selection of G_1 and G_2: Let p be a prime and set $q = p^n$. Let $E[\ell]$ be ℓ-torsion points on elliptic curve E defined over $\mathbb{F}_q$. Let $\phi(x, y) = (x^q, y^q)$ the q-th power

Frobenius morphism. Let $e : E[\ell] \times E[\ell] \to \mu_\ell$ be Weil pairing, where $\mu_\ell \subset \overline{\mathbb{F}}_q$ is the group of the ℓ-th root of 1.

We consider the eigenspace decomposition of $E[\ell]$ w.r.t. ϕ. For simplicity, we assume that ℓ is a prime, $\#E(\mathbb{F}_q)[\ell] = \ell$, m is minimal integer s.t. $\#E(\mathbb{F}_{q^m})[\ell] = \ell^2$. Since ϕ is the identity map on $E(\mathbb{F}_q)$, $E(\mathbb{F}_q)[\ell]$ is an eigenspace of ϕ with eigenvalue $\lambda_1 = 1$. Let λ_2 be another eigenvalue, and we have $\lambda_2 = \lambda_1 \lambda_2 = q$ mod ℓ. We also assume that the eigenvalues of ϕ are non-degenerative, i.e., $t = \lambda_1 + \lambda_2 \neq 2 \mod \ell$.

Then, there exists the eigenspace corresponding to λ_2 in $E[\ell] \setminus E(\mathbb{F}_q)[\ell]$. Let Q be its generator, and P be a generator of $E(\mathbb{F}_q)[\ell]$. Thus, we have $E[\ell] = \langle P \rangle \oplus \langle Q \rangle$, and we use $G_1 = \langle P \rangle \subset E(\mathbb{F}_q)[\ell], G_2 = \langle Q \rangle \subset E(\mathbb{F}_{q^m})[\ell], G_3 = \mu_\ell \subset \mathbb{F}_{q^m}^*$, and Weil pairing $e : \langle P \rangle \times \langle Q \rangle \to \mu_\ell$. We can use also $G_2 = \langle R \rangle$ where $R = \alpha P + \beta Q$, $\alpha, \beta \in \mathbb{Z}/\ell\mathbb{Z}$, and $\beta \neq 0$.

The (normalized) trace map $tr = 1/m \sum_{i=0,\dots,m-1} \phi^i : \mathbb{F}_{q^m} \to \mathbb{F}_q$ induces polynomial-time computable group isomorphism $tr : \langle R \rangle \to \langle P \rangle$.

The generator Q can be found by $Q = R - tr(R)$ from R. The generator R can be found with probability $(\ell - 1)/\ell$ if a point on $E[\ell]$ is chosen at random. A point on $E[\ell]$ can be found if a point on $E(\mathbb{F}_{q^m})$ is chosen at random, and multiplied by $\#E(\mathbb{F}_{q^m})/\ell^2$. A point on $E(\mathbb{F}_{q^m})$ can be found with probability of approximately $1/2$ if $x \in \mathbb{F}_{q^m}$ is chosen at random and $y \in \mathbb{F}_{q^m}$ is obtained by solving the curve equation.

The generator P of $E(\mathbb{F}_q)[\ell]$ can be found if a point on $E(\mathbb{F}_q)$ is chosen at random, and multiplied by $\#E(\mathbb{F}_q)/\ell$. A point on $E(\mathbb{F}_q)$ can be found with probability of approximately $1/2$ if $x \in \mathbb{F}_q$ is chosen at random and $y \in \mathbb{F}_q$ is obtained by solving the curve equation.

Selection of Elliptic Curve: To guarantee that our assumptions hold, we need to choose an elliptic curve that satisfies the following conditions.

On the the supersingular curve and the trace-2 curve, polynomial-time computable homomorphism from G_1 to G_2, called the distortion map, is known [JN01, Ver01]. Therefore, we should avoid the supersingular curve and the trace-2 curve.

On the other hand, it was shown that the distortion map on a non-supersingular non trace-2 curve cannot be described by any single rational map [Ver01]. Therefore, we choose the non-supersingular non trace-2 curve.

As in the case of a pairing-based cryptosystem, to guarantee that the CDH or DL problem is intractable, we need to choose an elliptic curve that satisfies the following two conditions.

- $\#E(\mathbb{F}_q)[\ell] = \ell$ is large enough s.t. the CDH or DL problem on $E(\mathbb{F}_q)[\ell]$ is intractable.
- $\#\mathbb{F}_{q^m}^*$ is large enough s.t. the CDH or DL problem on $\mathbb{F}_{q^m}^*$ is intractable.

We can choose in practice $\ell \sim q \geq 2^{160}$ and $q^m \geq 2^{1024}$.

Adding to the conditions above, we need to choose an elliptic curve with small q and m s.t. elliptic addition and pairing can be computed efficiently, i.e., polynomial-time computable.

Efficient methods to find non-supersingular non trace-2 pairing enabled secure curves are given in [MNT01, SB06, BLS02, DEM05].

C Proofs of Theorem 1

Anonymity: Let Σ be the proposed revocable ring signature scheme. Let D be a $(\tau, \epsilon, q_{SO}, q_H)$-adversary against Σ that requests signing oracle at most q_{SO} times and accesses random oracles at most q_H times in total and breaks the anonymity of Σ with advantage at least ϵ and running time at most τ. Let D' be a (τ', ϵ')-adversary against DDH assumption that breaks the assumption with advantage at least ϵ' and running time at most τ'. We construct adversary D' from adversary D as follows. Simulator maintains random oracle call lists H, H' and list Log, and performs followings.

At the beginning of the simulation, simulator D' is given instance of DDH problem $(g, g^\alpha, g^\beta, g^\gamma)$. Simulator selects random bit $b \in \{0, 1\}$ and random $v \in \mathbb{Z}_p$, sets $pk_b = y_b = (g^\alpha)^v$, generates secret, relink, and public keys $(sk_{1-b}, rk_{1-b}, pk_{1-b}) \leftarrow Gen(k)$, and gives pk_0, pk_1 to adversary D.

Random Oracle $H(r, m)$. If $H(r, m)$ is already defined, return defined value. Otherwise, select random $u \in_U \mathbb{Z}_p$, define $Log(g, h) = u$, and define $H(r, m) = h = g^u$ and return it.

Random Oracle $H'(L, h, w, (a_i)_{i \in L}, (b_i)_{i \in L})$. If $H'(L, h, w, (a_i)_{i \in L}, (b_i)_{i \in L})$ is already defined, return defined value. Otherwise, select random $c \in_U \mathbb{Z}_p$, and define $H'(L, h, w, (a_i)_{i \in L}, (b_i)_{i \in L}) = c$ and return it.

Key Registration Oracle $KO(i, \hat{y}, y)$. If y_i is already defined, reject. If $e(g, \hat{y}) \neq e(y, \hat{g})$, reject. Otherwise, define $\hat{y}_i = \hat{y}$ and $y_i = y$.

Signing Oracle $SO(i, L, m)$. Select random $r \in_U \{0, 1\}^k$. Call random oracle $h = H(r, m)$, if $Log(g, h)$ is not defined, abort, otherwise set $u = Log(g, h)$ and $w = y_j^u$. Create simulated proof $((c_i)_{i \in L}, (\hat{z}_i)_{i \in L})$ by setting $H'(L, h, w, (a_i)_{i \in L}, (b_i)_{i \in L}) = \sum_{i \in L} c_i$. Return $(r, w, (c_i)_{i \in L}, (\hat{z}_i)_{i \in L})$.

Relink Oracle $RO(L, L', m, \sigma)$. Verify signature $\sigma = (r, w, (c_i)_{i \in L}, (\hat{z}_i)_{i \in L})$, otherwise reject. Call random oracle $h = H(r, m)$. Create simulated proof $((c_i')_{i \in L'}, (\hat{z}_i')_{i \in L'})$ by setting $H'(L, h, w, (a_i')_{i \in L'}, (b_i')_{i \in L'}) = \sum_{i \in L'} c_i'$. Return $(r, w, (c_i')_{i \in L'}, (\hat{z}_i')_{i \in L'})$.

Challenge Oracle $CO(L^*, m^*)$. Select random $r \in_U \{0, 1\}^k$. If $H(r, m)$ is already defined, abort, otherwise select random $u \in_U \mathbb{Z}_p$, define $H(r, m) = h = (g^\beta)^u$ ($Log(g, h)$ can not be defined), and set $w = (g^\gamma)^{uv}$. Create simulated proof $((c_i)_{i \in L}, (\hat{z}_i)_{i \in L})$ by setting $H'(L, h, w, (a_i)_{i \in L}, (b_i)_{i \in L}) = \sum_{i \in L} c_i$. Return $(r, w, (c_i)_{i \in L}, (\hat{z}_i)_{i \in L})$.

Finally, D outputs bit $b' \in \{0,1\}$. Simulator D' outputs random bit if the simulation abort, bit 1 if $b = b'$, and random bit if $b \neq b'$.

The probability that the simulation does not abort is $Pr[\neg abort] \geq (1 - (q_H + q_S)/2^k)(1 - 1/2^k)^{q_S}$, since CO does not abort with probability at least $1 - (q_H + q_S)/2^k$ and SO does not abort with probability at least $(1 - 1/2^k)^{q_S}$.

In the case that the simulation aborts, we have $Pr[D'\ wins|abort] = 1/2$. Since the view of adversary is independent from b if the instance is not DDH-tuple, i.e., $\gamma \neq \alpha\beta$, we have $Pr[D'\ wins|\gamma \neq \alpha\beta|\neg abort] = 1/2$. Since the simulation is perfect if the instance is DDH-tuple, i.e., $\gamma = \alpha\beta$, we have $Pr[D'\ wins|\gamma = \alpha\beta|\neg abort] = Pr[D\ wins|\gamma = \alpha\beta|\neg abort]$.

Thus, we have $\epsilon' = |Pr[D'\ wins] - 1/2| = |Pr[D'\ wins|\neg abort] - 1/2| \cdot Pr[\neg abort] = |Pr[D'\ wins|\gamma = \alpha\beta|\neg abort] - 1/2| \cdot Pr[\gamma = \alpha\beta|\neg abort] \cdot Pr[\neg abort] = |Pr[D\ wins|\gamma = \alpha\beta|\neg abort] - 1/2| \cdot Pr[\gamma = \alpha\beta|\neg abort] \cdot Pr[\neg abort] \geq \epsilon \cdot 1/2 \cdot (1 - (q_H + q_S)/2^k)(1 - 1/2^k)^{q_S} \geq \epsilon \cdot 1/2 \cdot (1 - (q_H + 2q_S)/2^k)$.

Traceability: Let Σ be the proposed revocable ring signature scheme. Let F be a $(\tau, \epsilon, q_{SO}, q_H)$-adversary against Σ that requests signing oracle at most q_{SO} times and accesses random oracles at most q_H times in total and breaks the traceability of Σ with advantage at least ϵ and running time at most τ. Simulator maintains random oracle call lists H, H' and performs followings.

Random Oracle $H(r, m)$. If $H(r, m)$ is already defined, return defined value. Otherwise, select random $u \in_U \mathbb{Z}_p$, and define $H(r, m) = h = g^u$ and return it.

Random Oracle $H'(L, h, w, (a_i)_{i \in L}, (b_i)_{i \in L})$. If $H'(L, h, w, (a_i)_{i \in L}, (b_i)_{i \in L})$ is already defined, return defined value. Otherwise, select random $c \in_U \mathbb{Z}_p$, and define $H'(L, h, w, (a_i)_{i \in L}, (b_i)_{i \in L}) = c$ and return it.

Key Registration Oracle $KO(i, \hat{y}, y)$. If y_i is already defined, reject. If $e(g, \hat{y}) \neq e(y, \hat{g})$, reject. Otherwise, define $\hat{y}_i = \hat{y}$ and $y_i = y$.

Finally, F outputs L^*, m^*, σ^*. we write $\sigma^* = (r^*, w^*, (c_i)_{i \in L^*}, (\hat{z}_i)_{i \in L^*})$ and $h^* = H(r^*, m^*)$. By the second winning condition of adversary, $\forall i \in L^*$, (g, y_i, h^*, w^*) is not DDH-tuple, i.e., $x_i = \log_g(y_i) \neq x_i' = \log_{h^*}(w^*)$. Thus, the first winning condition of adversary holds with negligible probability $\epsilon \leq 1 - (1 - 1/p)^{q_H} \leq q_H/p$, since language soundness of zero-knowledge proof.

Unforgeability: Let Σ be the proposed revocable ring signature scheme. Let F be a $(\tau, \epsilon, q_{SO}, q_H)$-adversary against Σ that requests signing oracle at most q_{SO} times and accesses random oracles at most q_H times in total and breaks the unforgeability of Σ with advantage at least ϵ and running time at most τ. Let F' be a (τ', ϵ')-adversary against skew CDH assumption that breaks the assumption with advantage at least ϵ' and running time at most τ'. We construct adversary F' from adversary F as follows. Simulator maintains random oracle call lists H, H' and list Log, and performs followings.

At the beginning of the simulation, simulator F' is given instance of skew CDH problem $(g, g^\alpha, \hat{g})$. Simulator selects random $v_i \in \mathbb{Z}_p$, sets $pk_i = y_i = (g^\alpha)^{v_i}$ for $i = 0, ...n - 1$, and gives them to adversary F.

Random Oracle $H(r, m)$. If $H(r, m)$ is already defined, return defined value. Otherwise, select random $u \in_U \mathbb{Z}_p$, define $Log(g, h) = u$, and define $H(r, m) = h = g^u$ and return it.

Random Oracle $H'(L, h, w, (a_i)_{i \in L}, (b_i)_{i \in L})$. If $H'(L, h, w, (a_i)_{i \in L}, (b_i)_{i \in L})$ is already defined, return defined value. Otherwise, select random $c \in_U \mathbb{Z}_p$, and define $H'(L, h, w, (a_i)_{i \in L}, (b_i)_{i \in L}) = c$ and return it.

Key Registration Oracle $KO(i, \hat{y}, y)$. If y_i is already defined, reject. If $e(g, \hat{y}) \neq e(y, \hat{g})$, reject. Otherwise, define $\hat{y}_i = \hat{y}$ and $y_i = y$.

Signing Oracle $SO(i, L, m)$. Select random $r \in_U \{0, 1\}^k$. Call random oracle $h = H(r, m)$, set $u = Log(g, h)$ and $w = y_j^u$. Create simulated proof $((c_i)_{i \in L}, (\hat{z}_i)_{i \in L})$ by setting $H'(L, h, w, (a_i)_{i \in L}, (b_i)_{i \in L}) = \sum_{i \in L} c_i$. Return $(r, w, (c_i)_{i \in L}, (\hat{z}_i)_{i \in L})$.

Relink Oracle $RO(L, L', m, \sigma)$. Verify signature $\sigma = (r, w, (c_i)_{i \in L}, (\hat{z}_i)_{i \in L})$, otherwise reject. Call random oracle $h = H(r, m)$. Create simulated proof $((c'_i)_{i \in L'}, (\hat{z}'_i)_{i \in L'})$ by setting $H'(L, h, w, (a'_i)_{i \in L'}, (b'_i)_{i \in L'}) = \sum_{i \in L'} c'_i$. Return $(r, w, (c'_i)_{i \in L'}, (\hat{z}'_i)_{i \in L'})$.

Finally, F outputs L^*, m^*, σ^*. we write $\sigma^* = (r^*, w^*, (c_i)_{i \in L^*}, (\hat{z}_i)_{i \in L^*})$ and $h^* = H(r^*, m^*)$. By rewinding adversary F, F outputs $L^*, m^*, \sigma^{*'}$ where $\sigma^{*'} = (r^*, w^*, (c'_i)_{i \in L^*}, (\hat{z}'_i)_{i \in L^*})$ s.t. $\sum_{i \in L} c_i \neq \sum_{i \in L} c'_i$ with probability $1/q_H$, since adversary F uses same H' query for forgery with probability $1/q_H$. Find i s.t. $c_i \neq c'_i$ and compute $\hat{y}_i = (\hat{z}_i/\hat{z}'_i)^{1/(c'_i - c_i)}$. Simulator F' outputs $(\hat{y}_i)^{1/v_i}$ as guess of $\hat{g}^\alpha$. Using forking lemma, we have $\epsilon' \geq \epsilon(\epsilon/q_H - 1/p)$.

Relinker Unforgeability: Let Σ be the proposed revocable ring signature scheme. Let F be a $(\tau, \epsilon, q_{SO}, q_H)$-adversary against Σ that requests signing oracle at most q_{SO} times and accesses random oracles at most q_H times in total and breaks the relinker unforgeability of Σ with advantage at least ϵ and running time at most τ. Let F' be a (τ', ϵ')-adversary against hinted CDH assumption that breaks the assumption with advantage at least ϵ' and running time at most τ'. We construct adversary F' from adversary F as follows. Simulator maintains random oracle call lists H, H' and list Log, and performs followings.

At the beginning of the simulation, simulator F' is given instance of hinted CDH problem $(g, g^\alpha, g^\beta, \hat{g}^\alpha, \hat{g}^\beta)$.

Simulator selects random $v_i \in_U \mathbb{Z}_p$ and random $\gamma_i \in_U \{0, 1\}$. Let $\hat{g}_i = (\hat{g}^\alpha)^{\gamma_i}(\hat{g}^\beta)^{1-\gamma_i}$ and $g_i = (g^\alpha)^{\gamma_i}(g^\beta)^{1-\gamma_i}$. Simulator sets $rk_i = \hat{y}_i = \hat{g}_i^{v_i}$ and $pk_i = y_i = g_i^{v_i}$, and give them to adversary F.

Random Oracle $H(r, m)$. If $H(r, m)$ is already defined, return defined value. Otherwise, select random $u \in_U \mathbb{Z}_p, \gamma \in_U \{0, 1\}$ define $H(r, m) = h = ((g^\alpha)^\gamma(g^\beta)^{1-\gamma})^u$ and define $Log((g^\alpha)^\gamma(g^\beta)^{1-\gamma}, h) = u$, and return h.

Random Oracle $H'(L, h, w, (a_i)_{i\in L}, (b_i)_{i\in L})$. If $H'(L, h, w, (a_i)_{i\in L}, (b_i)_{i\in L})$ is already defined, return defined value. Otherwise, select random $c \in_U \mathbb{Z}_p$, and define $H'(L, h, w, (a_i)_{i\in L}, (b_i)_{i\in L}) = c$ and return it.

Key Registration Oracle $KO(i, \hat{y}, y)$. If y_i is already defined, reject. If $e(g, \hat{y}) \neq e(y, \hat{g})$, reject. Otherwise, define $\hat{y}_i = \hat{y}$ and $y_i = y$.

Signing Oracle $SO(i, L, m)$. Select random $r \in_U \{0, 1\}^k$. If $H(r, m)$ is already defined, abort, otherwise select random $u \in_U \mathbb{Z}_p$, define $H(r, m) = h = g_i^u$, define $Log(g_i, h) = u$, and set $w = h^{v_i}$. Create simulated proof $((c_i)_{i\in L}, (\hat{z}_i)_{i\in L})$ by setting $H'(L, h, w, (a_i)_{i\in L}, (b_i)_{i\in L}) = \sum_{i\in L} c_i$. Return $(r, w, (c_i)_{i\in L}, (\hat{z}_i)_{i\in L})$.

Finally, F outputs L^*, m^*, σ^*. we write $\sigma^* = (r^*, w^*, (c_i)_{i\in L^*}, (\hat{z}_i)_{i\in L^*})$ and $h^* = H(r^*, m^*)$. Find j s.t. $e(h^*, \hat{y}_j) = e(w^*, \hat{g})$, and set $v = v_j$. If $Log(g^\alpha g^\beta / g_j, h^*)$ is not defined, abort, otherwise set $u = Log(g^\alpha g^\beta / g_j, h^*)$. Simulator F' outputs $(w^*)^{1/uv}$ as guess of $g^{\alpha\beta}$. Since the simulation successes if SO doesn't select random r that is already queried by adversary F and $Log(g^\alpha g^\beta / g_j, h^*)$ is defined at the final step, we have $\epsilon' \geq Pr[Log(g^\alpha g^\beta / g_j, h^*) \text{ is } defined | F' \text{ finds } j | F \text{ wins} | \neg abort \text{ in } SO] \cdot Pr[F' \text{ finds } j | F \text{ wins} | \neg abort \text{ in } SO] \cdot Pr[F \text{ wins} | \neg abort \text{ in } SO] \cdot Pr[\neg abort \text{ in } SO] \geq 1/2 \cdot (1 - 1/p)^{q_H} \cdot \epsilon \cdot (1 - (q_H + q_S)/2^k)^{q_S} \geq \epsilon/2 \cdot (1 - q_H/p - (q_H + q_S)q_S/2^k)$.

Author Index